LONDON MATHEMATICAL SOCIETY LECTURE NOTE SERIES 487

Managing Editor: Professor Endre Süli, Mathematical Institute, U ˙ ˙ ˙ ˙ Woodstock Road, Oxford OX2 6GG, United Kingdom

The titles below are available from booksellers, or from Cambridge

Algebraic Combinatorics and the Monster Group

Edited by

ALEXANDER A. IVANOV
Russian Academy of Sciences, Moscow

CAMBRIDGE
UNIVERSITY PRESS

Shaftesbury Road, Cambridge CB2 8EA, United Kingdom

One Liberty Plaza, 20th Floor, New York, NY 10006, USA

477 Williamstown Road, Port Melbourne, VIC 3207, Australia

314–321, 3rd Floor, Plot 3, Splendor Forum, Jasola District Centre,
New Delhi – 110025, India

103 Penang Road, #05–06/07, Visioncrest Commercial, Singapore 238467

Cambridge University Press is part of Cambridge University Press & Assessment,
a department of the University of Cambridge.

We share the University's mission to contribute to society through the pursuit of
education, learning and research at the highest international levels of excellence.

www.cambridge.org
Information on this title: www.cambridge.org/9781009338042
DOI: 10.1017/9781009338073

First published 2024

Printed in the United Kingdom by TJ Books Limited, Padstow Cornwall

A catalogue record for this publication is available from the British Library.

*A Cataloging-in-Publication data record for this book is available from the Library of
Congress*

ISBN 978-1-009-33804-2 Paperback

Contents

Contributors

Tatsiana S. Busel
Institute of Mathematics, National Academy of Sciences of Belarus

Ilaria Cardinali
University of Siena

Luca Giuzzi
University of Brescia

Takuya Ikuta
Kobe Gakuin University, Japan

Tatsuro Ito
Kanazawa University, Japan

Alexander A. Ivanov
Institute for System Analysis, FRC CSC RAN, Moscow, Russia

Jack H. Koolen
University of Science and Technology of China

Ching Hung Lam
Institute of Mathematics, Academia Sinica

Atsushi Matsuo
The University of Tokyo

Ulrich Meierfrankenfeld
Michigan State University

Akihiro Munemasa
Tohoku University, Japan

Jongyook Park
Kyungpook National University, South Korea

Antonio Pasini
University of Siena

Sergey Shpectorov
University of Birmingham

Leonard H. Soicher
Queen Mary University of London

Irina D. Suprunenko
Institute of Mathematics, National Academy of Sciences of Belarus

Paul Terwilliger
University of Wisconsin

Hendrik Van Maldeghem
Ghent University

Hiroshi Yamauchi
Tokyo Woman's Christian University

Qianqian Yang
Shanghai University

Preface

The content of this collection is best viewed via the relationship and the inter-action between examples, axiomatization, and theory. In [1] Simon Norton wrote: 'It is often said that Euclid's "Elements" was intended not so much as an introduction to geometry but to show how to construct the five Platonic Solids; these solids had (in modern terminology) the most elaborate symmetry groups that had been encountered at that time, and were known to be of special interest. On the way to constructing them it would have been necessary to expound most of the geometry that was known at the time.'

The wideness of this viewpoint might be arguable, but we clearly observe a similar situation in recent mathematics. Ernst Witt constructed his designs in [2] to remove forever any doubts about the existence of the Mathieu groups M_{11}, M_{12}, M_{22}, M_{23}, and M_{24} discovered by Émile Mathieu some 70 years earlier [3]. The future development of the theory of designs demonstrated that these are the pearls of design theory. The largest of the Witt designs is on 24 points and it is the unique S(5,8,24).

In the 1960s, when the Leech lattice came to John Conway's attention, it became clear that the 'purpose' of the Witt design $S(5,8,24)$ associated with the largest Mathieu group M_{24} was to serve through the Golay code as the frame of the Leech lattice. The fundamental monograph [4] of 1318 grams contains many lattices and their theories, but essentially it is to justify that the Leech lattice is the most special one. The most common characterization of the Leech lattice is as the unique 24-dimensional, even, unimodular lattice without roots. If the roots are allowed, we obtain in addition 23 Niemeier lattices, whose existence can be justified by 23 constructions of the Leech lattice, one from each Niemeier lattice.

In its turn the 'purpose' of the Leech lattice Λ and its automorphism group $Co_0 = 2.Co_1$ is the structure of the centralizer $C \cong 2_+^{1+24}.Co_1$ of an involution in the Monster group, where $O_2(C)/Z(C)$ is isomorphic to $\Lambda/2\Lambda$ as a module for $C/O_2(C) \cong Co_1$. However, originally the way to the Monster was opened through a different door.

In the late 1960s, Bernd Fischer [5] axiomatized the order product property of the transpositions of the symmetric groups to build up a theory of 3-transposition groups. The highest point of this theory was discovery of three sporadic 3-transposition groups Fi_{22}, Fi_{23}, and Fi_{24}. The former two and the index 2 commutator subgroup Fi'_{24} of the last one are now known as Fischer's sporadic simple groups. The maximal set of pairwise commuting transpositions in Fi_{24} contains precisely 24 transpositions and carries a structure of Witt design $S(5, 8, 24)$ with the Mathieu group M_{24} being the action induced on this set by its stabilizer (which is a non-split extension of the dual Golay code module 2^{12} by M_{24}). The 'purpose' of the largest Fischer's sporadic 3-transposition group Fi'_{24} extended by its Schur multiplier of order 3 is to serve as the centralizer of an order 3 subgroup in the Monster group. But this is still not the original path to the Monster.

The non-sporadic 3-transposition groups include besides the symmetric groups the classical symplectic, orthogonal, and unitary groups. Fischer relaxed the axioms by admitting products of order 4, which allowed in the exceptional Lie-type groups including $^2E_6(2)\!:2$. Further on, a $\{3, 4\}$-transposition group, having $2 \cdot^2 E_6(2)\!:2$ as the transposition centralizer, turned out to be another sporadic simple group now known as Fischer's Baby Monster and is denoted by BM.

The non-split central extension $2 \cdot BM$ of the Baby Monster is an example of a 6-transposition group (in which products of order 5 are absent) and it served as the centralizer of an involution in another 6-transposition group, which is the Monster group M. Bernd Fischer came to this evidence in 1973 and, independently, in the same year the existence of the Monster was suggested by Robert Griess.

In the 1970s, there was an intensive study of various properties of the then-hypothetical Monster M and some of the results were left unpublished. It was shown that the minimal faithful complex representation of M has degree at least 196 883; Simon Norton showed that if such a representation exists, then it carries a non-zero inner product and a non-associative algebra product, which are unique up to rescaling. John Thompson proved the uniqueness of M subject to the structure of involution centralizers and the existence of a 196 883 representation [6]. In 1980, Griess constructed the 196 883-dimensional algebra, and the Monster itself [7]. The construction was improved by John Conway [8] who in particular added the identity to the algebra, increasing its dimension by 1. The extended algebra is called Conway–Griess–Norton algebra or simply the Monster algebra.

The theory around the Monster was initiated by the observation made by John McKay that $196\,883 + 1 = 196\,884$, where the right-hand side is the linear coefficient of the most celebrated modular form $J(q)$. A remarkable sequence of events, which goes under the name of Monstrous Moonshine [9], culminated (but was not concluded) in the construction of the Vertex Operator Algebra (VOA) $(V^\natural, *_n)$ known as the Moonshine Module [10]. This is an infinite-dimensional integer-graded algebra with infinitely many products $*_n$ satisfying infinitely many Jordan-type relations. In this context the Monster algebra and the inner product are realized as on grade 2 operators: $(V_2, *_3, *_1)$.

The axiomatic for VOAs was designed by Richard Borcherds [11] with a close look at the Moonshine Module construction [10]. Thus the theory was designed for the construction of $V^\natural$ based on the Leech lattice, although it can also be applied to any Niemeier lattice, and even in two different ways: twisted and untwisted constructions. There are quite remarkable isomorphisms between the lattice VOAs, so in total one obtains 39 (rather than 48) lattice VOAs with central charge 24 (which inherits the dimension of the lattice). These VOAs are rational, meaning that their theta series are the modular invariant $J(q)$ with the constant term (properly normalized) being the dimension of the level one sector V_1. Within this class, the Moonshine module (twisted construction based on the Leech lattice) is characterized by the condition $V_1 = 0$ (which has a deep analogy with the rootless property of the Leech lattice).

The main features of the VOA theory are now understood to be implicit in the physical string theory, and in physics VOAs correspond to the 2-dimensional meromorphic conformal field theories. In 1993, A. N. Shellekens [12] gave physical evidence for the existence of exactly 71 strongly regular holomorphic VOAs with central charge 24, including the lattice constructions. Since then the attempts to obtain a rigorous mathematical proof of this evidence, including constructions of the predicted VOAs, became one of the most fundamental problems in the theory of VOA.

Thus, through moonshine VOA, the Monster was placed in the centre of a remarkable theory with deep connections with modern physics through infinite-dimensional algebras. The attempts to axiomatize the finite-dimensional Monster algebra itself led to success when Alexander Ivanov axiomatized in [13] certain properties of the 2A-axes of the Monster algebra from [8] under the name of Majorana axes. This led to Majorana theory, which enables identification of the isomorphism type of various subgroups in the Monster generated by a set of 2A-involutions.

In [1] Simon Norton wrote 'one may consider an "ideal" ATLAS whose culmination was a simple explanation of the existence of the Monster, with properties of many smaller groups being covered on the way'. We hope that in its completion, Majorana theory might constitute the material for such an ideal ATLAS, where the Monster is characterized as the automorphism group of the largest Majorana algebra, which is the Monster algebra, and all 2A-generated subgroups of the Monster classified through the subalgebras generated by the relevant set of 2A-axes.

In [14] Jonathan Hall, Felix Rehren, and Sergey Shpectorov have relaxed certain Majorana axioms to obtain a class of axial algebras, which contains more examples including Jordan and Matsuo algebras, and exhibits a richer theory allowing the universal algebra construction and more.

We can now review the papers in the first part of this collection. The chapter by Atsushi Matsuo provides an introduction to the theory of Vertex Algebras and Vertex Operator Algebras. It illustrates, in much details, the most important examples of the algebras, including free bosons and lattice constructions. The survey both enables a newcomer to join the subject and provides an outsider the logic and starting content of the subject.

The chapter by Hiroshi Yamauchi describes the role of the theory and examples of 3-transposition groups in vertex operator algebras. These groups particularly appear from Matsuo algebras, which can be viewed as Majorana algebras missing the $\frac{1}{4}$-eigenspaces of Majorana axes. An important feature is that 3-transposition groups appear in a wider context of VOAs than the particular moonshine module associated with the Monster.

Ching Hung Lam is one of the leading and most active players in the almost 30-year-long attack on the Schellekens conjecture. In his chapter, he gives an account of the present state of proving this conjecture, which is very close to completion. All the 70 anticipated analogues of the moonshine module have been constructed by now. The technique of these constructions based on orbit-folding is clearly explained.

The chapter by Ulrich Meierfrankenfeld and Sergey Shpectorov is an almost classical and ingenious work on maximal 2-local subgroups in the Monster and Baby Monster and gives a unique insight into the local structure of these groups. Although written at the turn of the millennium and highly circulated, it was previously unpublished and we are happy to present it in this volume.

The chapter by Alexander A. Ivanov gives the account of the current status of Majorana theory. The Majorana representations of small and not so small groups are now classified, the main achievement being the classification of the saturate Majorana representation of A_{12}. At least two working computer packages for calculating Majorana representations are now available. This suggests the current strategy comprising classifying small sub-representations by a computer and assembling them together by hand.

The machinery for developing Majorana theory and axial algebras lies in the scope of Algebraic Combinatorics to which the second part of this collection is devoted. In particular, the central combinatorial object for both the theories is the association scheme of the Monster group acting by conjugation on the class of its $2A$-involutions. It is precisely through this scheme that the methods of algebraic combinatorics were originally used by Simon Norton to establish the uniqueness of the Monster. The procedure of recovering the Monster algebra from the structure of the centralizers of various elements encoded in the structure constants of the association scheme brought the term 'Norton algebras' into algebraic combinatorics [15]. We discuss some aspects of algebraic combinatorics also through relationships between examples, axiomatics, and theory illustrated by the chapters in Part II.

The understanding of the finite simple groups of Lie type started in the nineteenth century with linear groups and moved on to the classical groups, symplectic, orthogonal, and unitary, summarized in [16]. In the first half of the twentieth century exceptional groups of types G_2, E_6, and F_4 were constructed by A. A. Albert as the automorphism groups of exceptional Jordan algebras (now reincarnated in the form of axial algebras). At the beginning of the 1960s, these and other exceptional groups were axiomatized by Jacques Tits [17], [18] under the names of buildings and BN-pairs. The leading chapter of Part II, that by Hendrik Van Maldeghem, is a brilliant survey of some recent developments in the theory of buildings, geometries, and varieties related to the exceptional groups. Various exceptional geometries can be arranged in Freudenthal–Tits Magic Squares so that different series can be read off in both the rows and the columns. This square and its remarkable properties is at the centre of the chapter.

When geometries are axiomatized as buildings or even as point–line systems, they are no longer subspaces in any vector space but just abstract sets with incidence relations satisfying certain axioms. It appears fruitful to introduce the vector space structure by considering representations of geometries. A survey of this approach is given in the chapter by Ilaria Cardinali, Luca Giuzzi, and Antonio Pasini.

Given a point–line incidence system, one can define subspaces and hyperplanes, which are crucial in understanding representations. An ovoid is a subspace and the question of when it is maximal is discussed in the chapter by Antonio Pasini and Hendrik Van Maldeghem.

The representation theory is a very important tool in Majorana theory through algebraic combinatorics. So far, mainly ordinary representations appear, although modular ones are expected to come through modular Moonshine or otherwise. We are pleased to introduce a chapter in the collection written by leading experts in representation theory, Tatsiana S. Busel and Irina D. Suprunenko. They discuss rather delicate features of highest weight representations of algebraic groups. With sadness we report that Irina Dmitrievna Suprunrnko passed away on 10 August 2022.

The axiomatics of distance-regular graphs were introduced by Norman Biggs in the late 1960s [19]. By that time the classical generalized polygons were known, as were some sporadic examples including the Livingstone graph for the sporadic group J_1 of Janko. The theory of distance-regular graphs was at the centre of algebraic combinatorics in the 1970s and 1980s, which led to the celebrated monograph [20]. Jack H. Koolen is a leading researcher in the modern theory of distance-regular graphs. The chapter he wrote jointly with Jongyook Park and Qianqian Yang in the collection is on distance-regular graphs with classical parameters.

In the terminology of [15] the distance-regular graphs are precisely the P-polynomial association schemes. If the scheme is also Q-polynomial, then the usual matrix and Hadamard products are tri-diagonalizable in suitable bases. This property was axiomatized by Paul Terwilliger under the name of Terwilliger algebras. These algebras were studied by Paul together with Tatsuro Ito, and a survey on these algebras is in their chapter in the collection.

The theory of non-symmetric (analogues of) association schemes is less developed. A nice result in the chapter by Takuya Ikuta and Akihiro Munemasa contributes to this theory.

These days computer programs designed for calculations in algebraic combinatorics are often included in packages of the GAP system [21]. The GAP packages – GRAPE for graph theory, DESIGN for design theory, and FinInG for finite incidence geometry – described by Leonard Soicher in the concluding chapter, are of crucial importance.

The authors of the collection belong to a dynamic mathematical community. The editor has organized a number of meetings for this community starting with the 1991 conference in Vladimir, Russia (jointly with Igor Faradjev and Mikhail Klin). The latest was a hybrid 2021 conference in Rogla, Slovenia (jointly with Elena Konstantinova). Below is a photo from a London 2013 conference that shows many contributors to this collection.

The Editor

Participants of the conference 'Majorana Theory: the Monster and Beyond,' London, September 2013 (left to right, top to bottom): Chien Sheng Lim, Alonso Castello Ramines, Masahiko Miyamoto, Clara Franchi, Mario Mainardis, Alexander A. Ivanov, Sanhan Khasraw, Felix Rehren, Oliver Gray, Hiroshi Yamauchi, Leonard H. Soicher, Michael Tuite, Sophie Decelle, Hendrik Van Maldeghem, Simon Norton, Jonathan Hall, Antonio Pasini, Atsushi Matsuo, Hiroki Shimakura, Igor Faradjev, Yasuyuki Kawahigashi, Sergey Shpectorov, Peter Bantay, Ching Hung Lam, Ben Fairbrian, Terry Gannon, David Ghatei

References

[1] S. P. Norton, The Monster is fabulous, in *Finite Simple Groups: Thirty Years of the Atlas and Beyond*, Contemp. Math., **694**, pp. 3–10, AMS, Providence, RI, 2017.

[2] E. Witt, Über Steiner Systems Systeme, *Abh. Math. Semi. Univ. Hamburg* **12** (1938), 265–275.

[3] É. Mathieu, Mémoire sur l'étude des fonctions de plusieurs quantité, sur la manière des les former et sur les substitutions qui les laissent invariables, *J de Math. et App.* **6** (1861), 241–323.

[4] J. H. Conway and N. J. A. Sloane, *Sphere Packings, Lattices and Groups*, Grundlehren Math. Wiss., **290**, Springer, Berlin, 1988.

[5] B. Fischer, Finite groups generated by 3-transpositions, *Invent. Math.* **13** (1971), 232–246.

[6] J. G. Thompson, Uniqueness of the Fischer–Griess Monster, *Bull. London Math. Soc.* **11** (1979), 340–346.

[7] R. L. Griess, The friendly giant, *Invent. Math.* **69** (1982), 1–102.

[8] J. H. Conway, A simple construction for the Fischer–Griess Monster group, *Invent. Math.* **79** (1985), 513–540.

[9] J. H. Conway and S. P. Norton, Monstrous moonshine, *Bull. London. Math. Soc.* **11** (1979), 308–339.

[10] I. B. Frenkel, J. Lepowsky and A. Meurman, *Vertex Operator Algebras and the Monster*, Acadademic Press, Boston, 1988.

[11] R. E. Borcherds, Vertex algebras, Kac–Moody algebras, and the Monster, *Proc. Natl. Acad. Sci. USA* **83** (1986), 3068–3071.

[12] A. N. Schellekens, Meromorphic $c = 24$ conformal field theories, *Comm. Math. Phys.* **153** (1993), 159–185.

[13] A. A. Ivanov, *The Monster Group and Majorana Involutions*, Cambridge University Press, Cambridge, 2009.

[14] J. I. Hall, F. Rehren and S. Shpectorov, Universal axial algebras and a theorem of Sakuma, *J. Algebra* **421** (2015), 394–424.

[15] E. Bannai and T. Ito, *Algebraic Combinatorics I: Association Schemes*, Benjamin, Menlo Park, CA, 1984.

[16] L. E. Dickson, *Linear Groups with an Exposition of the Galois Field Theory*, Dover, New York, 1901.

[17] J. Tits, Algebraic and abstract simple groups, *Annals of Mathematics, 2nd Ser.* **80** (1964), 313–329.

[18] J. Tits, *Buildings of Spherical Type and Finite BN-pairs*. Lecture Notes in Mathematics, **386**, Springer, Berlin, 1974.

[19] N. L. Biggs, *Finite Groups of Automorphisms*, LMS Lecture Notes Series, **6**, Cambridge University Press, Cambridge, 1971.

[20] A. E. Brouwer, A. M. Cohen and A. Neumaier, *Distance-Regular Graphs*, Springer, Berlin, 1989.

[21] The GAP Group, *GAP – Groups, Algorithms, and Programming, Version 4.11.1;* 2021. (www.gap-system.org)

Part I

The Monster

1

Lectures on Vertex Algebras

Atsushi Matsuo

Abstract

The purpose of the present chapter is to explain the basics of vertex algebras, as well as some more advanced topics on vertex operator algebras, to the reader mainly in the fields of group theory and algebraic combinatorics.

CONTENTS

Introduction

The Monster, the largest sporadic finite simple group of order

$$2^{46} \cdot 3^{20} \cdot 5^9 \cdot 7^6 \cdot 11^2 \cdot 13^3 \cdot 17 \cdot 19 \cdot 23 \cdot 29 \cdot 31 \cdot 41 \cdot 47 \cdot 59 \cdot 71$$
$$= \underbrace{808017424794512875886459904961710757005754368000000000}_{54\ \text{digits}},$$

is known to be realized as the automorphism group of the moonshine module $V^\natural$, a distinguished example of a vertex operator algebra, equipped with a grading of the shape

$$\mathbf{V}^{\natural} = \mathbb{C}\mathbf{1} \oplus 0 \oplus \mathbf{B}^{\natural} \oplus \mathbf{V}_3^{\natural} \oplus \mathbf{V}_4^{\natural} \oplus \cdots ,$$

dim	1	0	196884

of which the dimensions of the homogeneous subspaces satisfy

$$q^{-1} \sum_{n=0}^{\infty} \dim \mathbf{V}_n^{\natural} q^n = j(\tau) - 744 \tag{1}$$
$$= q^{-1} + 0 + 196884q + 21493760q^2 + \cdots ,$$

where $j(\tau)$ is the elliptic modular function and $q = e^{2\pi\sqrt{-1}\tau}$.

The 196884-dimensional subspace $\mathbf{B}^{\natural}$ of degree 2 inherits a structure of a commutative nonassociative algebra with unity equipped with a nondegenerate symmetric invariant bilinear form, which we call the *Griess–Conway algebra*, as suggested by S. P. Norton. The algebra $\mathbf{B}^{\natural}$ is a variant of the algebras constructed by R. L. Griess in [61] to prove the existence of the Monster, and it is indeed the same as the algebra constructed by J. H. Conway in [38].

The notion of vertex algebras was introduced by R. E. Borcherds in the seminal paper [32] in 1986 by axiomatizing properties of infinite sequences of operators constructed from even lattices that generalize those considered for the root lattices of ADE type in the famous *Frenkel–Kac construction,* achieved by I. B. Frenkel and V. G. Kac in [57], to realize representations of affine Kac–Moody algebras associated with simple Lie algebras of the corresponding type. Such sequences of operators are related to the *vertex operators* in string theory, whence the term *vertex algebra*. The vertex operator is actually not a single operator but an infinite series with operator coefficients. The concept of vertex algebras can be seen to be a mathematical formulation of what is called the *operator product algebra* or the *chiral algebra* in physics.

Borcherds then applied vertex algebras to the study of the Monster via the moonshine module $\mathbf{V}^{\natural}$, which was previously introduced by I. B. Frenkel, J. Lepowsky, and A. Meurman [59] as a vector space equipped with some structures, and achieved in [33], with numerous outstanding ideas and works, the proof of the Conway–Norton conjecture, the conjecture that states the famous moonshine phenomena relating representations of the Monster and certain modular functions, the simplest among which is (1).

The concepts of vertex operator algebras (VOA) and their modules, in turn, were formulated by I. B. Frenkel, J. Lepowsky, and A. Meurman in [1] in order to set up appropriate "algebras" and "modules" by modifying those for vertex algebras. More precisely, a VOA is not just a vertex algebra, but a pair consisting of a vertex algebra and its element generating a representation of the Virasoro algebra satisfying a number of conditions that would make it suitable for applications.

Table 1 Codes, lattices and VOAs

Doubly even codes	Postive-definite even lattices	VOAs
Length	Rank	Central charge
Weight enumerator	Theta function	Conformal character
Self-dual	Unimodular	Holomorphic
Extended Hamming code H_8	Gosset lattice E_8	Lattice VOA V_{E_8}
Extended Golay code G_{24}	Leech lattice Λ	Moonshine module $V^\natural$
Mathieu group M_{24}	Conway group C_0	Monster $M = F_1$

For example, VOAs are assumed to be graded by integers with the homogeneous subspaces being finite-dimensional, so that one may consider the conformal character, the generating series of dimensions such as (1).

In fact, important applications of vertex algebras are often based on the properties of the Virasoro algebra, thus justifying the definition of VOAs.

The moonshine module $V^\natural$ indeed carries a natural structure of a VOA. It possesses a distinguished position among VOAs when viewed through the famous analogies of binary codes, lattices, and VOAs as indicated in Table 1, although the uniqueness of $V^\natural$ conjectured in [1], which is an analogue of the uniqueness of the extended Golay code G_{24} and the Leech lattice Λ, is yet to be settled. Thus the concept of VOAs is as natural as those of binary codes and lattices. However, even constructing a single example of a VOA is not so easy.

In Section 1.1, we will describe the definition of vertex algebras after preliminary sections, and then proceed to realization of vertex algebras by formal series with operator coefficients in Section 1.2, where the concept of modules over vertex algebras will also be introduced. Such realization enables us to state and prove the existence of vertex algebra structures under certain circumstances. Standard examples of vertex algebras will be described in Section 1.3.

Section 1.4 is devoted to construction of the vertex algebras associated with even lattices, where commutation relations of vertex operators play fundamental roles. In Section 1.5, we will explain the definition and construction of what are called *twisted modules* over vertex algebras by repeating the arguments of the previous sections in slightly more general settings, which enables one to construct the moonshine module $V^\natural$ as a module over a fixed-point subalgebra of the Leech lattice vertex algebra by a lift of the (-1)-involution.

In Section 1.6, we will give brief accounts of theory of VOAs including fusion rules and modular invariance. We will then finish the sections by mentioning properties of the moonshine module and their variants that opened ways to new research directions.

The author is grateful to Professors Alexander A. Ivanov and Elena V. Konstantinova for inviting him to give the series of lectures in G2G2 2021 at Rogla. It was a hard task, to be honest, but very much fruitful indeed. The lectures were actually given online from Tokyo, and the author wishes to visit Rogla sometime in the future.

The author thanks Takuro Abe, Tomoyuki Arakawa, Hiroki Shimakura and Hiroshi Yamauchi for useful conversations in preparation of the manuscripts for the lectures and the referee for useful comments. The present sections are partly based on the author's past lectures at Nagoya Institute of Technology, National Taiwan University, University of the Ryukyus in 2003 etc.

1.1 Axioms for Vertex Algebras

A vertex algebra is a vector space equipped with countably many binary operations indexed by integers satisfying a number of axioms.

In Section 1.1, we start with preliminary sections on algebras and formal series and then describe the definition of vertex algebras and some consequences of the axioms. We will give a few examples: the commutative vertex algebras, the Heisenberg vertex algebra, and a Virasoro vertex algebra as a vertex subalgebra of the Heisenberg vertex algebra.

We will work over a field $\mathbb{F}$ of any characteristic not 2, thus vector spaces and linear maps are always over such a field $\mathbb{F}$, unless otherwise stated. We denote the set of integers by $\mathbb{Z}$ and that of nonnegative integers by $\mathbb{N}$.

1.1.1 Preliminaries on Algebras

For a vector space $\mathbf{M}$, consider the set End $\mathbf{M}$ of all operators (endomorphisms) acting on $\mathbf{M}$. The symbol $I = I_{\mathbf{M}}$ refers to the identity operator.

For an operator $A \in \text{End}\,\mathbf{M}$, we will denote the value of A at $v \in \mathbf{M}$ by juxtaposition:

$$A: \mathbf{M} \longrightarrow \mathbf{M}, \quad v \mapsto Av.$$

Compositions of operators, also written by juxtaposition, are taken from right to left unless specified by parentheses: for $A, B, C \in \text{End}\,\mathbf{M}$ and $v \in \mathbf{M}$,

$$ABC = A(BC), \quad ABCv = A(B(Cv)), \quad \text{etc.}$$

The commutator of operators is denoted by the bracket as

$$[A, B] = AB - BA$$

for $A, B \in \text{End}\,\mathbf{M}$.

1.1.1.1 Associative Algebras

Let us first recall the definition of associative algebras. We will always assume that associative algebras are unital.

An *associative algebra* is a vector space **A** equipped with a bilinear map

$$\mathbf{A} \times \mathbf{A} \longrightarrow \mathbf{A}, \quad (a,b) \mapsto ab,$$

called *multiplication* or the *product operation*, satisfying the following axioms:

(A1) Associativity. For all $a,b,c \in \mathbf{A}$:

$$(ab)c = a(bc).$$

(A2) Unity. There exists an element $\mathbf{1} \in \mathbf{A}$ such that for all $a \in \mathbf{A}$:

$$1a = a \quad \text{and} \quad a1 = a.$$

The element $\mathbf{1} \in \mathbf{A}$ in (A2) is uniquely determined by the conditions therein and called the *unity* of **A**,

For a vector space **M**, the set End **M** of all operators acting on **M** becomes an associative algebra by composition of operators, of which the unity is the identity operator.

1.1.1.2 Modules over Associative Algebras

A *module* over **A**, or an **A**-*module,* is a vector space **M** equipped with a bilinear map

$$\mathbf{A} \times \mathbf{M} \longrightarrow \mathbf{M}, \quad (a,v) \mapsto av,$$

called an *action* of **A** on **M**, satisfying

(AM1) Associativity. For all $a,b \in \mathbf{A}$ and $v \in \mathbf{M}$:

$$(ab)v = a(bv).$$

(AM2) Identity. For all $v \in \mathbf{M}$: $\quad 1v = v.$

For $a \in \mathbf{A}$, the operator on **M** sending v to av is called the *action* of a on **M**.

For an **A**-module **M**, consider the map assigning the action on **M** to each element of **A**:

$$\rho_{\mathbf{M}} : \mathbf{A} \longrightarrow \text{End}\,\mathbf{M}, \quad a \mapsto [v \mapsto av].$$

Then this map is a homomorphism of algebras. Such a homomorphism is called a *representation* of **A** on **M**. The concepts of modules over **A** and representations of **A** are essentially the same.

The algebra **A** itself becomes an **A**-module by the product operation, for which the *left action* of $a \in \mathbf{A}$ sending x to ax is called *left multiplication* by a. The corresponding representation

$$\rho_{\mathbf{A}} : \mathbf{A} \longrightarrow \operatorname{End} \mathbf{A}, \quad a \mapsto [x \mapsto ax]$$

is an isomorphism of algebras onto its image.

1.1.1.3 Lie Algebras

A *Lie algebra* is a vector space **L** equipped with a bilinear map

$$[\ ,\ ] : \mathbf{L} \times \mathbf{L} \longrightarrow \mathbf{L}, \quad (X,Y) \mapsto [X,Y],$$

called the *bracket operation*, satisfying

(1) For all $X,Y,Z \in \mathbf{L}$:

$$[X,[Y,Z]] + [Y,[Z,X]] + [Z,[X,Y]] = 0.$$

(2) For all $X \in \mathbf{L}$:

$$[X,X] = 0.$$

As the base field is assumed to be not of characteristic 2, the set of the two conditions is equivalently replaced by

(L1) Jacobi identity. For all $X,Y,Z \in \mathbf{L}$:

$$[[X,Y],Z] = [X,[Y,Z]] - [Y,[X,Z]].$$

(L2) Antisymmetry. For all $X,Y \in \mathbf{L}$:

$$[X,Y] = -[Y,X].$$

Throughout the sections, we will take the latter conditions (L1) and (L2) as the axioms for Lie algebras and call the identity in (L1) the *Jacobi identity*, although this term usually refers to (1) rather than (L1).

For a vector space **M**, the space $\operatorname{End} \mathbf{M}$ becomes a Lie algebra by the commutator of operators, for which the Jacobi identity

$$[[A,B],C] = [A,[B,C]] - [B,[A,C]], \quad A,B,C \in \operatorname{End} \mathbf{M}$$

trivially holds by cancellation of terms in

$$(ABC - BAC) - (CAB - CBA)$$
$$= ((ABC - ACB) - (BCA - CBA))$$
$$\quad - ((BAC - BCA) - (ACB - CAB)).$$

A variant of this simple observation will serve as a basis for the Borcherds identity, the main identity for vertex algebras, where A, B, C are replaced by series with operator coefficients. (See Subsection 1.2.3.1.)

Similarly, any associative algebra $\mathbf{A}$ is regarded as a Lie algebra by the commutator

$$[a, b] = ab - ba, \quad a, b \in \mathbf{A}.$$

We will denote this Lie algebra by $\mathbf{L}(\mathbf{A})$.

Note 1.1. A vector space $\mathbf{L}$ equipped with a bracket operation satisfying (L1) but not necessarily (L2) is called a (left) *Leibniz algebra* and the property (L1) is called the (left) *Leibniz identity*. Note that (L1) is equivalently written as

$$[X, [Y, Z]] = [[X, Y], Z] + [Y, [X, Z]],$$

which says that the operations of taking the brackets by elements of $\mathbf{L}$ are derivations with respect to the bracket operation itself.

1.1.1.4 Modules over Lie Algebras

An $\mathbf{L}$-*module*, or a *module* over $\mathbf{L}$, is a vector space $\mathbf{M}$ equipped with a bilinear map

$$\mathbf{L} \times \mathbf{M} \longrightarrow \mathbf{M}, \quad (X, v) \mapsto Xv,$$

satisfying

(LM) For all $X, Y \in \mathbf{L}$ and $v \in \mathbf{M}$:

$$[X, Y]v = X(Yv) - Y(Xv).$$

For an $\mathbf{L}$-module $\mathbf{M}$, consider the map assigning the corresponding action on $\mathbf{M}$ to each element of $\mathbf{L}$:

$$\rho_{\mathbf{M}} \colon \mathbf{L} \longrightarrow \operatorname{End} \mathbf{M}, \quad X \mapsto [v \mapsto Xv].$$

Then this map is a homomorphism of Lie algebras. Such a homomorphism is called a *representation* of $\mathbf{L}$ on $\mathbf{M}$. The concepts of modules over $\mathbf{L}$ and representations of $\mathbf{L}$ are essentially the same.

The Lie algebra $\mathbf{L}$ itself becomes an $\mathbf{L}$-module by the bracket operation, for which the action of $X \in \mathbf{L}$ sending Y to $[X, Y]$ is called the *adjoint action* of X, and the corresponding representation

$$\rho_{\mathbf{L}} \colon \mathbf{L} \longrightarrow \operatorname{End} \mathbf{L}, \quad X \mapsto [Y \mapsto [X, Y]],$$

is called the *adjoint representation* of $\mathbf{L}$.

Let ρ be a representation of **L** on a vector space **M**:

$$\rho : \mathbf{L} \longrightarrow \text{End }\mathbf{M}, \quad \rho(X) : v \mapsto Xv.$$

For $X_1, X_2, \ldots, X_d \in \mathbf{L}$, the product $X_1 X_2 \cdots X_d$ makes sense in End **M** as

$$(X_1 X_2 \cdots X_d)v = X_1 X_2 \cdots X_d v,$$

but such a product $X_1 X_2 \cdots X_d$ does not make sense in the Lie algebra **L**.

The universal enveloping algebra resolves this inconvenience by collecting expressions of the form $X_1 X_2 \cdots X_d$ subject to appropriate relations. We will give the precise formulation in the next subsection.

1.1.1.5 Universal Enveloping Algebras

Let **L** be a Lie algebra and consider the tensor algebra over **L**,

$$\mathbf{T}(\mathbf{L}) = \bigoplus_{d=0}^{\infty} \mathbf{T}^d(\mathbf{L}), \quad \mathbf{T}^d(\mathbf{L}) = \underbrace{\mathbf{L} \otimes \cdots \otimes \mathbf{L}}_{d \text{ times}}.$$

We will identify the elements of $\mathbf{T}^0(\mathbf{L})$ with the scalars.

Let $\mathbf{U}(\mathbf{L})$ be the quotient of $\mathbf{T}(\mathbf{L})$ by the two-sided ideal $\mathbf{J}(\mathbf{L})$ generated by the elements of the form

$$X \otimes Y - Y \otimes X - [X,Y], \quad X,Y \in \mathbf{L}.$$

We will denote the image of $X_1 \otimes \cdots \otimes X_d$ in $\mathbf{U}(\mathbf{L})$ by $X_1 \cdots X_d$.

Let j be the canonical map which sends $X \in \mathbf{L}$ to its image in $\mathbf{U}(\mathbf{L})$:

$$j : \mathbf{L} \longrightarrow \mathbf{U}(\mathbf{L}) = \mathbf{T}(\mathbf{L})/\mathbf{J}(\mathbf{L}).$$

The associative algebra $\mathbf{U}(\mathbf{L})$ equipped with the map j is called the *universal enveloping algebra* of the Lie algebra **L**, which is characterized by the following universal property:

For any associative algebra **A** and any homomorphism $\varphi : \mathbf{L} \to \mathbf{L}(\mathbf{A})$ of Lie algebras, there exists a unique homomorphism of associative algebras $\psi : \mathbf{U}(\mathbf{L}) \to \mathbf{A}$ such that the diagram

$$\begin{array}{ccc} \mathbf{L} & \xrightarrow{\varphi} & \mathbf{L}(\mathbf{A}) \\ {\scriptstyle j}\downarrow & & \| \\ \mathbf{U}(\mathbf{L}) & \xrightarrow{\psi} & \mathbf{A} \end{array}$$

commutes.

Considering the case when $A = \text{End } M$ for a vector space M, we see that giving an L-module structure on M is equivalent to giving a $U(L)$-module structure on M:

$$L\text{-modules} \longleftrightarrow U(L)\text{-modules}.$$

The structure of the universal enveloping algebra as a vector space is described by the following theorem, called *Poincaré–Birkhoff–Witt theorem*, or PBW for short.

Theorem 1.2 (PBW) *Let L be a Lie algebra and B a totally ordered basis of L. Then the elements of the set*

$$\{ X_1 \cdots X_k \mid k \in \mathbb{N},\ X_1, \ldots, X_k \in B,\ X_1 \leq \cdots \leq X_k \}$$

form a basis of $U(L)$.

In particular, it follows that the canonical map $j : L \rightarrow U(L)$ is injective, and the representation

$$L \longrightarrow \text{End}\, U(L),$$

given by left multiplication, is an isomorphism of Lie algebras onto its image.

When the Lie algebra L is commutative, the algebra $U(L)$ reduces to the symmetric algebra $S(L)$ over the vector space L.

1.1.2 Preliminaries on Formal Series

We will substantially work with formal series with operator coefficients. Let us summarize notations and basic properties of formal series in advance.

The formal series we will be dealing with are series consisting of infinitely many terms of both positive and negative degrees. We will simply call such a formal series a *series* for short.

1.1.2.1 Spaces of Formal Series

Let z be an indeterminate, V a vector space, and $v(z)$ a series with coefficients in V. Throughout the text, unless otherwise stated, the coefficients of a series $v(z)$ are indexed as in

$$v(z) = \sum_n v_n z^{-n-1},$$

where the summation is over all $n \in \mathbb{Z}$. The set of such series is denoted as

$$V[[z, z^{-1}]] = \left\{ \sum_n v_n z^{-n-1} \,\middle|\, v_n \in V \text{ for all } n \in \mathbb{Z} \right\}.$$

Recall the following spaces of series of specific types:

$$V[[z]] = \left\{ \sum_n v_n z^{-n-1} \,\middle|\, v_n = 0 \text{ for all } n \geq 0 \right\},$$

$$V((z)) = \left\{ \sum_n v_n z^{-n-1} \,\middle|\, \exists N \in \mathbb{N} \colon v_n = 0 \text{ for all } n \geq N \right\},$$

$$V[z, z^{-1}] = \left\{ \sum_n v_n z^{-n-1} \,\middle|\, \exists N \in \mathbb{N} \colon v_n = 0 \text{ unless } -N \leq n \leq N \right\},$$

$$V[z] = \left\{ \sum_n v_n z^{-n-1} \,\middle|\, \exists N \in \mathbb{N} \colon v_n = 0 \text{ unless } -N \leq n < 0 \right\}.$$

Their elements are, respectively, called *formal power series, formal Laurent series, Laurent polynomials,* and *polynomials.* We may write

$$V((z)) = V[[z]][z^{-1}], \quad V[z, z^{-1}] = V[z][z^{-1}], \quad \text{etc.}$$

A bilinear map $U \times V \longrightarrow W$ induces bilinear maps on series such as

$$U((z)) \times V((z)) \longrightarrow W((z)), \quad U[z, z^{-1}] \times V[[z, z^{-1}]] \longrightarrow W[[z, z^{-1}]],$$

by the product

$$\sum_m u_m z^{-m-1} \sum_n v_n z^{-n-1} = \sum_{m,n} u_m v_n z^{-m-n-2}.$$

We will also consider series in many indeterminates. For example,

$$V((y))((z)) = \left\{ \sum_n v_n(y) z^{-n-1} \,\middle|\, \begin{array}{l} v_n(y) \in V((y)) \text{ for all } n \text{ and} \\ \exists N \geq 0 \colon v_n(y) = 0 \text{ for all } n \geq N \end{array} \right\},$$

$$V((z))((y)) = \left\{ \sum_m v_m(z) y^{-m-1} \,\middle|\, \begin{array}{l} v_m(z) \in V((z)) \text{ for all } m \text{ and} \\ \exists M \geq 0 \colon v_m(z) = 0 \text{ for all } m \geq M \end{array} \right\}.$$

We will write

$$V((y, z)) = V((y))((z)) \cap V((z))((y)).$$

The three spaces do not agree unless $V = 0$.

Note 1.3. A series of the form $v(z) = \sum_n v_n z^{-n-1}$ is seen to be a formal *Fourier series* by substitution $z = e^{2\pi i t}$. In this regard, the coefficients are sometimes called the *Fourier coefficients* or even the *Fourier modes* of the series.

1.1.2.2 Binomial Expansions

Let x, y, z, etc. be indeterminates. The binomial theorem states that, for non-negative integer n,

$$(x + z)^n = \sum_{i=0}^n \binom{n}{i} x^i z^{n-i} = \sum_{i=0}^n \binom{n}{i} x^{n-i} z^i.$$

We may consider similar expansions for negative powers, but then the two expansions become different. We will distinguish them by writing

$$(x+z)^n\big|_{|x|>|z|} = \sum_{i=0}^{\infty} \binom{n}{i} x^{n-i} z^i,$$

$$(x+z)^n\big|_{|x|<|z|} = \sum_{i=0}^{\infty} \binom{n}{i} x^i z^{n-i}.$$

We understand that the left-hand sides denote the formal series given by the right-hand sides, respectively. We thus have

$$(x+z)^n\big|_{|x|>|z|} \in \mathbb{F}[x,x^{-1}][[z]] \subset \mathbb{F}((x))((z)),$$

$$(x+z)^n\big|_{|x|<|z|} \in \mathbb{F}[z,z^{-1}][[x]] \subset \mathbb{F}((z))((x)).$$

The regions attached signify where the series are convergent when working over $\mathbb{C}$. Similarly, we write

$$(y-z)^n\big|_{|y|>|z|} = \sum_{i=0}^{\infty} (-1)^i \binom{n}{i} y^{n-i} z^i,$$

$$(y-z)^n\big|_{|y|<|z|} = \sum_{i=0}^{\infty} (-1)^{n-i} \binom{n}{i} y^i z^{n-i},$$

which belong to $\mathbb{F}((y))((z))$ and $\mathbb{F}((z))((y))$, respectively.

Note 1.4.   These expansions are often written in the literatures as

$$\iota_{y,z}(y-z)^n = \sum_{i=0}^{\infty} (-1)^i \binom{n}{i} y^{n-i} z^i,$$

$$\iota_{z,y}(y-z)^n = \sum_{i=0}^{\infty} (-1)^{n-i} \binom{n}{i} y^i z^{n-i}.$$

It is also common to distinguish them by the order of the summands in the argument as $(y-z)^n$ and $(-z+y)^n$, although sometimes confusing.

1.1.2.3 Divided Derivatives of Series
Consider the operators $\partial_z^{(k)}$ acting on series in z defined for $k \in \mathbb{N}$ as

$$\partial_z^{(k)}: V[[z,z^{-1}]] \longrightarrow V[[z,z^{-1}]], \quad v(z) \mapsto \partial_z^{(k)} v(z),$$

where $\partial_z^{(k)} v(z)$ denotes the kth *divided derivative*:

$$\partial_z^{(k)} \sum_n v_n z^{-n-1} = \sum_n \binom{-n-1}{k} v_n z^{-n-k-1}$$

$$= \sum_n \binom{-n+k-1}{k} v_{n-k} z^{-n-1}.$$

We will omit the subscript z in $\partial_z^{(k)}$ if there is no danger of confusion.

The operators $\partial^{(k)}$ are iterative in the following sense:

$$\partial^{(i)}\partial^{(j)} = \binom{i+j}{i}\partial^{(i+j)} \quad \text{for all } i, j \in \mathbb{N}.$$

They annihilate constants:

$$\partial^{(k)}v = 0 \quad \text{for all } k \geq 1 \text{ and } v \in V.$$

They satisfy the Leibniz rule

$$\partial^{(k)}(u(z)v(z)) = \sum_{i+j=k} (\partial^{(i)}u(z))(\partial^{(j)}v(z)),$$

as long as the products make sense.

For an indeterminate x, define a formal power series $e^{x\partial_z}$ with operator co-efficients by

$$e^{x\partial_z} = \sum_{k=0}^{\infty} x^k \partial_z^{(k)}.$$

Then it acts on a series $v(z)$ in the following sense:

$$e^{x\partial_z}v(z) = \sum_{k=0}^{\infty} x^k \partial_z^{(k)}v(z) = v(x+z)\big|_{|x|<|z|} \in V[[z, z^{-1}]][[x]]. \tag{1.1}$$

This is seen to be a formal analogue of Taylor expansion of $v(y)$ at $y = z$ giving a power series in $x = y - z$. Note that the Leibniz rule can be restated as

$$e^{x\partial_z}(u(z)v(z)) = (e^{x\partial_z}u(z))(e^{x\partial_z}v(z)).$$

Over a field of characteristic zero, we have

$$e^{x\partial_z} = \sum_{k=0}^{\infty} \frac{(x\partial_z)^k}{k!}, \quad \text{where } \partial_z = \frac{\partial}{\partial z},$$

which justifies the notation.

1.1.2.4 Formal Delta Functions

Let us next consider the *formal delta function*, which is a series defined by

$$\delta(x) = \sum_{n} x^n.$$

We will often encounter the following series of two indeterminates:

$$\delta(y, z) = z^{-1}\delta(y/z) = \sum_{n} y^n z^{-n-1} = \sum_{n} y^{-n-1}z^n.$$

Notice the following formula:

$$\delta(y, z) = \sum_{i=0}^{\infty} y^{-i-1}z^i + \sum_{i=0}^{\infty} y^i z^{-1-i} = \frac{1}{y-z}\Big|_{|y|>|z|} - \frac{1}{y-z}\Big|_{|y|<|z|}.$$

Let $\delta^{(k)}(y,z)$ denote the kth divided derivative of $\delta(y,z)$ with respect to z:

$$\delta^{(k)}(y,z) = \partial_z^{(k)}\delta(y,z) = \left.\frac{1}{(y-z)^{k+1}}\right|_{|y|>|z|} - \left.\frac{1}{(y-z)^{k+1}}\right|_{|y|<|z|}.$$

The relation

$$(y-z)^{k+1}\delta^{(k)}(y,z) = 0$$

holds for all $k \in \mathbb{N}$.

Note 1.5. The formal delta function $\delta(x)$ is seen to be the Fourier series expansion of a periodic analogue of the Dirac delta function up to a scalar factor.

1.1.3 Vertex Algebras

There are many equivalent ways to define vertex algebras. Here we pick up the one given by Borcherds in [33] and include the identity property. The resulting set of axioms is, in the author's opinion, the most natural.

To begin with, let us recall that an associative algebra $\mathbf{C}$ is said to be *commutative* if the multiplication is *symmetric*, that is, $ab = ba$ holds for all $a, b \in \mathbf{C}$. Since associative algebras are unital by assumption, we may replace symmetry by commutativity of left multiplication, that is, $a(bc) = b(ac)$ for $a, b, c \in \mathbf{C}$.

We may therefore define a commutative associative algebra alternatively by saying that it is a vector space $\mathbf{C}$ equipped with a bilinear map

$$\mathbf{C} \times \mathbf{C} \longrightarrow \mathbf{C}, \quad (a,b) \mapsto ab,$$

satisfying

(C1) Commutativity and associativity. For all $a, b, c \in \mathbf{C}$:

$$a(bc) = b(ac) \quad \text{and} \quad (ab)c = a(bc).$$

(C2) Unity. There exists an element $\mathbf{1} \in \mathbf{C}$ such that, for all $a \in \mathbf{C}$,

$$\mathbf{1}a = a \quad \text{and} \quad a\mathbf{1} = a.$$

Note that associativity follows from commutativity under the presence of unity.

1.1.3.1 Definition of Vertex Algebras

Let $\mathbf{V}$ be a vector space equipped with countably many bilinear maps indexed by integers n as

$$\mathbf{V} \times \mathbf{V} \longrightarrow \mathbf{V}, \quad (a,b) \mapsto a_{(n)}b.$$

We will call $a_{(n)}b$ the *nth product* for each n.

A *vertex algebra* is a vector space $\mathbf{V}$ equipped with such product operations satisfying the following axioms:

(V0) Local truncation. For any $a, b \in \mathbf{V}$, there exists an $N \in \mathbb{N}$ such that

$$a_{(N+i)}b = 0 \text{ for all } i \geq 0.$$

(V1) Borcherds identity. For all $a, b, c \in \mathbf{V}$ and $p, q, r \in \mathbb{Z}$:

$$\sum_{i=0}^{\infty} \binom{p}{i} (a_{(r+i)}b)_{(p+q-i)}c$$

$$= \sum_{i=0}^{\infty} (-1)^i \binom{r}{i} a_{(p+r-i)}(b_{(q+i)}c) - \sum_{i=0}^{\infty} (-1)^{r-i} \binom{r}{i} b_{(q+r-i)}(a_{(p+i)}c).$$

(V2) Vacuum. There exists an element $\mathbf{1} \in \mathbf{V}$ satisfying

1. Identity. For any $a \in \mathbf{V}$ and $n \in \mathbb{Z}$:

$$\mathbf{1}_{(n)}a = \begin{cases} 0 & (n \neq -1), \\ a & (n = -1). \end{cases}$$

2. Creation. For any $a \in \mathbf{V}$ and $n \in \mathbb{Z}_{\geq -1}$:

$$a_{(n)}\mathbf{1} = \begin{cases} 0 & (n \geq 0), \\ a & (n = -1). \end{cases}$$

Here are remarks on the axioms.

1. The three sums in the Borcherds identity in (V1) are finite sums by (V0). We therefore assume (V0) without a mention in referring to (V1).
2. The element $\mathbf{1}$ in (V2) is called the *vacuum* of $\mathbf{V}$, which is uniquely determined as it is a unity with respect to the (-1)st product.
3. The products $a_{(n)}\mathbf{1}$ with $n \leq -2$ are not specified in (V2), but their properties are encoded in the operators $T^{(k)}: \mathbf{V} \to \mathbf{V}$ defined by $T^{(k)}a = a_{(-k-1)}\mathbf{1}$ for $k = 0, 1, 2, \ldots$, called the *translation operators*, as described later in Subsection 1.1.3.4.
4. The identity property in (V2) in fact follows from the other axioms.

The concepts of subalgebras, homomorphisms, isomorphisms, ideals, and quotients, etc. are defined in obvious ways. A subalgebra of a vertex algebra is called a *vertex subalgebra*. For modules over vertex algebras, see Section 1.2.5.

The axioms (V1) and (V2) can be seen to be modelled on properties of series with operator coefficients under some assumptions on the set of series. See Section 1.2.3 for details.

Note 1.6.   1. Local truncation is usually called *truncation* in the literatures.
2. For each $a \in \mathbf{V}$ and $n \in \mathbb{Z}$, consider the action $a_{(n)}: x \mapsto a_{(n)}x$. Then the countably many product operations are collectively treated by the generating series $Y(a, z) = \sum_{n} a_{(n)}z^{-n-1}$. See Section 1.1.5 for details.

1.1.3.2 Structure of the Borcherds Identity

Although the Borcherds identity looks extremely complicated, it includes important properties as special cases. Here are a few instances.

1. The Borcherds identity (V1) with $(p, q, r) = (0, 0, 0)$ reads

$$(a_{(0)}b)_{(0)}c = a_{(0)}(b_{(0)}c) - b_{(0)}(a_{(0)}c),$$

   which is the Jacobi identity (L1) for Lie algebras with respect to the bracket given by the 0th product as $[a, b] = a_{(0)}b$.

2. If $a_{(n)}b = 0$ and $a_{(n)}c = 0$ hold for all $n \geq 0$, then the Borcherds identities with $(p, q, r) = (0, -1, -1)$ and $(p, q, r) = (-1, -1, 0)$, respectively, read

$$(a_{(-1)}b)_{(-1)}c = a_{(-1)}(b_{(-1)}c), \quad a_{(-1)}(b_{(-1)}c) = b_{(-1)}(a_{(-1)}c),$$

   which is the axiom (C1) for commutative associative algebras with respect to the (-1)st product.

Thus the Borcherds identity can be viewed as an "enhancement" of the Jacobi identity for Lie algebras and an "extension" of associativity and commutativity for commutative associative algebras.

There are redundancies in the Borcherds identity. Let $B(p, q, r)$ be either of the three sums in (V1):

$$B(p, q, r) = \sum_{i=0}^{\infty} \binom{p}{i} (a_{(r+i)}b)_{(p+q-i)}c, \quad \sum_{i=0}^{\infty} (-1)^i \binom{r}{i} a_{(p+r-i)}(b_{(q+i)}c),$$

$$\text{or } \sum_{i=0}^{\infty} (-1)^{r-i} \binom{r}{i} b_{(q+r-i)}(a_{(p+i)}c).$$

Then the following recurrence relation holds for all $p, q, r \in \mathbb{Z}$:

$$B(p + 1, q, r) = B(p, q + 1, r) + B(p, q, r + 1). \tag{1.2}$$

This implies the following lemma.

Lemma 1.7 *If the Borcherds identity holds for some p and all q, r and for some r and all p, q, then it holds for all p, q, r.*

Note 1.8. A vertex algebra equipped with the 0th product need not be a Lie algebra since skew-symmetry (L2) may not hold (cf. Notes 1.1 and 1.10).

1.1.3.3 Commutator and Associativity Formulas

The Borcherds identity with $(p, q, r) = (m, n, 0)$ reads

$$\sum_{i=0}^{\infty} \binom{m}{i} (a_{(i)}b)_{(m+n-i)}c = a_{(m)}(b_{(n)}c) - b_{(n)}(a_{(m)}c).$$

Therefore, the following property holds in any vertex algebra **V**.

(VC) Commutator formula. For all $a, b \in \mathbf{V}$ and $m, n \in \mathbb{Z}$:

$$[a_{(m)}, b_{(n)}] = \sum_{i=0}^{\infty} \binom{m}{i} (a_{(i)}b)_{(m+n-i)} .$$

The Borcherds identity with $(p, q, r) = (0, n, m)$ is as follows.

(VA) Associativity formula. For all $a, b, c \in \mathbf{V}$ and $m, n \in \mathbb{Z}$:

$$(a_{(m)}b)_{(n)}c = \sum_{i=0}^{\infty}(-1)^i \binom{m}{i} a_{(m-i)}b_{(n+i)}c - \sum_{i=0}^{\infty}(-1)^{m-i}\binom{m}{i} b_{(m+n-i)}a_{(i)}c.$$

By Lemma 1.7, the Borcherds identity holds if and only if both the commutator formula and the associativity formula hold:

$$(V1) \iff (VC) + (VA).$$

As an application of the commutator formula, consider the subspace of End **V** spanned by the left actions of elements of a vertex algebra **V**:

$$\text{Span}\left\{a_{(n)} \,\middle|\, a \in \mathbf{V}, n \in \mathbb{Z}\right\} \subset \text{End}\,\mathbf{V}.$$

Then (VC) implies that this space is closed under taking commutators, thus forms a Lie subalgebra of End **V**.

As for the associativity formula, let $\langle \mathbf{S} \rangle_{\mathrm{VA}}$ denote the vertex subalgebra *generated* by a subset **S** of a vertex algebra, that is, the span of the elements obtained by repeatedly applying the product operations to elements of **S** in arbitrary order. Then, by (VA) and (V2), it is actually given by left actions as

$$\langle \mathbf{S} \rangle_{\mathrm{VA}} = \text{Span}\left\{a^1_{(n_1)} \cdots a^k_{(n_k)}\mathbf{1} \,\middle|\, k \in \mathbb{N}, a^1, \ldots, a^k \in \mathbf{S}, n_1, \ldots, n_k \in \mathbb{Z}\right\}.$$

We understand that application zero time gives the vacuum **1**.

Note 1.9. 1. The associativity formula in the sense above is called the *iterate formula* or the *associator formula* in the literatures. 2. The Lie algebra spanned by the left actions of a vertex algebra is often called the *Lie algebra of Fourier modes* (cf. Subsection 1.6.3.1).

1.1.3.4 Translation Operators

For any vertex algebra **V**, canonically associated are the operators

$$T^{(k)} : \mathbf{V} \longrightarrow \mathbf{V}, \quad k = 0, 1, 2, \cdots$$

defined by setting, for $a \in \mathbf{V}$,

$$T^{(k)}a = a_{(-k-1)}\mathbf{1}.$$

These operators are called the *translation operators* or the *derivations* of the vertex algebra **V**. Note that $T^{(0)} = I$ is just the identity operator.

By these operators, the creation property in (V2) is completed as

$$a_{(n)}\mathbf{1} = \begin{cases} 0 & (n \ge 0), \\ T^{(k)}a & (n = -k - 1 < 0). \end{cases}$$

On the other hand, the identity property in (V2) implies

$$T^{(k)}\mathbf{1} = \begin{cases} 0 & (k \ge 1), \\ 1 & (k = 0). \end{cases}$$

The following properties are consequences of the Borcherds identity.

(VT) Translation. For all $a, b \in \mathbf{V}$, $n \in \mathbb{Z}$ and $k \in \mathbb{N}$:

$$(T^{(k)}a)_{(n)}b = (-1)^k \binom{n}{k} a_{(n-k)}b.$$

(VL) Leibniz rule. For all $a, b \in \mathbf{V}$, $n \in \mathbb{Z}$ and $k \in \mathbb{N}$:

$$T^{(k)}(a_{(n)}b) = \sum_{i+j=k} (T^{(i)}a)_{(n)}(T^{(j)}b).$$

(VI) Iterativity. For all $i, j \in \mathbb{N}$:

$$T^{(i)}T^{(j)} = \binom{i+j}{i} T^{(i+j)}.$$

Over a field of characteristic zero, (VI) implies

$$T^{(k)} = \frac{T^k}{k!} \quad \text{for } T = T^{(1)},$$

and the properties (VT) and (VL) for all $k \in \mathbb{N}$ follow from those with $k = 1$.

1.1.3.5 Skew-Symmetry

The following property follows from the axioms (V1) and (V2).

(VS) Skew-symmetry. For all $a, b \in \mathbf{V}$ and $n \in \mathbb{Z}$:

$$a_{(n)}b = (-1)^{n+1} \sum_{i=0}^{\infty} (-1)^i T^{(i)}(b_{(n+i)}a).$$

We may view this property as a counterpart in vertex algebras of symmetry and antisymmetry for commutative associative algebras and Lie algebras, respectively, depending on the parity of n.

To be more precise, consider the following subspace of $\mathbf{V}$:

$$T^{(\geq 1)}\mathbf{V} = \sum_{k=1}^{\infty} T^{(k)}\mathbf{V}.$$

Then, picking up the term with $i = 0$ in (VS), we have

$$a_{(n)}b \equiv (-1)^{n+1} b_{(n)}a \mod T^{(\geq 1)}\mathbf{V}.$$

Therefore, modulo the subspace $T^{(\geq 1)}\mathbf{V}$, the nth product of a vertex algebra $\mathbf{V}$ is symmetric for odd n and antisymmetric for even n.

Note 1.10. 1. The 0th product of a vertex algebra $\mathbf{V}$ satisfies the Jacobi identity (L1) on $\mathbf{V}$, while the skew-symmetry (L2) holds modulo $T^{(\geq 1)}\mathbf{V}$, and the quotient $\mathbf{V}/T^{(\geq 1)}\mathbf{V}$ indeed becomes a Lie algebra. 2. If the base field is of characteristic zero, then the subspace $T^{(\geq 1)}\mathbf{V}$ agrees with the image of the translation operator $T = T^{(1)}$. Moreover, if T agrees with the left action of an element of $\mathbf{V}$, then, by skew-symmetry, left ideals of $\mathbf{V}$ become two-sided ideals. This remark indeed applies to vertex operator algebras. See Sections 1.6.1 and 1.6.2 for details.

1.1.4 A Few Examples

It is not at all easy to construct examples of vertex algebras. The easiest is the commutative vertex algebra, but it is not really a new object since it is just a commutative associative algebra with iterative derivations.

The second easiest, the simplest noncommutative example of a vertex algebra, is supplied by *free boson theory* in physics. It is called the *Heisenberg vertex algebra* and contains another example of a vertex algebra, a *Virasoro vertex algebra,* as a vertex subalgebra.

1.1.4.1 Commutative Vertex Algebras

In a vertex algebra $\mathbf{V}$, the following conditions for elements $a, b \in \mathbf{V}$ are equivalent to each other by (VC) and (V2).

(1) $a_{(k)}b = 0$ for all $k \geq 0$.   (2) $[a_{(m)}, b_{(n)}] = 0$ for all $m, n \in \mathbb{Z}$.

A vertex algebra $\mathbf{V}$ is said to be *commutative* if the equivalent conditions hold for all $a, b \in \mathbf{V}$.

Regard such a vertex algebra as a vector space and equip it with the product given by the (-1)st product:

$$ab = a_{(-1)}b.$$

Then it becomes a commutative associative algebra, for which the vacuum 1 is the unity. The translation operators $T^{(k)}$ act as iterative derivations with respect to the product, by which the nth products are written as

$$a_{(n)}b = \begin{cases} 0 & (n \geq 0), \\ (T^{(k)}a)b & (n = -k - 1 < 0). \end{cases}$$

Thus the commutative vertex algebras fall into the concept of commutative associative algebras with iterative derivations. In this regard, vertex algebras are essentially infinite-dimensional objects, as we see by the following proposition.

Proposition 1.11   *Any finite-dimensional vertex algebra is commutative.*

Proof.   Assume $a_{(n)}b \neq 0$ for some $n \geq 0$ and take the minimal n among such. Then the matrix with entries $(T^{(i)}a)_{(n+j)}b$ indexed by $i, j \in \mathbb{N}$ is upper triangular with the diagonals $(-1)^i \binom{n+i}{i} a_{(n)}b$, which are nonzero for infinitely many $i \in \mathbb{N}$. Therefore, the sequence $a, Ta, T^{(2)}a, \cdots$ contains infinitely many linearly independent elements of $\mathbf{V}$. $\square$

1.1.4.2   Heisenberg Vertex Algebra

Let h be an element of a vertex algebra $\mathbf{V}$ satisfying

$$h_{(n)}h = \begin{cases} 0 & (n \geq 2), \\ 1 & (n = 1). \end{cases} \tag{1.3}$$

Then $h_{(0)}h = 0$ follows by skew-symmetry (VS), and the commutator formula (VC) implies that the operators $a_n = h_{(n)}$ satisfy

$$[a_m, a_n] = m\delta_{m+n,0}, \tag{1.4}$$

the commutation relation for the *Heisenberg algebra*.

The Heisenberg commutation relation (1.4) can be realized in a vertex algebra. To see it, consider the polynomial ring $\mathbb{F}[x_1, x_2, \cdots]$ with countably many indeterminates. Identify the scalar multiple of the unity of the ring with the scalars and denote the multiplication operator for a polynomial by the same symbol.

Let a_n, $n \in \mathbb{Z}$ denote the operators acting on $\mathbb{F}[x_1, x_2, \cdots]$ defined by

$$a_n = \begin{cases} n\frac{\partial}{\partial x_n} & (n > 0), \\ 0 & (n = 0), \\ x_k & (n = -k < 0). \end{cases} \tag{1.5}$$

Then they satisfy the commutation relation (1.4).

The vertex algebra given in the following proposition is called the *Heisenberg vertex algebra*.

Proposition 1.12 *The vector space* $\mathbb{F}[x_1, x_2, \cdots]$ *carries a unique structure of a vertex algebra such that*

$$x_{1(n)} = a_n \text{ for all } n \in \mathbb{Z}, \tag{1.6}$$

with the vacuum $\mathbf{1}$ *being the unity of the polynomial ring.*

By the definition (1.5) of the actions a_n, the condition (1.6) implies, for example:

$$x_{1(n)}x_1 = \begin{cases} 0 & (n \geq 2), \\ 1 & (n = 1), \\ 0 & (n = 0), \\ x_k x_1 & (n = -k < 0). \end{cases}$$

Thus the element $h = x_1$ satisfies (1.3). Repeated use of the associativity formula (VA) allows us to calculate the nth products for all polynomials.

The difficulty lies in guaranteeing consistency of the elements arising from the use of the Borcherds identity. We will see in the Section 1.2 that we can avoid this difficulty by identifying the nth products with certain products defined on series with operator coefficients, for which the Borcherds identity automatically holds under certain circumstances.

We will describe the details of this example including the proof of the proposition in Section 1.3, where the underlying polynomial ring $\mathbb{F}[x_1, x_2, \cdots]$ will be naturally identified with what is called the *Fock module* of *charge* 0 over the Heisenberg algebra (cf. Proposition 3.1).

Note 1.13. The vertex algebra described above is the Heisenberg vertex algebra of *rank one*. It is also called the *free bosonic vertex algebra* in the literatures.

1.1.4.3 Virasoro Vectors

Let ω be an element of a vertex algebra V satisfying the following condition with a scalar c:

$$\omega_{(n)}\omega = \begin{cases} 0 & (n \geq 4), \\ (c/2)\mathbf{1} & (n = 3), \\ 2\omega & (n = 1). \end{cases} \tag{1.7}$$

Then $\omega_{(0)}\omega = T^{(1)}\omega$ and $\omega_{(2)}\omega = 0$ follow by skew-symmetry (VS).

Such an ω is called a *Virasoro vector* since the operators $L_n = \omega_{(n+1)}$ satisfy

$$[L_m, L_n] = (m - n)L_{m+n} + \frac{m^3 - m}{12}c\delta_{m+n,0},$$

the commutation relation for the *Virasoro algebra* of central charge c, where

$$\frac{m^3 - m}{12} = \frac{1}{2}\binom{m + 1}{3}$$

is a half integer.

The Heisenberg vertex algebra $\mathbb{F}[x_1, x_2, \cdots]$ actually contains a Virasoro vector. Indeed, consider the following element:

$$u = x_1^2 = x_{1(-1)}x_1 = a_{-1}a_{-1}\mathbf{1}.$$

By the associativity formula (VA) with $m = -1$, we have

$$u_{(n)}u = (x_{1(-1)}x_1)_{(n)}u = \sum_{i=0}^{\infty} a_{-i-1}a_{n+i}a_{-1}a_{-1}\mathbf{1} + \sum_{i=0}^{\infty} a_{n-1-i}a_i a_{-1}a_{-1}\mathbf{1}.$$

After some algebra,

$$u_{(1)}u = 2a_{-1}a_1 a_{-1}a_{-1}\mathbf{1} = 4u,$$
$$u_{(3)}u = 2a_1 a_{-1}\mathbf{1} \qquad = 2,$$
$$u_{(4)}u = 2a_2 a_{-1}\mathbf{1} \qquad = 0, \cdots.$$

Therefore, the following element satisfies (1.7) as desired with $c = 1$:

$$\omega = \frac{u}{2} = \frac{1}{2}x_1^2. \tag{1.8}$$

We will call it the *standard* Virasoro vector of the Heisenberg vertex algebra to distinguish it from other Virasoro vectors in the same vertex algebra.

We may consider the vertex subalgebra generated by ω, which is actually spanned by the elements of the form

$$L_{n_1} \cdots L_{n_k}\mathbf{1} \quad (k \in \mathbb{N}, \ n_1, \ldots, n_k \in \mathbb{Z}).$$

This is an example of what is called a *Virasoro vertex algebra*.

Note 1.14. 1. Over the field $\mathbb{C}$ of complex numbers, the Virasoro vertex algebra obtained above is actually isomorphic to the simple vertex algebra denoted $\mathbf{L}(1,0)$ See Subsection 1.3.3.3 for details. 2. The construction of ω as above is a particular case of the process called the *Sugawara construction* for affine Lie algebras (cf. Subsection 1.3.2.4). 3. For any scalar λ, the vector $\omega_\lambda = x_1^2/2 + \lambda x_2$ is a Virasoro vector of central charge $c = 1 - 12\lambda^2$. This construction is called the *Feigin–Fuchs construction*.

1.1.5 Description by Generating Series

Let $\mathbf{V}$ be a vertex algebra. For each $a \in \mathbf{V}$ and $n \in \mathbb{Z}$, consider the left actions with respect to the nth product:

$$a_{(n)} : \mathbf{V} \longrightarrow \mathbf{V}, \quad x \mapsto a_{(n)}x.$$

Then the countably many product operations are collectively expressed in the generating series

$$Y(a, z) = \sum_n a_{(n)} z^{-n-1},$$

and the axioms (V0) and (V2), for instance, are expressed as follows:

(V0) Local truncation. For all $a, b \in \mathbf{V}$: $Y(a, z)b \in \mathbf{V}((z))$.

(V2) Vacuum. There exists an element $\mathbf{1} \in \mathbf{V}$ satisfying, for all $a \in \mathbf{V}$,

$$Y(\mathbf{1}, z)a = a \quad \text{and} \quad Y(a, z)\mathbf{1} \in a + \mathbf{V}[[z]]z.$$

In this section, we will describe various properties of vertex algebras in terms of generating series. The expression of the Borcherds identity (V1) by generating series in its full form, called the *Cauchy–Jacobi identity*, will be given in Subsection 1.1.5.3.

1.1.5.1 Local Commutativity and Associativity

Recall that the conjunction of the commutator formula (VC) and the associativity formulas (VA) is equivalent to the Borcherds identity (V1). In this subsection, we will consider another pair of conditions whose conjunction is equivalent to the Borcherds identity.

By local truncation (V0), there exists an N such that $a_{(N+i)}b = 0$ for all $i \geq 0$. For such an N, the Borcherds identity with $(p, q, r) = (m, n, N)$ reads:

$$0 = \sum_{i=0}^{\infty} (-1)^i \binom{N}{i} a_{(m+N-i)}(b_{(n+i)}c)$$

$$- \sum_{i=0}^{\infty} (-1)^{N-i} \binom{N}{i} b_{(n+N-i)}(a_{(m+i)}c).$$

Similarly, there exists an L such that $a_{(L+i)}c = 0$ for all $i \geq 0$. For such an L, the Borcherds identity with $(p,q,r) = (L,n,m)$ reads:

$$\sum_{i=0}^{\infty} \binom{L}{i} (a_{(m+i)}b)_{(n+L-i)}c = \sum_{i=0}^{\infty} (-1)^i \binom{m}{i} a_{(m+L-i)}(b_{(n+i)}c).$$

These properties are better described by generating series as follows:

(VLC) Local commutativity. For any $a,b \in V$, there exists an $N \in \mathbb{N}$ such that

$$(y-z)^N Y(a,y)Y(b,z) = (y-z)^N Y(b,z)Y(a,y).$$

(VLA) Local associativity. For any $a,c \in V$, there exists an $L \in \mathbb{N}$ such that, for all $b \in V$,

$$(x+z)^L Y(Y(a,x)b,z)c = (x+z)^L Y(a,x+z)\big|_{|x|>|z|} Y(b,z)c.$$

By the recurrence relation (1.2), we have the following implications:

$$(\mathrm{VC}) \implies (\mathrm{VLC}), \quad (\mathrm{VA}) \implies (\mathrm{VLA}),$$

$$(\mathrm{V1}) \iff (\mathrm{VLC}) + (\mathrm{VLA}).$$

Note 1.15.   1. In the literatures, local commutativity is usually called *locality* or *weak commutativity*, while local associativity is called *weak associativity*. They are sometimes called *commutativity* and *associativity*, respectively.
2. The commutator formula (VC) and the associativity formula (VA) are written respectively in terms of generating series as

$$[Y(a,y),Y(a,z)] = \sum_{i=0}^{\infty} Y(a_{(i)}b,z)\delta^{(i)}(y,z),$$

$$Y(a_{(m)}b,z) = Y(a,z)_{(m)}Y(b,z) \quad (m \in \mathbb{Z}).$$

These formulas, as well as local commutativity and local associativity, have clear meanings in the language of *operator product expansion*. See Section 1.2 for details.

1.1.5.2 Translation Covariance

Let us next consider the translation operators defined for each $k \in \mathbb{N}$ by

$$T^{(k)} : V \longrightarrow V, \quad a \mapsto T^{(k)}a = a_{(-k-1)}\mathbf{1}.$$

For an indeterminate x, we formally write

$$e^{xT} = \sum_{k=0}^{\infty} x^k T^{(k)}.$$

Here are properties related to the translation operators.

(VT) Translation. For all $a \in \mathbf{V}$ and $k \in \mathbb{N}$:

$$Y(T^{(k)}a, z) = \partial_z^{(k)} Y(a, z).$$

(VL) Leibniz rule. For all $a \in \mathbf{V}$:

$$e^{xT} Y(a, z) = Y(e^{xT} a, z) e^{xT}.$$

(VS) Skew-symmetry. For all $a, b \in \mathbf{V}$:

$$Y(a, z)b = e^{zT} Y(b, -z)a.$$

The translation property (VT) implies

$$Y(e^{xT} a, z) = e^{x\partial_z} Y(a, z) = Y(a, x + z)\big|_{|x|<|z|}.$$

Therefore, the operators $T^{(k)}$ are seen to generate *translation*.
Combining it with (VL), we have the *translation covariance*,

$$e^{xT} Y(a, z) e^{-xT} = Y(a, x + z)\big|_{|x|<|z|},$$

from which $Y(a, z)\mathbf{1} = e^{zT} a$ follows by (V2).

Note 1.16. Over a field of characteristic zero, we have

$$e^{xT} = \sum_{k=0}^{\infty} \frac{(xT)^k}{k!}, \quad \text{where} \ T = T^{(1)}.$$

The properties (VT) and (VL) for all k follow from those with $k = 1$:

$$Y(Ta, z) = \partial_z Y(a, z) \ \text{and} \ [T, Y(a, z)] = Y(Ta, z),$$

whence $\partial_z Y(a, z) = [T, Y(a, z)]$, the "equation of motion" of $Y(a, z)$.

1.1.5.3 Cauchy–Jacobi Identity

For convenience of readers in consulting the literatures, we will describe the Borcherds identity (V1) in terms of generating series.

Recall the formal delta function $\delta(z) = \sum_n z^n$, where n runs over the integers, and consider the following expressions:

$$y^{-1}\delta\left(\frac{z + x}{y}\right) = \sum_n \sum_{i=0}^{\infty} \binom{n}{i} x^i y^{-n-1} z^{n-i} \qquad (|x| < |z|),$$

$$x^{-1}\delta\left(\frac{y - z}{x}\right) = \sum_n \sum_{i=0}^{\infty} (-1)^i \binom{n}{i} x^{-n-1} y^{n-i} z^i \qquad (|y| > |z|),$$

$$x^{-1}\delta\left(\frac{z - y}{-x}\right) = \sum_n \sum_{i=0}^{\infty} (-1)^{n-i} \binom{n}{i} x^{-n-1} y^i z^{n-i} \ (|y| < |z|).$$

The region of expansion is signified by the order of variables in the numerator of the argument of delta for each.

The Borcherds identity (V1) is now expressed in terms of generating series.

(V1) Cauchy–Jacobi identity. For all $a, b, c \in V$:

$$y^{-1}\delta\left(\frac{z+x}{y}\right)Y(Y(a,x)b,z)$$
$$= x^{-1}\delta\left(\frac{y-z}{x}\right)Y(a,y)Y(b,z) - x^{-1}\delta\left(\frac{z-y}{-x}\right)Y(b,z)Y(a,y). \tag{1.9}$$

Indeed, the coefficients to $x^{-r-1}y^{-p-1}z^{-q-1}$ in the Cauchy–Jacobi identity form the Borcherds identity as described in Subsection 1.1.3.1.

Note 1.17. 1. The Cauchy–Jacobi identity is usually called the *Jacobi identity* for vertex algebras in the literatures. 2. The left-hand side of (1.9) is equivalently rewritten by the relation

$$y^{-1}\delta\left(\frac{z+x}{y}\right) = z^{-1}\delta\left(\frac{y-x}{z}\right),$$

which can be easily verified by direct calculations.

1.1.5.4 Tensor Product of Vertex Algebras

As an application of the description by generating series, let us briefly explain the *tensor product* of vertex algebras, which produces a new vertex algebra from a pair of given vertex algebras.

Let V and W be vertex algebras with the vacuums 1_V and 1_W. Consider the tensor product $V \otimes W$ of vector spaces and set

$$Y(a \otimes b, z) = Y(a,z) \otimes Y(b,z). \tag{1.10}$$

Then it equips $V \otimes W$ with a structure of a vertex algebra, called the *tensor product* of vertex algebras, for local commutativity and local associativity for V and W imply those for $V \otimes W$, and the vacuum properties holds with $1_{V \otimes W} = 1_V \otimes 1_W$.

As the right-hand side of (1.10) equals

$$\sum_{i,j} a_{(i)}z^{-i-1} \otimes b_{(j)}z^{-j-1} = \sum_{i,j} a_{(i)} \otimes b_{(j)}z^{-i-j-2},$$

the nth product of the tensor product is given by

$$(a \otimes b)_{(n)} = \sum_{i+j+1=n} a_{(i)} \otimes b_{(j)}.$$

The same process works for constructing the tensor product of a finite number of vertex algebras.

Bibliographic Notes

The main reference for Section 1.1 is the monograph [7] by K. Nagatomo and the author, where we worked over a field of characteristic zero. It is more or less straightforward to describe most materials covered in Section 1.1 over commutative rings (cf. Borcherds [32], [34] and Borcherds and Ryba [35]). See Mason [78] for accounts over $\mathbb{Z}$ under the name "vertex ring," including the proof that the creation property implies the identity property under the Borcherds identity.

For more information on the Borcherds identity and its formulation by generating series, see Frenkel, Lepowsky, and Meurman [1], Feingold, Frenkel, and Ries [2], Frenkel, Huang, and Lepowsky [3], or Lepowsky and Li [10]. The formulation of local commutativity as in (VLA) is due to Dong and Lepowsky [4]. Some textbooks such as Kac [6] or Frenkel and Ben-Zvi [8] are based on an equivalent but apparently different formulation of vertex algebras, where the translation operator is taken as a part of the structure. See Rosellen [11] for various formulations and their relation to other algebraic concepts.

For geometric interpretation of vertex algebras, see Frenkel and Ben-Zvi [8]. See Bakalov and Kac [29], Etingof and Kazhdan [52], and Li [75], for *noncommutative* or *nonlocal* analogues of vertex algebras, that is, objects satisfying local associativity but not necessarily local commutativity in our terminology. For more general frameworks, see Beilinson and Drinfeld [9] and Borcherds [34].

1.2 Vertex Algebras of Series

In order to construct an example of a group, it is often convenient to realize it as a set of bijective transformations of a set. If such a set $\mathbf{G}$ of transformations is closed under composition and inversion and contains the identity transformation, then it becomes a group by composition of transformations. The advantage of such construction lies in that associativity automatically holds for composition of maps. The set on which $\mathbf{G}$ acts then carries a structure of a *permutation representation* of $\mathbf{G}$.

Analogously, we can construct a commutative associative algebra $\mathbf{C}$ by realizing it as a vector space consisting of commuting operators on a vector space $\mathbf{M}$, for which associativity is again automatic, and $\mathbf{M}$ becomes a *representation* of $\mathbf{C}$ or equivalently a *module* over $\mathbf{C}$.

In Section 1.2, we will explain a way to construct vertex algebras along the same line. That is, we will realize a vertex algebra $\mathbf{V}$ as a vector space consisting of series with operator coefficients equipped with product operations for which the Borcherds identity is automatic under certain conditions. We will also introduce the concepts of a *representation* of **V** or a *module* over **V**.

We will continue to work over a field $\mathbb{F}$ of characteristic not 2 unless otherwise stated.

1.2.1 Residue Products of Series

In this section, we introduce a sequence of operations

$$\mathrm{Hom}(\mathbf{M}, \mathbf{M}((z))) \times \mathrm{Hom}(\mathbf{M}, \mathbf{M}((z))) \longrightarrow \mathrm{Hom}(\mathbf{M}, \mathbf{M}((z)))$$
$$(A(z), B(z)) \longmapsto A(z)_{(m)}B(z),$$

indexed by $m \in \mathbb{Z}$, called the *residue products*, which associate a series to a pair of series for each m.

1.2.1.1 Expansions in Various Regions

Let V be a vector space and x, y, z indeterminates, and consider the space

$$V((x, y, z)) = V[[x, y, z]][x^{-1}, y^{-1}, z^{-1}],$$

whose elements are written in the following form with some $L, M, N \in \mathbb{N}$:

$$w(x, y, z) = \frac{w_0(x, y, z)}{x^N y^L z^M}, \quad w_0(x, y, z) \in V[[x, y, z]].$$

For such an element, substitute $x = y - z$, and apply the binomial expansions to each term. Then we obtain

$$w(y - z, y, z)\big|_{|y| > |z|} \in V((y))((z)),$$
$$w(y - z, y, z)\big|_{|y| < |z|} \in V((z))((y)).$$

Similarly, we obtain the following series in x and z:

$$w(x, x + z, z)\big|_{|x| > |z|} \in V((x))((z)),$$
$$w(x, x + z, z)\big|_{|x| < |z|} \in V((z))((x)).$$

Since $x^N w(x, y, z) \in V[[x]]((y, z))$ and $y^L w(x, y, z) \in V[[y]]((x, z))$, we have

$$(y - z)^N w(y - z, y, z)\big|_{|y| > |z|} = (y - z)^N w(y - z, y, z)\big|_{|y| < |z|}, \tag{2.1}$$

$$(x + z)^L w(x, x + z, z)\big|_{|x| > |z|} = (x + z)^L w(x, x + z, z)\big|_{|x| < |z|}. \tag{2.2}$$

The expansions fit in the diagram

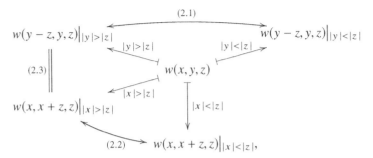

where the bent double-headed arrows indicate relations (2.1) and (2.2), and the vertical equality is given by the identification

$$w(y-z,y,z)\big|_{|y|>|z|} = w(x,x+z,z)\big|_{|x|>|z|} \qquad (2.3)$$

via the isomorphisms

$$V((y))((z)) \underset{\psi}{\overset{\phi}{\rightleftarrows}} V((x))((z))$$

inverse to each other, defined by

$$\phi\colon v(y,z) \mapsto v(x+z,z)\big|_{|x|>|z|}, \quad \psi\colon v(x,z) \mapsto v(y-z,z)\big|_{|y|>|z|}.$$

In this sense, we may think of the three series

$$w(y-z,y,z)\big|_{|y|>|z|}, \quad w(y-z,y,z)\big|_{|y|<|z|}, \quad w(x,x+z,z)\big|_{|x|<|z|}$$

as being "analytically continued" to each other under $x = y - z$.

1.2.1.2 Extracting Coefficients by Formal Residues

Let Res_z denote the operation of taking the formal residue in z:

$$\text{Res}_z\colon V[[z,z^{-1}]] \longrightarrow V, \quad v(z) = \sum_n v_n z^{-n-1} \mapsto \text{Res}_z\, v(z) = v_0.$$

By this operation, the coefficients of a series are extracted as

$$v_n = \text{Res}_z\, v(z) z^n.$$

If a series $v(y,z)$ in y and z belongs to $V((y))((z))$ or $V((z))((y))$, then $\text{Res}_y\, v(y,z)$ sits in $V((z))$, thus the operation Res_y gives rise to maps

$$\text{Res}_y\colon V((y))((z)) \longrightarrow V((z)), \quad \text{Res}_y\colon V((z))((y)) \longrightarrow V((z)).$$

Let $s(y,z)$ and $t(y,z)$ be series in $V((y))((z))$ and $V((z))((y))$, respectively, such that there exists a series $w(x,y,z)$ in $V((x,y,z))$ satisfying

$$s(y,z) = w(y-z,y,z)\big|_{|y|>|z|} \quad \text{and} \quad t(y,z) = w(y-z,y,z)\big|_{|y|<|z|}. \qquad (2.4)$$

We are interested in finding the expansion

$$u(x,z) = w(x,x+z,z)\big|_{|x|<|z|}$$

as in the following diagram:

$$s(y,z) \xleftarrow{\;|y|>|z|\;} \qquad \xrightarrow{\;|y|<|z|\;} t(y,z)$$
$$w(x,y,z)$$
$$\Big\downarrow {\scriptstyle |x|<|z|}$$
$$u(x,z).$$

Let us expand the series $u(x,z)$ as follows:

$$u(x,z) = \sum_m u_m(z) x^{-m-1} = \sum_{m,n} u_{m,n} x^{-m-1} z^{-n-1}.$$

Then the series $u_m(z)$ are expressed as

$$u_m(z) = \mathrm{Res}_y (y-z)^m\big|_{|y|>|z|} s(y,z) - \mathrm{Res}_y (y-z)^m\big|_{|y|<|z|} t(y,z), \qquad (2.5)$$

and their coefficients by

$$u_{m,n} = \sum_{i=0}^{\infty} (-1)^i \binom{m}{i} s_{m-i,n+i} - \sum_{i=0}^{\infty} (-1)^{m-i} \binom{m}{i} t_{i,m+n-i}.$$

In particular, $u(x,z)$ does not depend on the choice of $w(x,y,z)$ as in (2.4), although it is clear from the beginning.

The formula (2.5) is actually equivalent to the following identity valid for any series $v(x,y,z)$ in $V((x,y,z))$:

$$\mathrm{Res}_x\, v(x,x+z,z)\big|_{|x|<|z|} = \mathrm{Res}_y\, v(y-z,y,z)\big|_{|y|>|z|}$$
$$- \mathrm{Res}_y\, v(y-z,y,z)\big|_{|y|<|z|}. \qquad (2.6)$$

Indeed, (2.5) is obtained by substituting $x^m w(x,y,z)$ for $v(x,y,z)$ in (2.6).

Note 2.1.   1. Heuristically, take the series $s(y,z)$ and $t(y,z)$ as if they were expansions of a meromorphic function $v(y,z)$ of y with only poles at $y=0,z,\infty$. Then the coefficients to $(y-z)^{-m-1}$ in the expansion of $v(y,z)$ at $y=z$ are

$$\frac{1}{2\pi\sqrt{-1}} \oint_{C_z} (y-z)^m v(y,z)\,dy,$$

where C_z is a small circle surrounding z with $|y - z| < |z|$. As this becomes

$$\frac{1}{2\pi\sqrt{-1}} \oint_{C_{0,z}} (y - z)^m v(y,z)dy - \frac{1}{2\pi\sqrt{-1}} \oint_{C_0} (y - z)^m v(y,z)dy$$

by deformation of contour, where $C_{0,z}$ is a circle surrounding 0 and z with $|y| > |z|$ and C_0 surrounding 0 with $|y| < |z|$. Thus formula (2.5) is seen to describe $u_m(z)$ by the latter expression (cf. [8], [13], etc.). 2. Substituting $x^r y^p z^q w(x,y,z)$ for $v(x,y,z)$ in (2.6), we have

$$\sum_{i=0}^{\infty} \binom{p}{i} u_{r+i,p+q-i} = \sum_{i=0}^{\infty} (-1)^i \binom{r}{i} s_{p+r-i,q+i}$$

$$- \sum_{i=0}^{\infty} (-1)^{r-i} \binom{r}{i} t_{q+r-i,p+i}.$$

The resemblance with the Borcherds identity (V1) is not an accident, as we will see in the sequel.

1.2.1.3 Series Acting on Vector Spaces

Let **M** be a vector space and consider formal series in $(\text{End }\mathbf{M})[[z,z^{-1}]]$. We will call such a series a *series acting on* **M**, or just a *series on* **M** for short.

For such a series $A(z)$, set

$$A(z) = \sum_n A_n z^{-n-1},$$

where the summation is over all $n \in \mathbb{Z}$ and A_n are operators on **M**. For an element $v \in \mathbf{M}$, we write

$$A(z)v = \sum_n A_n v z^{-n-1}.$$

In particular, consider the series $I(z)$ such that the only nonzero term is the constant term being the identity operator I:

$$I(z) = \sum_n I_n z^{-n-1}, \quad \text{where} \quad I_n v = \begin{cases} 0 & (n \neq -1), \\ v & (n = -1). \end{cases}$$

We will call it the *identity series* and often identify it with the scalar 1.

For a series $A(z)$, split it into the sum of series with nonnegative and negative powers as

$$\sum_n A_n z^{-n-1} = \underbrace{\sum_{n<0} A_n z^{-n-1}}_{\substack{\text{nonnegative} \\ \text{powers in } z}} + \underbrace{\sum_{n\geq 0} A_n z^{-n-1}}_{\substack{\text{negative} \\ \text{powers in } z}}. \tag{2.7}$$

Let us denote the resulting series by $A(z)^{\geq 0}$ and $A(z)^{<0}$, respectively:

$$A(z)^{\geq 0} = \sum_{n<0} A_n z^{-n-1} = \sum_{n \geq 0} A_{-n-1} z^n,$$

$$A(z)^{<0} = \sum_{n \geq 0} A_n z^{-n-1} = \sum_{n<0} A_{-n-1} z^n.$$

We will also denote them by $A(z)_{<0}$ and $A(z)_{\geq 0}$, respectively.

1.2.1.4 Locally Truncated Series

We will say that a series $A(z)$ on a vector space $\mathbf{M}$ is *locally truncated* if $A(z)v \in \mathbf{M}((z))$ for all $v \in \mathbf{M}$, thus

$$A(z) \text{ is locally truncated} \iff A(x) \in \mathrm{Hom}(\mathbf{M}, \mathbf{M}((z))).$$

In other words, $A(z)$ is locally truncated if and only if for any $v \in \mathbf{M}$, there exists an $N \in \mathbb{N}$ such that $A_{N+i}v = 0$ for all $i \in \mathbb{N}$.

Consider series $A(z)$ and $B(z)$, split $A(y)$ as in (2.7), and set

$$\overset{\circ}{{}_{\circ}}A(y)B(z)\overset{\circ}{{}_{\circ}} = A(y)^{\geq 0}B(z) + B(z)A(y)^{<0}.$$

Assume that $A(z)$ and $B(z)$ are locally truncated. Then, for $v \in \mathbf{M}$,

$$\overset{\circ}{{}_{\circ}}A(y)B(z)\overset{\circ}{{}_{\circ}}v \in \mathbf{M}((y,z)).$$

Therefore, the following expression gives rise to a locally truncated series:

$$\overset{\circ}{{}_{\circ}}A(z)B(z)\overset{\circ}{{}_{\circ}} = A(z)^{\geq 0}B(z) + B(z)A(z)^{<0}. \tag{2.8}$$

Such an expression is called the *normally ordered product* of $A(z)$ and $B(z)$.

Note 2.2.   A series on a vector space is also called a *formal distribution* and a locally truncated series a *field* in the literatures following [6].

1.2.1.5 Residue products

Let $A(z)$ and $B(z)$ be series on a vector space $\mathbf{M}$ and m an integer:

$$A(z), B(z) \in (\mathrm{End}\,\mathbf{M})[[z, z^{-1}]], \quad m \in \mathbb{Z}.$$

If $m \geq 0$, then we may consider the following expression as a series with operator coefficients:

$$A(z)_{(m)}B(z) = \mathrm{Res}_y(y-z)^m[A(y), B(z)]$$
$$= \mathrm{Res}_y(y-z)^m A(y)B(z) - \mathrm{Res}_y(y-z)^m B(z)A(y).$$

For $m < 0$, assume that $A(z)$ and $B(z)$ are locally truncated. Then, for $v \in \mathbf{M}$,

$$A(y)B(z)v \in \mathbf{M}((y))((z)), \quad B(z)A(y)v \in \mathbf{M}((z))((y)),$$

and the following expressions make sense as series in y and z:

$$(y-z)^m\big|_{|y|>|z|}A(y)B(z)v, \quad (y-z)^m\big|_{|y|<|z|}B(z)A(y)v.$$

We may therefore consider the expression

$$A(z)_{(m)}B(z) = \mathrm{Res}_y(y-z)^m\big|_{|y|>|z|}A(y)B(z)$$
$$- \mathrm{Res}_y(y-z)^m\big|_{|y|<|z|}B(z)A(y)$$

as a series with operator coefficients.

We will call the series $A(z)_{(m)}B(z)$ thus obtained the mth *residue product* of $A(z)$ and $B(z)$ for each $m \in \mathbb{Z}$.

Lemma 2.3 *If series $A(z)$ and $B(z)$ on a vector space are locally truncated, then so is the residue product $A(z)_{(m)}B(z)$ for all $m \in \mathbb{Z}$.*

To describe the coefficients explicitly, set

$$A(z)_{(m)}B(z) = (A_{(m)}B)(z) = \sum_n (A_{(m)}B)_n z^{-n-1}.$$

Then we have

$$(A_{(m)}B)_n = \sum_{i=0}^\infty (-1)^m \binom{m}{i} A_{m-i}B_{n+i} - \sum_{i=0}^\infty (-1)^{m-i}\binom{m}{i} B_{m+n-i}A_i. \quad (2.9)$$

For $m = -1$, the residue product $A(z)_{(-1)}B(z)$ agrees with the normally ordered product ${}^\circ_\circ A(z)B(z){}^\circ_\circ$ defined by (2.8) and, for $k \in \mathbb{N}$,

$$A(z)_{(-k-1)}B(z) = {}^\circ_\circ \left(\partial^{(k)}A(z)\right)B(z){}^\circ_\circ.$$

In particular, for the identity series $I(z)$,

$$I(z)_{(m)}A(z) = \begin{cases} 0 & (m \neq -1), \\ A(z) & (m = -1), \end{cases}$$

$$A(z)_{(m)}I(z) = \begin{cases} 0 & (m \geq 0), \\ \partial^{(k)}A(z) & (m = -k-1 < 0), \end{cases}$$

which is the vacuum property (V2) completed by the divided derivatives.

1.2.2 Operator Product Expansions

In Section 1.2.2, we will explain a rigorous formulation in certain circumstances of what is called *operator product expansion* (OPE) in physics and its relation to residue products.

Let us briefly outline the concept by example. Let a_n, $n \in \mathbb{Z}$, be operators on a vector space $\mathbf{M}$ satisfying the Heisenberg commutation relation (1.4). By the generating series $a(z) = \sum_n a_n z^{-n-1}$, the relation is written as

$$[a(y), a(z)] = \sum_n n y^{-n-1} z^{n-1} \ (= \delta^{(1)}(y, z)). \tag{2.10}$$

Note that the equality $(y - z)^2 a(y)a(z) = (y - z)^2 a(z)a(y)$ follows.

Split (2.10) into two equalities by collecting terms with nonnegative and negative powers in y. Then, after some algebra, we arrive at

$$\begin{cases} a(y)a(z) = \dfrac{1}{(y-z)^2}\Big|_{|y|>|z|} + {}^{\circ}_{\circ}a(y)a(z)^{\circ}_{\circ}, \\ a(z)a(y) = \dfrac{1}{(y-z)^2}\Big|_{|y|<|z|} + {}^{\circ}_{\circ}a(y)a(z)^{\circ}_{\circ}. \end{cases} \tag{2.11}$$

The two equalities in (2.11) are written as

$$a(y)a(z) \simeq a(z)a(y) \sim \frac{1}{(y-z)^2} \tag{2.12}$$

and called the OPE. The mth residue products $a(z)_{(m)}a(z)$ for $m \geq 0$ are then read off from the OPE (2.12) as

$$a(z)_{(m)}a(z) = \begin{cases} 0 & (m \geq 2), \\ 1 & (m = 1), \\ 0 & (m = 0), \end{cases}$$

where 1 for $m = 1$ is the numerator in (2.12), that is, the identity series $I(z)$.

1.2.2.1 OPE of Locally Commutative Series

Let $A(z)$ and $B(z)$ be series on a vector space $\mathbf{M}$. We will say that they are *locally commutative* if the following holds for some $N \in \mathbb{N}$:

$$(y - z)^N A(y)B(z) = (y - z)^N B(z)A(y). \tag{2.13}$$

In other words,

$$\sum_{i=0}^{\infty}(-1)^i \binom{N}{i} A_{m+N-i}B_{n+i}v = \sum_{i=0}^{\infty}(-1)^{N-i}\binom{N}{i}B_{n+N-i}A_{m+i}v \tag{2.14}$$

for some $N \in \mathbb{N}$ and all $m, n \in \mathbb{Z}$.

Let $A(z)$ and $B(z)$ be locally commutative series on a vector space $\mathbf{M}$ and take $N \in \mathbb{N}$ such that (2.13) holds. Split $A(y)$ into the sum of series with nonnegative and negative powers as in (2.7). Then local commutativity becomes

$$(y - z)^N \big[A(y)_{\geq 0}, B(z)\big] = -(y - z)^N \big[A(y)_{<0}, B(z)\big].$$

Comparing the degrees in y, we see that there exist series $C_0(z), \ldots, C_{N-1}(z)$
in z such that

$$
\begin{cases}
(y - z)^N \left[A(y)^{<0}, B(z) \right] = \sum_{k=0}^{N-1} (y - z)^{N-1-k} C_k(z), \\
-(y - z)^N \left[A(y)^{\geq 0}, B(z) \right] = \sum_{k=0}^{N-1} (y - z)^{N-1-k} C_k(z).
\end{cases}
$$

Multiplying them by $(y - z)^{-N}|_{|y|>|z|}$ and $(y - z)^{-N}|_{|y|<|z|}$, respectively, and
adding ${}^\circ_\circ A(y)B(z){}^\circ_\circ = A(y)^{\geq 0}B(z) + B(z)A(y)^{<0}$, we have

$$
\begin{cases}
A(y)B(z) = \sum_{k=0}^{N-1} \dfrac{C_k(z)}{(y - z)^{k+1}}\Big|_{|y|>|z|} + {}^\circ_\circ A(y)B(z){}^\circ_\circ, \\
B(z)A(y) = \sum_{k=0}^{N-1} \dfrac{C_k(z)}{(y - z)^{k+1}}\Big|_{|y|<|z|} + {}^\circ_\circ A(y)B(z){}^\circ_\circ.
\end{cases} \tag{2.15}
$$

$$\underbrace{}_{\text{singular part}} \quad \underbrace{}_{\text{regular part}}$$

The two equalities are written at once as

$$
A(y)B(z) \simeq B(z)A(y) \sim \sum_{k=0}^{N-1} \frac{C_k(z)}{(y - z)^{k+1}}, \tag{2.16}
$$

and it is called (the singular part of) the OPE.

As the difference of the left-hand sides of (2.15) becomes the commutators $[A(y), B(z)]$, the commutation relations of the coefficients are encoded in the series $C_0(z), \ldots, C_{N-1}(z)$ in the OPE (2.16), which are related to the residue products as

$$
A(z)_{(m)}B(z) = \begin{cases} 0 & (N \leq m), \\ C_m(z) & (0 \leq m < N). \end{cases}
$$

The OPE as described above gives a rigorous formulation of what is called by the same term in physics.

1.2.2.2 Expansion of Regular Parts
Let us further expand the regular part ${}^\circ_\circ A(y)B(z){}^\circ_\circ$ of the equalities in (2.15) under the assumption that $A(z)$ and $B(z)$ are locally truncated. For $v \in \mathbf{M}$, set

$$
w(x, y, z) = \sum_{k=0}^{N-1} \frac{C_k(z)v}{x^{k+1}} + {}^\circ_\circ A(y)B(z){}^\circ_\circ v.
$$

By local truncation, we have ${}^\circ_\circ A(y)B(z){}^\circ_\circ v \in \mathbf{M}((y,z))$, thus

$$w(x,y,z) \in \mathbf{M}((y,z))[x^{-1}] \subset \mathbf{M}((x,y,z)).$$

The OPE (2.15) is now restated as

$$\begin{cases} A(y)B(z)v = w(y-z,y,z)\big|_{|y|>|z|}, \\ B(z)A(y)v = w(y-z,y,z)\big|_{|y|<|z|}. \end{cases}$$

Expand $w(x,y,z)$ in the region $|x| < |z|$ by substitution $y = x + z$ and denote the resulting series by $(A \circ B)(x,z)$. Then, by (1.1), we have

$$(A \circ B)(x,z)v = \sum_{k=0}^{N-1} \frac{C_k(z)v}{x^{k+1}} + \sum_{k=0}^{\infty} x^k {}^\circ_\circ \big(\partial^{(k)} A(z)\big) B(z) {}^\circ_\circ v,$$

or equivalently

$$(A \circ B)(x,z)v = \sum_m x^{-m-1} A(z)_{(m)} B(z).$$

The situation is summarized in the following diagram as in Subsection 1.2.1.1:

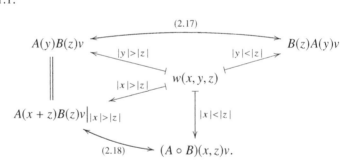

Here the bent double-headed arrows signify the following relations for sufficiently large L and N:

$$(y-z)^N A(y)B(z)v = (y-z)^N B(z)A(y)v, \qquad (2.17)$$

$$(x+z)^L A(x+z)B(z)v\big|_{|x|>|z|} = (x+z)^L (A \circ B)(x,z)v. \qquad (2.18)$$

The relation (2.17) is local commutativity which we have assumed, whereas (2.18) is a consequence, which is a form of local associativity.

In this sense, the series

$$A(y)B(z)v, \ B(z)A(y)v, \ (A \circ B)(x,z)v$$

are thought of as being "analytically continued" to each other, and the OPE (2.16) is formally completed as

$$A(y)B(z) \simeq B(z)A(y) \simeq \sum_{k=0}^{N-1} \frac{C_k(z)}{(y-z)^{k+1}} + \sum_{k=0}^{\infty} (y-z)^k \circ \left(\partial^{(k)} A(z) \right) B(z) \circ$$

by including the expansion of the regular part and formally substituting $y - z$ for x, although the result does not make sense in general as series in y and z.

Note 2.4. Let $A(z)$ and $B(z)$ be locally commutative and locally truncated. Then the formula in Note 2.1 implies the Borcherds identity in the form

$$\sum_{i=0}^{\infty} \binom{p}{i} (A_{(r+i)}B)_{p+q-i} = \sum_{i=0}^{\infty} (-1)^r \binom{r}{i} A_{p+r-i} B_{q+i}$$

$$- \sum_{i=0}^{\infty} (-1)^{r-i} \binom{r}{i} B_{q+r-i} A_{p+i}$$

for all $p, q, r \in \mathbb{Z}$, as noted by Tuite [22] in a different method, which is to be identified with the Borcherds identity (M1) for modules in Subsection 1.2.5.2.

1.2.2.3 Skew-Symmetry of Residue Products

Let us now consider the series $(B \circ A)(-x, y)$ obtained by switching the roles of $A(y)$ and $B(z)$ as in the following diagram:

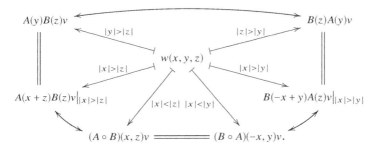

Then we have

$$(A \circ B)(x, z)v = (B \circ A)(-x, x + z)v \big|_{|x|<|z|}$$

$$= \sum_{m,n} (B_{(m)}A)_n (-x)^{-m-1} (x + z)^{-n-1} \big|_{|x|<|z|}$$

$$= e^{x\partial_z} \sum_{m,n} (B_{(m)}A)_n (-x)^{-m-1} z^{-n-1}$$

$$= e^{x\partial_z} (B \circ A)(-x, z)v.$$

Therefore,

$$A(z)_{(m)}B(z) = (-1)^{m+1} \sum_{i=0}^{\infty} (-1)^i \partial^{(i)} \left(B(z)_{(m+i)}A(z) \right),$$

which is the skew-symmetry for the residue products (cf. Subsection 1.1.3.5).

1.2.3 Vertex Algebras of Series

Recall that a linear space consisting of commuting operators on a vector space becomes a commutative associative algebra by composition of operators if it is closed under composition and contains the identity operator.

In this section, we will pursue such consideration for vertex algebras, where the analogue of operators is given by series with operator coefficients and that of composition is the residue products.

1.2.3.1 Borcherds Identity for Residue Products

Recall that the set of locally truncated series on **M** in an indeterminate z can be identified with the set $\mathrm{Hom}(\mathbf{M}, \mathbf{M}((z)))$, and the residue products of locally truncated series are again locally truncated. Therefore, the residue products equip $\mathrm{Hom}(\mathbf{M}, \mathbf{M}((z)))$ with a sequence of binary operations:

$$\mathrm{Hom}(\mathbf{M}, \mathbf{M}((z))) \times \mathrm{Hom}(\mathbf{M}, \mathbf{M}((z))) \longrightarrow \mathrm{Hom}(\mathbf{M}, \mathbf{M}((z)))$$

$$(A(z), B(z)) \longmapsto A(z)_{(n)}B(z).$$

Let $A(z)$, $B(z)$, and $C(z)$ be locally truncated series on a vector space **M**. Let us further assume that they are locally commutative with each other, that is, $A(z)$ and $B(z)$, $B(z)$ and $C(z)$, and $A(z)$ and $C(z)$ are locally commutative separately.

Theorem 2.5   *Let $A(z)$, $B(z)$, and $C(z)$ be locally truncated series on a vector space locally commutative with each other. Then the Borcherds identity*

$$\sum_{i=0}^{\infty} \binom{p}{i} (A(z)_{(r+i)}B(z))_{(p+q-i)}C(z)$$

$$= \sum_{i=0}^{\infty} (-1)^i \binom{r}{i} A(z)_{(p+r-i)}(B(z)_{(q+i)}C(z)) \qquad (2.19)$$

$$- \sum_{i=0}^{\infty} (-1)^{r-i} \binom{r}{i} B(z)_{(q+r-i)}(A(z)_{(p+i)}C(z))$$

holds for all $p, q, r \in \mathbb{Z}$ with respect to the residue products.

To see it, let us first consider the case when the integers p, q, r are nonnegative. The Jacobi identity for the commutators in the following form obviously holds by cancellation:

$$[[A(x), B(y)], C(z)] = [A(x), [B(y), C(z)]] - [B(y), [A(x), C(z)]]. \quad (2.20)$$

Multiply both sides by $D(x, y, z) = (x - z)^p (y - z)^q (x - y)^r$:

$$\underbrace{(x - z)^p (y - z)^q (x - y)^r} [[A(x), B(y)], C(z)]$$
$$= (x - z)^p (y - z)^q \underbrace{(x - y)^r} [A(x), [B(y), C(z)]]$$
$$- (x - z)^p (y - z)^q \underbrace{(x - y)^r} [B(y), [A(x), C(z)]].$$

To the underlined factors, apply the following expansions, respectively:

$$(x - z)^p = \big((x - y) + (y - z)\big)^p = \sum_{i=0}^{p} \binom{p}{i} (x - y)^i (y - z)^{p-i},$$

$$(x - y)^r = \big((x - z) - (y - z)\big)^r = \sum_{i=0}^{r} (-1)^i \binom{r}{i} (x - z)^{r-i} (y - z)^i,$$

$$(x - y)^r = \big((x - z) - (y - z)\big)^r = \sum_{i=0}^{r} (-1)^{r-i} \binom{r}{i} (x - z)^i (y - z)^{r-i}.$$

Then we have

$$\sum_{i=0}^{p} \binom{p}{i} (x - y)^{r+i} (y - z)^{p+q-i} [[A(x), B(y)], C(z)]$$

$$= \sum_{i=0}^{r} (-1)^i \binom{r}{i} (x - z)^{p+r-i} (y - z)^{q+i} [A(x), [B(y), C(z)]]$$

$$- \sum_{i=0}^{r} (-1)^{r-i} \binom{r}{i} (x - z)^{p+i} (y - z)^{q+r-i} [B(y), [A(x), C(z)]].$$

Taking $\text{Res}_x \text{Res}_y$, we arrive at the Borcherds identity (2.19).

For general p, q, r, the situation is much more complicated, for the factor $D(x, y, z)$ is no longer a polynomial.

In such a case, multiply the terms of the Jacobi identity (2.20) by the expansions of $D(x, y, z) = (x - z)^p (y - z)^q (x - y)^r$ in regions depending on the order of $A(x), B(y), C(z)$ as

$$Q_{123} = D(x,y,z)\big|_{|x|>|y|>|z|} A(x)B(y)C(z),$$
$$Q_{132} = D(x,y,z)\big|_{|x|>|z|>|y|} A(x)C(z)B(y),$$
$$Q_{213} = D(x,y,z)\big|_{|y|>|x|>|z|} B(y)A(x)C(z),$$
$$Q_{231} = D(x,y,z)\big|_{|y|>|z|>|x|} B(y)C(z)A(x),$$
$$Q_{312} = D(x,y,z)\big|_{|z|>|x|>|y|} C(z)A(x)B(y),$$
$$Q_{321} = D(x,y,z)\big|_{|z|>|y|>|x|} C(z)B(y)A(x).$$

Then, again by cancellation, we have

$$(Q_{123} - Q_{213}) - (Q_{312} - Q_{321})$$
$$= \big((Q_{123} - Q_{132}) - (Q_{231} - Q_{321})\big) - \big((Q_{213} - Q_{231}) - (Q_{132} - Q_{312})\big).$$

The result follows similarly by taking $\mathrm{Res}_x \, \mathrm{Res}_y$ after carefully manipulating series under local commutativity. See [7] and [11] for details.

Alternatively, for $A(z), B(z), C(z)$ as above and $v \in \mathbf{M}$, note that there exists an element $w(x,y,z,\xi,\eta,\zeta) \in \mathbf{M}((x,y,z,\xi,\eta,\zeta))$ such that

$$A(x)B(y)C(z)v = w(x,y,z,x-y,x-z,y-z)\big|_{|x|>|y|>|z|},$$
$$A(x)C(z)B(y)v = w(x,y,z,x-y,x-z,y-z)\big|_{|x|>|z|>|y|},$$
$$B(y)A(x)C(z)v = w(x,y,z,x-y,x-z,y-z)\big|_{|y|>|x|>|z|},$$
$$B(y)C(z)A(x)v = w(x,y,z,x-y,x-z,y-z)\big|_{|y|>|z|>|x|},$$
$$C(z)A(x)B(y)v = w(x,y,z,x-y,x-z,y-z)\big|_{|z|>|x|>|y|},$$
$$C(z)B(y)A(x)v = w(x,y,z,x-y,x-z,y-z)\big|_{|z|>|y|>|x|}.$$

Set $W(x,y,z,\xi,\eta,\zeta) = \xi^r \eta^p \zeta^q w(x,y,z,\xi,\eta,\zeta)$. Then the Borcherds identity (2.19) follows by taking Res_ζ of the identity

$$\mathrm{Res}_\xi \, W(\xi + \zeta + z, \zeta + z, z, \xi, \xi + \zeta, \zeta)\big|_{|\xi|<|\zeta+z|,|\xi|<|\zeta|<|z|}$$
$$= \mathrm{Res}_\eta \, W(\eta + z, \zeta + z, z, \eta - \zeta, \eta, \zeta)\big|_{|\zeta|<|\eta|<|z|}$$
$$- \mathrm{Res}_\eta \, W(\eta + z, \zeta + z, z, \eta - \zeta, \eta, \zeta)\big|_{|\eta|<|\zeta|<|z|},$$

which is in fact a variant of (2.6). Note that the region $|\xi| < |\zeta + z|$ attached in the left-hand side can be replaced by $|\xi + \zeta| < |z|$ without affecting the expansion.

1.2.3.2 Vertex Algebra of Series

We will say that a subset S of $\mathrm{Hom}(\mathbf{M}, \mathbf{M}((z)))$ is locally commutative if so are all the pairs of series belonging to S, including the pairs of the same series.

By combining the properties of the residue products obtained so far, we arrive at the following result due to H. S. Li.

Table 2 Vertex algebras, abstract versus realization

Abstract vertex algebra		Vertex algebra of series	
$\mathbf{V}$	abstract vector space	$\mathcal{V}$	subspace of $\mathrm{Hom}(\mathbf{M}, \mathbf{M}((z)))$
$a_{(n)}b$	abstract products	$A(z)_{(n)}B(z)$	residue products
$\mathbf{1}$	the vacuum	$I(z) = I$	the identity series
$T^{(k)}$	translation operators	$\partial_z^{(k)}$	divided derivatives

Corollary 2.6 *Let $\mathcal{V}$ be a vector space consisting of series on a vector space satisfying the following conditions:*

(1) $\mathcal{V}$ is locally truncated and locally commutative.

(2) $\mathcal{V}$ is closed under the residue products.

(3) $\mathcal{V}$ contains the identity series.

Then $\mathcal{V}$ becomes a vertex algebra by the residue products.

We will call the vertex algebra thus obtained a *vertex algebra of series* for short. See Table 2 for comparison of abstract vertex algebras and vertex algebras of series.

We have shown that a locally commutative subspace of $\mathrm{Hom}(\mathbf{M}, \mathbf{M}((z)))$ automatically becomes a vertex algebra by the residue products if it is closed under the residue products and contains $I(z)$. Conversely, any vertex algebra $\mathbf{V}$ is realized in this way by letting $\mathbf{M}$ be the vertex algebra $\mathbf{V}$ itself and considering the image $\mathcal{V}$ of the generating series map $Y(-, z)$. (cf. Section 1.2.4.)

In this regard, the axioms for vertex algebras are seen to be modelled on the properties of locally truncated locally commutative series with respect to the residue products in the same way as those for groups are modelled on properties of bijective transformations of a set with respect to composition of transformations.

1.2.3.3 Generation by a set of series

The following is called *Dong's Lemma*.

Lemma 2.7 *Let $A(z)$, $B(z)$, and $C(z)$ be locally truncated series on $\mathbf{V}$. If they are locally commutative with each other, then the residue products $A(z)_{(n)}B(z)$ and $C(z)$ are locally commutative for all $n \in \mathbb{Z}$.*

Let S be a locally commutative subset of $\mathrm{Hom}(\mathbf{M}, \mathbf{M}((z)))$. Then, by Lemma, we may construct a vertex algebra $\mathcal{V}$ by repeatedly applying residue products to the series belonging to S:

$$I(z), \; A(z), \; A(z)_{(m)}B(z), \; (A(z)_{(m)}B(z))_{(n)}C(z), \; A(z)_{(m)}(B(z)_{(n)}C(z)), \; \text{etc.}$$

We will denote the vertex algebra $\mathcal{V}$ thus defined by $\langle S \rangle_{RP}$ and call it the vertex algebra of series *generated by* S with respect to the residue products.

By the associativity formula (VA) for $\mathcal{V} = \langle S \rangle_{RP}$, the space is spanned by the elements of the following form:

$$A^1(z)_{(n_1)} \cdots A^k(z)_{(n_k)} I(z),$$

where $k \in \mathbb{N}$, $A^1(z), \ldots, A^k(z) \in S$ and $n_1, \ldots, n_k \in \mathbb{Z}$ and the operations are taken from right to left.

1.2.4 Identification of Vertex Algebras

Let $\mathbf{V}$ be a vertex algebra and $\mathcal{V}$ the image of the generating series map:

$$Y(-, z): \mathbf{V} \longrightarrow \mathrm{Hom}(\mathbf{V}, \mathbf{V}((z))).$$

Then $\mathcal{V}$ is a vertex algebra of series by Corollary 2.6, and the map $Y(-, z)$ is a homomorphism of vertex algebras by the associativity formula

$$Y(a_{(n)}b, z) = Y(a, z)_{(n)} Y(b, z).$$

Moreover, $Y(-, z)$ is an isomorphism onto $\mathcal{V}$ with the inverse given by

$$\sigma_{\mathcal{V}} : \mathcal{V} \longrightarrow \mathbf{V}, \quad Y(a, z) \mapsto Y(a, z)\mathbf{1}|_{z=0}.$$

Now, forget the vertex algebra structure on $\mathbf{V}$, but retain that Y is a linear isomorphism onto a vertex algebra $\mathcal{V}$ of series on $\mathbf{V}$ with the inverse given as above. Then the vertex algebra structure on $\mathbf{V}$ is reconstructed from $\mathcal{V}$ as

$$a_{(n)}b = \sigma_{\mathcal{V}}(Y(a, z)_{(n)} Y(b, z)).$$

In this section, we will construct a vertex algebra structure on a vector space $\mathbf{V}$ by identifying it with a vertex algebra of series along this line.

1.2.4.1 Creativity and the State Map

Let $\mathbf{V}$ be a vector space equipped with a candidate $\mathbf{1} \in \mathbf{V}$ of the vacuum. To relate a vertex algebra of series on $\mathbf{V}$ to a vertex algebra structure on $\mathbf{V}$, we introduce the concept of creativity for series.

A series $A(z)$ on $\mathbf{V}$ is said to be *creative* (with respect to $\mathbf{1} \in \mathbf{V}$) if

$$A_n \mathbf{1} = 0, \quad n \geq 0,$$

that is, $A(z)\mathbf{1} \in \mathbf{V}[[z]]$. We will say that a subset $\mathcal{V} \subset (\mathrm{End}\, \mathbf{V})[[z, z^{-1}]]$ is creative if so is every series in it.

Consider the following map, which we will call the *state map*:

$$\sigma : (\text{End } \mathbf{V})[[z, z^{-1}]] \longrightarrow \mathbf{V}, \quad A(z) \mapsto A_{-1}\mathbf{1}.$$

If $A(z)$ is creative, then $\sigma(A(z)) = A(z)\mathbf{1}\big|_{z=0}$.

Lemma 2.8 *Let $A(z)$ and $B(z)$ be locally truncated series on $\mathbf{V}$. If they are creative, then so are the residue products $A(z)_{(n)}B(z)$, for which*

$$\sigma(A(z)_{(n)}B(z)) = A_n B_{-1}\mathbf{1}$$

holds for all $n \in \mathbb{Z}$.

1.2.4.2 Identification by the State Map

Let $\mathcal{V}$ be a vertex algebra of series on $\mathbf{V}$. Assume that $\mathcal{V}$ is creative and the state map restricts to a bijection from $\mathcal{V}$ onto $\mathbf{V}$:

$$\sigma_{\mathcal{V}} = \sigma|_{\mathcal{V}} : \mathcal{V} \overset{\sim}{\longrightarrow} \mathbf{V}, \quad A(z) \mapsto A_{-1}\mathbf{1}.$$

Then we may transfer the vertex algebra structure on $\mathcal{V}$ to $\mathbf{V}$ via $\sigma_{\mathcal{V}}$. That is, for any $a, b \in \mathbf{V}$, choose $A(z), B(z) \in \mathcal{V}$ so that $\sigma(A(z)) = a$ and $\sigma(B(z)) = b$, and define the product as

$$a_{(n)}b = \sigma(A(z)_{(n)}B(z)).$$

By Lemma 2.8, we actually have

$$a_{(n)}b = A_n B_{-1}\mathbf{1} = A_n b.$$

As this holds for all $b \in \mathbf{V}$, we have

$$Y(a, z) = A(z).$$

Thus the map Y agrees with the inverse of $\sigma_{\mathcal{V}}$. Moreover, since the vacuum of $\mathcal{V}$ is the identity series $I(z)$, the vacuum of $\mathbf{V}$ is given by $\sigma(I(z)) = I_{-1}\mathbf{1} = \mathbf{1}$.

In summary, we have the following theorem.

Theorem 2.9 *Let $\mathcal{V}$ be a vertex algebra of series on $\mathbf{V}$ and assume that it is creative with respect to $\mathbf{1} \in \mathbf{V}$. If $\sigma_{\mathcal{V}} : \mathcal{V} \twoheadrightarrow \mathbf{V}$ is bijective, then $\mathbf{V}$ carries a unique structure of a vertex algebra such that*

$$Y(A_{-1}\mathbf{1}, z) = A(z) \text{ for all } A(z) \in \mathcal{V},$$

with vacuum $\mathbf{1}$.

Consider now the case when $\mathcal{V}$ is a vertex algebra of series generated by a locally commutative subset of $\text{Hom}(\mathbf{V}, \mathbf{V}((z)))$.

Corollary 2.10   *Let* **V** *be a vector space,* **1** *an element of* **V**, *and* $\mathcal{S}$ *a set of locally truncated series on* **V** *satisfying the following conditions:*

(1) $\mathcal{S}$ *is locally commutative.*
(2) $\mathcal{S}$ *is creative with respect to* **1**.
(3) $\sigma_{\mathcal{V}} : \mathcal{V} \longrightarrow$ **V** *is bijective for* $\mathcal{V} = \langle \mathcal{S} \rangle_{\mathrm{RP}}$.

Then **V** *carries a unique structure of a vertex algebra such that*

$$Y(S_{-1}\mathbf{1}, z) = S(z) \ \text{ for all } \ S(z) \in \mathcal{S}$$

with vacuum **1**.

1.2.4.3 Bijectivity by Translation Covariance

Let $\mathcal{V}$ be a vertex algebra of series on a vector space **V** and assume that $\mathcal{V}$ is creative with respect to $\mathbf{1} \in \mathbf{V}$. To transfer the vertex algebra structure to **V** by applying Theorem 2.9, we need know that the restriction $\sigma_{\mathcal{V}} = \sigma|_{\mathcal{V}}$ of the state map is bijective onto **V**.

For the convenience of readers, we will show that, under translation covariance, knowing surjectivity suffices, although we will not use this result in the rest of the sections.

Let $\mathbb{T} = (T^{(k)})$ be a sequence of operators on **V** indexed by positive integers:

$$T^{(k)} : \mathbf{V} \longrightarrow \mathbf{V} \ (k = 1, 2, \cdots).$$

We set $T^{(0)} = I$ and assume that the operators are iterative and annihilating **1** in the following senses, respectively:

$$T^{(i)}T^{(j)} = \binom{i+j}{i} T^{(i+j)} \ \text{ for } i, j \geq 0, \ \text{ and } \ T^{(k)}\mathbf{1} = 0 \ \text{ for } k \geq 1.$$

For an indeterminate x, define the formal exponentials $e^{\pm xT}$ by

$$e^{xT} = \sum_{k=0}^{\infty} x^k T^{(k)}, \ \ e^{-xT} = \sum_{k=0}^{\infty} (-x)^k T^{(k)}.$$

Then, by iterativity, they are inverse to each other:

$$e^{-xT} e^{xT} = 1 = e^{xT} e^{-xT}.$$

Now, let $A(z)$ be a series on **V**. It is said to be *translation covariant* (with respect to $\mathbb{T}$) if

$$e^{xT} A(z) e^{-xT} = e^{x\partial_z} A(z),$$

or equivalently,

$$e^{xT} A(z) e^{-xT} = A(x + z)\big|_{|x|<|z|}.$$

If $A(z)$ is translation covariant, then the coefficients of $v(z) = A(z)\mathbf{1}$ satisfy

$$T^{(k)}v_n = (-1)^k \binom{n}{k} v_{n-k}.$$

Therefore, $A(z)$ is creative and $v_{-k-1} = T^{(k)}v_{-1}$.

We will say that a subset $\mathcal{V} \subset (\text{End}\,\mathbf{V})[[z, z^{-1}]]$ is translation covariant if so is every series in it.

The following result is a variant of what is called *Goddard's uniqueness theorem*.

Proposition 2.11 *Let $\mathcal{V}$ be a locally commutative and translation covariant subspace of* $\text{Hom}(\mathbf{V}, \mathbf{V}((z)))$. *If $\sigma_{\mathcal{V}}$ is surjective, then it is bijective.*

Proof. Assume that $\sigma_{\mathcal{V}}$ is surjective. The result follows by showing injectivity of p and i in the diagram below:

$$
\begin{array}{ccccc}
\text{Hom}(\mathbf{V}, \mathbf{V}((z))) & \xrightarrow{\quad} & \mathbf{V}[[z]] & \xrightarrow{\quad} & \mathbf{V} \\
\cup & & \cup & & \| \\
\mathcal{V} & \xrightarrow{\;\;p\;\;} & \mathcal{V}\mathbf{1} & \xrightarrow{\;\;i\;\;} & \mathbf{V} \\
A(z) & \longmapsto & A(z)\mathbf{1} & \longmapsto & A_{-1}\mathbf{1}.
\end{array}
$$

Assume $A_{-1}\mathbf{1} = 0$. Then, by translation covariance, $A_{-k-1}\mathbf{1} = T^{(k)}A_{-1}\mathbf{1} = 0$ for all $k \geq 0$, thus $A(z)\mathbf{1} = 0$ since $A(z)$ is creative. Now, for any $b \in \mathbf{V}$, by surjectivity of $\sigma_{\mathcal{V}}$, there exists $B(z) \in \mathcal{V}$ such that $B_{-1}\mathbf{1} = b$. Since $A(z)$ and $B(z)$ are locally commutative, we have, for some $N \in \mathbb{N}$,

$$z^N A(z)b = (z-y)^N A(z)B(y)\mathbf{1}\big|_{y=0} = (z-y)^N B(y)A(z)\mathbf{1}\big|_{y=0} = 0.$$

Hence $A(z)b = 0$ holds for all $b \in \mathbf{V}$. Thus $A(z) = 0$. $\square$

Therefore, by Theorem 2.9, we have the following result.

Theorem 2.12 *Let $\mathbf{V}$ be a vector space, $\mathbb{T}$ a sequence of iterative operators annihilating $\mathbf{1} \in \mathbf{V}$, and $\mathcal{V}$ a vertex algebra of series on $\mathbf{V}$ satisfying:*

(1) *$\mathcal{V}$ is translation covariant with respect to $\mathbb{T}$.*
(2) *The state map σ restricts to a surjective map from $\mathcal{V}$ onto $\mathbf{V}$.*

Then $\mathbf{V}$ carries a unique structure of a vertex algebra such that

$$Y(A_{-1}\mathbf{1}, z) = A(z) \text{ for all } A(z) \in \mathcal{V}$$

with vacuum $\mathbf{1}$.

To deal with vertex algebras generated by series, we note the following.

Lemma 2.13   *Let $A(z)$ and $B(z)$ be locally truncated series on* **V***. If they are translation covariant, then so are the residue products $A(z)_{(n)}B(z)$ for all $n \in \mathbb{Z}$.*

Proof.   For $n \geq 0$, by $e^{x\partial_y} e^{x\partial_z} (y - z)^n = (y - z)^n$ and the Leibniz rule, we have

$$e^{x\partial_z} \operatorname{Res}_y (y - z)^n A(y)B(z) = \operatorname{Res}_y e^{x\partial_z} ((y - z)^n A(y)B(z))$$
$$= \operatorname{Res}_y e^{x\partial_y} e^{x\partial_z} ((y - z)^n A(y)B(z)) = \operatorname{Res}_y (y - z)^n (e^{x\partial_y} A(y))(e^{x\partial_z} B(z)),$$

as well as the one with the positions of $A(y)$ and $B(z)$ switched. Therefore, we have $e^{x\partial_z}(A(z)_{(n)}B(z)) = (e^{x\partial_z} A(z))_{(n)}(e^{x\partial_z} B(z))$, and the result is now clear by $e^{xT}(A(z)_{(n)}B(z))e^{-xT} = (e^{xT}A(z)e^{-xT})_{(n)}(e^{xT}B(z)e^{-xT})$. The same proof works for $n < 0$ as well by replacing $(y - z)^n$ with its expansions. $\square$

Therefore, we arrive at the following corollary, a variant of what is called the *existence theorem of Frenkel–Kac–Radul–Wang*.

Corollary 2.14   *Let* **V** *be a vector space,* $\mathbb{T}$ *a sequence of iterative operators annihilating* $\mathbf{1} \in$ **V***, and $\mathcal{S}$ a set of locally truncated series on* **V** *satisfying:*

(1) $\mathcal{S}$ *is locally commutative.*
(2) $\mathcal{S}$ *is translation covariant with respect to* $\mathbb{T}$.
(3) *The state map σ restricts to a surjective map from $\langle \mathcal{S} \rangle_{\mathrm{RP}}$ onto* **V**.

Then **V** *carries a unique structure of a vertex algebra such that*

$$Y(S_{-1}\mathbf{1}, z) = S(z) \quad \text{for all} \quad S(z) \in \mathcal{S}$$

with vacuum $\mathbf{1}$.

Note 2.15.   The equality $e^{x\partial_z}(A(z)_{(n)}B(z)) = (e^{x\partial_z} A(z))_{(n)}(e^{x\partial_z} B(z))$ achieved in the proof of Lemma 2.13 also follows from the Borcherds identity if $A(z)$ and $B(z)$ are not only locally truncated but also locally commutative.

1.2.5 Representations and Modules

Recall that a representation of a commutative associative algebra **C** is a vector space **M** equipped with a homomorphism $\rho \colon$ **C** $\longrightarrow$ End **M** of algebras. However, since End **M** is not commutative in general, let us alternatively define a representation of **C** to be a vector space **M** equipped with a map

$$\rho \colon \mathbf{C} \longrightarrow \operatorname{End} \mathbf{M}$$

satisfying the following properties:

(1) The image of ρ is a commutative associative algebra of operators on **M**.
(2) The map ρ gives a homomorphism of commutative associative algebras onto the image.

Then, letting the action of **C** on **M** be given by

$$\mathbf{C} \times \mathbf{M} \longrightarrow \mathbf{M}, \quad (a,v) \mapsto av = \rho(a)v,$$

the vector space **M** becomes a module over **C**.

In this section, we will follow this line to define the equivalent concepts of representations of a vertex algebra and modules over a vertex algebra.

1.2.5.1 Representations of Vertex Algebras

Let **V** be a vertex algebra and **M** a vector space. Consider a sequence (ρ_n) of countably many maps from **V** to End **M** indexed by integers $n \in \mathbb{Z}$,

$$\rho_n \colon \mathbf{V} \longrightarrow \text{End}\,\mathbf{M}, \quad n \in \mathbb{Z}.$$

For an indeterminate z, let us write the generating series as

$$\rho(a,z) = \sum_n \rho_n(a) z^{-n-1}, \quad a \in \mathbf{V},$$

which give rise to a map

$$\rho(-,z) \colon \mathbf{V} \longrightarrow \text{Hom}(\mathbf{M}, \mathbf{M}[[z,z^{-1}]]), \quad a \mapsto \rho(a,z).$$

The sequence (ρ_n) is said to be a *representation* of the vertex algebra **V** on the vector space **M** if the following conditions are satisfied:

(1) The image of $\rho(-,z)$ is a vertex algebra of series on **M**.
(2) The map $\rho(-,z)$ gives a homomorphism of vertex algebras onto the image.

Note that (1) in particular says that $\rho(a,z)$ is locally truncated for any $a \in \mathbf{V}$.

Let $\mathcal{V}$ be a vertex algebra of series on a vector space **M**. Then the obvious maps

$$\rho_n \colon \mathcal{V} \longrightarrow \text{End}\,\mathbf{M}, \quad A(z) \mapsto A_n$$

form a representation of $\mathcal{V}$ on **M**.

1.2.5.2 Modules over Vertex Algebras

Let $\rho = (\rho_n)$ be a representation of a vertex algebra **V** on a vector space **M**. Consider the corresponding actions of **V** on **M** given by

$$\mathbf{V} \times \mathbf{M} \longrightarrow \mathbf{M}, \quad (a,v) \mapsto a_n v = \rho_n(a)(v).$$

Then the conditions for representations imply the following properties:

(M0) Local truncation. For any $a \in \mathbf{V}$ and $v \in \mathbf{M}$, there exists an $N \in \mathbb{N}$ such that

$$a_{N+i}v = 0 \quad \text{for all } i \geq 0.$$

(M1) Borcherds identity. For all $a, b \in \mathbf{V}$, $v \in \mathbf{M}$, and $p, q, r \in \mathbb{Z}$:

$$\sum_{i=0}^{\infty} \binom{p}{i} (a_{(r+i)}b)_{p+q-i}v$$
$$= \sum_{i=0}^{\infty} (-1)^i \binom{r}{i} a_{p+r-i}b_{q+i}v - \sum_{i=0}^{\infty} (-1)^{r-i} \binom{r}{i} b_{q+r-i}a_{p+i}v.$$

(M2) Identity. For any $v \in \mathbf{M}$ and $n \in \mathbb{Z}$:

$$\mathbf{1}_n v = \begin{cases} 0 & (n \neq -1), \\ v & (n = -1). \end{cases}$$

A vector space $\mathbf{M}$ equipped with actions satisfying (M0)–(M2) is called a *module* over $\mathbf{V}$ or a $\mathbf{V}$-*module*. Conversely, any module $\mathbf{M}$ over $\mathbf{V}$ gives rise to a representation of $\mathbf{V}$ by letting $\rho_n(a) = a_n$ for $a \in \mathbf{V}$ and $n \in \mathbb{Z}$. Thus the concepts of modules and representations are essentially the same.

The generating series $\rho(a, z)$ for a $\mathbf{V}$-module $\mathbf{M}$ are usually denoted

$$Y_{\mathbf{M}}(a, z) = \sum_n a_n z^{-n-1}.$$

In particular, the vertex algebra $\mathbf{V}$ itself is thought of as a module over $\mathbf{V}$, called the *adjoint module*, for which $Y_{\mathbf{V}}(a, z) = Y(a, z)$ for all $a \in \mathbf{V}$.

1.2.5.3 Consequences of the Axioms

In the same way as in the case of the Borcherds identity for vertex algebras, we can derive various properties for modules. First note that (M0) means local truncation of the generating series $Y_{\mathbf{M}}(a, z)$:

$$Y_{\mathbf{M}}(a, z) \in \mathrm{Hom}(\mathbf{M}, \mathbf{M}((z))).$$

(MC) Commutator formula. For all $a, b \in \mathbf{V}$ and $m, n \in \mathbb{Z}$:

$$[a_m, b_n] = \sum_{i=0}^{\infty} \binom{m}{i} (a_{(i)}b)_{m+n-i}.$$

(MA) Associativity formula. For all $a, b \in \mathbf{V}$, and $m, n \in \mathbb{Z}$:

$$(a_{(m)}b)_n v = \sum_{i=0}^{\infty} (-1)^i \binom{m}{i} a_{m-i}b_{n+i}v - \sum_{i=0}^{\infty} (-1)^{m-i} \binom{m}{i} b_{m+n-i}a_i v.$$

In terms of generating series, they are written respectively as follows:

$$[Y_{\mathbf{M}}(a, y), Y_{\mathbf{M}}(b, z)] = \sum_{i=0}^{\infty} Y_{\mathbf{M}}(a_{(i)}b, z)\delta^{(i)}(y, z),$$

$$Y_{\mathbf{M}}(a_{(m)}b, z) = Y_{\mathbf{M}}(a, z)_{(m)}Y_{\mathbf{M}}(b, z) \quad (m \in \mathbb{Z}).$$

(MLC) Local commutativity. For any $a, b \in \mathbf{V}$, there exists an $N \in \mathbb{N}$ such that:

$$(y - z)^N Y_{\mathbf{M}}(a, y)Y_{\mathbf{M}}(b, z) = (y - z)^N Y_{\mathbf{M}}(b, z)Y_{\mathbf{M}}(a, y).$$

(MLA) Local associativity. For any $a \in \mathbf{V}$ and $v \in \mathbf{M}$, there exists an $L \in \mathbb{N}$ such that for all $b \in \mathbf{V}$:

$$(x + z)^L Y_{\mathbf{M}}(Y_{\mathbf{V}}(a, x)b, z)v = (x + z)^L Y_{\mathbf{M}}(a, x + z)\big|_{|x|>|z|} Y_{\mathbf{M}}(b, z)v.$$

Recall the translation operators:

$$T^{(k)} : \mathbf{V} \longrightarrow \mathbf{V}, \quad a \mapsto T^{(k)}a = a_{(-k-1)}\mathbf{1}.$$

The following property follows from (M1) and (M2):

(MT) Translation. For all $a \in \mathbf{V}$, $v \in \mathbf{M}$, $n \in \mathbb{Z}$ and $k \in \mathbb{N}$:

$$(T^{(k)}a)_n v = (-1)^k \binom{n}{k} a_{n-k} v.$$

Note 2.16. 1. The commutator formula (MC) is equivalent to the OPE

$$Y_{\mathbf{M}}(a, y)Y_{\mathbf{M}}(b, z) \simeq Y_{\mathbf{M}}(b, z)Y_{\mathbf{M}}(a, y) \sim \sum_{i=0}^{\infty} \frac{Y_{\mathbf{M}}(a_{(i)}b, z)}{(y - z)^{i+1}},$$

while the Borcherds identity (M1) is equivalent to

$$Y_{\mathbf{M}}(a, y)Y_{\mathbf{M}}(b, z) \simeq Y_{\mathbf{M}}(b, z)Y_{\mathbf{M}}(a, y) \simeq Y_{\mathbf{M}}(Y_{\mathbf{V}}(a, y - z)b, z).$$

Thus the Borcherds identity (M1) is characterized by the following condition: There exists an element $w(x, y, z) \in \mathbf{M}((x, y, z))$ such that

$$Y_{\mathbf{M}}(a, y)Y_{\mathbf{M}}(b, z)v = w(y - z, y, z)\big|_{|y|>|z|},$$

$$Y_{\mathbf{M}}(b, z)Y_{\mathbf{M}}(a, y)v = w(y - z, y, z)\big|_{|y|<|z|},$$

$$Y_{\mathbf{M}}(Y(a, x)b, z)v = w(x, x + z, z)\big|_{|x|<|y|}.$$

The Borcherds identity (V1) for vertex algebras, which is seen to be a particular case of (M1), is also formulated in the same way. 2. In the definition of modules, the Borcherds identity (M1) can be replaced by local associativity (MLA) thanks to skew-symmetry (VS) for the vertex algebra (cf. [8]).

Bibliographic Notes

General references for Section 1.2 are Kac [6], Matsuo and Nagatomo [7], Frenkel and Ben-Zvi [8], and Rosellen [11]. It is more or less straightforward to describe the materials covered in Section 1.2 over commutative rings. See Mason [78] for accounts over $\mathbb{Z}$.

For operator product expansions in various models in physics, consult [13] (cf. [19]). Formulation of local commutativity (locality) in terms of formal series is due to [4]. See [6] and [7] for formulation of operator product expansion in terms of formal series, and Tsuchiya and Kanie [96] for an earlier analytic formulation.

The statement of Theorem 2.5 is due to [7] (cf. [11]), although the result was implicitly given previously by Li in [70] in his proof of Corollary 2.6.

Construction of vertex algebras of series is due to [70], while properties of the residue products and their relation to vertex algebras are studied by Lian and Zuckerman [77]. See Goddard [20] for the original form of the uniqueness theorem and its applications, and Frenkel et al. [55] for the existence theorem.

1.3 Examples of Vertex Algebras

In Section 1.3, we will describe some standard examples of vertex algebras, the Heisenberg vertex algebra, the affine vertex algebras, and the Virasoro vertex algebras. Yet another class of standard examples, the lattice vertex algebras, will be described in Section 1.4.

For affine and Virasoro vertex algebras, we will first describe the *universal* ones as vector spaces and characterize the vertex algebra structure on them by conditions over a field of any characteristic not 2, and then describe some simple quotients over the field $\mathbb{C}$ of complex numbers.

1.3.1 Heisenberg Vertex Algebra

Let us start by constructing the Heisenberg vertex algebra of rank one briefly described in Subsection 1.1.4.2. Generalization to higher rank is straightforward (cf. Subsection 1.4.2.1).

Although they are particular cases of affine vertex algebras described in the next section, the Heisenberg vertex algebras have distinguished properties and play an important role in constructing the lattice vertex algebras in Section 1.4.

1.3.1.1 Heisenberg Algebra

Let a_n with $n \in \mathbb{Z}$ and ζ be indeterminates and set

$$\hat{\mathfrak{h}} = \bigoplus_{n \in \mathbb{Z}} \mathbb{F} a_n \oplus \mathbb{F}\zeta.$$

Then $\hat{\mathfrak{h}}$ becomes a Lie algebra by the bracket

$$[a_m, a_n] = m\delta_{m+n,0}\zeta, \quad [\zeta, a_n] = 0.$$

It is an infinite-dimensional *Heisenberg Lie algebra.*

Consider the quotient $\mathbf{U}(\hat{\mathfrak{h}}, 1)$ of the universal enveloping algebra $\mathbf{U}(\hat{\mathfrak{h}})$ by the two-sided ideal generated by $\zeta - 1$, where the scalar multiples of the unity of $\mathbf{U}(\hat{\mathfrak{h}})$ are identified with the scalars:

$$\mathbf{U}(\hat{\mathfrak{h}}, 1) = \mathbf{U}(\hat{\mathfrak{h}})/(\zeta - 1).$$

We will denote the images of the generators a_n by the same symbol. Then

$$[a_m, a_n] = m\delta_{m+n,0} \tag{3.1}$$

for $m, n \in \mathbb{Z}$, where the bracket denotes the commutator.

The algebra $\mathbf{U}(\hat{\mathfrak{h}}, 1)$ is the associative algebra generated by the symbols a_n, $n \in \mathbb{Z}$ subject to (3.1) as the fundamental relations. We will call it the *Heisenberg algebra.*

Let $\mathbf{M}$ be a $\mathbf{U}(\hat{\mathfrak{h}}, 1)$-module, regard the element a_n with $n \in \mathbb{Z}$ as operators acting on $\mathbf{M}$, and consider the generating series which we will call a *current* following physics terminology:

$$a(z) = \sum_n a_n z^{-n-1}.$$

As mentioned at the beginning of Section 1.2.2, we have

$$a(y)a(z) \simeq a(z)a(y) \sim \frac{1}{(y-z)^2}, \tag{3.2}$$

where the numerator 1 means the identity series $I(z)$. Note in particular that the current $a(z)$ is locally commutative with itself.

1.3.1.2 Fock Modules

Let $\hat{\mathfrak{h}}_{<0}$ and $\hat{\mathfrak{h}}_{\geq 0}$ be the Lie subalgebras of $\hat{\mathfrak{h}}$ spanned by a_n with $n < 0$ and $n \geq 0$, respectively. They are commutative, and generate subalgebras of $\mathbf{U}(\hat{\mathfrak{h}}, 1)$ isomorphic to the symmetric algebras $\mathbf{S}(\hat{\mathfrak{h}}_{<0})$ and $\mathbf{S}(\hat{\mathfrak{h}}_{\geq 0})$, respectively. By PBW for $\mathbf{U}(\hat{\mathfrak{h}})$, as vector spaces,

$$\mathbf{U}(\hat{\mathfrak{h}}, 1) = \mathbf{S}(\hat{\mathfrak{h}}_{<0}) \otimes \mathbf{S}(\hat{\mathfrak{h}}_{\geq 0}). \tag{3.3}$$

The element a_0 is central and, when acting on a module, its eigenvalue is called the *charge*.

Now, for each scalar λ, define a one-dimensional $S(\hat{\mathfrak{h}}_{\geq 0})$-module $\mathbb{F}\mathbf{v}_\lambda$ by

$$a_n \mathbf{v}_\lambda = \begin{cases} 0 & (n \geq 1), \\ \lambda \mathbf{v}_\lambda & (n = 0), \end{cases}$$

and consider the universal $U(\hat{\mathfrak{h}}, 1)$-module generated by $\mathbf{v}_\lambda$ given by

$$\mathbf{F}_\lambda = U(\hat{\mathfrak{h}}, 1) \otimes_{S(\hat{\mathfrak{h}}_{\geq 0})} \mathbb{F}\mathbf{v}_\lambda$$

on which $\hat{\mathfrak{h}}$ acts by left multiplication on $U(\hat{\mathfrak{h}}, 1)$. The $U(\hat{\mathfrak{h}}, 1)$-module $\mathbf{F}_\lambda$ thus obtained is called the *Fock module* of charge λ, also called the *Fock space* of charge λ.

Let us denote the element $1 \otimes \mathbf{v}_\lambda$ simply by $\mathbf{v}_\lambda$. By (3.3), we have

$$\mathbf{F}_\lambda = S(\hat{\mathfrak{h}}_{<0})\mathbf{v}_\lambda \simeq S(\hat{\mathfrak{h}}_{<0}).$$

The Fock module $\mathbf{F}_\lambda$ is characterized by the following universal property:

For any $U(\hat{\mathfrak{h}}, 1)$-module $\mathbf{M}$ and $\mathbf{w} \in \mathbf{M}$ satisfying

$$a_n \mathbf{w} = \begin{cases} 0 & (n \geq 1), \\ \lambda \mathbf{w} & (n = 0), \end{cases}$$

there exists a unique homomorphism $\psi : \mathbf{F}_\lambda \longrightarrow \mathbf{M}$ of $U(\hat{\mathfrak{h}}, 1)$-modules sending $\mathbf{v}_\lambda$ to $\mathbf{w}$.

Here the condition of ψ means that the following diagram commutes:

$$\begin{array}{ccc} \mathbb{F}\mathbf{v}_\lambda & \longrightarrow & \mathbb{F}\mathbf{w} \\ \downarrow & & \downarrow \\ \mathbf{F}_\lambda & \xrightarrow{\psi} & \mathbf{M}. \end{array}$$

Here the upper arrow sends $\mathbf{v}_\lambda$ to $\mathbf{w}$ and the vertical ones are inclusions.

We may further identify the Fock module with the polynomial ring via the linear isomorphism as in Table 3 defined by

$$\mathbb{F}[x_1, x_2, \cdots] \xrightarrow{\sim} \mathbf{F}_\lambda, \quad p(x_1, x_2, \cdots) \mapsto p(a_{-1}, a_{-2}, \cdots)\mathbf{v}_\lambda.$$

Then the actions of a_n for $n \neq 0$ turn out to be the same as (1.5) given by differential operators, while a_0 acts as multiplication by λ.

Among the Fock modules, the module $\mathbf{F}_0$ of charge 0 is called the *vacuum module* and the vector $\mathbf{v}_0$ the *vacuum*.

Table 3 Fock module of charge λ

0	1	2	3	4
1	x_1	x_2	x_3	x_4
		x_1^2	$x_1 x_2$	$x_1 x_3$
			x_1^3	$x_2 x_2$
				$x_1^2 x_2$
A basis of $\mathbb{F}[x_1, x_2, \cdots]$				x_1^4
v_λ	$a_{-1} v_\lambda$	$a_{-2} v_\lambda$	$a_{-3} v_\lambda$	$a_{-4} v_\lambda$
		$a_{-1} a_{-1} v_\lambda$	$a_{-1} a_{-2} v_\lambda$	$a_{-1} a_{-3} v_\lambda$
			$a_{-1} a_{-1} a_{-1} v_\lambda$	$a_{-2} a_{-2} v_\lambda$
				$a_{-1} a_{-1} a_{-2} v_\lambda$
A basis of $\mathbf{F}_\lambda$				$a_{-1} a_{-1} a_{-1} a_{-1} v_\lambda$

1.3.1.3 Vertex Algebra of Series on Fock Modules

Regard the current $a(z)$ as a series acting on $\mathbf{F}_\lambda$ for a $\lambda \in \mathfrak{h}^*$. Since it is locally truncated and locally commutative with itself, it generates a vertex algebra of series.

Recall that it is the span of the series obtained by repeatedly applying residue products of the current $a(z)$, which is actually spanned by the series of the form $a(z)_{(n_1)} \cdots a(z)_{(n_k)} I(z)$, where the residue products are taken from right to left. Let us denote it by

$$\mathcal{F}_0(\lambda) = \text{Span} \left\{ a(z)_{(n_1)} \cdots a(z)_{(n_k)} I(z) \,\middle|\, k \in \mathbb{N}, n_1, \ldots, n_k \in \mathbb{Z} \right\},$$

where λ signifies that the space consists of series acting on $\mathbf{F}_\lambda$.

Since $\mathcal{F}_0(\lambda)$ is a vertex algebra, the commutator formula is available. To describe it, consider the residue products of the current $a(z)$ with itself, which can be read off by extracting coefficients in the OPE (3.2) as

$$a(z)_{(n)} a(z) = \begin{cases} 0 & (n \geq 2), \\ 1 & (n = 1), \\ 0 & (n = 0). \end{cases}$$

Therefore, the commutator formula reads

$$[a(z)_{(m)}, a(z)_{(n)}] = \sum_{i=0}^{\infty} \binom{m}{i} (a(z)_{(i)} a(z))_{(m+n-i)} = m\delta_{m+n,0}.$$

Remarkably, this is the same as the Heisenberg commutation relation (3.1), and the space $\mathcal{F}_0(\lambda)$ becomes a $\mathbf{U}(\hat{\mathfrak{h}}, 1)$-module by

$$a_n : \mathcal{F}_0(\lambda) \longrightarrow \mathcal{F}_0(\lambda), \quad X(z) \mapsto a(z)_{(n)} X(z).$$

Table 4 Fock module of charge 0, abstract versus realization

0		2	3
$\mathbf{v}_\lambda$	$a_{-1}\mathbf{v}_\lambda$	$a_{-2}\mathbf{v}_\lambda$	$a_{-3}\mathbf{v}_\lambda$
		$a_{-1}a_{-1}\mathbf{v}_\lambda$	$a_{-1}a_{-2}\mathbf{v}_\lambda$
A basis of $\mathbf{F}_0$			$a_{-1}a_{-1}a_{-1}\mathbf{v}_\lambda$
$I(z)$	$a(z)_{(-1)}I(z)$	$a(z)_{(-2)}I(z)$	$a(z)_{(-3)}I(z)$
		$a(z)_{(-1)}a(z)_{(-1)}I(z)$	$a(z)_{(-1)}a(z)_{(-2)}I(z)$
A basis of $\mathcal{F}_0(\lambda)$			$a(z)_{(-1)}a(z)_{(-1)}a(z)_{(-1)}I(z)$

Moreover, since the identity series $I(z) = 1$ satisfies

$$a(z)_{(n)}I(z) = 0 \quad (n \geq 0),$$

the universal property of the Fock module implies that there exists a unique homomorphism of $\mathbf{U}(\hat{\mathfrak{h}}, 1)$-modules sending the vector $\mathbf{v}_0$ to the identity series $I(z)$ on $\mathbf{F}_\lambda$,

$$\psi_\lambda : \mathbf{F}_0 \longrightarrow \mathcal{F}_0(\lambda), \quad \mathbf{v}_0 \mapsto I(z), \qquad (3.4)$$

which is surjective since the $\mathbf{U}(\hat{\mathfrak{h}}, 1)$-module $\mathcal{F}_0(\lambda)$ is generated by $I(z)$ (cf. Table 4).

1.3.1.4 Identification of Heisenberg Vertex Algebra

For $\lambda = 0$, the current $a(z)$ is creative with respect to $\mathbf{v}_0$ and the map ψ_0 is inverse to the state map, hence the state map restricted to $\mathcal{F}_0(0)$ is bijective:

$$\sigma_{\mathcal{F}_0(0)} : \mathcal{F}_0(0) \xrightarrow{\ \sim\ } \mathbf{F}_0.$$

Therefore, we may transfer the vertex algebra structure on $\mathcal{F}_0(0)$ to $\mathbf{F}_0$ via the map $\sigma_{\mathcal{F}_0(0)}$, and general theory in Section 1.2 yields the following result, which restates Proposition 1.11.

Proposition 3.1 *The Fock module* $\mathbf{F}_0$ *of charge* 0 *carries a unique structure of a vertex algebra such that*

$$Y(a_{-1}\mathbf{v}_0, z) = a(z)$$

with vacuum $\mathbf{1} = \mathbf{v}_0$

For each $\lambda \in \mathfrak{h}^*$, the map (3.4) turns out to be a representation of $\mathbf{F}_0$, giving rise to a structure of a module over the Heisenberg vertex algebra $\mathbf{F}_0$ on the space $\mathbf{F}_\lambda$.

By considering the action of Heisenberg algebra, it is not difficult to show that, over a field of characteristic zero, the modules $\mathbf{F}_\lambda$ with $\lambda \in \mathfrak{h}^*$ are simple, as well as the Heisenberg vertex algebra $\mathbf{F}_0$ itself.

The construction of the Virasoro vector (1.8) is restated as

$$\omega = \frac{1}{2}a_{-1}a_{-1}\mathbf{1}. \tag{3.5}$$

The space $\mathbf{F}_0$ is given a grading by

$$\deg a_{-k_1-1}\cdots a_{-k_i-1}\mathbf{v}_0 = k_1 + \cdots + k_i,$$

which agrees with the eigenvalue for the action of L_0.

Let $\mathbf{F}_{0,d}$ denote the subspace of degree d. Then distribution of degrees is encoded in the *graded dimension*:

$$\sum_{d=0}^{\infty} q^d \dim \mathbf{F}_{0,d} = \prod_{k=1}^{\infty} \frac{1}{1-q^k} = \frac{1}{\phi(q)} = \frac{q^{1/24}}{\eta(\tau)},$$

where $\phi(q)$ is the Euler function and $\eta(\tau)$ the Dedekind eta function with $q = e^{2\pi\sqrt{-1}\tau}$.

1.3.2 Affine Vertex Algebras

The Heisenberg Lie algebra as in the preceding section is actually a particular example of an *affine Lie algebra*.

In this section, we will describe affine Lie algebras and the associated vertex algebras.

1.3.2.1 Affine Lie Algebras

A bilinear form $(\ |\ )$ on a Lie algebra $\mathfrak{g}$ is said to be *invariant* (with respect to the adjoint action of $\mathfrak{g}$) if, for all $X, Y, Z \in \mathfrak{g}$,

$$([X,Y]|Z) = (X|[Y,Z]).$$

Let $\mathbb{F}[t,t^{-1}]$ denote the ring of Laurent polynomials in t and let K be an indeterminate.

For a Lie algebra $\mathfrak{g}$, set

$$\hat{\mathfrak{g}} = \mathfrak{g} \otimes \mathbb{F}[t,t^{-1}] \oplus \mathbb{F}K.$$

Denote the element $X \otimes t^n$ for $X \in \mathfrak{g}$ and $n \in \mathbb{Z}$ as

$$X_n = X \otimes t^n.$$

Given a bilinear form $(\ |\ )$ on $\mathfrak{g}$, define a bracket operation by setting

$$[X_m, Y_n] = [X,Y]_{m+n} + m\delta_{m+n,0}(X|Y)K, \quad [K, X_n] = 0.$$

If the bilinear form is symmetric and invariant, then the bracket equips $\hat{\mathfrak{g}}$ with a structure of a Lie algebra. The Lie algebra $\hat{\mathfrak{g}}$ thus obtained is called the *affine Lie algebra* associated with $\mathfrak{g}$ and $(\ |\ )$.

A $\hat{\mathfrak{g}}$-module on which the central element K acts by a scalar k is said to be of *level* k, which is the same as a module over the quotient

$$\mathbf{U}(\hat{\mathfrak{g}}, k) = \mathbf{U}(\hat{\mathfrak{g}})/(K - k),$$

where the scalar k is identified with the multiple of the unity by k.

Let $\mathbf{M}$ be a $\hat{\mathfrak{g}}$-module of level k, regard the elements X_n for each $X \in \mathfrak{g}$ as operators on $\mathbf{M}$, and consider the generating series, again called a *current*:

$$X(z) = \sum_n X_n z^{-n-1}.$$

Then the commutation relation for $[X_m, Y_n]$ is equivalently described by the OPE:

$$X(y)Y(z) \simeq Y(z)X(y) \sim \frac{[X,Y](z)}{y - z} + \frac{k(X|Y)}{(y - z)^2}. \tag{3.6}$$

In particular, $X(z)$ and $Y(z)$ are locally commutative with each other.

1.3.2.2 Generalized Verma Modules

Consider the following Lie subalgebras of the affine Lie algebra $\hat{\mathfrak{g}}$:

$$\hat{\mathfrak{g}}_{<0} = \mathfrak{g} \otimes \mathbb{F}[t^{-1}]t^{-1}, \quad \hat{\mathfrak{g}}_{\geq 0} = \mathfrak{g} \otimes \mathbb{F}[t].$$

They generate subalgebras of $\mathbf{U}(\hat{\mathfrak{g}}, k)$ isomorphic to $\mathbf{U}(\hat{\mathfrak{g}}_{<0})$ and $\mathbf{U}(\hat{\mathfrak{g}}_{\geq 0})$, respectively, and PBW for $\mathbf{U}(\hat{\mathfrak{g}})$ implies

$$\mathbf{U}(\hat{\mathfrak{g}}, k) = \mathbf{U}(\hat{\mathfrak{g}}_{<0}) \otimes \mathbf{U}(\hat{\mathfrak{g}}_{\geq 0}).$$

Let V be a $\mathfrak{g}$-module and regard it as a $\mathbf{U}(\hat{\mathfrak{g}}_{\geq 0})$-module in the following way: for $X \in \mathfrak{g}$ and $v \in V$,

$$X_n v = \begin{cases} 0 & (n \geq 1), \\ Xv & (n = 0). \end{cases}$$

Define a $\hat{\mathfrak{g}}$-module $\mathbf{M}_k(V)$ by

$$\mathbf{M}_k(V) = \mathbf{U}(\hat{\mathfrak{g}}, k) \otimes_{\mathbf{U}(\hat{\mathfrak{g}}_{\geq 0})} V,$$

where the action of $\hat{\mathfrak{g}}$ is by left multiplication on $\mathbf{U}(\hat{\mathfrak{g}}, k)$. This is a particular type of what is called the *generalized Verma module.*

When V is the one-dimensional trivial $\mathfrak{g}$-module $\mathbb{F}v_0$, the module $\mathbf{M}_k(\mathbb{F}v_0)$ is called the *universal vacuum module,* and the vector v_0 the *vacuum vector.* We will often identify $\mathbb{F}v_0$ with $\mathbb{F}$ and denote the universal vacuum module by $\mathbf{M}_k(\mathbb{F})$, which has an obvious universal property by construction.

1.3.2.3 Affine Vertex Algebras

Let V be a $\mathfrak{g}$-module and consider the induced module $\mathbf{M}_k(V)$. Regard the currents $X(z)$ for $X \in \mathfrak{g}$ as series acting on $\mathbf{M}_k(V)$, which are locally truncated. Since they are locally commutative, they generate a vertex algebra of series:

$$\mathcal{M}_k(\mathbb{F})_V = \left\{ X^1(z)_{(n_1)} \cdots X^l(z)_{(n_l)} I(z) \,\middle|\, \begin{matrix} l \in \mathbb{N},\, X^1, \ldots, X^l \in \mathfrak{g} \\ n_1, \ldots, n_l \in \mathbb{Z} \end{matrix} \right\}.$$

By the OPE (3.6),

$$X(z)_{(n)} Y(z) = \begin{cases} 0 & (n \geq 2), \\ k(X \,|\, Y) & (n = 1), \\ [X, Y](z) & (n = 0). \end{cases}$$

The commutator formula implies

$$[X(z)_{(m)}, Y(z)_{(n)}] = [X, Y](z)_{(m+n)} + k(X \,|\, Y) m \delta_{m+n,0}.$$

The identity series satisfies

$$X(z)_{(n)} I(z) = 0 \quad (n \geq 0).$$

Therefore, by the universal property of $\mathbf{M}_k(\mathbb{F})$, there exists a unique homomorphism of $\mathbf{U}(\mathfrak{g}, k)$-modules sending the vector $\mathbf{v}_0$ to the identity series $I(z)$:

$$\psi_V : \mathbf{M}_k(\mathbb{F}) \longrightarrow \mathcal{M}_k(\mathbb{F})_V, \quad \mathbf{v}_0 \mapsto I(z),$$

which is surjective since the $\mathbf{U}(\mathfrak{g}, k)$-module $\mathcal{M}_k(\mathbb{F})_V$ is generated by $I(z)$.

When V is the one-dimensional trivial $\mathfrak{g}$-module $\mathbb{F} = \mathbb{F}\mathbf{v}_0$, the currents are creative with respect to $\mathbf{v}_0$, and the map $\psi = \psi_{\mathbb{F}}$ is inverse to the state map.

Proposition 3.2 *The universal vacuum module $\mathbf{M}_k(\mathbb{F})$ of level k over the affine Lie algebra $\hat{\mathfrak{g}}$ carries a unique structure of a vertex algebra such that*

$$Y(X_{-1}\mathbf{v}_0, z) = X(z), \quad X \in \mathfrak{g},$$

with vacuum $\mathbf{1} = \mathbf{v}_0$.

The vertex algebra thus obtained is called the *universal affine vertex algebra* associated with $\mathfrak{g}$ at level k. The module $\mathbf{M}_k(V)$ associated with a $\mathfrak{g}$-module V becomes a module over the vertex algebra $\mathbf{M}_k(\mathbb{F})$.

A quotient of $\mathbf{M}_k(\mathbb{F})$ is generally called an *affine vertex algebra* associated with $\mathfrak{g}$ at level k. The structure of such a quotient, including the simple one, heavily depends on the Lie algebra $\mathfrak{g}$ and the level k.

Note 3.3. Consider the subspace of $\mathbf{M}_k(\mathbb{F})$ spanned by $X_{(-1)}\mathbf{1}$ with $X \in \mathfrak{g}$:

$$\mathbf{M}_k(\mathbb{F})_1 = \mathrm{Span} \left\{ X_{(-1)}\mathbf{1} \,\middle|\, X \in \mathfrak{g} \right\}.$$

Then the 0th product equips it with a structure of a Lie algebra and the coefficient to $\mathbf{1}$ of the 1st product gives a symmetric bilinear form on it. The map

$$i\colon \mathfrak{g} \longrightarrow \mathbf{M}_k(\mathbb{F})_1, \quad X \mapsto X_{(-1)}\mathbf{1}$$

is an isomorphism of Lie algebras, which is an isometry multiplied by k.

1.3.2.4 Integrable Highest Weight Modules over $\widehat{\mathfrak{sl}}_2$

Let us consider the case when $\mathfrak{g}$ is the three-dimensional simple Lie algebra $\mathfrak{sl}_2$. We assume that the base field is $\mathbb{C}$.

The Lie algebra $\mathfrak{sl}_2$ is spanned by

$$E = \begin{bmatrix} 0 & 1 \\ 0 & 0 \end{bmatrix}, \quad H = \begin{bmatrix} 1 & 0 \\ 0 & -1 \end{bmatrix}, \quad F = \begin{bmatrix} 0 & 0 \\ 1 & 0 \end{bmatrix}$$

with the brackets

$$[H,E] = 2E, \quad [H,F] = -2F, \quad [E,F] = H.$$

An invariant bilinear form on $\mathfrak{sl}_2$ is a scalar multiple of the Killing form. We normalize it as

$$(H|H) = 2, \quad (E|F) = (F|E) = 1,$$
$$(E|E) = (F|F) = (H|E) = (H|F) = 0.$$

Let $\widehat{\mathfrak{sl}}_2$ denote the associated affine Lie algebra, which is (the derived algebra of) the affine Kac–Moody algebra of type $A_1^{(1)}$.

Finite-dimensional simple $\mathfrak{sl}_2$-modules are classified by their dimensions. We will denote the $(2j + 1)$-dimensional simple module by V_j, where j is a nonnegative half integer. The representation V_j is said to be of *spin j*.

1. The module V_0 corresponds to the one-dimensional *trivial representation.*
2. The module $V_{1/2}$ corresponds to the two-dimensional *vector representation,* the representation defining $\mathfrak{sl}_2$ by 2×2 matrices.
3. The module V_1 corresponds to the three-dimensional *adjoint representation,* the representation by the adjoint action of $\mathfrak{sl}_2$ on itself.

Let $\mathbf{M}(k, j)$ denote the module $\mathbf{M}_k(V_j)$ and $\mathbf{L}(k, j)$ its simple quotient. The image of an element of $\mathbf{M}(k, j)$ in the quotient $\mathbf{L}(k, j)$ will be denoted by the same symbol by abuse of notation.

For a positive integer k, consider the $k + 1$ simple quotients

$$\mathbf{L}(k, 0), \mathbf{L}(k, 1/2), \ldots, \mathbf{L}(k, k/2).$$

Table 5 Universal and simple $\widehat{sl}_2$-modules

0	1	2	0	1	2
		$E_{-1}E_{-1}v_0$			
		$E_{-2}v_0$			$E_{-2}v_0$
	$E_{-1}v_0$	$H_{-1}E_{-1}v_0$		$E_{-1}v_0$	
		$F_{-1}E_{-1}v_0$			
v_0	$H_{-1}v_0$	$H_{-2}v_0$	v_0	$H_{-1}v_0$	$H_{-2}v_0$
		$H_{-1}H_{-1}v_0$			$H_{-1}H_{-1}v_0$
	$F_{-1}v_0$	$F_{-1}H_{-1}v_0$		$F_{-1}v_0$	
		$F_{-2}v_0$			$F_{-2}v_0$
A basis of $\mathbf{M}(k,0)$		$F_{-1}F_{-1}v_0$	A basis of $\mathbf{L}(1,0)$		
$\underline{\mathbf{1}}$	$\underline{\mathbf{3}}$	$\underline{\mathbf{1}} \oplus \underline{\mathbf{3}} \oplus \underline{\mathbf{5}}$	$\underline{\mathbf{1}}$	$\underline{\mathbf{3}}$	$\underline{\mathbf{1}} \oplus \underline{\mathbf{3}}$

Then they form an important class of representations of the affine Kac–Moody Lie algebra $\widehat{sl}_2$, called the *integrable highest weight representations* of level k. Among them, $\mathbf{L}(k,0)$ is a simple vertex algebra, and the rest as well as itself are simple modules over it.

For example, when $k = 1$, the vector $E_{-1}E_{-1}v_0$ generates a maximal proper submodule of $\mathbf{M}(1,0)$ and the quotient $\mathbf{L}(1,0)$ is a simple vertex algebra (Table 5). The last rows exhibit the decomposition of each subspace under the action of sl_2 given by E_0, H_0, F_0, where $\underline{\mathbf{1}} = V_0$, $\underline{\mathbf{3}} = V_1$, and $\underline{\mathbf{5}} = V_2$.

For $k \neq -2$, we may consider the *Sugawara vector* defined by

$$\omega_k = \frac{1}{2(k+2)}\left(\frac{1}{2}H_{-1}H_{-1} + E_{-1}F_{-1} + F_{-1}E_{-1}\right)v_0. \qquad (3.7)$$

Then it becomes a Virasoro vector of central charge $c_k = 3k/(k+2)$. That is, the operators $L_n = \omega_{(n+1)}$ with $n \in \mathbb{Z}$ satisfy the commutation relation:

$$[L_m, L_n] = (m-n)L_{m+n} + \frac{m^3 - m}{12}c_k\delta_{m+n,0}, \quad c_k = \frac{3k}{k+2}.$$

We take the Virasoro vector ω_k defined by (3.7) as the standard choice for the affine vertex algebra associated with sl_2.

Note 3.4. 1. In this construction, the Lie algebra sl_2 can be replaced by any finite-dimensional simple Lie algebra $\mathfrak{g}$. Among the integrable highest weight modules over the affine Kac–Moody algebra $\hat{\mathfrak{g}}$ of level k, for which k is a positive integer, the vacuum module $\mathbf{L}(k,0)$ becomes a vertex algebra and the simple modules are classified as $\mathbf{L}(k,\lambda)$ where λ runs over the *dominant integral weights of level k*. For details, see [17]. 2. The construction (3.7) is called the *Sugawara construction*, which works for any finite-dimensional simple Lie algebra $\mathfrak{g}$ by means of the Casimir element of $\mathfrak{g}$ as long as $k + h^\vee \neq 0$, where $h^\vee$ the dual Coxeter number, that is, half the value of the Casimir action on the adjoint module, and the central charge of the resulting Virasoro action is given by $c_k = k \dim \mathfrak{g}/(k + h^\vee)$. For example, if $\mathfrak{g} = \mathfrak{sl}_2$, then $h^\vee = 2$, so $c_1 = 1$, $c_2 = 3/2$, $c_3 = 9/5$, etc.

1.3.3 Virasoro Vertex Algebras

In this section, we will describe the universal Virasoro vertex algebras and their simple quotients. There is a particularly nice family of such quotients, called the *Virasoro minimal models*.

1.3.3.1 Virasoro Algebras

Let L_n, $n \in \mathbb{Z}$ and C be indeterminates and set

$$\text{Vir} = \bigoplus_{n \in \mathbb{Z}} \mathbb{F}L_n \oplus \mathbb{F}C.$$

Then Vir becomes a Lie algebra, called the *Virasoro algebra*, by the bracket

$$[L_m, L_n] = (m - n)L_{m+n} + \frac{m^3 - m}{12}\delta_{m+n,0}C, \quad [C, L_n] = 0.$$

A Vir-module on which the central element C acts by a scalar c is said to be of *central charge c*. It is equivalently described as a module over the quotient

$$\text{U}(\text{Vir}, c) = \text{U}(\text{Vir})/(C - c),$$

where the scalar c is identified with the multiple of the unity by c.

Let **M** be a Vir-module of central charge c, regard the element L_n for $n \in \mathbb{Z}$ as operators acting on **M**, and consider the generating series which we denote, following physics, as follows:

$$T(z) = \sum_n L_n z^{-n-2}.$$

Then the Virasoro commutation relation is equivalently described by the OPE:

$$T(y)T(z) \simeq T(z)T(y) \sim \frac{\partial T(z)}{y - z} + \frac{2T(z)}{(y - z)^2} + \frac{c/2}{(y - z)^4}.$$

In particular, $T(z)$ is locally commutative with itself.

1.3.3.2 Verma Modules

Let $\text{Vir}_{<0}$ and $\text{Vir}_{\geq 0}$ be the subspaces of Vir spanned by L_n with $n < 0$ and $n \geq 0$, respectively, which form Lie subalgebras of Vir. They generate subalgebras of $\text{U}(\text{Vir}, c)$ isomorphic to $\text{U}(\text{Vir}_{<0})$ and $\text{U}(\text{Vir}_{\geq 0})$, respectively, and

$$\text{U}(\text{Vir}, c) = \text{U}(\text{Vir}_{<0}) \otimes \text{U}(\text{Vir}_{\geq 0}).$$

Table 6 Virasoro Verma module

0	1	2	3	4
$\mathbf{v}_h$	$L_{-1}\mathbf{v}_h$	$L_{-2}\mathbf{v}_h$	$L_{-3}\mathbf{v}_h$	$L_{-4}\mathbf{v}_h$
		$L_{-1}L_{-1}\mathbf{v}_h$	$L_{-2}L_{-1}\mathbf{v}_h$	$L_{-2}L_{-2}\mathbf{v}_h$
			$L_{-1}L_{-1}L_{-1}L_{-1}\mathbf{v}_h$	$L_{-3}L_{-1}\mathbf{v}_h$
				$L_{-2}L_{-1}L_{-1}\mathbf{v}_h$
A basis of $\mathbf{M}(c,h)$				$L_{-1}L_{-1}L_{-1}L_{-1}\mathbf{v}_h$

For each scalar h, define a one-dimensional $\mathbf{U}(\mathrm{Vir}_{\geq 0})$-module $\mathbb{F}\mathbf{v}_h$ by

$$L_n \mathbf{v}_h = \begin{cases} 0 & (n \geq 1), \\ h\mathbf{v}_h & (n = 0), \end{cases}$$

and a Vir-module $\mathbf{M}(c,h)$ by

$$\mathbf{M}(c,h) = \mathbf{U}(\mathrm{Vir},c) \otimes_{\mathbf{U}(\mathrm{Vir}_{\geq 0})} \mathbb{F}\mathbf{v}_h,$$

where the action of Vir is by left multiplication on $\mathbf{U}(\mathrm{Vir},c)$ (cf. Table 6). The resulting Vir-module is a highest weight module, called the *Verma module* of highest (conformal) weight h (although the value of the weight h is actually the lowest).

An element v of a Vir-module M is said to be a *singular vector* (for the Virasoro action) or a *primary vector* of M if the following condition holds:

$$L_n v = 0 \ (n \geq 1).$$

One often assumes that v is an eigenvector with respect to the action of L_0.

1.3.3.3 Virasoro Vertex Algebras

The module $\mathbf{M}(c,0)$ actually does not carry a natural structure of a vertex algebra. Indeed, if so with $\mathbf{1} = \mathbf{v}_0$, then, by the creation property, we must have

$$L_{-1}\mathbf{1} = \omega_{(0)}\mathbf{1} = 0.$$

We are thus led to consider a Virasoro module generated by a highest weight vector that is annihilated not only by L_n for $n \geq 0$ but also by L_{-1}.

To construct a universal one among such, let $\mathrm{Vir}_{\geq -1}$ be the subspace of Vir spanned by L_n with $n \geq -1$, which becomes a Lie subalgebra of Vir and generates a subalgebra of $\mathbf{U}(\mathrm{Vir},c)$ isomorphic to $\mathbf{U}(\mathrm{Vir}_{\geq -1})$.

Consider the one-dimensional trivial $\mathbf{U}(\mathrm{Vir}_{\geq -1})$-module with $\mathbf{v}_0$ a basis and define a Vir-module by

$$\mathbf{V}(c) = \mathbf{U}(\mathrm{Vir},c) \otimes_{\mathbf{U}(\mathrm{Vir}_{\geq -1})} \mathbb{F}\mathbf{v}_0.$$

Table 7 Universal vacuum module

0	1	2	3	4	5	6
v_0	$L_{-2}v_0$	$L_{-3}v_0$	$L_{-4}v_0$	$L_{-5}v_0$	$L_{-6}v_0$
			$L_{-2}L_{-2}v_0$	$L_{-2}L_{-3}v_0$	$L_{-2}L_{-4}v_0$
					$L_{-3}L_{-3}v_0$
A basis of $\mathbf{V}(c)$					$L_{-2}L_{-2}L_{-2}v_0$

Then $\mathbf{V}(c)$ is a highest weight Vir-module, which we will call the *universal vacuum module*. (See Table 7.)

The module $\mathbf{V}(c)$ is also described as a quotient of $\mathbf{M}(c,0)$ as

$$\mathbf{V}(c) = \mathbf{M}(c,0)/U(\mathrm{Vir},c)L_{-1}v_0,$$

where $U(\mathrm{Vir},c)L_{-1}v_0$ is the submodule generated by $L_{-1}v_0$. Note that the vector $L_{-1}v_0$ is a singular vector of weight 1 in $\mathbf{M}(c,0)$.

Proposition 3.5 *The universal vacuum module* $\mathbf{V}(c)$ *of central charge* c *over the Virasoro algebra carries a unique structure of a vertex algebra such that*

$$Y(L_{-2}v_0, z) = T(z)$$

with vacuum $\mathbf{1} = v_0$.

The vertex algebra $\mathbf{V}(c)$ thus obtained is called the *universal Virasoro vertex algebra* of central charge c. It is generated by the Virasoro vector $\omega = L_{-2}\mathbf{1}$. The quotients of $\mathbf{M}(c,h)$ are modules over $\mathbf{V}(c)$ for any $h \in \mathbb{F}$.

The simple quotient of $\mathbf{V}(c)$ is called the *simple Virasoro vertex algebra* and denoted $\mathbf{L}(c,0)$. For example, the Virasoro vector (3.5) generates a vertex subalgebra in the Heisenberg vertex algebra of rank one, which is isomorphic to $\mathbf{L}(1,0)$ over $\mathbb{C}$.

Theory of simple Virasoro vertex algebras $\mathbf{L}(c,0)$ heavily relies on representation theory of the Virasoro algebra, for its structure and properties seriously change by presence of singular vectors in $\mathbf{V}(c)$ for special values of c.

1.3.3.4 Virasoro Minimal Models

In this subsection, we will work over the field $\mathbb{C}$ of complex numbers.

Let p, q be a pair of coprime positive integers and consider the rational number $c_{p,q}$ defined by

$$c_{p,q} = 1 - \frac{6(p-q)^2}{pq}.$$

For positive integers r, s, consider the following numbers:

$$h_{r,s} = \frac{(pr - qs)^2 - (p - q)^2}{4pq}.$$

We have the following list of simple Virasoro modules of central charge $c_{p,q}$:

$$\mathbf{L}(c_{p,q}, h_{r,s}) \ (1 \le r \le q - 1, \ 1 \le s \le p - 1), \tag{3.8}$$

which constitute *Virasoro minimal models* in physics. Among them, $\mathbf{L}(c_{p,q}, 0)$ is a simple vertex algebra and the rest as well as itself are the simple modules over it. By $h_{r,s} = h_{q-r,p-s}$, we have $(p - 1)(q - 1)/2$ simple modules.

When $(p, q) = (m + 3, m + 2)$ for some $m = 1, 2, \cdots$, set

$$c_m = c_{m+3, m+2} = 1 - \frac{6}{(m + 2)(m + 3)}.$$

The corresponding Virasoro modules in (3.8) are precisely the unitarizable ones among the modules in the minimal models.

Here are some examples.

1. For $m = 1$, we have $(p, q) = (4, 3)$ and $c = 1/2$. The simple vertex algebra $\mathbf{L}(1/2, 0)$ is related to the *Ising model* in physics. The list of simple modules is as follows:
$$\mathbf{L}(1/2, 0), \ \mathbf{L}(1/2, 1/2), \ \mathbf{L}(1/2, 1/16).$$

2. For $m = 2$, we have $(p, q) = (5, 4)$ and $c_{p,q} = 7/10$. The simple vertex algebra $\mathbf{L}(7/10, 0)$ is related to the *tricritical Ising model* in physics. There are six simple modules.

3. For $m = 3$, we have $(p, q) = (6, 5)$ and $c_{p,q} = 4/5$. The simple vertex algebra $\mathbf{L}(4/5, 0)$ is related to the *three-state Potts model* in physics. There are ten simple modules.

The corresponding lists of highest weights $h_{r,s}$ are given in Table 8, respectively. Such a table is called the *Kac table* or the *conformal grid.*

Here is an example of a nonunitary minimal model.

For $(p, q) = (5, 2)$, the simple vertex algebra $\mathbf{L}(-22/5, 0)$ is related to the *Lee–Yang model* in physics. The list of simple modules is as follows:

$$\mathbf{L}(-22/5, 0), \ \mathbf{L}(-22/5, -1/5).$$

The last model is interesting in its relation to the Rogers–Ramanujan identities.

Table 8 Virasoro minimal models

3	$\frac{1}{2}$	0
2	$\frac{1}{16}$	$\frac{1}{16}$
1	0	$\frac{1}{2}$
s/r	1	2

$$c = 1/2$$

4	$\frac{3}{2}$	$\frac{7}{16}$	0
3	$\frac{3}{5}$	$\frac{3}{80}$	$\frac{1}{10}$
2	$\frac{1}{10}$	$\frac{3}{80}$	$\frac{3}{5}$
1	0	$\frac{7}{16}$	$\frac{3}{2}$
s/r	1	2	3

$$c = 7/10$$

5	3	$\frac{7}{5}$	$\frac{2}{5}$	0
4	$\frac{13}{8}$	$\frac{21}{40}$	$\frac{1}{40}$	$\frac{1}{8}$
3	$\frac{2}{3}$	$\frac{1}{15}$	$\frac{1}{15}$	$\frac{2}{3}$
2	$\frac{1}{8}$	$\frac{1}{40}$	$\frac{21}{40}$	$\frac{13}{8}$
1	0	$\frac{2}{5}$	$\frac{7}{5}$	3
s/r	1	2	3	4

$$c = 4/5$$

1.3.3.5 Ising Model and Majorana Fermions

Let us briefly describe an alternative construction of the simple Virasoro vertex algebra $\mathbf{L}(1/2,0)$ related to the Ising model over $\mathbb{C}$.

Recall that the simple Virasoro vertex algebra $\mathbf{L}(1,0)$ of central charge 1 is realized by the standard Virasoro vector in the Heisenberg vertex algebra $\mathbf{F}_0 = \mathbb{C}[x_1, x_2, \cdots]$, which is a mathematical formulation of free boson theory in physics.

In contrast, the simple Virasoro vertex algebra $\mathbf{L}(1/2,0)$ of central charge $1/2$, is realized by replacing the polynomial ring $\mathbb{C}[x_1, x_2, \cdots]$ by the exterior algebra $\Lambda(x_1, x_2, \cdots)$, resulting in the theory of *free Majorana fermions*, where the structure is described not by a vertex algebra but a vertex *superalgebra*, which is, roughly speaking, obtained by replacing the commutator by the anti-commutator when the operators are both from the odd subspace.

Let us consider the associative algebra generated by

$$\cdots, \psi_{-3/2}, \psi_{-1/2}, \psi_{1/2}, \cdots$$

subject to the fundamental relations

$$\psi_{m+1/2}\psi_{n+1/2} + \psi_{n+1/2}\psi_{m+1/2} = \delta_{m+n+1,0}.$$

Let us denote it by $\mathbf{A}^\psi$, which is the counterpart of $\mathrm{U}(\hat{\mathfrak{h}}, 1)$, and consider the generating series

$$\psi(z) = \sum_n \psi_{n+1/2} z^{-n-1}.$$

Then, we have $\psi(y)\psi(z) + \psi(z)\psi(y) = \delta(y, z)$ and, by considering local anti-commutativity instead of local commutativity, we have

$$\psi(y)\psi(z) \simeq -\psi(z)\psi(y) \sim \frac{1}{y-z}.$$

In particular, $\psi(z)$ is locally anticommutative with itself.

Let $\mathbf{A}^{\psi}_{>0}$ be the subalgebra generated by $\psi_{n+1/2}$ with $n \geq 0$ and consider the one-dimensional $\mathbf{A}^{\psi}_{>0}$-module $\mathbb{C}\mathbf{v}_0$ characterized by

$$\psi_{n+1/2}\mathbf{v}_0 = 0, \quad n \geq 0.$$

Define the fermionic Fock space $\mathbf{F}_{\psi}$ by setting

$$\mathbf{F}_{\psi} = \mathbf{A}^{\psi} \otimes_{\mathbf{A}^{\psi}_{>0}} \mathbb{C}\mathbf{v}_0 \simeq \Lambda(x_1, x_2, \cdots).$$

The space $\mathbf{F}_{\psi}$ carries a unique structure of a vertex superalgebra such that $Y(\psi_{-1/2}\mathbf{v}_0, z) = \psi(z)$ with vacuum $\mathbf{1} = \mathbf{v}_0$.

By analogy with the construction of the standard Virasoro vector (3.5) for the Heisenberg vertex algebra, consider the vector

$$\omega = \frac{1}{2}\psi_{-3/2}\psi_{-1/2}\mathbf{v}_0.$$

Then it generates a vertex subalgebra isomorphic to $\mathbf{L}(1/2, 0)$, and

$$\mathbf{F}_{\psi} \simeq \underbrace{\mathbf{L}(1/2, 0)}_{\mathbf{F}^+_{\psi}} \oplus \underbrace{\mathbf{L}(1/2, 1/2)}_{\mathbf{F}^-_{\psi}},$$

where $\mathbf{F}^{\pm}_{\psi}$ are the eigenspaces of the involution θ induced by the action $\psi \mapsto -\psi$ on the generator $\psi = \psi_{-1/2}\mathbf{v}_0$. We can then readily read off their graded dimensions as

$$\sum_{d=0}^{\infty} q^d \dim \mathbf{F}^{\pm}_{\psi,d} = \frac{1}{2}\left(\prod_{k=0}^{\infty}(1 + q^{k+1/2}) \pm \prod_{k=0}^{\infty}(1 - q^{k+1/2})\right),$$

where the degree is given by setting $\deg \psi_{-n-1/2} = n + 1/2$.

Note 3.6.   1. The fermionic construction as described is useful in constructing VOAs associated with binary even codes introduced by Miyamoto [84] (cf. [81]) and consequently in studying framed VOAs including the moonshine module (cf. [43]). See Subsection 1.4.4.3 for a related construction. 2. The module $\mathbf{L}(1/2, 1/16)$ can be constructed by considering the *twisted module* over vertex superalgebra $\mathbf{F}_{\psi}$. See Section 1.5 for the concept and examples of twisted modules over *vertex algebras*.

Bibliographic Notes

General references are Kac [6], Frenkel and Ben-Zvi [8], and Lepowsky and Li [10], for Section 1.3. For descriptions of algebras appearing in various models in physics, consult Di Francesco et al. [13] (cf. Ginsparg [19]).

Construction of affine and Virasoro vertex algebras are due to Frenkel and Zhu [60]. Our exposition follows Li [70]. See also Primc [88] for a related work.

See Kac [17] and Lepowsky and Li [10] (cf. Tsuchiya, Ueno, and Yamada [97]) for generalities on affine Kac–Moody algebras, and Kac, Raina, and Rozhkovskaya [18] and Iohara and Koga [15] for the Virasoro algebra.

For the construction of the Ising model $L(1/2, 0)$ by Majorana fermions, see [54], [13], and [18].

1.4 Lattice Vertex Algebras

From here on, we will work over the field $\mathbb{C}$ of complex numbers, although most of the results hold over a field of characteristic zero.

Recall that a (nondegenerate integral) *lattice* is a free $\mathbb{Z}$-module L of finite rank equipped with a nondegenerate symmetric bilinear form valued in $\mathbb{Z}$:

$$(\ | \ ) : L \times L \longrightarrow \mathbb{Z}.$$

Theory of lattices is important in many areas of mathematics.

For a lattice L, consider the Heisenberg Lie algebra $\hat{\mathfrak{h}}$; that is, the affine Lie algebra associated with the commutative Lie algebra $\mathfrak{h} = L \otimes_{\mathbb{Z}} \mathbb{C}$ and the bilinear form extending that of the lattice, and the direct sum of the Fock modules of charge belonging to the lattice:

$$\mathbf{V}_L = \bigoplus_{\lambda \in L} \mathbf{F}_\lambda.$$

Then the vertex algebra structure on the Heisenberg vertex algebra $\mathbf{F}_0$ can be extended to $\mathbf{V}_L$ by using the *vertex operators* in a natural but subtle way.

In Section 1.4, we will describe properties of vertex operators and the way how to construct a vertex algebra structure on $\mathbf{V}_L$.

Note that the Fock modules are written as $\mathbf{F}_\lambda = \mathbf{S}(\hat{\mathfrak{h}}_{<0}) \mathbf{v}_\lambda$, where $\mathbf{S}(\hat{\mathfrak{h}}_{<0})$ denotes the symmetric algebra over $\hat{\mathfrak{h}}_{<0}$. Thus the lattice vertex algebra is also written as

$$\mathbf{V}_L = \mathbf{S}(\hat{\mathfrak{h}}_{<0}) \otimes \mathbb{C}[L]$$

by identifying the vector $\mathbf{v}_\lambda$ with the basis vector e^λ of the group algebra.

1.4.1 Series with Homomorphism Coefficients

In the previous sections, we have considered series with operator coefficients, where the operators are endomorphisms of a vector space. In this section, we will consider slightly more general types of series, whose coefficients are homomorphisms (linear maps) between vector spaces.

1.4.1.1 Local Truncation and Residue Products

Let $\mathbf{M}$ and $\mathbf{N}$ be vector spaces and let $\Phi(z)$ be a series with coefficients in $\mathrm{Hom}(\mathbf{M}, \mathbf{N})$:

$$\Phi(z) = \sum_n \Phi_n z^{-n-1}, \quad \Phi_n : \mathbf{M} \longrightarrow \mathbf{N}.$$

Let $A(z)$ be a series on $\mathbf{M}$ and $\mathbf{N}$, that is, a series with coefficients in the set

$$\mathrm{Hom}((\mathbf{M}, \mathbf{N}), (\mathbf{M}, \mathbf{N})) = \mathrm{Hom}(\mathbf{M}, \mathbf{M}) \oplus \mathrm{Hom}(\mathbf{N}, \mathbf{N}).$$

Denote the actions of the coefficients on $\mathbf{M}$ and $\mathbf{N}$ by the same symbol:

$$A(z) = \sum_n A_n z^{-n-1}, \quad A_n : \mathbf{M} \longrightarrow \mathbf{M}, \ \mathbf{N} \longrightarrow \mathbf{N}.$$

We may then consider the compositions of the coefficients as in the diagram:

$$
\begin{array}{ccc}
\mathbf{M} & \xrightarrow{\Phi_n} & \mathbf{N} \\
{\scriptstyle A_m}\downarrow & & \downarrow{\scriptstyle A_m} \\
\mathbf{M} & \xrightarrow[\Phi_n]{} & \mathbf{N}.
\end{array}
$$

We say that $\Phi(z)$ is locally truncated if it belongs to $\mathrm{Hom}(\mathbf{M}, \mathbf{N}((z)))$. The concept of local truncation for $A(z)$ is defined in an obvious way, and denote the set of such series as

$$\mathrm{Hom}\left(((\mathbf{M}, \mathbf{N}), ((\mathbf{M}, \mathbf{N})((z)))\right) = \mathrm{Hom}(\mathbf{M}, \mathbf{M}((z))) \oplus \mathrm{Hom}(\mathbf{N}, \mathbf{N}((z))).$$

If $A(z)$ and $\Phi(z)$ are locally truncated, then the residue products make sense for all $m \in \mathbb{Z}$:

$$
\begin{aligned}
A(z)_{(m)}\Phi(z) &= \mathrm{Res}_y (y - z)^m \big|_{|y| > |z|} A(y)\Phi(z) \\
&\quad - \mathrm{Res}_y (y - z)^m \big|_{|y| > |z|} \Phi(z)A(y).
\end{aligned}
$$

Explicitly, the coefficients of $A(z)_{(m)}\Phi(z) = \sum_n (A_{(m)}\Phi)_n z^{-n-1}$ are given by

$$(A_{(m)}\Phi)_n = \sum_{i=0}^{\infty} (-1)^m \binom{m}{i} A_{m-i}\Phi_{n+i} - \sum_{i=0}^{\infty} (-1)^{m-i} \binom{m}{i} \Phi_{m+n-i}A_i,$$

as in the case of series acting on a vector space.

1.4.1.2 Local Commutativity and Borcherds Identity for Series

The series $A(z)$ and $\Phi(z)$ are *locally commutative* if the following holds for some $N \in \mathbb{N}$:

$$(y - z)^N A(y)\Phi(z) = (y - z)^N \Phi(z)A(y).$$

For such series, we have the OPE

$$A(y)\Phi(z) \simeq \Phi(z)A(y) \sim \sum_{k=0}^{N-1} \frac{\Psi_k(z)}{(y - z)^{k+1}},$$

by which the residue products are found as

$$A(z)_{(m)}\Phi(z) = \begin{cases} 0 & (N \le m), \\ \Psi_m(z) & (0 \le m < N). \end{cases}$$

Let $A(z), B(z)$ be locally truncated series acting on $\mathbf{M}$ and $\mathbf{N}$ and $\Phi(z)$ a locally truncated series with coefficients in $\mathrm{Hom}(\mathbf{M}, \mathbf{N})$:

$$A(z), B(z) \in \mathrm{Hom}((\mathbf{M}, \mathbf{N}), (\mathbf{M}, \mathbf{N})((z))), \quad \Phi(z) \in \mathrm{Hom}(\mathbf{M}, \mathbf{N}((z))).$$

If they are locally commutative with each other, then the Borcherds identity

$$\sum_{i=0}^{\infty} \binom{p}{i} (A(z)_{(r+i)}B(z))_{(p+q-i)}\Phi(z)$$

$$= \sum_{i=0}^{\infty} (-1)^i \binom{r}{i} A(z)_{(p+r-i)}(B(z)_{(q+i)}\Phi(z))$$

$$- \sum_{i=0}^{\infty} (-1)^{r-i} \binom{r}{i} B(z)_{(q+r-i)}(A(z)_{(p+i)}\Phi(z))$$

holds for all $p, q, r \in \mathbb{Z}$ with respect to the residue products.

1.4.1.3 OPEs in General Settings

Let $\mathbf{L}, \mathbf{M}_1, \mathbf{M}_2, \mathbf{N}$ be vector spaces and consider locally truncated series

$$\Psi(z) \in \mathrm{Hom}((\mathbf{L}, \mathbf{M}_1), (\mathbf{M}_2, \mathbf{N})((z))), \quad \Phi(z) \in \mathrm{Hom}((\mathbf{L}, \mathbf{M}_2), (\mathbf{M}_1, \mathbf{N})((z)))$$

and their compositions

$$\Phi(y)\Psi(z) \in \mathrm{Hom}(\mathbf{L}, \mathbf{N}((y))((z))), \quad \Psi(z)\Phi(y) \in \mathrm{Hom}(\mathbf{L}, \mathbf{N}((z))((y))),$$

where

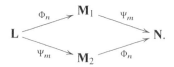

Consider the residue products

$$\Phi(z)_{(m)}\Psi(z) = \mathrm{Res}_y(y-z)^m\big|_{|y|>|z|}\Phi(y)\Psi(z)$$
$$- \mathrm{Res}_y(y-z)^m\big|_{|y|>|z|}\Psi(z)\Phi(y).$$

We will say that $\Phi(z)$ and $\Psi(z)$ are locally commutative if, for some $N \in \mathbb{N}$,

$$(y-z)^N\Phi(y)\Psi(z) = (y-z)^N\Psi(z)\Phi(y).$$

Then, for some series $\Gamma_0(z),\ldots,\Gamma_{N-1}(z) \in \mathrm{Hom}(\mathbf{L},\mathbf{N})((z))$, the OPE

$$\Phi(y)\Psi(z) \simeq \Psi(z)\Phi(y) \sim \sum_{k=0}^{N-1} \frac{\Gamma_k(z)}{(y-z)^{k+1}}$$

holds in the obvious sense, and the residue products are given by

$$\Phi(z)_{(m)}\Psi(z) = \begin{cases} 0 & (m \geq N), \\ \Gamma_m(z) & (0 \leq m < N) \end{cases}$$

for $m \in \mathbb{N}$.

1.4.1.4 Formal Taylor Expansion

For series $\Phi(z)$ and $\Psi(z)$ as in Subsection 1.4.1.3, assume that the composite $\Phi(y)$
$\Psi(z)$ is written in the following form with some $m_0 \in \mathbb{Z}$:

$$\Phi(y)\Psi(z) = (y-z)^{-m_0-1}\big|_{|y|>|z|}\Gamma(y,z), \quad \Gamma(y,z) \in \mathrm{Hom}(\mathbf{L},\mathbf{N}((y,z))).$$

Then Taylor expansion of $\Gamma(y,z)$ yields

$$\Gamma(x+z,z)\big|_{|x|<|z|} = \sum_{i=0}^{\infty} x^i \partial_y^{(i)}\Gamma(y,z)\big|_{y=z}.$$

Therefore, by substitution $x = y - z$,

$$\Phi(y)\Psi(z) = x^{-m_0-1}\big|_{|y|>|z|}\Gamma(y,z)$$
$$= \sum_{i=0}^{\infty} x^{-m_0+i-1}\big|_{|y|>|z|}\partial_y^{(i)}\Gamma(y,z)\big|_{y=z}.$$

Hence the residue products are determined as

$$\Phi(z)_{(m)}\Psi(z) = \begin{cases} 0 & (m > m_0), \\ \partial_y^{(i)}\Gamma(y,z)\big|_{y=z} & (m = m_0 - i \leq m_0). \end{cases}$$

1.4.2 Vertex Operators

Let us now describe *vertex operators*, which are the main ingredients in constructing lattice vertex algebras.

A *vertex operator* is a series of the form

$$V_\lambda(z) = \exp\left(-\sum_{n<0} \lambda_n \frac{z^{-n}}{n}\right) \exp\left(-\sum_{n>0} \lambda_n \frac{z^{-n}}{n}\right) e^\lambda z^{\lambda_0},$$

where λ_n with $n \in \mathbb{Z}$ are actions of the Heisenberg algebra (of higher rank in general). Thus, contrary to its name, a vertex operator is not a single operator, but a series with operator coefficients.

1.4.2.1 Heisenberg Vertex Algebras of Higher Rank

Let $\mathfrak{h}$ be a finite-dimensional vector space, and let $(\ |\ )$ be a symmetric bilinear form on $\mathfrak{h}$, which we assume to be nondegenerate.

Regard $\mathfrak{h}$ as a commutative Lie algebra and consider the affine Lie algebra associated with $\mathfrak{h}$ and $(\ |\ )$:

$$\hat{\mathfrak{h}} = \mathfrak{h} \otimes \mathbb{C}[t, t^{-1}] \oplus \mathbb{C}K.$$

Define the *Heisenberg algebra* associated with $\mathfrak{h}$ and $(\ |\ )$ by

$$\mathbf{U}(\hat{\mathfrak{h}}, 1) = \mathbf{U}(\hat{\mathfrak{h}})/(K - 1).$$

We will sometimes call a $\mathbf{U}(\hat{\mathfrak{h}}, 1)$-module a *Heisenberg module*.

Let $\hat{\mathfrak{h}}_{<0}$ and $\hat{\mathfrak{h}}_{\geq 0}$ be the commutative Lie subalgebra of $\hat{\mathfrak{h}}$ spanned by $\mathfrak{h} \otimes t^n$ with $n < 0$ and $n \geq 0$, respectively, for which

$$\mathbf{U}(\hat{\mathfrak{h}}, 1) = \mathbf{S}(\hat{\mathfrak{h}}_{<0}) \otimes \mathbf{S}(\hat{\mathfrak{h}}_{\geq 0})$$

as a vector space.

Let $\lambda \in \mathfrak{h}^*$ be a linear form on $\mathfrak{h}$, where $\mathfrak{h}^* = \mathrm{Hom}_\mathbb{C}(\mathfrak{h}, \mathbb{C})$. Consider the one-dimensional $\mathbf{S}(\hat{\mathfrak{h}}_{\geq 0})$-module $\mathbb{C}\mathbf{v}_\lambda$ given as follows for all $h \in \mathfrak{h}$ and $n \geq 0$:

$$h_n \mathbf{v}_\lambda = \begin{cases} 0 & (n \geq 1), \\ \lambda(h)\mathbf{v}_\lambda & (n = 0). \end{cases}$$

The *Fock module* of *charge* λ is the $\mathbf{U}(\hat{\mathfrak{h}}, 1)$-module:

$$\mathbf{F}_\lambda = \mathbf{U}(\hat{\mathfrak{h}}, 1) \otimes_{\mathbf{S}(\hat{\mathfrak{h}}_{\geq 0})} \mathbb{C}\mathbf{v}_\lambda.$$

It is isomorphic to $\mathbf{S}(\hat{\mathfrak{h}}_{<0}) \otimes_\mathbb{C} \mathbb{C}\mathbf{v}_\lambda$ as a vector space by PBW. Having this in mind, we often write

$$\mathbf{F}_\lambda = \mathbf{S}(\hat{\mathfrak{h}}_{<0})\mathbf{v}_\lambda.$$

The Fock module $\mathbf{F}_0$ carries a natural structure of a vertex algebra, and $\mathbf{F}_\lambda$ are simple modules over the vertex algebra $\mathbf{F}_0$ for all $\lambda \in \mathfrak{h}^*$.

The construction of the standard Virasoro vector (3.5) generalizes to higher rank cases by

$$\omega = \frac{1}{2} \sum_{i=1}^{d} a^i_{-1} a_{i,-1} \mathbf{v}_0,$$

where $(a^1, \ldots, a^d)$ and $(a_1, \ldots, a_d)$ are dual bases of $\mathfrak{h}$ with respect to the non-degenerate bilinear form $(-|-)$.

1.4.2.2 The Operators e^λ and z^{λ_0}

From here on, we identify the vector space $\mathfrak{h}$ with its dual $\mathfrak{h}^* = \mathrm{Hom}_{\mathbb{C}}(\mathfrak{h}, \mathbb{C})$ by the symmetric bilinear form $(\ |\ )$, which we have assumed to be nondegenerate, so that $\lambda(h) = (\lambda|h)$ for $\lambda, h \in \mathfrak{h}$.

For $\lambda \in \mathfrak{h}$, there exists a unique homomorphism of $\mathbf{S}(\hat{\mathfrak{h}}_{<0})$-modules sending $\mathbf{v}_\mu$ to $\mathbf{v}_{\lambda+\mu}$, which we denote by

$$e^\lambda : \mathbf{F}_\mu \longrightarrow \mathbf{F}_{\lambda+\mu}, \quad \mathbf{v}_\mu \mapsto \mathbf{v}_{\lambda+\mu}.$$

Next, for $\lambda, \mu \in \mathfrak{h}$ satisfying $(\lambda|\mu) \in \mathbb{Z}$, define

$$z^{\lambda_0} : \mathbf{F}_\mu \longrightarrow \mathbf{F}_\mu z^{(\lambda|\mu)}, \quad v \mapsto z^{\lambda_0} v = z^{(\lambda|\mu)} v.$$

Then, for $\lambda, h \in \mathfrak{h}$ and $n \in \mathbb{Z}$,

$$[h_n, e^\lambda] = (\lambda|h)\delta_{n,0} e^\lambda, \quad [h_n, z^{\lambda_0}] = 0.$$

The operators z^{λ_0} and e^μ do not commute, but satisfy

$$z^{\lambda_0} e^\mu = z^{(\lambda|\mu)} e^\mu z^{\lambda_0} \tag{4.1}$$

for $\lambda, \mu \in \mathfrak{h}$.

1.4.2.3 Vertex Operators

For $\lambda \in \mathfrak{h}$, consider the following expression:

$$V_\lambda(z) = \exp\left(\sum_{n<0} \lambda_n \frac{z^{-n}}{-n}\right) \exp\left(\sum_{n>0} \lambda_n \frac{z^{-n}}{-n}\right) e^\lambda z^{\lambda_0},$$

where the sums are over negative and positive integers, respectively, and the exponential of a series is defined as

$$\exp x(z) = \sum_{k=0}^{\infty} \frac{x(z)^k}{k!}.$$

To see the meaning and well-definedness of $V_\lambda(z)$, note the following structure:

$$\exp\left(\underbrace{\sum_{n<0}\lambda_n\frac{z^{-n}}{-n}}_{\substack{\text{positive powers}\\\text{coefficients in }\hat{\mathfrak{h}}_{<0}}}\right)\exp\left(\underbrace{\sum_{n>0}\lambda_n\frac{z^{-n}}{-n}}_{\substack{\text{negative powers}\\\text{coefficients in }\hat{\mathfrak{h}}_{>0}}}\right)\underbrace{e^\lambda}_{\substack{\text{shifting}\\\text{charge}}}\underbrace{z^{\lambda_0}}_{\substack{\text{shifting}\\\text{exponent}}}.$$

Let $\mu\in\mathfrak{h}$ satisfy $(\lambda|\mu)\in\mathbb{Z}$. Then, for any element $P\in S(\hat{\mathfrak{h}}_{<0})$,

$$z^{\lambda_0}P\mathbf{v}_\mu = P\mathbf{v}_\mu z^{(\lambda|\mu)},\quad e^\lambda z^{\lambda_0}P\mathbf{v}_\mu = P\mathbf{v}_{\lambda+\mu}z^{(\lambda|\mu)}.$$

Since if $n_1+\cdots+n_k$ is sufficiently large, then $\lambda_{n_1}\cdots\lambda_{n_k}P\mathbf{v}_{\lambda+\mu}=0$, we have

$$\exp\left(\underbrace{\sum_{n>0}\lambda_n\frac{z^{-n}}{-n}}_{\substack{\text{negative powers}\\\text{coefficients in }\hat{\mathfrak{h}}_{>0}}}\right)P\mathbf{v}_{\lambda+\mu}z^{(\lambda|\mu)}\in F_{\lambda+\mu}[z^{-1}]z^{(\lambda|\mu)},$$

thus

$$\exp\left(\underbrace{\sum_{n<0}\lambda_n\frac{z^{-n}}{-n}}_{\text{nonnegative powers}}\right)\exp\left(\underbrace{\sum_{n>0}\lambda_n\frac{z^{-n}}{-n}}_{\substack{\text{finitely many terms with}\\\text{negative powers}}}\right)P\mathbf{v}_{\lambda+\mu}z^{(\lambda|\mu)}\in F_{\lambda+\mu}((z))z^{(\lambda|\mu)}.$$

Therefore, for $\mu\in\mathfrak{h}^*$ with $(\lambda|\mu)\in\mathbb{Z}$, the expression $V_\lambda(z)$ gives rise to a locally truncated series with coefficients being maps from F_μ to $F_{\lambda+\mu}$:

$$(\lambda|\mu)\in\mathbb{Z}\implies V_\lambda(z)\in\mathrm{Hom}(F_\mu,F_{\lambda+\mu}((z))).$$

The series $V_\lambda(z)$ thus constructed is called the *vertex operator.*

Note 4.1.   1. Following the physics literatures, formally write

$$\phi_\lambda(z)=\phi_\lambda(z)_{<0}+\phi_\lambda(z)_{>0}+\lambda_0\log z+\lambda,$$

where

$$\phi_\lambda(z)_{<0}=\sum_{n<0}\lambda_n\frac{z^{-n}}{-n},\quad\phi_\lambda(z)_{>0}=\sum_{n>0}\lambda_n\frac{z^{-n}}{-n}.$$

Then we have $\partial\phi_\lambda(z)=\lambda(z)$ so that the expression $\phi_\lambda(z)$ is thought of as the "indefinite integral" of the series $\lambda(z)$. 2. The vertex operator $V_\lambda(z)$ as defined here is thought of as a regularization of the divergent expression $e^{\phi_\lambda(z)}$ by "normal ordering" and often denoted as

$$:e^{\phi_\lambda(z)}:\ =\exp\left(\phi_\lambda(z)_{<0}\right)\exp\left(\phi_\lambda(z)_{>0}\right)e^\lambda z^{\lambda_0}.$$

1.4.2.4 Commutation with Currents

Let X, Y be operators or series on a vector space such that $\exp Y = \sum_{k=0}^{\infty} Y^k/k!$ makes sense in an appropriate way. Then

$$[X, Y] \text{ commutes with } Y \implies [X, \exp Y] = [X, Y] \exp Y,$$

where the bracket refers to the commutator.

By using this, we have

$$[h(y), V_\lambda(z)] = \lambda(h) V_\lambda(z) \delta(y, z).$$

In particular, the current $h(z)$ and the vertex operator $V_\lambda(z)$ are locally commutative and their OPE is given by

$$h(y) V_\lambda(z) \simeq V_\lambda(z) h(y) \sim \frac{\lambda(h)}{y - z} V_\lambda(z).$$

We thus have

$$h(z)_{(n)} V_\lambda(z) = \begin{cases} 0 & (n \geq 1), \\ \lambda(h) V_\lambda(z) & (n = 0). \end{cases}$$

For each $\lambda \in \mathfrak{h}^*$, repeatedly apply the residue products by the currents to the vertex operator $V_\lambda(z)$, and let $\mathcal{F}_\lambda$ denote the span of such series:

$$\mathcal{F}_\lambda = \mathrm{Span}\left\{ h^1(z)_{(n_1)} \cdots h^k(z)_{(n_k)} V_\lambda(z) \,\middle|\, \begin{matrix} k \in \mathbb{N}, \, h^1, \ldots, h^k \in \mathfrak{h} \\ n_1, \ldots, n_k \in \mathbb{Z} \end{matrix} \right\}. \quad (4.2)$$

Then it becomes an $\mathbf{F}_0$-module by the residue products. By the OPE, it is isomorphic to the Fock module $\mathbf{F}_\lambda$ of charge λ as a Heisenberg module.

1.4.3 Residue Products of Vertex Operators

Assume that $\lambda, \mu, \nu \in L$ satisfy $(\lambda|\mu), (\mu|\nu), (\lambda|\nu) \in \mathbb{Z}$ and consider the vertex operators

$$V_\lambda(z) \in \mathrm{Hom}((\mathbf{F}_\nu, \mathbf{F}_{\mu+\nu}), (\mathbf{F}_{\lambda+\nu}, \mathbf{F}_{\lambda+\mu+\nu})((z))),$$

$$V_\mu(z) \in \mathrm{Hom}((\mathbf{F}_\nu, \mathbf{F}_{\lambda+\nu}), (\mathbf{F}_{\mu+\nu}, \mathbf{F}_{\lambda+\mu+\nu})((z))).$$

Their coefficients fit in

We are interested in commutation of $V_\lambda(y)$ and $V_\mu(z)$ for $\lambda, \mu \in L$.

1.4.3.1 Commutation of Vertex Operators

Let X, Y be operators or series on a vector space such that $\exp X$, $\exp Y$, and $\exp[X, Y]$ make sense in an appropriate way. Then

$$[X, Y] \text{ commutes with } X \text{ and } Y \implies \exp X \exp Y = \exp[X, Y] \exp Y \exp X.$$

Apply this to the following partial product of $V_\lambda(z)V_\mu(y)$:

$$\exp\Big(\underbrace{\sum_{n<0} \lambda_n \frac{y^{-n}}{-n}\Big) \exp\Big(\sum_{n>0} \lambda_n \frac{y^{-n}}{-n}}_{X}\Big) \exp\Big(\underbrace{\sum_{n<0} \mu_n \frac{z^{-n}}{-n}\Big) \exp\Big(\sum_{n>0} \mu_n \frac{z^{-n}}{-n}}_{Y}\Big).$$

Then, since

$$\Big[\underbrace{\sum_{n>0} \lambda_n \frac{y^{-n}}{-n}}_{X}, \underbrace{\sum_{n<0} \mu_n \frac{z^{-n}}{-n}}_{Y}\Big] = \sum_{m>0}\sum_{n<0} [\lambda_m, \mu_n] \frac{y^{-m}z^{-n}}{mn}$$

$$= (\lambda|\mu) \sum_{m>0} \frac{y^{-m}z^m}{-m} = (\lambda|\mu) \log\Big(1 - \frac{z}{y}\Big),$$

we have

$$\overbrace{\exp\Big(\sum_{n>0} \lambda_n \frac{y^{-n}}{-n}\Big)}^{\exp X} \overbrace{\exp\Big(\sum_{n<0} \mu_n \frac{z^{-n}}{-n}\Big)}^{\exp Y}$$

$$= \underbrace{\Big(1 - \frac{z}{y}\Big)^{(\lambda|\mu)}\Big|_{|y|>|z|}}_{\exp[X,Y]} \underbrace{\exp\Big(\sum_{n<0} \mu_n \frac{z^{-n}}{-n}\Big)}_{\exp Y} \underbrace{\exp\Big(\sum_{n>0} \lambda_n \frac{y^{-n}}{-n}\Big)}_{\exp X}.$$

On the other hand, by (4.1),

$$e^\lambda \underbrace{y^{\lambda_0} e^\mu}_{} z^{\mu_0} = y^{(\lambda|\mu)} e^\lambda \underbrace{e^\mu y^{\lambda_0}}_{} z^{\mu_0} = y^{(\lambda|\mu)} e^{\lambda+\mu} y^{\lambda_0} z^{\mu_0}.$$

Combining them together, we arrive at

$$V_\lambda(y)V_\mu(z) = (y - z)^{(\lambda|\mu)}\big|_{|y|>|z|} V_{\lambda,\mu}(y, z),$$
$$V_\mu(z)V_\lambda(y) = (z - y)^{(\mu|\lambda)}\big|_{|y|<|z|} V_{\lambda,\mu}(y, z),$$

$$(4.3)$$

where

$$V_{\lambda,\mu}(y, z) = \exp\Big(\sum_{n<0} \frac{\lambda_n y^{-n} + \mu_n z^{-n}}{-n}\Big)$$

$$\exp\Big(\sum_{n>0} \frac{\lambda_n y^{-n} + \mu_n z^{-n}}{-n}\Big) e^{\lambda+\mu} y^{\lambda_0} z^{\mu_0}.$$

Therefore, if $(\lambda|\mu), (\mu|\nu), (\lambda|\nu) \in \mathbb{Z}$, then we have the following equalities as series with coefficients in $\mathrm{Hom}(\mathbf{F}_\nu, \mathbf{F}_{\lambda+\mu+\nu})$.

1. If $(\lambda|\mu) \geq 0$, then

$$V_\lambda(y)V_\mu(z) = (-1)^{(\lambda|\mu)}V_\mu(z)V_\lambda(y).$$

2. If $(\lambda|\mu) < 0$, then, for $N = -(\lambda|\mu) \geq 0$,

$$(y - z)^N V_\lambda(y)V_\mu(z) = (-1)^{(\lambda|\mu)}(y - z)^N V_\mu(z)V_\lambda(y).$$

In particular, if $(\lambda|\mu) \in 2\mathbb{Z}$, then $V_\lambda(z)$ and $V_\mu(z)$ are locally commutative.

1.4.3.2 OPE of Vertex Operators

Consider the case with $(\lambda|\mu) \in \mathbb{Z}$. Then, by the last result, we have

$$V_\lambda(y)V_\mu(z) = (y - z)^{(\lambda|\mu)}\big|_{|y|>|z|}V_{\lambda,\mu}(y, z).$$

Applying Taylor expansion to $V_{\lambda,\mu}(y, z)$ in the first equality of (4.3), we have

$$V_{\lambda,\mu}(x + z, z) = e^\lambda \sum_{i=0}^\infty x^i \partial_y^{(i)}\Big(\exp\Big(\sum_{n<0}\frac{\lambda_n y^{-n}}{-n}\Big)V_\mu(z)$$
$$\exp\Big(\sum_{n>0}\frac{\lambda_n y^{-n}}{-n}\Big)y^{\lambda_0}\Big)\Big|_{y=z}.$$

The result can be written in a compact form as

$$V_{\lambda,\mu}(x + z, z)\big|_{|x|<|z|} = \exp\Big(\sum_{k=0}^\infty \frac{x^{k+1}}{k+1}\lambda(z)_{(-k-1)}\Big)V_{\lambda+\mu}(z).$$

When $(\lambda|\mu) \in 2\mathbb{Z}$, the residue products make sense and read

$$V_\lambda(z)_{(n)}V_\mu(z) = \begin{cases} 0 & (n \geq -(\lambda|\mu)), \\ V_{\lambda+\mu}(z) & (n = -(\lambda|\mu) - 1), \\ {}^\circ_\circ\lambda(z)V_{\lambda+\mu}(z)^\circ_\circ & (n = -(\lambda|\mu) - 2), \\ \quad\cdots\cdots\cdots. \end{cases}$$

In particular, $V_\lambda(z)_{(n)}V_\mu(z)$ belongs to $\mathcal{F}_{\lambda+\mu}$ for all $n \in \mathbb{Z}$.

1.4.4 Lattice Vertex Algebras for Rank One Even Lattices

Let L be an even lattice and set $\mathfrak{h} = L \otimes_\mathbb{Z} \mathbb{C}$. Consider the direct sum of vector spaces given by

$$\mathbf{V}_L = \bigoplus_{\lambda \in L}\mathbf{F}_\lambda,$$

where $\mathbf{F}_\lambda = \mathbf{S}(\hat{\mathfrak{h}}_{<0})\mathbf{v}_\lambda$ is the Fock module of charge λ.

If L is an even lattice of rank one, then the bilinear form takes values in $2\mathbb{Z}$. In such a case, the vertex operators $V_\lambda(z)$ with $\lambda \in L$ are locally commutative, and general theory in Section 1.2 is available.

In this section, we will describe the vertex algebra structure on $\mathbf{V}_L$.

1.4.4.1 Lattice Vertex Algebras of Rank One

Let L be an even lattice of rank one. Recall the currents $h(z)$ with $h \in \mathfrak{h}$ and the vertex operators $V_\lambda(z)$ with $\lambda \in L$, and regard them as series acting on $\mathbf{V}_L$. Since L is a lattice, the series $V_\lambda(z)$ has integral exponents by $z^{\lambda_0}\mathbf{v}_\mu = z^{(\lambda|\mu)}\mathbf{v}_\mu$ for $\lambda, \mu \in L$. Thus:

$$h(z), V_\lambda(z) \in \mathrm{Hom}(\mathbf{V}_L, \mathbf{V}_L((z))).$$

As we have already seen, the currents and the vertex operators are locally truncated and locally commutative with themselves.

Consider the vertex algebra of series generated by the currents and vertex operators:

$$\mathcal{V}_L = \langle h(z), V_\lambda(z) | h \in \mathfrak{h}, \lambda \in L \rangle_{\mathrm{RP}}.$$

Since the vertex operators, as well as the currents, are creative with respect to the vacuum vector $\mathbf{v}_0$, the state map σ restricts to a map

$$\sigma_{\mathcal{V}_L} : \mathcal{V}_L \longrightarrow \mathbf{V}_L.$$

The OPEs of vertex operators show

$$\mathcal{V}_L = \bigoplus_{\lambda \in L} \mathcal{F}_\lambda,$$

where $\mathcal{F}_\lambda$ is defined by (4.2) with the operators replaced by those acting on $\mathbf{V}_L$, which is isomorphic to the Fock module $\mathbf{F}_\lambda$ via the state map $\sigma_{\mathcal{V}_L}$. Therefore, we have the following result.

Proposition 4.2 *Let L be an even lattice of rank one. Then the vector space $\mathbf{V}_L$ carries a unique structure of a vertex algebra with vacuum $\mathbf{1} = \mathbf{v}_0$ such that*

$$Y(h_{-1}\mathbf{v}_0, z) = h(z) \text{ and } Y(\mathbf{v}_\lambda, z) = V_\lambda(z)$$

for all $h \in \mathfrak{h}$ and $\lambda \in L$.

The vertex algebra $\mathbf{V}_L$ thus obtained is called the *lattice vertex algebra* associated with L. It is not so difficult to show that it is a simple vertex algebra.

Consider the group algebra $\mathbb{C}[L]$ of the lattice L spanned by e^λ, $\lambda \in L$. We often identify the vector $\mathbf{v}_\lambda$ with $1 \otimes e^\lambda$ or e^λ as in the following diagram:

$$\mathbf{V}_L \;=\; \bigoplus_{\lambda \in L} \mathbf{F}_\lambda \;=\; S(\hat{\mathfrak{h}}_{<0}) \otimes \mathbb{C}[L] \longleftarrow \mathbb{C}[L]$$
$$\mathbf{v}_\lambda \longleftarrow 1 \otimes e^\lambda \longleftarrow e^\lambda.$$

Then the operator e^λ described in Subsection 1.4.2.2 is regarded as the multiplication by e^λ in $\mathbb{C}[L]$ and the conditions in Proposition 4.2 become

$$Y(h_{-1} \otimes e^0, z) = h(z) \quad \text{and} \quad Y(1 \otimes e^\lambda, z) = V_\lambda(z),$$

where e^0 is the unity of $\mathbb{C}[L]$.

The space $\mathbf{V}_L$ is given a grading by

$$\deg h_{-i_1} \cdots h_{-i_k} \mathbf{v}_\lambda = i_1 + \cdots + i_k + \frac{(\lambda | \lambda)}{2},$$

which agrees with the eigenvalue for the action of L_0 with respect to the Virasoro vector ω of the Heisenberg vertex subalgebra $\mathbf{F}_0$.

If L is positive-definite, then the degree takes values in nonnegative integers and the subspace of degree 0 is spanned by $\mathbf{v}_0$, and the graded dimension is given by

$$\sum_{d=0}^\infty q^d \dim \mathbf{V}_{L,d} = \sum_{\lambda \in L} \prod_{k=1}^\infty \frac{q^{(\lambda|\lambda)/2}}{1 - q^k} = \frac{\Theta_L(\tau)}{\phi(q)} = q^{1/24} \frac{\Theta_L(\tau)}{\eta(\tau)},$$

where $\Theta_L(\tau)$ is the theta constant associated with the lattice L.

1.4.4.2 The Lattice of Type A_1

Let us take L to be the root lattice $\sqrt{2}\mathbb{Z}$ of type A_1, a unique lattice generated by a root α; that is, an element of squared norm 2:

$$A_1 = \mathbb{Z}\alpha, \quad (\alpha | \alpha) = 2.$$

The structure of $\mathbf{V}_{A_1}$ looks as in Table 9. The graded dimension is given by

$$\sum_{d=0}^\infty q^d \dim \mathbf{V}_{A_1,d} = \frac{\sum_n q^{n^2}}{\prod_{k=1}^\infty (1 - q^k)} = \frac{\theta_3(\tau)}{\phi(q)} = q^{1/24} \frac{\theta_3(\tau)}{\eta(\tau)},$$

where $\theta_3(\tau)$ is the Jacobi theta constant.

Pick up the following elements of degree 1:

$$E = \mathbf{v}_\alpha, \quad H = \alpha_{-1}\mathbf{v}_\lambda, \quad F = \mathbf{v}_{-\alpha}.$$

Then the subspace of degree 1 becomes a Lie algebra isomorphic to $\mathfrak{sl}_2$ with respect to the bracket defined by $[X,Y] = X_{(0)}Y$:

$$[H,E] = 2E, \quad [H,F] = -2F, \quad [E,F] = H,$$
$$[H,H] = [F,F] = [E,E] = 0.$$

Table 9 Lattice vertex algebra of type A_1

0	1	2	3	4
				$v_{2\alpha}$
	v_α	$\alpha_{-1}v_\alpha$	$\alpha_{-2}v_\alpha$	$\alpha_{-3}v_\alpha$
			$\alpha_{-1}\alpha_{-1}v_\alpha$	$\alpha_{-1}\alpha_{-2}v_\alpha$
				$\alpha_{-1}\alpha_{-1}\alpha_{-1}v_\alpha$
v_0	$\alpha_{-1}v_0$	$\alpha_{-2}v_0$	$\alpha_{-3}v_0$	$\alpha_{-4}v_0$
		$\alpha_{-1}\alpha_{-1}v_0$	$\alpha_{-1}\alpha_{-2}v_0$	$\alpha_{-1}\alpha_{-3}v_0$
			$\alpha_{-1}\alpha_{-1}\alpha_{-1}v_0$	$\alpha_{-2}\alpha_{-2}v_0$
				$\alpha_{-1}\alpha_{-1}\alpha_{-2}v_0$
				$\alpha_{-1}\alpha_{-1}\alpha_{-1}\alpha_{-1}v_0$
	$v_{-\alpha}$	$\alpha_{-1}v_{-\alpha}$	$\alpha_{-2}v_{-\alpha}$	$\alpha_{-3}v_{-\alpha}$
			$\alpha_{-1}\alpha_{-1}v_{-\alpha}$	$\alpha_{-1}\alpha_{-2}v_{-\alpha}$
				$\alpha_{-1}\alpha_{-1}\alpha_{-1}v_{-\alpha}$
A basis of $V_{\sqrt{2}\mathbb{Z}}$				$v_{-2\alpha}$

Moreover, the bilinear form defined by $(X|Y)\mathbf{1} = X_{(1)}Y$ becomes

$$(H|H) = 2, \quad (E|F) = (F|E) = 1,$$
$$(E|E) = (F|F) = (H|E) = (H|F) = 0,$$

which is invariant with respect to the Lie bracket.

Therefore, there exists a homomorphism of vertex algebras from the universal affine vertex algebra $\mathbf{M}(1,0)$ associated with $\mathfrak{sl}_2$ at level $k = 1$. Since $\mathbf{V}_{A_1}$ is a simple vertex algebra generated by the degree 1 subspace, the map π induces an isomorphism of vertex algebras from $\mathbf{L}(1,0)$ onto $\mathbf{V}_{A_1}$:

$$\mathbf{L}(1,0) \xrightarrow{\ \sim\ } \mathbf{V}_{A_1}.$$

The $\mathbf{L}(1,0)$-module $\mathbf{L}(1,1/2)$ can be constructed as the space $\mathbf{V}_{A_1+1/\sqrt{2}}$ (cf. Subsection 1.4.5.3).

Note 4.3. 1. The Sugawara vector $\omega_k \in \mathbf{M}(k,0)$ given by (3.7) coincides for $k = 1$ with the Virasoro vector $\omega \in \mathbf{F}_0$ given by (3.5). 2. The construction of $\widehat{\mathfrak{sl}}_2$-modules as described above is a particular case of the famous *Frenkel–Kac construction* mentioned in the Introduction, which works for the root lattices of *ADE* type and realizes integrable highest weight representations of the corresponding affine Kac–Moody algebras of level 1.

1.4.4.3 The Lattice of Type D_1

Let L be the even lattice $2\mathbb{Z}$ generated by a norm 2 element as

$$2\mathbb{Z} = \sqrt{2}A_1 = \mathbb{Z}\beta, \quad (\beta|\beta) = 4,$$

Table 10 Lattice vertex algebra of type D_1

0	1	2	3	4
		v_β	$\beta_{-1}v_\beta$	$\beta_{-2}v_\beta$
				$\beta_{-1}\beta_{-1}v_\beta$
v_0	$\beta_{-1}v_0$	$\beta_{-2}v_0$	$\beta_{-3}v_0$	$\beta_{-4}v_0$
		$\beta_{-1}\beta_{-1}v_0$	$\beta_{-1}\beta_{-2}v_0$	$\beta_{-1}\beta_{-3}v_0$
			$\beta_{-1}\beta_{-1}\beta_{-1}v_0$	$\beta_{-2}\beta_{-2}v_0$
				$\beta_{-1}\beta_{-1}\beta_{-2}v_0$
				$\beta_{-1}\beta_{-1}\beta_{-1}\beta_{-1}v_0$
		$v_{-\beta}$	$\beta_{-1}v_{-\beta}$	$\beta_{-2}v_{-\beta}$
A basis of $V_{2\mathbb{Z}}$				$\beta_{-1}\beta_{-1}v_{-\beta}$

which can be thought of as the root lattice of type D_n with n formally set to 1.

The structure of $V_{2\mathbb{Z}}$ looks as in Table 10. Consider the elements

$$e^+ = \frac{\omega}{2} + \frac{v_\beta + v_{-\beta}}{4}, \quad e^- = \frac{\omega}{2} - \frac{v_\beta + v_{-\beta}}{4},$$

where ω is the Virasoro vector (3.5) with $a_n = (1/2)\beta_n$, $n \in \mathbb{Z}$. Then $e^\pm$ are Virasoro vectors of central charge $c = 1/2$ commuting with each other:

$$e^+_{(n)}e^- = 0 \ (n \geq 0), \quad e^\pm_{(n)}e^\pm = \begin{cases} 0 & (n \geq 4), \\ 1/4 & (n = 3), \\ 2e^\pm & (n = 1). \end{cases}$$

In fact, they generate vertex subalgebras isomorphic to the simple Virasoro vertex algebra of central charge $1/2$,

$$\langle e^+ \rangle_{\mathrm{VA}} \simeq \mathbf{L}(1/2, 0) \simeq \langle e^- \rangle_{\mathrm{VA}},$$

and we have a decomposition of the form

$$V_{2\mathbb{Z}} = \underbrace{\mathbf{L}(1/2, 0) \otimes \mathbf{L}(1/2, 0)}_{V^+_{2\mathbb{Z}}} \oplus \underbrace{\mathbf{L}(1/2, 1/2) \otimes \mathbf{L}(1/2, 1/2)}_{V^-_{2\mathbb{Z}}},$$

where $V^\pm_{2\mathbb{Z}}$ are the eigenspaces of an involution characterized by

$$\theta: V_{2\mathbb{Z}} \longrightarrow V_{2\mathbb{Z}}, \quad h \mapsto -h, \quad v_\beta \mapsto v_{-\beta}.$$

Note 4.4. 1. For more information on $V^+_{2\mathbb{Z}}$ and its applications, see [43]. 2. Many interesting and useful examples of vertex algebras are found as subalgebras of the vertex algebras (or vertex superalgebras) associated with a rank one lattice. For example, the vertex algebra $V^+_{\sqrt{6}\mathbb{Z}}$ is identified with what is called the minimal W_4 algebra of central charge 1, which can be used to describe the actions of 4A elements of the Monster on the moonshine module $\mathbf{V}^\natural$ (cf. [79] and [93]).

1.4.5 Lattice Vertex Algebras for General Even Lattices

Recall the commutation of vertex operators described by (4.3), which implies, for sufficiently large N,

$$(y - z)^N V_\lambda(y) V_\mu(z) = (-1)^{(\lambda|\mu)}(y - z)^N V_\mu(z) V_\lambda(y).$$

For an even lattice L of higher rank, the value $(\lambda|\mu)$ for $\lambda, \mu \in L$ can be odd and, for such a case, the series $V_\lambda(z)$ and $V_\mu(z)$ are not locally commutative.

By this reason, we wish to modify the vertex operators by multiplying it by a sign factor in such a way that the resulting series become locally commutative and creative. We will then describe the lattice vertex algebras $\mathbf{V}_L$ in the same way as in the rank one case, and give a brief account on the simple modules over $\mathbf{V}_L$.

1.4.5.1 Cocycle Factors and Central Extensions

For $\lambda, \mu \in L$, the vertex operator $V_\lambda(z)$ restricts to

$$V_\lambda(z)\big|_{\mathbf{F}_\mu} : \mathbf{F}_\mu \longrightarrow \mathbf{F}_{\lambda+\mu}((z)).$$

We will multiply it by a sign factor so that the resulting series become locally commutative. To be more precise, consider a function

$$\varepsilon\colon L \times L \longrightarrow \{\pm 1\},$$

that is to be called a *cocycle factor*, and set

$$V_{\lambda,\varepsilon}(z)\big|_{\mathbf{F}_\mu} = \varepsilon(\lambda,\mu) V_\lambda(z)\big|_{\mathbf{F}_\mu}. \tag{4.4}$$

Then the condition on the function ε so that $V_{\lambda,\varepsilon}(z)$ become locally commutative and creative with respect to $\mathbf{v}_0$ are stated, respectively, as follows:

(1) For all $\lambda, \mu, \nu \in L$: $\varepsilon(\lambda, \mu + \nu)\varepsilon(\mu, \nu) = (-1)^{(\lambda|\mu)}\varepsilon(\mu, \lambda + \nu)\varepsilon(\lambda, \nu)$.
(2) For all $\lambda \in L$: $\varepsilon(0, \lambda) = 1 = \varepsilon(\lambda, 0)$.

These conditions imply:

(3) For all $\lambda, \mu, \nu \in L$: $\varepsilon(\lambda, \mu)\varepsilon(\lambda + \mu, \nu) = \varepsilon(\lambda, \mu + \nu)\varepsilon(\mu, \nu)$.
(4) For all $\lambda, \mu \in L$: $\varepsilon(\lambda, \mu)\varepsilon(\mu, \lambda) = (-1)^{(\lambda|\mu)}$.

We actually have the equivalence

$$(1) + (2) \iff (2) + (3) + (4).$$

To see the meaning of the latter, consider a central extension of groups of the form

$$0 \longrightarrow \{\pm 1\} \longrightarrow \hat{L} \overset{\pi}{\longrightarrow} L \longrightarrow 0. \tag{4.5}$$

Choose a set-theoretical section $\iota \colon L \longrightarrow \hat{L}$ and denote its value at λ by e_λ. Assume that the group structure on $\hat{L}$ and the function $\varepsilon \colon L \times L \longrightarrow \{\pm 1\}$ are related by, for all $\lambda, \mu \in L$,

$$e_\lambda e_\mu = \varepsilon(\lambda, \mu) e_{\lambda + \mu}.$$

Then condition (2) says that e_0 is the identity element and (3) that the product is associative. In other words, the function $\varepsilon \colon L \times L \longrightarrow \{\pm 1\}$ satisfying (2) and (3) is a normalized 2-cocycle associated with the group extension (4.5).

Now the condition (4) means that the extension (4.5) is subject to

$$e_\lambda e_\mu = (-1)^{(\lambda | \mu)} e_\mu e_\lambda,$$

for which the correspondence $(e_\lambda, e_\mu) \mapsto e_\lambda e_\mu e_\lambda^{-1} e_\mu^{-1} = (-1)^{(\lambda | \mu)}$ is called the *commutator map*.

Note 4.5. As the vertex operators generate *intertwining operators* among the Fock modules (cf. Section 1.6.4), the problem of constructing a vertex algebra structure on $\mathbf{V}_L$ is a particular case of that for a direct sum of modules over a vertex algebra by intertwining operators. Such a problem is studied in detail for a good vertex algebra under certain conditions (cf. [37] and references therein), although $\mathbf{V}_L$ as a sum of Fock modules does not fulfill such conditions.

1.4.5.2 Lattice Vertex Algebras

Let us explicitly construct a cocycle factor as a bimultiplicative map:

$$\varepsilon(\lambda, \mu + \nu) = \varepsilon(\lambda, \mu)\varepsilon(\lambda, \nu), \quad \varepsilon(\lambda + \mu, \nu) = \varepsilon(\lambda, \nu)\varepsilon(\mu, \nu).$$

Let $\alpha_1, \ldots, \alpha_n$ be a basis of L, set the values of $\varepsilon(\alpha_i, \alpha_j)$ so that

$$\varepsilon(\alpha_i, \alpha_j) = (-1)^{(\alpha_i | \alpha_j)} \varepsilon(\alpha_j, \alpha_i),$$

and extend it bimultiplicatively to the whole L. Then it indeed satisfies the conditions for a cocycle factor and the following proposition holds for the series $V_{\lambda, \varepsilon}(z)$ defined by (4.4).

Proposition 4.6 *For an even lattice L and a cocycle factor ε, there exists a unique structure of a vertex algebra on the vector space $\mathbf{V}_L$ with vacuum $\mathbf{1} = \mathbf{v}_0$ such that*

$$Y(h_{-1}\mathbf{v}_0, z) = h(z) \quad and \quad Y(\mathbf{v}_\lambda, z) = V_{\lambda, \varepsilon}(z)$$

for $h \in \mathfrak{h}$ and $\lambda, \mu \in L$.

The vertex algebra $\mathbf{V}_L$ is called the *lattice vertex algebra* associated with the even lattice L, which is a simple vertex algebra. The isomorphism class of the resulting vertex algebra does not depend on the choice of the cocycle factor.

The space $\mathbf{V}_L$ is given a grading as in the rank one case. If L is positive-definite of rank n, the graded dimension is given by

$$\sum_{d=0}^{\infty} q^d \dim \mathbf{V}_{L,d} = \sum_{\lambda \in L} \prod_{k=1}^{\infty} \frac{q^{(\lambda|\lambda)/2}}{(1-q^k)^n} = \frac{\Theta_L(\tau)}{\phi(q)^n} = q^{n/24} \frac{\Theta_L(\tau)}{\eta(\tau)^n}.$$

Note 4.7.   1. The cocycle factor $\varepsilon \colon L \times L \longrightarrow \{\pm 1\}$ here can be chosen so that it is bimultiplicative and satisfies $\varepsilon(\lambda, \lambda) = (-1)^{(\lambda|\lambda)/2}$ for all $\lambda \in L$. 2. Let $\mathbf{F}_\lambda(d)$ be the subspace of $\mathbf{F}_\lambda$ of degree $(\lambda|\lambda)/2 + d$ for each $\lambda \in L$ and $d \in \mathbb{N}$. In particular, we have $\mathbf{F}_\lambda(0) = \mathbb{C}\mathbf{v}_\lambda$. Then the following properties hold:
(1) For all $h \in \mathfrak{h}$, $\lambda \in L$ and $k, m \in \mathbb{Z}$: $[h_k, \mathbf{v}_{\lambda(m)}] = \lambda(h)\mathbf{v}_{\lambda(k+m)}$.
(2) For all $\lambda, \mu \in L$ and $m \in \mathbb{Z}$: $\mathbf{v}_{\lambda(m)}\mathbf{v}_\mu \in \mathbf{F}_{\lambda+\mu}(-(\lambda|\mu) - m - 1)$.
(3) For all $\lambda, \mu \in L$: $\mathbf{v}_{\lambda(-(\lambda|\mu)-1)}\mathbf{v}_\mu = \pm\mathbf{v}_{\lambda+\mu}$.

These properties in fact characterize the vertex algebra structure of the lattice vertex algebra $\mathbf{V}_L$. (See the arguments in [76].)

1.4.5.3 Modules over Lattice Vertex Algebras
Let us briefly describe modules over the lattice vertex algebras. (See [40] for details.)

For a lattice L, the *dual lattice* is the set

$$L^\circ = \{\mu \in \mathfrak{h} \,|\, (\lambda|\mu) \in \mathbb{Z}\},$$

which is an additive subgroup of $\mathfrak{h}$ containing L. The values of the bilinear form on L° are rational numbers, and L° need not be a lattice in the sense we followed so far.

The isomorphism classes of simple $\mathbf{V}_L$-modules are in one-to-one correspondence with the cosets in L°/L. For each coset $M \in L^\circ/L$, it has the following shape:

$$\mathbf{V}_M = \bigoplus_{\lambda \in M} \mathbf{F}_\lambda.$$

In particular, the adjoint module $\mathbf{V}_L$ is the case with $M = L$.

Recall that a lattice is said to be *unimodular* if $L^\circ = L$. For a unimodular lattice L, the adjoint module is the only simple module over $\mathbf{V}_L$.

Here are examples of even unimodular positive-definite lattices. Note that the rank of such a lattice is a multiple of 8.

1. Rank 8. There is only one such lattice: the Gosset lattice E_8.
2. Rank 16. There are two such lattices: $E_8 \oplus E_8$ and D_{16}^+.
3. Rank 24. There are 24 such lattices, called *Niemeier lattices*. Among them, there is a distinguished one without roots called the *Leech lattice*.

The graded dimensions of the vertex algebras associated with the Gosset lattice E_8 and the Leech lattice Λ are given by

$$\sum_{d=0}^{\infty} q^d \dim \mathbf{V}_{E_8,d} = \frac{\Theta_{E_8}(\tau)}{\phi(q)^8} = q^{1/3} \frac{E_4(\tau)}{\eta(\tau)^8} \quad \text{and}$$

$$\sum_{d=0}^{\infty} q^d \dim \mathbf{V}_{\Lambda,d} = \frac{\Theta_\Lambda(\tau)}{\phi(q)^{24}} = q(j(\tau) - 720) = 1 + 24q + 196884q^2 + \cdots,$$

respectively, where $E_4(\tau)$ is the Eisenstein series of weight 4,

$$E_4(\tau) = 1 + 240 \sum_{n=1}^{\infty} \frac{n^3 q}{1 - q^n},$$

and $j(\tau)$ the elliptic modular function. The graded dimensions for the Niemeier lattices are all the same except for the coefficients to q.

Note 4.8. To construct the $\mathbf{V}_L$-module structures on $\mathbf{V}_M$ for cosets M in L°/L, we need to extend the factor ε to an appropriate one defined on $L \times L^\circ$ valued in a cyclic group containing $\{\pm 1\}$. See [40] for details.

Bibliographic Notes

Main references for Section 1.4 are Frenkel, Lepowsky, and Meurman [1], Dong and Lepowsky [4], and Dong [40]. See also Kac [6] and Lepowsky and Li [10]. For descriptions in models in physics, consult Di Francesco et al. [13]. For generalities on lattices, see Ebeling [14] and Conway and Sloan [12].

See Frenkel and Kac [57] and Kac [17] for Frenkel–Kac Construction (cf. (cf. Segal [92]), and Lepowsky and Wilson [69] for a slightly earlier construction of $\widehat{\mathfrak{sl}}_2$ by twisted vertex operators and Frenkel, Lepowsky, and Meurman [58] for generalizations.

Construction of the lattice vertex algebras, as well as introduction of the concept of vertex algebras, is due to Borcherds [32]. See Frenkel, Lepowsky, and Meurman [59] for an earlier attempt to construct and investigate the moonshine module $V^\natural$ by means of vertex operators.

There are a huge number of applications of vertex operators in various fields of mathematics and physics.

1.5 Twisted Modules

Let V be a vertex algebra and θ an involution of V, that is, an automorphism of order 2, and consider the subspace of fixed-points,

$$V^+ = \{a \in V \mid \theta a = a\},$$

which is a vertex subalgebra of V. Let $i\colon V^+ \longrightarrow V$ denote the inclusion.

For any representation $\rho(-,z)\colon V \longrightarrow \mathrm{Hom}(\mathbf{M}, \mathbf{M}((z)))$, we obtain a representation of V^+ by restriction as

$$V^+ \xrightarrow{\ i\ } V \xrightarrow{\ \rho(-,z)\ } \mathrm{Hom}(\mathbf{M}, \mathbf{M}((z))).$$

Representation of V^+ may also be obtained by restricting "generalized" representations of V in such a way that series in half-integral powers are involved but the restriction becomes valued in series with integral powers:

$$V^+ \xrightarrow{\ i\ } V \xrightarrow{\ \rho(-,z)\ } \mathrm{Hom}(\mathbf{M}, \mathbf{M}((z^{1/2})))$$
$$\mathrm{Hom}(\mathbf{M}, \mathbf{M}((z))).$$

The concept of θ-twisted modules over V corresponds to such a generalization of representations of V, achieved by appropriately generalizing the residue products to series with half-integral powers in z.

More generally, for an automorphism g of finite order N, the concept of g-twisted modules is defined by replacing $z^{1/2}$ with $z^{1/N}$:

$$V \xrightarrow{\ \rho(-,z)\ } \mathrm{Hom}(\mathbf{M}, \mathbf{M}((z^{1/N}))).$$

Although twisted modules are considered under presence of the action of an automorphism, generalization of the residue products works well for series with complex powers in z without actions of automorphisms.

In Section 1.5, we will first describe the general theory of twisted modules and then proceed to classical examples, the θ-twisted modules over the Heisenberg vertex algebras and the lattice vertex algebras, where θ is lift of the (-1)

-involution, the automorphism of order 2 induced by negation -1 of the generators of the Heisenberg algebra or the lattice.

We will work over the field $\mathbb{C}$ of complex numbers.

1.5.1 OPE of Shifted Series

In this section, we will generalize the residue products to those of the series whose exponents are integers shifted by a complex number for each. We will then appropriately adjust the concept of operator product expansion so that it fits such shifted series.

1.5.1.1 Preliminaries on Shifted Series

For a complex number α, consider the vector space $V[[z, z^{-1}]]z^{-\alpha}$ consisting of the formal expressions of the form

$$\sum_n v_{n+\alpha} z^{-n-\alpha-1},$$

where the summation is over the integers n. In this section, we will call such a series a series *shifted* by $\alpha \in \mathbb{C}$.

We will also consider subspaces such as $V((z))z^{-\alpha}$ or

$$V((y))((z))y^{-\alpha}z^{-\beta}, \quad V((z))((y))y^{-\alpha}z^{-\beta}, \quad V((y,z))y^{-\alpha}z^{-\beta}$$

for indeterminates y, z and complex numbers α, β.

Let x, y, z be indeterminates. We write

$$(x + z)^{n-\alpha}\big|_{|x|>|z|} = \sum_{i=0}^{\infty} \binom{n-\alpha}{i} x^{n-\alpha-i} z^i,$$

$$(x + z)^{n-\alpha}\big|_{|x|<|z|} = \sum_{i=0}^{\infty} \binom{n-\alpha}{i} x^i z^{n-\alpha-i}.$$

Note that

$$(x + z)^{n-\alpha}\big|_{|x|>|z|} \in \mathbb{C}[x, x^{-1}][[z]]x^{-\alpha} \subset \mathbb{C}((x))((z))x^{-\alpha},$$

$$(x + z)^{n-\alpha}\big|_{|x|<|z|} \in \mathbb{C}[z, z^{-1}][[x]]z^{-\alpha} \subset \mathbb{C}((z))((x))z^{-\alpha}.$$

Similarly, we write

$$(y - z)^{n-\alpha}\big|_{|y|>|z|} = \sum_{i=0}^{\infty} (-1)^i \binom{n-\alpha}{i} y^{n-\alpha-i} z^i \qquad \in \mathbb{C}((y))((z))y^{-\alpha},$$

$$(y - z)^{n-\alpha}\big|_{|y|<|z|} = \sum_{i=0}^{\infty} (-1)^{n-\alpha-i} \binom{n-\alpha}{i} y^i z^{n-\alpha-i} \in \mathbb{C}((z))((y))z^{-\alpha}.$$

1.5.1.2 Expansions of Shifted Series

Let x, y, z be indeterminates and α, β complex numbers. Consider the space

$$V((x, y, z))y^{-\alpha}z^{-\beta},$$

whose elements are written in the following form with some $L, M, N \in \mathbb{N}$:

$$w(x, y, z) = \frac{w_0(x, y, z)}{x^N y^{L+\alpha} z^{M+\beta}}, \quad w_0(x, y, z) \in V[[x, y, z]].$$

Since $x^N w(x, y, z) \in V[[x]]((y, z))y^{-\alpha}z^{-\beta}$,

$$(y - z)^N w(y - z, y, z)\big|_{|y|>|z|} = (y - z)^N w(y - z, y, z)\big|_{|y|<|z|}. \tag{5.1}$$

Similarly, $y^{L+\alpha}w(x, y, z) \in V[[y]]((x, z))z^{-\beta}$ implies

$$(x + z)^{L+\alpha}w(x, x + z, z)\big|_{|x|>|z|} = (x + z)^{L+\alpha}w(x, x + z, z)\big|_{|x|<|z|}.$$

Let $s(y, z)$ and $t(y, z)$ be series belonging to the spaces $V((y))((z))y^{-\alpha}z^{-\beta}$ and $V((z))((y))y^{-\alpha}z^{-\beta}$, respectively, and $w(x, y, z) \in V((x, y, z))y^{-\alpha}z^{-\beta}$ satisfy

$$s(y, z) = w(y - z, y, z)\big|_{|y|>|z|} \in V((y))((z))y^{-\alpha}z^{-\beta},$$
$$t(y, z) = w(y - z, y, z)\big|_{|y|<|z|} \in V((z))((y))y^{-\alpha}z^{-\beta}.$$

Consider the series $u(x, z) = w(x, x + z, z)\big|_{|x|<|z|}$. Then the coefficients $u_m(z)$ in $u(x, z) = \sum_m u_m(z)x^{-m-1}$ are determined by

$$u_m(z) = \sum_{i=0}^{\infty} \binom{-\alpha}{i} z^{-\alpha-i} \operatorname{Res}_y \left((y - z)^{m+i}\big|_{|y|>|z|} y^\alpha s(y, z)\right.$$
$$\left. - (y - z)^{m+i}\big|_{|y|<|z|} y^\alpha t(y, z)\right),$$

where the sum is a finite sum by (5.1). Note that $u_m(z)$ does not depend on the choice of α such that $y^\alpha w(x, y, z)$ is of integral powers in y.

Note 5.1. The right-hand side is obtained by formally calculating the expression

$$(y - z)^m s(y, z)\big|_{|y|>|z|} - (y - z)^m t(y, z)\big|_{|y|<|z|}.$$

Indeed, inserting $y^{-\alpha}y^\alpha$, we formally write it as

$$y^{-\alpha}(y - z)^m\big|_{|y|>|z|} y^\alpha s(y, z) - y^{-\alpha}(y - z)^m\big|_{|y|<|z|} y^\alpha t(y, z). \tag{5.2}$$

Replace the factor $y^{-\alpha}(y - z)^m$ by the expansion

$$(x + z)^{-\alpha}\big|_{|x|<|z|}x^m = \sum_{i=0}^{\infty} \binom{-\alpha}{i} z^{-\alpha-i} x^{m+i},$$

and then expand it by substitution $x = y - z$ in the respective regions. Although the result does not make sense as a series in y and z, the expression (5.2) turns out to give a well-defined expression by (5.1).

1.5.1.3 Residue Products of Shifted Series

Let **M** be a vector space and α a complex number. We will call such an element of $(\text{End }\mathbf{M})[[z, z^{-1}]]z^{-\alpha}$ a series on **M** *shifted* by α. For such a series $A(z)$, set

$$A(z) = \sum_n A_{n+\alpha} z^{-n-\alpha-1},$$

where the sum is over the integers n and $A_{n+\alpha}$ are operators acting on **M**.

We will say that $A(z)$ is *locally truncated* if $A(z)v \in \mathbf{M}((z))z^{-\alpha}$ for all $v \in \mathbf{M}$. The set of such series is identified with $\text{Hom}(\mathbf{M}, \mathbf{M}((z))z^{-\alpha})$.

Let $A(z)$ and $B(z)$ be series on a vector space **M** shifted by complex numbers. They are said to be *locally commutative* if the following holds for some $N \in \mathbb{N}$:

$$(y - z)^N A(y)B(z) = (y - z)^N B(z)A(y).$$

This is the same as the case of series with integral powers.

Let $A(z)$ and $B(z)$ be locally truncated and *locally commutative* series on a vector space **M** shifted by complex numbers α and β, respectively. Define the nth residue product of $A(z)$ and $B(z)$ by

$$
\begin{aligned}
A(z)_{(n)}B(z) = \sum_{i=0}^{\infty} \binom{-\alpha}{i} & z^{-\alpha-i} \left(\text{Res}_y (y - z)^{n+i} \big|_{|y|>|z|} y^{\alpha} A(y)B(z) \right. \\
& \left. - \text{Res}_y (y - z)^{n+i} \big|_{|y|<|z|} y^{\alpha} B(z)A(y) \right) \\
= \sum_{i=0}^{\infty} \binom{-\alpha}{i} & (z^{\alpha} A(z))_{(n+i)} B(z) \, z^{-\alpha-i}
\end{aligned}
$$

for each $n \in \mathbb{Z}$.

As the identity series $I(z)$ is unshifted, the identity property holds:

$$
I(z)_{(n)}A(z) = \begin{cases} 0 & (n \neq -1), \\ A(z) & (n = -1). \end{cases}
$$

The relation $A(z)_{(n)}I(z) = 0$ for $n \geq 0$ is clear and

$$
\begin{aligned}
A(z)_{(-1)}I(z) &= \sum_{i=0}^{\infty} \binom{-\alpha}{i} z^{-\alpha-i} (z^{\alpha} A(z))_{(-1+i)} I(z) \\
&= z^{-\alpha} (z^{\alpha} A(z))_{(-1)} I(z) = z^{-\alpha} z^{\alpha} A(z) = A(z).
\end{aligned}
$$

Therefore, the creation property also holds:

$$
A(z)_{(n)}I(z) = \begin{cases} 0 & (n \geq 0), \\ A(z) & (n = -1). \end{cases}
$$

It is not difficult to show the relation

$$A(z)_{(-k-1)}I(z) = \partial^{(k)} A(z)$$

for $n = -k - 1 < 0$ by $(z^{\alpha} A(z))_{(-k-1)} I(z) = \partial^{(k)}(z^{\alpha} A(z))$.

1.5.1.4 Modified OPE of Shifted Series

Let $A(z)$ and $B(z)$ be locally truncated locally commutative series on a vector space $\mathbf{M}$ shifted by α and β, respectively:

$$A(z) \in \mathrm{Hom}(\mathbf{M}, \mathbf{M}((z))z^{-\alpha}), \quad B(z) \in \mathrm{Hom}(\mathbf{M}, \mathbf{M}((z))z^{-\beta}).$$

Since $z^\alpha A(z)$ and $z^\beta B(z)$ are of integral powers, we have an OPE of the following form for some $N \in \mathbb{N}$ and series $C_0(z), \ldots, C_{N-1}(z)$ with integral powers:

$$y^\alpha A(y)z^\beta B(z) \simeq z^\beta B(z)y^\alpha A(y) \sim \sum_{k=0}^{N-1} \frac{C_k(z)}{(y-z)^{k+1}}.$$

Therefore,

$$A(y)B(z) \simeq B(z)A(y) \sim y^{-\alpha}z^{-\beta} \sum_{k=0}^{N-1} \frac{C_k(z)}{(y-z)^{k+1}}. \tag{5.3}$$

Let us further expand $y^{-\alpha}$ by substitution $y = x + z$ in $|x| < |z|$, connect by $\approx$ if the two sides are related by this process, and neglect the regular part. Then the right-hand side of (5.3) results in

$$y^{-\alpha}z^{-\beta} \sum_{k=0}^{N-1} \frac{C_k(z)}{(y-z)^{k+1}} \approx \sum_{k=0}^{N-1} \frac{D_k(z)}{(y-z)^{k+1}},$$

where

$$D_k(z) = \sum_{i=0}^{N-k} \binom{-\alpha}{i} C_{k+i}(z)z^{-\alpha-\beta-i} \quad (0 \le k < N).$$

We thus arrive at the following expression:

$$A(y)B(z) \simeq B(z)A(y) \approx \sum_{k=0}^{N-1} \frac{D_k(z)}{(y-z)^{k+1}}. \tag{5.4}$$

Let us call it (the singular part of) the modified OPE of $A(z)$ and $B(z)$.

The modified OPE (5.4) allows us to find the residue products for $m \in \mathbb{N}$ as

$$A(z)_{(m)}B(z) = \begin{cases} 0 & (N \le m), \\ D_m(z) & (0 \le m < N), \end{cases}$$

by the formula (5.4).

1.5.2 Shifted and Twisted Modules

In this section, we will generalize the concept of modules over a vertex algebra by replacing vertex algebra of series with vertex algebra of shifted series. Under presence of the action of an automorphism g of finite order, those fitting g give rise to the concept of g-twisted modules.

1.5.2.1 Vertex Algebras of Shifted Series

Let us now consider a finite sum of series shifted by complex numbers. Such a sum $v(z)$ is written in the following form with some $k \in \mathbb{N}$ and $\alpha_1, \ldots, \alpha_k \in \mathbb{C}$:

$$v(z) = \sum_{i=1}^{k} \sum_{n} v_{n+\alpha_i} z^{-n-\alpha_i-1}.$$

We will simply call such a series a *shifted series*. The concepts of local truncation, local commutativity, and residue products are generalized to shifted series in obvious ways.

The result shown Theorem 5.2 is a shifted analogue of Theorem 5.2 and can be proved in the same spirit.

Theorem 5.2   *Let $A(z), B(z), C(z)$ be locally truncated shifted series on a vector space. If they are locally commutative with each other, then the Borcherds identity*

$$\sum_{i=0}^{\infty} \binom{p}{i} (A(z)_{(r+i)}B(z))_{(p+q-i)}C(z)$$

$$= \sum_{i=0}^{\infty} (-1)^i \binom{r}{i} A(z)_{(p+r-i)}(B(z)_{(q+i)}C(z))$$

$$- \sum_{i=0}^{\infty} (-1)^{r+i} \binom{r}{i} B(z)_{(q+r-i)}(A(z)_{(p+i)}C(z))$$

holds for all $p, q, r \in \mathbb{Z}$.

As in the unshifted case, we readily obtain the following result due to Li and Roitman.

Corollary 5.3   *Let $\mathcal{V}$ be a vector space consisting of shifted series on a vector space $\mathbf{M}$ satisfying the following conditions.*

(1) *The space $\mathcal{V}$ is locally truncated and locally commutative.*
(2) *The space $\mathcal{V}$ is closed under the residue products.*
(3) *The space $\mathcal{V}$ contains the identity series.*

Then the space $\mathcal{V}$ becomes a vertex algebra by the residue products.

We will call the vertex algebra thus obtained a *vertex algebra of shifted series.*

1.5.2.2 Shifted Representations

Let us denote the sets of shifted series and shifted Laurent series with coefficients in a vector space V respectively by

$$V[[z,z^{-1}]]z^{\mathbb{C}} = \sum_{\alpha \in \mathbb{C}} V[[z,z^{-1}]]z^{-\alpha},$$

$$V((z))z^{\mathbb{C}} = \sum_{\alpha \in \mathbb{C}} V((z))z^{-\alpha} = \sum_{\alpha \in \mathbb{C}} V[[z]]z^{-\alpha}.$$

Let $\mathbf{V}$ be a vertex algebra and $\mathbf{M}$ a vector space, and consider a map of the
following form:

$$\rho(-,z) : \mathbf{V} \longrightarrow \mathrm{Hom}(\mathbf{M}, \mathbf{M}[[z,z^{-1}]]z^{\mathbb{C}}), \quad a \mapsto \rho(a,z).$$

Such a map is said to be a *shifted representation* of $\mathbf{V}$ if the following conditions
are satisfied:

(1) The image of $\rho(-,z)$ is a vertex algebra of shifted series on $\mathbf{M}$.
(2) The map $\rho(-,z)$ induces a homomorphism of vertex algebras onto its image.

Note that (1) in particular says that $\rho(a,z)$ is locally truncated for any $a \in \mathbf{V}$.

1.5.2.3 Twisted Modules

By abuse of notation, we will denote the image of a complex number $\alpha \in \mathbb{C}$ in
$\mathbb{C}/\mathbb{Z}$ by the same symbol α.

Let g be an automorphism of a vertex algebra $\mathbf{V}$ of finite order:

$$g : \mathbf{V} \longrightarrow \mathbf{V}.$$

Let $N \in \mathbb{N}$ be the order of g and set

$$\Gamma = ((1/N)\mathbb{Z})/\mathbb{Z} = \{0, 1/N, \ldots, (N-1)/N\}.$$

Then the eigenspace decomposition of $\mathbf{V}$ with respect to the action of g is written as follows:

$$\mathbf{V} = \bigoplus_{\alpha \in \Gamma} \mathbf{V}^{\alpha}, \quad \mathbf{V}^{\alpha} = \left\{ a \in \mathbf{V} \,\middle|\, ga = e^{2\pi\sqrt{-1}\alpha}a \right\}.$$

Since g is an automorphism of $\mathbf{V}$, we have

$$\mathbf{V}^{\alpha}{}_{(m)}\mathbf{V}^{\beta} \subset \mathbf{V}^{\alpha+\beta}$$

for all $\alpha, \beta \in \Gamma$, and $m \in \mathbb{Z}$.

In such a situation, a shifted representation given by a direct sum of maps

$$\rho_\alpha(-, z) \colon \mathbf{V}^\alpha \longrightarrow \mathrm{Hom}(\mathbf{M}, \mathbf{M}((z))z^{-\alpha})$$

is called a *g-twisted representation* of **V**.

For each $\alpha \in \Gamma$ and $a \in \mathbf{V}_\alpha$, let us write

$$\rho_\alpha(a, z) = \sum_n a_{n+\alpha} z^{-n-\alpha-1},$$

so that the coefficients give rise to maps

$$\rho_{\alpha,n} \colon \mathbf{V}^\alpha \times \mathbf{M} \longrightarrow \mathbf{M}, \quad (a, v) \mapsto a_{n+\alpha} v. \tag{5.5}$$

Then the direct sum of $\rho_\alpha(-, z)$ becomes a *g*-twisted representation of **V** if and only if the following properties holds:

(T0) Local truncation. For any $\alpha \in \Gamma$, $a \in \mathbf{V}^\alpha$ and $v \in \mathbf{M}$, there exists an $N \in \mathbb{N}$ such that

$$a_{N+\alpha+i} v = 0 \ \text{ for all } i \geq 0.$$

(T1) Borcherds identity. For all $\alpha, \beta \in \Gamma$, $a \in \mathbf{V}^\alpha$, $b \in \mathbf{V}^\beta$, $v \in \mathbf{M}$, and $p \in \mathbb{Z} + \alpha$, $q \in \mathbb{Z} + \beta$ and $r \in \mathbb{Z}$:

$$\sum_{i=0}^\infty \binom{p}{i} (a_{(r+i)}b)_{p+q-i} v = \sum_{i=0}^\infty (-1)^i \binom{r}{i} a_{p+r-i}(b_{q+i}v)$$
$$- \sum_{i=0}^\infty (-1)^{r-i} \binom{r}{i} b_{q+r-i}(a_{p+i}v).$$

(T2) Identity. For any $v \in \mathbf{M}$ and $n \in \mathbb{Z}$:

$$\mathbf{1}_n v = \begin{cases} 0 & (n \neq -1), \\ v & (n = -1). \end{cases}$$

A sequence of maps as in (5.5) satisfying the properties (T0)–(T2) is called a *g-twisted module* over **V** or a *g-twisted V-module*.

The generating series $\rho_\alpha(a, z)$ for $a \in \mathbf{V}^\alpha$ is usually written as

$$Y_\mathbf{M}(a, z) = \sum_n a_{n+\alpha} z^{-n-\alpha-1}.$$

A module over a vertex algebra in the ordinary sense is sometimes called an *untwisted module*. A *g*-twisted **V**-module **M** is an untwisted module over the subalgebra $\mathbf{V}^0$.

In later sections, we will be concerned with the case $N = 2$, where the eigenspaces $\mathbf{V}^0$ and $\mathbf{V}^{1/2}$ are denoted by $\mathbf{V}^+$ and $\mathbf{V}^-$, respectively.

Note 5.3.   In the physics literatures, the untwisted and twisted modules are often called the untwisted and twisted *sectors* (of the theory or the model under consideration), respectively.

1.5.3 Twisted Heisenberg Modules

In this section, we illustrate an example of a twisted module in the case of the Heisenberg vertex algebra with a particular involution θ; the one induced by negation of the standard generator, which is the most simple and classical example of twisted modules.

For simplicity, we will work with the Heisenberg vertex algebra of rank one following the description in Section 1.3.1. The higher rank cases are treated in the same way (cf. Subsection 1.5.4.1).

1.5.3.1 OPE of Twisted Current

Let $a_{n+1/2}$ with $n \in \mathbb{Z}$ and ζ be indeterminates and set

$$\hat{\mathfrak{h}}^{\mathrm{tw}} = \bigoplus_{n \in \mathbb{Z}} \mathbb{C}a_{n+1/2} \oplus \mathbb{C}\zeta.$$

Then $\hat{\mathfrak{h}}^{\mathrm{tw}}$ becomes a Lie algebra by the bracket

$$[a_{m+1/2}, a_{n+1/2}] = (m + 1/2)\delta_{m+n+1,0}\zeta, \quad [\zeta, a_{n+1/2}] = 0.$$

Consider the twisted Heisenberg algebra defined by

$$\mathbf{U}(\hat{\mathfrak{h}}^{\mathrm{tw}}, 1) = \mathbf{U}(\hat{\mathfrak{h}}^{\mathrm{tw}})/(\zeta - 1).$$

We will denote the images of the generators $a_{n+1/2}$ by the same symbol. Then

$$[a_{m+1/2}, a_{n+1/2}] = (m + 1/2)\delta_{m+n+1,0}$$

for $m, n \in \mathbb{Z}$, where the bracket denotes the commutator.

Consider the *twisted current*

$$a^{\mathrm{tw}}(z) = \sum_n a_{n+1/2} z^{-(n+1/2)-1} = \sum_n a_{n+1/2} z^{-n-3/2}.$$

Then the commutation relation turns out to be expressed as

$$[a^{\mathrm{tw}}(y), a^{\mathrm{tw}}(z)] - \sum_m (m + 1/2) y^{-m-1/2-1} z^{m-1/2}$$

$$= \partial_z \left(y^{-1/2} z^{1/2} \delta(y, z) \right)$$

$$= \frac{1}{2} y^{-1/2} z^{-1/2} \delta(y, z) + y^{-1/2} z^{1/2} \delta^{(1)}(y, z).$$

Table 11 Twisted Fock module

	0	1/2	1	3/2	2
$\mathbf{F}^{\mathrm{tw}}$:	$\mathbf{v}^{\mathrm{tw}}$	$a_{-1/2}\mathbf{v}^{\mathrm{tw}}$	$(a_{-1/2})^2\mathbf{v}^{\mathrm{tw}}$	$a_{-3/2}\mathbf{v}^{\mathrm{tw}}$	$a_{-1/2}a_{-3/2}\mathbf{v}^{\mathrm{tw}}$
				$(a_{-1/2})^3\mathbf{v}^{\mathrm{tw}}$	$(a_{-1/2})^4\mathbf{v}^{\mathrm{tw}}$

Therefore, we have the following OPE:

$$a^{\mathrm{tw}}(y)a^{\mathrm{tw}}(z) \simeq a^{\mathrm{tw}}(z)a^{\mathrm{tw}}(y) \sim \frac{1}{2}\frac{y^{-1/2}z^{-1/2}}{y-z} + \frac{y^{-1/2}z^{1/2}}{(y-z)^2}.$$

In particular, $a^{\mathrm{tw}}(z)$ is locally commutative with itself.
Expansion of the factor $y^{-1/2}$ by $y = x + z$ in $|x| < |z|$ yields

$$(x+z)^{-1/2}\big|_{|x|<|z|} = \sum_{k=0}^{\infty}\binom{-1/2}{k}x^k z^{-1/2-k}.$$

After some algebra, we find that the modified OPE is given by

$$a^{\mathrm{tw}}(y)a^{\mathrm{tw}}(z) \simeq a^{\mathrm{tw}}(z)a^{\mathrm{tw}}(y) \approx \frac{1}{(y-z)^2}.$$

Therefore, the shifted residue products for $n \in \mathbb{N}$ is determined as

$$a^{\mathrm{tw}}(z)_{(n)}a^{\mathrm{tw}}(z) = \begin{cases} 0 & (n \geq 2), \\ 1 & (n = 1), \\ 0 & (n = 0), \end{cases} \tag{5.6}$$

which is exactly of the same form as the untwisted case.

1.5.3.2 Twisted Fock Module

Consider the following subspaces of the Lie algebra $\hat{\mathfrak{h}}^{\mathrm{tw}}$:

$$\hat{\mathfrak{h}}^{\mathrm{tw}}_{<0} = \mathrm{Span}\left\{a_{n+1/2}\,\big|\,n < 0\right\}, \quad \hat{\mathfrak{h}}^{\mathrm{tw}}_{>0} = \mathrm{Span}\left\{a_{n+1/2}\,\big|\,n \geq 0\right\}.$$

They are commutative Lie subalgebras, and generate subalgebras of $\mathbf{U}(\hat{\mathfrak{h}}^{\mathrm{tw}}, 1)$ isomorphic to the symmetric algebras $\mathbf{S}(\hat{\mathfrak{h}}^{\mathrm{tw}}_{<0})$ and $\mathbf{S}(\hat{\mathfrak{h}}^{\mathrm{tw}}_{>0})$, respectively.
Let $\mathbb{C}\mathbf{v}^{\mathrm{tw}}$ be the one-dimensional trivial module over $\mathbf{S}(\hat{\mathfrak{h}}^{\mathrm{tw}}_{>0})$, for which

$$a_{n+1/2}\mathbf{v}^{\mathrm{tw}} = 0 \quad (n \geq 0).$$

Define the twisted Fock module by

$$\mathbf{F}^{\mathrm{tw}} = \mathbf{U}(\hat{\mathfrak{h}}^{\mathrm{tw}}, 1) \otimes_{\mathbf{S}(\hat{\mathfrak{h}}^{\mathrm{tw}}_{>0})} \mathbb{C}\mathbf{v}^{\mathrm{tw}} \simeq \mathbf{S}(\hat{\mathfrak{h}}^{\mathrm{tw}}_{<0}) \otimes \mathbb{C}\mathbf{v}^{\mathrm{tw}}.$$

See Table 11.

Unlike the untwisted case, there is no freedom of *charge,* for it is the eigen-value of the action of the central element a_0, which is present in $\mathbf{U}(\hat{\mathfrak{h}}, 1)$ but absent in $\mathbf{U}(\hat{\mathfrak{h}}^{tw}, 1)$.

1.5.3.3 Twisted $\mathbf{F}_0$-Module

Since the twisted current on $\mathbf{F}^{tw}$ is locally truncated and locally commutative with itself, the twisted currents generate a vertex algebra, which we denote as

$$\mathcal{F}_0^{tw} = \langle a^{tw}(z) \rangle_{RP}.$$

Now recall the (untwisted) Fock module $\mathbf{F}_0 = \mathbf{S}(\hat{\mathfrak{h}}_{<0})\mathbf{v}_0 = \mathbb{C}[x_1, x_2, \cdots]$ of charge 0. By the shifted OPE (5.6) and the universal property of the Fock module $\mathbf{F}_0$, there exists a unique homomorphism of $\mathbf{U}(\hat{\mathfrak{h}}, 1)$-modules sending the vacuum vector $\mathbf{v}_0$ to the identity series $I(z)$ on $\mathbf{F}^{tw}$,

$$\psi^{tw} : \mathbf{F}_0 \longrightarrow \mathcal{F}_0^{tw}, \quad \mathbf{v}_0 \mapsto I(z),$$

giving rise to a shifted representation of the Heisenberg vertex algebra $\mathbf{F}_0$ on the twisted Fock module $\mathbf{F}^{tw}$.

The corresponding shifted $\mathbf{F}_0$-module becomes a twisted module. To see it, consider the involution θ of the polynomial ring $\mathbf{F}_0 = \mathbb{C}[x_1, x_2, \cdots]$ which negates the indeterminates:

$$\theta : \mathbf{F}_0 \longrightarrow \mathbf{F}_0, \quad x_k \mapsto -x_k \quad (k = 1, 2, \cdots).$$

Then θ turns out to be an automorphism of a vertex algebra, actually determined by its action on the generator $x_1 = a_{-1}\mathbf{v}_0$, for which the eigenspace decomposition is of the form

$$\mathbf{F}_0 = \mathbf{F}_0^+ \oplus \mathbf{F}_0^-,$$

where

$$\mathbf{F}_0^+ = \{a \in \mathbf{F}_0 \mid \theta a = a\}, \quad \mathbf{F}_0^- = \{a \in \mathbf{F}_0 \mid \theta a = -a\}.$$

On the other hand, the space $\mathcal{F}_0^{tw}$ decomposes according to the shifts of the series as

$$\mathcal{F}_0^{tw} = \mathcal{F}_0^{tw,+} \oplus \mathcal{F}_0^{tw,-},$$

where

$$\mathcal{F}_0^{tw,+} = \mathcal{F}_0^{tw} \cap \mathrm{Hom}(\mathbf{F}^{tw}, \mathbf{F}^{tw}((z))),$$
$$\mathcal{F}_0^{tw,-} = \mathcal{F}_0^{tw} \cap \mathrm{Hom}(\mathbf{F}^{tw}, \mathbf{F}^{tw}((z))z^{-1/2}).$$

For example,

$$I(z) \in \mathcal{F}_0^{tw,+}, \quad a^{tw}(z) \in \mathcal{F}_0^{tw,-}, \quad a^{tw}(z)_{(n)}a^{tw}(z) \in \mathcal{F}_0^{tw,+} \quad (n \in \mathbb{Z}).$$

In general, the residue products of even numbers of $a^{\mathrm{tw}}(z)$ belong to $\mathcal{F}_0^{\mathrm{tw},+}$, whereas odd numbers to $\mathcal{F}_0^{\mathrm{tw},-}$.

The shifted representation of $\mathbf{F}_0$ now decomposes into the direct sum of

$$\psi^{\mathrm{tw},+} : \mathbf{F}_0^+ \longrightarrow \mathcal{F}_0^{\mathrm{tw},+}, \quad \psi^{\mathrm{tw},-} : \mathbf{F}_0^- \longrightarrow \mathcal{F}_0^{\mathrm{tw},-},$$

giving rise to a structure of a θ-twisted $\mathbf{F}_0$-module on the twisted Fock module $\mathbf{F}^{\mathrm{tw}}$.

As the space $\mathcal{F}_0^{\mathrm{tw},+}$ consists of series with integral powers and the fixed-point subspace $\mathbf{F}_0^+$ is a vertex subalgebra of $\mathbf{F}_0$, the map

$$\psi^{\mathrm{tw},+} : \mathbf{F}_0^+ \longrightarrow \mathcal{F}_0^{\mathrm{tw},+}$$

gives rise to an untwisted representation of $\mathbf{F}_0^+$ on $\mathbf{F}^{\mathrm{tw}}$.

1.5.3.4 Twisted Virasoro Actions

Recall the standard Virasoro vector for the Heisenberg vertex algebra given by (1.8) and (3.5):

$$\omega = \frac{1}{2}x_1^2 = \frac{1}{2}a_{-1}a_{-1}\mathbf{v}_0.$$

The corresponding twisted series on $\mathbf{F}^{\mathrm{tw}}$ turns out to be

$$T^{\mathrm{tw}}(z) = \frac{1}{2}a^{\mathrm{tw}}(z)_{(-1)}a^{\mathrm{tw}}(z)_{(-1)}I(z) = \frac{1}{2}a^{\mathrm{tw}}(z)_{(-1)}a^{\mathrm{tw}}(z).$$

After some algebra following the definition of the residue products,

$$T^{\mathrm{tw}}(z) = \frac{1}{2}z^{-1/2}\,{}^{\circ}_{\circ}(z^{1/2}a^{\mathrm{tw}}(z))a^{\mathrm{tw}}(z)^{\circ}_{\circ} + \frac{1}{16}z^{-2}.$$

The Fourier modes of $T^{\mathrm{tw}}(z)$ generate a representation of Virasoro algebra of central charge 1. For example,

$$L_0^{\mathrm{tw}} = \frac{1}{16} + a_{-1/2}a_{1/2} + a_{-3/2}a_{3/2} + a_{-5/2}a_{5/2} + \cdots.$$

Note that the Virasoro vector ω belongs to the subspace $\mathbf{F}_0^+$.

1.5.4 Twisted Vertex Operators

In this section, we will generalize and continue the consideration of the preceding section for higher-rank Heisenberg vertex algebras, and introduce the twisted version of the vertex operators. The twisted vertex operators will be used to construct twisted modules over lattice vertex algebras in the next section.

1.5.4.1 Twisted Currents of Higher Rank

Let $\mathfrak{h}$ be an n-dimensional vector space regarded as an abelian Lie algebra and $(\;|\;)$ a nondegenerate symmetric invariant bilinear form on $\mathfrak{h}$. Recall the Heisenberg algebra $\mathbf{U}(\hat{\mathfrak{h}}, 1)$ and related notations from Subsection 1.4.2.1.

Let us now consider the space

$$\hat{\mathfrak{h}}^{\mathrm{tw}} = \mathfrak{h} \otimes \mathbb{C}[t, t^{-1}]t^{1/2} \oplus \mathbb{C}K.$$

Denote $h \otimes t^s$ by h_s for $h \in \mathfrak{h}$ and $s \in \mathbb{Z} + 1/2$ and equip $\hat{\mathfrak{h}}^{\mathrm{tw}}$ with a structure of a Lie algebra by setting, for $g, h \in \mathfrak{h}$,

$$[g_r, h_s] = r(g\,|\,h)\delta_{r+s,0}K.$$

Define the *twisted Heisenberg algebra* associated with $\mathfrak{h}$ and $(\;|\;)$ by

$$\mathbf{U}(\hat{\mathfrak{h}}^{\mathrm{tw}}, 1) = \mathbf{U}(\hat{\mathfrak{h}}^{\mathrm{tw}})/(K - 1).$$

Define the spaces such as $\hat{\mathfrak{h}}^{\mathrm{tw}}_{<0}$ and the twisted module space $\mathbf{F}^{\mathrm{tw}}$ as before. By PBW,

$$\mathbf{F}^{\mathrm{tw}} = \mathbf{S}(\hat{\mathfrak{h}}^{\mathrm{tw}}_{<0}) \otimes \mathbb{C}\mathbf{v}^{\mathrm{tw}}$$

as vector spaces.

For $h \in \mathfrak{h}$ and $s \in \mathbb{Z} + 1/2$, denote the image of h_s in $\mathbf{U}(\hat{\mathfrak{h}}^{\mathrm{tw}}, 1)$ by the same symbol, and consider the *twisted current*

$$h^{\mathrm{tw}}(z) = \sum_s h_s z^{-s-1},$$

where the sum is over $s \in \mathbb{Z} + 1/2$. Then the twisted currents are locally truncated shifted series on the twisted Fock module, they are locally commutative, and their modified OPE is given by

$$g^{\mathrm{tw}}(z)_{(n)}h^{\mathrm{tw}}(z) = \begin{cases} 0 & (n \geq 2), \\ (g\,|\,h) & (n = 1), \\ 0 & (n = 0), \end{cases}$$

which is again of the same form as the untwisted case.

1.5.4.2 Twisted Vertex Operators

Regard $\lambda \in \mathfrak{h}^*$ as an element of $\mathfrak{h}$ via the identification $\mathfrak{h}^* \simeq \mathfrak{h}$ induced by $(\;|\;)$ and consider the actions of λ_s with $s \in \mathbb{Z} + 1/2$ on $\mathbf{F}^{\mathrm{tw}}$. The following expression is called the *twisted vertex operator*:

$$U_\lambda^{\mathrm{tw}}(z) = \exp\left(-\sum_{s<0} \lambda_s \frac{z^{-s}}{s}\right)\exp\left(-\sum_{s>0} \lambda_s \frac{z^{-s}}{s}\right).$$

Here s runs over the elements of $\mathbb{N} + 1/2$ and $-\mathbb{N} - 1/2$, respectively, and

$$U_\lambda^{\mathrm{tw}}(z) \in \mathrm{Hom}(\mathbf{F}^{\mathrm{tw}}, \mathbf{F}^{\mathrm{tw}}((z^{1/2}))).$$

Note that

$$U_\lambda^{\mathrm{tw}}(z) + U_{-\lambda}^{\mathrm{tw}}(z) \in \mathrm{Hom}(\mathbf{F}^{\mathrm{tw}}, \mathbf{F}^{\mathrm{tw}}((z))),$$
$$U_\lambda^{\mathrm{tw}}(z) - U_{-\lambda}^{\mathrm{tw}}(z) \in \mathrm{Hom}(\mathbf{F}^{\mathrm{tw}}, \mathbf{F}^{\mathrm{tw}}((z))z^{1/2}).$$

We will later modify $U_\lambda^{\mathrm{tw}}(z)$ in a suitable way and call the resulting expression by the same term.

For $h \in \mathfrak{h}$ and $\lambda \in \mathfrak{h}^*$, we have

$$\left[h^{\mathrm{tw}}(y), U_\lambda^{\mathrm{tw}}(z) \right] = y^{-1/2} z^{1/2} \lambda(h) \delta(y, z) U_\lambda^{\mathrm{tw}}(z).$$

In particular, the twisted currents are locally commutative with twisted vertex operators, and the modified OPE reads

$$h^{\mathrm{tw}}(y) U_\lambda^{\mathrm{tw}}(z) \simeq U_\lambda^{\mathrm{tw}}(z) h^{\mathrm{tw}}(y) \approx \frac{\lambda(h)}{y - z} U_\lambda^{\mathrm{tw}}(z).$$

Hence the residue products for nonnegative n become

$$h^{\mathrm{tw}}(z)_{(n)} U_\lambda^{\mathrm{tw}}(z) = \begin{cases} 0 & (n \geq 1), \\ \lambda(h) U_\lambda^{\mathrm{tw}}(z) & (n = 0), \end{cases}$$

which is of the same form as the untwisted case.

1.5.4.3 Commutation Relations

For $\lambda, \mu \in \mathfrak{h}^*$, consider the composite $U_\lambda^{\mathrm{tw}}(y) U_\mu^{\mathrm{tw}}(z)$ of twisted vertex operators,

$$\exp \left(\sum_{r<0} \lambda_r \frac{y^{-r}}{-r} \right) \underbrace{\exp \left(\sum_{r>0} \lambda_r \frac{y^{-r}}{-r} \right) \exp \left(\sum_{s<0} \mu_s \frac{z^{-s}}{-s} \right)} \exp \left(\sum_{s>0} \mu_s \frac{z^{-s}}{-s} \right),$$

and denote the expression obtained by switching the underlined factors by $U_{\lambda,\mu}^{\mathrm{tw}}(y, z)$, which is written as

$$U_{\lambda,\mu}^{\mathrm{tw}}(y, z) = \exp \left(\sum_{r<0} \lambda_r \frac{y^{-r}}{-r} + \sum_{s<0} \mu_s \frac{z^{-s}}{-s} \right) \exp \left(\sum_{r>0} \lambda_r \frac{y^{-r}}{-r} + \sum_{s>0} \mu_s \frac{z^{-s}}{-s} \right).$$

By the commutation relations of twisted Heisenberg algebra,

$$
\left[\sum_{r>0}\lambda_r\frac{y^{-r}}{-r}, \sum_{s<0}\mu_s\frac{z^{-s}}{-s}\right] = \sum_{r>0}\sum_{s<0}[\lambda_r,\mu_s]\frac{y^{-r}}{-r}\frac{z^{-s}}{-s}
$$

$$
= \sum_{r>0}\sum_{s<0}r(\lambda|\mu)\delta_{r+s,0}\frac{y^{-r}}{-r}\frac{z^{-s}}{-s}
$$

$$
= -(\lambda|\mu)\sum_{r>0}\frac{y^{-r}z^r}{r}.
$$

We have

$$
-\sum_{r>0}\frac{y^{-r}z^r}{r} = \log\left(\frac{y^{1/2}-z^{1/2}}{y^{1/2}+z^{1/2}}\right)\Bigg|_{|y|>|z|},
$$

where the region $|y| > |z|$ indicates that the expression is to be expanded in $|y^{1/2}| > |z^{1/2}|$. Therefore, we have the following equalities, with the latter obtained by switching the roles of $U_\lambda^{\mathrm{tw}}(y)$ and $U_\mu^{\mathrm{tw}}(z)$:

$$
\begin{aligned}
U_\lambda^{\mathrm{tw}}(y)U_\mu^{\mathrm{tw}}(z) &= \left(\frac{y^{1/2}-z^{1/2}}{y^{1/2}+z^{1/2}}\right)^{(\lambda|\mu)}\Bigg|_{|y|>|z|} U_{\lambda,\mu}^{\mathrm{tw}}(y,z),\\
U_\mu^{\mathrm{tw}}(z)U_\lambda^{\mathrm{tw}}(y) &= \left(\frac{z^{1/2}-y^{1/2}}{z^{1/2}+y^{1/2}}\right)^{(\lambda|\mu)}\Bigg|_{|z|>|y|} U_{\lambda,\mu}^{\mathrm{tw}}(y,z).
\end{aligned}
\tag{5.7}
$$

Therefore, it follows that, for sufficiently large N,

$$
(y-z)^N U_\lambda^{\mathrm{tw}}(y)U_\mu^{\mathrm{tw}}(z) = (-1)^{(\lambda|\mu)}(y-z)^N U_\mu^{\mathrm{tw}}(z)U_\lambda^{\mathrm{tw}}(y),
$$

since the numerators of the right-hand sides of

$$
\frac{y^{1/2}-z^{1/2}}{y^{1/2}+z^{1/2}} = \frac{\left(y^{1/2}-z^{1/2}\right)^2}{y-z}, \quad \frac{y^{1/2}+z^{1/2}}{y^{1/2}-z^{1/2}} = \frac{\left(y^{1/2}+z^{1/2}\right)^2}{y-z}
$$

are symmetric polynomials in $y^{1/2}$ and $z^{1/2}$.

1.5.4.4 Correction of Operators

Consider the following expansion:

$$
\left(y^{1/2}+z^{1/2}\right)^{-2(\lambda|\mu)} \approx 2^{-2(\lambda|\mu)}z^{-2(\lambda|\mu)/2} + \cdots.
$$

Then, by the first equality in (5.7),

$$
U_\lambda^{\mathrm{tw}}(y)U_\mu^{\mathrm{tw}}(z) \approx \underline{2^{-2(\lambda|\mu)}z^{-(\lambda|\mu)}}(y-z)^{(\lambda|\mu)}\big|_{|y|>|z|}\left(U_{\lambda+\mu}^{\mathrm{tw}}(z)+\cdots\right).
$$

To remove the unpleasant factors, define, for $\lambda \in \mathfrak{h}^*$:

$$
\tilde{U}_\lambda^{\mathrm{tw}}(z) = 2^{-(\lambda|\lambda)}z^{-(\lambda|\lambda)/2}U_\lambda^{\mathrm{tw}}(z).
\tag{5.8}
$$

Then we have

$$
\tilde{U}_\lambda^{\mathrm{tw}}(y)\tilde{U}_\mu^{\mathrm{tw}}(z) \approx (y-z)^{(\lambda|\mu)}\big|_{|y|>|z|}\left(\tilde{U}_{\lambda+\mu}^{\mathrm{tw}}(z)+\cdots\right).
$$

Note that the equalities in (5.7) become

$$\tilde{U}_\lambda^{\text{tw}}(y)\tilde{U}_\mu^{\text{tw}}(z) = \left.\left(\frac{y^{1/2} - z^{1/2}}{y^{1/2} + z^{1/2}}\right)^{(\lambda|\mu)}\right|_{|y|>|z|} \tilde{U}_{\lambda,\mu}^{\text{tw}}(y,z),$$

$$\tilde{U}_\mu^{\text{tw}}(z)\tilde{U}_\lambda^{\text{tw}}(y) = \left.\left(\frac{z^{1/2} - y^{1/2}}{z^{1/2} + y^{1/2}}\right)^{(\lambda|\mu)}\right|_{|z|>|y|} \tilde{U}_{\lambda,\mu}^{\text{tw}}(y,z),$$

where, for $\lambda, \mu \in \mathfrak{h}^*$,

$$\tilde{U}_{\lambda,\mu}^{\text{tw}}(y,z) = 2^{-(\lambda|\lambda)-(\mu|\mu)} y^{-(\lambda|\lambda)/2} z^{-(\mu|\mu)/2} U_{\lambda,\mu}^{\text{tw}}(y,z).$$

Therefore, by the same argument as in the preceding subsection, we have, for sufficiently large N,

$$(y-z)^N \tilde{U}_\lambda^{\text{tw}}(y)\tilde{U}_\mu^{\text{tw}}(z) = (-1)^{(\lambda|\mu)}(y-z)^N \tilde{U}_\mu^{\text{tw}}(z)\tilde{U}_\lambda^{\text{tw}}(y).$$

In particular, if $(\lambda|\mu)$ is even, then the twisted vertex operators $\tilde{U}_\lambda^{\text{tw}}(z)$ and $\tilde{U}_\mu^{\text{tw}}(z)$ are locally commutative.

1.5.5 Twisted Modules for Rank One Even Lattices

In this section, we will describe θ-twisted modules over the lattice vertex algebra $\mathbf{V}_L$ with respect to a lift θ of the (-1)-involution induced from the (-1)-involution of the lattice:

$$\theta : L \longrightarrow L, \quad \lambda \mapsto -\lambda.$$

We will first describe them for the rank 1 cases in this section, and then proceed to higher-rank cases in the next section.

1.5.5.1 Shifted $\mathbf{V}_L$-Module

Let L be an even lattice of rank one, set $\mathfrak{h} = L \otimes_{\mathbb{Z}} \mathbb{C}$, and extend the bilinear form on L to $\mathfrak{h}$. Recall the lattice vertex algebra:

$$\mathbf{V}_L = \bigoplus_{\lambda \in L} \mathbf{F}_\lambda$$

Since L is of rank 1, we have $(\lambda|\mu) \in 2\mathbb{Z}$ for all $\lambda, \mu \in L$.

Consider the twisted Fock module $\mathbf{F}^{\text{tw}}$ for the twisted Heisenberg algebra and the twisted currents $h^{\text{tw}}(z)$ for $h \in \mathfrak{h}$ acting on it. Let $V_\lambda^{\text{tw}}(z)$ denote the twisted vertex operator $\tilde{U}_\lambda^{\text{tw}}(z)$ with correction factors given by (5.8), and simply call it the twisted vertex operator from now on:

$$V_\lambda^{\text{tw}}(z) = \tilde{U}_\lambda^{\text{tw}}(z).$$

Then the twisted currents and the twisted vertex operators become locally commutative, and they generate a vertex algebra by the residue products, which we denote by $\mathcal{V}_L^{\text{tw}}$:

$$\mathcal{V}_L^{\mathrm{tw}} = \langle h^{\mathrm{tw}}(z), V_\lambda^{\mathrm{tw}}(z) \mid h \in \mathfrak{h}, \lambda \in L \rangle_{\mathrm{RP}}.$$

Moreover, for all $n \in \mathbb{Z}$,

$$V_\lambda^{\mathrm{tw}}(z)_{(n)} V_\mu^{\mathrm{tw}}(z) \in \mathcal{F}_\lambda^{\mathrm{tw}}(\lambda + \mu).$$

Therefore, we have

$$\mathcal{V}_L^{\mathrm{tw}} = \bigoplus_{\lambda \in L} \mathcal{F}_\lambda^{\mathrm{tw}},$$

where $\mathcal{F}_\lambda^{\mathrm{tw}}$ is the $\mathbf{F}_0$-submodule generated by $V_\lambda^{\mathrm{tw}}(z)$, which is isomorphic to $\mathbf{F}_\lambda$ as an $\mathbf{F}_0$-module.

It is therefore likely that there exists a homomorphism of vertex algebras satisfying

$$\rho^{\mathrm{tw}} \colon \mathbf{V}_L \longrightarrow \mathcal{V}_L^{\mathrm{tw}}, \quad \mathbf{v}_\lambda \mapsto V_\lambda^{\mathrm{tw}}(z).$$

This is indeed the case, and the map ρ^{tw} is actually an isomorphism of vertex algebras, since properties enough to characterize the lattice vertex algebra have already been verified. We thus have a shifted representation

$$\rho^{\mathrm{tw}} \colon \mathbf{V}_L \longrightarrow \mathrm{Hom}(\mathbf{F}^{\mathrm{tw}}, \mathbf{F}^{\mathrm{tw}}((z^{1/2})))$$

of the lattice vertex algebra $\mathbf{V}_L$ on the twisted Fock module $\mathbf{F}^{\mathrm{tw}}$.

1.5.5.2 Twisted $\mathbf{V}_L$-Module

The shifted $\mathbf{V}_L$-module $\mathbf{F}^{\mathrm{tw}}$ constructed in the preceding subsection can be given a structure of a twisted module with respect to a lift θ of the (-1)-involution of the lattice L.

Let θ denote the involution of $\mathfrak{h}$ induced from the (-1)-involution of L; that is,

$$\theta \colon \mathfrak{h} \longrightarrow \mathfrak{h}, \quad h \mapsto -h.$$

Then θ induces an involution of the Heisenberg algebra $\mathbf{U}(\hat{\mathfrak{h}}, 1)$, hence of $\mathbf{S}(\hat{\mathfrak{h}}_{<0})$, and of the lattice vertex algebra as

$$\theta \colon \mathbf{V}_L \longrightarrow \mathbf{V}_L, \quad P\mathbf{v}_\lambda \mapsto (\theta P)(\hat{\theta}\mathbf{v}_\lambda),$$

where $P \in \mathbf{S}(\hat{\mathfrak{h}}_{<0})$ and $\hat{\theta}\mathbf{v}_\lambda = \mathbf{v}_{-\lambda}$. The eigenspace decomposition with respect to the action of θ looks thus:

$$\mathbf{V}_L = \mathbf{V}_L^+ \oplus \mathbf{V}_L^-, \tag{5.9}$$

where $\mathbf{V}_L^\pm = \{ a \in \mathbf{V}_L \mid \theta a = \pm a \}$ are the eigenspaces.

Recall the vertex algebra $\mathcal{V}_L^{tw}$ of shifted series on the twisted Fock module $\mathbf{F}^{tw}$ generated by the twisted currents and the twisted vertex operators. Then it decomposes as

$$\mathcal{V}_L^{tw} = \mathcal{V}_L^{tw,+} \oplus \mathcal{V}_L^{tw,-},$$

where

$$\mathcal{V}_L^{tw,+} = \mathcal{V}_L^{tw} \cap \operatorname{Hom}(\mathbf{F}^{tw}, \mathbf{F}^{tw}((z))),$$
$$\mathcal{V}_L^{tw,-} = \mathcal{V}_L^{tw} \cap \operatorname{Hom}(\mathbf{F}^{tw}, \mathbf{F}^{tw}((z))z^{1/2}).$$

By the very definition of the twisted vertex operators, we have

$$V_\lambda^{tw}(z) + V_{-\lambda}^{tw}(z) \in \mathcal{V}_L^{tw,+} \quad \text{and} \quad V_\lambda^{tw}(z) - V_{-\lambda}^{tw}(z) \in \mathcal{V}_L^{tw,-}.$$

Therefore, it turns out that the shifted representation $\mathbf{F}^{tw}$ is actually a θ-twisted V_L-module by the decomposition (5.9).

Proposition 5.4 *Let L be an even lattice of rank 1. Then the twisted Fock module $\mathbf{F}^{tw}$ carries a unique structure of a θ-twisted V_L-module such that*

$$Y_{\mathbf{F}^{tw}}(h_{-1}\mathbf{1}, z) = h^{tw}(z) \quad and \quad Y_{\mathbf{F}^{tw}}(\mathbf{v}_\lambda, z) = V_\lambda^{tw}(z)$$

for $h \in \mathfrak{h}$ and $\lambda \in L$, respectively.

Note 5.5. 1. There is another θ-twisted V_L-module structure on the same space $\mathbf{F}^{tw}$ given by $V_\beta^{tw}(z) = -\tilde{U}_\beta^{tw}(z)$ for a generator β of L. In other words, for $\lambda = n\beta$,

$$V_{n\beta}^{tw}(z) = (-1)^n \tilde{U}_{n\beta}^{tw}(z).$$

The two θ-twisted modules correspond to the two 1-dimensional representations of L as a group, the trivial representation and the sign representation.
2. The θ-twisted V_L-modules for a rank one even lattice L are actually classified by representations of $L/2L \simeq \mathbb{Z}/2\mathbb{Z}$. See Subsection 1.5.6.3 for more details.

1.5.6 Twisted Modules for General Even Lattices

Let us now describe θ-twisted modules over lattice vertex algebras associated with general even lattices, where the cocycle factor has to be taken into account. We will first describe a lift θ of the (-1)-involution of the lattice L in terms of the central extension $\hat{L}$ of the lattice by $\{\pm 1\}$, and then classify $\hat{L}$-modules $\mathbf{T}$ such that the tensor products $\mathbf{F}^{tw} \otimes \mathbf{T}$ carry structures of θ-twisted modules.

1.5.6.1 Relation to Central Extensions

Let L be an even lattice, choose a cocycle factor $\varepsilon\colon L \times L \twoheadrightarrow \{\pm 1\}$, and consider the corresponding central extension as in Section 1.4.5:

$$0 \longrightarrow \{\pm 1\} \longrightarrow \hat{L} \overset{\pi}{\longrightarrow} L \longrightarrow 0.$$

Let $\lambda \mapsto e_\lambda$ be a set-theoretical section of π such that $e_\lambda e_\mu = \varepsilon(\lambda, \mu)e_{\lambda+\mu}$. Recall that the lattice vertex algebra can be written as

$$\mathbf{V}_L = \mathbf{S}(\hat{\mathfrak{h}}_{<0}) \otimes \mathbb{C}[L],$$

on which the generating series $Y(\mathbf{v}_\lambda, z)$ given by the series $V_{\lambda,\varepsilon}(z)$ defined as in (4.4) factors as

$$Y(\mathbf{v}_\lambda, z) = U_\lambda(z) \otimes e_\lambda z^{\lambda_0},$$

where $U_\lambda(z)$ is the part acting on $\mathbf{S}(\hat{\mathfrak{h}}_{<0})$ and e_λ sends e^μ to $\varepsilon(\lambda, \mu)e^{\lambda+\mu}$ so that $\mathbb{C}[L]$ becomes an $\hat{L}$-module on which the image of -1 in $\hat{L}$ acts by -1. We are to construct a shifted $\mathbf{V}_L$-module on the tensor product of the form

$$\mathbf{M} = \mathbf{S}(\hat{\mathfrak{h}}^{\mathrm{tw}}_{<0}) \otimes \mathbf{T}$$

so that the generating series for $\mathbf{v}_\lambda$ acts in the form

$$Y_{\mathbf{M}}(\mathbf{v}_\lambda, z) = \tilde{U}^{\mathrm{tw}}_\lambda(z) \otimes e^{\mathrm{tw}}_\lambda,$$

where $\tilde{U}^{\mathrm{tw}}_\lambda(z)$ refers to the series defined by (5.8) acting on $\mathbf{S}(\hat{\mathfrak{h}}^{\mathrm{tw}}_{<0})$ and e^{tw}_λ is an operator acting on $\mathbf{T}$.

Indeed, if $\mathbf{T}$ is an $\hat{L}$-module on which e^{tw}_λ operates by the action of e_λ on $\mathbf{T}$ in such a way that the image of -1 in $\hat{L}$ acts by -1, then

$$e^{\mathrm{tw}}_\lambda e^{\mathrm{tw}}_\mu = (-1)^{(\lambda\,|\,\mu)} e^{\mathrm{tw}}_\mu e^{\mathrm{tw}}_\lambda.$$

Hence the series $Y_{\mathbf{M}}(\mathbf{v}_\lambda, z)$ become locally commutative with each other, thus giving rise to a structure of a shifted $\mathbf{V}_L$-module on the space $\mathbf{M} = \mathbf{S}(\hat{\mathfrak{h}}^{\mathrm{tw}}_{<0}) \otimes \mathbf{T}$ by the same argument as in the rank one case.

1.5.6.2 Lifts of (-1)-Involutions

To generalize the construction of θ-*twisted* modules for rank one lattices to the higher rank cases, let us lift the (-1)-involution of the lattice L to an automorphism of the lattice vertex algebra $\mathbf{V}_L$.

To this end, first choose a lift $\hat{\theta}$ of $\theta = -1$ on L to an automorphism of $\hat{L}$.

$$
\begin{array}{ccccccccc}
0 & \longrightarrow & \{\pm 1\} & \longrightarrow & \hat{L} & \longrightarrow & L & \longrightarrow & 0 \\
  &   & \big\| &   & \big\downarrow{\hat{\theta}} & & \big\downarrow{\theta} & & \\
0 & \longrightarrow & \{\pm 1\} & \longrightarrow & \hat{L} & \longrightarrow & L & \longrightarrow & 0.
\end{array}
$$

Note that, by definition,

$$\hat\theta e_\lambda \in \{\pm e_{-\lambda}\} = \{\pm(e_\lambda)^{-1}\}.$$

Such an automorphism $\hat\theta$ automatically becomes an involution. There is actually a canonical choice defined by $\hat\theta e_\lambda = (-1)^{(\lambda|\lambda)/2}(e_\lambda)^{-1}$ for each $\lambda \in L$.

Now define the action of θ on $\mathbf{V}_L$ by

$$P\mathbf{v}_\lambda \mapsto (\theta P)(\theta \mathbf{v}_\lambda),$$

where $\theta \mathbf{v}_\lambda$ is given by $1 \otimes \hat\theta e_\lambda$ under the identification of $\mathbf{v}_\lambda$ with $1 \otimes e_\lambda$. Then the lattice vertex algebra decomposes as

$$\mathbf{V}_L = \mathbf{V}_L^+ \oplus \mathbf{V}_L^-, \quad \text{where} \quad \mathbf{V}_L^\pm = \{a \in \mathbf{V}_L \mid \theta a = \pm a\}$$

as in the rank one case. The isomorphism class of the vertex algebra $\mathbb{V}_L^+$ does not depend on the choice of the lift $\hat\theta$ (cf. [43]).

1.5.6.3 Construction of Twisted $\mathbf{V}_L$-Modules

In order for the shifted representation $\mathbf{M} = S(\hat{\mathfrak{h}}_{<0}^{\mathrm{tw}}) \otimes \mathbf{T}$ to become a θ-twisted $\mathbf{V}_L$-module, the generating series $Y_{\mathbf{M}}(\mathbf{v}_\lambda, z) = \tilde{U}_\lambda^{\mathrm{tw}}(z) \otimes e_\lambda^{\mathrm{tw}}$ must satisfy

$$Y_{\mathbf{M}}(\mathbf{v}_\lambda, z) + Y_{\mathbf{M}}(\theta\mathbf{v}_\lambda, z) \in \mathrm{Hom}(\mathbf{M}, \mathbf{M}((z))),$$
$$Y_{\mathbf{M}}(\mathbf{v}_\lambda, z) - Y_{\mathbf{M}}(\theta\mathbf{v}_\lambda, z) \in \mathrm{Hom}(\mathbf{M}, \mathbf{M}((z)))z^{1/2}.$$

Since $\tilde{U}_\lambda^{\mathrm{tw}}(z)$ already satisfy this property, it remains to impose the following condition on the $\hat{L}$-module $\mathbf{T}$: for all $\lambda \in L$,

$$(\hat\theta e_\lambda)\big|_{\mathbf{T}} = e_\lambda\big|_{\mathbf{T}}.$$

To describe it, consider the following set:

$$K = \{(\theta g)g^{-1} \mid g \in \hat{L}\} \subset \hat{L}.$$

Then the condition holds if and only if K acts on $\mathbf{T}$ by 1.

It is readily checked that K is a central subgroup of $\hat{L}$, which fits in the following commutative diagram of exact sequences:

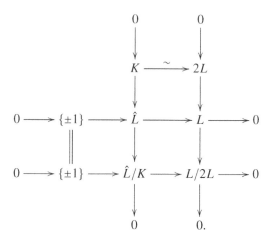

Therefore, the space $S(\hat{\mathfrak{h}}^{\mathrm{tw}}_{<0}) \otimes \mathbf{T}$ carries a structure of a θ-twisted $\mathbf{V}_L$-module if and only if $\mathbf{T}$ is an $\hat{L}/K$-module on which $(-1)K$ acts by -1.

To classify such $\hat{L}/K$-modules, let R be the radical of the commutator map modulo 2:

$$R = \{\lambda \in L \,|\, (\lambda|L) \subset 2\mathbb{Z}\}.$$

Then the inverse image of R in $\hat{L}$ agrees with the center of $\hat{L}$, and the number of central characters of $\hat{L}/K$ of which $(-1)K$ acts by -1 agrees with $|R/2L|$. For each such character χ, we have a unique $\hat{L}$-module $\mathbf{T}_\chi$ satisfying the desired properties, whose dimension is given by

$$\dim_{\mathbb{C}} \mathbf{T}_\chi = |L/R|^{1/2}.$$

See [1] for details.

Let us finally mention the case when L is the Leech lattice Λ, the unique even unimodular positive-definite lattice of rank 24 without roots. Being unimodular implies that the commutator map induces a nondegenerate bilinear form on $\Lambda/2\Lambda$, hence the radical R agrees with 2Λ. Therefore, there exists a unique $\hat{\Lambda}$-module $\mathbf{T}$ of which $(-1)K$ acts by -1, and its dimension is given by

$$\dim_{\mathbb{C}} \mathbf{T} = |\Lambda/2\Lambda|^{1/2} = \sqrt{2^{24}} = 2^{12}.$$

The central extension $\hat{\Lambda}/K$ is actually the extraspecial 2-group 2^{1+24}_+ for the canonical choice of the lift θ.

Bibliographic Notes

Main references for Section 1.5 are Frenkel et al. [1], Dong [41], and Li [71]. For descriptions of twisted modules in models in physics, consult Di Francesco et al. [13] (cf. Ginsparg [19]).

Our treatments by introducing the concepts of shifted series and shifted representation are more or less straightforward generalizations of those described in [71] along the line of [7]. See Roitman [89], Frenkel and Szczensny [56], and Doyon et al. [51] for related works.

Properties of twisted vertex operators are described in detail in [1] by explicit calculations. The idea of using characterization of lattice vertex algebras is stated in [89]. For earlier works on twisted vertex operators, see Lepowsky and Wilson [69], Frenkel, Lepowsky, and Meurman [58], and Lepowsky [68].

For twisted module for general finite-order automorphisms of lattice vertex algebras, see Dong and Lepowsky [45] and Bakalov and Kac [30].

Often useful in constructing twisted modules is an operator $\Delta(z)$ introduced by H. S. Li [71], called the *Delta operator*, which transforms a twisted module to another.

Under certain circumstances, existence of some twisted modules is guaranteed by general theory of modular invariance by Dong, Li, and Mason [46].

1.6 Vertex Operator Algebras

Recall that standard examples of vertex algebras, such as the lattice vertex algebra associated with a positive-definite even lattice, often admit an action of the Virasoro algebra that is *internal* in the sense that it is given by the actions of an element of the vertex algebra. Such a vector is called a *conformal vector* if the Virasoro action includes the translation operator. The presence of a conformal vector leads to additional features such as gradings and various transformation formulas.

A vertex algebra may have many conformal vectors, and it is natural to specify one and to impose appropriate conditions on it. The resulting concept is that of a *vertex operator algebra* (VOA). Thus a VOA is a pair $(\mathbf{V}, \omega)$ of a vertex algebra $\mathbf{V}$ and a conformal vector ω satisfying a number of conditions.

Section 1.6 is a brief introduction to theory of vertex operator algebras. We will start by describing consequences of the presence of a conformal vector and then give terminologies specific to VOAs. As a VOA is graded, we may talk about the category of $\mathbb{N}$-graded modules for which the simple objects are controlled by the top subspaces by means of *Zhu's algebra*, an associative algebra universally constructed by the VOA structure alone without addressing modules. Two more important topics, fusion rules among modules and modular invariance of conformal characters, are also included.

We will work over the field $\mathbb{C}$ of complex numbers.

1.6.1 Conformal Vectors

In this section, we will give the precise definition of conformal vectors of a vertex algebra and describe various consequences of the presence of a conformal vector. We will show that a grading is given to a vertex algebra by L_0 action for the choice of the conformal vector and that various transformation formulas hold by a part of the Virasoro action.

1.6.1.1 Virasoro Vectors in Vertex Algebras

Let $\mathbf{V}$ be a vertex algebra. Recall that an element $e \in \mathbf{V}$ is called a *Virasoro vector* if there exists a scalar c_e, called the *central charge*, such that

$$e_{(n)}e = \begin{cases} 0 & (n \geq 4), \\ (c_e/2)\mathbf{1} & (n = 3), \\ 2e & (n = 1), \end{cases}$$

from which the properties $e_{(0)}e = Te$ and $e_{(2)}e = 0$ follow, where $T = T^{(1)}$ is the translation operator.

Set $L_n^e = e_{(n+1)}$ so that

$$Y(e, z) = \sum_n L_n^e z^{-n-2}.$$

Then the condition above is equivalent to the OPE

$$Y(e, y)Y(e, z) \simeq Y(e, z)Y(e, y) \sim \frac{\partial Y(e, z)}{y - z} + \frac{2Y(e, z)}{(y - z)^2} + \frac{c_e/2}{(y - z)^4},$$

or to the Virasoro commutation relation

$$[L_m^e, L_n^e] = (m - n)L_{m+n}^e + \delta_{m+n,0}\frac{m^3 - m}{12}c_e.$$

Here we have identified a scalar with multiplication by it.

1.6.1.2 Conformal Vectors in Vertex Algebras

Let $\mathbf{V}$ be a vertex algebra and recall that the translation operator $T = T^{(1)}$ is given by

$$T: \mathbf{V} \longrightarrow \mathbf{V}, \quad a \mapsto Ta = a_{(-2)}\mathbf{1}.$$

Let ω be a Virasoro vector of **V**. Denote the actions L_n^ω simply by L_n and the central charge c_ω by c:

$$[L_m, L_n] = (m-n)L_{m+n} + \delta_{m+n,0}\frac{m^3-m}{12}c.$$

Such a vector ω is called a *conformal vector* of **V** if, for all $a \in$ **V**,

$$\omega_{(0)}a = a_{(-2)}\mathbf{1},$$

that is, L_{-1} agrees with the translation operator T as operators acting on **V**:

$$L_{-1} = T : \mathbf{V} \longrightarrow \mathbf{V}.$$

Then, for a conformal vector ω, the translation property (VT) with $k = 1$ gives

$$(L_{-1}a)_{(n)}b = -na_{(n-1)}b. \tag{6.1}$$

Namely:

$$Y(L_{-1}a, z) = \partial_z Y(a, z). \tag{6.2}$$

The standard Virasoro vectors for the Heisenberg vertex algebras, the lattice vertex algebras, the affine vertex algebras, and the Virasoro vertex algebras are conformal vectors of the vertex algebras under consideration.

Note 6.1. 1. A conformal vector in our sense is often called a *Virasoro element* in the literatures (cf. [1], [32]), and a Virasoro vector in our sense is sometimes called a conformal vector (cf. [83]). 2. The Virasoro vector obtained by Sugawara construction is a conformal vector of the affine vertex algebra as long as it makes sense. 3. Under the presence of a conformal vector ω, left ideals of the vertex algebra become two-sided ideals by skew-symmetry; for $T = \omega_{(0)}$ is one of the left actions.

1.6.1.3 Grading by Conformal Weights

Let ω be a conformal vector of a vertex algebra **V** and consider the action of L_0 on **V**. Then, for $a, b \in$ **V**,

$$L_0(a_{(n)}b) = (L_0a)_n b + a_{(n)}(L_0b) - (n+1)a_n b \tag{6.3}$$

by the Borcherds identity (V1) with $(p,q,r) = (1,n,0)$ and (6.1).

Now assume that the operator L_0 is semisimple on **V** with integral eigenvalues and consider the eigenspace decomposition:

$$\mathbf{V} = \bigoplus_d \mathbf{V}_d, \quad \text{where } \mathbf{V}_d = \text{Ker}(L_0 - d)|_{\mathbf{V}}. \tag{6.4}$$

The eigenvalue of a homogeneous element a is called the *conformal weight* of a with respect to ω, which we denote by $\Delta(a)$:

$$a \in \mathbf{V}_d \iff \Delta(a) = d.$$

For example, since $L_0\mathbf{1} = \omega_{(1)}\mathbf{1} = 0$ and $L_0\omega = \omega_{(1)}\omega = 2\omega$,

$$\Delta(\mathbf{1}) = 0 \text{ and } \Delta(\omega) = 2.$$

The relation (6.3) implies

$$\mathbf{V}_{d\,(n)}\mathbf{V}_e \subset \mathbf{V}_{d+e-n-1}. \tag{6.5}$$

Namely:

$$\Delta(a_{(n)}b) = \Delta(a) + \Delta(b) - n - 1.$$

Note 6.2. A direct sum decomposition of $\mathbf{V}$ satisfying the relation (6.5) in general is called a *grading* of the vertex algebra $\mathbf{V}$.

1.6.1.4 Projective Linear Transformations

Let ω be a conformal vector and consider the actions L_m with $m = -1, 0, 1$. Then, as a part of the Virasoro commutation relations, we have

$$[L_0, L_{-1}] = L_{-1}, \ [L_0, L_1] = -L_1, \ [L_1, L_{-1}] = 2L_0.$$

Thus the actions L_{-1}, L_0, L_1 give rise to a representation of $\mathfrak{sl}_2$ on $\mathbf{V}$ by identifying $L_{-1} = E$, $L_0 = (1/2)H$, and $L_1 = -F$, for which

$$\exp xE = \begin{bmatrix} 1 & x \\ 0 & 1 \end{bmatrix}, \ \exp xH = \begin{bmatrix} e^x & 0 \\ 0 & e^{-x} \end{bmatrix}, \ \exp xF = \begin{bmatrix} 1 & 0 \\ x & 1 \end{bmatrix}.$$

The corresponding projective linear transformations are given by the substitutions

$$\exp xL_{-1} : z \mapsto z + x, \ \exp xL_0 : z \mapsto e^x z, \ \exp xL_1 : z \mapsto \frac{z}{1 - xz}.$$

Moreover, the operators L_{-1}, L_0, L_1 generate formal actions on the space of generating series $Y(a, z)$ with $a \in \mathbf{V}$ corresponding to projective linear transformations of the variable z. Indeed, by the commutator formula (VC),

$$[L_m, a_{(n)}] = \sum_{i=0}^{\infty} \binom{m+1}{i} (L_{i-1}a)_{(m+n-i)}.$$

In particular, for $m = -1, 0, 1$, we have, by the translation property (6.2),

$$[L_{-1}, Y(a, z)] = \partial_z Y(a, z),$$
$$[L_0, Y(a, z)] = z\partial_z Y(a, z) + Y(L_0 a, z),$$
$$[L_1, Y(a, z)] = z^2 \partial_z Y(a, z) + 2z Y(L_0 a, z) + Y(L_1 a, z).$$

Then, by exponentiation,

$$
\begin{aligned}
e^{xL_{-1}} Y(a,z) e^{-xL_{-1}} &= Y(a, z+x), \\
x^{L_0} Y(a,z) x^{-L_0} &= Y(x^{L_0} a, xz), \\
e^{xL_1} Y(a,z) e^{-xL_1} &= Y\!\left(e^{x(1-xz)L_1}(1-xz)^{-2L_0} a, \frac{z}{1-xz}\right).
\end{aligned}
\tag{6.6}
$$

The first one is the translation covariance.

1.6.1.5 Formal Change of Coordinates

The transformations corresponding to L_0 and L_1, which send z to xz and to $z/(1-xz)$, respectively, are thought of as coordinate transformations that fix the origin, thus forming a part of more general transformations of the affine line $\mathbb{C}$ generated by vector fields in $z\mathbb{C}[[z]](d/dz)$ under the isomorphism

$$
z\mathbb{C}[[z]]\frac{d}{dz} \longrightarrow \bigoplus_{n\geq 0}\mathbb{C}L_n, \quad z^{n+1}\frac{d}{dz} \mapsto -L_n
\tag{6.7}
$$

of Lie algebra.

Assume that $\mathbf{V}$ is graded by integral conformal weights by L_0 for a conformal vector, and that the actions of $L_1, L_2, \cdots$ are locally nilpotent on $\mathbf{V}$.

Let $\phi : t \mapsto \phi(t)$ be a formal change of variables such that

$$
\phi(t) = \phi_1 t + \phi_2 t^2 + \cdots \quad \text{with } \phi_1 \neq 0.
$$

Let $c_0, c_1, c_2, \cdots$ be a sequence of complex numbers such that

$$
\phi(t) = \left(\exp \sum_{j=1}^{\infty} c_j t^{j+1} \partial_t\right) c_0^{t\partial_t} t.
$$

Accordingly, define an operator $R(\phi)$ acting on the vertex algebra $\mathbf{V}$ by the Virasoro action via the embedding (6.7) as

$$
R(\phi) = \exp\left(-\sum_{j=1}^{\infty} c_j L_j\right) c_0^{-L_0}.
$$

Here, by definition, $c_0^{t\partial_t} t^n = c_0^n t^n$ and $c_0^{-L_0} a = c_0^{-\Delta(a)} a$ for a homogeneous $a \in \mathbf{V}$.

Consider the series $\phi_z(t)$ in t with coefficients in $z\mathbb{C}[[z]]$ defined by

$$
\phi_z(t) = \phi(z+t) - \phi(z).
$$

Then, for $a \in \mathbf{V}$ and ϕ as above,

$$
R(\phi)^{-1} Y(a,z) R(\phi) = Y(R(\phi_z)^{-1} a, \phi(z)).
\tag{6.8}
$$

This formula is called *Huang's formula*.

1.6.2 Vertex Operator Algebras and their Modules

This section is devoted to giving the precise definition of a vertex operator algebra (VOA) and the concepts of modules of various types for it. We will be selecting the types of modules to work with according to the purpose.

1.6.2.1 Vertex Operator Algebras

A *vertex operator algebra* (VOA) is a pair $(\mathbf{V}, \omega)$ of a vertex algebra $\mathbf{V}$ and a conformal vector ω of $\mathbf{V}$ satisfying the following conditions:

(1) The action of L_0 is semisimple on $\mathbf{V}$ with integral eigenvalues.
(2) The eigenvalues of L_0 are bounded below.
(3) The eigenspaces for L_0 are finite-dimensional.

The central charge of ω is called the *central charge* of the VOA $(\mathbf{V}, \omega)$.

As usual, we alternatively say that a VOA is a vertex algebra $\mathbf{V}$ *equipped with* a conformal vector ω satisfying (1)–(3), and often refer to $(\mathbf{V}, \omega)$ by $\mathbf{V}$ without mentioning ω explicitly.

By (1), the underlying vertex algebra $\mathbf{V}$ is graded by the conformal weights as in (6.4). By (3), the eigenspaces are finite-dimensional:

$$\dim \mathbf{V}_d < \infty \quad \text{for all } d \in \mathbb{Z}.$$

By (2), the grading is of the following shape:

$$\mathbf{V} = \underbrace{\mathbf{V}_{-d_0} \oplus \cdots \oplus \mathbf{V}_{-1}}_{\text{negative degrees}} \oplus \mathbf{V}_0 \oplus \mathbf{V}_1 \oplus \cdots.$$

In other words, the conformal weights are bounded below.

A VOA $(\mathbf{V}, \omega)$ is said to be *of CFT type* if the grading satisfies $\mathbf{V}_d = 0$ for all $d < 0$ and $\mathbf{V}_0 = \mathbb{C}\mathbf{1}$:

$$\mathbf{V} = \mathbb{C}\mathbf{1} \oplus \mathbf{V}_1 \oplus \mathbf{V}_2 \oplus \cdots.$$

Many typical examples of VOAs, such as Heisenberg VOAs, lattice VOAs associated with positive-definite even lattices, Virasoro VOAs, affine VOAs associated with finite-dimensional simple Lie algebras at general levels, etc., are of CFT type under the standard choices of the conformal vectors.

Note 6.3. A pair $(\mathbf{V}, \omega)$ of a vertex algebra $\mathbf{V}$ and a conformal vector $\omega \in \mathbf{V}$ is called a *conformal vertex algebra* if it satisfies (1).

1.6.2.2 Weak Modules and Ordinary Modules

A module over the underlying vertex algebra $\mathbf{V}$ of a VOA $(\mathbf{V}, \omega)$ is a called a *weak module* for the VOA.

For a weak module $\mathbf{M}$, consider the action of L_0 on $\mathbf{M}$. Then, we have a counterpart of the property (6.3) for modules: for $a \in \mathbf{V}$ and $v \in \mathbf{M}$,

$$L_0(a_n v) = a_n(L_0 v) + (L_0 a)_n v - (n+1)a_n v. \tag{6.9}$$

Assume that $\mathbf{M}$ has an eigenspace decomposition

$$\mathbf{M} = \bigoplus_{\lambda \in \mathbb{C}} \mathbf{M}_\lambda, \quad \text{where } \mathbf{M}_\lambda = \operatorname{Ker}(L_0 - \lambda)\big|_{\mathbf{M}}.$$

Let us call such a module a *weight module* for $(\mathbf{V}, \omega)$ and the eigenspace $\mathbf{M}_\lambda$ the *weight space* of *conformal weight* λ.

Let us write $\Delta(v) = \lambda$ when $v \in \mathbf{M}_\lambda$. Then, by (6.9), we have

$$\Delta(a_n v) = \Delta(a) + \Delta(v) - n - 1. \tag{6.10}$$

In other words, for $d, n \in \mathbb{Z}$ and $\lambda \in \mathbb{C}$,

$$\rho_n(\mathbf{V}_d)\,\mathbf{M}_\lambda \subset \mathbf{M}_{\lambda + d - n - 1},$$

where ρ_n denotes the n-th action of $\mathbf{V}$ on $\mathbf{M}$.

A weight module $\mathbf{M}$ is called an *ordinary module* (or just a *module*) if the weight spaces are finite-dimensional and the real part of the conformal weight in every coset in $\mathbb{C}/\mathbb{Z}$ is bounded below. That is, for any coset in $\mathbb{C}/\mathbb{Z}$, there exists a representative λ such that

$$\bigoplus_{n \in \mathbb{Z}} \mathbf{M}_{\lambda + n} = \mathbf{M}_\lambda \oplus \mathbf{M}_{\lambda + 1} \oplus \cdots \quad \text{and} \quad \dim \mathbf{M}_{\lambda + n} < \infty \quad (n \in \mathbb{Z}).$$

The adjoint module $\mathbf{V}$ is an ordinary module by the conditions (1)–(3).

If $\mathbf{M}$ is a simple ordinary module, then there exists a complex number λ, called the *lowest conformal weight* of $\mathbf{M}$, such that

$$\mathbf{M} = \mathbf{M}_\lambda \oplus \mathbf{M}_{\lambda + 1} \oplus \cdots \quad \text{with } \mathbf{M}_\lambda \neq 0,$$

and the whole space $\mathbf{M}$ is generated by $\mathbf{M}_\lambda$ as a module over $\mathbf{V}$.

Note 6.4. 1. For ordinary modules, we may consider the *graded dimensions* and the *conformal characters* (cf. Section 1.6.5). 2. Assume that $\mathbf{M}$ has a generalized eigenspace decomposition with respect to L_0 and write $\Delta(v) = \lambda$ when v is in the generalized eigenspace of conformal weight λ. Then (6.10) also holds true for generalized eigenspaces.

1.6.2.3 ℕ-Graded Modules

Let us set up an appropriate category of modules for a VOA to work with in practice for representation theory.

An $\mathbb{N}$-*graded module* for a VOA $(\mathbf{V}, \omega)$ is a weak module $\mathbf{M}$ equipped with a grading of the form

$$\mathbf{M} = \bigoplus_{k=0}^{\infty} \mathbf{M}(k) = \underbrace{\mathbf{M}(0)}_{\text{top}} \oplus \mathbf{M}(1) \oplus \mathbf{M}(2) \oplus \cdots$$

satisfying the condition that, for all $k \in \mathbb{N}$ and $d, n \in \mathbb{Z}$,

$$\rho_n(\mathbf{V}_d)\,\mathbf{M}(k) \subset \mathbf{M}(k + d - n - 1).$$

A weak module is said to be $\mathbb{N}$-*gradable* if there exists a grading that makes it into an $\mathbb{N}$-graded module. Note that a simple $\mathbb{N}$-gradable module $\mathbf{M}$ can be given an $\mathbb{N}$-grading such that $\mathbf{M}$ is generated as a module over $\mathbf{V}$ by the top subspace $\mathbf{M}(0)$.

For example, a weight module $\mathbf{M}$ is $\mathbb{N}$-gradable if the real part of the conformal weight in every coset in $\mathbb{C}/\mathbb{Z}$ is bounded below. Indeed, such a module is $\mathbb{N}$-graded by

$$\mathbf{M}(k) = \bigoplus_{i \in I} \mathbf{M}_{\lambda_i + k}, \quad k = 0, 1, 2, \cdots,$$

where λ_i are the minimal representatives of the conformal weights modulo $\mathbb{Z}$. In particular, ordinary modules are $\mathbb{N}$-gradable.

The category of $\mathbb{N}$-graded modules is the category of which the objects are $\mathbb{N}$-graded modules and the morphisms are homomorphisms of modules over the underlying vertex algebra $\mathbf{V}$. Thus a morphism in this category need not respect the gradings.

Note 6.5. An $\mathbb{N}$-graded module is also called an *admissible module* in the literatures, although the same term is used in a different sense for affine Kac–Moody algebras.

1.6.2.4 Rationality, C_2-Cofiniteness, and Regularity

Let us briefly describe conditions on a VOA and their consequences that guarantee finiteness or semisimplicity of module categories and play prominent roles in representation theory of VOAs.

1. A VOA is said to be *rational* if any $\mathbb{N}$-graded module is a direct sum of simple $\mathbb{N}$-graded modules. In other words, the category of $\mathbb{N}$-graded module is semisimple,

   A rational VOA has only finitely many (isomorphism classes of) simple $\mathbb{N}$-gradable modules and they are ordinary.

2. A VOA $(\mathbf{V}, \omega)$ is said to be C_2-*cofinite*, if the following condition holds:

$$\dim \mathbf{V}/\mathbf{V}_{(-2)}\mathbf{V} < \infty.$$

A C_2-cofinite VOA *of CFT type* has only finitely many simple weak modules and they are ordinary. Moreover, any weak module is $\mathbb{N}$-gradable.
3. A VOA is said to be *regular* if any weak module is a direct sum of simple ordinary modules.

   A regular VOA is rational and C_2-cofinite. If a VOA *of CFT type* is rational and C_2-cofinite, then it is regular.

Since the Heisenberg VOA has infinitely many simple modules, it is neither rational nor C_2-cofinite, so not regular either. The lattice VOAs associated with positive-definite lattices, the affine VOAs associated with integrable representations of affine Kac–Moody algebras, and the VOAs associated with the Virasoro minimal models are known to be regular, hence rational and C_2-cofinite.

1.6.2.5 Contragredient Modules

Let $\mathbf{M}$ be an ordinary module for a VOA $(\mathbf{V}, \omega)$. By composition of the actions given by (6.6), the transformation corresponding to $z \mapsto 1/z$ is found as

$$Y_{\mathbf{M}}(a, z) \mapsto Y_{\mathbf{M}}(e^{zL_1}(-z^{-2})^{L_0}a, z^{-1}).$$

We will use this to define the concept of the contragredient module.

To this end, consider the restricted dual of $\mathbf{M}$ defined by

$$\mathbf{M}' = \bigoplus_{\lambda \in \mathbb{C}} \mathbf{M}_\lambda^*$$

where $\mathbf{M}_\lambda^* = \mathrm{Hom}_{\mathbb{C}}(\mathbf{M}_\lambda, \mathbb{C})$ is the dual space. Denote the canonical pairing by

$$\mathbf{M}' \times \mathbf{M} \longrightarrow \mathbb{C}, \quad (\varphi, v) \mapsto \langle \varphi, v \rangle,$$

and the action of $\mathbf{V}$ on $\mathbf{M}'$ by

$$\langle Y_{\mathbf{M}'}(a, z)\varphi, v \rangle = \langle \varphi, Y_{\mathbf{M}}(e^{zL_1}(-z^{-2})^{L_0}a, z^{-1})v \rangle.$$

Then $\mathbf{M}'$ becomes an ordinary module for $(\mathbf{V}, \omega)$, called the *contragredient module* or the *dual module* of $\mathbf{M}$.

1.6.3 Simple $\mathbb{N}$-Graded Modules

In this section, we will explain a way to develop representation theory of VOAs in the category of $\mathbb{N}$-graded modules. Recall that an $\mathbb{N}$-graded module has the shape

$$\mathbf{M} = \underbrace{\mathbf{M}(0)}_{\text{top}} \oplus \mathbf{M}(1) \oplus \mathbf{M}(2) \oplus \cdots,$$

where the grading is not necessarily given by the L_0 eigenvalues.

Since the actions of homogeneous elements of **V** raise, preserve, or lower the degrees, we can apply an analogue of "highest weight theory" to our categories.

Thus, to classify ℕ-graded simple **V**-modules, we will classify the top subspace as a module over a certain algebra that induces actions preserving the degree, called the *zero-mode actions*, and reconstruct the whole **M** as a simple quotient of the induced module.

Such an algebra can be universally constructed, denoted **A(V)**, and called *Zhu's algebra* associated with the VOA (**V**, ω).

1.6.3.1 Lie Algebra of Fourier Modes

Let **V** be a vertex algebra and **M** a module over it. It follows from the commutator formula (MC),

$$[a_m, b_n] = \sum_{i=0}^{\infty} \binom{m}{i} (a_{(i)}b)_{m+n-i},$$

that the actions of elements of **V** on **M** form a Lie subalgebra of End **M**, called the Lie algebra of *Fourier modes* on **M**.

Such a Lie algebra is actually the image of a universal one defined by taking the commutator formula as the defining relation. Indeed, consider the space

$$\hat{\mathbf{V}} = \mathbf{V} \otimes \mathbb{C}[t, t^{-1}],$$

and equip it with the bracket operation given by

$$[a \otimes t^m, b \otimes t^n] = \sum_{i=0}^{\infty} \binom{m}{i} (a_{(i)}b) \otimes t^{m+n-i}. \tag{6.11}$$

Let $\hat{T}$ denote the operator on $\hat{\mathbf{V}}$ defined by

$$\hat{T}(a \otimes t^n) = (Ta) \otimes t^n + na \otimes t^{n-1}.$$

Then the bracket operation on $\hat{\mathbf{V}}$ induces a Lie algebra structure on the quotient

$$\mathfrak{g}(\mathbf{V}) = \mathbf{V} \otimes \mathbb{C}[t, t^{-1}]/\hat{T}(\mathbf{V} \otimes \mathbb{C}[t, t^{-1}])$$

and any module **M** over **V** can be regarded as a $\mathfrak{g}(\mathbf{V})$-module by

$$(a \otimes t^m)v = a_m v \quad (a \in \mathbf{V}, \ v \in \mathbf{M}).$$

We will call the Lie algebra $\mathfrak{g}(\mathbf{V})$ the *Lie algebra of Fourier modes* associated with **V**, which covers the Lie algebra of Fourier modes on every module.

1.6.3.2 Triangular Decomposition

Let (V, ω) be a VOA. For a homogeneous element $a \in V$ and an integer n, let $J_n(a)$ denote the image of $a \otimes t^{n+\Delta(a)-1}$ in the quotient $\mathfrak{g}(V)$:

$$J_n(a) = \pi(a \otimes t^{n+\Delta(a)-1}).$$

The definition of the Lie bracket (6.11) turns out to be

$$[J_m(a), J_n(b)] = \sum_{i=0}^{\infty} \binom{m + \Delta(a) - 1}{i} J_{m+n}(a_{(i)}b).$$

The action of $J_n(a)$ on an $\mathbb{N}$-graded module lowers the degree by n:

$$J_n(a) \colon \mathbf{M}(k) \longrightarrow \mathbf{M}(k - n).$$

The Lie algebra $\mathfrak{g}(V)$ is $\mathbb{Z}$-graded by the degree as

$$\mathfrak{g}(V) = \bigoplus_n \mathfrak{g}_n(V),$$

where $\mathfrak{g}_n(V)$ is the subspace of degree n:

$$\mathfrak{g}_n(V) = \text{Span} \{J_n(a) \mid a \in V_d \ (d \in \mathbb{Z})\}.$$

We have the *triangular decomposition*

$$\mathfrak{g}(V) = \mathfrak{g}_{<0}(V) \oplus \mathfrak{g}_0(V) \oplus \mathfrak{g}_{>0}(V),$$

where $\mathfrak{g}_{<0}(V)$ and $\mathfrak{g}_{>0}(V)$ are the sums of subspaces of negative and positive degrees, respectively.

In particular, the action of an element of $\mathfrak{g}_0(V)$ is called the *zero-mode action*, usually denoted $o(a)$ for a homogeneous element $a \in V$. Thus

$$o(a) = J_0(a) = a_{\Delta(a)-1} \colon \mathbf{M}(k) \longrightarrow \mathbf{M}(k),$$

where we have identified $J_0(a)$ in $\mathfrak{g}_0(V)$ with its action on $\mathbf{M}$.

1.6.3.3 Zhu's Algebra

For a VOA (V, ω), the Borcherds identity (M1) for modules reads, for homogeneous elements $a, b \in V$,

$$\sum_{i=0}^{\infty} \binom{m + \Delta(a) - 1}{i} J_{m+n+r}(a_{(r+i)}b)$$

$$= \sum_{i=0}^{\infty} (-1)^i \binom{r}{i} J_{m+r-i}(a)J_{n+i}(b) \tag{6.12}$$

$$- \sum_{i=0}^{\infty} (-1)^{r+i} \binom{r}{i} J_{n+r-i}(b)J_{m+i}(a).$$

Since the top space $\mathbf{M}(0)$ of an $\mathbb{N}$-graded module $\mathbf{M}$ is annihilated by $\mathfrak{g}_{>0}(\mathbf{V})$ and preserved by $\mathfrak{g}_0(\mathbf{V})$, the relation (6.12) with $(m, n, r) = (1, 0, -1)$ reads

$$\sum_{i=0}^{\infty} \binom{\Delta(a)}{i} o(a_{(i-1)}b)\big|_{\mathbf{M}(0)} = o(a)o(b)\big|_{\mathbf{M}(0)}$$

and, with $(m, n, r) = (1, 1, -2)$,

$$\sum_{i=0}^{\infty} \binom{\Delta(a)}{i} o(a_{(i-2)}b)\big|_{\mathbf{M}(0)} = 0.$$

We take these relations as defining a universal algebra acting on the top space $\mathbf{M}(0)$ for every $\mathbb{N}$-graded module $\mathbf{M}$.

We therefore define, for homogeneous $a, b \in \mathbf{V}$, the elements $a * b$ and $a \circ b$ by setting

$$a * b = \sum_{i=0}^{\infty} \binom{\Delta(a)}{i} a_{(i-1)}b, \quad a \circ b = \sum_{i=0}^{\infty} \binom{\Delta(a)}{i} a_{(i-2)}b. \qquad (6.13)$$

Then it is not difficult to see that the operation $*$ induces a structure of an associative algebra on the quotient defined by the operation $\circ$ as

$$\mathbf{A}(\mathbf{V}) = \mathbf{V}/\mathbf{V} \circ \mathbf{V}.$$

The algebra $\mathbf{A}(\mathbf{V})$ thus obtained is called *Zhu's algebra*. By abuse of notation, we will denote the elements of $\mathbf{A}(\mathbf{V})$ by their representatives in $\mathbf{V}$.

The top space $\mathbf{M}(0)$ becomes an $\mathbf{A}(\mathbf{V})$-module by

$$\mathbf{A}(\mathbf{V}) \times \mathbf{M}(0) \longrightarrow \mathbf{M}(0), \quad (a, v) \mapsto o(a)v = J_0(a)v,$$

since $o(a * b)\big|_{\mathbf{M}(0)} = o(a)o(b)\big|_{\mathbf{M}(0)}$ and $o(a \circ b)\big|_{\mathbf{M}(0)} = 0$ holds by construction.

1.6.3.4 Zhu's One-to-One Correspondence

Let $\mathfrak{g}(\mathbf{V}) = \mathfrak{g}_{<0}(\mathbf{V}) \oplus \mathfrak{g}_0(\mathbf{V}) \oplus \mathfrak{g}_{>0}(\mathbf{V})$ be the Lie algebra of Fourier modes, and consider the Lie subalgebra $\mathfrak{g}_{\geq 0}(\mathbf{V}) = \mathfrak{g}_0(\mathbf{V}) \oplus \mathfrak{g}_{>0}(\mathbf{V})$. Note that the identity map induces a surjection

$$\pi_0 \colon \mathfrak{g}_0(\mathbf{V}) \longrightarrow \mathbf{A}(\mathbf{V}),$$

which sends the Lie bracket of $\mathfrak{g}_0(\mathbf{V})$ to the commutators in Zhu's algebra.

For a simple $\mathbb{N}$-graded $\mathbf{V}$-module $\mathbf{M} = \mathbf{M}(0) \oplus \mathbf{M}(1) \oplus \cdots$, the top $\mathbf{M}(0)$ is seen to be an $\mathbf{A}(\mathbf{V})$-module.

Conversely, for any simple $\mathbf{A}(\mathbf{V})$-module W, regard it as a $\mathbf{U}(\mathfrak{g}_0(\mathbf{V}))$-module via the surjection π_0 and further as a $\mathbf{U}(\mathfrak{g}_{\geq 0}(\mathbf{V}))$-module by letting $\mathfrak{g}_{>0}(\mathbf{V})$ act by 0, and consider the induced module:

$$\mathbf{M}(W) = \mathbf{U}(\mathfrak{g}(\mathbf{V})) \otimes_{\mathbf{U}(\mathfrak{g}_{\geq 0}(\mathbf{V}))} W.$$

Then a simple $\mathbb{N}$-graded module for $(\mathbf{V}, \omega)$ can be constructed as an appropriate quotient of $\mathbf{M}(W)$.

In this way, we arrive at the following result.

Theorem 6.6 (Zhu) *There exists a one-to-one correspondence between the isomorphism classes of simple $\mathbb{N}$-gradable modules for a VOA $(\mathbf{V}, \omega)$ and the isomorphism classes of simple $\mathbf{A}(\mathbf{V})$-modules.*

Note 6.7. Zhu's algebra plays an important role also in modular invariance of conformal characters (cf. Section 1.6.5).

1.6.4 Fusion Rules

Let $\mathbf{C}$ be a commutative associative algebra and let $\mathbf{L}, \mathbf{M}, \mathbf{N}$ be submodules of $\mathbf{C}$. Then the multiplication on $\mathbf{C}$ restricts to a map

$$\mathbf{L} \times \mathbf{M} \longrightarrow \mathbf{N}$$

satisfying properties coming from those of $\mathbf{C}$; that is, $(au)v = u(av) = a(uv)$ for $a \in \mathbf{C}$, $u \in \mathbf{L}$, and $v \in \mathbf{N}$. Let us now replace the submodules of $\mathbf{C}$ by arbitrary modules and consider a map $\mathcal{Y} : \mathbf{L} \times \mathbf{M} \longrightarrow \mathbf{N}$ satisfying

$$\mathcal{Y}(au)v = \mathcal{Y}(u)av = a\mathcal{Y}(u)v.$$

The first equality guarantees that $\mathcal{Y}$ induces a map $\mathbf{L} \otimes_{\mathbf{A}} \mathbf{M} \longrightarrow \mathbf{N}$, which becomes a homomorphism of modules over $\mathbf{C}$ by the second equality.

Such a map constructed for a triple of modules can be seen to motivate the concept of an *intertwining operator* for a VOA. For example, for the lattice VOA $\mathbf{V}_L$ associated with an even lattice L and $\lambda, \mu \in L$, consider the triple $\mathbf{F}_\lambda, \mathbf{F}_\mu, \mathbf{F}_{\lambda+\mu}$ of modules for the Heisenberg VOA $\mathbf{F}_0$. Then the generating series map $Y(-, z)$ of $\mathbf{V}_L$ restricts to a map

$$\mathbf{F}_\lambda \times \mathbf{F}_\mu \longrightarrow \mathbf{F}_{\lambda+\mu}((z)), \quad (u, v) \mapsto Y(u, z)v,$$

which is an intertwining operator for $\mathbf{F}_0$, and this generalizes to the case when λ, μ are replaced by arbitrary elements of $\mathfrak{h}^*$.

The dimension of the space of intertwining operators is in fact a mathematical formulation of the concept of *fusion rules* originally considered in physics, and the analogue of the tensor product is called the *fusion product*.

For a vector space V and an indeterminate z, consider series of the following form with coefficients in V:

$$v(z) = \sum_{\alpha \in \mathbb{C}} v_\alpha z^{-\alpha-1}.$$

The set of such series is denoted $V\{z\}$.

1.6.4.1 Intertwining Operators

Let $\mathbf{L}, \mathbf{M}, \mathbf{N}$ be ordinary modules for a VOA $(\mathbf{V}, \omega)$. An *intertwining operator* of type $\begin{pmatrix} \mathbf{N} \\ \mathbf{L}\,\mathbf{M} \end{pmatrix}$ is a linear map

$$\mathcal{Y}(-, z)\colon \mathbf{L} \longrightarrow \mathrm{Hom}(\mathbf{M}, \mathbf{N}\{z\}), \quad u \mapsto \left[v \mapsto \mathcal{Y}(u, z)v \right]$$

such that the coefficients in the expansion

$$\mathcal{Y}(u, z) = \sum_\alpha u_\alpha z^{-\alpha-1}$$

satisfy the following properties:

(10) Local truncation. For any $u \in \mathbf{L}$, $v \in \mathbf{M}$, and $\alpha \in \mathbb{C}$,

$$\sum_n z^{-\alpha-n-1} u_{\alpha+n} v \in \mathbf{N}((z)) z^{-\alpha}.$$

(11) Borcherds identity. For all $a \in \mathbf{V}$, $u \in \mathbf{L}$, $v \in \mathbf{M}$, $p, r \in \mathbb{Z}$, and $q \in \mathbb{C}$:

$$\sum_{i=0}^\infty \binom{p}{i} (a_{r+i}u)_{p+q-i}v = \sum_{i=0}^\infty (-1)^i \binom{r}{i} a_{p+r-i}(u_{q+i}v)$$
$$- \sum_{i=0}^\infty (-1)^{r+i} \binom{r}{i} u_{q+r-i}(a_{p+i}v).$$

(IT) Translation. For all $u \in \mathbf{L}$, $v \in \mathbf{M}$, and $\alpha \in \mathbb{C}$,

$$(L_{-1}u)_\alpha v = -\alpha u_{\alpha-1} v.$$

For example, for $\mathbf{L} = \mathbf{V}$, the generating series $Y_\mathbf{M}(-, z)$ giving a module structure on $\mathbf{M}$ is an intertwining operator of type $\begin{pmatrix} \mathbf{M} \\ \mathbf{V}\,\mathbf{M} \end{pmatrix}$.

By (11) with $a = \omega$, the property (IT) turns out to be given by

$$L_0(u_\alpha v) = (L_0 u)_\alpha v + u_\alpha(L_0 v) - (\alpha + 1)u_\alpha v.$$

Therefore, (IT) is equivalent to the following: for homogeneous $u \in \mathbf{L}$, $v \in \mathbf{M}$, and $\alpha \in \mathbb{C}$,

$$\Delta(u_\alpha v) = \Delta(u) + \Delta(v) - \alpha - 1.$$

The Borcherds identity (11) is stable under shifting the index q. Therefore, if the coefficients of $\mathcal{Y}(u, z)$ satisfy (11), then those of $\mathcal{Y}(u, z)z^{-\beta}$ also satisfy it for any scalar β, and this freedom is fixed by (IT).

By making use of this freedom, in turn, the series $\mathcal{Y}(u, z)v$ for simple ordinary modules is turned to a series with integral powers by setting

$$\mathcal{Y}^\circ(u, z)v = z^{\lambda+\mu-\nu}\mathcal{Y}(u, z)v, \tag{6.14}$$

where λ, μ, ν are the lowest conformal weights of $\mathbf{L, M, N}$, respectively. Expanding the modified series as

$$\mathcal{Y}^{\circ}(u, z)v = \sum_n \mathcal{Y}_n^{\circ}(u)v \, z^{-n-1},$$

we have $\mathcal{Y}_n^{\circ}(u)\mathbf{M}_{(m)} \subset \mathbf{N}(m + \Delta(u) - n - 1)$, where $\mathbf{M}(k) = \mathbf{M}_{\mu+k}$ and $\mathbf{N}(l) = \mathbf{N}_{\nu+l}$ are the homogeneous subspaces with respect to the associated $\mathbb{N}$-gradings.

Note 6.8. 1. The property (IT) is often called the $L(-1)$-*derivative property* in the literatures. 2. The symbol $\left(\begin{smallmatrix} \mathbf{N} \\ \mathbf{L\,M} \end{smallmatrix}\right)$ exhibits that the intertwining operator is contravariant in the lower entries $\mathbf{L}$ and $\mathbf{M}$ and covariant in the upper $\mathbf{N}$.

1.6.4.2 Fusion Rules

Let $\mathbf{L, M, N}$ be ordinary modules for a VOA $(\mathbf{V}, \omega)$. Then the set of intertwining operators of type $\left(\begin{smallmatrix} \mathbf{N} \\ \mathbf{L\,M} \end{smallmatrix}\right)$ forms a vector space denoted

$$I\left(\begin{smallmatrix} \mathbf{N} \\ \mathbf{L\,M} \end{smallmatrix}\right) = \left\{\text{intertwining operators of type } \left(\begin{smallmatrix} \mathbf{N} \\ \mathbf{L\,M} \end{smallmatrix}\right)\right\}.$$

The dimension of this space is called the *fusion rule*, and usually denoted.

$$N_{\mathbf{L,M}}^{\mathbf{N}} = \dim I\left(\begin{smallmatrix} \mathbf{N} \\ \mathbf{L\,M} \end{smallmatrix}\right).$$

The fusion rules satisfy the following properties:

$$N_{\mathbf{LM}}^{\mathbf{N}} = N_{\mathbf{ML}}^{\mathbf{N}} = N_{\mathbf{LN'}}^{\mathbf{M'}},$$

where $\mathbf{M'}$ and $\mathbf{N'}$ denote the contragredient modules.

1. The equality $N_{\mathbf{LM}}^{\mathbf{N}} = N_{\mathbf{ML}}^{\mathbf{N}}$ holds by the actions corresponding to the skew-symmetry. Indeed, we have an isomorphism $* : I\left(\begin{smallmatrix} \mathbf{N} \\ \mathbf{L\,M} \end{smallmatrix}\right) \longrightarrow I\left(\begin{smallmatrix} \mathbf{N} \\ \mathbf{M\,L} \end{smallmatrix}\right)$ sending $\mathcal{Y}$ to $\mathcal{Y}^*$ defined by

$$\mathcal{Y}^*(v, z)u = e^{zL_{-1}}\mathcal{Y}(u, -z)v$$

for $u \in \mathbf{L}$ and $v \in \mathbf{M}$,

2. The equality $N_{\mathbf{LM}}^{\mathbf{N}} = N_{\mathbf{LN'}}^{\mathbf{M'}}$ holds by the contragredient actions. Indeed, we have an isomorphism $' : I\left(\begin{smallmatrix} \mathbf{N} \\ \mathbf{L\,M} \end{smallmatrix}\right) \longrightarrow I\left(\begin{smallmatrix} \mathbf{M'} \\ \mathbf{L\,N'} \end{smallmatrix}\right)$ sending $\mathcal{Y}$ to $\mathcal{Y}'$ defined by

$$\langle \mathcal{Y}'(u, z)\varphi, v \rangle = \langle \varphi, \mathcal{Y}(e^{zL_1}(-z^2)^{-L_0}u, z^{-1})v \rangle$$

for $u \in \mathbf{L}$, $v \in \mathbf{M}$, and $\varphi \in \mathbf{N'}$.

1.6.4.3 Fusion Products

Let us first note that we may equivalently formulate intertwining operators as a map of the form

$$\mathcal{Y}: \mathbf{L} \otimes \mathbf{M} \longrightarrow \mathbf{N}\{z\}, \quad u \otimes v \mapsto \mathcal{Y}(u \otimes v, z) = \mathcal{Y}(u, z)v.$$

In the sequel, we will freely switch from one to the other.

For a pair $\mathbf{L}, \mathbf{M}$ of ordinary modules for a VOA $(\mathbf{V}, \omega)$, their *fusion product* is a pair $(\mathbf{L} \boxtimes \mathbf{M}, \mathcal{Y}_{\mathbf{L},\mathbf{M}}^{\mathbf{L} \boxtimes \mathbf{M}})$ of an ordinary module $\mathbf{L} \boxtimes \mathbf{M}$ and an intertwining operator

$$\mathcal{Y}_{\mathbf{L},\mathbf{M}}^{\mathbf{L} \boxtimes \mathbf{M}}: \mathbf{L} \otimes \mathbf{M} \longrightarrow \mathbf{L} \boxtimes \mathbf{M}\{z\},$$

satisfying the following universal property:

For any ordinary module $\mathbf{N}$ for $(\mathbf{V}, \omega)$ and any intertwining operator

$$\mathcal{Y}: \mathbf{L} \otimes \mathbf{M} \longrightarrow \mathbf{N}\{z\}$$

there exists a unique homomorphism $\phi: \mathbf{L} \boxtimes \mathbf{M} \twoheadrightarrow \mathbf{N}$ of modules such that the diagram

commutes, where $\phi: (\mathbf{L} \boxtimes \mathbf{M})\{z\} \twoheadrightarrow \mathbf{N}\{z\}$ is the obvious map induced by $\phi: \mathbf{L} \boxtimes \mathbf{M} \twoheadrightarrow \mathbf{N}$.

If the fusion product $\mathbf{L} \boxtimes \mathbf{M}$ exists, then, for any ordinary module $\mathbf{N}$, the fusion rule is given by

$$N_{\mathbf{L},\mathbf{M}}^{\mathbf{N}} = \dim \mathrm{Hom}_V(\mathbf{L} \boxtimes \mathbf{M}, \mathbf{N}),$$

hence, under semisimplicity of the module category,

$$\mathbf{L} \boxtimes \mathbf{M} = \bigoplus_{\mathbf{W}} N_{\mathbf{L},\mathbf{M}}^{\mathbf{W}} \mathbf{W},$$

where the direct sum runs over the isomorphism classes of simple modules, which we identify with their representatives.

Note 6.9. When the VOA is good enough, the fusion products satisfy *associativity* in an appropriate sense and, under semisimplicity of the module category, the free abelian group generated by the simple modules becomes a commutative ring by the fusion product, called the *fusion ring*, for which the fusion rules are the structure constants.

1.6.4.4 Determination of fusion rules

Let $\mathbf{L}$ be a simple ordinary module for a VOA $(\mathbf{V}, \omega)$. For homogeneous $a \in \mathbf{V}$ and $u \in \mathbf{L}$, consider the elements $a * u$ and $a \circ u$ given by the same formula as (6.13). That is,

$$a * u = \sum_{i=0}^{\infty} \binom{\Delta(a)}{i} a_{i-1} u, \quad a \circ u = \sum_{i=0}^{\infty} \binom{\Delta(a)}{i} a_{i-2} u.$$

Then, as in the case of Zhu's algebra, the operation $*$ induces an action of Zhu's algebra $\mathbf{A}(\mathbf{V})$ on the quotient defined by $\circ$ as

$$\mathbf{A}(\mathbf{L}) = \mathbf{L}/\mathbf{V} \circ \mathbf{L}.$$

Moreover, the space $\mathbf{A}(\mathbf{L})$ admits a right action of $\mathbf{A}(\mathbf{V})$ induced by

$$u * a = \sum_{i=0}^{\infty} \binom{\Delta(a) - 1}{i} a_{i-1} u.$$

It is not difficult to see that $\mathbf{A}(\mathbf{L})$ becomes an $\mathbf{A}(\mathbf{V})$-bimodule by the left and the right operations $*$, which is called the *Frenkel–Zhu bimodule*.

Let $\mathbf{L}, \mathbf{M}, \mathbf{N}$ be simple ordinary modules with the lowest conformal weights λ, μ, ν, respectively, and equip them with $\mathbb{N}$-grading by, for $k \in \mathbb{N}$,

$$\mathbf{L}(k) = \mathbf{L}_{\lambda+k}, \quad \mathbf{M}(k) = \mathbf{M}_{\mu+k}, \quad \mathbf{N}(k) = \mathbf{N}_{\nu+k}.$$

For an intertwining operator $\mathcal{Y}$ of type $\binom{\mathbf{N}}{\mathbf{L}\,\mathbf{M}}$, consider the following maps for $k \in \mathbb{N}$ by taking the series $\mathcal{Y}^{\circ}(u, z)$ given by (6.14):

$$\mathbf{L}(k) \times \mathbf{M}(0) \longrightarrow \mathbf{N}(0), \quad (u, v) \mapsto \mathcal{Y}^{\circ}_{k-1}(u)v.$$

Then it induces

$$\pi(\mathcal{Y}) \colon \mathbf{A}(\mathbf{L}) \otimes_{\mathbf{A}(\mathbf{V})} \mathbf{M}(0) \longrightarrow \mathbf{N}(0),$$

giving rise to a map

$$\pi \colon I\binom{\mathbf{N}}{\mathbf{L}\,\mathbf{M}} \longrightarrow \mathrm{Hom}_{\mathbf{A}(\mathbf{V})}(\mathbf{A}(\mathbf{L}) \otimes_{\mathbf{A}(\mathbf{V})} \mathbf{M}(0), \mathbf{N}(0)).$$

Theorem 6.10 (Frenkel–Zhu, Li) *Let* $\mathbf{L}, \mathbf{M}, \mathbf{N}$ *be simple ordinary modules for a VOA* $(\mathbf{V}, \omega)$. *Then*

$$\dim I\binom{\mathbf{N}}{\mathbf{L}\,\mathbf{M}} \leq \dim \mathrm{Hom}_{\mathbf{A}(\mathbf{V})}(\mathbf{A}(\mathbf{L}) \otimes_{\mathbf{A}(\mathbf{V})} \mathbf{M}(0), \mathbf{N}(0)).$$

Moreover, if the VOA is rational, then the equality holds.

1.6.4.5 Examples of Fusion Rules

Let us list some examples of fusion rules.

1. Heisenberg VOAs. Consider the Heisenberg algebra $\hat{\mathfrak{h}}$ and the associated vertex algebra $\mathbf{V} = \mathbf{F}_0$, which is a VOA with the standard conformal vector ω. Consider the Fock modules $\mathbf{F}_\lambda$ of charge $\lambda \in \mathfrak{h}^*$. The fusion rules among them are described as follows: for $\lambda, \mu, \nu \in \mathfrak{h}^*$,

$$\dim I\binom{\mathbf{F}_\nu}{\mathbf{F}_\lambda \, \mathbf{F}_\mu} = \begin{cases} 1 & if \, \lambda + \mu = \nu, \\ 0 & otherwise. \end{cases}$$

In terms of the fusion product, it is expressed as

$$\mathbf{F}_\lambda \boxtimes \mathbf{F}_\mu = \mathbf{F}_{\lambda+\mu}.$$

2. Lattice VOAs. For a positive-definite even lattice L, consider the lattice VOA $(\mathbf{V}_L, \omega)$ with ω the standard conformal vector. The simple ordinary modules are classified by the cosets in L° / L as

$$\mathbf{V}_{L+\lambda} = \bigoplus_{\mu \in L} \mathbf{F}_{\lambda+\mu}, \ (\lambda \in L^\circ),$$

and the fusion rules are described as, for representatives λ, μ of L° / L,

$$\mathbf{V}_{L+\lambda} \boxtimes \mathbf{V}_{L+\mu} = \mathbf{V}_{L+(\lambda+\mu)}.$$

3. Simple affine VOAs at integrable levels. Consider the simple affine VOA $(\mathbf{L}(k,0), \omega)$ associated with $\mathfrak{sl}_2$ at level $k = 1, 2, \cdots$. The simple ordinary modules are the module $\mathbf{L}(k,j)$ with spin $j = 0, 1/2, 1, \ldots, k/2$. The fusion rules are described as follows: for half integers i, j with $0 \leq i, j \leq k/2$,

$$\mathbf{L}(k,i) \boxtimes \mathbf{L}(k,j) = \bigoplus_l \mathbf{L}(k,l),$$

where the sum is over the half integers l with $1 \leq l \leq k/2$ satisfying

$(1)\, |i-j| \leq l \leq i+j$ and $i+j+l \in \mathbb{Z}$, and $(2)\, i+j+l \leq k$.

The condition (1) is just the ordinary Clebsch–Gordan rules of decomposition of tensor products.

For example, if $k = 1$, then

$$\begin{aligned} \mathbf{L}(1,0) \boxtimes \mathbf{L}(1,0) &= \mathbf{L}(1,0), \\ \mathbf{L}(1,0) \boxtimes \mathbf{L}(1,1/2) &= \mathbf{L}(1,1/2), \\ \mathbf{L}(1,1/2) \boxtimes \mathbf{L}(1,0) &= \mathbf{L}(1,1/2), \\ \mathbf{L}(1,1/2) \boxtimes \mathbf{L}(1,1/2) &= \mathbf{L}(1,0). \end{aligned}$$

Compare the result with that for the lattice VOA $\mathbf{V}_L$ associated with the one-dimensional lattice of type A_1, which is isomorphic to $\mathbf{L}(1,0)$.

The result is generalized to simple affine VOAs associated with integrable representations of affine Kac–Moody algebras. See [96] for $\mathfrak{sl}_2$, and [60] and [97] for the general case.

4. Virasoro minimal models. Consider the simple Virasoro VOA $\mathbf{L}(1/2,0)$ of central charge $1/2$. The list of simple modules is

$$\mathbf{L}(1/2,0), \mathbf{L}(1/2,1/2), \mathbf{L}(1/2,1/16),$$

and the fusion rules are described as

$$
\begin{aligned}
\mathbf{L}(1/2,0) \;\boxtimes\; \mathbf{L}(1/2,h) &= \mathbf{L}(1/2,h) \;\; (h = 0,1/2,1/16),\\
\mathbf{L}(1/2,1/2) \;\boxtimes\; \mathbf{L}(1/2,1/2) &= \mathbf{L}(1/2,0),\\
\mathbf{L}(1/2,1/2) \;\boxtimes\; \mathbf{L}(1/2,1/16) &= \mathbf{L}(1/2,1/16),\\
\mathbf{L}(1/2,1/16) \;\boxtimes\; \mathbf{L}(1/2,1/16) &= \mathbf{L}(1/2,0) \oplus \mathbf{L}(1/2,1/2).
\end{aligned}
$$

The other cases are obtained by the symmetry of fusion rules.

The results are generalized to the case of the Virasoro minimal models, where the fusion rules are given by

$$\mathbf{L}(c_{p,q},h_{k,l}) \boxtimes \mathbf{L}(c_{p,q},h_{m,n}) = \bigoplus_{(r,s)} \mathbf{L}(c_{p,q},h_{r,s}),$$

where the sum is over the pairs (r,s) of integers with $1 \le r \le q-1$ and $1 \le s \le p-1$, up to identification $(r,s) \sim (q-r,p-s)$, satisfying the following conditions:

(1) $|k-m| < r < k+m$ and $|l-n| < s < l+n$.
(2) $k+m+r \in 2\mathbb{Z}+1$ and $l+n+s \in 2\mathbb{Z}+1$.
(3) $k+m+r \le 2q$ and $l+n+s \le 2p$.

See Subsection 1.3.3.4 for the notations and [15] and [99] for details.

5. Fixed-point VOAs $\mathbf{V}_L^+$. Let $\mathbf{V}_L$ be the VOA associated with a positive-definite even lattice L and $\mathbf{V}_L^+$ the vertex subalgebra of fixed-points by the automorphism θ, a lift of (-1)-involution of L to $\mathbf{V}_L$.

For simplicity, consider the case when L is unimodular. Recall the twisted module $\mathbf{V}_L^{\mathrm{tw}}$, which becomes an (untwisted) module for $\mathbf{V}_L^+$ and decomposes into the direct sum of simple components $\mathbf{V}_L^{\mathrm{tw},\pm}$. Now the list of simple modules for $\mathbf{V}_L^+$ is

$$\mathbf{V}_L^+, \;\; \mathbf{V}_L^-, \;\; \mathbf{V}_L^{\mathrm{tw},+}, \;\; \mathbf{V}_L^{\mathrm{tw},-}.$$

The fusion rules are

$$\mathbf{V}_L^- \boxtimes \mathbf{V}_L^- = \mathbf{V}_L^+, \ \mathbf{V}_L^- \boxtimes \mathbf{V}_L^{\text{tw},\pm} = \mathbf{V}_L^{\text{tw},\mp},$$
$$\mathbf{V}_L^{\text{tw},\pm} \boxtimes \mathbf{V}_L^{\text{tw},\pm} = \mathbf{V}_L^+, \ \mathbf{V}_L^{\text{tw},\pm} \boxtimes \mathbf{V}_L^{\text{tw},\mp} = \mathbf{V}_L^-.$$

See [27] for details.

1.6.5 Modular Invariance

Let $\mathbf{M}$ be a simple ordinary module for a VOA $(\mathbf{V}, \omega)$ of central charge c. The graded dimension with respect to the grading by conformal weight multiplied by $q^{-c/24}$ is called the *conformal character* of $\mathbf{M}$ and denoted as.

$$\text{ch}_{\mathbf{M}}(q) = q^{-c/24} \text{Tr}_{\mathbf{M}} q^{L_0} = q^{-c/24} \sum_{n=0}^{\infty} q^{\lambda+n} \dim \mathbf{M}_{\lambda+n},$$

where λ is the lowest conformal weight of $\mathbf{M}$. We are interested in the behavior of the conformal characters as functions of τ, where $q = e^{2\pi\sqrt{-1}\tau}$, when $\mathbf{M}$ varies over the ordinary simple modules.

For simple affine VOAs $\mathbf{L}(k, 0)$ associated with integrable representations of affine Kac–Moody algebras, the conformal characters agree with specialization of the characters in the sense of Kac–Moody algebras, and they satisfy certain modular transformation properties.

On the other hand, for the moonshine module $\mathbf{V}^\natural$, the conformal character is given by

$$J(\tau) - 744 = q^{-1} + 196884q + \cdots,$$

where $J(\tau)$ is the elliptic modular function, which is certainly invariant under modular transformations.

Such nice modular transformation properties can be uniformly described as a consequence of a general fact, modular invariance of conformal characters, which holds not only for conformal characters, but also for functions called *torus one-point functions* under suitable conditions.

In this section, we will briefly describe the theory of modular invariance established by Y. C. Zhu by combining ideas from conformal field theory in physics with C_2-cofiniteness and Zhu's algebra.

1.6.5.1 Torus One-Point Functions

For a simple ordinary module $\mathbf{M}$ with the lowest conformal weight $\lambda \in \mathbb{C}$ for a VOA $(\mathbf{V}, \omega)$, consider the series in $q = e^{2\pi\sqrt{-1}\tau}$ defined for a homogeneous $a \in \mathbf{V}$ by

$$\chi_{\mathbf{M}}(a, \tau) = q^{-c/24} \text{Tr} \, o(a) q^{L_0} \big|_{\mathbf{M}} = q^{-c/24} \sum_{n=0}^{\infty} \text{Tr} \, o(a) \big|_{\mathbf{M}_{\lambda+n}} q^{\lambda+n},$$

where $o(a) = a_{\Delta(a)-1}$ is the zero-mode action on the weight spaces $\mathbf{M}_{\lambda+n}$. This series is called the *one-point function* (on the torus) associated with $\mathbf{M}$ at $a \in \mathbf{V}$.

In particular, when the element a is the vacuum $\mathbf{1}$, the one-point function agrees with the conformal character $\mathrm{ch}_{\mathbf{M}}(q)$:

$$\chi_{\mathbf{M}}(\mathbf{1}, \tau) = \mathrm{ch}_{\mathbf{M}}(q) = q^{-c/24} \sum_{\alpha \in \mathbb{C}} \dim \mathbf{M}_\alpha q^\alpha.$$

Now assume that $(\mathbf{V}, \omega)$ is rational; that is, the category of $\mathbb{N}$-graded modules is semisimple. Then there are only finitely many isomorphism classes of simple modules, say $\mathbf{M}^1, \ldots, \mathbf{M}^n$, which are ordinary. We are interested in the behavior of the one-point functions

$$\chi_{\mathbf{M}^1}(a, \tau), \ldots, \chi_{\mathbf{M}^n}(a, \tau).$$

under transformations of weight k by the full modular group $\mathrm{SL}_2(\mathbb{Z})$,

$$f(\tau) \mapsto (c\tau + d)^{-k} f\left(\frac{a\tau + b}{c\tau + d}\right),$$

where $a, b, c, d \in \mathbb{Z}$ and $ad - bc = 1$.

1.6.5.2 Eisenstein Series and Serre Derivative

In describing the properties of one-point functions as defined here, the Eisenstein series naturally arise.

For $k \in \mathbb{N}$, let $G_{2k}(\tau)$ denote the Eisenstein series in its q-expansion:

$$G_{2k}(\tau) = 2\zeta(2k) + \frac{2(2\pi\sqrt{-1})^{2k}}{(2k-1)!} \sum_{n=1}^\infty \sigma_{2k-1}(n)q^n,$$

where ζ is the Riemann zeta function and $\sigma_m(n)$ the sum of mth powers of divisors of n. For $k \geq 2$, the Eisenstein series is a modular form of weight $2k$:

$$G_{2k}\left(\frac{a\tau + b}{c\tau + d}\right) = (c\tau + d)^{2k} G_{2k}(\tau).$$

For $k = 2, 3$, we have

$$G_4(\tau) = \frac{\pi^4}{45}\left(1 + 240 \sum_{n=1}^\infty \sigma_3(n)q^n\right), \ G_6(\tau) = \frac{2\pi^6}{945}\left(1 - 504 \sum_{n=1}^\infty \sigma_5(n)q^n\right).$$

Consider the ring of modular forms $\mathbb{C}[G_4(\tau), G_6(\tau)]$, which is a Noetherian ring. For $k = 1$, we have

$$G_2(\tau) = \frac{\pi^2}{3}\left(1 - 24 \sum_{n=1}^\infty \sigma_1(n)q^n\right),$$

whose transformation property is

$$G_2\left(\frac{a\tau + b}{c\tau + d}\right) = (c\tau + d)^2 G_2(\tau) - 2\pi\sqrt{-1}c(c\tau + d).$$

Although it is not a modular form, for a modular form $f(\tau)$ of weight k, the Serre derivative

$$D_k f(\tau) = (2\pi\sqrt{-1})\frac{d}{d\tau}f(\tau) + kG_2(\tau)f(\tau)$$

becomes a modular form of weight $k + 2$.

1.6.5.3 Space of One-Point Functions

Let $\mathbf{M}$ be an ordinary module for a VOA $(\mathbf{V}, \omega)$ and consider the associated one-point function $\chi_{\mathbf{M}}(-, \tau)$. For homogeneous elements $a, b \in \mathbf{V}$, the following formulas hold by the axioms for vertex algebras and invariance of trace under cyclic permutation of entries: for $a, b \in \mathbf{V}$,

(1) $\chi_{\mathbf{M}}(a_{[0]}b, \tau) = 0$,

(2) $\chi_{\mathbf{M}}(a_{[-2]}b, \tau) + \sum_{k=2}^{\infty}(2k - 1)G_{2k}(\tau)\chi_{\mathbf{M}}(a_{[2k-2]}b, \tau)$,

(3) $\chi_{\mathbf{M}}(L_{[-2]}a, \tau) = D_{\Delta[a]}\chi_{\mathbf{M}}(a, \tau) + \sum_{k=2}^{\infty}G_{2k}(\tau)\chi_{\mathbf{M}}(L_{[2k-2]}a, \tau)$,

where $a_{[n]}b$ and $L_{[n]}$ are defined by

$$Y(e^{2\pi\sqrt{-1}zL_0}a, e^{2\pi\sqrt{-1}z} - 1)b = \sum_n a_{[n]}bz^{-n-1},$$

$$L_{[n]} = (2\pi\sqrt{-1})^2(\omega - (c/24)\mathbf{1})_{[n+1]},$$

respectively, and $D_{\Delta[a]}\chi_{\mathbf{M}}(a, \tau)$ is the Serre derivative with $\Delta[a]$ the conformal weight of a with respect to $L_{[0]}$.

Here the assignment $a \mapsto Y(e^{2\pi\sqrt{-1}zL_0}a, e^{2\pi\sqrt{-1}z} - 1)$ gives a new VOA structure on $\mathbf{V}$ with the conformal vector $\tilde{\omega} = (2\pi\sqrt{-1})^2(\omega - (c/24)\mathbf{1})$, which is seen to be achieved by a particular case of Huang's formula (6.8) corresponding to a coordinate on the cylinder, and this is where the factor $q^{-c/24}$ of the conformal character arises.

Now the properties (1), (2), and (3) do not depend on the choice of the module $\mathbf{M}$. We therefore take them as conditions on a functional $\chi(-, \tau)$ on $\mathbf{V} \otimes \mathbb{C}[G_4(\tau), G_6(\tau)]$ valued in series in q, and call such a functional an *abstract one-point function* if it satisfies the conditions.

Let $\mathbf{O}_q(\mathbf{V})$ be the $\mathbb{C}[G_4(\tau), G_6(\tau)]$-submodule of $\mathbf{V} \otimes \mathbb{C}[G_4(\tau), G_6(\tau)]$ generated by the elements of the following form with $a, b \in \mathbf{V}$:

$$a_{[0]}b \quad \text{or} \quad a_{[-2]}b + \sum_{k=2}^{\infty}(2k-1)G_{2k}(\tau)a_{[2k-2]}b. \tag{6.15}$$

Then an abstract one-point function $\chi(-, \tau)$ induces a functional on the quotient

$$\mathbf{V} \otimes \mathbb{C}[G_4(\tau), G_6(\tau)]/\mathbf{O}_q(\mathbf{V})$$

by (1) and (2).

1.6.5.4 Consequence of C_2-Cofiniteness

Recall that a VOA $(\mathbf{V}, \omega)$ is said to be C_2-*cofinite* if the quotient $\mathbf{V}/\mathbf{V}_{(-2)}\mathbf{V}$ is finite-dimensional.

If $(\mathbf{V}, \omega)$ is C_2-cofinite, $\mathbf{V} \otimes \mathbb{C}[G_4(\tau), G_6(\tau)]/\mathbf{O}_q(\mathbf{V})$ is a finitely generated $\mathbb{C}[G_4(\tau), G_6(\tau)]$-module, thus a Noetherian module since the ring $\mathbb{C}[G_4(\tau), G_6(\tau)]$ is Noetherian. Therefore, for any $a \in \mathbf{V}$, there exist $s \in \mathbb{N}$ and $g_i(\tau) \in \mathbb{C}[G_4(\tau), G_6(\tau)]$ such that

$$(L_{[-2]})^s a + \sum_{i=0}^{s-1} g_i(\tau)(L_{[-2]})^i a \in \mathbf{O}_q(\mathbf{V}).$$

After some algebra, the relation yields the following result.

Theorem 6.11 (Zhu) *Let $(\mathbf{V}, \omega)$ be a C_2-cofinite VOA, $\chi(-, \tau)$ a one-point function and $a \in V$ satisfy $L_{[n]}a = 0$ for $n > 0$. Then, the value of the one-point function at $a \in \mathbf{V}$ satisfies a differential equation of the form*

$$\left(q\frac{d}{dq}\right)^s \chi(a, \tau) + \sum_{i=0}^{s-1} h_i(\tau)\left(q\frac{d}{dq}\right)^i \chi(a, \tau) = 0,$$

where $h_i(\tau) \in \mathbb{C}[G_2(\tau), G_4(\tau), G_6(\tau)]$.

Note 6.12. 1. Theorem 6.11 only guarantees existence of a differential equation, and finding it explicitly is a separate problem. 2. By the property (1) in the preceding subsection, it follows that the result of Theorem 6.11 holds under a weaker assumption. See [28] and [85].

1.6.5.5 Modular Invariance

Assume that $(\mathbf{V}, \omega)$ is C_2-cofinite. Then the values of one-point functions $\chi(a, \tau)$ at any $a \in \mathbf{V}$ satisfy a linear ordinary differential equation of regular singular type for which $q = 0$ is a the only regular singular point, so the series solutions at $q = 0$ converge. Therefore, one-point functions can be viewed as taking values in holomorphic functions of τ on the upper half-plane, and the space of abstract one-point functions is invariant under the modular transformations since the conditions that characterize it are modular invariant. Moreover, the initial terms of one-point functions give rise to linear functionals F on Zhu's algebra $\mathbf{A(V)}$ satisfying $F(a * b) = F(b * a)$.

Assume further that $(\mathbf{V}, \omega)$ is rational. Then Zhu's algebra $\mathbf{A(V)}$ is semisimple, and the simple $\mathbf{A(V)}$-modules are in one-to-one correspondence with the top spaces of the simple ordinary modules. It then follows that the modular invariant space of abstract one-point functions is spanned by the one-point functions associated with simple ordinary modules.

Theorem 6.13 (Zhu, Dong-Li-Mason) *Let* $(\mathbf{V}, \omega)$ *be rational and* C_2*-cofinite and let* $\mathbf{M}^1, \ldots, \mathbf{M}^n$ *be the list of simple ordinary modules. Then, for any* $k \in \mathbb{N}$ *and any element* $a \in \mathbf{V}$ *of weight* k *with respect to* $L_{[0]}$, *the series*

$$\chi_{\mathbf{M}^1}(a, \tau), \ldots, \chi_{\mathbf{M}^n}(a, \tau) \tag{6.16}$$

define holomorphic functions on the upper half-plane that span a vector space invariant under the weight k *action of the full modular group* $SL_2(\mathbb{Z})$.

In other words, the transformation of $\chi_{\mathbf{M}^i}(a, \tau)$ under a modular transformation of weight k becomes a linear combination of the functions (6.16) with constant coefficients.

Note 6.14.   When the VOA and the module category are good enough, the matrix representing the modular transformation $\tau \mapsto -1/\tau$ is related to the fusion rules by a famous formula called the *Verlinde formula*.

Bibliographic Notes

General references are Frenkel, Huang, and Lepowsky [3], Lepowsky and Li [10], and Frenkel and Ben-Zvi [8]. For related models in physics, consult Di Francesco et al. [13] (cf. [19]).

For various transformation formulas, see Frenkel, Huang, and Lepowsky [3] and Huang [5] (cf. [8], [97], [100].).

Classification of $\mathbb{N}$-graded modules by Zhu's algebra is due to Y. C. Zhu in [24] (cf. [60], [100]). See Matsuo, Nagatomo, and Tsuchiya [82] for an alternative approach.

Interpretation of fusion rules by means of intertwining operators was initiated by Tsuchiya and Kanie [96] for affine Lie algebras and further developed geometrically in Tsuchiya, Ueno, and Yamada [97]. For general theory of fusion products, see [72] and references therein (cf. [65] and subsequent papers). For geometric interpretations, see Huang [5] and Nagatomo and Tsuchiya [86].

Fusion rules for lattice VOAs are described in Dong and Lepowsky [4]. For determination of fusion rules by Frenkel and Zhu bimodules, see [60] and [74]. Fusion rules for affine vertex algebras were treated in [60]. See [96] and [97] for early treatments. Fusion rules for Virasoro minimal models were determined by Wang [99] (cf. Dong et al. [49]) based on Feigin and Fuchs [53]. Fusion rules for V_L^+ are determined by Abe, Dong and Li [27].

General theory of modular invariance of conformal characters as well as the concept of C_2-cofiniteness of VOAs was established by Y. C. Zhu in [24] (cf. [100]), and generalized to twisted modules by Dong et al. [46]. See Mason and Tuite [21] for a good survey. For earlier studies on modular invariant characters for affine Kac–Moody algebras, see Kac and Peterson [66].

For regularity and rationality of VOAs, see Dong, Li, and Mason [47], Li [73], and Abe, Buhl, and Dong [26] (cf. [82]). For rationality of V_L^+ for positive-definite lattices L, see Dong, Jiang, and Lin [44].

Finding the automorphism group of a VOA is an interesting problem. See Dong and Nagatomo [50] for Aut V_L and Shimakura [94] for Aut V_L^+.

Epilogue

Let Λ be the Leech lattice, the unique even unimodular positive-definite lattice of rank 24 without roots. Consider the lattice VOA V_Λ with the standard conformal vector. It is a regular VOA of CFT type and the conformal character is given by

$$\mathrm{ch}_{V_\Lambda}(q) = j(\tau) - 720$$
$$= q^{-1} + \underline{24} + 196884q + 21493760q^2 + \cdots.$$

In particular, the dimension of the degree 1 subspace is 24, which is nonzero. Since Λ is unimodular, the VOA V_Λ is holomorphic, that is, the adjoint module V_Λ is the only simple module.

Consider the decomposition of V_Λ under a lift θ of the (-1)-involution of the lattice:

$$V_\Lambda = V_\Lambda^+ \oplus V_\Lambda^-.$$

Then the fixed-point subVOA V_Λ^+ is again of CFT type but with the degree 1 subspace now being 0. The VOA V_Λ^+ is still regular, but not holomorphic, and the list of simple modules

$$V_\Lambda^+, \quad V_\Lambda^-, \quad V_\Lambda^{\mathrm{tw},+}, \quad V_\Lambda^{\mathrm{tw},-},$$

where $V_\Lambda^{\mathrm{tw},\pm}$ are simple components of V_Λ^{tw} as a V_Λ^+-module and let $V_\Lambda^{\mathrm{tw},+}$ be the one with integral grading with respect to L_0.

Now the moonshine module $\mathbf{V}^\natural$ is constructed as a module over $\mathbf{V}_\Lambda^+$ as

$$\mathbf{V}^\natural = \mathbf{V}_\Lambda^+ \oplus \mathbf{V}_\Lambda^{\mathrm{tw},+},$$

of which the conformal character is given as described in the Introduction by

$$\begin{aligned}
\mathrm{ch}_{\mathbf{V}^\natural}(q) &= j(\tau) - 744 \\
&= q^{-1} + \underbrace{0} + 196884q + 21493760q^2 + \cdots.
\end{aligned}$$

The moonshine module $\mathbf{V}^\natural$ actually becomes a regular holomorphic VOA (cf. [42], [47]), which is of CFT type with the degree 1 subspace now being 0.

Existence of a vertex algebra structure on $\mathbf{V}^\natural$ that extends the structure of a module over $\mathbf{V}_\Lambda^+$ is stated in [32] and its detailed proof by hard calculations, heavily based on group theoretical consideration, is given in [1]. Alternatively, note that $\mathbf{V}_\Lambda^+$ and $\mathbf{V}_\Lambda^{\mathrm{tw},+}$ are closed under the fusion products as

$$\mathbf{V}_\Lambda^{\mathrm{tw},+} \boxtimes \mathbf{V}_\Lambda^{\mathrm{tw},+} = \mathbf{V}_\Lambda^+,$$

the intertwining operators are series with integral powers in z, and the fusion rules are at most one-dimensional. The vertex algebra structure on $\mathbf{V}^\natural$ can actually be obtained by intertwining operators among $\mathbf{V}_\Lambda^+$ and $\mathbf{V}_\Lambda^{\mathrm{tw},+}$, multiplied by appropriate constant scalar factors.

Nowadays, many constructions of $\mathbf{V}^\natural$ are known. See [95] for a nice construction from $(\mathbf{V}_{\sqrt{2}E_8}^+)^{\otimes 3}$ and [36] for constructions from $\mathbf{V}_\Lambda$. See [37] for general theory of constructing vertex algebras by intertwining operators.

Let $\mathbf{V}$ be a VOA of CFT type with the degree 1 subspace being 0, such as the moonshine module.

$$\mathbf{V} = \mathbb{C}\mathbf{1} \oplus 0 \oplus \mathbf{B} \oplus \cdots.$$

Then the degree 2 subspace $\mathbf{B}$ satisfies

$$\mathbf{B}_{(1)}\mathbf{B} \subset \mathbf{B}, \quad \mathbf{B}_{(3)}\mathbf{B} \subset \mathbf{V}_0 = \mathbb{C}\mathbf{1}.$$

Let us equip $\mathbf{B}$ with a product and a bilinear form:

$$\begin{aligned}
\cdot \;\; &: \mathbf{B} \times \mathbf{B} \longrightarrow \mathbf{B}, \;\; (a,b) \mapsto a \cdot b, \\
(\,|\,) &: \mathbf{B} \times \mathbf{B} \longrightarrow \mathbb{C}, \;\; (a,b) \mapsto (a|b),
\end{aligned}$$

in such a way that

$$a \cdot b = a_{(1)}b \;\; \text{and} \;\; (a|b)\mathbf{1} = a_{(3)}b.$$

It is easy to see that the axioms for vertex algebras and their consequences imply, for all $a, b, c \in \mathbf{B}$,

$$a \cdot b = b \cdot a, \quad (a|b) = (b|a), \quad \text{and} \quad (a \cdot b|c) = (a|b \cdot c).$$

In other words, $\mathbf{B}$ is a commutative algebra with symmetric invariant bilinear form, often called the *Griess algebra* of $\mathbf{V}$. The Griess algebra $\mathbf{B}^\natural$ of the moonshine module $\mathbf{V}^\natural$ is the Griess–Conway algebra mentioned in the Introduction.

One of the specific features of $\mathbf{V}^\natural$ is the existence of an *Ising frame*, often called a *Virasoro frame*; that is, a full subVOA of a VOA isomorphic to a tensor product of the simple Virasoro VOA $\mathbf{L}(1/2, 0)$ of central charge $1/2$ (cf. [49], [43]). Here a subVOA of a VOA is said to be *full* if their conformal vectors agree. The moonshine module $\mathbf{V}^\natural$ inherits such a frame of length 48 from $\mathbf{V}_\Lambda^+ \subset \mathbf{V}_\Lambda$:

$$\underbrace{\mathbf{L}(1/2, 0) \otimes \cdots \otimes \mathbf{L}(1/2, 0)}_{48 \text{ times}} \subset \mathbf{V}_\Lambda^+.$$

In other words, the Griess–Conway algebra $\mathbf{B}^\natural$ has a set of 48 Virasoro vectors of central charge $1/2$, orthogonal to each other in the sense that the corresponding Virasoro actions commute, such that each generates a subVOA isomorphic to $\mathbf{L}(1/2, 0)$.

In general, let $\mathbf{V}$ be a VOA of CFT type with $\mathbf{V}_1 = 0$. Let e be a Virasoro vector in the Griess algebra $\mathbf{B}$ of central charge $1/2$; that is,

$$\omega \cdot e = e \cdot e = 2e, \quad (e|e) = \frac{1}{4}.$$

Such an e is called an *Ising vector* if it generates a subVOA $\mathbf{V}_e$ isomorphic to $\mathbf{L}(1/2, 0)$. Since $\mathbf{L}(1/2, 0)$ is regular, $\mathbf{V}$ decomposes as

$$\mathbf{V} = \mathbf{V}(e, 0) \oplus \mathbf{V}(e, 1/2) \oplus \mathbf{V}(e, 1/16), \tag{2}$$

where $\mathbf{V}(e, h)$ denotes the sum of components isomorphic to $\mathbf{L}(1/2, h)$ for each $h = 0, 1/2, 1/16$. Note that $\mathbf{V}_e$ itself is one of the simple components in $\mathbf{V}(e, 0)$.

Consider the map

$$Y(-, z) \colon \mathbf{V} \longrightarrow \operatorname{Hom}(\mathbf{V}, \mathbf{V}((z))), \quad a \mapsto Y(a, z).$$

Then it induces an intertwining operator for the VOA $\mathbf{L}(1/2, 0)$ for each triple of appropriate simple components of $\mathbf{V}$ when multiplied by a rational power of z depending on the triple. Consequently, the fusion rules for the Ising model imply that the product operations of the vertex algebra $\mathbf{V}$ satisfy, for $n \in \mathbb{Z}$,

$$V(e,0)_{(n)}V(e,h) \quad \subset V(e,h) \ (h = 0, 1/2, 1/16),$$
$$V(e,1/2)_{(n)}V(e,1/2) \subset V(e,0),$$
$$V(e,1/2)_{(n)}V(e,1/16) \subset V(e,1/16),$$
$$V(e,1/16)_{(n)}V(e,1/16) \subset V(e,0) \oplus V(e,1/2),$$

which imply that the following map τ_e gives rise to an automorphism of the whole VOA V (cf. [19] and [83]):

$$\tau_e : V \longrightarrow V, \quad \tau_e v = \begin{cases} v & (v \in V(e,0) \oplus V(e,1/2)), \\ -v & (v \in V(e,1/16)). \end{cases}$$

This automorphism is called the *Miyamoto involution* in the literatures.

Now let a be the half of an Ising vector e in B. Then a is an idempotent of B with squared norm $1/16$:

$$a \cdot a = a, \quad (a|a) = \frac{1}{16}.$$

Consider the eigenspace with eigenvalue λ for the action of the idempotent a on B by multiplication and write

$$B(a,\lambda) = \{ x \in B \,|\, a \cdot x = \lambda x \}.$$

Then, since V is of CFT type with $V_1 = 0$ and the highest weight vector of V_e is the vacuum $1 \in V$, we have $V_e \cap B = \mathbb{C}e$, and it follows that the other simple components of $V(e,0)$ can be chosen so that they intersect B with the span of the highest weight vectors. Therefore, we have $B(a,1) = \mathbb{C}a$ and

$$B = B(a,1) \oplus B(a,0) \oplus B(a,1/4) \oplus B(a,1/32), \tag{3}$$

where the eigenspaces are related to the decomposition (2) by

$$B(a,1) \oplus B(a,0) = B \cap V(e,0),$$
$$B(a,1/4) = B \cap V(e,1/2), \quad B(a,1/32) = B \cap V(e,1/16).$$

Regarding the decomposition (3), the fusion rules turn out to be described by Table 12. The information given to the idempotent a is strong enough to ensure that the structures of subalgebras generated by a pair of such idempotents actually fall into nine types ([90], [62]) corresponding to the conjugacy classes of the products of pairs of 2A involutions of the Monster (see [16]).

The properties of the idempotents a in the Griess algebra as here were axiomatized in [16], and a general framework in dealing with such algebras was formulated in [62] under the term *axial algebras* (cf. [80], [63] and [23]).

Let A be a commutative nonassociative algebra and consider a set of distinguished idempotents, called the *axes*. Assume that the actions of axes a by multiplication are semisimple with eigenvalues from a fixed set Λ as

Table 12 Ising fusion rules for axes

	1	0	1/4	1/32
1	1	∅	1/4	1/32
0	∅	0	1/4	1/32
1/4	1/4	1/4	1,0	1/32
1/32	1/32	1/32	1/32	1,0,1/4

$$\mathbf{A} = \bigoplus_{\lambda \in \Lambda} \mathbf{A}(a, \lambda),$$

for which the multiplication of $\mathbf{A}$ obeys a prescribed set of fusion rules, recently called a *fusion law*, which sends a pair (λ, μ) of eigenvalues in Λ to a subset $\lambda * \mu$ of Λ, in such a way that

$$\mathbf{A}(a, \lambda)\mathbf{A}(a, \mu) \subset \sum_{\nu \in \lambda * \mu} \mathbf{A}(a, \nu).$$

Since $\mathbf{A}$ is commutative, we may and do assume that the fusion rules are symmetric; that is, $\lambda * \mu = \mu * \lambda$ for all $\lambda, \mu \in \Lambda$. An axial algebra is an algebra equipped with axes that satisfy the properties discussed here and generate the algebra.

The concept of axial algebras was further generalized in [39] to a class of algebras called *axial decomposition algebras.* These algebras have been extensively studied in recent years with fruitful outcomes.

Let us finally recall that the moonshine module $\mathbf{V}^\natural$ is of CFT type, regular, hence C_2-cofinite, and holomorphic of central charge $c = 24$. Note that the lattice VOAs associated with Niemeier lattices and many other VOAs also satisfy the same properties. For recent progress on the classification of such VOAs, initiated by Schellekens [91], see [98].

Let $\mathbf{V}$ be a VOA satisfying the properties listed. By Zhu's theory of modular invariance, the one-dimensional vector space spanned by the conformal character $\mathrm{ch}_{\mathbf{V}}(q) = \chi_{\mathbf{V}}(\mathbf{1}, \tau)$ is invariant under modular transformations. After inspecting the possible characters of the full modular group for the transformation, it turns out that the conformal character itself is modular invariant and it must be identical to the elliptic modular function $j(\tau)$ up to shifting the constant term (cf. [25], [21]). In particular, the dimension of the degree 2 subspace $\mathbf{V}_2$ of such a VOA must be 196884.

Let a be any element of $\mathbf{V}$ that is homogeneous of conformal weight k with respect to $L_{[0]}$, and consider the one-point function:

$$\chi_{\mathbf{V}}(a,\tau) = q^{-1} \sum_{n=0}^{\infty} \mathrm{Tr}\, o(a)\big|_{\mathbf{V}_n}\, q^n.$$

Then it is also invariant under modular transformations, but now of weight k. Moreover, if a is a singular vector with respect to the Virasoro actions, then $\chi_{\mathbf{V}}(a,\tau)$ is a cusp form (cf. [48]), which is 0 if the weight is less than 12.

Let us apply this observation to the case with $\mathbf{V}_1 = 0$, when the conformal character equals $j(\tau)-744$ and the shape of $\mathbf{V}$ agrees with that of the moonshine module:

$$\mathbf{V} = \mathbb{C}\mathbf{1} \oplus 0 \oplus \mathbf{B} \oplus \mathbf{V}_3 \oplus \mathbf{V}_4 \oplus \cdots.$$

$$\text{dim} \quad 1 \quad 0 \quad 196884$$

Then the absence of cusp forms, as mentioned, combined with general properties of the VOA such as the Borcherds identity and their consequences, implies the following trace formulae for the multiplication operators on the Griess algebra $\mathbf{B}$ for up to five elements of $\mathbf{B}$:

$$\mathrm{Tr}\, o(a_1) = 32814\,(a_1|\omega),$$
$$\mathrm{Tr}\, o(a_1)o(a_2) = 4620\,(a_1|a_2) + 5084\,(a_1|\omega)(a_2|\omega),$$
$$\mathrm{Tr}\, o(a_1)o(a_2)o(a_3) = 900\,(a_1|a_2|a_3) + 620\,\mathrm{Cyc}\,(a_1|a_2)(a_3|\omega)$$
$$+ 744\,(a_1|\omega)(a_2|\omega)(a_3|\omega),$$
$$\mathrm{Tr}\, o(a_1)o(a_2)o(a_3)o(a_4)$$
$$= 166\,(a_1a_2|a_3a_4) - 116\,(a_1a_3|a_2a_4) + 166\,(a_1a_4|a_2a_3)$$
$$+ 114\,\mathrm{Sym}\,(a_1|a_2|a_3)(a_4|\omega) + 52\,\mathrm{Sym}\,(a_1|a_2)(a_3|a_4)$$
$$+ 80\,\mathrm{Sym}\,(a_1|a_2)(a_3|\omega)(a_4|\omega) + 104\,(a_1|\omega)(a_2|\omega)(a_3|\omega)(a_4|\omega),$$
$$\mathrm{Tr}\, o(a_1)o(a_2)o(a_3)o(a_4)o(a_5) = 30\,\mathrm{Cyc}\,(a_1a_2|a_3|a_4a_5) + \cdots.$$

Note that for $a \in \mathbf{B}$, the zero-mode action $o(a)$ restricted to $\mathbf{B}$ agrees with multiplication by a. See [79] for details and [67] for a recent application.

These formulae were originally obtained by S. P. Norton for the Griess–Conway algebra $\mathbf{B}^\natural$ in [87] by investigating detailed structures of the algebra (cf. [38]).

The derivation of the formulae by modular invariance as described remarkably shows that such detailed properties of the algebra $\mathbf{B}^\natural$ are revealed as a consequence of general properties of the whole $\mathbf{V}^\natural$ endowed with the structure of a VOA, which is in accordance with the uniqueness of $\mathbf{V}^\natural$ conjectured by I. B. Frenkel et al. in [1] as mentioned in the Introduction.

Bibliography

Textbooks and Monographs

[1] I. B. Frenkel, J. Lepowsky, and A. Meurman: *Vertex operator algebras and the Monster.* Academic Press, Boston, 1988.

[2] A. J. Feingold, I. B. Frenkel, and J. F. X Ries: *Spinor construction of vertex operator algebras, triality, and $E_8^{(1)}$.* Contemporary Mathematics, 121. Amer. Math. Soc., Providence, 1991.

[3] I. B. Frenkel, Y. Huang, and J. Lepowsky: *On axiomatic approaches to vertex operator algebras and modules.* Mem. Amer. Math. Soc. 104, Amer. Math. Soc., Providence, 1993.

[4] C. Y. Dong and J. Lepowsky: *Generalized vertex algebras and relative vertex operators.* Birkhäuser, Boston, 1993.

[5] Y. Z. Huang: *Two-dimensional conformal geometry and vertex operator algebras.* Progr. Math., 148, Birkhäuser, Boston, 1997.

[6] V. G. Kac: *Vertex algebras for beginners.* Second edition. Amer. Math. Soc., Providence, 1998.

[7] A. Matsuo and K. Nagatomo: *Axioms for a vertex algebra and the locality of quantum fields.* MSJ Memoirs 4. Math. Soc. Japan, Tokyo, 1999.

[8] E. Frenkel and D. Ben-Zvi: *Vertex algebras and algebraic curves.* Second edition. Amer. Math. Soc., Providence, 2004.

[9] A. Beilinson and V. Drinfeld: *Chiral algebras.* Amer. Math. Soc. Colloquium Publications 51. Amer. Math. Soc., Providence, 2004.

[10] J. Lepowsky and H. S. Li: *Introduction to vertex operator algebras and their representations.* Progr. Math., 227. Birkhäuser, Boston, 2004.

[11] M. Rosellen: *A course in vertex algebra.* Preprint, arXiv:math/0607270.

Related Books

[12] J. H. Conway and N. J. A. Sloane: *Sphere packings, lattices and groups.* Third edition. Grundlehren der mathematischen Wissenschaften 290. Springer-Verlag, New York, 1999.

[13] P. Di Francesco, P. Mathieu, and D. Sénéchal: *Conformal field theory.* Springer, New York, 1997.

[14] W. Ebeling: *Lattices and codes. A course partially based on lectures by Friedrich Hirzebruch.* Third edition. Advanced Lectures in Mathematics. Springer Spektrum, Wiesbaden, 2013.

[15] K. Iohara and Y. Koga: *Representation theory of the Virasoro algebra.* Springer Monographs in Mathematics. Springer, London, 2011.

[16] A. A. Ivanov: *The Monster group and Majorana involutions.* Cambridge Tracts in Mathematics 176. Cambridge University Press, Cambridge, 2009.

[17] V. G. Kac: *Infinite-dimensional Lie algebras.* Third edition. Cambridge University Press, Cambridge, 1990.

[18] V. G. Kac, A. K. Raina, and N. Rozhkovskaya: *Bombay lectures on highest weight representations of infinite dimensional Lie algebras.* Second edition. Advanced Series in Mathematical Physics 29. World Scientific, Hackensack, 2013.

Lecture Notes and Surveys

[19] P. Ginsparg: Applied conformal field theory. In: *Champs, cordes et phénomènes critiques* (Les Houches, 1988), 1–168, North-Holland, Amsterdam, 1990.

[20] P. Goddard: Meromorphic conformal field theory. In: *Infinite-dimensional Lie algebras and groups*, 556–587, Adv. Ser. Math. Phys., 7, World Scientific, Teaneck, 1989.

[21] G. Mason and M. Tuite: Vertex operators and modular forms. In: *A window into zeta and modular physics*, 183–278, Math. Sci. Res. Inst. Publ., 57, Cambridge University Press, Cambridge, 2010.

[22] M. Tuite: Vertex algebras according to Isaac Newton. *J. Phys. A*, **50**, (2017), no. 41.

[23] J. I. Hall: Transposition algebras. *Bull. Inst. Math. Acad. Sin. (N.S.)*, **14**, (2019), 155–187.

Theses

[24] Y. C. Zhu: Vertex operator algebras, elliptic functions and modular forms. Yale University, 1990.

[25] G. Höhn: Selbstduale Vertexoperatorsuperalgebren und das Babymonster. 1995, Bonn. arXiv:0706.0236.

Articles

[26] T. Abe, G. Buhl, and C. Dong: Rationality, regularity and C_2-cofiniteness. *Trans. Amer. Math. Soc.*, **356**, (2004), 3391–3402.

[27] T. Abe, C.Y. Dong, and H.S. Li: Fusion rules for the vertex operator algebras $M(1)^+$ and V_L^+. *Comm. Math. Phys.*, **253**, (2005), 171–219.

[28] T. Arakawa and K. Kawasetsu: Quasi-lisse vertex algebras and modular linear differential equations. In: *Lie groups, geometry, and representation theory*, 41–57, Progr. Math., 326, Birkhäuser/Springer, Cham, 2018.

[29] B. Bakalov and V. G. Kac: Field algebras. *Int. Math. Res. Not.*, **2003**, 123–159.

[30] B. Bakalov and V. G. Kac: Twisted modules over lattice vertex algebras. In: *Lie theory and its applications in physics V*, 3–26, World Scientific, River Edge, 2004.

[31] A. A. Belavin, A. M. Polyakov, and A. B. Zamolodchikov: Infinite conformal symmetry in two-dimensional quantum field theory. *Nucl. Phys.*, **B241**, (1984), 333–380.

[32] R. B. Borcherds: Vertex algebras, Kac–Moody algebras, and the Monster. *Proc. Natl. Acad. Sci. USA*, **83**, (1986), 3068–3071.

[33] R. E. Borcherds: Monstrous moonshine and monstrous Lie superalgebras. *Invent. Math.*, **109**, (1992), 405–444.

[34] R. E. Borcherds: Vertex algebras. In: *Topological field theory, primitive forms and related topics* (Kyoto, 1996), 35–77, Progr. Math., 160, Birkhäuser, Boston, 1998.

[35] R. E. Borcherds and A. J. E. Ryba: Modular moonshine. II. *Duke Math. J.*, **83**, (1996), 435–459.

[36] S. Carnahan: 51 constructions of the moonshine module. *Comm. Number Theory Phys.*, **12**, (2018), 305–334.

[37] S. Carnahan: Building vertex algebras from parts. *Comm. Math. Phys.*, **373**, (2020), 1–43.

[38] J. H. Conway: A simple construction for the Fischer-Griess monster group. *Invent. Math.*, **79**, (1985), 513–540.

[39] T. De Medts, S. F. Peacock, S. Shpectorov, and M. Van Couwenberghe: Decomposition algebras and axial algebras. *J. Algebra*, **556**, (2020), 287–314.

[40] C. Y. Dong: Vertex algebras associated with even lattices. *J. Algebra*, **160**, (1993), 245–265.

[41] C. Y. Dong: Twisted modules for vertex algebras associated with even lattices *J. Algebra*, **165**, (1994), 91–112.

[42] C. Y. Dong: Representations of the moonshine module vertex operator algebra. In: *Mathematical aspects of conformal and topological field theories and quantum groups* (South Hadley, MA, 1992), 27–36, Contemp. Math., 175, Amer. Math. Soc., Providence, 1994.

[43] C. Y. Dong, R. L. Griess, and G. Höhn: Framed vertex operator algebras, codes and Moonshine module. *Comm. Math. Phys.*, **193**, (1998), 407–448.

[44] C. Y. Dong, C. Jiang, and X. Lin: Rationality of vertex operator algebra V_L^+: higher rank. *Proc. London Math. Soc.*, (3), **104**, (2012), 799–826.

[45] C. Dong and J. Lepowsky: The algebraic structure of relative twisted vertex operators. *J. Pure Appl. Algebra*, **110**, (3), (1996), 259–295.

[46] C. Y. Dong, H. S. Li, and G. Mason: Modular-invariance of trace functions in orbifold theory and generalized moonshine. *Comm. Math. Phys.*, **214**, (1), (2000), 1–56.

[47] C. Y. Dong, H. S. Li, and G. Mason: Regularity of rational vertex operator algebras. *Adv. Math.*, **132**, (1997), 148–166.

[48] C. Y. Dong and G. Mason: Monstrous moonshine of higher weight. *Acta Math.*, **185**, (2000), 101–121.

[49] C. Y. Dong, G. Mason, and Y.C. Zhu: Discrete series of the Virasoro algebra and the moonshine module. In: *Algebraic groups and their generalizations: quantum and infinite-dimensional methods* (University Park, PA, 1991), 295–16, Proc. Sympos. Pure Math., 56, Part 2, Amer. Math. Soc., Providence, 1994.

[50] C. Y. Dong and K. Nagatomo: Automorphism groups and twisted modules for lattice vertex operator algebras. In: *Recent developments in quantum affine algebras and related topics* (Raleigh, NC, 1998), 117–133, Contemp. Math., 248, Amer. Math. Soc., Providence, 1999.

[51] B. Doyon, J. Lepowsky, and A. Milas: Twisted vertex operators and Bernoulli polynomials. *Comm. Contemp. Math.*, **8**, (2006), 247–307.

[52] P. Etingof and D. Kazhdan: Quantization of Lie bialgebras. V. Quantum vertex operator algebras. *Selecta Math.*, (N.S.), **6**, (2000), 105–130.

[53] B. L. Feigin and D. B. Fuchs: Cohomology of some nilpotent subalgebras of the Virasoro and Kac-Moody Lie algebras. *J. Geom. Phys.*, **5**, (1988), 209–235.

[54] A. J. Feingold, J. F. Ries, and M. D. Weiner: Spinor construction of the c=1/2 minimal model. In: *Moonshine, the Monster, and related topics* (South Hadley, MA, 1994), 45–92, Contemp. Math., 193, Amer. Math. Soc., Providence, 1996.

[55] E. Frenkel, V. Kac, A. Radul, and W. Wang: $\mathcal{W}_{1+\infty}$ and $\mathcal{W}(gl_\infty)$ with central charge N. *Comm. Math. Phys.*, **170**, (1995), 337–357

[56] E. Frenkel and M. Szczesny: Twisted modules over vertex algebras on algebraic curves. *Adv. Math.*, **187**, (2004), 195–227.

[57] I. B. Frenkel and V. G. Kac: Basic representations of affine Lie algebras and dual resonance models. *Inventiones Math.*, **62**, (1980), 23–66.

[58] I. B. Frenkel, J. Lepowsky, and A. Meurman: An E_8-approach to F_1. In: *Finite groups—coming of age* (Montreal, Que., 1982), 99–120, Contemp. Math., 45, Amer. Math. Soc., Providence, 1985.

[59] I. B. Frenkel, J. Lepowsky, and A. Meurman: A natural representation of the Fischer-Griess Monster with the modular function J as character. *Proc. Nati. Acad. Sci. USA*, **81**, (1984), 3256–3260.

[60] I. B. Frenkel and Y. C. Zhu: Vertex operator algebras associated to representations of affine and Virasoro algebras. *Duke Math. J.*, **66**, (1992), 123–168.

[61] R. L. Griess, Jr.: The friendly giant. *Invent. Math.*, **69**, (1982), 1–102.

[62] J. I. Hall, F. Rehren, and S. Shpectorov: Universal axial algebras and a theorem of Sakuma. *J. Algebra*, **421**, (2015), 394–424.

[63] J. I. Hall, F. Rehren, and S. Shpectorov: Primitive axial algebras of Jordan type. *J. Algebra*, **437**, (2015), 79–115.

[64] Y. Z. Huang: Differential equations and intertwining operators. *Comm. Contemp. Math.*, **7**, (2005), 375–400.

[65] Y. Z. Huang and J. Lepowsky: A theory of tensor products for module categories for a vertex operator algebra. *Selecta Math.* (N.S.), **1**, (1995), 699–756.

[66] V. G. Kac and D. H. Peterson: Infinite-dimensional Lie algebras, theta functions and modular forms. *Adv. Math.*, **53**, (1984), 125–264

[67] C. H. Lam and H. Yamauchi: The Conway–Miyamoto correspondences for the Fischer 3-transposition groups. *Trans. Amer. Math. Soc.*, **375**, (2022), 2025–2067.

[68] J. Lepowsky: Calculus of twisted vertex operators, *Proc. Natl. Acad. Sci. USA*, **82**, (1985), 8295–8299,

[69] J. Lepowsky and R. L. Wilson: Construction of the affine Lie algebra $A_1^{(1)}$. *Comm. Math. Phys.*, **62**, (1978), 43–53.

[70] H. S. Li: Local systems of vertex operators, vertex superalgebras and modules. *J. Pure Appl. Algebra*, **109**, (1996), 143–195.

[71] H. S. Li: Local systems of twisted vertex operators, vertex operator superalgebras and twisted modules. In: *Moonshine, the Monster, and related topics* (South Hadley, MA, 1994), 203–236, Contemp. Math., 193, Amer. Math. Soc., Providence, 1996.

[72] H. S. Li: An analogue of the Hom functor and a generalized nuclear democracy theorem. *Duke Math. J.*, **93**, (1998), 73–114.

[73] H. S. Li: Some finiteness properties of regular vertex operator algebras. *J. Algebra*, **212**, (1999), 495–514.

[74] H. S. Li: Determining fusion rules by $A(V)$-modules and bimodules. *J. Algebra*, **212**, (1999), 515–556.

[75] H. S. Li: Axiomatic G_1-vertex algebras. *Comm. Contemporary Mathematics*, **5**, (2003), 281–327.

[76] H. S. Li and X. P. Xu: A characterization of vertex algebras associated to even lattices. *J. Algebra*, **173**, (1995), 253–270.

[77] B. H. Lian and G. J. Zuckerman: Some classical and quantum algebras. In: *Lie theory and geometry*, 509–529, Progr. Math., 123, Birkhäuser, Boston, 1994.

[78] G. Mason: Vertex rings and their Pierce bundles. In: *Vertex algebras and geometry*, 45–104, Contemp. Math., 711, Amer. Math. Soc., Providence, 2018.

[79] A. Matsuo: Norton's trace formulae for the Griess algebra of a vertex operator algebra with larger symmetry. *Comm. Math. Phys.*, **224**, (2001), 565–591.

[80] A. Matsuo: 3-transposition groups of symplectic type and vertex operator algebras. *J. Math. Soc. Japan*, **57**, (2005), 639–649.

[81] A. Matsuo and M. Matsuo: The automorphism group of the Hamming code vertex operator algebra. *J. Algebra*, **228**, (2000), 204–226.

[82] A. Matsuo, K. Nagatomo, and A. Tsuchiya: Quasi-finite algebras graded by Hamiltonian and vertex operator algebras. In: *Moonshine: the first quarter century and beyond*, 282–329, London Math. Soc. Lecture Note Ser., 372, Cambridge Univ. Press, Cambridge, 2010.

[83] M. Miyamoto: Griess algebras and conformal vectors in vertex operator algebras. *J. Algebra*, **179**, (1996), 523–548.

[84] M. Miyamoto: Binary codes and vertex operator (super)algebras. *J. Algebra*, **181**, (1996), 207–222.

[85] M. Miyamoto: Vertex operator algebras and modular invariance. In: *Vertex operator algebras, number theory and related topics*, 233–250, Contemp. Math., 753, Amer. Math. Soc., Providence, 2020.

[86] K. Nagatomo and A. Tsuchiya: Conformal field theories associated to regular chiral vertex operator algebras. I. Theories over the projective line. *Duke Math. J.*, **128**, (2005), 393–471.

[87] S. P. Norton: The Monster algebra: some new formulae. In: *Moonshine, the Monster, and related topics* (South Hadley, MA, 1994), 297–306, Contemp. Math., 193, Amer. Math. Soc., Providence, 1996.

[88] M. Primc: Vertex algebras generated by Lie algebras. *J. Pure Appl. Algebra*, **135**, (1999), 253–293.

[89] M. Roitman: On twisted representations of vertex algebras. *Adv. Math.*, **176**, (2003), 53–88.

[90] S. Sakuma: 6-transposition property of τ-involutions of vertex operator algebras. *Int. Math. Res. Not.*, **2007**, Art. ID rnm 030, 19 pp.

[91] A. N. Schellekens: Meromorphic $c = 24$ conformal field theories. *Comm. Math. Phys.*, **153**, (1993), 159–185.

[92] G. Segal: Unitary representations of some infinite-dimensional groups. *Comm. Math. Phys.*, **80**, (1981), 301–342.

[93] H. Shimakura: Decompositions of the Moonshine module with respect to sub-VOAs associated to codes over $\mathbb{Z}_{2k}$. *J. Algebra*, **251**, (2002), 308–322.

[94] H. Shimakura: The automorphism groups of the vertex operator algebras V_L^+: general case. *Math. Z.*, **252**, (2006), 849–862.

[95] H. Shimakura: An E_8-approach to the moonshine vertex operator algebra *J. London Math. Soc.*, (2) **83**, (2011), 493–516.

[96] A. Tsuchiya and Y. Kanie: Vertex operators in conformal field theory on $\mathbf{P}^1$ and monodromy representations of braid group. In: *Conformal field theory and solvable lattice models* (Kyoto, 1986), 297–372, Adv. Stud. Pure Math., 16, Academic Press, Boston, 1988.

[97] A. Tsuchiya, K. Ueno, and Y. Yamada: Conformal field theory on universal family of stable curves with gauge symmetries. In: *Integrable systems in quantum field theory and statistical mechanics*, 459–566, Adv. Stud. Pure Math., 19, Academic Press, Boston, 1989.

[98] J. van Ekeren, C. H. Lam, S. Möller, and H. Shimakura: Schellekens' list and the very strange formula. *Adv. Math.*, **380**, (2021).

[99] W. Q. Wang: Rationality of Virasoro vertex operator algebras. *Int. Math. Res. Not.*, **1993**, (1993), 197–211.

[100] Y. C. Zhu: Modular invariance of characters of vertex operator algebras. *J. Amer. Math. Soc.*, **9**, (1996), 237–302.

2

3-Transposition Groups Arising in Vertex Operator Algebra Theory

Hiroshi Yamauchi[*]

Abstract

We review 3-transposition groups arising in vertex operator algebra (VOA) theory. One can construct a commutative algebra called the Matsuo algebra out of a 3-transposition group. Some 3-transposition groups arise as automorphism groups of vertex operator algebras via Matsuo algebras but there exist some 3-transposition groups that do not arise through Matsuo algebras. We will exhibit examples of those groups together with VOAs.

CONTENTS

* Partially supported by Grant-in-Aid for Scientific Research (C) 19K03409

2.1 Introduction

An *n-transposition group* is a pair (G, I) of a group G and its subset I of involutions such that I is a union of conjugacy classes of G and the product of any two involutions in I has order at most n. Three-transposition groups are first considered by Fischer [Fi71] and three sporadic groups called Fischer groups are discovered on the way of classification of finite 3-transposition groups. Cuypers and Hall [CH95] have investigated the classification program further.

Given a 3-transposition group (G, I), one can define a graph structure on I by adjoining two distinct involutions in I if and only if their product has order 3. The permutation representation of G on the graph I has rank 3 under the assumptions that $Z(G) = O_2(G) = O_3(G) = 1$ and $G' = G''$ (cf. [As97, CH95, Fi71]). The study of center-free 3-transposition groups is actually equivalent to that of the graph structures defined as above. As a linearization of the graph structure, one can define a commutative algebra called the Matsuo algebra associated with a 3-transposition group which is linearly spanned by idempotents indexed by transpositions (cf. Definition 3.1). Therefore, if we realize the Matsuo algebra associated with a 3-transposition group as a substructure of some algebraic structures, then we may obtain a natural action of the 3-transposition group on the algebraic structures under consideration. This is one way 3-transposition groups arise in the theory of vertex operator algebras.

A vertex operator algebra (VOA) is an infinite dimensional vector space equipped with infinitely many bilinear operations indexed by integers satisfying a certain set of axioms (cf. Definition 4.1). Given a VOA V and its sub VOA U, one can decompose V as a U-module and, if U is rational, then V is a direct sum of irreducible U-submodules. In this case, fusion rules among U-modules constrain binary operations in V so that if the fusion rules among irreducible U-modules have a cyclic symmetry then this symmetry gives rise to define an automorphism of V through its decomposition as a U-module. This is a typical way to obtain automorphisms of VOAs and in practice U is often chosen to be a subalgebra with a small number of generators such as a Virasoro VOA or a W_3-algebra, since they are easy to handle. If U is a simple Virasoro VOA in the BPZ series [BPZ84] then its fusion rules have a $\mathbb{Z}_2$-symmetry and the involutive automorphism defined by U is called the *Miyamoto involution* (cf. [Mi96]). The most fundamental example is the Virasoro VOA with central charge $1/2$. A Virasoro vector is called an *Ising vector* if it generates a simple Virasoro sub VOA with central charge $1/2$. The sub VOAs generated by two Ising vectors are studied in [S07] and it is shown that the product of Miyamoto involutions associated with two Ising vectors is bounded by 6, which is nowadays known as Sakuma's theorem. Thus in general we cannot directly obtain 3-transposition groups if we consider automorphism groups generated by Miyamoto involutions associated with Ising vectors.

This article is an exposition of 3-transposition groups arising in VOA theory. We will mainly consider VOAs of OZ-type over $\mathbb{R}$ with positive definite invariant bilinear forms. The degree-2 subspace of a VOA of OZ-type carries a structure of a commutative but in general nonassociative algebra called the Griess algebra with symmetric invariant bilinear form (cf. [FLM88, G82]). Let V be a VOA of OZ-type over $\mathbb{R}$. It is known that if $x \in V_2$ is an idempotent of the Griess algebra of V then $2x$ is a Virasoro vector of V with central charge $8(x \mid x)$ (cf. [Mi96]). Sakuma's theorem [S07] describes subalgebras of the Griess algebra generated by two Ising vectors that correspond to idempotents with squared norm $1/16$. The notions of Majorana algebras [I09] and axial algebras [HRS15] are introduced by axiomatizing the role of Ising vectors in the Griess algebra. Recall that a Matsuo algebra is a commutative algebra linearly spanned by idempotents of fixed squared norm. Those idempotents are called axial vectors. If the Griess algebra V_2 contains a subalgebra isomorphic to (the nondegenerate quotient of) a Matsuo algebra and each axial vector generates a rational Virasoro sub VOA, then Miyamoto involutions associated with axial vectors generate a 3-transposition group. In this chapter we will review some families of Matsuo algebras that arise as Griess algebras of VOAs and related topics like the classification problems and the uniqueness problems. On the other hand, there are constructions of 3-transposition groups as automorphism groups of VOAs based on Ising vectors that do not arise from Matsuo algebras. We will also review such examples.

The organization of this chapter is as follows. In Section 2.2, we will review 3-transposition groups based on [CH95, K00]. In Section 2.3, we will review Matsuo algebras based on [Ma05]. Note that the published version of [Ma05] and the one posted to the arXiv are different, and we will refer to the arXiv version in this chapter. In Section 2.4, we will review the notion of vertex operator algebras and Miyamoto involutions. Section 2.5 is the main body of this article and we will review several constructions of 3-transposition groups in VOA theory, except for Section 2.5.5 where a construction of 4-transposition groups will be explained. We will close this chapter by proposing some open questions in Section 2.6.

2.2 3-Transposition Groups

Recall the notion of 3-transposition groups (cf. [As97, CH95, Fi71, K00]).

Definition 2.1 A *3-transposition group* is a pair (G, I) of a group G and its subset I of involutions called *transpositions* such that I is closed under the conjugation and the product of any two involutions in I has order at most 3. A subgroup of G generated by a subset of I is called an *I-subgroup*, which is, again, a 3-transposition group.

In this definition, some authors indeed require I to be a single conjugacy class of G. To include some subgroups such as $\mathfrak{S}_k \times \mathfrak{S}_{n-k}$ of $\mathfrak{S}_n$, we admit I to be a union of conjugacy classes.

Example 2.2   Here are some typical examples of 3-transposition groups.

(1) The symmetric group $G = \mathfrak{S}_n$ with $I = \{(i\ j) \mid 1 \le i < j \le n\}$.
(2) Weyl groups of ADE types.
(3) Orthogonal groups over $\mathbb{F}_2$ where the 3-transpositions are the transvections.
(4) Symplectic groups over $\mathbb{F}_2$ where the 3-transpositions are the transvections.
(5) Orthogonal groups over $\mathbb{F}_3$ where the 3-transpositions are the reflections.
(6) Unitary groups over $\mathbb{F}_4$ where the 3-transpositions are the transvections.

We have the following isomorphisms between several 3-transposition groups with different base fields:

$$U_4(2) \cong O_5(3) < O_5(3){:}2 \cong O_6^-(2){:}2 \cong W(E_6),\ \ S_6(2) \cong W(E_7)/\{\pm 1\},$$

$$O_8^+(2) \cong W(E_8)/\{\pm 1\},\ \ O_6^+(2){:}2 \cong \mathfrak{S}_8,\ \ O_4^-(3){:}2 \cong S_4(2) \cong \mathfrak{S}_6,$$
$$O_4^-(2){:}2 \cong \mathfrak{S}_5.$$

Here we use the notation as in [ATLAS] for simple groups and $W(X)$ denotes the Weyl group of type X.

The study of 3-transposition groups was initiated by Fischer and, under certain assumptions, he classified finite 3-transposition groups up to centers as follows.

Theorem 2.3 ([CH95, Fi71])   *Let G be a finite 3-transposition group generated by a single conjugacy class of 3-transpositions. If $O_2(G)O_3(G) < Z(G)$ then up to the center, G is isomorphic to one of the following:*

(1) *Symmetric groups $\mathfrak{S}_n$, $n \ge 5$.*
(2) *Symplectic groups $S_{2n}(2)$, $n \ge 2$.*
(3) *Orthogonal groups $O_{2n}^\varepsilon(2){:}2$ where $n \ge 3$ if $\varepsilon = +$ and $n \ge 2$ if $\varepsilon = -$.*
(4) *Orthogonal groups $O_{2n}^\varepsilon(3){:}2$ where $n \ge 3$ if $\varepsilon = +$ and $n \ge 2$ if $\varepsilon = -$.*
(5) *Orthogonal groups $O_{2n+1}^\pm(3)$ or $O_{2n+1}^\pm(3){:}2$ with $n \ge 2$ and $\varepsilon = \pm$.*
(6) *Unitary groups $U_n(2)$, $n \ge 4$.*
(7) *One of 5 exceptionals: Fi_{22}, Fi_{23}, Fi_{24}, $O_8^+(2){:}\mathfrak{S}_3$, or $O_8^+(3){:}\mathfrak{S}_3$.*

Cupyers and Hall classified the structure $G/Z(G)$ of a 3-transposition group G without the assumptions that G is finite and $O_2(G)O_3(G) < Z(G)$ in [CH95]. In their classification, 3-transposition groups are divided into four classes as follows. Set

$$H := \langle a, b, c \mid a^2 = b^2 = c^2 = (ab)^3 = (bc)^3 = (ca)^3 = (abac)^3 = 1 \rangle. \quad (2.1)$$

Then $Z(H) = \langle (abc)^2 \rangle \cong 3$ and $H \cong 3^{1+2}{:}2$. The group H is denoted by $\mathrm{SU}_3(2)'$ in [CH95]. A 3-transposition group (G, I) is called of *symplectic type* if G does not contain an I-subgroup isomorphic to H, and G is of *orthogonal type* if it contains an I-subgroup isomorphic to H but no I-subgroup isomorphic to $2^{1+6}{:}H$, and G is of *unitary type*, if it contains an I-subgroup isomorphic to $2^{1+6}{:}H$ but no $\mathrm{P}\Omega_8^+(2){:}\mathfrak{S}_3$, and G is of *sporadic type* if G does contain an I-subgroup isomorphic to $\mathrm{P}\Omega_8^+(2){:}\mathfrak{S}_3$ (cf. [CH]). Among the groups in Theorem 2.3, those in (1), (2) and (3) are of symplectic type, those in (4) and (5) are of orthogonal type, those in (6) are of unitary type and those in (7) are of sporadic type. Note that $O_{2n}^{\mp}(2) < S_{2n}(2) < O_{2(n+1)}^{\pm}(2)$, $\mathfrak{S}_{2n+1} < S_{2n}(2)$, and $\mathfrak{S}_{2n+2} < S_{2n}(2)$ so that the groups in (1), (2), (3) of Theorem 2.3 are called of symplectic type.

Remark 2.4 In the definition of H in (2.1), if we replace the relation $(abac)^3 = 1$ by $abac = 1$, then we obtain $\mathfrak{S}_3$, and if we replace the relation $(abac)^3 = 1$ by $(abac)^2 = 1$, then we obtain $\mathfrak{S}_4$.

2.3 Matsuo Algebras

Let (G, I) be a 3-transposition group. We define a binary relation on I such that $a \sim b$ for $a, b \in I$ if and only if ab has order 3; that is, we have $a^b = b^a$ in I, and in this case we set $a \circ b := a^b = b^a \in I$. This binary relation amounts to defining an irreflexive undirected graph structure on I. We say (G, I) (or just G) is *connected* or *indecomposable* if I is connected with respect to the adjacency relation defined by $a \sim b$. This is equivalent to saying that I is a single conjugacy class of involutions.

Based on the graph structure on I, we define the Matsuo algebra associated with a 3-transposition group as follows.

Definition 3.1 ([Ma05]) For $\alpha, \beta \in \mathbb{R}$, set $B_{\alpha,\beta}(G) := \oplus_{i \in I} \mathbb{R}x^i$ where $\{x^i \mid i \in I\}$ is a formal basis and define the product and the bilinear form as follows.

$$x^i x^j := \begin{cases} 2x^i & \text{if } i = j, \\ \dfrac{\alpha}{2}(x^i + x^j - x^{i \circ j}) & \text{if } i \sim j, \\ 0 & \text{otherwise,} \end{cases} \qquad (x^i \mid x^j) := \begin{cases} \dfrac{\beta}{2} & \text{if } i = j, \\ \dfrac{\alpha\beta}{8} & \text{if } i \sim j, \\ 0 & \text{otherwise.} \end{cases}$$

(3.1)

Then $B_{\alpha,\beta}(G)$ becomes a commutative algebra with symmetric invariant bilinear form. We call $B_{\alpha,\beta}(G)$ the *Matsuo algebra* associated with (G, I), and we also call each x^i with $i \in I$ an *axial vector*. By the invariance, the radical of the bilinear form becomes an ideal of $B_{\alpha,\beta}(G)$. We call the quotient algebra of $B_{\alpha,\beta}(G)$ by the radical of the bilinear form the *nondegenerate quotient*.

Remark 3.2 By definition, $x^i/2$ is an idempotent of $B_{\alpha,\beta}(G)$ so that a Matsuo algebra is linearly spanned by idempotents.

Unity Suppose the set I of 3-transpositions is not a single conjugacy class, for example, $I = I_1 \sqcup I_2$ where I_1 and I_2 are disconnected with respect to the adjacency relation. Then G is isomorphic to a central product $G_1 * G_2$ of $G_1 = \langle I_1 \rangle$ and $G_2 = \langle I_2 \rangle$. Correspondingly, we have a decomposition into a direct sum

$$B_{\alpha,\beta}(G) = B_{\alpha,\beta}(G_1) \oplus B_{\alpha,\beta}(G_2)$$

of ideals, which is also an orthogonal sum. Therefore, in the study of the Matsuo algebra $B_{\alpha,\beta}(G)$, we may assume that G is indecomposable; that is, G is generated by a single conjugacy class of 3-transpositions. In particular, $B_{\alpha,\beta}(G)$ is unital if and only if the Matsuo algebra associated with each indecomposable component of G has the unity.

Suppose G is indecomposable. Then $k := \#\{j \in I \mid i \sim j\}$ is independent of $i \in I$ and the vector

$$\omega := \frac{4}{k\alpha + 4} \sum_{i \in I} x^i \tag{3.2}$$

satisfies $\omega x^i = 2x^i$ and $(\omega \mid x^i) = (x^i \mid x^i) = \beta/2$ for all $i \in I$ so that $\omega/2$ gives the unity of $B_{\alpha,\beta}(G)$ if $k\alpha + 4 \neq 0$.

Remark 3.3 When the Matsuo algebra $B_{\alpha,\beta}(G)$ (or its quotient) is realized by a VOA as its Griess algebra, then twice the unity of $B_{\alpha,\beta}(G)$ corresponds to the conformal vector of the VOA. In this case, each axis x^i gives a Virasoro vector with central charge β.

Involutions It is obvious that G acts on $B_{\alpha,\beta}(G)$ by conjugation. Namely, there exists a unique group homomorphism

$$
\begin{aligned}
\sigma: \quad G \quad &\longrightarrow \quad \operatorname{Aut} B_{\alpha,\beta}(G) \\
a \quad &\longmapsto \quad \sigma_a: x^i \mapsto x^{aia^{-1}},
\end{aligned} \tag{3.3}
$$

such that $\sigma_i(x^j) = x^{i\circ j}$ if $i \sim j$ and $\sigma_i(x^j) = x^j$ otherwise for $i, j \in I$. Clearly $\operatorname{Ker} \sigma = Z(G)$ and σ is injective if G is center-free.

Suppose $\alpha \neq 2$. Then the adjoint action $\operatorname{ad} x^i$ of x^i is semisimple and has three eigenvalues, 0, 2, and α, on $B_{\alpha,\beta}(G)$. Clearly x^i is an eigenvector of $\operatorname{ad} x^i$ with eigenvalue 2. Let $j \in I$ be distinct from i. If $i \nsim j$, then $(\operatorname{ad} x^i) x^j = x^i x^j = 0$ so that x^j belongs to the eigenspace of eigenvalue 0. If $i \sim j$, then $i \circ (i \circ j) - j$ and one has

$$
x^i(x^j + x^{i\circ j}) = \alpha x^i, \quad x^i(x^j - x^{i\circ j}) = \alpha(x^j - x^{i\circ j}),
$$

from which it follows that $\operatorname{ad} x^i$ is semisimple and has eigenvalues, 0, 2, and α. One can also verify that $\operatorname{Ker}(\operatorname{ad} x^i - 2) = \mathbb{R} x^i$ and σ_i acts by the identity on $\operatorname{Ker} \operatorname{ad} x^i$ and by -1 on $\operatorname{Ker}(\operatorname{ad} x^i - \alpha)$. Therefore, we have a decomposition of $B_{\alpha,\beta}(G)$ such that

$$
\begin{array}{ccccccc}
B_{\alpha,\beta}(G) & = & \mathbb{R} x^i & \oplus & \operatorname{Ker}(\operatorname{ad} x^i) & \oplus & \operatorname{Ker}(\operatorname{ad} x^i - \alpha), \\
\operatorname{ad} x^i & : & 2 & & 0 & & \alpha, \\
\sigma_i & : & 1 & & 1 & & -1.
\end{array} \tag{3.4}
$$

If we set $B^{i,\pm}_{\alpha,\beta}(G) := \{v \in B_{\alpha,\beta}(G) \mid \sigma_i(v) = \pm v\}$ then we have $B_{\alpha,\beta}(G) = B^{i,+}_{\alpha,\beta}(G) \oplus B^{i,-}_{\alpha,\beta}(G)$ with

$$
B^{i,+}_{\alpha,\beta}(G) = \mathbb{R} x^i \oplus \operatorname{Ker}(\operatorname{ad} x^i), \quad B^{i,-}_{\alpha,\beta}(G) = \operatorname{Ker}(\operatorname{ad} x^i - \alpha).
$$

Therefore, the action of G on $B_{\alpha,\beta}(G)$ can be described by the adjoint action of $B_{\alpha,\beta}(G)$. The involution σ_i defined as in (3.4) is called the *Miyamoto involution* associated with the axial vector x^i of $B_{\alpha,\beta}(G)$.

2.4 Vertex Operator Algebras

Recall the notion of vertex operator algebras (VOAs).

Definition 4.1 A vertex operator algebra is a $\mathbb{Z}_{\geq 0}$-graded vector space $V = \bigoplus_{n \geq 0} V_n$ over $\mathbb{C}$ equipped with bilinear products $V \otimes V \ni a \otimes b \mapsto a_{(n)} b \in V$ indexed by $n \in \mathbb{Z}$ satisfying the following axioms.

(1) For any $a, b \in V$, there exists $N \in \mathbb{Z}$ such that $a_{(i)} b = 0$ for $i \geq N$.

(2) There exists a special element $\mathbb{1} \in V_0$ called the *vacuum vector* of V such that for any $a \in V$, one has $a_{(n)}\mathbb{1} = \delta_{n,-1}a$ for $n \geq -1$.

(3) There exists a special element $\omega \in V_2$ called the *conformal vector* of V such that if we set $L(n) = \omega_{(n+1)} \in \mathrm{End}(V)$ for $n \in \mathbb{Z}$, then we have

$$[L(m), L(n)] = (m-n)L(m+n) + \delta_{m+n,0}\frac{m^3-m}{12}c_V, \qquad (4.1)$$

where $c_V \in \mathbb{C}$ is called the *central charge* of V. We also have

$$[L(-1), a_{(m)}] = -m\,a_{(m-1)}$$

for any $a \in V$ and $m \in \mathbb{Z}$ and $V_n = \mathrm{Ker}\,(L(0) - n)$ with $\dim V_n < \infty$.

(4) The following Borcherds identity holds for any $a, b, c \in V$ and $k, m, n \in \mathbb{Z}$:

$$\sum_{i=0}^{\infty}\binom{m}{i}\left(a_{(k+i)}b\right)_{(m+n-i)}c = \sum_{i=0}^{\infty}(-1)^i\binom{k}{i}\left\{a_{(m+k-i)}b_{(n+i)}\right.$$
$$\left. -(-1)^k b_{(n+k-i)}a_{(m+i)}\right\}c.$$

There are various equivalent but apparently different formulations of the vertex operator algebra. The set of axioms here is an apparently minimum one (cf. [MaN99]). We refer the readers to [FHL93, FLM88, Ka90, LL04, MaN99] for more details.

If $a \in V_n$ is homogeneous with respect to the grading of V, then we define $\mathrm{wt}(a) := n$ and call it the *weight* of a. It follows from our axioms that for homogeneous $a, b \in V$, the product $a_{(n)}b$ is also homogeneous with weight $\mathrm{wt}(a) + \mathrm{wt}(b) - n - 1$ for $n \in \mathbb{Z}$. In particular, the *zero-mode operator* $\mathrm{o}(a) = a_{(\mathrm{wt}(a)-1)} \in \mathrm{End}(V)$ preserves each graded subspace of V.

Griess Algebra In this chapter, we will mainly consider VOAs of OZ-type defined over $\mathbb{R}$. A VOA $V = \bigoplus_{n\geq 0} V_n$ over $\mathbb{R}$ is called of *OZ-type*[*] if $V_0 = \mathbb{R}\mathbb{1}$ and $V_1 = 0$. In this case, we can define a product and a bilinear form on the degree-2 subspace V_2 as follows. Let $a, b \in V_2$. We define

$$ab := \mathrm{o}(a)b = a_{(1)}b \in V_2, \qquad (a\,|\,b)\mathbb{1} := a_{(3)}b \in V_0 = \mathbb{R}\mathbb{1}. \qquad (4.2)$$

Then V_2 forms a commutative algebra with symmetric invariant bilinear form (cf. proposition 8.2.1 of [MaN99]; see also section 8.9 of [FLM88]). The algebra V_2 is called the *Griess algebra* of V.

---

[*] OZ stands for "One-Zero." This notion is introduced by Bob Griess.

Remark 4.2 A VOA of OZ-type has a unique symmetric invariant bilinear form defined by

$$(\mathbb{1}\,|\,\mathbb{1}) := 1 \quad \text{and} \quad (a\,|\,b) := \sum_{i \geq 0} \frac{(-1)^{\mathrm{wt}(a)}}{i!} (\mathbb{1}\,|\,\underbrace{(L(1)^i a)_{(2\mathrm{wt}(a)-i-1)}b}_{\in V_0 = \mathbb{R}\mathbb{1}}).$$

For details, see [FHL93, L96]. The bilinear form on V_2 defined in (4.2) is a restriction of the form above to the weight-2 subspace of V.

In the rest of the chapter, we always assume that **our VOA V is defined over $\mathbb{R}$, of OZ-type, and the invariant bilinear form on V is positive definite** unless otherwise stated. Due to the positivity, it follows that every sub VOA of V is always simple.

A vector $a \in V_2$ is called a *Virasoro vector* with central charge $c_a \in \mathbb{R}$ if the modes $L^a(n) = a_{(n+1)} \in \mathrm{End}(V)$, $n \in \mathbb{Z}$, satisfy the commutation relation

$$[L^a(m), L^a(n)] = (m-n)L^a(m+n) + \delta_{m+n,0}\frac{m^3-m}{12}c_a \qquad (4.3)$$

of the Virasoro algebra. By (4.1), the conformal vector of V is a Virasoro vector (but not vice versa, in general). For a VOA of OZ-type, Virasoro vectors corresponds to idempotents of the Griess algebra.

Lemma 4.3 ([Mi96]) *A vector $a \in V_2$ is an idempotent of the Griess algebra of V if and only if $2a$ is a Virasoro vector with central charge $8(a\,|\,a)$.*

A sub VOA $\langle a \rangle$ generated by a Virasoro vector $a \in V_2$ is called a *Virasoro VOA*. Thanks to the assumption of the positivity, a Virasoro sub VOA is always simple in our setting.

Unitary series Recall the unitary series of the Virasoro algebra.

$$c_m := 1 - \frac{6}{(m+2)(m+3)} \quad (m = 1, 2, 3, \dots)$$

$$h_{r,s}^{(m)} := \frac{(r(m+3) - s(m+2))^2 - 1}{4(m+2)(m+3)}, \quad 1 \leq r \leq m+1, \ 1 \leq s \leq m+2$$

$$\hspace{12cm} (4.4)$$

In this chapter, $L(c_m, h_{r,s}^{(m)})$ denotes the irreducible highest weight modules over the Virasoro algebra with central charge c_m and highest weight $h_{r,s}^{(m)}$. It was shown in [FZ92] that $L(c_m, 0)$ forms a simple and rational Virasoro VOA and its irreducible $L(c_m, 0)$-modules are provided by highest weight representations $L(c_m, h_{r,s}^{(m)})$, $1 \leq s \leq r \leq m+1$. Note that we have redundancy

$h_{r,s}^{(m)} = h_{m+2-r,m+3-s}^{(m)}$. The fusion rules among irreducible $L(c_m, 0)$-modules are computed in [W93] as follows:

$$L(c_m, h_{r,s}^{(m)}) \underset{L(c_m,0)}{\boxtimes} L(c_m, h_{r',s'}^{(m)}) = \bigoplus_{\substack{1 \leq i \leq I \\ 1 \leq j \leq J}} L(c_m, h_{|r-r'|+2i-1, |s-s'|+2j-1}), \quad (4.5)$$

where $I = \min\{r, r', m+2-r, m+2-r'\}$ and $J = \min\{s, s', m+3-s, m+3-s'\}$.

Miyamoto Involutions Let $x \in V$ be a Virasoro vector with central charge c_m. Since $\langle x \rangle \cong L(c_m, 0)$ is rational, we have the isotypical decomposition

$$V = \bigoplus_{1 \leq s \leq r \leq m+1} V[x; h_{r,s}^{(m)}], \quad (4.6)$$

$$V[x; h_{r,s}^{(m)}] \cong L(c_m, h_{r,s}^{(m)}) \otimes \mathrm{Hom}_{\langle x \rangle}(L(c_m, h_{r,s}^{(m)}, h_{r,s}^{(m)}), V).$$

Based on the decomposition (4.6), we can define a linear automorphism τ_x of V as follows:

$$\tau_x := (-1)^{4(m+2)(m+3)\mathrm{o}(x)} = \begin{cases} (-1)^{r+1} & \text{on } V[x; h_{r,s}^{(m)}] & \text{if } m \text{ is even,} \\ (-1)^{s+1} & \text{on } V[x; h_{r,s}^{(m)}] & \text{if } m \text{ is odd.} \end{cases} \quad (4.7)$$

Then it follows from the fusion rules in (4.5) that τ_x preserve the VOA structure.

Theorem 4.4 ([Mi96]) $\tau_x \in \mathrm{Aut}\, V$.

For instance, we consider the case $m = 1$ in the unitary series (4.4). Suppose $x \in V$ is a Virasoro vector with central charge $c_1 = 1/2$. Then $\langle x \rangle \cong L(1/2, 0)$ have three irreducible representations $L(1/2, 0)$, $L(1/2, 1/2)$, and $L(1/2, 1/16)$, so that we have the following isotypical decomposition as in (4.6):

$$\begin{array}{ccccccc} V & = & V[x; 0] & \oplus & V[x; 1/2] & \oplus & V[x; 1/16] \\ \tau_x & : & 1 & & 1 & & -1. \end{array} \quad (4.8)$$

Since $V_2[x; 0] = \mathbb{R}x \oplus \mathrm{Ker}_{V_2}\mathrm{o}(x)$ and $V_2[x; h] = \mathrm{Ker}_{V_2}(\mathrm{o}(x) - h)$ for $h = 1/2$ and $1/16$, the corresponding decomposition of the Griess algebra looks as follows:

V_2	$=$	$\mathbb{R}x$	$\oplus$	$\mathrm{Ker}_{V_2}o(x)$	$\oplus$	$\mathrm{Ker}_{V_2}(o(x)-1/2)$	$\oplus$	$\mathrm{Ker}_{V_2}(o(x)-1/16)$
$o(x)$	:	2		0		$1/2$		$1/16$
τ_x	:	1		1		1		$-1.$

$$(4.9)$$

By (4.8), we see that the fixed point subalgebra $V^{\langle \tau_x \rangle}$ has a decomposition as follows:

$$
\begin{aligned}
V^{\langle \tau_x \rangle} &= V[x;0] \oplus V[x;1/2] \\
\sigma_x \quad : &\quad 1 \qquad\qquad -1
\end{aligned}
$$
$$(4.10)$$

If we define the linear automorphism σ_x as in (4.10), then again by the fusion rules in (4.5) it follows that $\sigma_x \in \mathrm{Aut}\, V^{\langle \tau_x \rangle}$ (cf. [Mi96]). By (4.9), the action of σ_x on the Griess algebra looks as follows:

V_2	$=$	$\mathbb{R}x$	$\oplus$	$\mathrm{Ker}_{V_2}o(x)$	$\oplus$	$\mathrm{Ker}_{V_2}(o(x)-1/2)$
$o(x)$	:	2		0		$1/2$
σ_x	:	1		1		-1

$$(4.11)$$

We can generalize the construction of the second automorphism as follows.

Let x be a $c = c_m$ Virasoro vector of V as before. Set

$$
P_m := \begin{cases} \{h_{1,s}^{(m)} \mid 1 \leq s \leq m+2\} & \text{if } m \text{ is even,} \\ \{h_{r,1}^{(m)} \mid 1 \leq r \leq m+1\} & \text{if } m \text{ is odd,} \end{cases}
$$
$$(4.12)$$

and define $V[x;P_m] := \bigoplus_{h \in P_m} V[x;h]$. Then it follows from the fusion rules of $L(c_m,0)$-modules in (4.5) that the subspace $V[x;P_m]$ forms a subalgebra of the fixed point subalgebra $V^{\langle \tau_x \rangle}$. We define a linear automorphism σ_x of $V[x;P_m]$ as follows:

$$
\sigma_x := \begin{cases} (-1)^{s+1} \text{ on } V[x;h_{1,s}^{(m)}] & \text{if } m \text{ is even,} \\ (-1)^{r+1} \text{ on } V[x;h_{r,1}^{(m)}] & \text{if } m \text{ is odd.} \end{cases}
$$
$$(4.13)$$

Theorem 4.5 ([Mi96]) $\quad \sigma_x \in \mathrm{Aut}\, V[x;P_m]$.

The involutions $\tau_x \in \mathrm{Aut}\, V$ and $\sigma_x \in \mathrm{Aut}\, V[x;P_m]$ are called *Miyamoto involutions*. Note that τ_x is always well-defined on the whole space V whereas σ_x is locally defined on the subspace $V[x;P_m]$. We call a $c = c_m$ Virasoro vector $x \in V$ of σ-*type on V* if $V = V[x;P_m]$. Note that if x is of σ-type, then $V = V^{\langle \tau_x \rangle}$ so that τ_x is trivial on V and σ_x is globally defined on V.

Sakuma's theorem A $c = 1/2$ Virasoro vector $x \in V$ is called an *Ising vector* of V. The structures of a subalgebra generated by two Ising vectors as well as the dihedral groups generated by associated Miyamoto involutions are almost classified by Sakuma.

Theorem 4.6 ([S07]) *Let e and f be Ising vectors of V. Then the Griess algebra of the subalgebra $\langle e, f \rangle$ has nine possible structures and it follows that $|\tau_e \tau_f| \leq 6$ on V.*

Type of $\langle e, f \rangle$	1A	2A	3A	4A	5A	6A	4B	2B	3C
max $|\tau_e \tau_f|$	1	2	3	4	5	6	4	2	3
$2^{10}(e \mid f)$	2^8	2^5	13	2^3	6	5	2^2	0	2^2
dim$\langle e, f \rangle_2$	1	3	4	5	6	8	5	2	3
# of Ising vectors	1	3	3	4	5	7	5	2	3
Miyamoto type	σ	σ	τ	τ	τ	τ	τ	σ	τ

In this table, max $|\tau_e \tau_f|$ denotes the possible maximum order of $\tau_e \tau_f$ on V and Miyamoto type denotes the type of Ising vectors e and f on the subalgebra $\langle e, f \rangle$.

If $\langle e, f \rangle$ is the subalgebra of type nX in the first row of the table, then we will call $\langle e, f \rangle$, which is also denoted by U_{nX}, the *dihedral subalgebra* of type nX or just the nX-*algebra*. We refer readers to [LYY05] for more information about dihedral subalgebras. In this chapter, we mainly treat dihedral subalgebras of types 2A and 3A. Here we list some properties of the 2A- and 3A-algebras.

Lemma 4.7 *Let e and f be Ising vectors of a VOA V.*

(1) $\langle e, f \rangle$ *is a 2A-algebra if and only if $(e \mid f) = 2^{-5}$.*
(2) $\langle e, f \rangle$ *is a 3A-algebra if and only if $(e \mid f) = 13 \cdot 2^{-10}$.*

Theorem 4.8 ([Mi96]) *Let $\langle e, f \rangle \cong U_{2A}$ be a 2A-subalgebra of a VOA V.*

(1) U_{2A} *has three Ising vectors e, f and $\sigma_e f = \sigma_f e$, all of them are of σ-type on U_{2A}.*
(2) $\tau_e \tau_f = \tau_{\sigma_e f}$ *on V.*
(3) $\langle \tau_e, \tau_f \rangle$ *is an elementary abelian 2-group of order at most 4 (and possibly trivial).*
(4) $|\sigma_e \sigma_f| = 3$ *on $\langle e, f \rangle$ and Aut $U_{2A} = \langle \sigma_e, \sigma_f \rangle \cong \mathfrak{S}_3$.*

Theorem 4.9 ([Mi03, SY03])   *Let $\langle e, f \rangle \cong U_{3A}$ be a 3A-subalgebra of a VOA V.*

(1) *U_{3A} has 3 Ising vectors e, f and $\tau_e f = \tau_f e$.*
(2) *$|\tau_e \tau_f| = 3$ and Aut $U_{3A} = \langle \tau_e, \tau_f \rangle \cong \mathfrak{S}_3$.*
(3) *The 3A-algebra U_{3A} has a full subalgebra isomorphic to $L(4/5, 0) \otimes L(6/7, 0)$ where the central charges $c_3 = 4/5$ and $c_4 = 6/7$ are in the unitary series (4.4).*

2.5 3-Transposition Groups and VOAs

In this section, we review some constructions of 3-transposition groups as automorphisms of vertex operator algebras.

2.5.1 3-Transposition Groups and Ising Vectors

Let E_V be the set of Ising vectors of a VOA V of OZ-type and $E_V(\sigma)$ the subset of E_V consisting of Ising vectors of σ-type on V.

Proposition 5.1 ([Mi96])   *If e, $f \in E_V(\sigma)$ then the type of the dihedral subalgebra $\langle e, f \rangle$ is either 1A, 2A or 2B.*

Let e, $f \in E_V(\sigma)$. Then their product in the Griess algebra looks as follows (cf. [Mi96, Ma05]):

$$
ef = \begin{cases} 2e & \text{if } e = f, \text{ i.e., } \langle e, f \rangle \text{ is of 1A-type,} \\ 0 & \text{if } (e \mid f) = 0, \text{ i.e., } \langle e, f \rangle \text{ is of 2B-type,} \\ \dfrac{1}{4}(e + f - \sigma_e f) & \text{if } (e \mid f) = 2^{-8}, \text{ i.e., } \langle e, f \rangle \text{ is of 2A-type.} \end{cases} \tag{5.1}
$$

Note that it follows from (4) of Theorem 4.8 that $|\sigma_e \sigma_f| = 3$ if $\langle e, f \rangle$ is of 2A-type, whereas $|\sigma_e \sigma_f| \leq 2$ if $\langle e, f \rangle$ is of 2B-type (cf. [Mi96]). We find that the relation (5.1) is the same as the local structure in (3.1) of the Matsuo algebra with $\alpha = \beta = 1/2$. Therefore, if we collect all the Ising vectors of σ-type, then we obtain a realization of a 3-transposition group as well as the associated Matsuo algebra in the Griess algebra.

Theorem 5.2 ([Ma05, Mi96])   *Let $E_V(\sigma)$ be the set of Ising vectors of a VOA V of OZ-type and let $G_V = \langle \sigma_e \in \text{Aut} V \mid e \in E_V(\sigma) \rangle$. Then G_V is a 3-transposition group. The Griess algebra of the subalgebra $\langle E_V(\sigma) \rangle$ of V is isomorphic to the nondegenerate quotient of the Matsuo algebra $B_{1/2,1/2}(G_V)$.*

Historically, the Matsuo algebra was formulated based on Theorem 5.2. Matsuo has classified the 3-transposition groups realized by Ising vectors of σ-type.

Theorem 5.3 ([JLY19, Ma05])   *Let $E_V(\sigma)$ be the set of Ising vectors of a VOA V of OZ-type. The 3-transposition group $G_V = \langle \sigma_e \in \operatorname{Aut} V \mid e \in E_V(\sigma) \rangle$ is of symplectic type and a nontrivial connected component of G_V is isomorphic to one of the following:*

$$\mathfrak{S}_{n\geq 3}, \quad F^{\leq 2}{:}\mathfrak{S}_{n\geq 4}, \quad O^+_{8\,\text{or}\,10}(2), \quad 2^8{:}O^+_8(2), \quad O^-_{6\,\text{or}\,8}(2),$$
$$2^6{:}O^-_6(2), \quad \mathrm{Sp}_{6\,\text{or}\,8}(2), \quad 2^6{:}\mathrm{Sp}_6(2),$$

where $F \cong 2^{2n}$ is the natural module for $\mathrm{Sp}_{2n}(2) > \mathfrak{S}_{2n+2\,\text{or}\,2n+1}$. Moreover, if V is generated by $E_V(\sigma)$ as a VOA and V is simple then V is uniquely determined by its Griess algebra, the nondegenerate quotient of the Matsuo algebra $B_{1/2,1/2}(G_V)$ associated with G_V.

In [Ma05], the 3-transposition groups realizable in this manner were classified but the uniqueness of structures of VOAs realizing each 3-transposition group in Matsuo's list was open, and this uniqueness was established in [JLY19] under the assumption that the VOA is simple. It turns out that all Matsuo algebras appearing in Theorem 5.2 are isomorphic to some subalgebras of the Griess algebras of $V^+_{\sqrt{2}R}$ associated with the root lattice R of ADE-type. Recall that a 3-transposition group (G, I) is of symplectic type if and only if there is no I-subgroup isomorphic to $H \cong 3^{1+2}{:}2$ or its quotient $H/Z(H) \cong 3^2{:}2$ (cf. Eq. (2.1)). In Matsuo's classification, it is crucial to prove the absence of the I-subgroup isomorphic to $3^2{:}2$. The proof of this part goes as follows. Suppose we have a subalgebra B isomorphic to $B_{1/2,1/2}(H)$ in the Griess algebra of V. Then one can find a $c = c_2$ Virasoro sub VOA $L(7/10, 0)$ and its module isomorphic to $L(7/10, 7/10)$ in the sub VOA $\langle B \rangle$ of V. However, by the list of possible highest weights in (4.4), there is no irreducible highest weight representation of highest weight $7/10$ with central charge $c_2 = 7/10$. Thus 3-transposition groups realizable by Ising vectors of σ-type are only of symplectic type.

2.5.2 W_3-Algebra and $\mathbb{Z}_3$-Symmetry

We explain another example of Matsuo algebras arising in VOA theory. Here we consider the third value $c_3 = 4/5$ of the unitary series (4.4). The Virasoro VOA $L(4/5, 0)$ can be extended to a larger VOA $W_3(4/5) = L(4/5, 0) \oplus L(4/5, 3)$ called the W_3-*algebra* at $c = 4/5$ (cf. [FaZa87]). It is shown in (loc. cit.) (see also [KMY00, Mi01]) that $W_3(4/5)$ is rational and has six irreducible representations over $\mathbb{C}$ as follows:

$$W_3(^4/_5) = L(^4/_5, 0 \oplus 3), \quad L(^4/_5, ^2/_5 \oplus ^7/_5), \quad L(^4/_5, ^2/_3)^\pm, \quad L(^4/_5, ^1/_{15})^\pm$$

$$\xi : \qquad\qquad 1 \qquad\qquad\qquad 1 \qquad\qquad\qquad \zeta^{\pm 1} \qquad\quad \zeta^{\pm 1} \tag{5.2}$$

Here $L(c, h \oplus h')$ denotes $L(c, h) \oplus L(c, h')$ and $L(^4/_5, h)^\pm$ denotes inequivalent irreducible $W_3(^4/_5)$-modules that are isomorphic to $L(^4/_5, h)$ as $L(^4/_5, 0)$-modules. Let ξ be the assignment of the powers of the cubic root $\zeta = e^{2\pi\sqrt{-1}/3}$ of unity as in (5.2). Then it is known that ξ preserves the fusion rules (cf. [FaZa87, Mi01]). Therefore, fusion rules of $W_3(4/5)$-modules have a $\mathbb{Z}_3$-symmetry.

Let V be a VOA of OZ-type over $\mathbb{R}$. A $c = 4/5$ Virasoro vector $x \in V$ is called *extendable* in V if $\langle x \rangle \cong L(^4/_5, 0)$ can be extended to a larger sub VOA $W_3(^4/_5) = L(^4/_5, 0 \oplus 3)$ inside V (such an extension is known to be unique if it exists). Based on the $\mathbb{Z}_3$-symmetry of the fusion rules of $W_3(^4/_5)$-modules, one can define an automorphism ξ_x of $\mathbb{C}V = \mathbb{C} \otimes_{\mathbb{R}} V$ associated to an extendable $c = 4/5$ Virasoro vector $x \in V$ in the same manner as Miyamoto involutions.

Lemma 5.4 ([Mi96]) *Let $x \in V$ be an extendable $c = 4/5$ Virasoro vector and ξ_x the automorphism of the complexified VOA $\mathbb{C}V$ defined as in (5.2).*

(1) *The map ξ_x keeps the real space V invariant and defines an element of Aut V.*

(2) *The $c = 4/5$ Virasoro vector x is of σ-type on the fixed point subalgebra $V^{\langle \xi_x \rangle}$.*

Let F_V be the set of extendable $c = 4/5$ Virasoro vectors of V and $F_V(\sigma)$ the subset of F_V consisting of those of σ-type. It follows from (4.4), (5.2), and (2) of Lemma 5.4 that $x \in F_V(\sigma)$ if and only if $\xi_x = 1$ on V. Based on $c = 4/5$ Virasoro vectors, we obtain another family of Matsuo algebras with $\alpha = 2/5$ and $\beta = 4/5$ as follows.

Theorem 5.5 ([LY14]) *Let V be a simple VOA of OZ-type over $\mathbb{C}$ and let $G_V = \langle \sigma_x \in \mathrm{Aut}\, V \mid x \in F_V(\sigma) \rangle$. Then G_V is a 3-transposition group. The Griess algebra of the subalgebra $\langle F_V(\sigma) \rangle$ of V is isomorphic to the nondegenerate quotient of the Matsuo algebra $B_{2/5,4/5}(G_V)$ associated with G_V.*

The classification of 3-transposition groups based on extendable $c = 4/5$ Virasoro vectors of σ-type is open. Some systematic examples are constructed in [LY14] and it turns out that, contrary to the case of Ising vectors of σ-type, there exists an example of a VOA realizing a 3-transposition group of orthogonal type based on $c = 4/5$ Virasoro vectors.

Example 5.6 ([LY14])   Let K_{12} be the Coxeter–Todd lattice of rank 12 and consider the lattice VOA $V_{K_{12}}$. There exists an order-3 automorphism ν of $V_{K_{12}}$ such that $X = V_{K_{12}}^{\langle \nu \rangle}$ is of OZ-type. Let $F_X(\sigma)$ be the set of extendable $c = 4/5$ Virasoro vectors of X which are of σ-type on X and take $G_X = \langle \sigma_x \mid x \in F_X(\sigma) \rangle$. Then $G_X \cong {}^+\Omega_8^-(3)$. It is worth mentioning that K_{12} can be defined over the Eisenstein ring and its automorphism group preserving this complex structure is $3.{}^+\Omega_6^-(3)$ (cf. [CS99]) so that there is a canonical subgroup $3^6{:}{}^+\Omega_6^-(3)$ in Aut X. The interesting point here is that if we consider the automorphism group of the VOA $X = V_{K_{12}}^{\langle \nu \rangle}$ then the canonical symmetry $3^6{:}{}^+\Omega_6^-(3)$ can be extended to a larger symmetry ${}^+\Omega_8^-(3)$ which is highly nontrivial. This is due to the fact that $X = V_{K_{12}}^{\langle \nu \rangle}$ has some extra $c = 4/5$ extendable Virasoro vectors of σ-type. For details, see section 5.5 of [LY14].

Question 5.7   Is there an example of a VOA V such that $G_V = \langle \sigma_x \in \text{Aut } V \mid x \in F_V(\sigma) \rangle$ is of unitary type? If exists, then $B_{2/5,4/5}(2^{1+6}{:}H)$ appears as a Griess subalgebra of such a VOA.

2.5.3 3-Transposition Groups Not Arising from Matsuo Algebras

As we have seen in Theorem 5.2, the 3-transposition groups generated by σ-involutions associated with Ising vectors of σ-type are always of symplectic type so that we cannot obtain 3-transposition groups other than of symplectic type as long as we consider Ising vectors of σ-type. On the other hand, by the 6-transposition property in Sakuma's theorem, we cannot directly obtain 3-transposition groups if we use τ-involutions associated with Ising vectors without any restriction. However, if we collect suitable subsets of Ising vectors, then we can construct 3-transposition groups generated by τ-involutions. Here we explain a general recipe of a construction of 3-transposition groups from Ising vectors that are not arising from Matsuo algebras.

Let V be a VOA of OZ-type over $\mathbb{R}$ and let E_V be the set of Ising vectors of V. Suppose we have a pair a, $b \in E_V$ such that $(a \mid b) = 13 \cdot 2^{-10}$, which is equivalent to that $\langle a, b \rangle$ is a 3A-algebra by (2) of Lemma 4.7. We call such a pair a *3A-pair*. Fix a 3A-pair a, $b \in E_V$ and define

$$I_{a,b} := \{ x \in E_V \mid (a \mid x) = (b \mid x) = 2^{-5} \}. \tag{5.3}$$

By Lemma 4.7, $x \in I_{a,b}$ if and only if $\langle a, x \rangle \cong \langle b, x \rangle$ are 2A-algebras. In this case one can also show that $\langle a, b, x \rangle = \langle \sigma_x a, b \rangle$ is a 6A-algebra (cf. [LY16]).

Theorem 5.8 ([LY16])   *Let a, $b \in E_V$ be a 3A-pair and define $I_{a,b} \subset E_V$ as in* (5.3).

(1) *If x, $y \in I_{a,b}$ then the type of $\langle x, y \rangle$ is either 1A, 2A, or 3A. In particular, one has $|\tau_x \tau_y| \leq 3$ on V.*

(2) *Set $G_V := \langle \tau_x \mid x \in I_{a,b} \rangle$. Then G_V is a 3-transposition group acting on the commutant subalgebra $\mathrm{Com}_V \langle a, b \rangle \subset V$ of $\langle a, b \rangle$ in V whereas G_V acts on $\langle a, b \rangle$ trivially.*

Hereafter, $\mathrm{Com}_V A$ denotes the commutant subalgebra of A in V (cf. [FZ92]). It follows from (2) of Theorem 4.8 that G_V in Theorem 5.8 centralizes $\langle \tau_a, \tau_b \rangle \cong \mathfrak{S}_3$.

Example 5.9 Let $V^\natural$ be the moonshine VOA [FLM88]. It is known that all the 3A-pairs in $E_{V^\natural}$ are mutually conjugate under the Monster $\mathbb{M} = \mathrm{Aut}\, V^\natural$ (cf. [C85, Mi96]). Let a, $b \in V^\natural$ be Ising vectors such that $\langle a, b \rangle$ is a 3A-subalgebra and take $I_{a,b} \subset E_{V^\natural}$ as in (5.3). Then we obtain $G_{V^\natural} = \langle \tau_x \mid x \in I_{a,b} \rangle \cong \mathrm{Fi}_{23}$ (cf. [LY16]). Indeed, Theorem 5.8 is hinted at by the fact that $\mathfrak{S}_3 \times \mathrm{Fi}_{23} < \mathbb{M}$ where $\mathfrak{S}_3 = \langle \tau_a, \tau_b \rangle$.

As in Example 5.9, we see that 3-transposition groups obtained by Theorem 5.8 include those of sporadic type. However, this example strongly relies upon the Monster and there is no conceptual understanding of what kind of 3-transposition groups we can obtain from Theorem 5.8. Apart from this extreme example, we can simply say that the 3-transposition groups obtained by Theorem 5.8 are not arising from Matsuo algebras with $\beta = 1/2$ as follows. If x, $y \in I_{a,b}$ satisfy $|\tau_x \tau_y| = 3$ then x and y is also a 3A-pair. In this case, the Griess algebra of the dihedral subalgebra $\langle x, y \rangle$ is four-dimensional by Theorem 4.6 whereas subalgebras generated by two axes in Matsuo algebras are at most three-dimensional. Therefore, 3-transposition groups obtained by Theorem 5.8 are beyond those arising from Matsuo algebras with $\beta = 1/2$.

2.5.4 3-Transposition Groups as Homomorphic Images

We have seen the construction of a 3-transposition group based on τ-involutions associated with Ising vectors in Theorem 5.8. Here we give another slightly different construction of 3-transposition groups.

Let V be a VOA of OZ-type over $\mathbb{R}$ and E_V the set of Ising vectors of V. We consider all the 3A-pairs in V. Set

$$T_V := \{ (x, y) \in E_V^2 \mid (x \mid y) = 13 \cdot 2^{-10} \} \tag{5.4}$$

and define

$$u_{x,y} := \frac{448}{135}(x + y + \tau_x y) - \frac{512}{405}(x + y + \tau_x y)_{(1)}(x + y + \tau_x y) \in \langle x, y \rangle \cong U_{3A}. \tag{5.5}$$

The vector $u_{x,y}$ in (5.5) provides the $c = 4/5$ Virasoro vector described in (3) of Theorem 4.9.

Lemma 5.10 *If $(x, y) \in T_V$, then $u_{x,y}$ is an extendable $c = 4/5$ Virasoro vector of $\langle x, y \rangle$.*

Fix a 3A-pair $(a, b) \in T_V$ and set

$$J_{a,b} := \left\{ x \in V \mid \exists (x, y) \in T_V \text{ s.t. } u_{x,y} = u_{a,b} \right\}, \quad G := \langle \tau_x \mid x \in J_{a,b} \rangle. \tag{5.6}$$

By Theorem 4.6, G itself is not a 3-transposition group but a 6-transposition group in general but we can extract a 3-transposition group from G as follows.

Theorem 5.11 ([HLY12b]) *For a 3A-pair $(a, b) \in T_V$, define $J_{a,b}$ and G as in (5.6). Then the following restriction map is a group homomorphism:*

$$\begin{aligned} \psi_{a,b} : \quad G \quad &\longrightarrow \quad \mathrm{Aut}\,\mathrm{Com}_V\langle u_{a,b} \rangle \\ g \quad &\longmapsto \quad g|_{\mathrm{Com}_V\langle u_{a,b} \rangle} \end{aligned}.$$

The homomorphic image $\psi_{a,b}(G) = \langle \psi_{a,b}(\tau_x) \mid x \in J_{a,b} \rangle$ forms a 3-transposition group acting on the commutant subalgebra $\mathrm{Com}_V\langle u_{a,b} \rangle$ of the Virasoro sub VOA $\langle u_{a,b} \rangle \cong L(4/5, 0)$ of V.

Example 5.12 If we apply Theorem 5.11 to $V^{\natural}$ then we obtain $G \cong 3.\mathrm{Fi}_{24}$ and $\psi_{a,b}(G) \cong \mathrm{Fi}_{24}$ for any $(a, b) \in T_{V^{\natural}}$ since $\mathbb{M} = \mathrm{Aut}\,V^{\natural}$ acts on $T_{V^{\natural}}$ transitively (cf. [C85, Mi96]). Note that $3.\mathrm{Fi}_{24} < \mathbb{M}$ but $\mathrm{Fi}_{24} \not< \mathbb{M}$ and $3.\mathrm{Fi}_{24}$ is a 6-transposition group but not a 3-transposition group so that we need to take a homomorphic image to obtain a 3-transposition group in Theorem 5.11.

2.5.5 4-Transposition Groups as Homomorphic Images

In this subsection, we explain a construction of 4-transposition groups similar to the one in the previous subsection.

Let V be a VOA of OZ-type over $\mathbb{R}$ and E_V the set of Ising vectors of V as before. Fix $e \in E_V$ and set

$$I_e := \{ x \in E_V \mid (e \mid x) = 2^{-5} \}, \quad G := \langle \tau_x \in \mathrm{Aut}(V) \mid x \in I_e \rangle. \tag{5.7}$$

By Theorem 4.6, G itself is a 6-transposition group in general. Similar to Theorem 5.11, we can extract a 4-transposition group from G as follows.

Theorem 5.13 ([HLY12a]) *Take $e \in E_V$ and define I_e and G as in (5.7). Then the following restriction map is a group homomorphism:*

$$\begin{aligned} \varphi_e : \quad G \quad &\longrightarrow \quad \mathrm{Aut}\,\mathrm{Com}_V\langle e \rangle \\ g \quad &\longmapsto \quad g|_{\mathrm{Com}_V\langle e \rangle} \end{aligned}.$$

The homomorphic image $\varphi_e(G) = \langle \varphi_e(\tau_x) \mid x \in I_e \rangle$ is a 4-transposition group acting on the commutant subalgebra $\mathrm{Com}_V \langle e \rangle$ of the Virasoro sub VOA $\langle e \rangle \cong L(1/2, 0)$ of V.

Example 5.14 If we apply Theorem 5.13 to $V^\natural$, then we obtain $G \cong 2.\mathbb{B}$ and $\varphi_e(G) \cong \mathbb{B}$ for any $e \in E_{V^\natural}$ since $\mathbb{M} = \mathrm{Aut}\, V^\natural$ acts on $E_{V^\natural}$ transitively (cf. [C85, Mi96]). Note that $2.\mathbb{B} < \mathbb{M}$ but $\mathbb{B} \not< \mathbb{M}$ and $2.\mathbb{B}$ is a 6-transposition group but not a 4-transposition group so that we need to take a homomorphic image to obtain a 4-transposition group in Theorem 5.13.

2.6 Open Questions

We list a few open questions about 3-transposition groups, Matsuo algebras, and vertex operator algebras.

Problem 6.1 Classify 3-transposition groups arising from Matsuo algebra $B_{2/5,4/5}(G)$ based on $W_3(4/5)$. Are they all of orthogonal type? To determine this, one needs to analyze subalgebras of the Matsuo algebra $B_{2/5,4/5}(2^{1+6}{:}H)$ and determine whether there exists a subalgebra realizable by a VOA or not.

Problem 6.2 Other than $(\alpha, \beta) = (1/2, 1/2)$ and $(2/5, 4/5)$, it is possible to realize Matsuo algebras with $(\alpha, \beta) = (1/2, 1/16)$ as Griess algebras of VOAs [CL14a, CL14b, Ma05]. It is shown in [CL14a] that the Matsuo algebra $B_{1/2,1/16}(H)$ for H in (2.1) has a realization by a VOA so that 3-transposition groups of nonsymplectic type can be realized as automorphism groups of VOAs via Matsuo algebras with $(\alpha, \beta) = (1/2, 1/16)$. Some examples are constructed in [CL14b]. Give more examples of VOAs having Matsuo algebras with $(\alpha, \beta) = (1/2, 1/16)$ as Griess algebras and classify the 3-transposition groups realizable in this way, if possible.

Problem 6.3 One may consider Majorana algebras [I09] and axial algebras [HRS15] as axiomatizations of Griess algebras generated by Ising vectors. Can we generalize the construction of 3-transposition groups in Theorem 5.8 to arbitrary axial algebras? Also, can we generalize Theorems 5.11 and 5.13 to general axial algebras?

Problem 6.4 It seems possible to generalize Matsuo algebras to "nonsimply laced" cases. In the original definition, all idempotents have the same spectrum and squared norm but we can mix two or more different idempotents. For instance, when we consider the 3-transposition groups obtained by Theorems 5.8 and 5.11 we observe some commutative algebras spanned by two kinds of idempotents corresponding to $c_1 = 1/2$ and $c_4 = 6/7$ Virasoro vectors. We do not include explicit examples here but we refer interested readers to section 4 of [LY16].

Acknowledgment

This article is based on the author's talk at the seminar on "Majorana, Axial, Vertex Algebras and the Monster (MAVAM)" held online on June 4th, 2021. The author wishes to thank the organizers of MAVAM seminar for the opportunity to give a talk there.

References

[ATLAS] J. H. Conway, R. T. Curtis, S. P. Norton, R. A. Parker and R. A. Wilson, *ATLAS of finite groups.* Clarendon Press, Oxford, 1985.

[As97] M. Aschbacher, 3-transposition groups. Cambridge Tracts in Mathematics, 124. *Cambridge University Press,* Cambridge, 1997.

[BPZ84] A. A. Belavin, A. M. Polyakov, and A. B. Zamolodchikov, Infinite conformal symmetries in two-dimensional quantum field theory. *Nucl. Phys.* B **241** (1984), 333–380.

[CL14a] H. Y. Chen and C. H. Lam, An explicit Majorana representation of the group $3^2{:}2$ of $3C$-pure type. *Pacific J. Math.* **271** (2014), no. 1, 25–51.

[CL14b] H. Y. Chen and C. H. Lam, Weyl groups and vertex operator algebras generated by Ising vectors satisfying the $(2B, 3C)$ condition. *Rev. Math. Phys.* **26** (2014), no. 6, 1450011.

[C85] J. H. Conway, A simple construction for the Fischer–Griess Monster group. *Invent. Math.* **79** (1985), 513–540.

[CS99] J. H. Conway and N. J. A. Sloane, Sphere packings, lattices and groups, 3rd edition, Springer, New York, 1999.

[CH95] H. Cuypers and J. I. Hall, The 3-transposition groups with trivial center. *J. Algebra* **178** (1995), 149–193.

[FaZa87] V. A. Fateev and A. B. Zamolodchikov, Conformal quantum field theory models in two dimensions having Z_3 symmetry. *Nuclear Phys. B* **280** (1987), 644–660.

[Fi71] B. Fischer, Finite groups generated by 3-transpositions. I. *Invent. Math.* **13**, 232–246.

[FHL93] I. B. Frenkel, Y.-Z. Huang and J. Lepowsky, On axiomatic approaches to vertex operator algebras and modules. *Memoirs Amer. Math. Soc.* **104**, 1993.

[FLM88] I. B. Frenkel, J. Lepowsky and A. Meurman, *Vertex Operator Algebras and the Monster.* Academic Press, New York, 1988.

[FZ92] I. B. Frenkel and Y. Zhu, Vertex operator algebras associated to representation of affine and Virasoro algebras. *Duke Math. J.* **66** (1992), 123–168.

[G82] R. L. Griess, The friendly giant. *Invent. Math.* **69** (1982), 1–102.

[HLY12a] G. Höhn, C. H. Lam and H. Yamauchi, McKay's E_7 observation on the Baby Monster. *Internat. Math. Res. Notices* (2012), doi:10.1093/imrn/rnr009.

[HLY12b] G. Höhn, C. H. Lam and H. Yamauchi, McKay's E_6 observation on the largest Fischer group. *Comm. Math. Physics* **310** Vol. 2 (2012), 329–365.

[HRS15] J. I. Hall, F. Rehren and S. Shpectorov, Universal axial algebras and a theorem of Sakuma. *J. Algebra* **421** (2015), 394–424.

[I09] A. A. Ivanov, The Monster group and Majorana involutions. *Cambridge Tracts in Mathematics* **176**, 2009.

[Ka90] V. G. Kac, *Vertex algebras for beginners*, 2nd edition, Cambridge University Press, Cambridge, 1998.

[K00] M. Kitazume, 3-transposition groups (Japanese). *Group theory and related topics–summary and prospects, Surikaisekikenkyusho Kokyuroku No.* **1214** (2001), 62–75.

[KMY00] M. Kitazume, M. Miyamoto and H. Yamada, Ternary codes and vertex operator algebras. *J. Algebra* **223** (2000), 379–395.

[JLY19] C. Jiang, C. H. Lam and H. Yamauchi, Vertex operator algebras generated by Ising vectors of σ-type. *Math. Z.* **293** (2019), 425–442.

[LYY05] C. H. Lam, H. Yamada and H. Yamauchi, McKay's observation and vertex operator algebras generated by two conformal vectors of central charge 1/2. *Internat. Math. Res. Papers* **3** (2005), 117–181.

[LY14] C. H. Lam and H. Yamauchi, On 3-transposition groups generated by σ-involutions associated to $c = 4/5$ Virasoro vectors. *J. Algebra* **416** (2014), 84–121.

[LY16] C. H. Lam and H. Yamauchi, The Conway-Miyamoto correspondences for the Fischer 3-transposition groups. To appear in *Trans. Amer. Math. Soc.*, `arXiv:1604.04989`

[LL04] J. Lepowsky and H. Li, Introduction to vertex operator algebras and their representations. *Progress in Mathematics*, Vol. 227, Birkhäser Boston, Boston, MA, 2004.

[L96] H. Li, Symmetric invariant bilinear forms on vertex operator algebras. *J. Pure Appl. Algebra* **96** (1994), 279–297.

[Ma05] A. Matsuo, 3-transposition groups of symplectic type and vertex operator algebras. *J. Math. Soc. Japan* **57** (2005), no. 3, 639–649. `arXiv:math/0311400`.

[MaN99] A. Matsuo and K. Nagatomo, Axioms for a vertex algebra and the locality of quantum fields. *MSJ Memoirs* 4 (1999), Mathematical Society of Japan.

[Mi96] M. Miyamoto, Griess algebras and conformal vectors in vertex operator algebras. *J. Algebra* **179** (1996), 528–548.

[Mi01] M. Miyamoto, 3-state Potts model and automorphisms of vertex operator algebras of order 3. *J. Algebra* **239** (2001), 56–76.

[Mi03] M. Miyamoto, VOAs generated by two conformal vectors whose τ-involutions generate S_3. *J. Algebra* **268** (2003), 653–671.

[S07] S. Sakuma, 6-transposition property of τ-involutions of vertex operator algebras. *Internat. Math. Res. Notices* **2007**, no. **9**, Art. ID rnm 030, 19 pp.

[SY03] S. Sakuma and H. Yamauchi, Vertex operator algebra with two Miyamoto involutions generating S_3. *J. Algebra* **267** (2003), 272–297.

[W93] W. Wang, Rationality of Virasoro vertex operator algebras. *Internat. Math. Res. Notices* **71** (1993), 197–211.

3

On Holomorphic Vertex Operator Algebras of Central Charge 24

Ching Hung Lam

Abstract

This chapter is a survey on the recent progress towards the classification of strongly regular holomorphic vertex operator algebras of central charge 24. We review the construction of all holomorphic vertex operator algebras that realize the 71 Lie algebras in Schellekens' list. We also discuss the techniques that we used to prove the uniqueness conjecture that the isomorphism class of a holomorphic vertex operator algebra of central charge 24 is uniquely determined by the Lie algebra structure of the weight one subspace V_1 if $V_1 \neq 0$. We will also discuss several important observations by G. Höhn and their connections to the Leech lattice VOA and generalized deep holes.

CONTENTS

3.1 Introduction

The classification of strongly regular holomorphic vertex operator algebras (VOAs) of central charge 24 is probably one of the fundamental problems in vertex operator algebras. The weight one subspace of a VOA of CFT-type has a Lie algebra structure via the 0th mode [Bo86]. For a strongly regular holomorphic VOA V of central charge 24, it was proved in [DM04b] that the weight one Lie algebra V_1 is zero, abelian of rank 24 or semisimple; if V_1 is abelian of rank 24, then V is isomorphic to the Leech lattice VOA V_Λ; if V_1 is zero, then it is conjectured that V is isomorphic to the moonshine VOA ([FLM88]). In 1993, Schellekens [Sc93] (see also [EMS20]) showed that there are at most 71 possibilities for this Lie algebra (see Table 2) by exploiting the modular invariance of the character of V.[1]

By the combined efforts of many people, the following theorem is now proved.

[1] 2010 *Mathematics Subject Classification.* Primary 17B69. Partially supported by grant AS-IA -107-M02 of Academia Sinica and MoST grant 110-2115-M-001- 011-MY3 of Taiwan.

Theorem 1.1   *Each potential Lie algebra on Schellekens' list is realized by a strongly regular holomorphic VOA of central charge* 24 *and this VOA is uniquely determined by its V_1-structure if $V_1 \neq \{0\}$.*

The classification of holomorphic VOAs of central charge 24 may be viewed as a VOA analogue of the classification of (positive-definite) even unimodular lattices of rank 24. It is a famous theorem by Niemeier [Ni73] (cf. [Ve78]) that there are exactly 24 even unimodular lattices of rank 24 and their isometry types are determined by their root systems, a combinatorial structure of norm 2 vectors. It seems that the weight one Lie algebra of a holomorphic VOA of central charge 24 plays a similar role as the root system of the even unimodular lattices of rank 24. In particular, there exists a unique even unimodular lattice of rank 24 that has no vectors of norm 2 and it is isometric to the Leech lattice [Co69]. The characterization of the moonshine VOA mentioned previously would be a VOA analogue of the uniqueness of the Leech lattice.

In this chapter, we will give a survey on the recent progress towards the classification of strongly regular holomorphic vertex operator algebras of central charge 24. We review the construction of all holomorphic vertex operator algebras that realize the 71 Lie algebras in Schellekens' list. We also discuss the techniques that we used to prove the uniqueness conjecture that the isomorphism class of a holomorphic vertex operator algebra of central charge 24 is uniquely determined by the Lie algebra structure of the weight one subspace V_1 if $V_1 \neq 0$. We will also discuss several important observations by G. Höhn and their connections to the Leech lattice VOA and a notion called "generalized deep holes."

3.2 Preliminary

Throughout this chapter, all VOAs are defined over the field $\mathbb{C}$ of complex numbers.

A *vertex operator algebra* (VOA) $(V, Y, \mathbb{1}, \omega)$ is a $\mathbb{Z}$-graded vector space $V = \bigoplus_{m \in \mathbb{Z}} V_m$ equipped with a linear map

$$Y(a, z) = \sum_{i \in \mathbb{Z}} a_{(i)} z^{-i-1} \in (\mathrm{End}(V))[[z, z^{-1}]], \quad a \in V,$$

the *vacuum vector* $\mathbb{1}$ and the *conformal vector* ω satisfying a number of conditions ([Bo86, FLM88]). For $a \in V$ and $n \in \mathbb{Z}$, the operator $a_{(n)}$ is called the *nth mode* of a. Note that $L(n) = \omega_{(n+1)}$ satisfy the Virasoro relation:

$$[L_{(m)}, L_{(n)}] = (m - n)L_{(m+n)} + \frac{1}{12}(m^3 - m)\delta_{m+n,0} c \, \mathrm{id}_V,$$

where c is a complex number, called the *central charge* of V.

A linear automorphism g of V is called a *(VOA) automorphism* of V if

$$g\omega = \omega \quad \text{and} \quad gY(v, z) = Y(gv, z)g \quad \text{for all } v \in V.$$

A *vertex operator subalgebra* (or a *subVOA*) is a graded subspace of V that has a structure of a VOA such that the operations and its grading agree with the restriction of those of V and that they share the vacuum vector. When they also share the conformal vector, we will call it a *full subVOA*. For an automorphism g of a VOA V, the set of fixed points of g is denoted by V^g and V^g is a full subVOA of V.

A VOA is said to be *rational* if its admissible module category is semisimple, and a rational VOA is said to be *holomorphic* if it itself is the only irreducible module up to isomorphism. A VOA is said to be *of CFT-type* if $V_0 = \mathbb{C}\mathbb{1}$ (note that $V_n = 0$ for all $n < 0$ if $V_0 = \mathbb{C}\mathbb{1}$ [DM06b, lemma 5.2]), and is said to be C_2-*cofinite* if the codimension in V of the subspace spanned by the vectors of form $u_{(-2)}v$, $u, v \in V$, is finite. A module is said to be *self-contragredient* if its contragredient module is isomorphic to itself. A VOA is said to be *strongly regular* if it is rational, C_2-cofinite, self-contragredient, and of CFT-type. Note that a strongly regular VOA is simple.

Let g be an automorphism of a VOA V of order $n \in \mathbb{Z}_{>0}$ and let M be an irreducible g-twisted V-module. For the definition of (twisted) modules, see [FHL93, DLM00]. Note that if $g = 1$, then M is just an irreducible V-module. Then the twisted module M can be decomposed as $M = \bigoplus_{i \in (1/n)\mathbb{Z}_{\geq 0}} M_{i+w}$, where $w \in \mathbb{C}$, $M_w \neq 0$ and $M_s = \{v \in M \mid L(0)v = sv\}$. The number w is called the *conformal weight* of M.

Let V be a VOA of CFT-type. Then, the weight one space V_1 has a Lie algebra structure via the 0th mode ([Bo86, FLM88]), which we often call the *weight one Lie algebra* of V. Moreover, the nth modes, $v_{(n)}$, $v \in V_1$, $n \in \mathbb{Z}$, define an affine representation of the Lie algebra V_1 on V. For a simple Lie subalgebra $\mathfrak{a}$ of V_1, the *level* of $\mathfrak{a}$ is defined to be the scalar by which the canonical central element acts on V as the affine representation. When the type of the root system of $\mathfrak{a}$ is X_n and the level of $\mathfrak{a}$ is k, we denote the type of $\mathfrak{a}$ by $X_{n,k}$.

3.3 Weight One Lie Algebras and Schellekens' List

First we recall a theorem from [DM06a]. For the explicit construction of affine VOAs, we refer to [FZ92] for details.

Proposition 3.1 ([DM06a, theorem 1.1, corollary 4.3]) *Let V be a strongly regular VOA. Then V_1 is reductive. Let $\mathfrak{s}$ be a simple Lie subalgebra of V_1. Then V is an integrable module for the affine representation of $\mathfrak{s}$ on V, and the subVOA generated by $\mathfrak{s}$ is isomorphic to the simple affine VOA associated with $\mathfrak{s}$ at the positive integral level.*

If we further assume that V is holomorphic and has central charge 24, then one has the following result.

Proposition 3.2 ([DM04a, (1.1), theorem 3, and proposition 4.1]) *Let V be a strongly regular holomorphic VOA of $c = 24$. If the Lie algebra V_1 is neither $\{0\}$ nor abelian of rank 24, then V_1 is semisimple, and the subVOA generated by V_1 is full; that is, its conformal vector coincides with that of V. In addition, for any simple ideal of V_1 at level k, the identity*

$$\frac{h^\vee}{k} = \frac{\dim V_1 - 24}{24} \tag{3.1}$$

holds, where $h^\vee$ is the dual Coxeter number of the simple ideal.

This proposition restricts the structure of the semisimple weight one Lie algebra of a holomorphic VOA of central charge 24. Indeed, there are only 221 possibilities for them ([Sc93, EMS20]) but some cases are clearly not realizable. In order to further restrict the possibility for V_1, one needs to consider some higher trace formulas.

Suppose V_1 is semisimple and $L(V_1) = L_{\mathfrak{g}_1}(k_1, 0) \otimes \cdots \otimes L_{\mathfrak{g}_r}(k_r, 0)$. Since $L(V_1)$ is rational and full, V decomposes into a direct sum of finitely many irreducible $L_{\mathfrak{g}_1}(k_1, 0) \otimes \cdots \otimes L_{\mathfrak{g}_r}(k_r, 0)$-modules and

$$V \cong \bigoplus_\lambda m_\lambda L_{\hat{\mathfrak{g}}_1}(k_1, \lambda_1) \otimes \ldots \otimes L_{\hat{\mathfrak{g}}_r}(k_r, \lambda_r)$$

with $m_\lambda \in \mathbb{N}$, and the sum runs over finitely many $\lambda = \lambda_1 + \cdots + \lambda_r$ with dominant integral weights $\lambda_i \in P_+^{k_i}(\mathfrak{g}_i)$; that is, of level at most k_i. The character $\mathrm{ch}\, V : \mathcal{H} \times \mathfrak{h} \to \mathbb{C}$ is defined by

$$\mathrm{ch}\, V(\tau, z) = tr_V e^{2\pi i z} q^{L_0 - 1},$$

where $q = e^{2\pi i \tau}$, τ is in upper half plane $\mathcal{H}$, and $z \in \mathfrak{h}$. It is proved [KM12, theorem 2] that $\mathrm{ch}\, V$ is holomorphic on $\mathcal{H} \times \mathfrak{h}$ and transforms as a Jacobi form of weight 0 and index 1 (see also [Mi00]). Let M be a finite-dimensional $\mathfrak{g}$-module. Then M decomposes into weight spaces and one can define a function

$$S_M^j(z) = \sum_{\mu \in \prod(M)} m_\mu \mu(z)^j,$$

where $\prod(M)$ denotes the set of weights of M and m_μ the multiplicity of μ on M. By using the transformation rules of ch_V, the following theorem is proved in [EMS20].

Theorem 3.3   *Let V be strongly regular holomorphic VOA of central charge* 24. *Then*

$$S^0_{V_2}(z) = \dim(V_2) = 196884;$$

$$S^2_{V_2}(z) = 32808\langle z, z\rangle - 2\dim(V_1)\langle z, z\rangle;$$

$$S^4_{V_2}(z) = 240S^4_{V_1}(z) + 15264\langle z, z\rangle^2 - \dim(V_1)\langle z, z\rangle^2;$$

$$S^6_{V_2}(z) = -504S^6_{V_1}(z) + 900S^4_{V_1}(z)\langle z, z\rangle + 11160\langle z, z\rangle^3$$
$$- 15\dim(V_1)\langle z, z\rangle^3;$$

$$S^8_{V_2}(z) = 480S^8_{V_1}(z) - 2352S^6_{V_1}(z)\langle z, z\rangle + 2520S^4_{V_1}(z)\langle z, z\rangle^2$$
$$+ 10920\langle z, z\rangle^4 - 35\dim(V_1)\langle z, z\rangle^4;$$

$$S^{10}_{V_2}(z) = -264S^{10}_{V_1}(z) + 2700S^8_{V_1}(z)\langle z, z\rangle - 7560S^6_{V_1}(z)\langle z, z\rangle^2$$
$$+ 6300S^4_{V_1}(z)\langle z, z\rangle^3 + 13230\langle z, z\rangle^5 - \frac{315}{4}\dim(V_1)\langle z, z\rangle^5;$$

$$\vdots$$

For each of the 221 solutions in Proposition 3.2, V_1 is explicitly known and one can determine $S^j_{V_1}(z)$. The equations in Theorem 3.3 can be reduced to a system of linear equations with variables m_μ and $m_\mu \in \mathbb{N}$.

By using some computer packages in linear programming, it is possible to reduce the number of solutions to 69 [EMS20].

Theorem 3.4 ([EMS20, Sc93])   *Let V be a strongly regular holomorphic VOA of central charge* 24. *If* $V_1 \neq 0$ *is semisimple, then* V_1 *is isomorphic to one of the Lie algebras listed in Table 2.*

By this theorem, the classification problem of $c = 24$ can be divided into the following two steps:

(1) For each Lie algebra $\mathfrak{g}$ in the list, construct a (strongly regular) holomorphic VOA of $c = 24$ whose weight one Lie algebra is isomorphic to $\mathfrak{g}$.

(2) Show that the isomorphism class of a holomorphic VOA of $c = 24$ is uniquely determined by the Lie algebra structure of the weight one space.

Step (1) has been completed and there are usually several different constructions; see Sections 3.4 and 3.5 for details. Step (2) has also been completed for cases where $V_1 \neq 0$; see Section 3.6. Therefore, the only unsolved case is the case when $V_1 = 0$.

3.4 Orbifold Construction, Li's Δ-Operator and Dimension Formulas

In this section, we review several main techniques for constructions of holomorphic VOAs of $c = 24$.

3.4.1 $\mathbb{Z}_n$-Orbifold Construction

In this subsection, we will review the $\mathbb{Z}_n$-orbifold construction associated with a holomorphic VOA and an automorphism of arbitrary finite order from [EMS20, Mö16].

Let V be a strongly regular holomorphic VOA. Let g be an automorphism of V with a finite order n. For $0 \leq i \leq n - 1$, let $V[g^i]$ be the unique irreducible g^i-twisted V-module (cf. [DLM00, theorem 1.2]). For $i = 0$, we have $V[g^0] = V$. It is known that the group $\langle g \rangle$ acts on the twisted module $V[g^j]$ for each $1 \leq j \leq n - 1$. More precisely, there exists $\phi_j \colon \langle g \rangle \to \mathrm{Aut}_{\mathbb{C}}(V[g^j])$ such that for all $v \in V$ and $i \in \mathbb{Z}$,

$$\phi_j(g^i) Y_{V[g^j]}(v, z) \phi_j(g^i)^{-1} = Y_{V[g^j]}(g^i v, z).$$

Note that such an action is unique up to a multiplication of an nth root of unity. Set $\phi_0(g) = g \in \mathrm{Aut}(V)$. For $0 \leq j, k \leq n - 1$, we denote

$$W^{(j,k)} = \{ w \in V[g^j] \mid \phi_j(g)w = e^{(2\pi k\sqrt{-1})/n} w \}.$$

The fixed-point subspace V^g of g is a full subVOA of V. Note that $W^{(0,0)} = V^g$ and all $W^{(j,k)}$'s are irreducible V^g-modules ([MT04, theorem 2]). It was also shown recently by [CM, Mi15] that V^g is strongly regular. Moreover, any irreducible V^g-module is a submodule of $V[g^i]$ for some i, and there exist exactly n^2 nonisomorphic irreducible V^g-modules ([DRX17]), which are represented by $\{ W^{(j,k)} \mid 0 \leq j, k \leq n - 1 \}$. By calculating the S-matrix of V^g ([Zh96, DLM00]), it was proved in [EMS20, Mö16] that all irreducible V^g-modules $W^{(j,k)}$ are simple current modules. It implies that the set of isomorphism classes of irreducible V^g-modules, denoted by $\mathrm{Irr}(V^g)$, forms an abelian group of order n^2 under the fusion product. We often identify an element in $\mathrm{Irr}(V^g)$ with its representative irreducible V^g-module. Note that $\mathrm{Irr}(V^g) = \{ W^{(j,k)} \mid 0 \leq j, k \leq n - 1 \}$.

Now we assume the following:

(I) For $1 \leq i \leq n - 1$, the conformal weight of $V[g^i]$ is positive.

(II) The conformal weight of $V[g]$ belongs to $(1/n)\mathbb{Z}_{>0}$.

Then by [EMS20, Mö16], the abelian group $\mathrm{Irr}(V^g)$ is isomorphic to $\mathbb{Z}_n \times \mathbb{Z}_n$ and for any $M \in \mathrm{Irr}(V^g)$, the conformal weight $q_\Delta(M)$ of M belongs to $(1/n)\mathbb{Z}$. Moreover, the map $q \colon \mathrm{Irr}(V^g) \to \mathbb{Z}_n$ given by

$$q(M) = n \cdot q_\Delta(M) \quad (\mathrm{mod}\ n)$$

defines a nonsingular quadratic form on $\mathrm{Irr}(V^g)$ [EMS20, Mö16].

Theorem 4.1 ([EMS20, Mö16]) *Let H be a totally isotropic subgroup of* $\mathrm{Irr}(V^g)$; *that is, $q(M) = 0$ for all $M \in H$. Then, the V^g-module $\bigoplus_{M \in H} M$ has a strongly regular VOA structure as an H-graded simple current extension of V^g. Moreover, if H is maximal, then $\bigoplus_{M \in H} M$ is holomorphic.*

Under assumptions (I) and (II), it is proved in [EMS20] that one can choose the ϕ_is such that

- $W^{(i,j)} \boxtimes_{V^g} W^{(k,\ell)} \cong W^{(i+k,j+\ell)}$, where $\boxtimes_{V^g}$ is the fusion product of V^g-modules;
- $W^{(i,j)}$ has the conformal weight $q_\Delta(W^{(i,j)}) = ji/n \mod \mathbb{Z}$.

Then $I = \{W^{(i,0)} \mid 0 \le i \le n - 1\}$ forms a maximal totally isotropic subgroup of $\mathrm{Irr}(V^g)$ and by Theorem 4.1, the V^g-module

$$\widetilde{V}_g = \bigoplus_{M \in I} M$$

is a strongly regular holomorphic VOA. Since I is isomorphic to $\mathbb{Z}_n$, $\widetilde{V}_g$ is a $\mathbb{Z}_n$-graded simple current extension of V^g. The construction of $\widetilde{V}_g$ is often called the $\mathbb{Z}_n$-orbifold construction associated with V and g. We should note that $\widetilde{V}_g$ is uniquely determined by V and g, up to isomorphism.

Remark 4.2   Let g' be an automorphism of V which is conjugate to g. Then g' also satisfies the conditions (I) and (II), and $\widetilde{V}_{g'}$ is isomorphic to $\widetilde{V}_g$ as a VOA.

3.4.2 Li's Δ-Operator and Associated Twisted Modules

In this subsection, we review Li's Δ-operator from [Li96], which gives an explicit construction of the irreducible twisted modules for inner automorphisms.

Let V be a self-contragredient VOA of CFT-type. Let $\langle \cdot | \cdot \rangle$ be the invariant bilinear form on V such that $\langle \mathbb{1} | \mathbb{1} \rangle = -1$ ([Li94]). Let $u \in V_1$ such that $u_{(0)}$ acts semisimply on V. Let $\sigma_u = \exp(-2\pi\sqrt{-1}u_{(0)})$ be the (inner) automorphism of V associated with u. We assume that there exists a positive integer n such that the spectrum of $u_{(0)}$ on V belongs to $(1/n)\mathbb{Z}$. Then we have $\sigma_u^n = 1$ on V.

Conversely, if $\sigma_u^n = 1$, then the spectrum of $u_{(0)}$ on V belongs to $(1/n)\mathbb{Z}$. Let $\Delta(u, z)$ be Li's Δ-operator defined in [Li96]:

$$\Delta(u, z) = z^{u_{(0)}} \exp\left(\sum_{i=1}^{\infty} \frac{u_{(i)}}{-i}(-z)^{-i}\right).$$

Proposition 4.3 ([Li96, proposition 5.4]) *Let σ be a finite order automorphism of V and let $u \in V_1$ such that $u_{(0)}$ acts semisimply on V and $\sigma(u) = u$. Let (M, Y_M) be a σ-twisted V-module and define $(M^{(u)}, Y_{M^{(u)}}(\cdot, z))$ as follows:*

$$M^{(u)} = M \quad \text{as a vector space;}$$
$$Y_{M^{(u)}}(a, z) = Y_M(\Delta(u, z)a, z) \quad \text{for any } a \in V.$$

Then $(M^{(u)}, Y_{M^{(u)}}(\cdot, z))$ is a $\sigma_u\sigma$-twisted V-module. Furthermore, if M is irreducible, then so is $M^{(u)}$.

We will discuss the orbifold construction associated with inner automorphisms in a later section. For this purpose, we will discuss the conditions (I) and (II), also. By the definition of the Δ-operator, the weight operator acts on $V^{(u)}$ by

$$\omega_{(1)} + u_{(0)} + \frac{\langle u|u\rangle}{2}\text{id}.$$

Since the spectrum of $u_{(0)}$ belongs to $(1/n)\mathbb{Z}$, we obtain the following.

Lemma 4.4 *If $\langle u|u\rangle \in (2/n)\mathbb{Z}$, then the weights of $V^{(u)}$ belong to $(1/n)\mathbb{Z}$.*

Next, we consider the conformal weight of $V^{(u)}$. We now assume the following.

- V is strongly regular.
- V_1 is semisimple; let $V_1 = \bigoplus_{i=1}^{t} \mathfrak{g}_i$ be the decomposition of V_1 into the direct sum of simple ideals.
- The subVOA U generated by V_1 is full; that is, its conformal vector coincides with $\omega \in V$.

Note that the latter two assumptions hold for any (strongly regular) holomorphic VOA of $c = 24$ if V_1 is neither $\{0\}$ nor abelian of rank 24 (see Proposition 3.2). By Proposition 3.1, U is the tensor products of the simple affine VOAs $L_{\mathfrak{g}_i}(k_i, 0)$ associated with $\mathfrak{g}_i$ at positive level k_i. Then any irreducible U-module M is isomorphic to $\bigotimes_{i=1}^{t} L_{\mathfrak{g}_i}(k_i, \lambda_i)$, where $L_{\mathfrak{g}_i}(k_i, \lambda_i)$'s are irreducible $L_{\mathfrak{g}_i}(k_i, 0)$-modules with dominant integral weights λ_i of $\mathfrak{g}_i$ at level k_i. Computing the conformal weight of the irreducible twisted module $M^{(u)}$, the following lemma is proved in [LS16a].

Lemma 4.5 (cf. [LS16a])   *If u does not belong to the weight lattice of V_1, then the conformal weight of $V^{(u)}$ is positive.*

Even if u belongs to the weight lattice of V_1, one can still obtain a lower bound of the conformal weight of $V^{(u)}$ as follows:

(Step 1)   Find all irreducible U-modules M whose conformal weight is in $\mathbb{Z}_{\geq 2} \cup \{0\}$.
(Step 2)   Calculate the conformal weight of $M^{(u)}$ for all M in (Step 1).

Then the minimum of the conformal weights in (Step 2) gives a lower bound. By using this argument, it is confirmed in [LS16a, LS19] that the conformal weight of $V^{(u)}$ is at least 1 for some cases.

3.4.3   Dimension Formulas and Weight One Lie Algebras

In this subsection, we review certain dimension formulas for the weight one Lie algebras (see [Mo94, Mö16, MS19]. This is useful for determining the weight one Lie algebra structure of a holomorphic VOA of $c = 24$.

The following theorem is a variant of the dimension formula mentioned in [Mo94].

Theorem 4.6 ([Mö16, (4.10)])   *Let V be a strongly regular holomorphic VOA of $c = 24$. Let g be an automorphism of order n satisfying the conditions (I) and (II) in Section 3.4.1. Assume that $n \in \{2, 3, 5, 7, 13\}$ and that the conformal weight of $V[g^i]$ is at least 1 for all $1 \leq i \leq n - 1$. Then the following equation holds:*

$$\dim V_1 + \dim(\widetilde{V}_g)_1 = (n + 1) \dim V_1^g + 24.$$

In [MS19], the following general formula is proved (see also [Mo94, Mö16, EMS20b]).

Theorem 4.7 (Dimension Formula)   *The dimension of the weight one subspace of $\widetilde{V}_g$ is given by*

$$\dim(\widetilde{V}_g)_1 = 24 + \sum_{d \mid n} c_n(d) \dim(V^{g^d})_1 - R(g)$$

where the $c_n(d) \in \mathbb{Q}$ are defined by $\sum_{d \mid n} c_n(d)(t, d) = n/t$ for all $t \mid n$ and the rest term $R(g)$ is nonnegative.

The rest term $R(g)$ can be described explicitly by the dimensions of the weight spaces of the irreducible V^g-modules of weight less than 1.

Let V be a (strongly regular) holomorphic VOA of $c = 24$ and g an order n automorphism of V satisfying the conditions (I) and (II). Then we can compute $\dim(\widetilde{V}_g)_1$ by using the dimension formula. There are often only few semisimple Lie algebra structures that satisfy the equation in Proposition 3.2 with positive integral levels. We also note that $\widetilde{V}_g$ has an automorphism z of order n such that $(\widetilde{V}_g)_1^z = (V^g)_1$. This is related to the reverse orbifold construction, which will be discussed in Section 3.6.

3.5 Constructions of 71 Holomorphic VOAs of Central Charge 24

We will give a brief review on the construction of holomorphic VOAs of central charge 24 that realize the Lie algebras in Schellekens' list. The main idea is to try to construct a new VOA from a known VOA by using the "orbifold construction."

3.5.1 Lattice Vertex Operator Algebras

We first recall the notion of lattice VOAs and review some of their properties. We use the standard notation for the lattice vertex operator algebra,

$$V_L = M(1) \otimes \mathbb{C}\{L\},$$

associated with a positive-definite even lattice L of rank d [FLM88]. Let $\mathfrak{h} = \mathbb{C} \otimes_{\mathbb{Z}} L$ be an abelian Lie algebra. We extend the bilinear form $(\cdot|\cdot)$ to $\mathfrak{h}$ by $\mathbb{C}$-linearity. Let $\hat{\mathfrak{h}} = \mathfrak{h} \otimes \mathbb{C}[t, t^{-1}] \oplus \mathbb{C}k$ be the corresponding affine algebra, where $\mathbb{C}k$ is the 1-dimensional center of $\hat{\mathfrak{h}}$. The subspace $M(1) = \mathbb{C}[\alpha_i(n)|1 \leq i \leq d, n < 0]$ for a basis $\{\alpha_1, \ldots, \alpha_d\}$ of $\mathfrak{h}$, where $\alpha(n) = \alpha \otimes t^n$, is the unique irreducible $\hat{\mathfrak{h}}$-module such that $\alpha(n) \cdot 1 = 0$ for all $\alpha \in \mathfrak{h}$ and n nonnegative, and k acts as the scalar 1. Also, $\mathbb{C}\{L\} = Span\{e^\beta \mid \beta \in L\}$ is the twisted group algebra of the additive group L such that $e^\beta e^\alpha = (-1)^{(\alpha|\beta)} e^\alpha e^\beta$ for any $\alpha, \beta \in L$. The vacuum vector $\mathbb{1}$ of V_L is $1 \otimes e^0$ and the conformal vector ω is $\frac{1}{2} \sum_{i=1}^d \beta_i(-1)^2 \cdot \mathbb{1}$, where $\{\beta_1, \ldots, \beta_d\}$ is an orthonormal basis of $\mathfrak{h}$. For the explicit definition of the corresponding vertex operators, we will refer to [FLM88] for details. We also note that V_L is strongly regular and the central charge of V_L is equal to d, the rank of L.

Let $L^* = \{v \in \mathbb{R} \otimes_{\mathbb{Z}} L \mid (v|L) \subset \mathbb{Z}\}$ be the dual lattice of L. For $\alpha + L \in L^*/L$, denote $V_{\alpha+L} = M(1) \otimes \mathbb{C}\{\alpha + L\}$, where $\mathbb{C}\{\alpha + L\} = Span\{e^\beta \mid \beta \in \alpha + L\} \subset \mathbb{C}\{L^*\}$. Then $V_{\alpha+L}$ is an irreducible V_L-module [FLM88]. It was proved in [Do93] that any irreducible V_L-module is isomorphic to $V_{\alpha+L}$ for some $\alpha + L \in L^*/L$. In particular, we have the following result.

Theorem 5.1   *Let L be an even lattice of rank d. If L is unimodular, that is, $L^* = L$, then V_L is a strongly regular holomorphic VOA of central charge d.*

There are exactly 24 even unimodular lattices of rank 24 and their isometry types are determined by the structures of their norm 2 vectors, which are also known as the root systems in the literature [Ni73]. Therefore, there are also 24 holomorphic lattice VOA of central charge 24.

Proposition 5.2   *There exist holomorphic VOAs of central charge* 24 *whose weight one Lie algebras have the following* 24 *types:*

$$A_{1,1}^{24}, \qquad A_{2,1}^{12}, \qquad A_{3,1}^8, \quad A_{4,1}^6, \quad A_{5,1}^4 D_{4,1}, \qquad D_{4,1}^6, \quad A_{6,1}^4, \quad A_{7,1}^2 D_{5,1}^2,$$
$$A_{8,1}^3, \qquad A_{9,1}^2 D_{6,1}, \quad D_{6,1}^4, \quad E_{6,1}^4, \quad A_{11,1} D_{7,1} E_{6,1}, \quad A_{12,1}^2, \quad D_{8,1}^3, \quad A_{15,1} D_{9,1},$$
$$A_{17,1} E_{7,1}, \quad D_{10,1} E_{7,1}^2, \quad D_{12,1}^2, \quad A_{24,1}, \quad D_{16,1} E_{8,1}, \qquad E_{8,1}^3, \quad D_{24,1}, \quad U(1)^{24},$$

where $U(1)$ *is a* 1-*dimensional abelian Lie algebra.*

3.5.2 Irreducible Twisted Modules for Lattice VOAs

In this subsection, we review a construction of irreducible twisted V_L-modules, which will be used to apply the $\mathbb{Z}_n$-orbifold construction to V_L.

Let L be an even unimodular lattice and $O(L)$ the isometry group of L. For $g \in O(L)$, set $L^g = \{v \in L \mid g(v) = v\}$ and let P_0^g be the orthogonal projection from $\mathbb{Q} \otimes_{\mathbb{Z}} L$ to $\mathbb{Q} \otimes_{\mathbb{Z}} L^g$. Let $\hat{L} = \{\pm e^\alpha \mid \alpha \in L\}$ be a central extension of L by $\langle -1 \rangle$ with the commutator relation $e^\beta e^\alpha = (-1)^{(\alpha|\beta)} e^\alpha e^\beta$. Let $\mathrm{Aut}\,(\hat{L})$ be the set of all group automorphisms of $\hat{L}$. For $\varphi \in \mathrm{Aut}\,(\hat{L})$, we define the group automorphism $\bar{\varphi}$ of L by $\varphi(e^\alpha) \in \{\pm e^{\bar{\varphi}(\alpha)}\}$, $\alpha \in L$. Set

$$O(\hat{L}) = \{\varphi \in \mathrm{Aut}\,(\hat{L}) \mid \bar{\varphi} \in O(L)\}.$$

Then $O(\hat{L})$ acts on V_L as an automorphism group ([FLM88]).

Let $g \in O(L)$ be of order n. We call $\varphi \in O(\hat{L})$ a *lift* of g if $\bar{\varphi} = g$. A lift $\phi_g \in O(\hat{L})$ of g is called *standard* if $\phi_g(e^\alpha) = e^\alpha$ for $\alpha \in L^g$. Note that a standard lift ϕ_g always exists ([Le85]) and its order is n or $2n$ (see [EMS20, Mö16] for details). Then V_L has a unique irreducible ϕ_g-twisted V_L-module, up to isomorphism ([DLM00]). Such a module $V_L[\phi_g]$ was constructed in [Le85, DL96] explicitly; as a vector space,

$$V_L[\phi_g] \cong M(1)[g] \otimes \mathbb{C}[P_0^g(L)] \otimes T,$$

where $M(1)[g]$ is the "g-twisted" free bosonic space, $\mathbb{C}[P_0^g(L)]$ is the group algebra of $P_0^g(L)$ and T is an irreducible module for a certain "g-twisted" central extension of L. (see [Le85, propositions 6.1 and 6.2] and [DL96, remark 4.2] for details). Recall that

$$\dim T = |L_g/(1-g)L|^{1/2}$$

and that the weight ρ_g of T is given by

$$\rho_g := \frac{1}{4n^2} \sum_{j=1}^{n-1} j(n-j) \dim \mathfrak{h}_{(j)}, \qquad (5.1)$$

where $L_g = \{v \in L \mid (v|L^g) = 0\}$ and $\mathfrak{h}_{(j)} = \{v \in \mathfrak{h} \mid g(v) = e^{(2j\pi\sqrt{-1})/n}v\}$. The weight of an element $v_1(-n_1)\dots v_s(-n_s) \otimes e^\alpha \otimes t \in V_L[\phi_g]$ is given by

$$\sum_{i=1}^{s} n_i + \frac{(\alpha|\alpha)}{2} + \rho_g, \qquad (5.2)$$

where $v_1(-n_1)\dots v_s(-n_s) \in M(1)[g]$, $e^\alpha \in \mathbb{C}[P_0^g(L)]$ and $t \in T$. Note that $n_i \in (1/n)\mathbb{Z}_{>0}$ and that the conformal weight of $V_L[\phi_g]$ is ρ_g.

3.5.3 $\mathbb{Z}_2$-Orbifold Construction Associated with a Lift of the -1-Isometry

The original construction of the moonshine VOA $V^\natural$ is done by applying the $\mathbb{Z}_2$-orbifold construction to the Leech lattice VOA and an order 2 lift of the -1-isometry [FLM88]. This construction was later generalized to other Niemeier lattice VOAs by [DGM96]. Now let us recall the construction.

Let N be a Niemeier lattice, a positive-definite even unimodular lattice of rank 24, and V_N the lattice VOA associated with N. Let θ be an order 2 automorphism of V_N such that $\theta(e^\alpha) = \lambda_\alpha e^{-\alpha}$, $\lambda_\alpha \in \{\pm 1\}$ for all $\alpha \in N$; that is, θ is a lift of the -1-isometry of N. Let $V_N[\theta]$ be the irreducible θ-twisted V_L-module. Then, by (5.1), the conformal weight of $V_N[\theta]$ is $3/2$. Hence $V_N[\theta]$ satisfies the conditions (I) and (II) for the orbifold construction in Section 3.4.1. Applying the $\mathbb{Z}_2$-orbifold construction to V_N and θ, we obtain a strongly regular holomorphic VOA $(\widetilde{V_N})_\theta$ of central charge 24.

Since the conformal weight of $V_N[\theta]$ is $3/2$, we have $((\widetilde{V_N})_\theta)_1 = \{v \in (V_N)_1 \mid \theta(v) = v\}$. The weight one Lie algebra of $(\widetilde{V_N})_\theta$ is determined in [FLM88, DGM96], and there are 9 cases that the holomorphic VOA $(\widetilde{V_N})_\theta$ is again isomorphic to a Niemeier lattice VOA. The remaining 15 holomorphic VOAs are nonisomorphic to each other since their weight one Lie algebras are nonisomorphic.

Proposition 5.3 ([FLM88, DGM96]) *There exist holomorphic VOAs of central charge* 24 *whose weight one Lie algebras have the following* 15 *types:*

$A_{1,4}^{12}$, $A_{1,2}^{16}$, $\qquad$ $C_{2,2}^6$, $A_{3,2}^4 A_{1,1}^4$, $B_{3,2}^4$, $\qquad$ $D_{4,2}^2 C_{2,1}^4$, $B_{4,2}^3$, $D_{5,2}^2 A_{3,1}^2$,
$C_{4,1}^4$, $D_{6,2} C_{4,1} B_{3,1}^2$, $B_{6,2}^2$, $D_{8,2} B_{4,1}^2$, $D_{9,2} A_{7,1}$, $B_{12,2}$, $\qquad$ 0.

Remark 5.4   Let C be a doubly even self-dual binary code of length 24. Define

$$A(C) = \frac{1}{\sqrt{2}}\{(x_1,\ldots,x_{24}) \in \mathbb{Z}^{24} \mid (x_i \pmod 2)) \in C\},$$

$$C(C) = \{v \in A(C) \mid (v|(1,1,\ldots,1)) \in \sqrt{2}\mathbb{Z}\} \oplus \mathbb{Z}\frac{1}{2\sqrt{2}}(-3,1,1,\ldots,1).$$

Then $A(C)$ and $C(C)$ are even unimodular [CS99] and their rank are 24. It was shown in [DGM96, proposition 7.3] that $V_{C(C)} \cong \widetilde{(V_{A(C)})_\theta}$. Note that there exist exactly 9 doubly even self-dual binary codes of length 24, up to equivalence.

3.5.4 Framed vertex operator algebras

Next we review the constructions for 17 holomorphic VOAs of central charge 24 as framed VOAs from [La11, LS12]. For the fundamental results about framed VOAs, see [DGH98, Mi04, LY08].

Let $L(\frac{1}{2},0)$ be the simple Virasoro VOA of central charge $\frac{1}{2}$ and let $L(\frac{1}{2},h)$, $h \in \{\frac{1}{2},\frac{1}{16}\}$, be the irreducible $L(\frac{1}{2},0)$-module with conformal weight h. It is well-known that $L(\frac{1}{2},0)$ is strongly regular and has exactly three irreducible modules, namely, $L(\frac{1}{2},0)$, $L(\frac{1}{2},\frac{1}{2})$, and $L(\frac{1}{2},\frac{1}{16})$. The fusion products are also known [DMZ94]:

$$L(\tfrac{1}{2},\tfrac{1}{2}) \boxtimes L(\tfrac{1}{2},\tfrac{1}{2}) = L(\tfrac{1}{2},0), \quad L(\tfrac{1}{2},\tfrac{1}{2}) \boxtimes L(\tfrac{1}{2},\tfrac{1}{16}) = L(\tfrac{1}{2},\tfrac{1}{16}),$$
$$L(\tfrac{1}{2},\tfrac{1}{16}) \boxtimes L(\tfrac{1}{2},\tfrac{1}{16}) = L(\tfrac{1}{2},0) \oplus L(\tfrac{1}{2},\tfrac{1}{2}). \tag{5.3}$$

Definition 5.5 ([DGH98])   A simple VOA V of central charge $n \in \mathbb{Z}/2$ is said to be *framed* if it contains a full subVOA isomorphic to the tensor product VOA $L(\frac{1}{2},0)^{\otimes 2n}$. Such a full subVOA is called a *Virasoro frame* of V.

Given a framed VOA V of central charge $n/2$ with a Virasoro frame F, one can associate two binary codes C and D of length n with V and F as follows: Let $F(\cong L(\frac{1}{2},0)^{\otimes n})$. Then

$$V \cong \bigoplus_{h_i \in \{0,\frac{1}{2},\frac{1}{16}\}} m_{h_1,\ldots,h_n} L(\tfrac{1}{2},h_1) \otimes \cdots \otimes L(\tfrac{1}{2},h_n),$$

where the nonnegative integer $m_{h_1,\ldots,h_n}$ is the multiplicity of $L(\frac{1}{2},h_1) \otimes \cdots \otimes L(\frac{1}{2},h_n)$ in V and all instances of $m_{h_1,\ldots,h_n}$ are finite.

Definition 5.6   Let $M \cong L(\frac{1}{2},h_1)\otimes\cdots\otimes L(\frac{1}{2},h_n)$ be an irreducible F-module. We define $\tau(M)$ of M as the binary word $\beta = (\beta_1,\ldots,\beta_n) \in \mathbb{Z}_2^n$ such that

$$\beta_i = \begin{cases} 0 & \text{if } h_i = 0 \text{ or } 1/2, \\ 1 & \text{if } h_i = 1/16. \end{cases} \tag{5.4}$$

For any $\beta \in \mathbb{Z}_2^n$, let V^β be the sum of all irreducible F-submodules M of V such that $\tau(M) = \beta$.

Definition 5.7   Let $D := \{\beta \in \mathbb{Z}_2^n \mid V^\beta \neq 0\}$. Then D is a linear subcode of $\mathbb{Z}_2^n$, and it is called the $1/16$-*code* of V with respect to F.

For $\alpha = (\alpha_1, \ldots, \alpha_n) \in \mathbb{Z}_2^n$, define

$$m_\alpha = m_{\alpha_1/2, \ldots, \alpha_n/2} \quad \text{and} \quad M_\alpha = L(1/2, \alpha_1/2) \otimes \cdots \otimes L(1/2, \alpha_n/2).$$

Here we identify $\{0, 1\}$ with $\mathbb{Z}_2$. Then V^0 is given by

$$V^0 = \bigoplus_{\alpha \in \mathbb{Z}_2^n} m_\alpha M_\alpha.$$

It is proved in [DMZ94] that $m_{h_1, \ldots, h_n}$ is at most 1 if all h_i are different from $1/16$.

Definition 5.8   Let $C := \{\alpha \in \mathbb{Z}_2^n \mid m_\alpha \neq 0\}$. Then C is also a linear code and is called the $1/2$-*code* of V with respect to F.

Summarizing, there exists a pair of binary codes (C, D) of length n such that

$$V = \bigoplus_{\beta \in D} V^\beta \quad \text{and} \quad V^0 = \bigoplus_{\alpha \in C} M_\alpha.$$

The next lemma follows from the integral condition on the weights of V.

Lemma 5.9   *The $1/16$-code D is triply even:* $\text{wt}(\alpha) \equiv 0 \mod 8$ *for all $\alpha \in D$ and the $1/2$-code C is even. Moreover, $D < C^\perp = \{\beta \in \mathbb{Z}_2^n \mid (\alpha, \beta) = 0$ for all $\alpha \in C\}$, where $(\cdot, \cdot)$ is the standard inner product of $\mathbb{Z}_2^n$.*

The following theorems are also well-known.

Theorem 5.10 ([DGH98, theorem 2.9] and [Mi04, theorem 6.1])   *Let V be a framed VOA with the structure codes (C, D). Then, V is holomorphic if and only if $C = D^\perp$.*

Theorem 5.11 ([LY08, theorem 10])   *Let D be a linear binary code of length $16k, k \in \mathbb{Z}_{>0}$. Assume that D is triply even and contains the all-one vector. Then there exists a holomorphic framed VOA of central charge $8k$ with the structure codes $(D^\perp, D)$.*

Theorem 5.11 suggests a method for classifying all holomorphic framed VOAs of central charge 24 as follows:

(i) Classify all triply even codes of length 48 containing the all-one vector.
(ii) For each code D in (i), classify all possible holomorphic framed VOA structures with structure codes $(D^\perp, D)$.

A classification of all triply even codes of length 48 has been obtained in [BM12]. By the results in [BM12], all possible Lie algebra structures for the weight one Lie algebras of holomorphic framed VOAs of central charge 24 were determined in [La11, LS12]. In particular, we have the following theorem.

Theorem 5.12 ([La11, LS12]) *There exist holomorphic framed VOAs of central charge* 24 *whose weight one Lie algebras have the following* 17 *types:*

$$A_{3,4}^3 A_{1,2}, \qquad D_{5,8} A_{1,2}, \qquad D_{4,4} A_{2,2}^4, \qquad A_{7,4} A_{1,1}^3, \quad D_{5,4} C_{3,2} A_{1,1}^2, \quad C_{4,2} A_{4,2}^2,$$
$$A_{5,2}^2 C_{2,1} A_{2,1}^2, \quad A_{7,2} C_{3,1}^2 A_{3,1}, \quad C_{7,2} A_{3,1}, \qquad A_{8,2} F_{4,2}, \quad A_{9,2} A_{4,1} B_{3,1}, \quad E_{6,2} C_{5,1} A_{5,1},$$
$$C_{6,1} B_{4,1}, \qquad C_{8,1} F_{4,1}^2, \qquad E_{7,2} B_{5,1} F_{4,1}, \quad C_{10,1} B_{6,1}, \quad E_{8,2} B_{8,2}.$$

A complete answer to (ii) was also established in [LS15] as follows:

Theorem 5.13 ([LS15]) *The isomorphism class of a holomorphic framed VOA of central charge* 24 *is uniquely determined by the Lie algebra structure of its weight one subspace. In particular, there exist exactly* 56 *holomorphic framed VOAs of central charge* 24, *up to isomorphism.*

Remark 5.14 In [HS14], holomorphic VOAs of central charge 24 whose weight one Lie algebra has type $A_{8,2} F_{4,2}$, $C_{4,2} A_{4,2}^2$, and $D_{4,4} A_{2,2}^4$ were constructed as simple current extensions of $V_{\sqrt{2}E_8 \oplus \sqrt{2}D_4} \otimes V_{\sqrt{2}D_{12}^+}^+$. In [DJX14], a holomorphic VOA of central charge 24 whose weight one Lie algebra has type $A_{9,2} A_{4,1} B_{3,1}$ was constructed by using mirror extension.

3.5.5 $\mathbb{Z}_n$-Orbifold Constructions Associated with the Niemeier Lattice VOAs

In this subsection, we review $\mathbb{Z}_n$-orbifold constructions associated with the Niemeier lattice VOAs, which provides explicit construction of many holomorphic VOAs of central charge 24.

3.5.5.1 $\mathbb{Z}_3$-Orbifold Construction

Let N be a Niemeier lattice. Let g be an order 3 isometry of N such that the rank of N^g is a multiple of 6. Since the order of g is odd, a standard lift ϕ_g has order 3 ([EMS20, proposition 7.3]). The conformal weight ρ_g can be determined by the rank of N^g (see (5.1)). In [Mi13], the $\mathbb{Z}_3$-orbifold construction associated with V_N and ϕ_g was established, and, as an application, a holomorphic VOA of central charge 24 whose weight one Lie algebra has type $E_{6,3}G_{2,1}^3$ was constructed. By considering all Niemeier lattices and suitable order 3 isometries of the Niemeier lattices, two more holomorphic VOAs of central charge 24 were constructed in [SS16] (cf. [ISS15]); their weight one Lie algebras have the type $A_{2,3}^6$ and $A_{5,3}D_{4,3}A_{1,1}^3$.

Theorem 5.15 ([Mi13, SS16])   *There exist holomorphic VOAs of central charge 24 whose weight one Lie algebras have the following 3 types: $A_{2,3}^6$, $A_{5,3}D_{4,3}A_{1,1}^3$, and $E_{6,3}G_{2,1}^3$.*

Remark 5.16   In [Mi13], a holomorphic VOA of central charge 24 with trivial weight one space was constructed by applying the $\mathbb{Z}_3$-orbifold construction to the Leech lattice VOA and a fixed-point free isometry of order 3. It was first announced in [DM94] that it is isomorphic to $V^\natural$, which was recently verified in [CLS18].

3.5.5.2 $\mathbb{Z}_n$-Orbifold Construction for $n \geq 4$

As we mentioned in Section 3.4.1, a $\mathbb{Z}_n$-orbifold construction was established in [EMS20, Mö16]. Applying the $\mathbb{Z}_n$-orbifold construction to Niemeier lattice VOAs and some finite order isometries of the Niemeier lattices, the following result is obtained.

Theorem 5.17 ([EMS20, Mö16])   *There exist holomorphic VOAs of central charge 24 whose weight one Lie algebras have the following five types: $A_{2,1}B_{2,1}$, $E_{6,4}$, $A_{4,5}^2$, $A_{2,6}D_{4,12}$, $A_{1,1}C_{5,3}G_{2,2}$, and $C_{4,10}$.*

Remark 5.18   Many holomorphic framed VOAs in Theorem 5.12 can be constructed by $\mathbb{Z}_2$, $\mathbb{Z}_4$, or $\mathbb{Z}_8$-orbifold construction associated with the Niemeier lattice VOAs, also.

3.5.6 Orbifold Construction Associated with Inner Automorphisms

In [LS16a], the orbifold construction associated with inner automorphisms was studied. In particular, five holomorphic VOAs of central charge 24 were constructed. In this subsection, we review the constructions.

In [LS16a], the orbifold construction associated with an inner automorphism σ_u was studied based on the analysis in [Li96] and [DLM96]. An advantage for using an inner automorphism is that σ_u is defined globally on the whole VOA V. An explicit construction of the irreducible σ_u-twisted module was also obtained in [Li96] by using the Δ-operator (see Section 3.4.2). Therefore, we can apply the orbifold construction to nonlattice type VOA. By the definition

Table 1 Lie algebra structures of V_1, $(V^{\sigma_u})_1$ and $\widetilde{V}_1$

(Original) Lie algebra V_1	(Fixed point) Lie subalgebra $(V^{\sigma_u})_1$	(New) Lie algebra $\widetilde{V}_1$
$E_{6,3}G_{2,1}^3$	$D_{5,3}A_{1,1}^2A_{1,3}^2G_{2,1}U(1)$	$D_{7,3}A_{3,1}G_{2,1}$
$D_{7,3}A_{3,1}G_{2,1}$	$D_{6,3}A_{3,1}A_{1,1}A_{1,3}U(1)$	$E_{7,3}A_{5,1}$
$E_{7,3}A_{5,1}$	$A_{7,3}A_{2,1}^2U(1)$	$A_{8,3}A_{2,1}^2$
$C_{5,3}G_{2,2}A_{1,1}$	$A_{4,6}A_{1,6}A_{1,2}U(1)^2$	$A_{5,6}C_{2,3}A_{1,2}$
$A_{4,5}^2$	$A_{3,5}^2U(1)^2$	$D_{6,5}A_{1,1}^2$

of the twisted modules, it is relatively easy to obtain a necessary and sufficient condition on u so that the conformal weights of the irreducible twisted σ_u^i-module belong to $(1/n)\mathbb{Z}_{>0}$ (see Lemmas 4.4 and 4.5).

By choosing the holomorphic VOA V and its inner automorphism σ_u of order 2 carefully, several new holomorphic VOAs of central charge 24 were constructed in [LS16a] by applying the $\mathbb{Z}_2$-orbifold construction to V and σ_u. The Lie algebra structures of V_1, $(V^{\sigma_u})_1$ and $\widetilde{V}_1$ were summarized in Table 1, where $\widetilde{V}$ denotes the resulting VOA $\widetilde{V}_{\sigma_u}$ obtained by the orbifold construction and V^{σ_u} is the set of fixed-points of σ_u.

Theorem 5.19 ([LS16a]) *There exist holomorphic VOAs of central charge 24 whose weight one Lie algebras have the following 5 types: $D_{7,3}A_{3,1}G_{2,1}$, $E_{7,3}A_{5,1}$, $A_{8,3}A_{2,1}^2$, $D_{6,5}A_{1,1}^2$, and $A_{5,6}C_{2,3}A_{1,2}$.*

In the determination of the Lie algebra structure of the resulting VOA, the dimension formula plays a crucial role (cf. Section 3.4.3).

3.5.7 Orbifold Construction Associated with Nonstandard Lifts

In this subsection, we discuss a construction of a holomorphic VOA of central charge 24 whose weight one Lie algebra has the type $A_{6,7}$ from [LS16b].

It was shown in [LS16b] that a holomorphic VOA V of central charge 24 with $V_1 = A_{6,7}$ can be constructed by applying the $\mathbb{Z}_7$-orbifold construction to the Leech lattice VOA V_Λ and an order 7 automorphism of V_Λ but the desired automorphism is somewhat tricky, being product of a standard lift of an order 7 isometry of Λ and an order 7 inner automorphism of V_Λ. An explicit construction of the irreducible twisted V_Λ-module is also obtained in [LS16b] by combining the explicit construction of the twisted V_Λ-modules for an isometry of Λ (see Section 3.5.2) and the modification by Li's Δ-operator (see Section 3.4.2). Moreover, it was proved that the weight one subspace of the resulting orbifold VOA has dimension 48 and it is a simple Lie algebra of type A_6.

Theorem 5.20 ([LS16b])   *There exists a holomorphic VOA of central charge 24 whose weight one Lie algebra has the type $A_{6,7}$.*

3.5.8 A Holomorphic VOA V with $V_1 = F_{4,6}A_{2,2}$

Finally, we discuss a construction of a strongly regular holomorphic VOA of central charge 24 whose weight one Lie algebra has the type $F_{4,6}A_{2,2}$. Such a VOA is constructed in [LL20] by applying a $\mathbb{Z}_2$-orbifold construction to a holomorphic VOA U with the weight one Lie algebra $A_{8,3}A_{2,1}^2$ and a suitable automorphism g of order 2. In this construction, the fixed points of the automorphism g on $U_1 = A_{8,3}A_{2,1}^2$ have the type $B_{4,6}A_{2,2}$. Therefore, $g|_{U_1}$ is an outer automorphism of the Lie algebra U_1. In general, it is very difficult to determine if an outer automorphism of Lie algebra U_1 can be extended to an automorphism of the whole VOA. An explicit construction of the corresponding twisted module is also missing for such an automorphism.

The main trick in [LL20] is to use an alternative construction a holomorphic VOA U of central with $U_1 = A_{8,3}A_{2,1}^2$ based on mirror extensions of VOAs [DJX14, Xu07]. This alternative construction is first obtained in [Xu07] in terms of conformal nets and is proposed in [DJX14] in the VOA setting. Using mirror extensions and the theory of modular invariants, it was shown in [LL20] that the VOA structure of a holomorphic VOA U of central charge 24 with $U_1 = A_{8,3}A_{2,1}^2$ is unique, up to isomorphism. Moreover, by generalizing a result of [Sh04], a sufficient condition for extending an automorphism of a subVOA to the whole VOA is obtained. By these two facts, it was shown that if U is a strongly regular holomorphic VOA of central charge 24 such that $U_1 = A_{8,3}A_{2,1}^2$, then there exists an involution $g \in \mathrm{Aut}(U)$ such that U_1^g is a Lie algebra of type $B_{4,6}A_{2,2}$ [LL20, corollary 4.20]. The conformal weight of the unique irreducible g-twisted U-module $U[g]$ is also determined using the explicit action of g on U_1. As a consequence, the following result is proved in [LL20].

Theorem 5.21 ([LL20])   *There exists a holomorphic VOA of central charge 24 whose weight one Lie algebra has the type $F_{4,6}A_{2,2}$.*

Remark 5.22   A holomorphic VOA with $V_1 = F_{4,6}A_{2,2}$ is also constructed in [La20a] by a completely different method.

By the results in Section 3.5, it is known that all 71 Lie algebras in Schellekens' list can be realized as the weight one Lie algebra of some holomorphic VOA of central charge 24.

3.6 Uniqueness of Holomorphic VOAs of Central Charge 24

In this section, we deal with the uniqueness problem. In particular, we review a technique which we call "reverse orbifold construction" and discuss its application towards the uniqueness problem.

First, we recall a characterization of Niemeier lattice VOAs.

Theorem 6.1 ([DM04b, corollary 1.4]) *Let V be a strongly regular holomorphic VOA of central charge 24. If the Lie rank of V_1 is 24, then V is isomorphic to a Niemeier lattice VOA.*

By this theorem, the uniqueness of holomorphic VOAs are established for the 24 types of Lie algebras in Theorem 5.2.

It is also well-known that the simple affine VOAs of type $A_{1,2}$, $B_{8,1}$, and $E_{8,2}$ are framed (cf. [La11]). Hence, as a corollary of Theorem 5.13, we obtain the following.

Corollary 6.2 *Let V be a strongly regular holomorphic VOA of central charge 24. Assume that the weight one Lie algebra V_1 has type $A_{1,2}^{16}$ or $B_{8,1}E_{8,2}$. Then V is framed. In particular, the VOA structure of V is unique up to isomorphism.*

3.6.1 Reverse Orbifold Construction and Its Application

In this subsection, we review a technique which we call "reverse orbifold construction" from [LS19] (cf. [EMS20, Mö16]) and discuss its application.

Let V be a holomorphic VOA and let g be an order n automorphism of V satisfying the conditions (I) and (II) in Section 3.4.1. Let $W = \widetilde{V}_g$ be the resulting holomorphic VOA by applying $\mathbb{Z}_n$-orbifold construction to V and g. Let h be an order n automorphism of W associated with the $\mathbb{Z}_n$-grading of W. Then $W^h = V^g$ and, for $1 \leq i \leq n-1$, the irreducible h^i-twisted W-module is a direct sum of irreducible V^g-modules. Hence h also satisfies (I) and (II). By the uniqueness of the resulting holomorphic VOA by the orbifold construction, we obtain the following.

Corollary 6.3 ([LS19], see also [EMS20, Mö16]) *The VOA $\widetilde{W}_h$ is isomorphic to V.*

We call this procedure the *reverse orbifold construction*, which is called the inverse orbifold in [EMS20, Mö16]. By using this corollary, we can prove the following theorem about the uniqueness of holomorphic VOAs.

Theorem 6.4 *Let $\mathfrak{g}$ be a Lie algebra and $\mathfrak{p}$ a subalgebra of $\mathfrak{g}$. Let $n \in \mathbb{Z}_{>0}$ and let W be a strongly regular holomorphic VOA of central charge c. Assume the following hypotheses:*

> (i) *For any holomorphic VOA V of central charge c with $V_1 \cong \mathfrak{g}$, there exists an order n automorphism σ of V such that $\widetilde{V}_\sigma \cong W$ and the fixed point Lie algebra $V_1^\sigma \cong \mathfrak{p}$, where W is another holomorphic VOA of central charge c whose structure is known.*
>
> (ii) *Any order n automorphism φ of W belongs to a unique conjugacy class if $(W^\varphi)_1 \cong \mathfrak{p}$ and $(\widetilde{W}_\varphi)_1 \cong \mathfrak{g}$.*

Then any strongly regular holomorphic VOA of central charge c with weight one Lie algebra $\mathfrak{g}$ is isomorphic to $\widetilde{W}_\varphi$. In particular, such a holomorphic VOA is unique up to isomorphism.

Therefore, the main strategy for proving the uniqueness is to find suitable σ and W such that (i) and (ii) hold. For (i), it is actually quite easy to find an inner automorphism $\sigma \in \mathrm{Aut}(V)$ such that $V_1^\sigma \cong \mathfrak{p}$ and $\widetilde{V}_\sigma \cong W$ and there are usually several different choices for σ and W. The difficult part is to show (ii); that is, any automorphism of W satisfying the conditions $(W^\varphi)_1 \cong \mathfrak{p}$ and $(\widetilde{W}_\varphi)_1 \cong \mathfrak{g}$ belongs to a unique conjugacy class. It often requires careful case-by-case analysis.

Remark 6.5   In Theorem 6.4, the restriction of g to V_1 may have order less than n. Indeed, the case $\mathfrak{p} = \mathfrak{g}$ and $n = 2$ was considered in [KLL18], and clearly, $g = id$ on V_1.

There are several different cases.

3.6.2 Case: W is a lattice VOA V_N with $N_2 \neq \emptyset$

Since the automorphism group of V_N is determined in [DN99], the theorem was first applied to a holomorphic VOA U of central charge 24 such that

- $U \cong (\widetilde{V_N})_g$ for some Niemeier lattice N and an order n automorphism g of V_N;
- an order n automorphism of U associated with the $\mathbb{Z}_n$-grading of U as a simple current extension of $(V_N)^g$ is inner.

We should note that many holomorphic VOAs of central charge 24 can be constructed by an orbifold construction from Niemeier lattice VOAs (see also

[HM] for general constructions). The main strategy is to try to reverse the construction and to find a suitable semisimple element $u \in \mathfrak{g}$ such that $\mathfrak{g}^{\sigma_u} = \mathfrak{p}$. By using the representation theory of simple affine VOAs at positive levels and the dimension formula, one needs to verify that (1) σ_u has order n on V and (2) σ_u satisfies the condition (i) for the orbifold construction. Moreover, $\tilde{V}_{\sigma_u}$ is isomorphic to V_N.

Finally, one needs to show that the order n automorphism φ of V_N associated with the simple current extension of V^{σ_u} is uniquely determined by the condition (ii), up to conjugation. The proof for this step often requires some explicit information of the full automorphism group $\mathrm{Aut}(V_N)$ and some case by case analysis.

In [KLL18, LS19], the uniqueness for 17 cases has been proved using the reverse orbifold construction and the argument in 3.6.2.

Theorem 6.6 ([KLL18, LS19]) *A strongly regular holomorphic VOA of central charge* 24 *is unique up to isomorphism if the weight one Lie algebra is one of the 17 types:*
$$A_{1,4}^{12}, \quad A_{1,2}^{16}, \quad C_{2,2}^{6}, \quad A_{3,2}^4 A_{1,1}^4, \quad B_{3,2}^4, \quad D_{4,2}^2 C_{2,1}^4, \quad B_{4,2}^3,$$
$$D_{5,2}^2 A_{3,1}^2, \quad C_{4,1}^4, \quad D_{6,2} C_{4,1} B_{3,1}^2, \quad B_{6,2}^2, \quad D_{8,2} B_{4,1}^2, \quad D_{9,2} A_{7,1}, \quad B_{12,2},$$
$$A_{2,3}^6, \quad A_{5,3} D_{4,3} A_{1,1}^3, \quad E_{6,3} G_{2,1}^3.$$

By using a similar strategy, the uniqueness for 14 other cases was also proved in [EMS20b].

Theorem 6.7 ([EMS20b]) *A strongly rational, holomorphic vertex operator algebra V of central charge* 24 *is uniquely determined by the Lie algebra structure of its weight one space, up to isomorphism if V_1 is isomorphic to one of the following Lie algebras:*
$$A_{5,1} C_{5,1} E_{6,2}, \quad A_{3,1} A_{7,2} C_{3,1}^2, \quad A_{8,2} F_{4,2}, \quad \mathbf{B_{8,1} E_{8,2}}, \quad A_{2,1}^2 A_{5,2}^2 B_{2,1}, \quad C_{8,1} F_{4,1}^2,$$
$$A_{4,2}^2 C_{4,2}, \quad A_{2,2}^4 D_{4,4}, \quad B_{5,1} E_{7,2} F_{4,1}, \quad B_{4,1} C_{6,1}^2, \quad A_{4,5}^2, \quad A_{4,1} A_{9,2} B_{3,1}, \quad B_{6,1} C_{10},$$
$$A_{1,1} C_{5,3} G_{2,2} \text{ and } A_{1,2} A_{3,4}^3.$$

Note that the cases $A_{1,2}^{16}$ and $B_{8,1} E_{8,2}$ were also established in [LS15] (cf. Corollary 6.2).

3.6.3 Case: W is the Leech Lattice VOA V_Λ

Next we discuss the cases when $W \cong V_\Lambda$ is the Leech lattice VOA. This approach is motivated by the construction of a holomorphic VOA with $V_1 \cong A_{6,7}$ [LS16b]. The case $A_{6,7}$ is somewhat special because it is the only case in Schellekens' list that contains an affine VOA of level 7. If we try to apply an orbifold construction to a holomorphic VOA V with $V_1 = A_{6,7}$ and a suitable automorphism of finite order, it seems that the weight one Lie algebra of the resulting VOA is either abelian or isomorphic to $A_{6,7}$ (cf. [LS16a, proposition 5.5]). This observation suggests that one may probably prove the uniqueness for $V_1 \cong A_{6,7}$ by "reversing" the original orbifold construction associated with the Leech lattice VOA. One advantage of using Leech lattice VOA is that $(V_\Lambda)_1$ is an abelian Lie algebra and the subgroup generated by inner automorphisms is also abelian. Then Aut V_Λ is an extension of the isometry group $O(\Lambda)$ of the Leech lattice Λ by an abelian group $(\mathbb{C}^\times)^{24}$ [DN99]. Therefore, Aut V_Λ is easier to handle than the automorphism group of the other Niemeier lattice VOA. Indeed, the uniqueness of conjugacy classes can be verified by calculations on the Leech lattice and its isometry group, Co_0. The technical details on finite order automorphisms of (semi)simple Lie algebras in [Ka90] (cf. [EMS20b, LS19]) are not necessary.

In order to apply the method in [LS19], one should define an automorphism σ of a holomorphic VOA V with $V_1 = A_{6,7}$ so that the Leech lattice VOA is obtained by applying the orbifold construction to V and σ. Since the weight one Lie algebra of the Leech lattice VOA is abelian, the restriction of σ to the weight one Lie algebra V_1 must be regular; that is, the fixed-point subalgebra is abelian. There is a natural choice for such a regular automorphism, which is given by the exponential automorphism associated with the Weyl vector divided by the (dual) Coxeter number. One can confirm the necessary conditions on σ for the orbifold construction and the orbifold construction associated with V and σ actually gives the Leech lattice VOA.

The remaining task is to prove the uniqueness of the conjugacy class of the automorphism φ in the automorphism group Aut V_Λ of the Leech lattice VOA V_Λ under the assumption that the orbifold construction associated with V_Λ and φ gives the original holomorphic VOA V. The assumption on φ gives some constraints, such as dimension or weights for $(V_\Lambda^\varphi)_1$-modules, on the weight one subspaces of the fixed-point subalgebra and the irreducible φ-twisted V_Λ-module. These constraints turn out to be sufficient to determine the conjugacy class of g in $O(\Lambda)$ uniquely. In addition, under these constraints, $\sigma \in (\mathbb{C}^\times)^{24}$ is unique up to conjugation by the centralizer $C_{O(\Lambda)}(g)$ of g in $O(\Lambda)$. Thus φ belongs to the unique conjugacy class in Aut V_Λ.

Using these methods, the following theorem is proved in [LS20a].

Theorem 6.8 *The structure of a strongly regular holomorphic vertex operator algebra of central charge 24 is uniquely determined by its weight one Lie algebra if the Lie algebra has the type $A_{6,7}$, $A_{3,4}^3 A_{1,2}$, $A_{4,5}^2$, $D_{4,12}A_{2,6}$, $A_{6,7}$, $A_{7,4}A_{1,1}^3$, $D_{5,8}A_{1,2}$, or $D_{6,5}A_{1,1}^2$.*

This result also implies that there are construction of holomorphic VOAs whose weight one Lie algebras have the type $A_{3,4}^3 A_{1,2}$, $A_{4,5}^2$, $D_{4,12} A_{2,6}$, $A_{7,4} A_{1,1}^3$, $D_{5,8} A_{1,2}$, and $D_{6,5} A_{1,1}^2$ directly from the Leech lattice VOA. We also remark that the uniqueness for the cases $A_{3,4}^3 A_{1,2}$ and $A_{4,5}^2$ are proved in [EMS20b] by using the orbifold construction from the Niemeier lattices with root lattice D_4^6 and A_4^6, respectively.

This theorem indeed suggested much deeper relations between holomorphic VOAs of central charge 24 and the Leech lattice VOA V_Λ. We will discuss these relations in Section 3.7.

3.6.4 Case: W is a Nonlattice Type VOA

Next we discuss the uniqueness for remaining cases when $V_1 \neq 0$. The strategy is similar to that in Sections 3.6.2 and 3.6.3 but we will also consider the cases when W is not a lattice VOA. The automorphism groups of W are more complicated for these cases. Therefore, one needs to use the theory of Kac about the classification of finite automorphisms of (finite-dimensional) simple Lie algebras [Ka90, chapter 8] (cf. [He78]). It is indeed quite straightforward to show that $\varphi_1|_{W_1}$ is conjugate to $\varphi_2|_{W_1}$ by an inner automorphism in $\mathrm{Aut}\,(W_1)$ as Lie algebra automorphisms if φ_1 and φ_2 are involutions and $(W^{\varphi_1})_1 \cong (W^{\varphi_2})_1$. It means that there is an inner automorphism x of W such that $\varphi_1 x \varphi_2^{-1} x^{-1}$ acts trivially on W_1. Therefore, the main step is to determine the subgroup of $\mathrm{Aut}\,(W)$ that acts trivially on W_1. We call such a subgroup of $\mathrm{Aut}\,(W)$ the inertia group of W and denote it by $I(W)$. In [LS20b], the inertia groups for four holomorphic VOAs of central charge 24 are computed.

Theorem 6.9 *Let W be a holomorphic VOA of central charge* 24 *such that the weight one Lie algebra W_1 has the type $E_{6,3} G_{2,1}^3$, $A_{4,5}^2$, $C_{5,3} G_{2,2} A_{1,1}$, or $A_{7,4} A_{1,1}^3$. Then*

$$I(W) = \begin{cases} 1 & \text{if } W_1 = E_{6,3} G_{2,1}^3 \text{ or } A_{4,5}^2, \\ \langle \sigma_{(0,0,\Lambda_1)} \rangle \cong \mathbb{Z}_2 & \text{if } W_1 = C_{5,3} G_{2,2} A_{1,1}, \\ \langle \sigma_{(0,\Lambda_1,0,0)}, \sigma_{(0,0,\Lambda_1,0)}, \sigma_{(0,0,0,\Lambda_1)} \rangle \cong \mathbb{Z}_2^3 & \text{if } W_1 = A_{7,4} A_{1,1}^3, \end{cases}$$

where Λ_1 denotes the fundamental weight of a root system of type A_1.

By using this information, the uniqueness for six more cases was established.

Theorem 6.10 (see [LS20b]) *Up to isomorphism, there exists a unique strongly regular holomorphic VOA of central charge* 24 *if its weight one Lie algebra has the type $C_{4,10}$, $D_{7,3} A_{3,1} G_{2,1}$, $A_{5,6} C_{2,3} A_{1,2}$, $A_{3,1} C_{7,2}$, $D_{5,4} C_{3,2} A_{1,1}^2$, or $E_{6,4} C_{2,1} A_{2,1}$.*

3.6.5 $A_{8,3}A_{2,1}^2$, $F_{4,6}A_{2,2}$, and $E_{7,3}A_{5,1}$

As we discussed in Section 3.5.8, there is an alternative construction of a holo-morphic VOA U of central charge 24 with $U_1 = A_{8,3}A_{2,1}^2$ based on mirror ex-tensions. Such a construction was first obtained by [Xu07] in terms of confor-mal nets and is proposed in [DJX14] in VOA setting. Using mirror extensions and the theory of modular invariants, the following proposition was shown in [LL20].

Proposition 6.11 *A strongly regular holomorphic VOA structure of central charge* 24 *is unique if the weight one Lie algebra has the type* $A_{8,3}A_{2,1}^2$.

In addition, some information about the automorphism group of U is also determined in the process. By similar arguments as before, the uniqueness for two more cases was also established.

Theorem 6.12 *A holomorphic VOA V of central charge* 24 *is uniquely deter-mined by its weight one Lie algebra if* V_1 *has the type* $F_{4,6}A_{2,2}$, *or* $E_{7,3}A_{5,1}$.

As a consequence of these results, the following theorem has been estab-lished.

Theorem 6.13 *The isomorphism class of a strongly regular holomorphic VOA of central charge* 24 *is determined by its weight one Lie algebra if the weight one subspace is nonzero.*

The remaining case is a famous conjecture in VOA theory [FLM88], namely, a strongly regular holomorphic VOA of central charge 24 with trivial weight one subspace is isomorphic to Frenkel–Lepowsky–Meurman's moonshine VOA. Unfortunately, the technique used there cannot be applied to this case since the VOA has no inner automorphisms if the weight one subspace is zero.

3.7 Leech Lattice and Schellekens' List

As we discussed in Section 3.6.3, there are some nontrivial relations between holomorphic VOAs of central charge 24 and the Leech lattice VOA. In this section, we will discuss these relations in detail.

Let V be a strongly regular holomorphic VOA of central charge 24. Assume that $V_1 \cong \oplus_{j=1}^t \mathcal{G}_j \neq 0$ is semisimple, where $\mathcal{G}_j, j = 1, \ldots, t$, are simple Lie algebras. As we have mentioned in Section 3.6.3, there is a natural regular automorphism on V_1, which can be defined by $\sigma = \exp(2\pi i \alpha(0))$, where $\alpha = \sum_{j=1}^t \rho_j / h_j^\vee \in V_1$ is the sum of Weyl vectors ρ_j of $\mathcal{G}_j$ divided by the dual Coxeter numbers $h_j^\vee$.

By combining the dimension formula in [MS19] (see also Theorem 4.7) with the "very strange formula" in [Ka90], the following theorem is proved in [ELMS21].

Theorem 7.1 *Let V be a strongly regular holomorphic VOA of central charge 24 whose weight-one Lie algebra $V_1 \cong \oplus_{j=1}^t \mathcal{G}_j$ is semisimple. Let $\sigma_\alpha = e^{2\pi i \alpha(0)} \in$ $\mathrm{Aut}\,(V)$, where $\alpha = \sum_{j=1}^t \rho_j / h_j^\vee$. Then the VOA $\widetilde{V}_{\sigma_\alpha}$ obtained from a orbifold construction associated with V and σ_α is isomorphic to the Leech lattice VOA V_Λ.*

The reverse orbifold construction [EMS20, LS19] then implies the following.

Theorem 7.2 *Let V be a strongly regular holomorphic VOA of central charge 24 with $V_1 \neq \{0\}$. Then V is isomorphic to $(\widetilde{V_\Lambda})_g$ for some automorphism $g \in$ $\mathrm{Aut}\,(V_\Lambda)$.*

As an application, a novel proof for Schellekens' list is also provided in [ELMS21].

Theorem 7.3 *Let V be a strongly regular holomorphic VOA of central charge 24 with $V_1 \neq \{0\}$. Then V_1 is isomorphic to one of the 70 nonzero Lie algebras in table 1 of [Sc93].*

This new proof uses only equation (3.1) but none of the higher-order trace identities listed in Theorem 3.3. It is also much simpler than the original proof [EMS20, Sc93] and for the most part does not rely on computer calculations.

Remark 7.4   An alternative proof for Theorem 7.1 is also given in [CLM], which does not require the dimension formula in [MS19].

3.7.1 Generalized Deep Holes

By Theorem 7.1, we know that for any holomorphic VOA of central charge 24, the VOA $\widetilde{V}_{\sigma_\alpha}$ obtained from a orbifold construction associated with V and σ_α is isomorphic to the Leech lattice VOA V_Λ for $\alpha = \sum_{j=1}^t \rho_j / h_j^\vee$. When we consider the reverse orbifold construction, then one can obtain an automorphism $g \in \mathrm{Aut}\,(V_\Lambda)$ such that $\widetilde{(V_\Lambda)}_g \cong V$. By the structure of $\mathrm{Aut}\,(V_\Lambda)$, $g = \phi_\tau \sigma_\beta$, where $\tau \in O(\Lambda)$, $\beta \in \mathbb{C}\Lambda^\tau$ and ϕ_τ is a standard lift of τ. Indeed such an automorphism arising from the reverse orbifold construction is quite special and it may be taken to be a generalized deep hole as introduced in [MS19]. In some sense, the classification of holomorphic VOA of central charge 24 is almost equivalent to the classification of generalized deep holes.

First let us recall the definition of generalized deep holes from [MS19]. By Theorem 4.7, we know that

$$\dim(\widetilde{V}_g)_1 = 24 + \sum_{m|n} c_n(m)\dim(V_1^{g^m}) - R(g),$$

where the rest term $R(g)$ is nonnegative. In particular,

$$\dim(\widetilde{V}_g)_1 \leq 24 + \sum_{m|n} c_n(m)\dim(V_1^{g^m}).$$

Möller and Scheithauer define an automorphism g as a generalized deep hole if the above upper bound is achieved. More precisely, let V be a strongly regular holomorphic VOA of central charge 24 and g an automorphism of V of finite order $n > 1$. Suppose g is of type 0 and V^g satisfies the positivity condition. Then g is called a *generalized deep hole* of V if

(1) $\dim(\widetilde{V}_g)_1 = 24 + \sum_{m|n} c_n(m)\dim(V_1^{g^m})$, and
(2) the Cartan subalgebra of $(\widetilde{V}_g)_1$ has no contributions from the twisted modules; that is, $\operatorname{rank}(\widetilde{V}_g)_1 = \operatorname{rank}(V_1^g)$.

By giving generalized deep holes explicitly, Möller and Scheithauer proved the following theorem.

Theorem 7.5 *Let $\mathfrak{g}$ be one of the 70 nonzero Lie algebras on Schellekens' list (table 1 in [Sc93]). Then there exists a generalized deep hole $g \in \operatorname{Aut}(V_\Lambda)$ such that $((\widetilde{V_\Lambda})_g)_1 \cong \mathfrak{g}$.*

By an averaged version of Kac's very strange formula [Ka90], they also prove the following theorem.

Theorem 7.6 *The cyclic orbifold construction $g \mapsto (\widetilde{V_\Lambda})_g$ defines a bijection between the algebraic conjugacy classes of generalized deep holes g in $\operatorname{Aut}(V_\Lambda)$ with $\operatorname{rank}(V_\Lambda^g)_1 > 0$ and the isomorphism classes of strongly regular holomorphic VOAs V of central charge 24 with $V_1 \neq \{0\}$.*

In addition, they classified the generalized deep holes for the Leech lattice VOA V_Λ and observed the following.

Theorem 7.7 *Under the natural projection $\operatorname{Aut}(V_\Lambda) \to O(\Lambda)$, the 70 algebraic conjugacy classes of generalized deep holes g with $\operatorname{rank}(V_\Lambda^g)_1 > 0$ map to the 11 algebraic conjugacy classes in $O(\Lambda)$ with cycle shapes 1^{24}, $1^8 2^8$, $1^6 3^6$, 2^{12}, $1^4 2^2 4^4$, $1^4 5^4$, $1^2 2^2 3^2 6^2$, $1^3 7^3$, $1^2 2^1 4^1 8^2$, $2^3 6^3$, and $2^2 10^2$.*

This result recovers an observation described in [Hö2].

3.8 Höhn's Observation and ℓ-Duality

Next we discuss several observations and ideas proposed by G. Höhn [Hö2]. His idea provides a completely new view point towards our understanding of the classification. The results that we discussed in Section 3.7 are indeed inspired by his article [Hö2].

Höhn's idea [Hö2] may be viewed as a generalization of the theory of Cartan subalgebras to VOA. The main idea is to study the commutant subVOA of the subVOA generated by a Cartan subalgebra of V_1 and try to construct a holomorphic VOA using certain simple current extensions of lattice VOAs and some orbifold subVOAs in the Leech lattice VOA. Let us first explain his ideas. Let $\mathfrak{g}$ be a Lie algebra in Schellekens' list and let V be a strongly regular holomorphic VOA of central charge 24 that realized $\mathfrak{g}$. Suppose that

$$\mathfrak{g} = \mathfrak{g}_{1,k_1} \oplus \cdots \oplus \mathfrak{g}_{r,k_r}$$

is semisimple, where $\mathfrak{g}_{i,k_i}$ are simple ideals of $\mathfrak{g}$ at level k_i. Then the subVOA U generated by V_1 is isomorphic to the tensor of simple affine VOAs

$$L_{\widehat{\mathfrak{g}_1}}(k_1,0) \otimes \cdots \otimes L_{\widehat{\mathfrak{g}_r}}(k_r,0)$$

and U is a full subVOA of V [DM04b]. Let $\mathfrak{H}$ be a Cartan subalgebra of V_1 and $M(\mathfrak{H})$ the vertex subalgebra generated by $\mathfrak{H}$. Set $W = \mathrm{Com}_V(M(\mathfrak{H}))$ and $X = \mathrm{Com}_V(W)$.

It was shown [DW] that for each $1 \leq i \leq r$, $L_{\mathfrak{g}_i}(k_i,0)$ contains a lattice VOA $V_{\sqrt{k_i}Q_L^i}$, where Q_L^i is the lattice spanned by the long roots of $\mathfrak{g}_i$. Set $Q_{\mathfrak{g}} = \sqrt{k_1}Q_L^1 \oplus \cdots \oplus \sqrt{k_r}Q_L^r$. Then it is clear that $X \supset V_{Q_{\mathfrak{g}}}$ and $\mathrm{Com}_V(X) = W$ and $\mathrm{Com}_V(W) = X$. In this case, $X \cong V_L$ for some even lattice $L > Q_{\mathfrak{g}}$. The set of all irreducible modules $\mathrm{Irr}(V_L)$ of V_L forms an abelian group with respect to the fusion product. Indeed, $\mathrm{Irr}(V_L)$ has a quadratic form $q\colon \mathrm{Irr}(V_L) \to \mathbb{Q}/\mathbb{Z}$ defined by

$$q(V_{\alpha+L}) = \mathrm{wt}(V_{\alpha+L}) = \frac{(\alpha|\alpha)}{2} \quad \mathrm{mod}\ \mathbb{Z},$$

where $\mathrm{wt}(\cdot)$ denotes the conformal weight of the module. Moreover, $\mathrm{Irr}(V_L)$ is isomorphic to $\mathcal{D}(L) = L^*/L$ as a quadratic space.

By some recent results on coset constructions (see [KMi15] and [Lin17]), the set of all irreducible modules $\mathrm{Irr}(W)$ of W also forms a quadratic space and is isomorphic to $(\mathrm{Irr}(V_L), -q)$, where the quadratic form is defined by conformal weights modulo $\mathbb{Z}$. Then the holomorphic VOA V defines a maximal totally singular subspace of $\mathrm{Irr}(V_L) \times \mathrm{Irr}(W)$; hence it induces an anti-isomorphism of quadratic spaces $\varphi\colon (\mathrm{Irr}(V_L), q) \to (\mathrm{Irr}(W), q')$ such that $q(M) + q'(\varphi(M)) = 0$ for all $M \in irr(V_L)$.

Conversely, let $\varphi \colon (\mathrm{Irr}(V_L), q) \to (\mathrm{Irr}(W), -q')$ be an isomorphism of quadratic spaces. Then the set $\{(M, \varphi(M)) \mid M \in \mathrm{Irr}(V_L)\}$ is a maximal totally singular subspace of $\mathrm{Irr}(V_L) \times \mathrm{Irr}(W)$ and hence $U = \bigoplus_{M \in \mathrm{Irr}(V_{L_3})} M \otimes \varphi(M)$ has a structure of a holomorphic VOA.

Under this setting, for $f \in \mathrm{Aut}\,(U)$, the map $\mathrm{Irr}(U) \to \mathrm{Irr}(U)$, $W \mapsto W \circ f$, preserves q_U. Recall that for a VOA V and $f \in \mathrm{Aut}\,(V)$, the f-conjugate $(W \circ f, Y_{W \circ f}(\cdot, z))$ of W is defined as follows:

$$
\begin{aligned}
W \circ f &= W \quad \text{as a vector space;} \\
Y_{W \circ f}(a, z) &= Y_W(fa, z) \quad \text{for any } a \in V.
\end{aligned}
\tag{8.1}
$$

Hence we obtain the canonical group homomorphism

$$
\mu_U \colon \mathrm{Aut}\,(U) \to O(\mathrm{Irr}(U), q_U),
\tag{8.2}
$$

where

$$
O(\mathrm{Irr}(U), q_U) = \{h \in \mathrm{Aut}\,(\mathrm{Irr}(U)) \mid q_U(W) = q_U(h(W)) \text{ for all } W \in \mathrm{Irr}(U)\}
$$

is the orthogonal group of the quadratic space $(\mathrm{Irr}(U), q_U)$.

Let $q_L \colon \mathcal{D}(L) \to \mathbb{Q}/\mathbb{Z}$ be the quadratic form on $\mathcal{D}(L)$ defined by $q_L(v + L) = \langle v \mid v \rangle / 2 + \mathbb{Z}$, $v + L \in \mathcal{D}(L)$. Then $O(L)$ acts on $\mathcal{D}(L)$ and let

$$
\mu_L \colon O(L) \to O(\mathcal{D}(L), q_L)
$$

be the canonical group homomorphism. It is proved in [BLS2] that

$$
\mathrm{Im}\,\mu_{V_L} \cong O(L)/\mathrm{Ker}\,\mu_L \cong \mathrm{Im}\,\mu_L.
$$

Let O be the set of all isometries from $(\mathcal{D}(L), q_L)$ to $(\mathrm{Irr}(W), -q_W)$. For any $\psi \in O$,

$$
V_\psi = \bigoplus_{\lambda + L \in \mathcal{D}(L)} V_{\lambda + L} \otimes \psi(\lambda + L)
$$

has a holomorphic VOA structure as a simple current extension of $V_L \otimes W$ ([EMS20, theorem 4.2]).

Let $f \in \mathrm{Im}\,\mu_W$ and $h \in \mathrm{Im}\,\mu_L$. Then $f \circ \psi \circ h$ is also in O and $(h, f^{-1}) S_{f \circ \psi \circ h} = S_\psi$. Hence (h, f^{-1}) induces an isomorphism between the holomorphic VOAs V_ψ and $V_{f \circ \psi \circ h}$. Conversely, we assume that $\psi, \psi' \in O$ satisfies $V_\psi \cong V_{\psi'}$ as simple current extensions of $V_L \otimes W$. Then there exists an isomorphism $\xi \colon V_\psi \to V_{\psi'}$ such that $\psi(V_L \otimes W) = V_L \otimes W$. Hence S_ψ and $S_{\psi'}$ are conjugate by the restriction of ψ to $V_L \otimes W$. Therefore, the number of inequivalent extensions is equal to the number of double cosets in

$$
\mathrm{Im}\,\mu_W \backslash O / \mathrm{Im}\,\mu_L.
$$

Note that, in general, inequivalent extensions may still be isomorphic as VOAs.

Now fix an isometry $i \in O$. Then $i^*(h) = i^{-1} \circ h \circ i \in O(\mathcal{D}(L), q_L)$ for any $h \in O(\mathrm{Irr}(W), -q_W)$. We consider the double cosets in $i^*(\mathrm{Im}\mu_W)\backslash O(\mathcal{D}(L), q_L)/\mathrm{Im}\mu_L$. Notice that $i \circ f \in O$ for any $f \in O(\mathcal{D}(L), q_L)$. Conversely, $i^{-1} \circ \psi \in O(\mathcal{D}(L), q_L)$ for any $\psi \in O$. Therefore, i induces a bijective map between O and $O(\mathcal{D}(L), q_L)$.

Proposition 8.1 [Hö2, theorem 2.7] *Let $\psi, \psi' \in O$. Then ψ and ψ' are in the same double coset of $\mathrm{Im}\mu_W\backslash O/\mathrm{Im}\mu_L$ if and only if $i^{-1}\psi$ and $i^{-1}\psi'$ are in the same double coset of $i^*(\mathrm{Im}\mu_W)\backslash O(\mathcal{D}(L), q_L)/\mathrm{Im}\mu_L$. In particular, the number of inequivalent extensions in $\{V_\psi \mid \psi \in O\}$ is equal to $|i^*(\mathrm{Im}\mu_W)\backslash O(\mathcal{D}(L), q_L)/\mathrm{Im}\mu_L|$.*

Höhn noticed that the VOA W seems to be related to a certain coinvariant sublattice of the Leech lattice Λ. In particular, he proposed the following conjecture.

Conjecture 8.2 For each semisimple case in Schellekens' list, there exists an isometry $g \in O(\Lambda)$ such that $(\mathrm{Irr}(V_{\Lambda_g}^{\phi_g}), q) \cong (\mathrm{Irr}(V_L), -q)$ as quadratic spaces.

The isometry g for each case has also been described by Höhn (see [Hö2, table 4]). Höhn's conjecture has been confirmed in [LS17] and [La20a]. The automorphism groups for the orbifold VOA $V_{\Lambda_g}^{\phi_g}$ are also determined in [BLS1]. Moreover, the construction proposed by [Hö2] has been confirmed in [BLS2]. Inequivalent extensions of $V_L \otimes V_{\Lambda_g}^{\phi_g}$ are also determined. This provides an alternative proof for the uniqueness of holomorphic vertex operator algebras of central charge 24 with nontrivial weight one Lie algebras.

3.8.1 ℓ-Duality

In this subsection, we discuss more relations among generalized deep holes, Leech lattice and other Niemeier lattices.

Let V be a strongly regular holomorphic VOA of central charge 24 and let $V_1 \cong \oplus_{j=1}^t \mathcal{G}_j$ be semisimple. , where $\mathcal{G}_j, j = 1, \ldots, t$, are simple Lie algebras. Set $\alpha = \sum_{j=1}^t \rho_j/h_j^\vee \in V_1$, where ρ_j is a Weyl vector of $\mathcal{G}_j$ and $h_j^\vee$ is the dual Coxeter number. Then we know that

$$\widetilde{V}_{\sigma_\alpha} \cong V_\Lambda.$$

By considering the reverse orbifold construction, there is a $g \in \mathrm{Aut}(V_\Lambda)$ such that

$$g = \phi_\tau \exp(2\pi i \beta) \quad \text{and} \quad \widetilde{(V_\Lambda)}_g \cong V, \tag{8.3}$$

where ϕ_τ is a standard lift of $\tau \in O(\Lambda)$ and β is fixed by τ.

By some elementary calculations (along with Proposition 3.2), it is possible to show that $\tau \in O(\Lambda)$ must be one of the 11 conjugacy classes mentioned in [Hö2, table 4]. In other words, one can prove that there are only 11 possible conjugacy classes for τ without assuming Schellekens' list, the dimension formula in [MS19], nor the higher trace formulas in Theorem 3.3 (see [LM]).

It turns out that the isometries in [Hö2, table 4] are very special. They all have positive frame shapes and their fixed point sublattices are kind of "self-dual" in the following sense.

Let $\mathcal{P}_0$ be the set of conjugacy classes of $Co.0$ listed in [Hö2, table 4]; that is, $\mathcal{P}_0 = \{1A, 2A, 2C, 3B, 4C, 5B, 6E, 6G, 7B, 8E, 10F\}$. For $\tau \in \mathcal{P}_0$, let ϕ_τ be a standard lift of τ and set $\ell = |\phi_\tau|$. The first observation is the following lemma.

Lemma 8.3   *For $\tau \in \mathcal{P}_0$, we have* $\det(\Lambda^\tau) = \ell^{\mathrm{rank}(\Lambda^\tau)/2}$.

Indeed, one can prove the following theorem.

Theorem 8.4 ([LM, HL90])   *For $\tau \in \mathcal{P}_0$, there is an isometry:*

$$\varphi = \varphi_\tau : \sqrt{\ell}(\Lambda^\tau)^* \to \Lambda^\tau,$$

where $\ell = |\phi_\tau|$.

In other words, Λ^τ is self-dual with respect to the ℓ-dual $\sqrt{\ell}(\Lambda^\tau)^*$. We call such a phenomenon ℓ-*duality*.

The isometry φ can be extended to $\mathbb{C}\Lambda^\tau$ and one can define $\tilde{\beta} = \sqrt{\ell}\varphi(\beta)$, where β is as in (8.3). It is easy to show that $\tilde{\beta}$ has an even norm since the irreducible g-twisted $V_\Lambda[g]$ has elements of integral weights. Then one can construct a neighbor lattice of Λ by using $\tilde{\beta}$. Namely, the lattice $N = \Lambda^{[\tilde{\beta}]} = \Lambda_{\tilde{\beta}} + \mathbb{Z}\tilde{\beta}$, where $\Lambda_{\tilde{\beta}} = \{x \in \Lambda \mid \langle x, \tilde{\beta}\rangle\}$.

The following theorem is proved in [LM].

Theorem 8.5   *Let $\tilde{\beta}$ be as above. Then $N = \Lambda^{[\tilde{\beta}]} \not\cong \Lambda$. Moreover, $\tilde{\beta} = \sqrt{\ell}\varphi(\beta)$ is a deep hole of Λ.*

Via ℓ-duality, we also have:

Theorem 8.6   *Let $N = \Lambda^{[\tilde{\beta}]}$. Then*

$$N^{<\tau>} \cong \sqrt{\ell}\varphi_\tau(L^*).$$

Remark 8.7   (1) The lattice $N^{<\tau>}$ is related to the orbit lattice discussed by Höhn in [Hö2, Section 3].

(2) By Theorems 8.5 and 8.6, a generalized deep hole for the Leech lattice VOA indeed defines a "true" deep hole of the Leech lattice satisfying some extra properties. It provides a new combinatorial approach towards the classification of holomorphic VOAs of central charge 24. This phenomenon

is related to another mysterious observation by Höhn [Hö2, table 3], who noticed that the Lie algebras in Schellekens' list may be described by using a Niemeier lattice and a diagram automorphism of the corresponding affine root system.

References

[BM12] K. Betsumiya and A. Munemasa, On triply even binary codes, *J. London Math. Soc.* **86** (2012), 1–16.

[BLS1] K. Betsumiya, C. H. Lam and H. Shimakura, Automorphism groups of cyclic orbifold vertex operator algebras associated with the Leech lattice and some non-prime isometries, to appear in *Israel J. Math.*

[BLS2] K. Betsumiya, C. H. Lam and H. Shimakura, Automorphism groups and uniqueness of holomorphic vertex operator algebras of central charge 24, Comm. Math. Phys. 399 (2023), 1773–1810; arXiv:2203.15992.

[Bo86] R. E. Borcherds, Vertex algebras, Kac–Moody algebras, and the Monster, *Proc. Nat'l. Acad. Sci. U.S.A.* **83** (1986), 3068–3071.

[CM] S. Carnahan and M. Miyamoto, Regularity of fixed-point vertex operator subalgebras, arXiv:1603.05645.

[CLS18] H. Y. Chen, C. H. Lam and H. Shimakura, $\mathbb{Z}_3$ orbifold construction of the Moonshine vertex operator algebra and some maximal 3-local subgroups of the Monster, *Math. Z.* **288** (2018), 75–100.

[CLM] N. Chigira, C. H. Lam and M. Miyamoto, Orbifold construction and Lorentzian construction of Leech lattice vertex operator algebra, *J. Algebra* , **593**(2022), 26–71.

[Co69] J. H. Conway, A characterisation of Leech's lattice, *Invent. Math.* **7** (1969) 137–142.

[CS99] J. H. Conway and N. J. A. Sloane, Sphere Packings, Lattices and Groups, 3rd edition, Springer, New York, 1999.

[DGM96] L. Dolan, P. Goddard and P. Montague, Conformal field theories, representations and lattice constructions, *Comm. Math. Phys.* **179** (1996), 61–120.

[Do93] C. Dong, Vertex algebras associated with even lattices, *J. Algebra* **161** (1993), 245–265.

[DGH98] C. Dong, R. L. Griess, and G. Höhn, Framed vertex operator algebras, codes and Moonshine module, *Comm. Math. Phys.* **193** (1998), 407–448.

[DJX14] C. Dong, X. Jiao, and F. Xu, Mirror extensions of vertex operator algebras, *Comm. Math. Phys.* **329**, (2014), 263–294

[DL96] C. Dong and J. Lepowsky, The algebraic structure of relative twisted vertex operators, *J. Pure Appl. Algebra* **110** (1996), 259–295.

[DLM96] C. Dong, H. Li, and G. Mason, Simple currents and extensions of vertex operator algebras, *Comm. Math. Phys.* **180** (1996), 671–707.

[DLM00] C. Dong, H. Li, and G. Mason, Modular-invariance of trace functions in orbifold theory and generalized Moonshine, *Comm. Math. Phys.* **214** (2000), 1–56.

[DM94] C. Dong and G. Mason, The construction of the moonshine module as a Z_p-orbifold, in *Mathematical Aspects of Conformal and Topological Field Theories and Quantum Groups* (South Hadley, MA, 1992), *Contemp. Math.*, **175**, Amer. Math. Soc., Providence, RI, 1994, 37–52.

[DM04a] C. Dong and G. Mason, Holomorphic vertex operator algebras of small central charge, *Pacific J. Math.* **213** (2004), 253–266.

[DM04b] C. Dong and G. Mason, Rational vertex operator algebras and the effective central charge, *Int. Math. Res. Not.* (2004), 2989–3008.

[DM06a] C. Dong and G. Mason, Integrability of C_2-cofinite vertex operator algebras, *Int. Math. Res. Not.* (2006), Art. ID 80468, 15 pp.

[DM06b] C. Dong and G. Mason, Shifted vertex operator algebras, *Math. Proc. Cambridge Philos. Soc.* **141** (2006), 67–80.

[DMZ94] C. Dong, G. Mason and Y. Zhu, Discrete series of the Virasoro algebra and the moonshine module, *Proc. Sympos. Pure Math.* **56** (1994), 295–316.

[DN99] C. Dong and K. Nagatomo, Automorphism groups and twisted modules for lattice vertex operator algebras, *in* Recent Developments in Quantum Affine Algebras and Related Topics (Raleigh, NC, 1998), 117–133, *Contemp. Math.* **248**, Amer. Math. Soc., Providence, RI, 1999.

[DR17] C. Dong and L. Ren, Representations of the parafermion vertex operator algebras, *Adv. Math.* **315** (2017) 88–101.

[DRX17] C. Dong, L. Ren and F. Xu, On orbifold theory, *Adv. Math.* **321** (2017), 1–30.

[DW] C. Dong and Q. Wang, The structure of parafermion vertex operator algebras: General case, *Comm. Math. Phys.* 299 (2010), no. 3, 783–792.

[ELMS21] J. van Ekeren, C. H. Lam, S. Moller and H. Shimakura, Schellekens' list and the very strange formula, *Adv. Math.*, **380** (2021), 107567.

[EMS20] J. van Ekeren, S. Möller and N. Scheithauer, Construction and classification of holomorphic vertex operator algebras, *J. Reine Angew. Math.*, **759** (2020), 61–99.

[EMS20b] J. van Ekeren, S. Möller and N. Scheithauer, Dimension formulae in genus zero and uniqueness of vertex operator algebras, *Internat. Math. Res. Notices* 2020(7):2145–2204, 2020.

[FHL93] I. B. Frenkel, Y. Huang and J. Lepowsky, On axiomatic approaches to vertex operator algebras and modules, *Mem. Amer. Math. Soc.* **104** (1993), viii+64 pp.

[FLM88] I. Frenkel, J. Lepowsky and A. Meurman, Vertex operator algebras and the Monster, Pure and Appl. Math., Vol. 134, Academic Press, Boston, 1988.

[FZ92] I. Frenkel and Y. Zhu, Vertex operator algebras associated to representations of affine and Virasoro algebras, *Duke Math. J.* **66** (1992), 123–168.

[HL90] K. Harada and M. L. Lang, On some sublattices of the Leech lattice, *Hokkaido Mathematical Journal*, **19** (1990), 435–446.

[He78] S. Helgason, Differential geometry, Lie groups, and symmetric spaces, Pure and Appl. Math., Vol. **80**, Academic Press, New York, 1978.

[Hö] G. Höhn, Selbstduale Vertexoperatorsuperalgebren und das Babymonster, Dissertation, Bonn, 1995.

[Hö2] G. Höhn, On the genus of the Moonshine module, arXiv:1708.05990.

[HM] G. Höhn and Sven Möller, Systematic orbifold constructions of Schellekens' vertex operator algebras from Niemeier lattices; arXiv: 2010.00849.

[HS14] G. Höhn and N. R. Scheithauer, A generalized Kac–Moody algebra of rank 14, *J. Algebra* **404**, (2014), 222–239.

[ISS15] M. Ishii, D. Sagaki and H. Shimakura, Automorphisms of Niemeier lattices for Miyamoto's $\mathbb{Z}_3$-orbifold construction, *Math. Z.* **280** (2015), 55–83.

[Ka90] V. G. Kac, Infinite-dimensional Lie algebras, 3rd edition, Cambridge University Press, Cambridge, 1990.

[KM12] M. Krauel, G. Mason, Vertex operator algebras and weak Jacobi forms, *Internat. J. Math.* **23** (2012), 1250024.

[KMi15] M. Krauel and M. Miyamoto, A modular invariance property of multivariable trace functions for regular vertex operator algebras, *J. Algebra* **444** (2015), 124–142.

[KLL18] K. Kawasetsu, C. H. Lam and X. Lin, $\mathbb{Z}_2$-orbifold construction associated with (-1)-isometry and uniqueness of holomorphic vertex operator algebras of central charge 24, *Proc. Amer. Math. Soc.* **146** (2018), 1937–1950.

[La11] C. H. Lam, On the constructions of holomorphic vertex operator algebras of central charge 24, *Comm. Math. Phys.* **305** (2011), 153–198.

[La20a] C. H. Lam, Cyclic orbifolds of lattice vertex operator algebras having group-like fusions, *Lett. Math. Phys.* **110** (2020), 1081–1112.

[La20b] C. H. Lam, Automorphism group of an orbifold vertex operator algebra associated with the Leech lattice, *in* Vertex Operator Algebras, Number Theory and Related Topics, 127–138, *Contemp. Math.*, **753**, Amer. Math. Soc., Providence, RI, 2020.

[La] C. H. Lam, Some observations about the automorphism groups of certain orbifold vertex operator algebras, to appear in RIMS Kôkyûroku Bessatsu.

[Lin17] X. Lin, Mirror extensions of rational vertex operator algebras, *Trans. Amer. Math. Soc.* **369** (2017) 3821–3840.

[LL20] C. H. Lam and X. Lin, A holomorphic vertex operator algebra of central charge 24 with weight one Lie algebra $F_{4,6}A_{2,2}$, *J. Pure Appl. Algebra*, **224** (2020), 1241–1279.

[LM] C. H. Lam and M. Miyamoto, A lattice theoretical interpretation of generalized deep holes of the Leech lattice vertex operator algebra, arXiv:2205.04681.

[LS12] C. H. Lam and H. Shimakura, Quadratic spaces and holomorphic framed vertex operator algebras of central charge 24, *Proc. Lond. Math. Soc.* **104** (2012), 540–576.

[LS15] C. H. Lam and H. Shimakura, Classification of holomorphic framed vertex operator algebras of central charge 24, *Amer. J. Math.* **137** (2015), 111–137.

[LS16a] C. H. Lam and H. Shimakura, Orbifold construction of holomorphic vertex operator algebras associated to inner automorphisms, *Comm. Math. Phys.* **342** (2016), 803–841.

[LS16b] C. H. Lam and H. Shimakura, A holomorphic vertex operator algebra of central charge 24 whose weight one Lie algebra has type $A_{6,7}$, *Lett. Math. Phys.* **106** (2016), 1575–1585.

[LS17] C. H. Lam and H. Shimakura, Construction of holomorphic vertex operator algebras of central charge 24 using the Leech lattice and level p lattices, *Bull. Inst. Math. Acad. Sin. (N. S.),* **12** (2017) No. 1, 39– 70.

[LS19] C. H. Lam and H. Shimakura, Reverse orbifold construction and uniqueness of holomorphic vertex operator algebras, *Trans. Amer. Math. Soc.* , **372** (2019), 7001–7024.

[LS20a] C. H. Lam and H. Shimakura, On orbifold constructions associated with the Leech lattice vertex operator algebra, *Math. Proc. Cambridge Phil. Soc.*, **168** (2020), 261–285.

[LS20b] C. H. Lam and H. Shimakura, Inertia subgroups and uniqueness of holomorphic vertex operator algebras, *Transformation Groups*, **25** (2020), 1223–1268.

[LY08] C. H Lam and H. Yamauchi, On the structure of framed vertex operator algebras and their pointwise frame stabilizers, *Comm. Math. Phys.* **277** (2008), 237–285.

[Le85] J. Lepowsky, Calculus of twisted vertex operators, *Proc. Natl. Acad. Sci. USA* **82** (1985), 8295–8299.

[Li94] H. Li, Symmetric invariant bilinear forms on vertex operator algebras, *J. Pure Appl. Algebra* **96** (1994), 279–297.

[Li96] H. Li, Local systems of twisted vertex operators, vertex operator superalgebras and twisted modules, *in* Moonshine, the Monster, and Related Topics, 203–236, *Contemp. Math.* **193**, Amer. Math. Soc., Providence, RI, 1996.

[Mi96] M. Miyamoto, Griess algebras and conformal vectors in vertex operator algebras, *J. Algebra*, **179** (1996), 523–548.

[Mi00] M. Miyamoto, A modular invariance on the theta functions defined on vertex operator algebras, *Duke Math. J.* **101** (2000), 221–236.

[Mi04] M. Miyamoto, A new construction of the Moonshine vertex operator algebra over the real number field, *Ann. of Math.* **159** (2004), 535–596.

[Mi13] M. Miyamoto, A $\mathbb{Z}_3$-orbifold theory of lattice vertex operator algebra and $\mathbb{Z}_3$-orbifold constructions, *in* Symmetries, Integrable Systems and Representations, 319–344, *Springer Proc. Math. Stat.* **40**, Springer, Heidelberg, 2013.

[Mi15] M. Miyamoto, C_2-cofiniteness of cyclic-orbifold models, *Comm. Math. Phys.* **335** (2015), 1279–1286.

[MT04] M. Miyamoto and K. Tanabe, Uniform product of $A_{g,n}(V)$ for an orbifold model V and G-twisted Zhu algebra, *J. Algebra* **274** (2004), 80–96.

[Mö16] S. Möller, A cyclic orbifold theory for holomorphic vertex operator algebras and applications, Dissertation, Darmstadt, 2016; arXiv:1611.09843.

[MS19] S. Moller, N. R. Scheithauer, Dimension formulae and generalised deep holes of the Leech lattice vertex operator algebra; arXiv:1910.04947.

[Mo94] P. S. Montague, Orbifold constructions and the classification of self-dual $c = 24$ conformal field theories, *Nuclear Phys.* B **428** (1994), 233–258.

[Ni73] H. V. Niemeier, Definite quadratische Formen der Dimension 24 und Diskriminante 1, *J. Number Theory* **5** (1973), 142–178.

[SS16] D. Sagaki and H. Shimakura, Application of a $\mathbb{Z}_3$-orbifold construction to the lattice vertex operator algebras associated to Niemeier lattices, *Trans. Amer. Math. Soc.* **368** (2016), 1621–1646.

[Sc93] A. N. Schellekens, Meromorphic $c = 24$ conformal field theories, *Comm. Math. Phys.* **153** (1993), 159–185.

[Sh04] H. Shimakura, The automorphism group of the vertex operator algebra V_L^+ for an even lattice L without roots, *J. Algebra* **280** (2004), 29–57.

[Ve78] B. B. Venkov, On the classification of integral even unimodular 24-dimensional quadratic forms, Algebra, number theory and their applications, *Trudy Mat. Inst. Steklov* **148** (1978), 65–76, 273.

[Xu07] F. Xu, Mirror extensions of local nets, *Comm. Math. Phys.* **270** (2007), 835–847.

[Zh96] Y. Zhu, Modular invariance of characters of vertex operator algebras, *J. Amer. Math. Soc.* **9** (1996), 237–302.

Appendix 3.A Table of 71 holomorphic VOAs of central charge 24

dim(V_1)	Lie algebra	rank	Exist.	Uniq.	dim(V_1)	Lie algebra	rank	Exist.	Uniq.
0	0	0	[FLM88]	??	24	$U(1)^{24}$	24	[FLM88]	[DM04a]
36	$C_{4,10}$	4	[EMS20]	[LS20b]	36	$A_{2,6}D_{4,12}$	6	[EMS20]	[LS20a]
36	$A_{1,4}^{12}$	12	[DGM96]	[KLL18]	48	$A_{6,7}$	6	[LS16b]	[LS20a]
48	$A_{4,5}^2$	8	[EMS20]	[EMS20b]	48	$A_{2,3}^6$	12	[SS16]	[LS19]
48	$A_{1,2}D_{5,8}$	6	[La11]	[LS20a]	48	$A_{1,2}A_{5,6}C_{2,3}$	8	[LS16a]	[LS20b]
48	$A_{1,2}A_{3,4}^3$	10	[La11]	[EMS20b]	48	$A_{1,2}^{16}$	16	[DGM96]	[LS15]
60	$C_{2,2}^6$	12	[DGM96]	[KLL18]	60	$A_{2,2}F_{4,6}$	6	[LL20]	[LL20]
60	$A_{2,2}^4 D_{4,4}$	12	[EMS20]	[EMS20b]	72	$A_{1,1}C_{5,3}G_{2,2}$	8	[EMS20]	[EMS20b]
72	$A_{1,1}^2 D_{6,5}$	8	[LS16a]	[LS20a]	72	$A_{1,1}^2 C_{3,2}D_{5,4}$	10	[La11]	[LS20b]
72	$A_{1,1}^3 A_{7,4}$	10	[La11]	[LS20a]	72	$A_{1,1}^3 A_{5,3}D_{4,3}$	12	[SS16]	[LS19]
72	$A_{1,1}^4 A_{3,2}^4$	16	[DGM96]	[KLL18]	72	$A_{1,1}^{24}$	24	[FLM88]	[DM04a]
84	$B_{3,2}^4$	12	[DGM96]	[KLL18]	84	$A_{4,2}^2 C_{4,2}$	12	[LS12]	[EMS20b]
96	$C_{2,1}^4 D_{4,2}^2$	16	[DGM96]	[KLL18]	96	$A_{2,1}C_{2,1}E_{6,4}$	10	[EMS20]	[LS20b]
96	$A_{2,1}^2 A_{8,3}$	12	[LS16a]	[LL20]	96	$A_{2,1}^2 A_{5,2}^2 C_{2,1}$	16	[La11]	[EMS20b]
96	$A_{2,1}^{12}$	24	[FLM88]	[DM04b]	108	$B_{4,2}^3$	12	[DGM96]	[KLL18]
120	$E_{6,3}G_{2,1}^3$	12	[Mi13]	[LS19]	120	$A_{3,1}D_{7,3}G_{2,1}$	8	[LS16a]	[LS20b]
120	$A_{3,1}C_{7,2}$	10	[La11]	[LS20b]	120	$A_{3,1}A_{7,2}C_{3,1}^2$	16	[La11]	[EMS20b]
120	$A_{3,1}^2 D_{5,2}^2$	16	[DGM96]	[KLL18]	120	$A_{3,1}^8$	24	[FLM88]	[DM04a]
132	$A_{8,2}F_{4,2}$	12	[LS12]	[EMS20b]	144	$C_{4,1}^4$	16	[DGM96]	[KLL18]
144	$B_{3,1}^2 C_{4,1}D_{6,2}$	16	[DGM96]	[KLL18]	144	$A_{4,1}A_{9,2}B_{3,1}$	16	[LS12]	[EMS20b]
144	$A_{4,1}^6$	24	[FLM88]	[DM04a]	156	$B_{6,2}^2$	12	[DGM96]	[KLL18]
168	$D_{4,1}^6$	24	[FLM88]	[DM04a]	168	$A_{5,1}E_{7,3}$	12	[LS16a]	[LS20b]
168	$A_{5,1}C_{5,1}E_{6,2}$	16	[LS12]	[EMS20b]	168	$A_{5,1}^4 D_{4,1}$	24	[FLM88]	[DM04a]
192	$B_{4,1}C_{6,1}^2$	16	[La11]	[EMS20b]	192	$B_{4,1}^2 D_{8,2}$	16	[DGM96]	[KLL18]
192	$A_{6,1}^4$	24	[FLM88]	[DM04a]	216	$A_{7,1}D_{9,2}$	16	[DGM96]	[KLL18]
216	$A_{7,1}^2 D_{5,1}^2$	24	[FLM88]	[DM04a]	240	$C_{8,1}F_{4,1}^2$	16	[LS12]	[EMS20b]
240	$B_{5,1}E_{7,2}F_{4,1}$	16	[LS12]	[EMS20b]	240	$A_{8,1}^3$	24	[FLM88]	[DM04a]
264	$D_{6,1}^4$	24	[FLM88]	[DM04a]	264	$A_{9,1}^2 D_{6,1}$	24	[FLM88]	[DM04a]
288	$B_{6,1}C_{10,1}$	16	[LS12]	[EMS20b]	300	$B_{12,2}$	12	[DGM96]	[KLL18]
312	$E_{6,1}^4$	24	[FLM88]	[DM04a]	312	$A_{11,1}D_{7,1}E_{6,1}$	24	[FLM88]	[DM04a]
336	$A_{12,1}^2$	24	[FLM88]	[DM04a]	360	$D_{8,1}^3$	24	[FLM88]	[DM04a]
384	$B_{8,1}E_{8,2}$	16	[LS12]	[LS15]	408	$A_{15,1}D_{9,1}$	24	[FLM88]	[DM04a]
456	$D_{10,1}E_{7,1}^2$	24	[FLM88]	[DM04a]	456	$A_{17,1}E_{7,1}$	24	[FLM88]	[DM04a]
552	$D_{12,1}^2$	24	[FLM88]	[DM04a]	624	$A_{24,1}$	24	[FLM88]	[DM04a]
744	$E_{8,1}^3$	24	[FLM88]	[DM04a]	744	$D_{16,1}E_{8,1}$	24	[FLM88]	[DM04a]
1128	$D_{24,1}$	24	[FLM88]	[DM04a]					

4

Maximal 2-Local Subgroups of the Monster and Baby Monster

Ulrich Meierfrankenfeld and Sergey Shpectorov

Abstract

The lists of the maximal 2-local subgroups of the Monster and Baby Monster simple groups in the Atlas are complete.

CONTENTS

4.1 Introduction

The Monster and the Baby Monster are the two largest groups among the 26 sporadic finite simple groups. After the classification of finite simple groups was announced in 1981, the focus of research in the area of finite simple groups moved toward the study of properties of the known groups. One of the most important pieces of information about a simple group G is its list of maximal subgroups, taken up to conjugation in G.

Methods used to classify maximal subgroups H of G differ significantly depending on whether or not H is p-local. A subgroup H of G is *p-local*, where p is a prime number dividing the order of G, if H is the normalizer of a nontrivial p-subgroup. We say that H is a maximal p-local subgroup if H is maximal by inclusion among p-local subgroups of G. Notice that a maximal p-local subgroup H may or may not be maximal in G. Nevertheless, the classification of all maximal p-local subgroups of G for each p is an important step toward a complete determination of all maximal subgroups of G.

In the case where G is one of the 26 sporadic finite simple groups, the lists of maximal p-locals subgroups – as well as lists of non–p-local maximal subgroups – have been compiled and proven complete for almost all G by the work of many people, but most notably, R. A. Wilson. One significant omission to date has been the lists of maximal 2-local subgroups of M and BM. *The Atlas of Finite Groups* [ATLAS] provides lists of the *known* maximal 2-local subgroups of M and BM. What was missing was a proof that these lists are in fact complete. In this chapter and its sequel [M] we bridge this gap by supplying necessary proofs.

Let us now review the lists from [ATLAS]. Seven conjugacy classes of maximal 2-local subgroups are known for $G = M$. The corresponding structures are as follows (see [ATLAS] for the exact meaning of these structures; however, notice that [ATLAS] uses the notation B for the Baby Monster group; also we use the good old "Ω" and "Sp" where [ATLAS] uses "O" and "S"):

(1) $2 \cdot BM$;
(2) $2^2 \cdot (^2E_6(2)) : S_3$;
(3) $2^{1+24}_+ . Co_1$;
(4) $2^2 . 2^{11} . 2^{22} . (S_3 \times M_{24})$;
(5) $2^3 . 2^6 . 2^{12} . 2^{18} . (L_3(2) \times 3 \cdot S_6)$;
(6) $2^5 . 2^{10} . 2^{20} . (S_3 \times L_5(2))$; and
(7) $2^{10+16} \cdot \Omega^+_{10}(2)$.

Eight classes of maximal 2-local subgroups are known for $G = BM$. Their structures are shown in [ATLAS] as follows.

(1) $2 \cdot (^2E_6(2)) : S_3$;
(2) $(2^2 \times F_4(2)) : 2$;
(3) $S_4 \times {}^2F_4(2)$;
(4) $2^{1+22}_+ \cdot Co_2$;
(5) $2^2 . 2^{10} . 2^{20} . (M_{22} : 2 \times S_3)$;

(6) $2^3.[2^{32}].(S_5 \times L_3(2))$;

(7) $2^5.[2^{25}].L_5(2)$; and

(8) $2^9.2^{16}.Sp_8(2)$.

Recall that $O_2(H)$ denotes the largest normal 2-subgroup of H. We say that H is of *characteristic* 2 if $C_H(Q) \leq Q$, where $Q = O_2(H)$. We split the work as follows. In this chapter, we determine all maximal 2-local subgroups of M and BM that are of characteristic 2. The sequel [M] deals with the remaining classes. In the aforementioned lists, the partition into the characteristic 2 type and noncharacteristic 2 type is as follows: for M, classes (1) and (2) are not of characteristic 2, while classes (3)–(7) are of characteristic 2. For BM, classes (1)–(3) are not of characteristic 2, while classes (4)–(8) are of characteristic 2. Notice that classifying the maximal 2-local subgroups that are not of characteristic 2 is rather more simple and some may even say that this part of the lists has been known to be complete. However, as we are unaware of any published proof, we include this subcase in our work. Needless to say, we believe that our result on the maximal 2-local subgroups of characteristic 2 is entirely new.

Our approach was in part motivated by the work on the geometries of the groups M and BM, by A. A. Ivanov and the second author (see [IS]). We noticed that the known maximal 2-local subgroups of characteristic 2 are either the normalizers of certain very special elementary abelian subgroups that we call *singular subgroups*, or the normalizers of yet another type of elementary abelian subgroups (of order 2^{10}) that we call *arks*. The above-mentioned geometries of M and BM consist of singular subgroups, while arks also have a geometrical meaning, namely, they correspond to certain natural subgeometries.

We introduce singular subgroups in Section 4.4, in which we also classify them up to conjugation. Arks are introduced in Section 4.5. They form a single conjugacy class of subgroups. Our choice of the word "ark" was motivated by the fact that an ark contains representatives of all "species" (i.e., conjugacy classes) of singular subgroups.

We will now formally state the principal results of this chapter. We start with the following definitions of the groups M and BM: the *Monster M* is a finite simple group with a large extraspecial 2-subgroup Q whose normalizer C has the following structure: $C \sim 2^{1+24}.Co_1$. (Here and in what follows we use $\sim$ as shorthand for "has structure," while $\cong$ as usual stands for "is isomorphic.") Recall that a group G is said to have a *large extraspecial subgroup* if for some involution $z \in G$ its centralizer $C = C_G(z)$ contains a normal extraspecial 2-subgroup Q such that $C_G(Q) \leq Q$. It easily follows from this that z is *2-central* in G, that is, the centralizer of Z contains a Sylow 2-subgroup of G. It also follows that G has a unique conjugacy class of 2-central involutions. Returning to the Monster group M, in addition to the structure of C, we have that the Co_1-module arising in the action of C on $Q/\langle z \rangle$ is isomorphic to the module on $\Lambda/2\Lambda$, where Λ is the Leech lattice. This due to R. L. Griess who showed in [Gr2] that $\Lambda/2\Lambda$ is the only faithful Co_1-module in dimension 24.

For the purposes of this chapter, the *Baby Monster BM* is simply the group $H/\langle t \rangle$, where t is a non 2-central involution in the Monster M and $H = C_M(t)$. All the work in this chapter takes place in M; singular subgroups, arks, and the Baby Monster BM "live" in M.

Under these definitions, we prove the following.

Theorem 1.1 *The Monster M contains exactly five conjugacy classes of maximal 2-local subgroups of characteristic 2. They are* (a) *the normalizers of singular subgroups of types 2^1, 2^2, 2^3, and 2^5_2; and* (b) *the normalizers of arks.*

Theorem 1.2 *The Baby Monster BM contains exactly five conjugacy classes of maximal 2-local subgroups of characteristic 2. Their preimages in H are* (a) *the normalizers in H of special singular subgroups U of types 2^1, 2^2, 2^3, and 2^5_1; and* (b) *the normalizers of arks containing t.*

The exact meaning of the word "special" in this last theorem is as follows: With each singular subgroup U we associate in Section 4.4 a second subgroup Q_U. The special singular subgroups U are those for which $t \in Q_U$.

In simple words, Theorems 1.1 and 1.2 state that the lists of maximal 2-local subgroups of M and BM given in [ATLAS] are complete in their characteristic 2 part.

Finally, we need to explain our policy with respect to citing vs. proving. A lot of information is available about the two monsters. However, much of it exists as a sort of finite group theory lore, that is, with no proper published proof known. In particular, at least some of the information given in [ATLAS] should be considered as semi-lore because there are no proofs there and very little by way of citation. Of course, we cannot prove everything in a single chapter, and so we decided to take an "inductive" approach. We use some "lore" information about the smaller simple groups involved in M (mostly, Co_1 and $\Omega_{10}^+(2)$). At the same time, we prove everything we need as far as the properties of M and BM themselves are concerned. Likely, some of the facts that we prove can be found in the available sources such as [As1], [AsSe], [Gr1], [Se], and many others. However, we believe that the bulk of the detailed information that we need cannot be covered by citation. Notice also that M. Aschbacher in [As1] determines all maximal subgroups of M containing a Sylow 2-subgroup.

4.2 Classes of Involutions, I

In this section we classify conjugacy classes of involutions of a group H satisfying the following conditions:

(H1) $O_2(H) \sim 2^{24}$;
(H2) $H/O_2(H) \cong Co_1$; and
(H3) the action of $H/O_2(H)$ on $O_2(H)$ is equivalent to the action of Co_1 on $\hat{\Lambda} = \Lambda/2\Lambda$, the Leech lattice taken modulo 2.

If C is the centralizer of a 2-central involution z in the Monster group M, then the group $H = C/Z$, where $Z = \langle z \rangle$, satisfies (H1)–(H3) and so the results of this section give us some insight into the structure of C.

We refer to [ATLAS], page 180, for a description of the Leech lattice Λ, terminology and notation related to Λ, and a summary of properties of Λ. A whole wealth of information about the Leech lattice can be found in [CS]. Let $(x, y) = \frac{1}{8} \sum_{i=1}^{24} x_i y_i$ be the integral inner product that exists on Λ. Let $\Lambda_n = \{x \in \Lambda \mid (x, x) = 2n\}$. One useful fact is the following.

Lemma 2.1   *The orbits of Co_1 on $W = \hat{\Lambda}^{\#}$ are the sets $\hat{\Lambda}_2$, $\hat{\Lambda}_3$ and $\hat{\Lambda}_4$.*

For $i = 2$, 3, and 4, let $w_i \in \hat{\Lambda}_i$. The structure of the stabilizer of w_i in Co_1 is also well-known.

Lemma 2.2   *The following hold:*

(1) $C_{Co_1}(w_2) \cong Co_2$;
(2) $C_{Co_1}(w_3) \cong Co_3$;
(3) $C_{Co_1}(w_4) \cong 2^{11} : M_{24}$.

Recall that in its action on $W = \hat{\Lambda}$, the group Co_1 preserves a nondegenerate quadratic form, q, defined as follows: if $u = \hat{x}$ for some $x \in \Lambda$ then $q(u) = \frac{1}{2}(x, x)$ (mod 2). Let Φ denote the symmetric bilinear form that corresponds to q: for $u = \hat{x}$ and $v = \hat{y}$, we have $\Phi(u, v) = q(u + v) + q(u) + q(v) = (x, y)$ (mod 2). It follows from the definition of q that w_2 and w_4 are singular, while w_3 is nonsingular.

Before we go on, let us record the following property that can be verified, say, using the description of Λ from [ATLAS].

Lemma 2.3   *There is no $\hat{\Lambda}_2$-pure subgroups 2^3 in $W = \hat{\Lambda}$.*

According to [ATLAS], Co_1 contains three conjugacy classes of involutions. We will need to know how the involutions and their centralizers act on W.

Let t be an involution in Co_1 and let $C_t = C_{Co_1}(t)$. Define $U_t = C_W(t)$ and $V_t = [W, t]$. Since t is an involution, $V_t \subseteq U_t$. Furthermore, $\dim W/U_t = \dim V_t$.

In fact, U_t is the orthogonal complement (with respect to Φ) of V_t, and so the C_t-modules W/U_t and V_t are dual to each other.

First, let t be an involution of type $2A$. Then C_t is an extension of an extraspecial group 2_+^{1+8} by $\Omega_8^+(2)$. The action on the 8-dimensional quotient of $O_2(C_t)$ provides an irreducible module for $C_t/O_2(C_t) \cong \Omega_8^+(2)$. We will refer to this module as to the *natural* module. Notice that $\Omega_8^+(2)$ has two more irreducible 8-dimensional modules, and we will refer to those as to the two *halfspin* modules. Notice also that the natural module and the halfspin modules are all self-dual.

Lemma 2.4 *If t is of type $2A$, then*

(1) V_t *has dimension 8 and U_t has dimension 16;*
(2) C_t *acts irreducibly on each of W/U_t, U_t/V_t, and V_t; furthermore, V_t and U_t/V_t are two non-isomorphic halfspin modules and $W/U_t \cong V_t$.*

Proof Notice that t (or rather, its preimage in Co_0) can be chosen to act on the standard frame, inverting signs in an octad. This allows to establish (1) by direct computation. Let $x \in C_t$ be an element of order three, such that x has a 6-dimensional centralizer in the natural module. Then 2^{13} divides the order of the centralizer of x in C_t, and hence x is of type $3A$ (a Suzuki 3-element). According to [ATLAS], x acts fixed-point-freely on W, which implies that W/U_t, U_t/V_t, and V_t are halfspin modules for $C_t/O_2(C_t) \cong O_8^+(2)$. Since the halfspin modules are self-dual, we have $W/U_t \cong V_t$. Finally, if $U_t/V_t \cong V_t$ then C_t contains a 3-element with an 18-dimensional centralizer in W, which contradicts the information from [ATLAS]. $\qquad\square$

Let t be an involution of type $2B$. Then $C_t \sim (2^2 \times G_2(4)).2$.

Lemma 2.5 *If t is of type $2B$, then*

(1) $U_t = V_t$ *and so they are of dimension 12;*
(2) C_t *acts transitively on $V_t^\#$; furthermore, $V_t^\# \subseteq \hat{\Lambda}_4$.*

Proof Notice that 13 divides the order of $G_2(4)$, and it does not divide the orders of Co_2, Co_3, and $2^{11} : M_{24}$. Therefore, $G_2(4)$ fixes no nonzero vector in W. Again, since 13 divides the order of $G_2(4)$, the latter group has no nontrivial $GF(2)$-modules in dimensions less than 12. It follows that V_t has dimension at least 12. Hence, $U_t = V_t$ and they are both of dimension exactly 12. According to [MOD], V_t must be the natural module for $G_2(4)$. In particular, $G_2(4)$ is transitive on $V_t^\#$. Since $\hat{\Lambda}_4$ has odd length, t fixes a vector in $\hat{\Lambda}_4$. Now the transitivity implies that $V_t^\# \subseteq \hat{\Lambda}_4$. $\qquad\square$

Finally, let t be of type $2C$. Then $C_t \sim 2^{11} : \operatorname{Aut} M_{12}$.

Lemma 2.6 *If t is of type 2C then*

(1) $U_t = V_t$, *and so they are of dimension* 12;
(2) *as a C_t-module, U_t is uniserial with submodules of dimension* 1 *and* 11; *furthermore, the nonzero vector fixed by C_t is from $\hat{\Lambda}_4$, and $\hat{\Lambda}_3 \cap V_t$ coincides with the setwise complement of the* 11*-dimensional submodule.*

Proof   In this case, again, t can be chosen inside the diagonal subgroup stabilizing the standard frame. Namely, t inverts signs in a dodecad. This allows to compute all vectors in W fixed by t and thus establish (1) and also that V_t contains some elements from $\hat{\Lambda}_3$. Since $U_t = V_t$, we have that V_t is totally isotropic (with regard to Φ). Since the vectors in $\hat{\Lambda}_3$ are nonsingular, we obtain that $V_t \cap \hat{\Lambda}_3$ coincides with the complement of a hyperplane. So V_t contains an 11-dimensional subspace V_0 left invariant by C_t. Observe that C_t is fully contained in the stabilizer of the standard frame. So C_t stabilizes a vector $v \in \hat{\Lambda}_4$, the image of the standard frame. Clearly, $v \in V_0$. Since 11 divides the order of M_{12}, C_t acts irreducibly on $V_0/\langle v \rangle$. It remains to notice that v is the only vector in W fixed by C_t (indeed, already the diagonal group 2^{11} fixes no other vector in $W^{\#}$). The uniseriality now follows.   □

We can now determine the conjugacy classes of involutions in a group H satisfying conditions (H1)–(H3). Let $E = O_2(H)$ and $\bar{H} = H/E$. First of all, Lemma 2.1 implies the following.

Lemma 2.7   *The group H has exactly three classes of involutions contained in E. If e_2, e_3, and e_4 are representatives of those classes then $C_H(e_2) \sim 2^{24}.Co_2$, $C_H(e_3) \sim 2^{24}.Co_3$, and $C_H(e_4) \sim 2^{24}.(2^{11} : M_{24})$.*

We will classify the classes of involutions outside E case by case, depending on whether $\bar{x}$ is of type 2A, 2B, or 2C. We start with a general lemma. Let $U = C_E(\bar{x})$ and $V = [E, \bar{x}]$. Let $X = \langle x, U \rangle$ and $\tilde{X} = X/V$. Let C be the full preimage in H of $C_{\bar{H}}(\bar{x})$.

Lemma 2.8   *Suppose $x \in H$ is an involution, and $\bar{x} \neq 1$. Then the following hold:*

(1) *An element $y \in xE$ is an involution if and only if $y \in xU$.*
(2) *The subgroups X and V are invariant under C; furthermore, E acts trivially on $\tilde{X}$.*
(3) *If $y, z \in xU$, then y and z are conjugate in H if and only if $\tilde{y}$ and $\tilde{z}$ are in the same $\bar{C}$-orbit.*

Remark.   Part (2) contends that the action of $\bar{C} = C_{\bar{H}}(\bar{x})$ on $\tilde{X}$ is well defined, which allows us to view $\tilde{X}$ as a $\bar{C}$-module. Since $U = X \cap E$, $\tilde{U}$ is invariant under $\bar{C}$, and so $\bar{C}$ permutes the vectors in $\tilde{X} \setminus \tilde{U}$. The meaning of part (3) is that the orbits of $\bar{C}$ on $\tilde{X} \setminus \tilde{U}$ bijectively correspond to those conjugacy classes of involutions in H that map onto $\bar{x}^{\bar{H}}$.

Proof If $e \in E$, then $(xe)^2 = [x, e]$ since both x and e are involutions. Part (1) follows. Clearly, V is invariant under C. Since X is generated by all the involutions from the coset $\bar{x} = xU$, X is C-invariant, too. Clearly, E acts trivially on $\tilde{U}$. Furthermore, E fixes $\tilde{x}$, because $V = [E, x]$. This proves (2). For (3), let $y, z \in xU$. If $z = y^h$ for some $h \in H$ then $\bar{h} \in \bar{C}$, since $\bar{y} = \bar{x} = \bar{z}$. So $\tilde{y}$ and $\tilde{z}$ are in the same $\bar{C}$-orbit. Conversely, suppose that $\tilde{y}^{\bar{c}} = \tilde{z}$ for some $c \in C$. Then $y^c = zv$ for some $v \in V$. Since $V = [E, x]$, there exists an element $e \in E$ such that $v = [e, x]$. However, $[e, x] = [e, z]$, since $z \in xU$. Therefore, $z^e = vz = zv$, implying that y and z are conjugate. $\qquad\square$

We will first classify those involutions x for which $\bar{x}$ is of type $2A$. We will need the following fact proved in [Po].

Lemma 2.9 *Suppose Y is a $GF(2)$-module for $\Omega_8^+(2)$ that is an extension of an irreducible 8-dimensional submodule Y_0 by a 1-dimensional module. Then Y splits.*

Lemma 2.10 *The group H has exactly three classes of involutions whose images in $\bar{H}$ are of type $2A$. If a_1, a_2, and a_3 are representatives of these classes, then $C_H(a_i)$ has the structure $2^{16}.2^{1+8}.\Omega_8^+(2)$, $2^{16}.2^{1+8}.Sp_6(2)$, and $2^{16}.2^{1+8}.(2^6 : L_4(2))$, for $i = 1$, 2, and 3, respectively.*

Proof Let x be an element of H such that $\bar{x}$ is of type $2A$. We will first show that the coset $\bar{x}$ contains an involution, and so x can be chosen to be an involution. Let $R = \langle x, E \rangle$. Then R is a normal subgroup of C, where C is defined, as previously, as the full preimage in H of $C_{\bar{H}}(\bar{x})$. Let $U = C_E(x)$ and $V = [E, x]$. Clearly, $U = Z(R)$. Consider $X = R/U$. According to Lemma 2.4, C has two chief factors within X, of dimensions 8 and 1. This implies that X is an elementary abelian group, which we can view as a module for $\bar{C}$. Furthermore, by the same Lemma 2.4, the 8-dimension chief factor in X is not a natural module for $\bar{C}/O_2(\bar{C}) \cong \Omega_8^+(2)$. Therefore, $O_2(\bar{C})$ acts trivially on X. Now Lemma 2.9 implies that X contains a 1-dimensional subspace T invariant under C. Let R_0 be the full preimage in R of T. Clearly, R_0 is normal in C. Next, define $X_0 = R_0/V$. Again, X_0 is an extension of an 8-dimensional chief factor U/V by a 1-dimensional R_0/U. We conclude again that X_0 is elementary abelian. Clearly, R_0 acts trivially on X_0. Furthermore, by Lemma 2.4 the 8-dimensional chief factor in X_0 differs, as a module, from the chief factors in $O_2(C)/R_0$, which means that $O_2(C)$ acts trivially on X_0. Applying Lemma 2.9

again, we obtain a C-invariant 1-dimensional subspace T_0 in X_0. Let R_1 be the full preimage of T_0 in R. Setting $X_1 = R_1$, we observe for the third time that X_1 is an extension of an 8-dimensional chief factor V by a 1-dimensional one, R_1/V. Hence R_1 is elementary abelian. Since $R_1 \not\leq E$, we finally conclude that the coset $\bar{x}$ contains some involutions. Without loss of generality, we can now assume that x is itself an involution and so Lemma 2.8 applies. In the notation introduced before Lemma 2.8, $\tilde{X}$ (which has already appeared as $X_0 = R_0/V$) is the direct sum of a halfspin module and a 1-dimensional module. Therefore, C has three orbits on $\tilde{X} \setminus \tilde{U}$, of sizes 1, 120, and 135, and this immediately leads to the conclusion as in the lemma. □

The classification of involutions x with $\bar{x}$ of type $2B$ or $2C$ is an easy corollary of Lemmas 2.5 and 2.6.

Lemma 2.11 *For $L = B$ or C, H has a unique conjugacy class of involutions whose images in $\bar{H}$ are of type $2L$. If b and c are representatives of those two classes then $C_H(b) \sim 2^{12}.(2^2 \times G_2(4)).2$ and $C_H(c) \sim 2^{12}.(2^{11} :$ Aut $M_{12})$.*

Proof Let x be an element of H such that $\bar{x}$ is of type $2L$. Then according to Lemmas 2.5 and 2.6, we have that $C_E(x) = [E, x]$. In particular, $x^2 = [e, x]$ for some $e \in E$. It follows that $(xe)^2 = x^2[x, e] = 1$; that is, the coset xE contains involutions. Furthermore, in the notation of Lemma 2.8 we have that $\tilde{X}$ is 1-dimensional, and so the claim follows. □

This completes the classification of conjugacy classes of involutions in H. According to Lemmas 2.7, 2.10 and 2.11, the group H contains eight classes of involutions. We will refer to these classes as to the classes $2e_i$, $2 \leq i \leq 4$, $2a_i$, $1 \leq i \leq 3$, $2b$, and $2c$.

4.3 A Fusion Lemma and an Application

In the first part of this section, G is an arbitrary group having a large extraspecial subgroup. This means that for some involution $z \in G$, the centralizer $C = C_G(z)$ contains a normal extraspecial 2-subgroup Q and, furthermore, $C_G(Q) \leq Q$. This implies that G contains a unique class of 2-central involutions (recall that a 2-central involution is an involution in the center of some Sylow 2-subgroup of G), and that z is itself 2-central. We let S denote the class of 2-central involutions in G. For $x = z^g \in S$, we denote $C_x = C_G(x) = C^g$ and $Q_x = Q^g$. Thus, $C = C_z$ and $Q = Q_z$.

We will assume throughout this section that

$$(*) \quad S \cap C \neq \{z\}.$$

Indeed, the principal case of interest for us is where G is simple. However, in that case the Z^*-theorem of Glauberman makes $S \cap C = \{z\}$ impossible. In this section, we prove that, modulo some small configurations, $(*)$ implies the following stronger condition:

$$(**) \quad S \cap Q \neq \{z\}.$$

Let, as above, $Q = Q_z$ and let $\bar{Q}$ denote Q/Z where $Z = \langle z \rangle$.

Lemma 3.1 *Suppose* $S \cap Q = \{z\}$ *and let* $x \in S \cap C$, $x \neq z$. *Denote* $E = Q \cap Q_x$. *Then one of the following holds:*

(1) $E = 1$; *or*
(2) $|E| = 2$ *and either*

    (a) $\bar{E} \not\leq [\bar{Q}, x]$, *or*
    (b) $z \neq [x, y]$ *for all* $y \in Q$; *or*

(3) $|E| = 4$ *and furthermore, for* $W = \langle E, z, x \rangle$,

    (a) $N_G(W)$ *induces on* $W \cong 2^4$ *either* $O_4^-(2)$, *or* $\Omega_4^-(2)$ *acting as on the natural module;*
    (b) $|W \cap S| = 5$ *and, under the identification of* W *with the orthogonal module, the involutions in* $W \cap S$ *are the singular vertors; moreover, for each* $w \in W \cap S$, $W \cap Q_w$ *is the perp of* w.

Proof Suppose that $E \neq 1$. (Otherwise, (1) holds.) If $e \in E$, then $e^2 \in Z \cap \langle x \rangle = 1$, since E is contained in both Q and Q_x. Hence E is elementary abelian. Let U and V be defined as the full preimages in Q of $C_{\bar{Q}}(x)$ and $[\bar{Q}, x]$, respectively. Observe that since Q is extraspecial we have that $V = C_Q(U) = Z(U)$ and that $C_Q(x)$ is either equal to U or $[U : C_Q(x)] = 2$. In the latter case, $[x, y] = z$ for all $y \in U \setminus C_Q(x)$.

Notice that $[C_Q(x), E] \leq [Q, Q] \cap [C_Q(x), Q_x] \leq Z \cap Q_x$. By assumption, $S \cap Q = \{z\}$ and hence $S \cap Q_x = \{x\}$. We conclude that $z \notin Q_x$ and hence $[C_Q(x), E] = 1$.

Let $e \in E^\#$ and suppose $e = [x, y]$ for some $y \in Q$. Then $ex = x^y$ is a conjugate of x contained in Q_x, i.e., $S \cap Q_x \neq \{x\}$. This contradiction shows that no nontrivial element from E is an elementary commutator $[x, y]$, for $y \in Q$.

If $C_Q(x) \neq U$ then $[U : C_Q(x)] = 2$, which implies that $[C_Q(C_Q(x)) : V] = 2$. Hence $[E : E \cap V] \leq 2$. If $E \cap V = 1$, then $|E| = 2$, implying (2a). So let us assume that $E \cap V \neq 1$ and let $e \in (E \cap V)^\#$. Since $e \in V$, we have that either $e = [x, y]$ or $ez = [x, y]$ for some $y \in Q$. However, by the preceding paragraph, $e \neq [x,y]$. So $ez = [x,y]$. Furthermore, $z = [x,t]$ for $t \in U \setminus C_Q(x)$. Thus, $[x,ty] = [x,t]^y[x,y] = zez = e$; a contradiction.

Now assume that $C_Q(x) = U$ and so $z \neq [x,y]$ for all $y \in Q$. Also, $E \leq V$, since $[C_Q(x), E] = 1$. If $|E| = 2$, then we obtain (2b). Hence we may assume that $|E| \geq 4$. Set $W = \langle E, z, x \rangle$. Our next step is to determine which involutions from W are in S. First of all, the involutions in W fall into the following types: z, x, e, ze, xe, zx, and zxe, where e denotes an arbitrary involution from $E^{\#}$. Clearly, $z, x \in S$. By assumption, no other involution in $Q \cup Q_x$ is in S. Hence the involutions e, ze, and xe are not in S. Since $E \leq V$, we have that $\bar{e} = [\bar{y}, x]$ for some $y \in Q$. We have shown that $e \neq [x,y]$. Hence $ze = [x,y]$. It follows that $zxe = xze = x[x,y] = x^y$. Thus, all elements zxe are in S. The element zx may or may not be in S.

Observe also that the element y normalizes W. Indeed, $\langle E, z \rangle$ is normal in Q, so y leaves it invariant. Also $x^y = zxe \in W$. Thus, $W^y = W$. We conclude that x and all the elements zxe are conjugate under $N_G(W)$. Symmetrically, z is conjugate under $N_G(W)$ to the elements zxe and so also to x. Notice that $W_x = \langle E, x \rangle$ is an index two subgroup of W such that $|W_x \cap S| = \{x\}$. Pick an element $e \in E^{\#}$. By transitivity, there exists a subgroup W_{zxe} of index two in E such that $W_{zxe} \cap S = \{zxe\}$. Since z and x are not in W_{zxe}, we have that $zx \in W_{zxe}$. Thus, $zx \notin S$, which completes the enumeration of the elements in $W \cap S$.

If $|E| > 4$, then for every $e' \in E^{\#}$ there exist elements e_1 and $e_2 = e_1 e'$ such that $e_1 \neq e \neq e_2$. Since both zxe_1 and zxe_2 are not in W_{zxe}, we conclude that $e' = (zxe_1)(zxe_2)$ is in W_{zxe}, i.e., $E \leq W_{zxe}$. However, in that case all elements zxe', $e' \in E^{\#}$, are in $W_{zxe} \cap S$, a contradiction. This establishes that $|E| = 4$ and, consequently, $|W \cap S| = 5$. Observe that the elements y from the preceeding paragraph stabilize z. This proves that $N_G(W)$ induces a 2-transitive group on $W \cap S$. Also $y^2 \in Z$ for all those elements y, which rules out the Frobenius group F_5^4. Hence $N_G(W)$ induces on $W \cap S$ one of the groups $S_5 \cong O_4^-(2)$ or $A_5 \cong \Omega_4^-(2)$ and the claim (2) follows. $\square$

In the remainder of this section, $G = M$, the Monster, z is a 2-central involution in G, $Z = \langle z \rangle$, $C = C_z = C_G(z)$, and $Q = Q_z = O_2(C_z)$. Since by assumption M is simple, Glauberman's Z^* theorem [Gl] shows that $S \cap C \neq \{z\}$. Recall that S denotes the conjugacy class of 2-central involutions, z^M. Recall also that for $x = z^g$ we set $C_x = C^g$ and $Q_x = Q^g$. Since M has a large extraspecial subgroup, Lemma 3.1 applies to it. We use that lemma to prove the following.

Proposition 3.2 $Q \cap S \neq \{z\}$.

Proof   Suppose $Q \cap S = \{z\}$. Since $C \cap S \neq \{z\}$, we can choose $x \in C \cap S$, $x \neq z$. According to Lemma 3.1, one of the exceptional cases (2a), (2b), or (3) must hold. In particular, $E = Q \cap Q_x$ has size at most four.

Observe now that the group $H = \bar{C} = C/Z$ satisfies the conditions (H1)–(H3) from Section 2. In particular, we can use the classification of conjugacy classes of involutions obtained in that section. Let $D = C \cap C_x$ and $R = Q_x \cap C$. Clearly, R is normal in D, and $\bar{D}$ is of index two or one in $C_{\bar{C}}(\bar{x})$ depending on whether or not x and xz are conjugate in C.

Suppose first that $\bar{x}$ is in the class $2a_i$ for some i. Then also $z\langle x \rangle$ is in $2a_i$ in $C_x/\langle x \rangle$. In particular, R is of order at least 2^{16} (cf., Lemma 2.4). Consider $\tilde{C} = C/Q$. Since $E = Q \cap R$ is of order at most four, we obtain that $\tilde{D}$ contains a normal 2-subgroup $\tilde{R}$ of order at least 2^{14}. Comparing with Lemma 2.1, we see that $i = 3$ must hold. However, $i = 3$ also leads to a contradiction. Indeed, let $\tilde{Y}$ be the normal extraspecial subgroup 2^{1+8} of $C_{\tilde{C}}(\tilde{x}) \sim 2^{1+8}.\Omega_8^+(2)$. We have that $[\tilde{Y} : \tilde{Y} \cap \tilde{R}] \leq 2$ and $[\tilde{R}, \tilde{R}] \leq \langle \tilde{x} \rangle = Z(\tilde{Y})$. Therefore, all elements of $\tilde{R}$ centralize a hyperplane in the 8-dimensional quotient of $\tilde{Y}$, which is impossible.

Suppose next that $\bar{x}$ is in $2b$. Then also $z\langle x \rangle$ is in the class $2b$ in $C_x/\langle x \rangle$. Hence $|R| \geq 2^{12}$. Considering again $\tilde{C} = C/Q$ and taking into account that $|E| \leq 4$, we see that $\tilde{D}$ contains a normal 2-group of size at least 2^{10}, clearly contradicting Lemma 2.11.

Finally, suppose $\bar{x}$ is in the class $2c$. In this case our argument must be slightly more subtle. Let U be the full preimage in Q of $\bar{U} = C_{\bar{Q}}(\bar{x})$. Then U is a subgroup of order 2^{13}. We claim that $C_Q(x)$ is a proper subgroup of U. Indeed, according to Lemma 2.4, $\bar{U}$ contains elements from the class $2e_3$. Observe that the mapping $\bar{e} \mapsto e^2$ defines a nondegenerate quadratic form g on $\bar{Q}$. Since, as a module for $C/Q \cong Co_1$, $\bar{Q}$ is absolutely irreducible, this quadratic form is unique, and hence g is equivalent to the form q (cf. Section 2). In particular, if $\bar{e}$ is in $2e_3$ then e is of order four. Since $\bar{U} = [\bar{Q}, \bar{x}]$, we have that $\bar{e} = [\bar{q}, \bar{x}]$ for some $q \in Q$. Therefore, $[q, x] = e$ or e^3. Since, clearly, q can be chosen to be an involution, we obtain that x inverts e, i.e., $e \notin C_Q(x)$.

This has two consequences. First, $C_Q(x)$ is of size 2^{12}, and symmetrically, also $|R| = 2^{12}$. (Clearly, $z\langle x \rangle$ must also be in the class $2c$ in $C_x/\langle x \rangle$.) Secondly, we record for further use that $z = [q, x]$ for some $q \in Q$.

Since $|E| \leq 4$, we have that $\tilde{R}$ (where, as above, $\tilde{C} = C/Q$) has size 2^{10}, 2^{11}, or 2^{12}. Comparing with Lemma 2.11 and using that $\tilde{R}$ is normal in $\tilde{D}$, we obtain that $|\tilde{R}| = 2^{11}$, and hence $|E| = 2$. This means that either (2a) or (2b) of Lemma 3.1 must hold. Previously we recorded that $z = [q, x]$ for some $q \in Q$. Hence, in fact, it must be the case (2a). To obtain a contradiction in this last case, it remains to see that $\bar{E} \leq [\bar{Q}, x]$. However, this is clear because $\bar{E} \leq C_{\bar{Q}}(\bar{x}) = [\bar{Q}, \bar{x}] = [\bar{Q}, x]$. □

4.4 Singular Subgroups

First, let G again be a group with a large extraspecial subgroup, that is, let there be an involution $z \in G$ and an extraspecial 2-subgroup Q normal in $C = C_G(z)$ such that $C_G(Q) \leq Q$. Adopt the notation from Section 4.3; that is, let $S = \{z^G\}$ be the class of 2-central involutions in G and, for $x = z^g \in S$, let $C_x = C_G(x) = C^g$ and $Q_x = Q^g$.

Let x and y be two 2-central involutions. We will say that x is *perpendicular* to y if and only if $y \in Q_x$. The following important lemma is a slight improvement on [As0], lemma 8.7 (3).

Lemma 4.1   *The perpendicularity relation is symmetric.*

Proof   If $|Q| > 2^3$, then this is proven in [As0], lemma 8.7 (3). So suppose $Q \sim 2^{1+2}$. Suppose that the relation is not symmetric so that for some $x \in S$ we have that $x \in Q$, but $z \notin Q_x$. In particular, there is no $g \in G$ such that $z^g = x$ and $x^g = z$. Since $x \in Q$, Q must be isomorphic to D_8. Since $C_G(Q_x) = \langle x \rangle$ and since $\mathrm{Out}\, D_8$ is of order two, we have that $C_x = Q_x \langle z \rangle \cong D_{16}$. It remains to notice that an element from $N_{C_x}(U)$, where $U = \langle z, x \rangle$, permutes z and zx and, likewise, an element from $N_C(U)$ permutes x and zx. Hence the normalizer of U induces on it the full group S_3. Thus, there exists an element $g \in G$ such that $z^g = x$ and $x^g = z$; a contradiction. $\square$

Let U be a purely 2-central (i.e., all involutions in U are in S) elementary abelian 2-subgroup of G. We will say that U is *singular* if $U \leq Q_u$ for every $u \in U^{\#}$. If U is singular, define $Q_U = \cap_{u \in U^{\#}} Q_u$ and $L_U = \langle Q_u \mid u \in U^{\#} \rangle$. Clearly, $U \leq Q_U \leq L_U$.

Lemma 4.2   *Let U be singular. Then the following hold:*

(1) *U and Q_U are normal in L_U and L_U acts trivially on Q_U / U;*

(2) *if $W \leq Q_U$ and $W \cap U = 1$ then $C_{L_U}(W)$ induces on U the full group $L_n(2)$ (where n is the rank of U); in particular (for $W = 1$), $N_G(U) = L_U C_G(U)$;*

(3) *if $W_1, W_2 \leq Q_U$, $W_1 \cap U = W_2 \cap U = 1$, and $W_1 U = W_2 U$, then there is an element $x \in C_{L_U}(U)$ such that $W_1^x = W_2$; in particular, $C_{L_U}(U)$ acts transitively on every coset qU, $q \in Q \setminus U$;*

(4) *if $|U| > 2$, then Q_U is elementary abelian.*

Proof   Let $U \leq U' \leq Q_U$. If $u \in U^{\#}$ then $U' \leq Q_u$. Since Q_u is extraspecial, U' is normal in Q_u and hence U' is normal in L_U. This proves (1).

For W as in (2), take $U' = WU$. Clearly, U' is elementary abelian and so we can view it as a $GF(2)$-vector space. Notice that Q_u induces on U' all transvections with center $\langle u \rangle$. Since $W \cap U = 1$, $C_{Q_u}(W)$ induces on U all transvections with center $\langle u \rangle$. This implies (2), since the group generated by all transvections of U is $L_n(2)$.

Let W_1 and W_2 be as in (3). We will use induction on the rank of W_1. If $W_1 = 1$ then there is nothing to prove. Otherwise, choose $W_1' \leq W_1$ such that $|W_1/W_1'| = 2$, and let $W_2' = W_2 \cap W_1'U$. By induction, there is an element $x' \in C_{L_U}(U)$ such that $(W_1')^{x'} = W_2'$. Let $w_1 \in W_1 \setminus W_1'$ and let $\{w_2\} = W_2 \cap w_1U$. Notice that, by (1), $w_1^x \in w_1U = w_2U$ and hence $w_1^x = w_2u$ for some $u \in U$. If $u = 1$ then take $x = x'$. Otherwise, let y be an element of Q_u that induces on W_1U the transvection with center $\langle u \rangle$ and axis $W_2'U$. Clearly, $y \in C_{L_U}(U)$ and $W_1^x = W_2$, where $x = x'y$. This proves (3).

Finally, if $q \in Q_U$ then $q^2 \in \Phi(Q_u) = \langle u \rangle$ for every $u \in U^\#$. This proves (4). □

Lemma 4.3 *A subgroup U is singular if and only if it is generated by a set of pairwise perpendicular 2-central involutions. Furthermore, if $U = \langle u_1, \ldots, u_k \rangle$ ($u_i \neq 1$ for all i) is singular then $Q_U = \cap_{i=1}^{k} Q_{u_i}$.*

Proof We only need to prove the 'if' part of the first claim. Suppose $U = \langle u_1, \ldots, u_k \rangle$, where $u_1, \ldots, u_k$ are 2-central and pairwise perpendicular. By induction, $U' = \langle u_2, \ldots, u_k \rangle$ is singular. Since u_1 is perpendicular to $u_2, \ldots, u_k$, we have that $U' \leq Q_{u_1}$ which by Lemma 4.1 implies that $u_1 \in Q_{U'}$. By Lemma 4.2 (2), all involutions in $U \setminus U' = u_1U'$ are 2-central, since u_1 is 2-central. Finally, let $u \in U^\#$. Then $u \in U \leq Q_{u_i}$ for every i, since the involutions u_i are pairwise perpendicular. By Lemma 4.1, $U = \langle u_1, \ldots, u_k \rangle \leq Q_u$.

Suppose now that $U = \langle u_1, \ldots, u_k \rangle$ is singular. Clearly, $Q_0 = \cap_{i=1}^{k} Q_{u_i}$ contains Q_U. So it remains to see that $Q_0 \leq Q_U$. Let us use induction on k. The claim is obviously true if $k = 1$. Consider now the case $k > 1$ and set $U' = \langle u_2, \ldots, u_k \rangle$. By induction, $Q_{U'} = \cap_{i=2}^{k} Q_{u_i}$ and hence $Q_0 \leq Q_u$ for all $u \in U^\#$. However, this means that Q_0 is normal in Q_u. In particular, Q_0 is invariant under an element $x \in Q_u$, which induces on U a transvection taking u_1 to u_1u. Hence $Q_0 = Q_0^x \leq Q_{u_1}^x = Q_{u_1u}$. Thus $Q_0 \leq Q_u$ for all $u \in U^\#$. □

We now switch back to the case $G = M$. Our goal is to classify all singular subgroups in M up to conjugation. Notice that Proposition 3.2 means that the perpendicularity relation on 2-central involutions in M is nontrivial; that is, there exist singular subgroups of size more than two. We start by getting the details of the perpendicularity relation in M. For that, we need to know the fusion of involutions in Q. Let $\bar{C} = C/Z$, where $Z = \langle z \rangle$. Recall that the classes of involutions in $\bar{C}$ were determined in Section 4.2.

Lemma 4.4 *The group C has exactly two classes of involutions $x \neq z$, contained in Q. If q_2 and q_4 are representatives of those classes, then $C_C(q_2) \sim 2^{1+23}.Co_2$ and $C_C(q_4) \sim 2^{1+23}.(2^{11} : M_{24})$. Furthermore, q_4 is 2-central and q_2 is not.*

Proof   For $x \in Q \setminus Z$, the mapping $\bar{x} \mapsto x^2$ defines a nondegenerate quadratic form g on $\bar{Q}$. Since the action of $C/Q \cong Co_1$ on $\bar{Q}$ is absolutely irreducible, g is unique and hence g is equivalent to the form q existing on $\hat{\Lambda} = \Lambda/2\Lambda$ (cf. Section 4.2). The form q is zero on $\hat{\Lambda}_2$ and $\hat{\Lambda}_4$, and it is non-zero on $\hat{\Lambda}_3$. This means that x is an involution if and only if $\bar{x}$ belongs to the class $2e_2$ or $2e_4$. The involutions x and xz are conjugate in Q, because Q is extraspecial. Combined with Lemma 2.7, this establishes the first two claims of the lemma.

According to Proposition 3.2, at least one of q_2 and q_4 is conjugate to z in G. So, to complete the proof of the lemma, it suffices to show that $x = q_2$ is not 2-central. Suppose that $x \in S$. Let $D = C_C(x) = C \cap C_x$ and $R = Q \cap Q_x$. From the structure of D, it is clear that $R = O_2(D) \sim 2^{1+23}$. Since $R \leq Q$, we have that $[R, R] = Z$. Symmetrically, since $R \leq Q_x$, we have that $[R, R] = \langle x \rangle$, implying that $z = x$, a contradiction.   □

In particular, if a 2-central involution $y \neq z$ is perpendicular to z, then y is conjugate in C to $x = q_4$. This lemma implies that every singular subgroup $U \sim 2^2$ in M is conjugate to $\langle z, q_4 \rangle$. So there is only one conjugacy class of such subgroups.

Lemma 4.5 *Let $U \sim 2^2$ be singular. Then $W = Q_U/U \sim 2^{11}$ and $C_M(U)$ induces on W a group M_{24} acting as on the Todd module. Under the identification of W with the Todd module, the images of 2-central involutions from $Q_U \setminus U$ correspond to sextets, while the images of non 2-central involutions correspond to pairs.*

Proof   Without loss of generality, $U = \langle z, x \rangle$, where $x = q_4$. Let $D = C \cap C_x = C_M(U)$ and $R = Q \cap Q_x$. Notice that by Lemma 4.3 we have $R = Q_U$. Recall that $\bar{Q} = Q/Z$ affords a quadratic form g defined by $\bar{y} \mapsto y^2$. By Lemma 4.2 (4), Q_U is elementary abelian. In particular, $\bar{R}$ is a totally singular subspace with respect to g. This implies that $|\bar{R}| \leq 2^{12}$, and hence $|R| \leq 2^{13}$. On the other hand, both $C \cap Q_x$ and $C_x \cap Q$ have order 2^{24} and they are normal in D. Since $(C \cap Q_x) \cap (C_x \cap Q) = R$, the order of $(C \cap Q_x)(C_x \cap Q)$ is at least $2^{24+24-13} = 2^{35} = |O_2(D)|$ (see Lemma 4.4 for the structure of D). Hence $|R| = 2^{13}$.

It follows that $\bar{R}$ is a 12-dimensional subspace in $\bar{Q}$ invariant under the monomial group $D/Q \sim 2^{11} : M_{24}$. Such a subspace is known to be unique. Identifying $\bar{Q}$ with $\hat{\Lambda}$ and assuming that $\bar{x}$ is the image of the standard frame, we get that $\bar{R}^{\#}$ consists of the images of the vectors of the shape $\pm 8^1 0^{23}$ ($\bar{x}$), $\pm 4^2 0^{22}$

($\hat{\Lambda}_2$, non 2-central), and $\pm 4^4 0^{20}$ ($\hat{\Lambda}_4$, 2-central). Each pair of coordinates gives four vectors of the second kind, mapping onto two elements in $\bar{R} \cap \hat{\Lambda}_2$. These two elements of $\bar{R}$ sum up to $\bar{x}$. Similarly, every sextet produces 96 vectors of the third kind (two frames), mapping onto two elements in $\bar{R} \cap \hat{\Lambda}_4$. These two elements of $\bar{R}$ again sum up to $\bar{x}$. Thus, in $R/U \cong \bar{R}/\langle \bar{x} \rangle$, the nonidentity elements correspond simply to pairs and sextets. By Lemma 4.4, the elements from $R/U^{\#}$ corresponding to pairs (respectively, sextets) are the images of non 2-central (respectively, 2-central) involutions from $R \setminus U$. □

For the record, the normalizer of a singular subgroup $U \sim 2^2$ is now known to be an extension of a normal 2-subgroup of order 2^{35} by $S_3 \times M_{24}$. (The latter being the action of $N_M(U)$ on $U \times Q_U/U$.)

In this proof, if we do not assume that $\bar{x}$ is the image of the standard frame, then the condition for $\bar{y}$ to be in $\bar{R}$ looks as follows: Let $\{v_i\}$ be the frame corresponding to $\bar{x}$ (i.e., $\hat{v}_i = \bar{x}$ and $v_i \in \Lambda_4$ for all i) and let u be a short vector in Λ (i.e., a vector from $\Lambda_2 \cup \Lambda_3 \cup \Lambda_4$) such that $\bar{v} = \hat{y}$. Then $\bar{y} \in \bar{R}$ if and only if $(v_i, y) \in \{0, \pm 4, \pm 8\}$ for all i. When $\{v_i\}$ is the standard frame, this corresponds to the statement in the proof about the shapes of the vectors mapping into $\bar{R}$.

Combining Lemma 4.2 (3) with the fact that M_{24} acts transitively on pairs and on sextets, we obtain the following.

Corollary 4.6 *If $U \cong 2^2$ is singular then $C_M(U)$ has exactly two conjugacy classes in $Q_U \setminus U$, one consisting of non 2-central involutions, and one other consisting of 2-central involutions.*

Since $Q_U \setminus U$ contains a unique class of 2-central involutions, M has exactly one conjugacy class of singular subgroups 2^3.

Before we proceed further we need to understand better the perpendicularity relation among the elements in $Q_U \setminus U$, where U is a singular subgroup 2^2. For a non 2-central (respectively, 2-central) involution $y \in Q_U \setminus U$, let $P(y)$ (respectively, $S(y)$) be the pair (respectively, sextet) corresponding to $yU \in Q_U/U$.

We say that two sextets S_1 and S_2 intersect *evenly* if $|T_1 \cap T_2|$ is even for all tetrads $T_1 \in S_1$ and $T_2 \in S_2$. Suppose T_1 and T_2 are two tetrads and suppose $|T_1 \cap T_2| = 2$. Then the sextets defined by T_1 and T_2 intersect evenly if and only if $T_1 \cup T_2$ is contained in an octad. This allows us to compute that every sextet evenly intersects exactly 90 other sextets.

Lemma 4.7 *Let $U \sim 2^2$ be singular. Suppose $y, t \in Q_U \setminus U$, and suppose y is 2-central. Then*

(1) *if t is non 2-central, then $t \in Q_y$ if and only if $P(t)$ is contained in one of the tetrads from $S(y)$; and*

(2) *if t is 2-central, then $t \in Q_y$ if and only if $S(t)$ and $S(y)$ intersect evenly.*

Proof Assume again that $U = \langle z, x \rangle$, where $x = q_4$. Let $\{v_i\}$ be the frame in Λ that corresponds to $\bar{y}$ and let u be a short vector in Λ such that $\hat{u}$ corresponds to $\bar{t}$. Then the vectors v_i are of the shape $\pm 4^4 0^{20}$, where the nonzero coordinates appear in a tetrad from the sextet $S(y)$. Similarly, u is of shape $\pm 4^2 0^{22}$ (respectively, $\pm 4^4 0^{20}$) with the nonzero coordinates appearing in the pair $P(t)$ (respectively, sextet $S(t)$) if t is non 2-central (respectively, 2-central). According to the remark after the proof of Lemma 4.5 we have $t \in Q_y$ if and only if $(v_i, u) \in \{0, \pm 4, \pm 8\}$ for all i. The claim of the lemma follows. □

One implication of Corollary 4.6 is that M contains exactly one conjugacy class of singular subgroups 2^3. Indeed, pick a 2-central involution $y \in Q_{\langle z, x \rangle} \setminus \langle z, x \rangle$. Then every singular subgroup 2^3 is conjugate to $\langle z, x, y \rangle$.

Lemma 4.8 *Let $U \sim 2^3$ be singular. Then $W = Q_U / U \sim 2^6$ and $C_M(U)$ induces on W a group $3 \cdot S_6$ that acts on W irreducibly. Furthermore, $N_M(U)$ has two orbits on $W^\#$: an orbit of length 18 (images of non 2-central involutions from $Q_U \setminus U$) and an orbit of length 45 (images of 2-central involutions).*

Proof Without loss of generality, $U = \langle z, x, y \rangle$. We set $U_0 = \langle z, x \rangle$ and $V = \tilde{Q}_{U_0} / U_0$. According to Lemma 4.5, $C_M(U_0)$ induces on V a group M_{24} acting on V as on the Todd module. Let $S = S(y)$ be the sextet corresponding to $\tilde{y}$ under the identification of V with the Todd module. According to Lemma 4.7, $\tilde{Q}_U^\#$ consists of elements corresponding to pairs contained in the tetrads of S and to sextets evenly intersecting S. By counting, $\tilde{Q}_U^\#$ consists of 91 sextets (including S) and 36 pairs. Hence $|\tilde{Q}_U| = 2^7$. This means that $|Q_U / U| = 2^6$. Furthermore, $(Q_U / U)^\#$ contains 45 (respectively, 18) elements that are images of 2-central (respectively, non 2-central) involutions.

Recall that $C_M(U_0)$ induces on V a group M_{24}. The stabilizer of S in the latter group is a subgroup $2^6 : 3 \cdot S_6$. Let $D = N_M(U) \cap C_M(U_0)$ be the full preimage in $C_M(U_0)$ of the stabilizer of S. According to Lemma 4.2 (2), $N_M(U) \cap N_M(U_0)$ induces on $U \setminus U_0$ a group S_4. Since D is normal in $N_M(U) \cap N_M(U_0)$ and since $C_M(U)$ is the kernel of the action of D on $U \setminus U_0$, we conclude that $C_M(U)$ induces the whole sextet stabilizer $2^6 : 3 \cdot S_6$ in its action on V. Thus, it induces a quotient of $2^6 : 3 \cdot S_6$ on Q_U / U. Let $a \in C_M(U)$ be a 3-element mapping into the normal 3-subgroup of the quotient $3 \cdot S_6$. Consider the action of a on V. Clearly, a stabilizes every tetrad in the sextet S. Let S' be a sextet evenly intersecting S. Observe that every tetrad from S meets exactly two tetrads from S'. Being a 3-element, if a stabilizes S', then it must stabilize it tetradwise.

However, in that case, a stabilizes every part of a partition of $\{1,\ldots,24\}$ into 12 pairs (intersections of tetrads from S with tetrads from S'), which makes a to act on $\{1,\ldots,24\}$ trivially. This contradiction shows that a cannot stabilize S' and hence a acts nontrivially on Q_U/U. Therefore, $C_M(U)$ induces on Q_U/U either $2^6 : 3 \cdot S_6$ or $3 \cdot S_6$.

It is easy to see that the stabilizer of S in M_{24} acts transitively on pairs contained in tetrads from S and on sextets evenly intersecting S. Consequently, $C_M(U)$ has orbits of size 18 and 45 on Q_U/U. This makes the action on Q_U/U irreducible, implying that the group induced by $C_M(U)$ is in fact $3 \cdot S_6$. □

For the record, this lemma and Lemma 4.2 (2) imply that $N_M(U)$, where U is a singular subgroup 2^3, is an extension of a normal subgroup of order 2^{39} by $L_3(2) \times 3 \cdot S_6$.

Also, let us record what we proved about the classes of 2-central and non 2-central involutions in Q_U.

Corollary 4.9 *If $U \cong 2^3$ is singular then $C_M(U)$ has exactly two conjugacy classes in $Q_U \setminus U$, one consisting of non 2-central involutions, and one other consisting of 2-central involutions.*

Proof Follows from Lemma 4.2 (3). □

In particular, M contains a unique conjugacy class of singular subgroups 2^4.

Lemma 4.10 *Let $U \sim 2^4$ be singular. Then $W = Q_U/U \sim 2^3$. Furthermore, $W^{\#}$ contains exactly three elements that are images of non 2-central involutions, and these three elements generate W. The group $C_M(U)$ induces on W a group S_3.*

Proof Without loss of generality, $U \geq U_0 = \langle z, x \rangle$, say, $U = \langle z, x, y, t \rangle$. We will work with the Todd module $V = \tilde{Q}_{U_0} = Q_{U_0}/U_0$. Let $S = S(y)$ and $S' = S(t)$. If $s \in Q_U \setminus U$ is a non 2-central involution, then $P(s)$ is contained in a tetrad from S and in a tetrad from S'. Hence $T(s)$ must be one of the twelve pairs $P_1,\ldots,P_{12}$ (partitioning $\{1,\ldots,24\}$) that are intersections of tetrads from S with tetrads from S'. This proves that Q_U/U contains exactly three involutions that are images of non 2-central involutions. (Indeed, the twelve involutions in $\tilde{Q}_U$ merge into three involutions in Q_U/U.)

Let us be more specific. Since S and S' intersect evenly, there is a unique trio $T : = \{O_1, O_2, O_3\}$ of which both S and S' are refinements (we view trios and sextets as partitions of $\{1,\ldots,24\}$). Then every O_i is a union of some four pairs P_j. It is easy to see that pairs P_j and P_k produce the same element in Q_U/U if and only if they are contained in the same octad O_i. Thus, the octads O_i correspond to the "non 2-central" elements $a_i \in Q_U/U$. Clearly, the stabilizer of S and S' in M_{24} induces an S_3 on the trio T. Hence also $N_M(U)$ induces an S_3 on the three involutions a_i. Furthermore, since $N_M(U)$ induces a simple group $L_4(2)$ on U, we also have that $C_M(U)$ induces an S_3 on the a_i's. It remains to see that they are linearly independent and generate Q_U/U.

Observe that if P_j and P_k belong to distinct octads O_i, then the sum (we switch to the additive notation in V and Q_U/U) of the elements from V corresponding to P_j and P_k is of sextet type and, furthermore, that sextet is not a refinement of T. This means that the sum of two distinct involutions a_i is nontrivial and "2-central." This implies the linear independence. Let b be an arbitrary "2-central" element from $(Q_U/U)^{\#}$; say, it is the image of an element of V that corresponds to a sextet $S'' = \{R_1, \ldots, R_6\}$. Observe that S'' evenly intersects both S and S'. In particular, $|O_i \cap R_j|$ is even for all i and j. Suppose for some i and j, we have $|O_i \cap R_j| = 2$. (We will say that such an S'' is of the *first kind*.) Observe that O_i is a union of some four pairs P_k. If R_j meets two of these pairs, then R_j meets a tetrad from S or from S' in just one point, a contradiction. Hence, $O_i \cap R_j$ coincides with some P_k. Similarly, considering a nontrivial intersection of R_j with some other $O_{i'}$ we obtain that R_j contains a second pair $P_{k'}$ and hence b is the sum of two of the a_i's. It remains to consider the case where $|O_i \cap R_j| \in \{0, 4\}$ for all i and j, that is, every R_j is fully contained in some O_i. (Then we will say that S'' is of the *second kind*.) Fix O_i and R_j with $R_j \subset O_j$. If $P_k \subset R_j$ then $b + a_i$ is "non 2-central," which means that $R_j \setminus P_k = P_{k'}$. However, since P_k and $P_{k'}$ are both in O_i, we get $b = a_i + a_i = 0$, a contradiction. Therefore, R_i meets each of the four pairs P_k partitioning O_i in one point. Fix $P_k \subset O_i$ and consider $c = b + a_i$. Then one of the preimages of c in V will correspond to the sextet S''' containing the tetrad $R_j \triangle P_k$ ($\triangle$ denotes symmetric difference of sets). If S''' is of the second kind then S''' contains a tetrad contained in $O_{i'} \neq O_i$. That tetrad of S''' will meet some tetrad of S'' in at least two points. This gives us two octads meeting in five points, a contradiction. Therefore, S''' is of first kind. By the above, c is in the span of a_i's and hence so is also b.   □

We will continue using the notation a_i for the three "non 2-central" elements from Q_U/U. According to Lemma 4.10, $C_M(U)$ has three orbits on $(Q_U/U)^{\#}$: $\{a_1, a_2, a_3\}$ ("non 2-central"), $\{a_1 + a_2, a_1 + a_3, a_2 + a_3\}$ ("2-central"; sextets of the first kind), and $\{a_1 + a_2 + a_3\}$ ("2-central"; sextets of the second kind).

We record this as the following.

Corollary 4.11 *If $U \cong 2^4$ is singular, then $C_M(U)$ has exactly three conjugacy classes in $Q_U \setminus U$, two consisting of 2-central involutions, and one other consisting of non 2-central involutions.*

For the record, the normalizer of a singular subgroup $U \sim 2^4$ is an extension of a normal subgroup of order 2^{39} by $L_4(2) \times S_3$.

It follows from Corollary 4.11, that M contains two conjugacy classes of singular subgroups 2^5. One of these two classes is represented by $\langle z, x, y, t, s \rangle$ with the image of s in $Q_{\langle z,x,y,t \rangle}/\langle z, x, y, t \rangle$ being $a_1 + a_2$, while for the other the image of s can be chosen as $a_1 + a_2 + a_3$. We will write "a singular subgroup 2_1^5" (respectively, 2_2^5) for the two types of singular subgroups 2^5.

We will need the following corollary of Lemma 4.11.

Corollary 4.12 *Every singular subgroup 2^4 is contained in exactly three singular subgroups 2_1^5 and a unique singular subgroup 2_2^5.*

To complete the classification of singular subgroups of M, we need to discuss perpendicularity between the elements of $Q_U \setminus U$.

Lemma 4.13 *Suppose $U \sim 2^4$ is singular. Let s and r be two elements from $Q_U \setminus U$, whose images in Q_U/U are distinct. If s is 2-central and $r \in Q_s$, then the image of s is $a_i + a_j$ for some i and j. Furthermore, the image of r is either a_i or a_j.*

Proof Without loss of generality, $U = \langle z, x, y, t \rangle$ as in Lemma 4.10. Since $r \in Q_s$, we have that r and rs are both 2-central or both non 2-central. This implies that the image of s cannot be $a_1 + a_2 + a_3$. Hence the image of s coincides with some $a_i + a_j$. Next, it is easy to see that if r' maps onto a_i or a_j, then $r' \in Q_s$. Since no element mapping onto $a_1 + a_2 + a_3$ can be in Q_s, we conclude that the image of $Q_s \cap Q_U$ in Q_U/U coincides with $\langle a_i, a_j \rangle$. $\qquad\square$

The information in Lemma 4.13 allows us to determine Q_U for singular subgroups $U \sim 2^5$.

Lemma 4.14 *The following hold.*

(1) *If U is singular 2_1^5 then Q_U/U is of order two. Furthermore, all involutions in $Q_U \setminus U$ are non 2-central.*
(2) *If U is singular 2_2^5 then $Q_U = U$.*

Proof Follows from Lemma 4.13. $\qquad\square$

For the record, the normalizer of a singular 2_1^5 is an extension of a subgroup of order 2^{36} by $L_5(2)$, while the normalizer of a singular 2_2^5 is an extension of a subgroup of order $2^{36}3$ by $L_5(2)$.

Lemma 4.14 means that M contains no singular subgroups of order more than 2^5 and so we have completed the classification of the singular subgroups in M.

Proposition 4.15   *The Monster group M contains exactly 6 classes of nontrivial singular subgroups. The corresponding orders are 2, 2^2, 2^3, 2^4, 2^5 and 2^5.*

Let S_i, $1 \le i \le 4$, denote the conjugacy class of all singular subgroups 2^i of M. For $i = 5$, we will use the notation $S_{5,1}$ and $S_{5,2}$ for the conjugacy classes of singular subgroups 2^5_1 and 2^5_2, respectively.

Notice that in this section we only indicated the order of the normalizers of singular subgroups and their action on $U \times Q_U/U$. A more detailed information about the structure of these 2-local subgroups can be found elsewhere.

4.5 Arks

From this section on, $G = M$, the Monster simple group. In this section, we construct and study a class of subgroups 2^{10} of M, associated with singular subgroups.

Let U be a singular subgroup 2^5_1. According to Lemma 4.12, every index two subgroup of U is contained in a unique singular 2^5_2. Let $\mathcal{A} = \{U' \in S_{5,2} | [U : U \cap U']| = 2\}$ and let $A(U)$, the *ark* defined by U, be the subgroup of M generated by all $U' \in \mathcal{A}$. Clearly, $A(U)$ is invariant under $N_M(U)$.

Lemma 5.1   *The ark $A(U)$ is elementary abelian of order 2^{10}. Furthermore, U and $A(U)/U$ are dual to each other as modules for $N_M(U)$.*

Proof   Suppose $U', U'' \in \mathcal{A}$ with $U' \ne U''$. Then $W = U' \cap U'' \cap U$ is a singular subgroup 2^3. Since $U', U'' \le Q_W$ and since Q_W is elementary abelian by Lemma 4.2 (4), we have that U' and U'' commute elementwise and, therefore, $A = A(U)$ is elementary abelian.

Consider $\bar{A} = A/U$. If $U' \in \mathcal{A}$ then $\bar{U}'$ is of order two. This yields a mapping $V \mapsto \bar{a}_V$ from the set of index two subgroups $V < U$ to $\bar{A}^\#$. Namely, $\langle \bar{a}_V \rangle = \bar{U}'$, where $U' \in \mathcal{A}$ is the only singular 2^5_1 containing V. Clearly, the elements $\bar{a}_V$ generate $\bar{A}$. Furthermore, the subgroups V correspond to the elements in $(U^*)^\#$, where U^* is the dual of U. Therefore, in order to complete the proof of this lemma, it suffices to establish the three-term relations: $\bar{a}_{V_1} \bar{a}_{V_2} \bar{a}_{V_3} = 1$ whenever V_1, V_2, and V_3 are three index two subgroups of U, containing a given index four subgroup $W < U$.

Consider $\hat{Q}_W = Q_W/W$. According to Lemma 4.8, $\hat{Q}_W$ is 6-dimensional (as a vector space over $GF(2)$) and $C_M(W)$ induces on $\hat{Q}_W$ a group $3 \cdot S_6$. Let $x \in C_M(W)$ be a 3-element that maps onto a nontrivial element in the center of that action. Let V be a singular subgroup 2^4 containing W. Then $\hat{V}$ is of order two, and we claim that if U' is the unique singular 2^5_2 containing V, then

$\hat{U}' = \hat{V}\hat{V}^x$. Indeed, on the one hand, each of the 45 (cf. Lemma 4.8) subgroups V is contained in a unique U'. On the other hand, each U' contains three subgroups V. Therefore, W is contained in exactly 15 singular subgroups 2_2^5. It follows that each of them is invariant under x, since S_6 cannot nontrivially act on $15/3 = 5$ points. This proves our claim.

We can now finish the proof of the lemma. Suppose V_1, V_2, and V_3 are the three index two subgroups of U, containing W. Let U'_i, $i = 1, 2, 3$, be the unique singular 2_2^5 containing V_i. Working again in $\hat{Q}_W = Q_W/W$, we obtain that the image of $\langle U'_1, U'_2, U'_3 \rangle$ in $\hat{Q}_W$ coincides with $\hat{U}\hat{U}^x$, since $\langle V_1, V_2, V_3 \rangle = U$. Thus, $\langle \bar{a}_{V_1}, \bar{a}_{V_2}, \bar{a}_{V_3} \rangle = \langle U'_1, U'_2, U'_3 \rangle / U$ is of order four and hence $\bar{a}_{V_1} \bar{a}_{V_2} \bar{a}_{V_3} = 1$ holds. $\qquad\square$

Let $U \in \mathcal{S}_{5,1}$ and let $\mathcal{A}$, $A = A(U)$ and $\bar{A} = A/U$ be as discussed.

Lemma 5.2 *The following hold.*

(1) *If $a \in A \setminus U$, then $\langle U, a \rangle = \langle U, U' \rangle$ for some $U' \in \mathcal{A}$. In particular, every coset of U in A contains a 2-central involution.*

(2) *If $a \in A \setminus U$ is 2-central, then $U \cap Q_a$ is of index two in U. Furthermore, au (where $u \in U$) is 2-central if and only if $u \in U \cap Q_a$.*

Proof Part (1) follows directly from Lemma 5.1. Let $U' \in \mathcal{A}$ be such that $\langle U, a \rangle = \langle U, U' \rangle$. Then $\langle U, a \rangle \leq Q_W$, where $W = U \cap U'$. Since perpendicularity is symmetric, we have that $W \leq Q_a$. On the other hand, $U \not\leq Q_a$, because otherwise $\langle U, a \rangle$ must be singular in view of Lemma 4.3. Thus, $U \cap Q_a = W$ is of index two in U. Clearly, $\langle W, a \rangle$ is a singular subgroup; in particular, au is singular if $u \in W$. Comparing now with Lemma 4.10, we see that all elements in $\langle U, a \rangle \setminus (U \cup U')$ are non 2-central. Consequently, $a \in U'$ (and hence $U' = \langle W, a \rangle$) and au is non 2-central for all $u \in U \setminus W$. $\qquad\square$

It follows from this lemma that A contains exactly $31 \cdot 16 = 496$ non 2-central and $31 + 31 * 16 = 527$ 2-central involutions. Moreover, all non 2-central involutions in A are conjugate to q_2. Also notice that both non 2-central and 2-central involutions generate A.

Next, we analize the embedding of the ark $A = A(U)$ in C_u for $u \in U^\#$. First of all, we claim that Lemma 5.2 implies that $A \cap Q_u$ has index two in A. Indeed, u is contained in 15 subgroups $W = U \cap U'$, $U' \in \mathcal{A}$, and hence $\overline{A \cap Q_u}$ is of order 16. Let $a \in A \setminus (A \cap Q_u)$. Since A is generated by 2-central involutions, we can choose a to be 2-central.

Lemma 5.3 *We have $A \cap Q_u = [Q_u, a]$ and the image of a in $C_u/Q_u \cong Co_1$ is a 2A-involution.*

Proof Notice that Q_u normalizes U and hence it also normalizes A. Therefore, $[Q_u, A] \le A \cap Q_u$. If the image of a in Co_1 is of type $2B$ or $2C$, then Lemmas 2.5 and 2.6 imply that $|[Q_u, a]| \ge 2^{12}$, a contradiction. Hence, the image of a is of type $2A$. Furthermore, it follows from Lemma 2.4 that $[Q_u, a]\langle u \rangle / \langle u \rangle$ is of order 2^8, implying that $[Q_u, a]\langle u \rangle = A \cap Q_u$. Since $[Q_u, a]$ is normal in Q_u, it contains u and hence $[Q_u, a] = A \cap Q_u$. $\square$

Let $D = N_{C_u}(Q_u \langle a \rangle) \sim 2^{1+24}.2^{1+8}.\Omega_8^+(2)$. Let $\bar{C}_u = C_u / \langle u \rangle$.

Lemma 5.4   *A is normal in D.*

Proof   Let $R = \langle Q_u, a \rangle$. Then $\bar{R}$ is normal in $\bar{D}$. Observe that $\bar{R}$ has exactly two maximal elementary abelian subgroups: $\bar{Q}_u$ and $\bar{R}_0 = \langle C_{\bar{Q}_u}(\bar{a}), \bar{a} \rangle$. Since Q_u is normal in D, we conclude that R_0 (defined as the full preimage of $\bar{R}_0$ in D) is also normal in D. We claim that $A = Z(R_0)$. Indeed, clearly, $A \cap Q_u = [Q_u, a]$ is the center of $R_0 \cap Q_u$, because Q_u is extraspecial and because $\overline{R_0 \cap Q_u} = C_{\bar{Q}_u}(\bar{a})$. Hence, it remains to see that $[R_0 \cap Q_u, a] = 1$. However, this is clear: since $\bar{R}_0$ is abelian, we have that $[R_0 \cap Q_u, a] \le \langle u \rangle$; on the other hand, by Lemma 5.2 (2), the involution au is not 2-central. Hence, u cannot be written as a commutator $[r, a]$ for $r \in R$. Since R_0 is normal in D and $A = Z(R_0)$, we finally obtain that A is normal in D. $\square$

Let $N = N_M(A)$.

Corollary 5.5   *The action of N on A is irreducible. In particular, $A = \langle u^N \rangle$.*

Proof   According to Lemma 5.1, A has two 5-dimensional composition factors as a module for $N_M(U) \le N$. On the other hand, it follows from Lemmas 5.4, 5.3, and 2.4 that A has composition factors of dimensions 1, 8, and 1 as a module for $D \le N$. $\square$

In view of this lemma, we can assume that a is conjugate to u in N. Let $A_0 = A \cap Q_a \cap Q_u$. Notice that $A_0 \sim 2^8$.

Lemma 5.6   *We have $C_D(\langle a, u \rangle) \sim 2^{10}.2^{16}.\Omega_8^+(2)$. In particular, $\bar{a}$ is of type $2a_1$ in $\bar{C}_u$ (cf. Section 4.2) and $C_D(\langle a, u \rangle) = C_a \cap C_u$. Furthermore, $C_a \cap C_u$ induces on A_0 a group $\Omega_8^+(2)$ acting as on a halfspin module.*

Proof   Since $A \cap Q_u = [Q_u, a]$, the orbit of a under Q_u consists of at least 2^8 elements. On the other hand, if a' is a 2-central involution in $A \setminus Q_u$, then $a'u$ is non 2-central by Lemma 5.2. Therefore, $A \setminus (A \cap Q_u)$ consists of exactly 2^8 2-central and 2^8 non 2-central involutions. Furthermore, all 2-central (respectively, non 2-central) involutions in $A \setminus (A \cap Q_u)$ are conjugate by Q_u.

This shows that $Q_u C_D(\langle a, u \rangle) = D$. Comparing with Lemma 2.4 we obtain that $C_D(\langle a, u \rangle) \sim 2^{10}.2^8.2^8.\Omega_8^+(2)$. Notice that the two 8-dimensional chief factors (again, see Lemma 2.4) provide nonisomorphic modules for the quotient $\Omega_8^+(2)$. Therefore, $O_2(C_D(\langle a, u \rangle))/A$ is elementary abelian and so we can record the structure of $C_D(\langle a, u \rangle)$ as $2^{10}.2^{16}.\Omega_8^+(2)$.

Comparing with Lemmas 2.10 and 2.11, we see that $\bar{a}$ must be of type $2a_1$ and that $C_D(\langle a, u \rangle) = C_a \cap C_u$. Clearly, A_0 is invariant under $C_a \cap C_u$. Since $x \mapsto \bar{x}$ establishes an isomorphism between A_0 and $[\bar{Q}_u, a]$, the last claim follows from Lemma 2.4 (2). $\qquad\square$

Define a mapping $f \colon A \longrightarrow GF(2)$ as follows: for $x \in A$, $f(x) = 0$ if and only if x is the identity or a 2-central involution.

Lemma 5.7 *The mapping f is a nondegenerate quadratic form of plus type.*

Proof We will switch to the additive notation in A. Decompose A as $A = \langle a, u \rangle \oplus A_0$. Then the restriction of f on $\langle a, u \rangle$ is a plus type form, because au is non 2-central. It was shown in Lemma 5.6 that $C_a \cap C_u$ induces on A_0 a group $\Omega_8^+(2)$ acting as on a halfspin module (which is a triality conjugate of the natural module). In particular, $C_a \cap C_u$ has two orbits on $A_0^\#$, of length 120 and 135. Thus, in order to show that the restriction of f to A_0 is a quadratic form of plus type, it suffices to show that A_0 contains exactly 120 non 2-central involutions. However, this is clear. Indeed, by Lemma 5.2, each of the 15 cosets $a' + (U \cap A_0) = a' + (U \cap Q_u)$, with $a' \in A_0 \setminus (U \cap Q_u)$, contains exactly eight 2-central and eight non 2-central involutions. We have shown that the restriction of f on A_0 is also a plus type form.

It remains to verify the values of f on the elements $x + y$, $x \in \langle a, u \rangle^\#$, and $y \in A_0$. If $x = a$ or u then $f(x + y) = f(y)$ because y and $x + y$ are conjugate in Q_x. In view of Lemma 5.6, $C_a \cap C_u$ has orbits of length 120 and 135 on the set $a + u + A_0^\#$. Since the total number of non 2-central involutions in A is known to be 496, we compute that among the elements in $a + u + A_0^\#$ there are exactly 135 non 2-central involutions and 120 2-central involutions. Hence, $f(a + u + y) = 1 + f(y)$ for all $y \in A_0^\#$. $\qquad\square$

We can now pin down the structure of $N = N_M(A)$. Let $P_A = O_2(N)$.

Lemma 5.8 *We have $N \sim 2^{10+16}.\Omega_{10}^+(2)$. In particular, $P_A = C_M(A) \sim 2^{10+16}$.*

Proof First of all, Lemma 5.6 yields that $C_M(A)$ is an extension of A by a group 2^{16}, i.e., $C_M(A) \sim 2^{10+16}$. Consider now the action of N on A. Clearly, N leaves the form f invariant. So $N/C_M(A)$ is isomorphic to a subgroup of $O_{10}^+(2)$. We claim that it is isomorphic to $\Omega_{10}^+(2)$. Indeed, observe that D and $N_M(U)$ share a Sylow 2-subgroup T (indeed, the 2-parts of the orders of D and

$N_M(U)$ coincide and hence as T we can take a Sylow 2-subgroup of $N_M(U)$ centralizing u). Consider an index two subgroup in U invariant under T and the unique singular 2_2^5, say U', containing that subgroup. Both U and U' are maximal totally singular with respect to f and T leaves invariant both U and U'. This yields that the image of T lies in $\Omega_{10}^+(2)$ and, moreover, the images of D and $N_M(U)$ lie in $\Omega_{10}^+(2)$, too. Comparing the orders, we obtain that they are two maximal parabolics in $\Omega_{10}^+(2)$. Therefore, $N/C_M(A)$ is either $\Omega_{10}^+(2)$ or $O_{10}^+(2)$. It remains to notice that $N/C_M(A) \cong O_{10}^+(2)$ is impossible, because U and U' are not conjugate. □

For $x \in A^\#$, let $x^\perp$ be the orthogonal complement of $\langle x \rangle$ with respect to the symplectic form $(x_1, x_2) = f(x_1 + x_2) - f(x_1) - f(x_2)$ on A. (We continue using the additive notation in A.)

Corollary 5.9 *If $x \in A^\#$ is 2-central then $x^\perp = A \cap Q_x$.*

Proof First of all, by the preceding lemma, N is transitive on 2-central involutions in A. Hence $A \cap Q_x$ has index two in A. Furthermore, since y and $x + y$ have the same type whenever $y \in Q_x$, we have that $f(y) = f(x + y)$ for all $y \in A \cap Q_x$. This proves that $A \cap Q_x \leq x^\perp$. □

This shows that a subgroup of A is singular if and only if it is totally singular with respect to f. Notice that A contains both a singular 2_1^5 and a singular 2_2^5 and so, indeed, an ark contains all species of singular subgroups. Furthermore, all singular subgroups of A of the same kind are conjugate in N. This implies, in particular, the following

Lemma 5.10 *If $U \in S_{5,1}$, then $A(U)$ is the only ark containing U. If $U \in S_{5,2}$, then U is contained in exactly three arks. Furthermore, those three arks are conjugate under $N_M(U)$.*

Proof The first claim follows since $N_M(U) \leq N$ if $U \in S_{5,1}$. If $U \in S_{5,2}$ and $U \leq A$, we compute that $N_N(U)$ has index three in $N_M(U)$. □

4.6 Elementary Abelian Subgroups in P_A

Let A be an ark and $N = N_M(A)$. We first produce an inventory of the elements from $P_A \setminus A$. Since $A \leq Z(P_A)$, every coset xA with $x \in P_A \setminus A$ consists entirely of involutions or entirely of elements of order four.

Let $\tilde{N} = N/P_A \cong \Omega_{10}^+(2)$.

Lemma 6.1 *If u is a 2-central involution from A, then $R = P_A \cap Q_u$ is of order 2^{17}. In particular, P_A is nonabelian.*

Proof Notice that Q_u normalizes any singular 2_1^5 subgroup U such that $u \in U \leq A$, and hence Q_u normalizes $A = A(U)$. Thus, $Q_u \leq N$. Notice further that Q_u cannot be fully contained in P_A. Indeed, if $Q_u \leq P_A$ then Q_u has index two in P_A, which implies that Q_u must have a center of size at least 2^5; clearly a contradiction. Thus, $Q_u \not\leq P_A$, which means that $\tilde{Q}_u$ is a nontrivial normal subgroup of $\tilde{D}$, where $D = N \cap C_u$. Since $\tilde{D}$ is a maximal parabolic (the stabilizer of a singular vector from the natural module), we get that $\tilde{Q}_u \sim 2^8$. Hence $|R| = 2^{25-8} = 2^{17}$. Being a subgroup of an extraspecial group 2^{1+24}, R must be nonabelian. Hence, P_A is also nonabelian. $\qquad\square$

We will now classify the cosets xA with $x \in P_A \setminus A$. It turns out that the cosets consisting of involutions correspond to singular subgroups 2_1^5 from A.

Lemma 6.2 *Suppose $U \leq A$, $U \in \mathcal{S}_{5,1}$. Then*

(1) *$Q_U \leq P_A$ and $Q_U \not\leq A$; hence, $X = Q_U A \setminus A$ is a coset from $P_A \setminus A$, consisting of involutions; if $x \in X$, then $U = [P_A, x]$; and*

(2) *$K = N_M(U)$ has exactly two orbits on X; one of the orbits is $Q_U \setminus U$, and it consists of non 2-central involutions (conjugate to q_2); the other orbit is $X \setminus Q_U$, and it consists of 2-central involutions.*

Proof Let $K = N_M(U)$. Notice that $\tilde{K}$ is a maximal parabolic in $\tilde{N} \cong \Omega_{10}^+(2)$; namely, it is the stabilizer of a maximal totally singular subspace U from the natural module A. Clearly, Q_U is invariant under K. Since K has two 5-dimensional chief factors in A and since Q_U/U has order two, we conclude that $Q_U \not\leq A$. If $y \in A \setminus U$ is 2-central then $W = U \cap y^\perp$ is a singular 2^4. Since $\langle y, Q_U \rangle \leq Q_W$ (which is abelian), the subgroup Q_U centralizes every y and hence $Q_U \leq P_A$.

Recall that $K = N_M(U)$ is contained in N, because $A = A(U)$. Clearly, K acts on $X = Q_U A \setminus A$. Notice that $[P_A, Q_U] \leq Q_U \cap A = U$ and hence $[P_A, Q_U] = U$, because K acts on U irreducibly. This implies that for $x \in Q_U \setminus U$ we have $[P_A, x] = U$. Since $A = Z(P_A)$, the same must be true for all $x \in X$. In particular, for all $x \in X$, all elements in xU are conjugate under P_A. Let $T = Q_U A$ and let $\hat{T} = T/U$. Clearly, $\hat{T}$ is the product of $\hat{A} \sim 2^5$ and $\hat{Q}_U \sim 2$. Furthermore, K stabilizes both $\hat{A}$ and $\hat{Q}_U$, and it acts transitively on $\hat{A}^\#$. Thus, K indeed has exactly two orbits on $xA = T \setminus A$.

We already know from Lemma 4.14 (1) that the involutions from $Q_U \setminus U$ are non 2-central. Since those involutions are contained in Q_u for $u \in U^\#$, they are conjugate to q_2. To see that the involutions in $X \setminus Q_U$ are 2-central, consider $W \leq U, W \sim 2^4$. Let U' be the unique singular subgroup 2_2^5 containing W. Then, by definition of $A = A(U)$, we have $U' \leq A$. Let $x \in Q_U \setminus U$ and $s \in U' \setminus W$. Comparing with Corollary 4.11 and with the definition of singular subgroups 2_2^5 (following Corollary 4.11), we see that xs is 2-central. Clearly, $xs \in X \setminus Q_U$, and so the claim follows. □

In particular, this lemma shows that Q_U and hence also U can be recognized from the coset $X = Q_U A \setminus A$. Therefore, such cosets are in a natural bijection with the set of all singular subgroups 2_1^5 from A. The latter set has size 2295, which means that at least 2295 cosets of A in $P_A \setminus A$ consist of involutions. We claim that the remaining $(2^{16} - 1) - 2295$ cosets of A in $P_A \setminus A$ consist of elements of order four.

Suppose $x \in P_A$ is of order four. Then $s = x^2$ is an element of A and furthermore $s = y^2$ for each $y \in xA$.

Lemma 6.3 *If u is a 2-central involution from A, then $(P_A \cap Q_u)A \setminus A$ contains exactly 120 cosets xA such that $u = x^2$. The group $K = N \cap C_u$ transitively permutes those cosets.*

Proof Consider $R = P_A \cap Q_u$. By Lemma 6.1, $|R| = 2^{17}$. Let a be a 2-central involution from $A \setminus u^\perp$. Then $R \leq C_a$, the image of a in $C_u/Q_u \cong Co_1$ is of type 2A, and, comparing with Lemma 2.4, we see that $R = Q_u \cap C_a$. Since $Q_u \leq N$, we have that $[Q_u, a] \leq Q_u \cap A = u^\perp$. Since $F = N \cap C_a \cap C_u$ involves $\Omega_8^+(2)$ (indeed, if we view A as the natural module for $\tilde{N} = N/P_A \cong \Omega_{10}^+(2)$, then a and u span in A a nondegenerate subspace of plus type), Lemma 2.4 gives us that $\bar{R} \sim 2^8$ and F induces on $\bar{R}$ a group $\Omega_8^+(2)$ acting as on a halfspin module. Here the bar indicates the image in $\bar{P}_A = P_A/A$.

Define a mapping $q: \bar{R} \longrightarrow GF(2)$ by $q(xA) = 0$ if $x^2 = 1$, and $q(xA) = 1$ if $x^2 = u$. Then q is a quadratic form on $\bar{R}$ and this form is invariant under K. Since the halfspin module for $\Omega_8^*(2)$ is triality conjugate to the natural module and since the latter admits a unique invariant quadratic form, the claims of the lemma follow. □

Since A contains 527 2-central involutions u, Lemma 6.3 accounts for $527 \cdot 120$ cosets xA consisting of elements of order four. Since $527 \cdot 120 = (2^{16} - 1) - 2295$, all the cosets of A in $P_A \setminus A$ have been accounted for. Thus, we obtain the following.

Lemma 6.4 *The group N has exactly two orbits on the nonidentity elements of $\bar{P}_A = P_A/A$. The smaller orbit has length 2295 and it consists of cosets containing involutions. The longer orbit has length $527 \cdot 120$ and it consists of cosets containing elements of order four.*

In particular, $\bar{P}_A$ is irreducible as a module for $\tilde{N} = N/P_A \cong \Omega_{10}^+(2)$. We remark that this module is isomorphic to the halfspin module. Indeed, this follows from the fact that the stabilizer of maximal totally singular subspace $U \leq A$, $U \in \mathcal{S}_{5,1}$, fixes a vector in $\bar{P}_A$.

Additionally, Lemma 6.2 gives us the following.

Corollary 6.5 *The group N has exactly two conjugacy classes of involutions in $P_A \setminus A$. One class has length $2295 \cdot 32$, and it consists of non 2-central involutions conjugate to q_2. The other class has length $2295 \cdot (1024 - 32)$ and it consists of 2-central involutions.*

We will not classify the classes of elements of order four in P_A. However, we will need the following fact.

Lemma 6.6 *If $x \in P_A$ is of order four, then $x \in (P_A \cap Q_u)A$ where $u = x^2$.*

Proof We have seen in Lemma 6.1 that $P_A \cap Q_u$ is nonabelian and hence it contains an element y of order four. Clearly, $y^2 = u$. Since N has just one orbit on cosets from P_A/A that consist of elements of order four, there is a conjugate y^n, $n \in N$, of y which lies in the coset xA. Then $(y^n)^2 = x^2 = u$ and hence, without loss of generality, we may assume that $y = y^n$ lies in xA. Now since $y \in (P_A \cap Q_u)A$, we have that $x \in yA \leq (P_A \cap Q_u)A$. $\square$

Next, we need to know when two involutions from $P_A \setminus A$ commute. For an involution $x \in P_A \setminus A$, let $U(x, A)$ (or simply $U(x)$, if A is clear from the context) be the singular subgroup 2_1^5 from A, that corresponds to the coset xA. Recall that $U(x) = [x, P_A]$ (cf. Lemma 6.2 (1)).

Lemma 6.7 *Involutions x and y from $P_A \setminus A$ commute if and only if $U(x) \cap U(y)$ has size at least eight.*

Proof Notice first of all that $[x, y] = [x', y']$ for arbitrary $x' \in xA$ and $y' \in yA$. Hence, commutation of x and y depends solely on $U' = U(x)$ and $U'' = U(y)$. In particular, we may assume that $x \in Q_{U'}$ and $y \in Q_{U''}$. Secondly, observe that U' and U'' meet in a subgroup of order 2, 2^3, or 2^5. Since $N = N_M(A)$ acts transitively on pairs (U', U'') with $U' \cap U''$ of a given size and since P_A is nonabelian, it suffices to show that x and y commute if $W = U' \cap U''$ has size eight. However, this is clear: both x and y are contained in Q_W, which is abelian. $\square$

This lemma allows us to determine now all maximal elementary abelian subgroups of P_A.

Lemma 6.8 *With respect to conjugation by $N = N_M(A)$, the group P_A has exactly two classes of maximal elementary abelian subgroups Y:*

(1) *for $W \leq A$, $W \in S_2$, Y consists of A and all involutions $y \in P_A \setminus A$ such that $W < U(y)$; and*

(2) *for $V \leq A$, $V \in S_{5,2}$, Y consists of A and all involutions $y \in P_A \setminus A$ such that $V \cap U(y)$ is of order 2^4.*

Proof First of all, it follows from Lemma 6.7 that the subgroups Y from (1) and (2) are elementary abelian. (Indeed, in (1) if $U(y_1)$ and $U(y_2)$ both contain W, then $U(y_1) \cap U(y_2)$ is of order at least 2^3; the other case is even easier.)

Let E now be an elementary abelian subgroup of P_A. Let $\mathcal{E} = \{U(x)|x \in E \setminus A\}$. It follows from Lemma 6.7 that if U and U' are distinct elements of $\mathcal{E}$, then $U \cap U'$ has order 2^3. Let us now show that if $U, U', U'' \in \mathcal{E}$ are pairwise distinct, then $U \cap U' \cap U''$ is of order at least 2^2. Suppose not. Then $U \cap U' \cap U''$ has order two. Observe that by Lemma 4.3 the subgroup $W = \langle U \cap U', U \cap U,'' U' \cap U'' \rangle$ is singular. Furthermore, since $(U \cap U') \cap (U \cap U'')$ has order two, we have $U = \langle U \cap U', U \cap U'' \rangle$, which means that $U \leq W$. Similarly, $U', U'' \leq W$; clearly, a contradiction, since U, U', and U'' are maximal singular. Thus, indeed, $U \cap U' \cap U''$ has order at least 2^2.

Fix $U \in \mathcal{E}$ and let $\mathcal{T} = \{U \cap U'|U' \in \mathcal{E}, U' \neq U\}$. This is a set of subgroups 2^3 from U, such that any two of them meet in a subgroup 2^2. We claim that one of the following two possibilities holds: (a) there is a subgroup $W \leq U$ of order four such that every $T \in \mathcal{T}$ contains W; or (b) there is a subgroup $W \leq U$ of order 16, such that every $T \in \mathcal{T}$ is contained in W. Let T_1, T_2 be distinct elements from $\mathcal{T}$, and let $W_1 = T_1 \cap T_2$ and $W_2 = \langle T_1, T_2 \rangle$. Then clearly W_1 is of order four and W_2 is of order 16. If every $T \in \mathcal{T}$ is contained in W_2 then we have case (b) with $W = W_2$. So suppose $T_3 \in T$ and $T_3 \not\leq W_2$. Observe that $T_3 \cap W_2$ has order four and hence $T_3 \cap W_2 = T_3 \cap T_1 = T_3 \cap T_2 = W_1$. Finally, consider an arbitrary $T \in \mathcal{T}$. If $T \not\geq W_1$ then $T \cap T_1 \neq T \cap T_2$ and hence $T = \langle T \cap T_1, T \cap T_2 \rangle \leq W_2$. However, this means that $T \cap T_3 = T_3 \cap W_2 = W_1$, that is, $T \geq W_1$, a contradiction. We proved that case (a) holds with $W = W_1$.

We can now complete the proof of the lemma. If $\mathcal{T}$ satisfies the condition in (a) then E is contained in the subgroup from (1) defined by W. If, on the other hand, $\mathcal{T}$ satisfies the condition from (b), then E is contained in the subgroup from (2), where V is defined as the unique singular 2_2^5 containing W. Indeed, every U' from $\mathcal{E}$ meets V in a subgroup of size at least eight. Since U' and V are nonconjugate maximal totally singular subgroups from A, we must have that $V \cap U'$ has size 16. $\qquad\square$

We will use the following notation. For an ark A and a singular $W \leq A$, $W \in S_2$, let $\mathrm{Ab}_2(A, W)$ (or simply, $\mathrm{Ab}_2(W)$) be the maximal elementary abelian subgroup of P_A defined by W as in (1). Similarly, if $V \leq A$ and $V \in S_{5,2}$ then let $\mathrm{Ab}_5(A, V)$ (or just $\mathrm{Ab}_5(V)$) denote the maximal elementary abelian subgroup defined by V as in (2). Notice that $|\mathrm{Ab}_2(W)| = 2^{14}$ and $|\mathrm{Ab}_5(V)| = 2^{15}$.

Recall that $N = N_M(A)$ and let $\bar{P}_A = P_A / A$.

Lemma 6.9 *Suppose $W, V \leq A$ with $W \in S_2$ and $V \in S_{5,2}$. Let $Y = \mathrm{Ab}_2(W)$ and $Y' = \mathrm{Ab}_5(V)$. The following hold:*

(1) *$N_N(W)$ induces on $\bar{Y} \sim 2^4$ the group $L_4(2)$; and*
(2) *$N_N(V)$ induces on $\bar{Y}' \sim 2^5$ the group $L_5(2)$.*

Proof The involutions $\bar{y} \in \bar{Y}$ bijectively correspond to the 2_1^5 subgroups U in A, that contain W. Since $W^\perp / W$ is a 6-dimensional orthogonal space of plus type, $N_N(W)$ induces on $\bar{Y}$ the group $\Omega_6^+(2) \cong L_4(2)$. Similarly, the involutions in $\bar{y}' \in \bar{Y}'$ bijectively correspond to index two subgroups in V. So $N_N(V)$ induces on $\bar{Y}'$ the group $L_5(2)$. $\qquad\square$

Lemma 6.10 *Suppose $B \leq P_A$ is elementary abelian and $\bar{B} \sim 2^3$. Then there exist unique $W, V \leq A$ with $W \in S_2$ and $V \in S_{5,2}$, such that $\mathrm{Ab}_2(W)$ and $\mathrm{Ab}_5(V)$ contain B.*

Proof Let $b_1, b_2, b_3 \in B$ and $\bar{B} = \langle \bar{b}_1, \bar{b}_2, \bar{b}_3 \rangle$. Then $W = U(b_1) \cap U(b_2) \cap U(b_3)$ and V is the unique singular 2_2^5 in A containing $(U(b_1) \cap U(b_2))(U(b_1) \cap U(b_3)) \sim 2^4$. $\qquad\square$

We complete this section with a different construction of the maximal elementary abelian subgroups $\mathrm{Ab}_5(A, V)$. Suppose V is a singular subgroup 2_2^5 and let A_1, A_2, and A_3 be the three arks containing V (cf. Lemma 5.10). Furthermore, let $P_i = P_{A_i}$ for all i.

Lemma 6.11 *Let $\{i, j, k\} = \{1, 2, 3\}$ and $R = A_i A_j$. Then the following hold:*

(1) *the subgroups A_i and A_j commute elementwise and $A_i \cap A_j = V$; in particular, $R \sim 2^{15}$; furthermore, $R = \mathrm{Ab}_5(A_i, V) = \mathrm{Ab}_5(A_j, V)$;*
(2) *$P_i \cap P_j = R$; in particular, R is maximal abelian;*
(3) *R contains A_k; in particular, $R = A_i A_k = A_j A_k$.*

Proof If $a \in A_i$ and $b \in A_j$, then a and b lie in Q_W for some subgroup $W \cong 2^3$ of V. Since Q_W is abelian, we have that a and b commute. In particular, $A_j \leq P_i$ and hence also $R \leq P_i$. The stabilizer K of V in $N_M(A_i)$ involves $L_5(2)$ acting irreducibly on both V and A_i / V (see Lemma 5.1). Since the stabilizer of A_j in

K is of index at most two, we obtain that $A_i \cap A_j = V$. If $y \in A_j \setminus V$ then $U(y)$ contains an index two subgroup from V. Hence $A_j \le \mathrm{Ab}_5(A_i, V)$, proving (1).

For (2), let $N = N_M(A_j)$, $K = N_N(V)$, and $S = P_i \cap N$. Since K acts on $\{A_i, A_k\}$, the index of $K_0 = N_M(A_i) \cap N$ in K is at most two. Notice that the image of K in $N/P_j \cong \Omega_{10}^+(2)$ is a maximal parabolic $2^{10}.L_5(2)$ with an irreducible action of $L_5(2)$ on the normal 2^{10}. This means that S, being normal in K_0, is either contained in P_j and so $S = P_i \cap P_j$, or R has index 2^{10} in S. In its turn, S has index at most two in P_i. Thus, in the first case, $P_i \cap P_j$ has index two in P_i, which is clearly impossible. In the second case, $P_i \cap P_j$ has order 2^{15} or 2^{16}, implying that R has index one or two in $P_i \cap P_j$. Suppose $R \ne P_i \cap P_j$. Since R is in the center of $P_i \cap P_j$ and since R is a maximal elementary abelian subgroup by (1), all elements in $(P_i \cap P_j) \setminus R$ are of order four and they all square to the same involution in $A_i \cap A_j = V$. Since K_0 involves $L_5(2)$ acting transitively on the involutions from V, we obtain a contradiction, proving that $R = P_i \cap P_j$. This shows that R is self-centralized, which means that R is maximal abelian. So (2) is proven.

Since A_k commutes with both A_i and A_j, it is contained in $C_M(R) = R$. $\square$

In particular, this lemma shows that the subgroup $\mathrm{Ab}_5(A, V)$ depends, in fact, only on V. So we will use the notation $\mathrm{Ab}_5(V)$.

4.7 Classes of Involutions, II

In this section, z is a 2-central involution in M, the Monster, $Z = \langle z \rangle$, $C = C_z$, and $Q = Q_z$. We classify conjugacy classes of involution in C and determine the fusion of these classes in M.

Let $\bar{C} = C/Z$. If $x \ne z$ is an involution from C then $\bar{c}$ is again an involution. Notice that the group $\bar{C}$ satisfies the conditions (H1)–(H3) from Section 4.2. The results from that section tell us that $\bar{C}$ has exactly eight classes of involutions. If $\bar{x}$ is an involution, then x is either an involution or an element of order four, having the property that $x^2 = z$. Suppose x is an involution. Then $y = xz$ is also an involution, and either x and y are conjugate in C, or they are not. As a result, each conjugacy class of involutions from $\bar{C}$ leads to zero, one, or two conjugacy classes of involutions in C. We now have to decide which case takes place for each of the eight classes of involutions from $\bar{C}$.

The three classes contained in $\bar{Q}$, namely, $2e_2$, $2e_3$, and $2e_4$, were discussed in Lemma 4.4. There we proved that $2e_3$ produces a class of elements of order four, while each of $2e_2$ and $2e_4$ leads to a class of involutions. We denoted by q_2 and q_4 representatives of those classes and noted that q_4 is 2-central (i.e., conjugate to z) and q_2 is non 2-central. Thus, in this section we only need to discuss the classes $2a_i$, $2b$, and $2c$.

Table 1 $(\hat{\Lambda}_2 \cup \hat{\Lambda}_4)$-pure subgroups 2^2 in $\hat{\Lambda}$

Orbit	Type	Stabilizer	Induced group
1	222	$U_6(2)$	S_3
2	224	$2^{10} : \operatorname{Aut} M_{22}$	2
3	244	$2^{1+8} A_8$	2
4	444	$2^{4+12}.3 \cdot S_6$	S_3
5	444	$\operatorname{Aut} M_{12}$	S_3
6	444	$[2^{11}].L_3(2)$	S_3

Let $q = q_4$, $D = C \cap C_q$, and $R = Q \cap Q_q \sim 2^{1+23}.(2^{11} : M_{24})$. According to Lemma 4.5, $Q \cap Q_q$ has order 2^{13} and hence $(C \cap Q_q)(C_q \cap Q) = O_2(D)$. That is, the subgroup $C \cap Q_q$ maps in $\tilde{C} = C/Q \cong Co_1$ onto the diagonal subgroup 2^{11}. Since the latter contains representatives of the conjugacy classes $2A$ and $2C$ from Co_1, we may be able to find elements y with $\bar{y}$ in the classes $2a_1$, $2a_2$, $2a_3$, and $2c$ by looking at $C \cap Q_q$.

If x is an involution in $C \cap Q_q$ and $x \notin \langle z, q \rangle$, then $\langle z, x \rangle$ maps onto a size four subgroup of $Q_q/\langle q \rangle$. Under the identification of the latter with $\hat{\Lambda} = \Lambda/2\Lambda$, the Leech lattice modulo two, the images of z, x, and xz belong to $\hat{\Lambda}_2 \cup \hat{\Lambda}_4$, because these elements are involutions (cf. Lemma 4.4).

To proceed, we will need some information about the subgroups of order four in $\hat{\Lambda}$. According to [ATLAS], Co_1 has exactly fifteen orbits on such subgroups. Representatives of nine of those orbits contain elements from $\hat{\Lambda}_3$. The remaining six orbits are shown in Table 1. Let $\hat{U} \sim 2^2$ be a representative of one of these orbits. Then the second column contains the types of the three involutions from $\hat{U}$. For example, if the entry there is 244, then one involution lies in $\hat{\Lambda}_2$, while two involutions lie in $\hat{\Lambda}_4$. The third column shows the structure of the elementwise stabilizer (centralizer) of $\hat{U}$ in Co_1. The fourth column shows the group induced on $\hat{U}$ by the setwise stabilizer (normalizer) of $\hat{U}$ in Co_1. Notice that [ATLAS] shows a different structure (namely, $[2^{12}].L_3(2)$) of the elementwise stabilizer of $\hat{U}$ for $\hat{U}$ from orbit 6. That structure is incorrect.

Lemma 7.1 *Suppose $x \in C$. Then the following hold:*

(1) *if $\bar{x}$ is of type $2a_1$, then x and xz are involutions; one of them is non 2-central (fused with q_2) and the other one is 2-central;*

(2) *if $\bar{x}$ is of type $2a_3$, then x and xz are 2-central involutions and they are conjugate in C;*

(3) *if $\bar{x}$ is of type $2c$, then x and xz are 2-central involutions and they are conjugate in C.*

Proof Let x be an involution from $C \cap Q_q \setminus \langle z, q \rangle$. Let $U = \langle z, x \rangle$ and $\hat{U}$ be the image of U in $\hat{\Lambda}$. Notice that z maps onto an element from $\hat{\Lambda}_4$ (cf. Lemma 4.4). Thus, $\hat{U}$ cannot be in orbit 1 from Table 1. If $x \in Q$ and x is 2-central, then $\langle z, q, x \rangle$ is singular. Comparing with with Lemma 4.8, we see that in this case $\hat{U}$ is in orbit 4. If $x \in Q$ and x is non 2-central, then xz is also non 2-central and hence $\hat{U}$ is of type 224; that is, $\hat{U}$ is in orbit 2. Thus, orbits 2 and 4 correspond to x's from Q.

Suppose next that we choose x so that $\hat{U}$ is in orbit 3. Then, according to Table 1 and Lemma 4.4, one of the elements x and xz is 2-central and the other one is non 2-central. In particular, $x \notin Q$. Let A be an ark containing z and q and let $a \in A$ be a 2-central involution contained in $q^{\perp}$, but not in $z^{\perp}$. Then az is non 2-central and hence $\langle z, a \rangle$ is conjugate to U in C_q. Furthermore, since the normalizer of $\hat{U}$ in Co_1 permutes the two elements from $\hat{\Lambda}_4 \cap \hat{U}$ and since U is contained in the extraspecial group Q_q, we get that $\langle z, a \rangle$ is conjugate to U in $C \cap C_q$. It follows from Lemma 5.8 that $C \cap C_a$ involves $\Omega_8^+(2)$. Comparing with Lemmas 2.10 and 2.11, we obtain that $\bar{a}$ (and hence also $\bar{x}$) is of type $2a_1$.

Next choose x so that $\hat{U}$ is in orbit 5. Then $C \cap C_x$ involves M_{12}, whose order is divisible by 11. Comparing with Lemmas 2.10 and 2.11, we immediately obtain that $\bar{x}$ is of type $2c$.

Finally, let x be such that $\hat{U}$ is in orbit 6. In this case, both x and xz are 2-central, which rules out the possibility that $\bar{x}$ is of type $2a_1$. It follows from Table 1 that 2^{38} divides $|C \cap C_x|$. Comparing again with Lemmas 2.10 and 2.11, we see that $\bar{x}$ can only be of type $2a_3$. $\square$

In fact, we have proved a bit more.

Lemma 7.2 *Suppose x is an involution in C. Then there exists a 2-central involution $q \notin \langle z, x \rangle$, such that $z, x \in Q_q$, if and only if either $x \in Q$ or $\bar{x}$ is of type $2a_1$, $2a_3$, or $2c$. Furthermore, if such q's exist, then $C \cap C_x$ permutes them transitively.*

Proof Only the transitivity claim requires proof. Let q and q' be 2-central involutions such that $z, x \in Q_q \cap Q_{q'}$. Since $q, q' \in Q$, there exists $c \in C$ such that $(q')^c = q$. Then $\overline{x^c}$ is of the same type as $\bar{x}$. It follows then that $\langle z, x^c \rangle$ and U are conjugate in C_q and moreover there is an element $s \in C \cap C_q$ such that $(x^c)^s = x$. Clearly, $cs \in C \cap C_x$ and $(q')^{cs} = q$. $\square$

Our next goal is to show that the classes $2a_2$ and $2b$ do not lead to involutions. We start with $2a_2$.

Lemma 7.3 *If $x \in C$ and $\bar{x}$ is of type $2a_2$, then x is of order four.*

Proof Let R be the full preimage in Q of $C_{\bar{Q}}(\bar{x})$ and let $F = \langle R, x \rangle$. Then $|R| = 2^{17}$ and $|F| = 2^{18}$. The coset $F \setminus R$ consists entirely of elements y with $\bar{y}$ of type $2a_i$ for some i. According to Lemma 7.1, y is an involution if $i = 1$ or 3. Suppose x is also an involution. Then all elements in $F \setminus R$ are involutions. This implies that $C_R(x)$ contains no elements of order four, implying that $C_R(x)$ is elementary abelian. Since $C_R(x) \leq Q$ and $|C_R(x)| \geq 2^{16}$, we get a contradiction. □

It remains to consider the class $2b$.

Lemma 7.4 *If $x \in C$ and $\bar{x}$ is of type $2b$ then x is of order four.*

Proof Suppose by contradiction that x is an involution. We first show that x and xz are not conjugate in C. Indeed, according to Lemma 2.11, $C \cap C_x$ involves $G_2(4)$. Let D be the Sylow 13-subgroup from $C \cap C_x$. Clearly, if x and xz are conjugate in C, then they are also conjugate in $N_C(D)$. Let $\tilde{C} = C/Q \cong Co_1$. According to [ATLAS], $C_{\tilde{C}}(\tilde{D}) \cong 13 \times A_4$. Notice also that $C_Q(D) = Z$. This means that $F = C_C(D)$ is an extension of Z by a group $13 \times A_4$. Now the fact that x is an involution yields that $F \cong 2 \times 13 \times A_4$. (Here Z is the direct factor of order two.) Consequently, one of x and xz is contained in the commutator subgroup of F and the other is not. Thus, x and xz cannot be conjugate in $N_C(D)$, and hence they are not conjugate in C.

Choose an involution $a \in Q$ such that $\bar{a} \notin C_{\bar{Q}}(\bar{x})$. Then for $b = [a, x]$ we have that $\bar{b} \in C_{\bar{Q}}(\bar{x})$. It follows from Lemma 2.5 that b is a 2-central involution. In particular, x commutes with b and, furthermore, x and xb are conjugate in $\langle a, x \rangle \leq C_b$. Shifting our attention now to C_b we see that the image of x in $C_b/\langle b \rangle$ cannot be of type $2a_1$ or $2b$ because x and xb are conjugate in C_b (cf. Lemma 7.1 for $2a_1$). Also, by Lemma 7.3, it cannot be of type $2a_2$. Thus, the image of x in $C_b/\langle b \rangle$ must be of type $2a_3$ or $2c$. Lemma 7.1 forces now that x is 2-central.

Clearly, the image of z in $C_x/\langle x \rangle$ must also be of type $2b$. Observe that $Q \cap C_x$ is of index at most four in $O_2(C \cap C_x)$. Symmetrically, $Q_x \cap C$ is of index at most four in $O_2(C \cap C_x)$. This shows that $Q \cap Q_x \neq 1$; that is, there exists a 2-central involution q such that $z, x \in Q_q$. Now Lemma 7.2 provides a contradiction. □

We summarize all as follows.

Proposition 7.5 *The group C has exactly seven conjugacy classes of involutions, three in Q and four in $C \setminus Q$.*

One important corollary of Lemmas 7.1, 7.3, and 7.4 is that the group M contains exactly two classes of involutions. Indeed, we have shown that every involution is either 2-central and hence fused with z, or non 2-central, fused with q_2. We record this as the following.

Proposition 7.6   *Every involution in M is either conjugate with z or with q_2.*

A second important corollary is that we now know all pairs of commuting 2-central involutions in M.

Lemma 7.7   *Let a and b be two commuting 2-central involutions, $a \neq b$, and let $R = Q_a \cap Q_b$. Let bar indicate the image in $C_a / \langle a \rangle$. Then one of the following is true:*

(1) $\bar{b}$ *is of type $2e_4$; that is, $b \in Q_a$ and $\langle a, b \rangle$ is singular;*
(2) $\bar{b}$ *is of type $2a_1$; moreover, $\langle a, b \rangle$ is contained in a unique ark A and $R = a^\perp \cap b^\perp$; in particular, $A = \langle a, b \rangle R$;*
(3) $\bar{b}$ *is of type $2a_3$ and $R \sim 2^5$; R contains a singular 2^4 subgroup W and every involution in $R \setminus W$ is non 2-central; subgroups $W_a = W\langle a \rangle$ and $W_b = W\langle b \rangle$ are singular 2_1^5; we have that $R\langle a \rangle = Q_{W_a}$ and $R\langle b \rangle = Q_{W_b}$; finally, $Q_W = R\langle a, b \rangle$; or*
(4) $\bar{b}$ *is of type $2c$, $R \cong 2$, and the involution in R is 2-central.*

Proof   According to Lemmas 4.4, 7.1, 7.3, and 7.4, $\bar{b}$ is of type $2e_4$, $2a_1$, $2a_3$, or $2c$ in $\bar{C}_a$. By Lemma 7.2, $C_a \cap C_b$ is transitive on the (nonempty) set of 2-central involutions $q \notin \langle a, b \rangle$ such that a and b are in Q_q. Dividing $|C_a \cap C_b|$ by $|C_a \cap C_b \cap C_q|$ (the latter can be found using Table 1) we obtain that the number of involutions q is equal to 7084, 135, 15, and 1, depending on whether the type of $\bar{b}$ is $2e_4$, $2a_1$, $2a_3$, or $2c$. We now turn to the concrete cases.

If $\bar{b}$ is of type $2e_4$, then, clearly, (1) holds. Suppose $\bar{b}$ is of type $2a_1$. Then a and b are contained in an ark A. Clearly, $a^\perp \cap b^\perp \sim 2^8$ lies in R. On the other hand, R is abelian and invariant under the action of $C_a \cap C_b$. It follows from Lemma 2.4, that R cannot have size more than 2^8 and hence it coincides with $a^\perp \cap b^\perp$. Now all claims of (2) follow.

Consider next the case where $\bar{b}$ is of type $2a_3$. Choose a singular 2^4 subgroup W and let A and B be two singular 2_1^5 containing W. Let $a' \in A \setminus W$ and $b' \in B \setminus W$. If the image of b' is of type $2a_1$ in $C_{a'} / \langle a' \rangle$, then both A and B are contained in the unique ark containing a' and b'. This is impossible because in an ark every singular 2^4 is contained in a unique singular 2_1^5. Also the image of b' cannot be of type $2c$ because all the involutions $q \in W$ have the property that $a', b' \in Q_q$. Hence, the image of b' is of type $2a_3$ and without loss of generality we may assume that $a' = a$ and $b' = b$. Clearly, $W \leq R$ and W contains all 15

2-central involutions from R. Hence all involutions in $R \setminus W$ are non 2-central. In particular, if $c \in R \setminus W$ and $w \in W$, then cw is non 2-central. According to Lemmas 7.1, 7.3, and 7.4, this means that $c \in Q_w$. That is, $R \leq Q_W$. Since also $a, b \in Q_W$, the claim (3) follows from Lemmas 4.10 and 4.13.

Finally, suppose that $\bar{b}$ is of type $2c$. Let q be the only 2-central involution in R. Then it follows from Lemma 2.3 that $[R : \langle q \rangle] \leq 4$. Comparing with Lemmas 2.6 and 2.11, we see that $R = \langle q \rangle$. □

4.8 More on P_A

In this section, A is an arc. For an involution $x \in P_A \setminus A$, let $W(x, A)$ (or simply $W(x)$) be the subgroup of A generated by all 2-central involutions in $a \in A$ such that $x \in Q_a$. Recall from Section 4.6 that $U(x, A)$ (or simply $U(x)$) is the singular subgroup 2_1^5 in A that corresponds to the coset xA.

Lemma 8.1   *We have $W(x) \leq U(x)$. If x is non 2-central, then $W(x) = U(x)$; otherwise, $W(x)$ is an index two subgroup in $U(x)$.*

Proof   We first notice that $W(x)$ is a singular subgroup. Indeed, suppose $a, b \in A$ are 2-central involutions such that $x \in Q_a \cap Q_b$. If a and b are not perpendicular, then $\langle a, b \rangle$ is as in Lemma 7.7 (2). However, in this case, $Q_a \cap Q_b \leq A$, which means that $x \in A$, a contradiction. Hence $W(x)$ is singular.

If x is non 2-central, then $x \in Q_U$, where $U = U(x)$. Clearly, this means that $U \leq W(x)$. Since U is maximal singular, we obtain that $W(x) = U$. Now suppose x is 2-central. Then for some $a \in A$, we have that $t = xa$ is non 2-central. Let $U = U(x) = U(t)$ and $W = U \cap a^\perp$. Clearly, the index of W in U is at most two. If $w \in W$, then $a \in Q_w$ and $t \in Q_w$. Hence also $x \in Q_w$, i.e., $W \leq W(x)$. On the other hand, $W(x)\langle x \rangle$ is singular, and hence $|W(x)| \leq 2^4$. Thus, $W(x) = W$. □

The next result adds to Lemma 6.1.

Lemma 8.2   *Suppose u is a 2-central involution in A and x is an involution in $P_A \setminus A$. Then $x \in (P_A \cap Q_u)A$ if and only if $u \in U(x)$.*

Proof   Suppose first that $x \in (P_A \cap Q_u)A$; that is, for some $a \in A$, the element $t = xa$ is in Q_u. By Lemma 8.1, this means that $u \in U(t) = U(x)$. Reversely, suppose $u \in U(x)$. Let $a \in A$ be such that $t = xa$ is non 2-central. Since $u \in U(x) = U(t)$, Lemma 8.1 implies that $t \in Q_u$. Therefore, $x \in (P_A \cap Q_u)A$. □

4.9 Heart of the Proof

Suppose $t \in M$, where either $t = 1$, or t is a non 2-central involution. We call a 2-central involution u *marked* if $t \in Q_u$. In this section, we are going to prove the following result.

Proposition 9.1   *Let Q be a 2-subgroup of M such that $C_M(Q) = Z(Q)$ and $t \in Z(Q)$. Let $E = \Omega_1 Z(Q)$. Let $\mathcal{J}$ be the set of all those marked involutions $u \in E$ for which $|E \cap Q_u|$ reaches maximum. Then either $\langle \mathcal{J} \rangle$ is singular, or E is an ark.*

The proposition will be proven in a sequence of lemmas. We first show that $\mathcal{J}$ is nonempty.

Lemma 9.2   *There exists a marked 2-central involution in E. In particular, $\mathcal{J}$ is nonempty.*

Proof   If $t = 1$, then E contains a 2-central involution, since if T is a Sylow 2-subgroup of M containing Q, then $Z(T) \le Z(Q)$. So suppose now that $t \neq 1$. Suppose T_0 is a Sylow 2-subgroup of $C_M(t)$ containing Q and T is a Sylow 2-subgroup of M containing T_0. Suppose z is the 2-central involution in the center of T. Then $z \in C_M(Q)$ and hence $z \in E$. Also, since C_z contains a full Sylow 2-subgroup of $C_M(t)$, it follows that $\bar{t}$ cannot be of type $2a_1$ in $\bar{C}_z = C_z/\langle z \rangle$. Hence $t \in Q_z$, i.e., z is marked.   □

Thus, $\mathcal{J}$ is nonempty. We need to show that either $\langle \mathcal{J} \rangle$ is singular, or E is an ark. Notice that in both cases $\mathcal{J}$ is fully contained in an ark. Inside an ark, every pair of 2-central involutions (in particular, two involutions from $\mathcal{J}$) is either as in case (1), or as in case (2) of Lemma 7.7. Therefore, we first show that the other two cases are impossible. Fix $a, b \in \mathcal{J}$, $a \neq b$. Let $U = \langle a, b \rangle$ and $R = Q_a \cap Q_b$. Notice that $t \in R$ because a and b are marked. Let $\bar{Q}_a = Q_a/\langle a \rangle$.

Lemma 9.3   *The pair (a, b) is not in case (4) of Lemma 7.7.*

Proof   Indeed, suppose the pair (a, b) is in case (4). Since the only nontrivial element q from R is 2-central, we have $t = 1$. In particular, q is marked. According to Lemma 7.7, $\bar{b}$ is of type $2c$, which means that, with respect to the identification of $Q_q/\langle q \rangle$ with Λ, we have that $\hat{U}$ is in orbit 5 from Table 1. In particular, the image of $C_a \cap C_b$ in $C_q/Q_q \cong Co_1$ is isomorphic to M_{12} or Aut M_{12}. We obtain that $C_a \cap Q_b$ (which is normal in $C_a \cap C_b$!) is contained in Q_q. Similarly, $C_b \cap Q_a \le Q_q$.

We claim that $q \in E$. Indeed, Q centralizes a and b, hence also q, as it is the only nontrivial element in $R = Q_a \cap Q_b$. Therefore, $q \in C_M(Q) = Z(Q)$; that is, $q \in E$, as claimed. On the other hand, we have $E \cap Q_b \le C_a \cap Q_b \le Q_q$.

This means that $E \cap Q_b \leq E \cap Q_q$, hence, by maximality of $|E \cap Q_b|$, we have $q \in \mathcal{J}$ and $E \cap Q_b = E \cap Q_q$. Symmetrically, $E \cap Q_a = E \cap Q_q$. However, as $b \notin Q_a$, we have $E \cap Q_b \neq E \cap Q_a$; a contradiction. $\qquad\square$

The second case, where (a, b) is as in Lemma 7.7 (3), is harder and requires several lemmas. Let us start with some additional notation. By Lemma 7.7, $R = Q_a \cap Q_b \cong 2^5$ and it contains a singular 2^4 subgroup W. All elements in $R \setminus W$ are non 2-central. Furthermore, $W_a = \langle a, W \rangle$ and $W_b = \langle b, W \rangle$ are singular subgroups 2_1^5. By Lemma 5.10, W_a is contained in a unique ark $A_a = A(W_a)$. Similarly, let $A_b = A(W_b)$ be the only ark containing W_b. We have $A_a \neq A_b$, since (a, b) is not as in case (1) or (2) of Lemma 7.7. Let also $P_a = P_{A_a}$ and $P_b = P_{A_b}$.

By Lemma 4.12, W is contained in a unique singular 2_2^5 subgroup T. By the definition of an ark, $T \leq A_a$ and $T \leq A_b$. It follows from Lemma 6.11 that $A_a \cap A_b = T$ and that $S = P_a \cap P_b$ coincides with $A_a A_b$ and is elementary abelian. (It coincides with $\text{Ab}_5(T)$.) In particular, $a \in P_b$ and $b \in P_a$. Notice that $W = T \cap W_a = T \cap W_b$. Also $W = W(b, A_a) < U(b, A_a)$ (cf. Lemma 8.1) and similarly $W = W(a, A_b) < U(a, A_b)$.

First of all, we note the following.

Lemma 9.4 *For $x, y \in \{a, b\}$, $x \neq y$, we have that $W_x = U(y, A_x)$. Also, $1 \neq C_{W_x}(Q) \leq E$ and all involutions in W_x are marked.*

Proof By Lemma 8.1, $W = W(y, A_x) < U(y, A_x)$. Since W is contained in A_x in a unique singular 2_1^5, it follows that $W_x = U(y, A_x)$. The next claim follows from the fact that $E = C_M(Q)$. Since $t \in R$ (because a and b are marked) and since $R \leq Q_W$ by Lemma 7.7 (3), we conclude that $t \in Q_W$. Taking now in the account that also $t \in Q_x$ and using Lemma 4.3, we obtain that $t \in Q_{W_x}$, because $W_x = \langle x, W \rangle$. Thus, every involution in W_x is marked. $\qquad\square$

Let $E_a = E \cap P_a$ and $E_b = E \cap P_b$. Since $a \in P_b$ and $b \in P_a$, we have that $a \in E_b$ and $b \in E_a$.

Lemma 9.5 *We have $E \cap Q_a \leq P_a$ and hence $E \cap Q_a \leq E_a$. Symmetrically, $E \cap Q_b \leq E_b$.*

Proof Since $E \leq C_b$, it suffices to show that $C_b \cap Q_a \leq P_a$. Let $c \in T \setminus W$. Clearly, A_a is the only ark containing a and c. Since $b, c \in Q_W$ and since $Q_a \cap Q_W = Q_{W_a}$ has index two in Q_W, we have that $bQ_a = cQ_a$. It follows from Lemma 7.1 that $C_c \cap Q_a$ has order 2^{17}, namely; it is the full preimage in Q_a of $C_{\bar{Q}_a}(c)$. Since $C_{\bar{Q}_a}(b) = C_{\bar{Q}_a}(c)$, we obtain that $C_b \cap Q_a \leq C_c \cap Q_a$. On the other hand, $P_a \cap Q_a \leq C_c$ and according to Lemma 6.1 the size of $P_a \cap Q_a$ is exactly 2^{17}. Therefore, $Q_a \cap C_c = P_a \cap Q_a \leq Q_a$. $\qquad\square$

Since $E \cap Q_a$ is fully contained in E_a, the maximality property of $a \in \mathcal{J}$ implies that for every marked 2-central involution $s \in E_a$, we have $|E_a \cap Q_s| \leq |E_a \cap Q_a|$. Symmetrically, for every marked 2-central involution $s \in E_b$, we have that $|E_b \cap Q_s| \leq |E_b \cap Q_b|$. Since $b \notin Q_a$, it follows that $E_a \cap Q_a \neq E_a$ and similarly $E_b \cap Q_b \neq E_b$.

Our argument depends on how E_a and E_b embed into P_a and P_b, respectively. Since E_a and E_b are elementary abelian we can make use of the classification from Section 4.6.

Lemma 9.6   *For $x \in \{a, b\}$, if $E_x \cap Q_x$ has index more than two in E_x then $E_x A_x = \mathrm{Ab}_5(V)$ for some singular 2_2^5 subgroup $V_x \leq A_x$, and the index of $E_x \cap Q_x$ in E_x is four.*

Proof   We may assume that $x = a$. Suppose $[E_a : E_a \cap Q_a] > 2$. Notice that $E_a A_a$ is elementary abelian. It follows from Lemma 6.8 that either $E_a A_a = \mathrm{Ab}_5(V_a)$ for some singular 2_2^5 subgroup $V_a \leq A_a$, or the intersection F of all $U(s)$, $s \in E_a \setminus A_a$, is nontrivial. Suppose the latter. Since $b \in E_a \setminus A_a$, we get $F \leq W_a$. Therefore Q centralizes some $1 \neq e \in F$, as Q clearly normalizes F. It follows from Lemma 9.4 that $e \in E$ and e is marked. If $s \in E_a$, then $e \in U(s)$, which by Lemma 8.2 means that $E_a \leq (P_a \cap Q_a)A_a$. The latter group contains $P_a \cap Q_a$ as an index two subgroup. It follows that $[E_a : E_a \cap Q_e] \leq 2$. By the maximality property of a, we now have that $[E_a : E_a \cap Q_a] = 2$, since $E_a \not\leq Q_a$. This is a contradiction. Thus, if $[E_a : E_a \cap Q_a] > 2$, then $E_a A_a = \mathrm{Ab}_5(V_a)$ for some singular 2_2^5 subgroup $V_a \leq A_a$. Let $K = \mathrm{Ab}_5(V_a)$. Observe that Q acts trivially on V_a, since it acts trivially on K/A_a. Hence $V_a \leq E$. Since $b \in E_a$, we have that $V_a \cap W_a \sim 2^4$. Let $e \in V_a \cap W_a$. Then e is a marked 2-central involution in E. Using Lemma 8.2, we see that $[K : K \cap Q_e] = 4$ and hence $[E_a : E_a \cap Q_e] \leq 4$. By the maximality of a, we now must conclude that $[E_a : E_a \cap Q_a] \leq 4$.   □

Out of the two options given by this lemma, we will first dispose of the possibility that $E_x A_x = \mathrm{Ab}_5(V_x)$ for some singular 2_2^5 subgroup $V_x \leq A_x$.

Lemma 9.7   *For $x \in \{a, b\}$, we have that $E_x A_x / A_x \not\sim 2^5$. In particular, $[E_x : E_x \cap Q_x] = 2$.*

Proof   Suppose by contradiction that $K = E_x A_x = \mathrm{Ab}_5(V_x)$ for some singular 2_2^5 subgroup $V_x \leq A_x$. Then again Q acts trivially on V_x because it acts trivially on K/A_x. Suppose $x \notin V_x$. If $s \in E_x \cap Q_x$ then $U(s)$ contains x (cf. Lemma 8.2) and also $U(s)$ meets V_x in a subgroup of index two (because $s \in K = \mathrm{Ab}_5(V_x)$). Since $U(s)$ is singular, we have that $U(s) \cap V_x = V_x \cap Q_x$. This shows that $U(s) = \langle x, V_x \cap Q_x \rangle$ is unique, which means that $[E_x \cap Q_x : E_x \cap Q_x \cap A_x] = 2$.

However, in that case the index of $E_x \cap Q_x$ in E_x is at least 16, a contradiction with Lemma 9.6. Thus, $x \in V_x$. In particular, $V_x \neq T$. Now let $y \in \{a, b\}$, $y \neq x$. Observe that $V_x \cap Q_y \leq W(y) = W \leq T$. Consequently, $V_x \cap Q_y \leq V_x \cap T$, which gives us that $[V_x : V_x \cap Q_y] \geq 4$. (Here we use that both V_x and T are singular subgroups 2_2^5 and hence $|V_x \cap T| \leq 2^3$.) As $V_x \leq E \cap P_y = E_y$, it follows that $[E_y : E_y \cap Q_y] \geq 4$. By Lemma 9.6, $E_y A_y = \mathrm{Ab}_5(V_y)$ for some singular 2_2^5 subgroup $V_y \leq A_y$. Repeating this argument with y in place of x, we obtain that $y \in V_y$ and hence $V_y \neq T$, and also that $[E_x : E_x \cap Q_x] = 4$.

Since $U(y, A_x)$ meets both V_x and T in a subgroup of order 16, we conclude that that $|V_x \cap T| \geq 8$. Symmetrically, $|V_y \cap T| \geq 8$. As a result, $F = V_x \cap V_y \cap T \neq 1$. Clearly, Q normalizes F and so we can choose $e \in C_F(Q)$. Then e is a marked 2-central involution and $e \in E$. Since $e \in V_x$, we have that $[K : K \cap Q_e] = 4$ and therefore $[E_x : E_x \cap Q_e] \geq 4$. Because of the maximality of x, we must have that $E \cap Q_e = E_x \cap Q_e$ and that this subgroup has index four in E_x. In particular, $E \cap Q_x \leq E_x$ and, symmetrically, $E \cap Q_e \leq E_y$. Hence $E \cap Q_e \leq E_x \cap E_y] \leq P_x \cap P_y = S = \mathrm{Ab}_5(T)$. This shows that $E \cap Q_e$ is contained in $K \cap S$. By Lemma 6.10, $[E \cap Q_e : E \cap Q_e \cap A_x] \leq 4$. However, this means that $[E_x : E \cap Q_e] \geq 8$, a contradiction. $\qquad\square$

Thus, we now know that $|E_x A_x / A_x| \leq 2^4$ and $[E_x : E_x \cap Q_x] = 2$ for $x = a$ and b. We will obtain the final contradiction by showing that $[E_b : E_b \cap Q_b]$ must at the same time be equal to four. However, the proof of this claim will be different for the following two cases: (a) $|E_a A_a / A_a| \leq 4$, and (b) $|E_a A_a / A_a| \geq 8$. We first consider the case (a).

Lemma 9.8 *If $|E_a A_a / A_a| \leq 4$, then the index of $E_b \cap Q_b$ in E_b is at least four.*

Proof By assumption, $E_a = \langle E_a \cap A_a, b, c \rangle$ for some $c \in E_a$. Notice that since $b \notin Q_a$ and since $[E_a : E_a \cap Q_a] = 2$, we can choose $c \in Q_a$. Since b and c commute, Lemma 6.7 tells us that $U(b) \cap U(c)$ has size at least eight. Since b is 2-central, $W(b) = W$ is of index two in $U(b)$. Depending on the type of c, the subgroup $W(c)$ either coincides with $U(c)$ or is an index two subgroup in it. In any case, $F = W(b) \cap W(c)$ is nontrivial. In particular, we can select $1 \neq e \in C_F(Q)$. This e is a marked 2-central involution and $e \in E$. Since $b, c \in Q_e$, we have $[\langle A_a, b, c \rangle : \langle A_a, b, c \rangle \cap Q_e] \leq 2$, which gives us that $[E_a : E_a \cap Q_e] \leq 2$. By the maximality of a, we have that $e \in \mathcal{J}$ and, furthermore, that $E \cap Q_e$ is an index two subgroup in E_x.

Since $b, c \in Q_e$ and $E_a = \langle E_a \cap A_a, b, c \rangle$, we can choose $s \in E_a \cap A_a$ such that $s \notin Q_e$. Consider the subgroup $X = \langle a, s \rangle = \{1, a, s, as\}$. Clearly, $X \leq E_b$, because $X \leq E$ and $X \leq A_a \leq P_b$. We have $a \notin Q_b$. Also, both s and as are not contained in Q_e, because $a \in Q_e$ (the latter holds since both a and e are in $U(b)$). We claim that s and as do not belong to Q_b. For that, view $A_a \cap Q_b$ as a subspace of the orthogonal space A_a (cf. Lemma 5.7). All singular vectors in $A_a \cap Q_b$ form the subspace W. This implies that W is in the radical of $A_a \cap Q_b$. So if s or as is in Q_b, then it must be also perpendicular to e, because $e \in W$.

We have shown that X trivially intersects Q_b. Since $X \leq E_b$, this means that $[E_b : E_b \cap Q_b] \geq 4$. $\qquad\square$

This lemma and Lemma 9.7 rule out case (a). So it remains to deal with case (b); that is, we now assume that $|E_a A_a / A_a| \geq 8$. We borrow ideas for our argument from the proof of Lemma 9.7.

First a preparatory lemma.

Lemma 9.9 *There is a subgroup $D \leq E_a$ such that*

(1) $|D A_a / A_a| \geq 2^3$ *and* $b \in D$;
(2) D *is contained in* $\mathrm{Ab}_5(V)$ *for some singular* 2^5_2 *subgroup* $V \leq A_a$;
(3) *for every 2-central involution* $e \in A_a$ *we have that* $D \not\leq Q_e$.

Proof We consider two cases: either (a) $E_a A_a = \mathrm{Ab}_2(F)$ for some singular 2^2 subgroup $F \leq A_a$, or (b) $E_a A_a \neq \mathrm{Ab}_2(F)$ for all such F. Suppose we are in case (a); that is, $E_a A_a = \mathrm{Ab}_2(F)$ for some F. Let $\bar{P}_a = P_a / A_a$. Notice that since $b \in E_a$ we have that $F \leq U(b) = W_a$. Let $F = \{1, f_1, f_2, f_3\}$. Lemma 9.4 yields that each f_i is marked. Suppose $E_a \leq Q_{f_i}$ for some i. The set of all such f_i together with 1 forms a Q-invariant subspace F_0 of F. Hence Q centralizes some f_i with $E_a \leq Q_{f_i}$; a contradiction to the maximality of a. Thus, $E_a \not\leq Q_{f_i}$ for all i. In particular, there exist $x_1, x_2,$ and $x_3 \in E_a$ such that $x_i \notin Q_{f_i}$. Moreover, we can choose x_1 and x_2 equal, because no group can be fully covered by two proper subgroups. Let D be the full preimage in E_a of a subgroup 2^3 from $\bar{E}_a$ that contains the three elements $\bar{x}_1 = \bar{x}_2, \bar{x}_3,$ and $\bar{b}$. Clearly, (1) is satisfied for this D. Also, (2) follows from Lemma 6.10. In case (b), we simply take $D = E_a$. Then (1) is trivially satisfied, while (2) follows from Lemmas 6.8 and 6.10.

It remains to show that (3) holds in both cases. If $e \in A_a$ is a 2-central involution such that $D \leq Q_e$, then e lies in the intersection X of all $W(s)$, $s \in D \setminus A_a$. Conversely, if $e \in X$, then e is 2-central and $D \leq Q_e$. Suppose $X \neq 1$. In case (a), we have that F is the intersection of all $U(s)$, $s \in D \setminus A_a$. Therefore, $X \leq F$. However, no f_i can be in X because the corresponding x_i is in D, a contradiction. Suppose now we are in case (b). Since $b \in D$, we have that $X \leq W(b) = W$, and Lemma 9.4 implies that all $e \in X$ are marked. Finally, since Q centralizes D, it normalizes X, and hence it centralizes some $1 \neq e \in X$. That e is a marked 2-central involution in E_a with the property that $D = E_a \leq Q_e$. However, we know that no such e exists. $\qquad\square$

Choose D as in this lemma and let $V \leq A_a$ be the singular subgroup 2_2^5 such that $D \leq \mathrm{Ab}_5(V)$. In view of Lemma 6.10, this V is unique. Since V is unique, Q normalizes V.

Lemma 9.10   *The group Q centralizes V and, in particular, $V \leq E$.*

Proof If $s \in D \setminus A_a$ then we set $B(s) = V \cap U(s)$. If $s \in (D \cap A_a) \setminus V$ then we set $B(s) = V \cap s^\perp$. In both cases, $B(s)$ is a hyperplane of V. Let F be the intersection of subgroups $B(s)$ for all $s \in D \setminus V$. Since Q normalizes every $B(s)$, it normalizes F and acts trivially on V/F. That is, $[V, Q] \leq F$. If $F = 1$, then there is nothing else to prove, so we assume that $F \neq 1$. Notice that since $|DA_a/A_a| \geq 2^3$ and since $F \leq U(s)$ for all $s \in D \setminus A_a$, the size of F is at most four. Notice also that every $e \in F$ is perpendicular to all of $D \cap A_a$. Finally, since $F \leq U(s)$ for all $s \in D \setminus A_a$, we have that $D \leq A_a Q_e$ for every $e \in F$, $e \neq 1$. Therefore, $[D : D \cap Q_e] = 2$ for every such e.

Set $X = D \cap Q_F$ and for a hyperplane H of F set $Y = Y(H)$ to be $D \cap Q_H$. (Here if $H = 1$ then Q_H is the entire group M and $Y = D$.) Since $[D : D \cap Q_e] = 2$ for every $e \in F$, we have $|D/Y| \leq 2$, $|Y/X| \leq 2$, and so $|XA_a/A_a| \geq 2$.

We claim that Y is never equal to X. Indeed, suppose $Y \leq Q_F$. If $|F| = 2$, we have $Y = D$ and so $D \leq Q_F$, a contradiction with the definition of D. Thus $|F| = 4$. Pick $x \in D \setminus Y$. Since $W(x)$ is of index at most two in $U(x)$ and since $F \leq U(x)$, there exists $1 \neq e \in F$ with $e \in W(x)$. For that e, we have $D = \langle Y, x \rangle \leq Q_e$, since $Y = X \leq Q_e$. This is a contradiction, proving that $Y \neq X$.

Thus, $D \cap A_a < X < Y$, and there exist $s_1, s_2 \in Y \setminus X$ with $s_1 A_a \neq s_2 A_a$. Then $U(s_1) \neq U(s_2)$, and since every singular subgroup 2^4 in A_a lies in a unique singular 2_2^5, we also get $B(s_1) \neq B(s_2)$. Since $s_i \notin Q_F$ but $F \leq U(s_i)$, the involution s_i is 2-central and $U(s_i) = FW(s_i)$. Notice that $F \cap W(s_i) = H$ for $i = 1, 2$. As Q centralies s_i, Q normalizes $W(s_i)$, F, and H. Observe that $[F, Q] \leq H$ and also $[W(s_i), Q] \leq F \cap W(s_i) = H$. Thus, $[U(s_i), Q] \leq H$. Finally, since $V = U(s_1)U(s_2)$, we have $[V, Q] \leq H$. As H was an arbitrary hyperplane of F we conclude $[V, Q] = 1$ and $V \leq E$. $\square$

The next question is whether $V = T$ or not.

Lemma 9.11   *We have $V \neq T$. In particular, $[E_b : E_b \cap Q_b] \geq 4$.*

Proof Since $a \notin T$, it suffices to show that $a \in V$. Suppose $a \notin V$. If $s \in \mathrm{Ab}_5(V) \cap Q_a$, then $U(s) = \langle a, V \cap U(s) \rangle$. Since $U(s)$ is singular, $V \cap U(s)$ must coincide with $V \cap a^\perp$. Hence $U(s)$ is unique. This shows that $\mathrm{Ab}_5(V) \cap Q_a \cap A_a$ has index at most two in $\mathrm{Ab}_5(V) \cap Q_a$. Therefore, also $|(D \cap Q_a)A_a/A_a| \leq 2$. Since DA_a/A_a has size at least eight, we must have that $[D : D \cap Q_a] \geq 4$, yielding $[E_a : E_a \cap Q_a] \geq 4$, which is a contradition with Lemma 9.7. Hence $V \neq T$. Finally, notice that $V \leq E_b$ and $V \cap Q_b \leq V \cap W(b) \leq V \cap T$. Since $[V : V \cap T] \geq 4$, we conclude that $[E_b : E_b \cap Q_b] \geq 4$. □

Manifestly, the conclusion of this lemma contradicts Lemma 9.7, ruling out case (2) and thus showing that $\bar{b}$ cannot be of type $2a_3$ in $\bar{C}_a = C_a/\langle a \rangle$.

Corollary 9.12　　*The pair (a, b) is not in case (3) of Lemma 7.7.*

According to Lemma 9.3 and Corollary 9.12, any two involutions $a, b \in \mathcal{J}$ are either perpendicular, or (a, b) is as in case (2) of Lemma 7.7. If any two involutions in $\mathcal{J}$ are perpendicular then the subgroup generated by $\mathcal{J}$ is singular. Thus, it order to complete the proof of Proposition 9.1 all we need is to prove the following lemma.

Lemma 9.13　　*If the pair (a, b) is as in case (2) of Lemma 7.7, then E is an ark and $\langle \mathcal{J} \rangle = E \cap t^{\perp}$.*

Proof　　In this case, $\bar{b}$ is of type $2a_1$ in $\bar{C}_a = C_a/\langle a \rangle$. By Lemma 7.7, $A = \langle a, b \rangle$ $(Q_a \cap Q_b)$ is the unique ark containing a and b. Let $P = P_A$. Notice that $Q_a \cap C_b$ has order 2^{17} by Lemma 7.1 and that $P \cap Q_a$ also has size 2^{17} by Lemma 6.1. Since $P \leq C_b$, we conclude that $Q_a \cap C_b = P \cap Q_a$. In particular, $E \cap Q_a \leq P$. Symmetrically, $E \cap Q_b \leq P$. Let $D = (E \cap Q_a)(E \cap Q_b)$. Then $D \leq P$ and $E \cap Q_x = D \cap Q_x$ for $x \in \{a, b\}$. The maximality of x now shows that $|D \cap Q_e| \leq |D \cap Q_x|$ for all marked 2-central involutions from E. Let $y \in \{a, b\}$, $y \neq x$. Since $y \notin Q_x$, we have that $E \cap A \cap Q_x$ is a hyperplane of $E \cap A$. This implies that $|E \cap A \cap Q_x| = |E \cap A \cap Q_y|$ and hence also $|(E \cap Q_x)A/A| = |(E \cap Q_y)A/A|$. We will denote this latter number by r. We intend to prove that $r = 1$; that is, $E \cap Q_x \leq A$.

If $s \in P \setminus A$, then $U(s) = U(s, A)$ is singular, and hence it cannot contain both x and y. This means that $(E \cap Q_x)A/A$ and $(E \cap Q_y)A/A$ meet trivially in P/A. Therefore, $|DA/A| = r^2$. Furthermore, this shows that $[D : D \cap Q_x] = [DA/A : (D \cap Q_x)A/A] \cdot [D \cap A : D \cap Q_x \cap A] = 2r$. Since D is abelian, we have that $|DA/A| \leq 2^5$ and hence $r \leq 4$.

Suppose first that $r = 2$ and let F be the intersection of all $U(s)$, $s \in D \setminus A$. Then F is a singular subgroup 2^3. Note that $t \in Q_a \cap Q_b \leq A$. This means that $t^{\perp}$ is a hyperplane in A and hence $F_0 = F \cap t^{\perp}$ is nontrivial. Clearly, Q normalizes F_0 so we can select $1 \neq e \in C_{F_0}(Q)$. This e is a marked 2-central involution in E. Since $e \in F$, we have that $[D : D \cap Q_e] \leq 2$. In view of maximality of x, we must have that $2 \geq 2r = 4$, a contradiction.

Suppose now that $r = 4$. Pick a hyperplane H in D such that $D \cap A \leq H$. Let F be the intersection of all $U(s)$, $s \in H \setminus A$. Then $F \sim 2^2$ and hence again

$F_0 = F \cap t^\perp \neq 1$. Choosing $1 \neq e \in C_{F_0}(Q)$, we see that e is a marked 2-central involution from E and $[D : D \cap Q_e] \leq 2[H : H \cap Q_e] \leq 4 < 2r = 8$. So again we have a contradiction with the maximality of x.

Thus, $r = 1$; that is, $E \cap Q_x \leq A$ for $x \in \{a, b\}$. Let e be any 2-central involution in $E \cap A$ that is perpendicular to t. Since $[E \cap A : E \cap A \cap Q_a] \leq 2$, the maximality of x implies that $e \in \mathcal{J}$, $E \cap Q_e \leq A$, and $[E \cap A : E \cap A \cap Q_e] = 2$. Consider $F = (E \cap A)^\perp$. We would like to show that $A \leq E$; that is, we need to prove that $F = 1$. Suppose the contrary. We first remark that $t \in E \cap A$. Indeed, $t \in A$ and $t \in E = \Omega_1 Z(Q)$ by assumption. Since $t \in E \cap A$, every 2-central involution in F is marked. Recall now that the 2-central involutions in A are simply the singular vectors with respect to the quadratic form f on A. Since Q normalizes F, if the number of singular vectors in F is odd, then Q centralizes a 2-central involution $e \in F$. Then e is marked and 2-central, $e \in E$ and $E \cap A \leq Q_e$, a contradiction with the maximality of x. Hence the number of singular vectors in F is even. This means that that either F is 1-dimensional containing a nonsingular vector, or 2-dimensional nondegenerate. Suppose F is 2-dimensional. Then F contains an odd number of nonsingular vectors and hence Q centralizes one of them, say c. Since $E = C_M(Q)$, we have that $c \in E$. Hence c is in the radical of the symplectic form on F, a contradiction, since that form is nondegenerate. If F is 1-dimensional, then Q centralizes a hyperplane in A. Since $N_M(A)$ induces on A the group $\Omega_{10}^+(2)$, no element of Q can act on A as a transvection. This means that Q centralizes A; that is, $A \leq E$ and $W = 1$, a contradiction. Thus, $F = 1$ and $A \leq E$.

Now, $A \leq E$ implies that $Q \leq C_M(A) = P$. In particular, $E \leq P$. Suppose $s \in E \setminus A$. Then $W(s) \cap t^\perp \neq 1$. By this, every involution $e \in W(s) \cap t^\perp$ is in $\mathcal{J}$ and $E \cap Q_e \leq A$. This contradicts the fact that $s \in Q_e$. □

The proof of Proposition 9.1 is now complete.

4.10 Proof of the Theorems

In this section, we derive Theorems 1.1 and 1.2. We start with the Monster group M.

Proof of Theorem 1.1   Suppose N is a maximal 2-local subgroup in M such that $C_N(Q) \leq Q$, where $Q = O_2(N)$. Let $E = \Omega_1 Z(Q)$. If E is an ark, then N is the normalizer of an ark, which agrees with Theorem 1.1. So now suppose that E is not an ark. Set $t = 1$ and let $\mathcal{J}$ be the set of all 2-central involutions $e \in E$ for which $|E \cap Q_e|$ reaches maximum. According to Proposition 9.1, $U = \langle \mathcal{J} \rangle$ is singular, since E is not an ark. Thus, N coincides with the normalizer of a singular subgroup. The singular subgroups of M have been classified in

Section 4.4 (cf. Proposition 4.15). The normalizer of a singular 2^4 is not maximal because it is contained in the larger normalizer of a singular subgroup 2^5_2 (cf. Lemma 4.12). Also the normalizer of a singular subgroup 2^5_1 is not maximal, because it is contained in the normalizer of an ark (cf. Lemma 5.10). The remaining possibilities agree with Theorem 1.1.   □

We now turn to the case of the Baby Monster BM. Recall that for us BM is defined as the group $\bar{H}$, where $H = C_M(t)$, t is a non 2-central involution in M, and $\bar{H} = H/\langle t \rangle$.

Proof of Theorem 1.2   Let $\bar{N}$ be a maximal 2-local subgroup of $\bar{H}$ and suppose $C_{\bar{H}}(\bar{Q}) \le \bar{Q}$, where $\bar{Q} = O_2(\bar{H})$. Let N and Q be the full preimages of $\bar{N}$ and $\bar{Q}$ in H. Then $C_H(Q) \le Q$ and $t \in Z(Q)$. Thus Proposition 9.1 applies to Q. Let $E = \Omega_1 Z(Q)$ and let $\mathcal{J}$ be the set of all marked 2-central involutions $e \in E$, for which $|E \cap Q_e|$ reaches maximum. According to Proposition 9.1, either E is an ark, or $U = \langle \mathcal{J} \rangle$ is singular. Suppose first that E is an ark. Notice that $t \in E$ and that t is a nonsingular vector in the orthogonal space E. The group $\Omega^+_{10}(2)$ induced on E acts transitively on nonsingular vectors. Thus, H has a unique conjugacy class of arks containing t. This leads to one of the cases from Theorem 1.2. Next suppose that E is not an ark and hence U is singular. Observe that U is generated by marked 2-central involutions which means that $t \in Q_U$. It follows from the results of Section 4.4 (namely, Lemma 4.4, Corollaries 4.6, 4.9 and 4.11, and Lemma 4.14; see also Lemma 4.2 (3)) that U cannot be a singular 2^5_2 and that in every other case the normalizer of U has a unique conjugacy class of non 2-central involutions in Q_U. Thus, H contains exactly five classes of singular subgroups U' having the property that $t \in Q_{U'}$. For the case $U' \sim 2^4$, we claim that in fact $N_H(U')$ is not a maximal subgroup (so $U = \langle \mathcal{J} \rangle$ can never be a singular 2^4). Namely, we claim that U' and t are contained together in a unique ark A and hence $N_H(U') \le N_H(A)$. First notice that U' and t are contained in some ark. Indeed, if A is an ark containing U' then $A \cap Q_{U'}$ does contain some non 2-central involutions. Since $N_M(U')$ is transitive on non 2-central involutions from $Q_{U'}$, there must be an ark containing U' and t. It follows from Lemma 4.12 that U' is contained in exactly three singular 2^5_1 subgroups V, each of which is in turn contained in a unique ark A. By Lemma 4.13, t belongs to Q_V for two of these V. If $t \in Q_V$, then by Lemma 6.2 we have for the corresponding ark A that $t \in P_A \setminus A$. Thus, there is at most one ark containing U' and t. We have shown that $U \nsim 2^4$. The remaining four classes of possible singular subgroups U appear in Theorem 1.2.   □

References

[As0]   M. Aschbacher, *Sporadic Groups*, Cambridge University Press, 1994.

[As1]   M. Aschbacher, *Overgroups of Sylow Subgroups in Sporadic Groups*, Memoirs of the AMS, **60** (1986).

[AsSe] M. Aschbacher and Y. Segev, Extending morphisms of groups and graphs, *Ann. of Math.* (2) **135** (1992), 297–323.

[ATLAS] J. H. Conway, R. T. Curtis, S. P. Norton, R. A. Perkel, and R. A. Wilson, *Atlas of Finite Groups*, Clarendon Press, Oxford, 1985.

[CS]   J. H. Conway and N. J. A. Sloane *Sphere Packings, Lattices and Groups* Springer Verlag, New York, 1988.

[Gl]   G. Glauberman, Central elements in core-free groups, *J. Algebra* **4** (1966), 403–420.

[Gr1]   R. L. Griess, The Friendly Giant, *Invent. Math.* **69** (1982), 1–102.

[Gr2]   R. L. Griess, A remark about representations of .1, *Comm. Alg.* **13** (1985), 835–844.

[IS]   A. A. Ivanov and S. Shpectorov, The flag-transitive tilde and Petersen type geometries are all known, *Bull. Amer. Math. Soc.* (N. S.) **31** (1994), 173–184.

[M]   U. Meierfrankenfeld, The maximal 2-locals of the Monster and Baby Monster, II, preprint.

[MOD]   C. Jansen, K. Lux, R. Parker, and R. Wilson, *An Atlas of Brauer Characters*, LMS monographs N. S. bf. 11, Clarendon Press (1995).

[Po]   H. Pollatsek, First cohomology groups of some orthogonal groups, *J. Algebra* **28** (1974), 477–483.

[Se]   Y. Segev, On the uniqueness of Fischer's Babymonster, *Proc. LMS* (3) **62** (1991), 509–536.

5

The Future of Majorana Theory II

Alexander A. Ivanov

To the memory of Igor Alexandrovich Faradjev

Abstract

The Monster group is the largest among the 26 sporadic simple groups, 21 of which can be found in the Monster as factors of subgroups. The discovery of the Monster by B. Fischer and R. Griess around 1973 and its construction by R. Griess in 1980 are believed to be the most important outcome of the whole Classification of Finite Simple Groups. It was coined as a "Twenty-First Century Mathematics found in the Twentieth Century," that is as an alien bolide that hit our modern science and sprayed its traces all over it, from number theory to physical string theory. Various properties of the Monster and its 196 884-dimensional Conway–Griess–Norton algebra were axiomatised by the author in 2009 under the name of *Majorana algebra*, which is a real vector space V endowed with a positive-definite inner $(,)$ and a non-associative algebra $\cdot$ products that associate with each other. The space V contains a generating set A of idempotents of length one, called *Majorana axes*. For $a \in A$, the adjoint operator

$$\mathrm{ad}(a) \colon v \mapsto a \cdot v$$

is semi-simple with spectrum $S = \{1, 0, \frac{1}{4}, \frac{1}{32}\}$. The 1-eigenspace is 1-dimensional (spanned by a). The linear transformation $\tau(a)$ (called the *Majorana involution*) which negates the $\frac{1}{32}$-eigenspace of $\mathrm{ad}(a)$ and centralises the other eigenspaces is an automorphism of $(V, (,), \cdot)$. The involution $\sigma(a)$ which negates the $\frac{1}{4}$-eigenspace and centralises the 1- and 0-eigenspaces is an automorphism of the centraliser of $\tau(a)$ in V with restricted inner and algebra products. This is the whole axiomatic besides the beautiful Norton inequality:

$$(u \cdot u, v \cdot v) \geq (u \cdot v, u \cdot v),$$

which holds for all $u, v \in V$, and ensures that the angle between the eigenvectors is acute. Surprisingly, the Norton inequality did not prove to be very restrictive on the top of the other conditions. The Monster is the Majorana group of the Conway–Griess–Norton algebra with 2A-axes as the Majorana axes. The launch of Majorana Theory was inspired by Sakuma's theorem of 2007, which can be phrased as an identification of the 2-generated Majorana algebras with the eight such subalgebras in the Monster algebra indexed by the Monster classes 2A, 2B, 3A, 3C, 4A, 4B, 5A and 6A, where the numerical part indicates the order of the product of the corresponding Majorana involutions. Subsequent development of Majorana Theory demonstrated strong evidence for the *Straight Flush Conjecture* posed by the author in 2012:

Every Majorana algebra which (1) is not a direct sum of proper pairwise annihilating and orthogonal Majorana subalgebras, and (2) contains a 2-generated Majorana subalgebra for every possible numerical value 2, 3, 4, 5 and 6, is a Majorana subalgebra in the Monster algebra. In particular, the Monster algebra is the universal Majorana algebra subject to these conditions.

The Majorana axiomatic is inductive so that 2A-generated subgroups of the Monster are also involved in the game. The conjecture, when proved, will provide an ideal explanation for the universe of the Monster and its subgroups in the spirit of Richard Feynman:

"I'm just looking to find out more about the world and if it turns out there is a simple ultimate law which explains everything, so be it; that would be very nice to discover."

In this survey we discuss the current state of the art and some perspectives of Majorana Theory.

CONTENTS

5.1 Origins

Our original interest in the Monster group came from the classification project of Petersen and Tilde geometries [Iv98], [IS02]. In [RS84] a number of remarkable 2-local geometries for sporadic groups were constructed, which show striking similarity to the geometries of the classical groups defined over $GF(2)$. Among them, the Tilde geometry $\mathcal{G}(M)$ for the Monster group M is described by the following diagram, where the structures of maximal parabolic subgroups are indicated below the corresponding nodes on the diagram:

$$\underset{\substack{Co_1 \\ 2^{1+24}_+}}{\overset{4}{\circ}} \text{---} \underset{\substack{S_3\times M_{24} \\ 2^2_{\otimes}2^{11} \\ 2^{2+11}}}{\overset{3}{\circ}} \text{---} \underset{\substack{L_3(2)\times 3\cdot S_6 \\ 2^3_{\otimes}2^6 \\ 2^3_{\otimes}2^4 \\ 2^6 \\ 2^3}}{\overset{2}{\circ}} \text{---} \underset{\substack{L_4(2)\times S_3 \\ [2^{39}]}}{\overset{1}{\circ}} \overset{\sim}{\text{---}} \underset{\substack{L_5(2) \\ 2^5 \\ 2^{10} \\ 2^{10} \\ 2^5 \\ 2^{5+1}}}{\overset{0}{\circ}} \qquad \mathcal{G}(M)$$

The nearest classical geometry to $\mathcal{G}(M)$ is the C_5-geometry of $Sp_{10}(2)$ (also of rank 5 with string diagram), where the rightmost edge is the classical generalised quadrangle of $Sp_4(2) \cong S_6$ with diagram $\underset{2}{\circ}\!=\!=\!\underset{2}{\circ}$. The corresponding residue in $\mathcal{G}(M)$ is the triple cover of that generalised quadrangle associated with the non-split triple cover $3\cdot S_6$ with diagram $\underset{2}{\circ}\overset{\sim}{=\!=}\underset{2}{\circ}$. This is the only difference between $\mathcal{G}(Sp_{10}(2))$ and $\mathcal{G}(M)$ in terms of the local structure encoded in the diagrams.

The difference in global structures of classical and sporadic geometries are usually dramatic. Nonetheless, in the most important examples, they share the simple connectedness feature, which is very important for the classical geometries (which are buildings). This is reflected in the validity of the Steinberg presentations for the relevant groups, because all the relations can be checked inside the rank 2 residues. It turned out that the 2-simple connectedness feature of the classical geometries is shared by their sporadic analogues. A standard way of proving simple connectedness of geometries is to establish triangulability of their collinearity graphs.

The triangulation procedure works perfectly well for the classical geometries where the collinearity graph has small permutation rank (which is just three for C_n-geometries). It is also applicable to geometries of small sporadic groups, such as Mathieu and Conway. On the other hand, this procedure becomes problematic for geometries of large sporadic groups including the Monster. The rank of the collinearity graph (with point stabiliser being the centraliser of a $2B$-involution, isomorphic to $2^{1+24}_+.Co_1$) is in the hundreds. Some crucial breakthrough ideas were needed. It came through consideration the chain of subgeometries:

$$\mathcal{G}(^2E_6(2)) \subset \mathcal{G}(BM) \subset \mathcal{G}(M),$$

where BM is the Baby Monster group and the classical geometry of $^2E_6(2)$ is simply connected. An immediate application is that a cycle contained in a simply connected subgeometry is contractible. But one can do better. The $\mathcal{G}(BM)$-subgeometry in $\mathcal{G}(M)$ is on the elements fixed by a $2A$-involution with centraliser $2\cdot BM$. This is a Petersen geometry with diagram

$$\underset{\substack{Co_2 \\ 2^{1+22}_+}}{\overset{4}{\circ}} \text{-----} \underset{\substack{S_3\times \mathrm{Aut}(M_{22}) \\ 2^2_{\otimes}2^{10} \\ 2^{2+10}}}{\overset{3}{\circ}} \text{-----} \underset{\substack{L_3(2)\times S_5 \\ [2^{32}] \\ 2^3}}{\overset{2}{\circ}} \text{-----} \underset{\substack{L_4(2)\times S_3 \\ [2^{34}]}}{\overset{1}{\circ}} \overset{P}{\text{-----}} \underset{\substack{L_5(2) \\ 2^{10} \\ 2^{10} \\ 2^5 \\ 2^5}}{\overset{0}{\circ}} \qquad \mathcal{G}(BM).$$

This geometry is simply but not 2-simply connected (its universal 2-cover is associated with the non-split extension $3^{4371} \cdot BM$). This enables us to reduce the simply connectedness task for $\mathcal{G}(M)$ to the triangulation of the Monster graph $\Gamma(M)$ on the set of $2A$-involutions in M, where two of them are adjacent whenever their product is again a $2A$-involution. Recently the embedding of $\mathcal{G}(BM)$ in $\mathcal{G}(M)$ was put into a general setting [Iv21] of densely embedded subgraphs. For the classical C_5-geometry of $Sp_{10}(2)$ this will be the embedding of the orthogonal geometry of $O_{10}^+(2)$, and this way the simply connectedness problem will be reduced to the triangulability of the complete graph (with doubly transitive action of $Sp_{10}(2)$), which is obvious. For the Monster we arrive with the following.

Problem T Prove that the fundamental group of the Monster graph $\Gamma(M)$ is generated by homotopy classes of triangles. Here $\Gamma(M)$ is the graph on the set of $2A$-involutions in the Monster group M, where two vertex-involutions are adjacent whenever their product is again a $2A$-involution.

The Monster graph $\Gamma(M)$ is still pretty large; the number of vertices is of magnitude of 10^{20}, but the permutation rank of the action of the Monster on its vertex-set is only 9. The information on $\Gamma(M)$ contained in [N85] enabled the author to deduce the affirmative answer to the above problem in [Iv92]. Our goal is to deduce the information (and much more) from a single principle within the Majorana Theory discussed in the present chapter.

In this section, we first discuss the importance of Problem T for understanding the Monster and how our attempts to give a self-contained proof of this problem led to the formulation of Majorana Theory and its more general version, the theory of axial algebras.

According to a general principle, the 2-simply connectedness of $\mathcal{G}(M)$ implies the validity of a Steinberg-type presentation based on the amalgam of rank 2 parabolics of the action of M on the geometry. Such presentation itself is not very practical: it is rather hard to write it down and not clear how to use it, since the parabolics are large and complicated. During the Durham symposium in 1990, the author and S. P. Norton proved the equivalence of such presentation to the famous Y-presentation for the Monster, thus solving a long-standing problem attacked by B. Fischer, J. Tits, J. H. Conway, L. H. Soicher, and others about its validity. In [Iv93] the justification of the Y-presentation was directly reduced to Problem T.

Through the discussions that follow, we will single out a few properties of the Monster graph and of the relative algebra. These properties both led us to the launch of Majorana Theory and have a natural explanation within it.

I. The action on the Monster M on the vertex set of $\Gamma(M)$ has rank 9. The orbit containing a pair (a, b) of involution-vertices is uniquely determined by the conjugacy class containing the group product ab and this class is one of the following nine:

$$1A, \ 2A, \ 2B, \ 3A, \ 3C, \ 4A, \ 4B, \ 5A, \ 6A.$$

The cycle structure of $\Gamma(M)$ was described in [N85] in the following form. By the defining property, every edge $\{a, b\}$ of $\Gamma(M)$ is contained in a (unique) triangle $\{a, b, ab\}$ and for every other vertex c, the M-orbit of the triple $\{a, b, c\}$ is determined by (1) the classes of the elements ac, bc and abc; (2) the isomorphism type of the subgroup in M generated by a, b, c; (3) the centraliser in M of the subgroup generated by a, b, c. In particular:

II. There are 37 M-orbits of $\{a, b, c\}$'s as in table 3 in [N85].

An inspection of the 37 orbits shows the following property, which plays an important role in Majorana Theory.

III. There are no subgroups of order 2^3 in M all of whose involutions are of type $2A$.

The description of the $\{a, b, c\}$-orbits carries very detailed information about the action of the Monster M on its graph $\Gamma(M)$. In fact, the structure constants

$$p^{2A}_{KL} = \#\{c \in 2A \mid ac \in K, bc \in L\}$$

for $a, b, c, ac \in 2A$, where a, b, c are $2A$-involutions, while L and K are conjugacy classes (from the list of nine above) are sums of the orbit lengths in II. On the other hand, these structure constants are sufficient to calculate the complete set of the parameters of the Bose–Mesner algebra associated with the action of M on the vertex set of $\Gamma(M)$, as it was done in [N85]. In particular, these parameters expose the primitive idempotent whose rank is the famous number 196 883, which is one less the linear coefficient of the celebrated J-modular invariant. At this point, Simon Norton was able to cite J. Thompson's paper [Th79] to deduce the uniqueness of the Monster.

An independent uniqueness proof for the Monster was achieved in [GMS89]. The central part of the proof there amounts to solving Problem T under the assumption that the vertex stabiliser (isomorphic to $2 \cdot BM$) is still the centraliser of an involution in the covering group.

So we emphasise again that understanding the cycle structure of the Monster graph is a very important task.

The Monster graph is a very important example of a multiplicity-free permutation action which is not metric (that is, not distance-transitive). When we were characterising the Hermitian form graphs, we learned about the great advantage of realising vertices of the graph in question by norm-one vectors in a finite-dimensional real inner product space in such a way that the inner product of

Table 1 Inner products between Majorana axes

	1A	2A	2B	3A	3C	4A	4B	5A	6A
196 883	752	80	−16	23	−4	8	−4	2	−1
196 884	1	$\frac{1}{2^3}$	0	$\frac{13}{2^8}$	$\frac{1}{2^6}$	$\frac{1}{2^5}$	$\frac{1}{2^6}$	$\frac{3}{2^7}$	$\frac{5}{2^8}$

the vertex-vectors uniquely determines the adjacency and, possibly, the type of the relation they belong to.

There is a standard procedure of achieving the *geometric representation* for a multiplicity-free permutation action: by projecting the (characteristic functions of) vertices onto a primitive idempotent. Then the projection of the Hadamard product of the characteristic functions becomes a (commonly non-associative) algebra product in the idempotent space. The algebra might turn out to be zero, but this is a very special situation ruled by the Krein parameter q_{ii}^i to be non-zero.

In the case of the Monster graph, it is most rewarding to project onto the minimal non-trivial idempotent of rank 196 883, where the relevant Krein parameter is non-zero. Because of this particular example, the whole class of algebras obtained in this way is called *Norton algebras* [B184]. This is the algebra used by R. L. Griess for his first construction of the Monster in [Gr82], and it was studied in detail in [C85] by J. H. Conway, who extended it to a 196 884-dimensional algebra by adjoining a unit. Therefore, this 196 884-dimensional algebra is called the Conway–Griess–Norton algebra or simply the *Monster algebra.*

IV. The inner products of the vertex-vectors of the Monster graph when projected onto the 196 883-rank idempotent depending on their relation, as calculated in [N85], p. 279 and normalised to be integers, are given in the second row of Table 1 below. These numbers, shifted by a constant (which is 16) making use of the identity in the 196 884-dimensional algebra to become non-negative, and normalised to be 1 in relation 1A (dividing by $768 = 2^8 \cdot 3$), are given in the third row. All the denominators in the third row are 2-powers due to the fact that all numbers in the second row are congruent to 2 modulo 3, as noted in the last paragraph on p. 279 in [N85].

Notice that the inner products in row two of the table above are a column of the eigenmatrix of the corresponding Bose–Mesner algebra and usually one needs the whole matrix to calculate it. This was the case for calculations in [N85]. Only very few incidence algebras of very special type, such as those from *BN*-pairs, reserve the privilege that the inner products are determined by a much smaller configuration, like the Weyl group. We will see that in the Monster, case there is a similar situation.

Let $(V_M, (\,,\,), *)$ be the 194 884-dimensional Monster algebra equipped with the invariant inner product $(\,,\,)$. Notice that the algebra product $*$ is non-associative but commutative. For a 2A-involution τ in the Monster, let $a(\tau)$ denote the idempotent in the algebra corresponding to the inner products in the third row of Table 1. Then $a(\tau)$ is called a 2A-axis.

V. The 2A-axes $a(\tau)$ for $\tau \in 2A$ are idempotents with $(a(\tau), a(\tau)) = 1$, and the inner and algebra products associate in the sense that

$$(u, v * w) = (u * v, w)$$

for all $u, v, w \in V_M$. The algebra $(V_M, (\,,\,))$ is known [C85] to satisfy also the *Norton inequality*:

$$(u * u, v * v) \geq (u * v, u * v)$$

for all $u, v \in V_M$. This implies that the angle between any two idempotents is right or acute, which is reflected in the feature that the inner products in the third row of Table 1 are all non-negative.

For $a = a(\tau)$, a 2A-axis as above, consider the adjoint operator

$$\mathrm{ad}(a) : u \mapsto u * a$$

for all $u \in V_M$. The spectrum of $\mathrm{ad}(a)$ described below is Lemma 1 in [N96], subject to rescaling.

VI. $\mathrm{ad}(a)$ has eigenvalues

$$1,\ 0,\ \frac{1}{4},\ \frac{1}{32}.$$

The corresponding eigenspaces, as modules for $C_M(a) = C_M(\tau) \cong 2 \cdot BM$, have dimensions

$$1,\ 1 + 96\,255,\ 4371,\ 96\,256,$$

respectively, with the first 1-entry corresponding to a and the second to $\mathbf{1} - a$, where $\mathbf{1}$ is the identity of the algebra, properly rescaled. The $\frac{1}{32}$-eigenspace, of dimension 96 256, is the only faithful component (on which it acts by negation).

Let τ_0 and τ_1 be a pair of 2A-involutions in M and let $a_0 = a(\tau_0)$ and $a_1 = a(\tau_1)$ be the corresponding 2A-axes. One can consider the algebra product $a_0 * a_1$, aiming to express it in terms of the other 2A-axes (and something else), and even describe the whole subalgebra $\langle\langle a_0, a_1 \rangle\rangle$ generated by these two 2A-axes. These products and subalgebras were computed by Simon Norton, cf. table 1 in [N96] which is reproduced as Table 2 in our scaling.

VII. We make a few remarks.

Table 2 The Norton–Sakuma algebras

Type	Basis	Products and angles
2A	a_0, a_1, a_ρ	$a_0 \cdot a_1 = \frac{1}{2^3}(a_0 + a_1 - a_\rho)$, $a_0 \cdot a_\rho = \frac{1}{2^3}(a_0 + a_\rho - a_1)$
		$(a_0, a_1) = (a_0, a_\rho) = (a_1, a_\rho) = \frac{1}{2^3}$
2B	a_0, a_1	$a_0 \cdot a_1 = 0$, $(a_0, a_1) = 0$
		$a_0 \cdot a_1 = \frac{1}{2^5}(2a_0 + 2a_1 + a_{-1}) - \frac{3^3 \cdot 5}{2^{11}} u_\rho$
3A	$a_{-1}, a_0, a_1,$	$a_0 \cdot u_\rho = \frac{1}{3^2}(2a_0 - a_1 - a_{-1}) + \frac{5}{2^5} u_\rho$
	u_ρ	$u_\rho \cdot u_\rho = u_\rho$
		$(a_0, a_1) = \frac{13}{2^8}$, $(a_0, u_\rho) = \frac{1}{2^2}$, $(u_\rho, u_\rho) = \frac{2^3}{5}$
3C	a_{-1}, a_0, a_1	$a_0 \cdot a_1 = \frac{1}{2^6}(a_0 + a_1 - a_{-1})$, $(a_0, a_1) = \frac{1}{2^6}$
		$a_0 \cdot a_1 = \frac{1}{2^6}(3a_0 + 3a_1 + a_2 + a_{-1} - 3v_\rho)$
4A	$a_{-1}, a_0, a_1,$	$a_0 \cdot v_\rho = \frac{1}{2^4}(5a_0 - 2a_1 - a_2 - 2a_{-1} + 3v_\rho)$
	a_2, v_ρ	$v_\rho \cdot v_\rho = v_\rho$, $a_0 \cdot a_2 = 0$
		$(a_0, a_1) = \frac{1}{2^5}$, $(a_0, a_2) = 0$, $(a_0, v_\rho) = \frac{3}{2^3}$, $(v_\rho, v_\rho) = 2$
4B	$a_{-1}, a_0, a_1,$	$a_0 \cdot a_1 = \frac{1}{2^6}(a_0 + a_1 - a_{-1} - a_2 + a_{\rho^2})$
	a_2, a_{ρ^2}	$a_0 \cdot a_2 = \frac{1}{2^3}(a_0 + a_2 - a_{\rho^2})$
		$(a_0, a_1) = \frac{1}{2^6}$, $(a_0, a_2) = (a_0, a_\rho) = \frac{1}{2^3}$
		$a_0 \cdot a_1 = \frac{1}{2^7}(3a_0 + 3a_1 - a_2 - a_{-1} - a_{-2}) + w_\rho$
5A	$a_{-2}, a_{-1}, a_0,$	$a_0 \cdot a_2 = \frac{1}{2^7}(3a_0 + 3a_2 - a_1 - a_{-1} - a_{-2}) - w_\rho$
	a_1, a_2, w_ρ	$a_0 \cdot w_\rho = \frac{7}{2^{12}}(a_1 + a_{-1} - a_2 - a_{-2}) + \frac{7}{2^5} w_\rho$
		$w_\rho \cdot w_\rho = \frac{5^2 \cdot 7}{2^{19}}(a_{-2} + a_{-1} + a_0 + a_1 + a_2)$
		$(a_0, a_1) = \frac{3}{2^7}$, $(a_0, w_\rho) = 0$, $(w_\rho, w_\rho) = \frac{5^3 \cdot 7}{2^{19}}$
		$a_0 \cdot a_1 = \frac{1}{2^6}(a_0 + a_1 - a_{-2} - a_{-1} - a_2 - a_3 + a_{\rho^3}) + \frac{3^2 \cdot 5}{2^{11}} u_{\rho^2}$
6A	$a_{-2}, a_{-1}, a_0,$	$a_0 \cdot a_2 = \frac{1}{2^5}(2a_0 + 2a_2 + a_{-2}) - \frac{3^3 \cdot 5}{2^{11}} u_{\rho^2}$
	a_1, a_2, a_3	$a_0 \cdot u_{\rho^2} = \frac{1}{3^2}(2a_0 - a_2 - a_{-2}) + \frac{5}{2^5} u_{\rho^2}$
	a_{ρ^3}, u_{ρ^2}	$a_0 \cdot a_3 = \frac{1}{2^3}(a_0 + a_3 - a_{\rho^3})$, $a_{\rho^3} \cdot u_{\rho^2} = 0$, $(a_{\rho^3}, u_{\rho^2}) = 0$
		$(a_0, a_1) = \frac{5}{2^8}$, $(a_0, a_2) = \frac{13}{2^8}$, $(a_0, a_3) = \frac{1}{2^3}$

VII A There are the following inclusions among the algebras in the Table 2, ruled by the inclusions among the relevant dihedral groups:

$$2A \subset 4B, \ 2B \subset 4A, \ 2A \subset 6A, \ 3A \subset 6A.$$

VII B The vectors a_ρ and a_{ρ^2}, in $2A$ and $4B$ algebras, respectively, are $2A$-axes.
VII C The algebras $3A$, $4A$, $5A$, $6A$ are not closed on the linear span of the $2A$-axes contained in these subalgebras. The additional vectors u_ρ, v_ρ, w_ρ, in the algebras $3A$, $4A$, and $5A$, respectively, can be chosen to depend exclusively on the element $\rho = \tau_0 \tau_1$. These are called $3A$-, $4A$-, and $5A$-axes and play a very important role in the subsequent exposition. Furthermore,

$$u_\rho = u_{\rho^{-1}}, \ v_\rho = v_{\rho^{-1}}, \ w_\rho = -w_{\rho^2} = -w_{\rho^3} = w_{\rho^{-1}}.$$

The $6A$-algebra is closed on the linear span of its $2A$-axes and the $3A$-axes from the $3A$-subalgebra.

By VI, given $(V_M,(,),*)$ and a $2A$-axis a, the involutory automorphism τ such that $a = a(\tau)$ can be recovered from the algebra product by the following rule:

VIII. Every $\frac{1}{32}$-eigenvector is negated and all other eigenvectors are centralised.

It was noticed in section 7 of [N96], and stated without a proof, that the subalgebra on $C_{V_M}(\tau)$ is preserved by the involution $\sigma(a)$ which negates the 4371-irreducible of $C_M(\tau)$ and centralises the other irreducibles. It was shown how $\sigma(a)$ acts on the $2A$- and $3A$-subalgebras containing a: by negating the $\frac{1}{4}$-eigenspace and centralising the 1- and 0-eigenspaces. It is an easy check that such an action works for all other algebras in Table 2. This gives a proof of the existence of $\sigma(a)$.

IX. The transformation of $C_{V_M}(\tau)$, which negates the $\frac{1}{4}$ eigenspace of $\mathrm{ad}(a)$ and centralises the 1- and 0-eigenspaces, is an automorphism of the restrictions to $C_{V_M}(\tau)$ of the inner $(,)$ and algebra $*$ products. This automorphism does not extend to an automorphism of the whole of $(V_M,(,),*)$.

X. The Monster group M is known to contain a single conjugacy class of A_5-subgroups whose elements of order 2, 3, and 5 are contained in the conjugacy classes $2A$, $3A$, and $5A$, respectively. If $A \cong A_5$ is a representative of this conjugacy class, then

$$N_M(A) \cong (S_5 \times S_{12})^+,$$

which is the intersection of the group in the brackets acting naturally on $5+12 = 17$ points with A_{17}. Notice that this normaliser is a very important group in the Y-story of the Monster. By fusing the conjugacy classes of $N_M(A)$ into that of the Monster (compare Table 7) one can calculate that

$$\dim C_{V_M}(C_M(A)) = 27.$$

This means that the fifteen $2A$-axes corresponding to the involutions in A generate a subalgebra of dimension at most 27. This subalgebra clearly contains the $3A$-axes (10 of them) and $5A$-axes (6 of them) corresponding to the relevant subgroups in A. The Gram matrix of the totality of $31 = 15 + 10 + 6$ axes was computed in [N96], p. 300. The rank of this matrix turned out to be 26, with the $5A$-axes being all equal (up to a sign) modulo the subspace spanned by the $2A$- and $3A$-axes. The determination of the exact structure of the subalgebra on this subspace was not attempted [N96].

The subalgebra in the Monster algebra $(V_M,(,),*)$ generated by the fifteen $2A$-axes contained in an $(2A, 3A, 5A)$-subgroup isomorphic to A_5 has dimension at most 26.

As we will see in the next section, this discussion contains enough prophecy for Majorana Theory, although historically it was introduced through Vertex

Table 3 Fusion rules of Virasoro algebra

	0	$\frac{1}{2}$	$\frac{1}{16}$
0	0	$\frac{1}{2}$	$\frac{1}{16}$
$\frac{1}{2}$	$\frac{1}{2}$	0	$\frac{1}{16}$
$\frac{1}{16}$	$\frac{1}{16}$	$\frac{1}{16}$	$0, \frac{1}{2}$

Operator Algebras (VOAs), particularly through the example $V^\natural$ of VOA possessing the Monster group as the automorphism group. This VOA is known as the Moonshine VOA.

A VOA (of the type we are interested in) is a $\mathbb{Z}$-graded vector space:

$$V^\infty = \bigoplus_{i=0}^{\infty} V_i,$$

on which there are infinitely many products $*_n : (v, u) \mapsto v *_n u$, $n = 0, 1, \ldots$ satisfying infinitely many Jacobi-type relations. The products respect the grading, in the sense that for $v, u \in V_m$ we have $v *_n u$ is contained in V_k, where $k = 2m - n - 1$. The VOA $V^\natural$ was constructed in [FLM88] with $\mathrm{Aut}(V^\natural) \cong M$, $V_0 = \mathbb{R}$, $V_1 = 0$, $\dim V_2 = 196\,884$; furthermore,

$$(V_2, *_3, *_1) \cong (V_M, (\,,\,), *).$$

Notice that the products in $V^\natural$ are not subject to any rescaling, and the fact that Conway's version on the right-hand side of the isomorphism exactly fits the VOA version on the left-hand side is the ingenious "coincidence." In terms of $V^\natural$, the 2A-axes, when divided by 2, become *conformal vectors of central charge* $\frac{1}{2}$. The components of such a vector generate a Virasoro algebra of that central charge. Such an algebra is known [BPZ84] to possess only three irreducible representations with highest weights 0, $\frac{1}{2}$, and $\frac{1}{16}$ whose fusion rules are described in Table 3:

It was M. Miyamoto who noticed in [Miy96] that these fusion rules possess an obvious $\mathbb{Z}_2$-grading, and this resulted in the following:

XI. Let V^∞ be a VOA of the above specified class, which includes the Moonshine VOA $V^\natural$; let $c \in V_2$ be a conformal vector of central charge $\frac{1}{2}$; and let $Vir(c)$ be the corresponding Virasoro algebra. Then the transformation $\tau(c)$, which negates the irreducibles of $Vir(c)$ with highest weight $\frac{1}{16}$ and centralises the irreducibles with highest weight 0 and $\frac{1}{2}$, is an automorphism of V^∞. In the theory of VOAs, the automorphism $\tau(c)$ is called τ-*involution* associated with the conformal vector c of central charge $\frac{1}{2}$.

In the case of $V^\natural$, the automorphism $\tau(c)$ is the $2A$-involution in the Monster group identified with the automorphism group of $V^\natural$. Since $a(\tau(c)) = 2c$, the numbers in the spectrum in VI are halves of the corresponding highest weights.

In [Miy04] Miyamoto used his epoch-making observation to reconstruct in the automorphism group of $V^\natural$ the maximal parabolics of the Tilde $\mathcal{G}(M)$, thus closing the cycle of our discussions so far.

The final nail is the following result by S. Sakuma [Sak07], based on earlier work by M. Miyamoto [Miy03].

XII. Let c_0 and c_1 be conformal vectors of central charge $\frac{1}{2}$ in a VOA V^∞ from the above specified class. Then there are at most nine possibilities for the isomorphism type of the subalgebra generated by c_0 and c_1.

XIII. In [N17], Simon Norton wrote, "One may consider an 'ideal' ATLAS whose culmination was a simple explanation of the existence of the Monster, with properties of many smaller groups being covered on the way."

5.2 Axiomatic

While writing [Iv09], the author noticed that Sakuma's theorem in [Sak07] can be stated and proved without leaving the finite-dimensional domain under the following axiomatic.

First we define *Majorana algebras* and introduce some relevant terms. Let V be a real vector space equipped with a bilinear form $(\,,\,)$ and an algebra product $\cdot$.

(M1) $(\,,\,)$ is a symmetric positive-definite bilinear form on V that associates with $\cdot$ in the sense that $(u, v \cdot w) = (u \cdot v, w)$ for all $u, v, w \in V$, and $\cdot$ is a bilinear commutative non-associative algebra product on V;

(M2) the Norton inequality holds, so that $(u \cdot u, v \cdot v) \geq (u \cdot v, u \cdot v)$ for all $u, v \in V$.

A vector $a \in V$ is said to be a *Majorana axis* if it satisfies the following five conditions (M3) to (M7), where $\mathrm{ad}(a) : v \mapsto a \cdot v$ is the adjoint operator of a on V.

(M3) $(a, a) = 1$ and $a \cdot a = a$, so that a is an idempotent of length 1;

(M4) $\mathrm{ad}(a)$ is semi-simple with spectrum $Sp = \{1, 0, \frac{1}{4}, \frac{1}{32}\}$:

$$V = V_1^{(a)} \oplus V_0^{(a)} \oplus V_{\frac{1}{4}}^{(a)} \oplus V_{\frac{1}{32}}^{(a)},$$

where $V_\mu^{(a)} = \{v \mid v \in V, a \cdot v = \mu v\}$ is the set of μ-eigenvectors of $\mathrm{ad}(a)$ on V;

Table 4 Fusion rules for Majorana axes

Sp	1	0	$\frac{1}{4}$	$\frac{1}{32}$
1	1	0	$\frac{1}{4}$	$\frac{1}{32}$
0	0	0	$\frac{1}{4}$	$\frac{1}{32}$
$\frac{1}{4}$	$\frac{1}{4}$	$\frac{1}{4}$	$1,0$	$\frac{1}{32}$
$\frac{1}{32}$	$\frac{1}{32}$	$\frac{1}{32}$	$\frac{1}{32}$	$1,0,\frac{1}{4}$

(M5) $V_1^{(a)} = \{\lambda a \mid \lambda \in \mathbb{R}\}$;

(M6) the linear transformation $\tau(a)$ of V defined via

$$\tau(a) : u \mapsto (-1)^{32\mu} u$$

for $u \in V_\mu^{(a)}$ with $\mu = 1, 0, \frac{1}{4}, \frac{1}{32}$, preserves the algebra product (i.e., $u^{\tau(a)} \cdot v^{\tau(a)} = (u \cdot v)^{\tau(a)}$ for all $u, v \in V$). The automorphism $\tau(a)$ is called the *Majorana involution* associated with the Majorana axis a;

(M7) if $V_+^{(a)}$ is the centraliser of $\tau(a)$ in V, so that $V_+^{(a)} = V_1^{(a)} \oplus V_0^{(a)} \oplus V_{\frac{1}{4}}^{(a)}$,

then the linear transformation $\sigma(a)$ of $V_+^{(a)}$ defined via

$$\sigma(a) : u \mapsto (-1)^{4\mu} u$$

for $u \in V_\mu^{(a)}$ with $\mu = 1, 0, \frac{1}{4}$, preserves the restriction of the algebra product to the subalgebra $V_+^{(a)}$.

The conditions (M1) to (M7) imply that the eigenspaces $V_\mu^{(a)}$ of the adjoint action of a satisfy the fusion rules described in Table 4.

The meaning of the fusion rules is the inclusion

$$V_\lambda^{(a)} \cdot V_\mu^{(a)} \subseteq \bigoplus_{\nu \in Sp(\lambda,\mu)} V_\nu^{(a)},$$

where $\lambda, \mu \in Sp$ and $Sp(\mu, \lambda)$ is the (λ, μ)-entry in Table 4, which is essentially the fusion table for conformal vectors of central charge $\frac{1}{2}$ subject to dividing by 2 and adjoining the 1-eigenvalue.

Definition 2.1 Let $(V, (,), \cdot)$ be a triple satisfying $(M1)$ and $(M2)$, let A be a set of Majorana axes in V satisfying $(M3)$ to $(M7)$, and let G be the subgroup in the automorphism group of $(V, (,), \cdot)$ generated by the Majorana involutions associated with the axes in A. Then the quintet $(V, (,), \cdot, A, G)$ is said to be a Majorana algebra, which is a *Majorana representation* of G.

Notice that in this definition we do not exclude the possibility that there are further Majorana axes in V besides those in A. It is tempting to restrict V to the subalgebra generated by the axes in A (as was done in the original definition in [Iv09]), although this hides a problem since G might act unfaithfully on the subalgebra generated by A. For example, if different axes correspond to different involutions, then the centre of G acts trivially on the subalgebra generated by A.

The way a Majorana representation was defined in [Iv09] and followed thereafter actually specifies a way of constructing such a representation. In fact, if a_0, a_1 are Majorana axes, then the image $(a_0)^{\tau(a_1)}$ of a_0 under the automorphism $\tau(a_1)$ is also a Majorana axis. Therefore, A can be assumed to be closed under this sort of conjugation, in which case

$$T := \{\tau(a) \mid a \in A\}$$

is a union of conjugacy classes of involutions in G. Thus to construct a Majorana representation of G, one can start by choosing T and assigning a vector in V for every involution in T and proceed to construct the whole of V with inner and algebra products. In a more general setting, we might be given only partial information about G and A, as in Sakuma's theorem.

Comparing items **V** through **VIII** in the previous section, we deduce the expected result.

Theorem 2.2 *The quintet $(V_M, (\ ,\ ), *, A, M)$, where A is the set of 2A-axes in the Monster algebra V_M, is a Majorana representation of M.*

In the new terminology the finite-dimensional version of Sakuma's theorem [Sak07], as proved in [IPSS10], reads as follows.

Theorem 2.3 *Let $B = \{a_0, a_1\}$ for $a_0 \neq a_1$ and let G be the dihedral subgroup D_{2n} generated by $\tau(a_0)$ and $\tau(a_1)$, where n is the order of the product of the generators. Then $n \leq 6$ and there are at most eight possibilities for the isomorphism type of the subalgebra generated by A in a Majorana representation of G with A containing B.*

Comparing Theorems 2.2, 2.3, and Table 2 in the previous section, we obtain the following.

Theorem 2.4 *There are exactly eight Majorana algebras generated by pairs of distinct Majorana axes as in Table 2. Each of them is a subalgebra in the Monster algebra. They are called Norton–Sakuma algebras.*

The Majorana axioms are local, therefore we have the following.

Theorem 2.5 *Let A be a set of 2A-axes in V_M, let G be the subgroup of M generated by the 2A-involutions whose axes are in A, and let V be the subalgebra generated by A; then*

$$(V, (\ ,\ )|_V, *|_V, A, G)$$

*is a Majorana representation of G (which is said to be based on the embedding
of G into the Monster). The dimension of the span of A is the dimension of the
representation.*

By Theorem 2.5, the complete list of Majorana representations will include
the list (so far unknown) of the $2A$-generated subgroups of the Monster and the
non-embeddable examples.

The remarkable recent development of Majorana Theory demonstrated a
strong tendency of Majorana representations to be based on embeddings in the
Monster (excluding "small" cases), and this led to posing the following conjecture [Iv17], [Iv18].

Straight Flush Conjecture Suppose that $(V, (\,,\,), \cdot, A, G)$ is an indecomposable Majorana algebra in which, for every $i \in \{2, 3, 4, 5, 6\}$, there exists a pair
of Majorana involutions t_1 and t_2 in G, such that the order of the product $t_1 t_2$ is
i. Then A embeds into the Monster algebra.

A Majorana representation is *irreducible* if the set of Majorana axes does
not possess a partition into proper subsets such that axes from different subsets,
say a_0 and a_1, generate a $2B$-algebra, meaning that a_0 and a_1 are perpendicular
and annihilate each other. It is easy to see that every Majorana algebra is an
orthogonal direct sum of indecomposable Majorana algebras, so that the group
is the direct product of the subgroups of the summands.

The group G in Theorem 2.5 might or might not act faithfully on V, but
this should not cause much of a problem. This already happens with Norton–
Sakuma algebras associated with dihedral groups of double even order.

When someone is primarily interested in the subalgebra structure of the Monster algebra – that is, in representations based on embeddings into the Monster – in
view of **VII B** and **VII C** it might be reasonable to expand the axiomatic by a
further axiom:

(M8) The vectors a_ρ and a_{ρ^2}, in the $2A$ and $4B$ algebras respectively, are Majo-
       rana axes on the whole space. The vectors u_ρ, v_ρ, w_ρ, in the algebras $3A$,
       $4A$, and $5A$, respectively, depend solely on the group element ρ (rather
       than on the whole dihedral group containing it).

5.3 Representations of A_5 as an Example

After Sakuma's theorem, one can start thinking about algebras generated by
more than two Majorana axes. It is natural to start with three.

Let $V = (V, \cdot, (\,,\,))$ be a Majorana algebra, and let a_0, a_1, a_2 be distinct
Majorana axes in V. By Sakuma's theorem, for $0 \le i < j \le 2$ the subal-
gebra $\langle\langle a_i, a_j \rangle\rangle$ is one of the eight non-trivial Norton–Sakuma algebras, say

$N_{ij}X$, where $N_{ij} \in \{2,3,4,5,6\}$ and $X \in \{A,B,C\}$ depending on N_{ij}. Then if $\tau_i = \tau(a_i)$, the subgroup G in $\mathrm{Aut}(V)$ generated by the τ_i's is a quotient of the Coxeter group C, whose diagram is a triangle with edges N_{ij}. We can start with the situation when the Coxeter group C itself is finite.

Suppose that the diagram of C is

$$H_3 : \circ\!\!-\!\!-\!\!-\!\!-\!\!-\!\!-\!\!-\!\!-\!\!\circ\overset{5}{-\!\!-\!\!-\!\!-\!\!-\!\!-\!\!-}\circ,$$

and G is either $C \cong A_5 \times 2$, or the A_5-quotient of C over its centre of order 2.

In view of the following obvious general statement,

Theorem 3.1 *The image of a Majorana axis under an automorphism of V is a Majorana axis;*

we have the following setting. The subalgebra $U := \langle\!\langle a_0, a_1, a_2 \rangle\!\rangle$ is generated by a set

$$A = \{a_t \mid t \in C/Z(C) \cong A_5, t^2 = 1 \neq t\}$$

of 15 Majorana axes indexed by the involutions in A_5, so that $\tau(a_t) \in G$ acts on A as t acts on the set T of involutions in A_5 by conjugation.

If we write

$$\varphi : G \to GL(V), \quad \psi : T \to V \setminus \{0\},$$

where φ is determined by the action of the generating involutions on the generating set A of the algebra, and $\psi(t) = a_t$, we obtain the tuple

$$\mathcal{R} = (G, T, V, \; \cdot\,, (\,,\,), \varphi, \psi)$$

called the *Majorana representation* of G (which is A_5 or $A_5 \times 2$ in the considered situation) with respect to the set T of involutions in G. In plain words, such a representation is a realisation of G as an automorphism group of a Majorana algebra generated by the Majorana axes corresponding to the involutions in the set T.

Thus we consider a Majorana representation $\mathcal{R}$ of A_5 or $A_5 \times 2$. For the original triple of generators $\{a_0, a_1, a_2\}$ we assume that

$$S_{01} = \langle\!\langle a_0, a_1 \rangle\!\rangle \cong 3A \text{ or } 3C;$$

$$S_{12} = \langle\!\langle a_1, a_2 \rangle\!\rangle \cong 5A;$$

$$S_{02} = \langle\!\langle a_0, a_2 \rangle\!\rangle \cong 2A \text{ or } 2B.$$

The particular choice of the algebras S_{01} and S_{02} is said to be the *shape* of the representation. By the obvious symmetry if $t, s \in T$ are distinct involutions of A_5, then

$$\langle\!\langle a_t, a_s \rangle\!\rangle \cong S_{ij},$$

Table 5 $2A$-generated A_5-subgroups of M

classes	$\bar{G}$	$C_M(G)$	dim(C)
$(2A, 3A, 5A)$	A_5	A_{12}	26
$(2A, 3C, 5A)$	A_5	$U_3(8).3$	21
$(2B, 3A, 5A)$	$A_5 \times 2$	$2.M_{22}.2$	?
$(2B, 3C, 5A)$	$A_5 \times 2$	$2^{1+4}(A_4 \times A_5)$	?

where ij is 01, 12, or 02 when the order of ts is 3, 5, or 2, respectively. Therefore, the shape determines the isomorphism type of the Norton–Sakuma algebras generated by every pair of the symmetric generators.

Let us now discuss the subgroups A_5 and $A_5 \times 2$ in the Monster group M which are generated by $2A$-involutions and the corresponding subalgebras in the Monster algebra. These were determined by Simon Norton in [N98].

Here $\bar{G} = C_M(C_M(G))$, $C = C_{V_M}(C_M(G))$. It would be very useful to fill up the questioned entries in Table 5.

Table 5 shows particularly that for each of the possible four shapes of Majorana representation of $A_5 \times 2$ (possibly with non-injective φ) there exists a representation into a subalgebra in the Monster algebra. We say that these four representations are *based on embeddings into the Monster*.

Let $\mathcal{R}_{Sh}$ be a representation of $A_5 \times 2$ of shape Sh, where the latter is understood as a map from the set of pairs of Majorana generators into the set of Norton–Sakuma algebras.

The dimension calculation amounts to determination of the rank of the Gram matrix of the symmetric generators, which is the matrix

$$\Gamma = \|(a_t, a_s)\|_{15 \times 15}.$$

The action of A_5 on the generators allows us to apply the theory of association schemes and to reduce the task to hand calculations. The calculations can be performed with the association scheme corresponding to the action of A_5 on the set of its involutions by conjugation. It is sufficient to calculate the eigenmatrix of the association scheme. Let T be the set of involutions in A_5, and let A_1, A_2, A_3, A_5 be the relations on T defined by the products order 1, 2, 3, 5, respectively. If we consider the A_i's as matrices, then

$$\Gamma = A_0 + p_2 A_2 + p_3 A_3 + p_5 A_5,$$

where $p_i = (a_t, a_s)$ with $o(st) = i$. Then $a_2 = \frac{1}{8}$ or 0 for 2-shapes $2A$ and $2B$, $a_3 = \frac{13}{256}$ or $\frac{1}{64}$ for 3-shapes $3A$ and $3C$, and $a_5 = \frac{5}{256}$ for only 5-shape $5A$.

The matrix

$$P = \begin{pmatrix} 1 & 4 & 2 & 8 \\ 1 & 1 & 1/2 & -4 \\ 1 & 2 & -1 & -2 \\ 1 & -2 & -8 & 2 \end{pmatrix}$$

simultaneously diagonalises the structure relations and therefore the diagonalised version of Γ is

$$\bar{\Gamma} = P(1, p_2, p_3, p_5)^T,$$

and to accomplish the exercise you should check that the eigenvalues of $\bar{\Gamma}$ are never zero, and hence Γ is of full rank 15.

Next we turn to the 2-closure. The algebra product is not closed on the space V_1, which is the 1-closure; that is, the linear span of the generators. When we multiply axes a_t and a_s for commuting involutions, we assume that the product is closed on $Sp\{a_t, a_s, a_{st}\}$, but when we multiply generators for noncommuting involutions t and s, we need to adjoin u_ρ for the 3A-shape and w_ρ for the 5A-shape, where $\rho = ts$. These are (potentially) new vectors, not contained in V_1. By adjoining them and taking the linear span, we obtain the 2-closure V_2. In V_2, all the Majorana generators can be multiplied, but the products involving 3A-axes u_ρ and 5A-axes w_ρ can only be multiplied by the generators used in their definitions. Recall that

$$u_\rho = u_{\rho^{-1}}, \quad \text{while} \quad w_\rho = -w_{\rho^2} = -w_{\rho^3} = w_{\rho^{-1}}.$$

As a next step, we might decide to calculate the rank of the 2-closure; that is, the dimension of V_2, or better to describe $(V_2, (\,,\,))$ up to isomorphism as an inner product space. Towards this end we have to determine the inner products between 2A-, 3A-, and 5A-axes. For this purpose, it is usually sufficient to multiply the eigenvectors with different eigenvalues of a given Majorana generator a_t and from different Norton–Sakuma algebras containing a_t. These pairs of eigenvectors will also be used later to recover the algebra multiplication law, mostly through the resurrection principle.

The rank for the $(2A, 3A, 5A)$-shape turned out to be 26, and for the $(2A, 3C, 5A)$-shape it is 20. In the former case, we have fifteen linearly independent 2A-axes, ten 3A-axes and six 5A-axes; since $15 + 10 + 6 - 26 = 5$, we have five relations. In the latter case we again have 15 linearly independent 2A-axes and six 5A-axes, $15+6-20 = 1$, thus a single relation (up to scalar multiples). Let F be a set of six elements of order 5 in A_5 which generate pairwise different cyclic subgroups and are contained in the same conjugacy class. Then in $(2A, 3C, 5A)$ we have

$$\sum_{f \in F} w_f = 0,$$

while in $(2A, 3A, 5A)$ the six differences

$$w_g - \sum_{f \in F} w_f$$

are explicitly given expressions in terms of $2A$- and $3A$-axes, whose particular form depends on g (the sum is independent on the choice of F subject to the conditions).

The final list of Majorana representations of A_5 is given in the following.

Theorem 3.1 *The group A_5 possesses precisely four Majorana representations, all based on embeddings into the Monster group, whose shape and dimensions are as follows:*

(i) *shape $(2A, 3A, 5A)$ of dimension 26;*

(ii) *shape $(2A, 3C, 5A)$ of dimension 20;*

(iii) *shape $(2B, 3C, 5A)$ of dimension 21;*

(iv) *shape $(2B, 3A, 5A)$ of dimension 46.*

The representations in (i) to (iii) are 2-closed and the one in (iv) is not. □

The representations in (i) to (iii) have been classified in [IS12], and that in (iv) was constructed and characterised by Ákos Seress [S12].

5.4 A Standard Procedure

In the process of understanding Majorana representations, the following procedure has become rather standard.

Identifying G. The group G might not be given at the first place. Instead, we can be given a generating set B of Majorana axes together with information on the isomorphism type of the algebras generated by pairs of axes from B. Then, for the group G, we have a generating set $\{\tau(a) \mid a \in B\}$ of involutions together with the orders of pairwise products. This defines the obvious Coxeter group G^c of which our group G is a homomorphic image. The group G must be a 6-transposition group with respect to the normal closure of the generating involutions. This condition (hopefully) reduces the situation to a limited number of cases. Of course that situation is much easier when G^c is already finite.

Generating Involutions. Choose a generating union T of conjugacy classes in G. We assume that G is generated by involutions, since only such a group might possess a Majorana representation. The involutions in T will be the

Majorana involutions of the axes in A and we start building V by producing a vector $a(t)$ for every $t \in T$; that is, by building $A = \{a(t) \mid t \in A\}$ just as a set of vectors in V.

Shape. In order to define the inner products between vectors in A and to start multiplying them, we define the *shape* of the representation to be constructed. The shape is a map sh from the set of dihedral subgroups generated by pairs of distinct involutions in T to the set of the eight non-trivial Norton–Sakuma algebras, so that the algebra generated by $a(t)$ and $a(s)$ for $t, s \in T$ is isomorphic to $sh(< t, s >)$. In order to construct at least one shape, T should be a 6-transposition set; that is,

$$o(ts) \leq 6 \text{ for all } t, s \in T.$$

The shape should be invariant under the action on G on T by conjugation and also should map the inclusions of the dihedral subgroups to the inclusions between Norton–Sakuma algebras as in VII A. The shape determines the restriction of the bilinear form (,) to A. This restriction should be a positive-definite bilinear form, otherwise our choice (of sh, T or G) should be adjusted. If there is a vector of negative length in the span of A, the situation is not repairable. If a vector has zero length, then it must be perpendicular to all other vectors. Then these vectors must be zero vectors forming the radical of the form to be factored out.

2-closure. The shape also shows the additional set of vectors to be adjoined to V: the $3A$-, $4A$- and $5A$-axes of the relevant Norton–Sakuma algebras. Not all of these vectors will be linearly independent of the Majorana axes and among themselves, but this will be discussed later.

The multiplication heavily depends on the eigenvectors of Majorana axes. Such eigenvectors inside the Norton–Sakuma algebras are well known since [Iv09], and can be seen in Table 6.

5.5 Dimensions of Majorana (sub)Representations

Let $(V, (,), \cdot, A, G)$ be a Majorana representation of a group G, and let W be a subspace of V (which could be the whole of V) of which we have some information. On various stages of the construction and classification of Majorana representations, it is important to know the dimension of W. In this section, we discuss various tools to determine, or at least estimate the dimension of W.

5.5.1 Rank of Gram Matrix on Axes

In the previous paragraph, consider the situation when W is spanned by a set X of vectors such that the pairwise inner products (x, y) are known. Since the form $(\, , \,)$ is assumed to be positive-definite,

$$\dim(W) = \text{rank}(\widehat{W}, (\, , \,)),$$

where the right-hand side is the codimension of the radical of the space $\widehat{W}$ freely spanned by X with respect to the inner product values inherited from W. This in turn is equal to the rank of the Gram matrix

$$\Gamma = ||(x, y)||_{X \times X}$$

of X. It is convenient to consider Γ as the Gram matrix of vectors in $\widehat{W}$. Since Γ is a real symmetric matrix, it can be diagonalised via conjugation by a unitary matrix U, so that

$$U^{-1} \Gamma U = \text{diag}(\lambda_1, \ldots, \lambda_{|X|}),$$

where the λ_i's are the eigenvalues of Γ. These eigenvalues must be non-negative; if they are positive, then Γ is of full rank and $\dim(W) = |X|$. On the other hand, if there is a zero eigenvalue, then the zero-eigenspace is the radical and the codimension is the rank of Γ equal to the dimension of W. An expression of a zero-eigenvector in terms of the spanning vectors from X is the zero vector in W, so that it is perpendicular to any other vector in W and annihilates it. No wonder that such vectors can help build up the aimed Majorana representation.

The search for the zero-eigenspace of the Gram matrix is particularly handy (at least on the theoretical level) when the following hold:

 (i) the group G acts transitively on the set X;
(ii) the transitive permutation group (G, X) is multiplicity-free.

The condition (i) holds, for instance, when X is a G-orbit on axes in V (which can be Majorana, $3A$-, $4A$-, or $5A$-axes). The condition (ii) means that the irreducible constituents of the $\mathbb{C}$-permutation module $\mathbb{C}^X$ of (G, X) are pairwise non-isomorphic. This is much more restrictive, although it holds for some G-orbits on Majorana axes (in some representations).

The situation under (i) and (ii) is controlled by the theory of association schemes as in [BI84], which goes as follows. Let $O_0, O_1, \ldots, O_{r-1}$ be the orbits of G on $X \times X$ (known as the orbitals), where $O_0 = \{(x, x) \mid x \in X\}$ is the diagonal. Let $A_0, A_1, \ldots, A_{r-1}$ be the incidence matrices of the orbitals. This means that A_i is a $\{0, 1\}$-matrix, where the (x, y)-entry is 1 if and only if $(x, y) \in O_i$. Then $A_0 = I$ and every $(X \times X)$-matrix invariant under the action

Table 6 Eigenvectors of Norton–Sakuma algebras

Type	0	$\frac{1}{4}$	$\frac{1}{32}$
2A	$a_1 + a_\rho - \frac{1}{2^2}a_0$	$a_1 - a_\rho$	
2B	a_1		$a_1 - a_{-1}$
3A	$u_\rho - \frac{2\cdot5}{3^3}a_0 + \frac{2^5}{3^3}(a_1 + a_{-1})$	$u_\rho - \frac{2^3}{3^2\cdot5}a_0 - \frac{2^5}{3^2\cdot5}a(a_1 + a_{-1})$	$a_1 - a_{-1}$
3C	$a_1 + a_{-1} - \frac{1}{2^5}a_0$		$a_1 - a_{-1}$
4A	$v_\rho - \frac{1}{2}a_0 + 2(a_1 + a_{-1}) + a_2,\ a_2$	$v_\rho - \frac{1}{3} - \frac{2}{3}(a_1 + a_{-1}) - \frac{1}{5}a_2$	$a_1 - a_{-1}$
4B	$a_1 + a_{-1} - \frac{1}{2^5}a_0 - \frac{1}{2^3}(a_{\rho^2} - a_2),$ $a_2 + a_{\rho^2} - \frac{1}{2^2}a_0$	$a_2 - a_{\rho^2}$	$a_1 - a_{-1}$
5A	$w_\rho + \frac{3}{2^9}a_0 - \frac{3\cdot5}{2^7}(a_1 + a_{-1}) - \frac{1}{2^7}(a_2 + a_{-2}),$ $w_\rho - \frac{3}{2^9}a_0 + \frac{1}{2^7}(a_1 + a_{-1}) + \frac{3\cdot5}{2^7}(a_2 + a_{-2})$	$w_\rho + \frac{1}{2^7}(a_1 + a_{-1} - a_2 - a_{-2})$	$a_1 - a_{-1}$ $a_2 - a_{-2}$
6A	$u_{\rho^2} + \frac{2}{3^2\cdot5}a_0 - \frac{2^4}{3^2\cdot5}(a_1 + a_{-1}) - \frac{2^3}{3^2\cdot5}(a_2 + a_{-2} + a_3 - a_{\rho^3}),$ $a_3 + a_{\rho^3} - \frac{1}{2^2}a_0,\ u_{\rho^2} - \frac{2\cdot5}{3^3}a_0 + \frac{2^5}{3^3}(a_2 + a_{-2})$	$u_{\rho^2} - \frac{2^3}{3^2\cdot5}(a_2 + a_{-2} + a_3 - a_{\rho^3}),\ a_1 - a_{-1}$ $a_3 - a_{\rho^3}$	$a_2 - a_{-2}$

of G (including the Gram matrix Γ) is a linear combination of the A_i's. More specifically,

$$\Gamma = \sum_{i=0}^{r-1} d_i A_i,$$

where d_i is the inner product (x,y) for $(x,y) \in O_i$. Then (ii) is equivalent to the commutativity of the matrix algebra $\mathcal{A}$ generated by the incidence matrices. Then the whole $\mathcal{A}$ can be diagonalised with common eigenspaces, which are precisely the irreducible constituents of G on $\mathbb{C}^X$. The number of these irreducibles is r, the permutation rank of (G,X). Let E_j be the orthogonal projection matrix onto the jth eigenspace.

Then

$$A_i = \sum_{j=0}^{r-1} p_i(j) E_j,$$

where $p_i(j)$ is the jth eigenvalue of A_i. These eigenvalues are arranged in the first eigenmatrix:

$$P = ||p_i(j)||_{r \times r},$$

whose columns are indexed by the orbitals and rows by the irreducibles. It is commonly assumed that E_0 projects onto the trivial 1-dimensional irreducible, so that E_0 is the all-ones matrix. Then the jth eigenvalue of Γ is equal to the component-wise inner product of the vectors

$$(d_0, d_1, \ldots, d_{r-1}) \text{ and } (p_0(j), p_1(j), \ldots, p_{r-1}(j)),$$

so that we have the following.

Lemma 5.1 *Under (i) and (ii) the jth eigenvalue of Γ is zero if and only if the distribution vector $(d_0, d_1, \ldots, d_{r-1})$ is perpendicular to the jth row of the first eigenmatrix P.*

In light of this, the following question is of a practical interest. Given a transitive multiplicity-free permutation group (G,X), a Gram matrix Γ, and an irreducible $G\mathbb{C}$-module U, isomorphic to a constituent $U_j = E_j X$ of $X^{\mathbb{C}}$, is the orthogonal projection of Γ onto U_j non-zero? The answer is again given by the theory of association schemes.

Let $x \in X$, and let H be the stabilizer of x in G. Since (G,X) is multiplicity-free and since $U_j \cong U$ is a constituent of $X^{\mathbb{C}}$, by Frobenius reciprocity, $C_U(H)$ is 1-dimensional spanned by a vector u, say, and there is a G-invariant mapping

$$\varphi : X \to U$$

which is unique up to a scalar multiple in the class of such mappings, and $\varphi(x) = u$. It is clear that U carries a non-zero symmetric form $[\ ,\ ]_U$ (which is the projection onto U_j of the pointwise inner product on X). By Schur's lemma, this form is unique up to scalar multiples. Assume that we can calculate the vector:

$$e := (e_0, e_1, \ldots, e_{r-1}),$$

where $e_i = [\varphi(x), \varphi(y)]$ and $(x, y) \in O_i$. Rescaling e to achieve $e_0 = \frac{m_j}{|X|}$, where $m_j = \text{rank}(E_j) = \dim(U_j)$, we have the following lemma.

Lemma 5.2 *In the above terms e_i is equal to the entry $q_j(i)$ of the second eigenmatrix Q of (G, X), where*

$$q_j(i) = p_i(j) \frac{m_j}{k_i},$$

for m_j being the dimension of U_j and k_i being the valency of O_i.

Proof Let $x, y \in O_i$ and let $\widehat{x}, \widehat{y}$ be the corresponding vectors of $X^{\mathbb{C}}$. Then (up to a global scalar) we have

$$[\varphi(x), \varphi(y)] = (E_j\widehat{x}, E_j\widehat{y}),$$

where here $(\ ,\ )$ is the pointwise inner product on X. Since different eigenspaces U_i are pairwise orthogonal,

$$(E_j\widehat{x}, E_j\widehat{y}) = \left((\sum_{i=0}^{r-1} E_i\,\widehat{x}, E_j\widehat{y}) = (\widehat{x}, E_j\widehat{y}) \right)$$

is the (x, y)-entry of E_j. On the other hand,

$$E_j = \frac{1}{|X|} \sum_{i=0}^{r-1} q_j(i) A_i.$$

Since $p_0(j) = 1$ for all j and $q_j(i) = p_i(j)\frac{m_j}{k_i}$, the result follows. $\square$

5.5.2 An Upper Bound for Embedded Representations

Suppose now that the considered Majorana representation $(V, (\ ,\ ), \cdot, A, G)$ is based on an embedding of G into the Monster group. This means that A is a set of $2A$-axes in the Monster algebra V_M, G is the subgroup of M generated by the $2A$-involutions of the axes in A, V is the subalgebra in V_M generated by A, while $(\ ,\ )$ and $\cdot$ are, respectively, the inner and algebra products of the Monster algebra restricted to V. Then A is centralised by $C_M(G)$ and so is the subalgebra V generated by A. Therefore we have the following.

Lemma 5.3 *For a representation $(V, (\ ,\ ), \cdot, A, G)$ based on an embedding of G into the Monster*

$$V \leq C_{V_M}(C_M(G)), \text{ in particular } dim(V) \leq dim(C_{V_M}(C_M(G))). \qquad \square$$

The dimension of $C_{V_M}(C_M(G))$ is clearly the multiplicity of the trivial 1-dimensional representation in the decomposition of V_M into $G\mathbb{C}$-irreducibles. This in turn can be calculated based on the fusion pattern of the conjugacy classes of G into that of the Monster group and the Monster character on V_M, the latter being the direct sum of 1-dimensional trivial and the 196 883-dimensional irreducible (whose character is known [CCNPW85]).

5.5.3 Harada–Norton Representation

We illustrate the aforementioned material through analysis of Majorana representations of the Harada–Norton group HN. It is known [CCNPW85] that HN is the centraliser in M of a D_{10}-subgroup:

$$HN = C_M(D_{10}).$$

It is also known that involutions τ in HN with centraliser of the form

$$C_{HN}(\tau) \cong 2.HS.2,$$

(where HS stands for the Higman–Sims sporadic simple group) are $2A$-involutions in the Monster. Therefore, HN possesses a (Monster-based) Majorana representation with the generating set of Majorana axes naturally corresponding to the conjugacy class

$$T := \{\tau^g \mid g \in HN\}$$

of involutions in HN of size $1\,539\,000$. The permutation action of HN on T by conjugation was analysed in [ILLSS95]. In that paper, it was shown that the action is multiplicity-free of rank 9, the irreducible constituents were identified by their dimensions, and the intersection parameters of an adjacency matrix of the corresponding association scheme were computed. In [FIM13], based on this information, the first eigenmatrix of the association scheme was reconstructed, as given in Table 7, where the last column gives the dimension of the irreducibles and the rows are ordered so that the values in the last column increase from the top to the bottom.

Let us first calculate the character-induced upper bound on the dimension of the representation. The Monster classes of $D_{10} = C_M(HN)$ are $1A$, $2A$, and $5A$, so that

$$C_{V_M}(C_M(HN)) = 1 + \frac{1}{10}(196\,883 + 5 \cdot 4371 + 4 \cdot 133) = 21\,928.$$

Table 7 Eigenmatix for HN

1	1408	2200	35200	123200	354816	739200	277200	5775	1
1	208	−50	2200	−2800	2016	4200	−6300	525	3344
1	−112	300	1000	−2200	−864	−1800	3600	75	8910
1	208	100	1000	1400	2016	−4200	0	−525	9405
1	128	200	0	1600	−2304	0	0	375	16929
1	−47	−50	250	350	−504	0	0	0	214016
1	28	−50	−50	−100	396	−750	450	75	267520
1	−32	40	−80	80	576	−240	−360	15	365759
1	16	4	−56	−136	−288	504	0	−45	653120

This bound shows that the 1-closure of the representation involves up to two non-trivial irreducibles from the last column and rows 2, 3, 4, and 5 of the extended first eigenmatrix (clearly it involves the trivial 1-dimensional representation). The exact module structure of the 1-closure comes from the distribution of the inner product vector.

The shape of the Monster-based representation of HN, as recovered in [FIM13], is

$$(1A, 5A, 2A, 3A, 4B, 5A, 6A, 4A, 2B)$$

with respect to the ordering of the classes of dihedral subgroups by the columns of the first eigenmatrix. Therefore, the inner product distribution vector is

$$d = \left(1, \frac{3}{2^7}, \frac{1}{2^3}, \frac{13}{2^8}, \frac{1}{2^6}, \frac{3}{2^7}, \frac{5}{2^8}, \frac{1}{2^5}, 0\right).$$

Now by direct calculations one observes that the inner product of d and a row of the first eigenmatrix is non-zero for rows 1, 3, and 4 only. The orthogonality with rows 6, 7, 8, 9 follows from the above upper bound on the dimension of the representation. Thus the dimension of the 1-closure is

$$1 + 8\,910 + 9\,405 = 18\,316.$$

This is how the dimension analysis works. As for the representation, it was eventually shown in [FIM22] that it is 1-closed and unique in the class of all representations with the given generating set.

Theorem 5.4 *The Harada–Norton group HN has a unique Majorana representation with the generating set of involutions being the conjugacy class of involutions with centraliser of the form 2.HS.2. This representation is based on an embedding into the Monster; it is 1-closed and 18 316-dimensional.* $\quad\square$

Table 8 Majorana representations of S_4

shape	$\bar{G}$	$C_M(G)$	dim(U)
$(2B,3A)$	S_4	$2^{11}.M_{23}$	13
$(2A,3A)$	S_4	$Sp_8(2)$	13
$(2B,3C)$	S_4	$^3D_4(2).3$	6
$(2A,3C)$	$S_4 \times 2$	$2^{1+8}_+.A_8$	9

Table 9 Majorana representations of $L_3(2)$

classes	$\bar{G}$	$C_M(G)$	dim(U)
$(2A,3A,7A)$	$L_3(2)$	$Sp_4(4).2$	49
$(2A,3C,7B)$	$L_3(2)$	$7^2.6A_4$	21
$(2B,3A,7A)$	$L_3(2)$	$2^2.M_{21}.S_3$	80
$(2B,3C,7X)$	$L_3(2)$	?	57

5.6 Representations of S_4

Historically, the first group for which all the Majorana representations were classified is the symmetric group S_4 [IPSS10]. The classes of $2A$-generated S_4-subgroups in the Monster are given in Table 8, where U is the Monster-embedded representation.

The algebras are 2-closed except for the one of shape $(2B,3A)$, which, besides six $2A$-axes and four $3A$-axes, contains three remarkable vectors:

$$\delta_{(ij)(kl)} = \sigma_i \cdot a_{(ij)} - \frac{1}{32}\sigma_i + \frac{1}{1024}a_{(ij)},$$

indexed by the cyclic subgroups of order 4 in S_4, and could be viewed as *shadows* of $4A$-axes in a larger algebra.

5.7 Representations of $L_3(2)$

Consider the second smallest non-Abelian simple group $L_3(2)$ of order 168. The Majorana representations of $L_3(2)$ are summarised in Table 9.

Important information on $L_3(2)$-subgroups in the Monster can be found in the relevant section of [N98]. The representations in the first and second rows were constructed and characterised in [IS10]. The representation in the fourth row was constructed by Justin McInroy and Sergey Shpectorov in [MS20]. The

Table 10 Fano plane labeling

a_{11}	a_{12}	a_{13}	u_{14}	u_{15}	u_{16}	u_{17}
a_{12}	u_{22}	u_{23}	a_{24}	a_{25}	u_{26}	u_{27}
a_{13}	u_{32}	u_{33}	u_{34}	u_{35}	a_{36}	a_{37}
u_{14}	a_{42}	u_{43}	a_{44}	u_{45}	a_{46}	u_{47}
u_{15}	a_{52}	u_{53}	a_{54}	a_{55}	u_{56}	a_{57}
u_{16}	u_{62}	a_{63}	a_{64}	u_{65}	u_{66}	a_{67}
u_{17}	u_{72}	a_{73}	u_{74}	a_{75}	u_{76}	u_{77}

representation in the third row was found by Clara Franchi and Mario Mainardis (private communication) inside the A_{12}-algebra.

Here we discuss some results and methods from [IS10]. The group $L_3(2)$ is the automorphism group of the Fano plane Π. The involutions a and the subgroups u of order 3 in $L_3(2)$ and naturally indexed by the flags and by the anti-flags of Π: $N_{L_3(2)}(a) \cong D_8$, $N_{L_3(2)}(u) \cong S_3$, shown in Table 10.

The representation of shape $(2A, 3C, 4B)$ is 1-closed (since all the 2-generated algebras involved are such). It is 21-dimensional since the Gram matrix of the generators is non-singular. This can be proved similarly to the A_5-case above within the association scheme of the flag-graph (distance-regular) of the Fano plane. The algebra product is closed on the linear span of the generators and is determined by the Norton-Sakuma algebra ($2A$, $3C$ or $4B$) generated by pairs of the generators.

In the representation of shape $(2A, 3A, 4B)$ the following lemma plays an important role.

Lemma 7.1 If $\alpha_1 = a_{24} + a_{25} - \frac{1}{4}a_{21}$, $\alpha_2 = a_{42} + a_{25} - \frac{1}{4}a_{12}$, then $a_{11} \cdot (\alpha_1 \cdot \alpha_2) = 0$.

Proof The two α's are the 0-eigenvectors of a_{11} inside two $4B$ algebras. By the fusion rules, their product is again a 0-eigenvector: □

$$\alpha_1 \cdot \alpha_2 = \frac{1}{16}(a_{24} + a_{42} + a_{25} + a_{52}) - \frac{1}{64}(a_{64} + a_{46} + a_{57} - a_{44} - a_{55}) +$$

$$\frac{1}{1024}(a_{11} - a_{13} - a_{31} - 31a_{21} - 31a_{12}) + \frac{1}{32}(a_{67} + a_{76}) -$$

$$\frac{1}{128}(a_{36} + a_{63} + a_{37} + a_{73}) + \frac{135}{8192}(u_{41} + u_{14} + u_{51} + u_{15})$$

$$- \frac{135}{2028}(u_{45} + u_{54}).$$

Since point and line stabilisers are isomorphic to S_4, whose Majorana representation (of the shape $(2A, 3A)$) is known [IPSS10], we know the products of a_{11} with every $2A$- or $3A$-axis in the first row or in the first column.

Since t_{11} swaps u_{45} and u_{54}, we also have

$$a_{11}(u_{45} - u_{54}) = \frac{1}{32}(u_{45} - u_{54}).$$

Summarising, we obtain:

$$a_{11} \cdot u_{45} = \frac{1}{64 \cdot 45} \begin{pmatrix} 128 & -32 & 32 & 45 & 45 & -45 & -45 \\ -32 & -45 & 45 & 0 & 0 & 0 & 0 \\ 32 & 45 & -90 & 0 & 0 & 0 & 0 \\ 45 & 0 & 0 & 32 & 45 & -32 & 0 \\ 45 & 0 & 0 & -45 & 32 & 0 & -32 \\ -45 & 0 & 0 & -32 & 0 & 0 & 64 \\ -45 & 0 & 0 & 0 & 32 & 64 & 0 \end{pmatrix}.$$

The product $u_{22} \cdot u_{33}$, coming by a resurrection principle, is described by an analogous matrix.

5.8 $\{3,4\}^+$-Transpositions

The following lemma was one of the first examples of application of Majorana Theory. It proves the property III in the introduction.

Theorem 8.1 *There is no Majorana representation of shape 2A of the elementary abelian group of order 8.*

Proof Let $a_0, \ldots, a_6$ be the Majorana axes, such that $\{a_0, a_i, a_{i+1}\}$ are lines of the Fano plane for $i = 1, 3, 5$. Then $a_i - a_{i+1}$ are $\frac{1}{4}$-eigenvectors of $\mathrm{ad}(a_0)$ by Table 6. Then, by the structure of the $2A$-algebra, we have

$$(a_1 - a_2) \cdot (a_3 - a_4) = 2(a_6 - a_5),$$

which is a $\frac{1}{4}$-eigenvector, contradicting the fusion rules. □

Theorem 8.2 *Let G be one of the groups $L_3(2)$, $G_2(2)'$, or $^3D_4(2)$. Then the following assertions hold:*

 (i) *G possesses a unique Majorana representation of shape $(2A, 3C, 4B)$;*
 (ii) *the Majorana axes are indexed by the points of the generalised hexagon $GH(2,t)$ associated with G, where $t = 1, 2$, and 8, respectively;*
(iii) *the lines of the $GH(2,t)$ are the 2A-subalgebras in the representation;*
(iv) *the dimension of the representation is 21, 63, and 793, respectively;*
 (v) *the representation of $L_3(2)$ is based on an embedding into the Monster and the other two are not.* □

Theorem 8.3 (F. Timmesfeld, 1973) *Let G be a finite group generated by a class D of $\{3,4\}^{+}$-transpositions. Let $Z_{*}(G)$ be the maximal coimage of $Z(G/O_2$ (G)). Then $G/Z_{*}(G)$ is isomorphic to $A_n(2)$; $n \geq 2$, $B_n(2)$; $n \geq 3$, $D_n(2)$, $^{2}D_n(2)$; $n \geq 4$, $G_2(2)'$, $^{3}D_4(2)$, $F_4(2)$, $^{2}E_6(2)$, $E_6(2)$, $E_7(2)$ or $E_8(2)$. Moreover, except in $F_4(2)$, D is the class of 'root elements' according to the long roots of the corresponding Lie algebra. In $F_4(2)$, the root elements for the short and long roots are $\{3,4\}^{+}$-transpositions. Furthermore, either G contains a D-closed subgroup of order 8 or $G \cong A_2(2) \cong L_3(2)$, $G_2(2)'$, or $^{3}D_4(3)$.*

Theorem 8.4 *Let G be a group whose maximal coimage of $Z(G/O_2(G))$ is trivial and suppose that G possesses a Majorana representation of shape $(2A, 3C, 4B)$ with respect to a conjugacy class of involutions in G. Then G is $L_3(2)$, $G_2(2)'$, or $^{3}D_4(2)$ and the representation is as in Theorem 8.2.* $\qquad\square$

5.9 3*A*-Axes

The 3*A*-axes have already appeared many times in our discussion as the vectors u_ρ in the Norton–Sakuma algebra of type 3*A*. It is appropriate to give the following explicit definition.

Definition 9.1 *Let $(V,(\ ,\ ),\cdot)$ be a Majorana algebra, let a_0, a_1 be Majorana axes such that $U = \langle\!\langle a_0, a_1 \rangle\!\rangle$ is of type 3A. Then*

$$u := \frac{64}{135}(2a_0 + 2a_1 + a_{-1} - 32\, a_0 \cdot a_1)$$

is a 3A-axis, where

$$a_{-1} = -\frac{78}{7}a_0 - \frac{512}{7}a_0 \cdot a_1 + \frac{2048}{7}a_0 \cdot (a_0 \cdot a_1) = a_1^{\tau(a_0)}$$

is the third Majorana axis in U.

The automorphism group of the 3*A*-algebra *U* is

$$S_3 \cong \langle \tau(a_0), \tau(a_1) \rangle,$$

acting naturally on $\{a_0, a_1, a_{-1}\}$ and stabilising *u*, so that any pair of Majorana axes can be taken as generators. The 3*A*-axis *u* is an idempotent of length $\frac{8}{5}$:

$$u \cdot u = u, \quad (u, u) = \frac{8}{5}.$$

In this section, we discuss the properties of 3*A*-axes in the Monster algebra and the possibility of deducing them from the Majorana axioms following [N96] and [C17].

In the action of the Monster M on its algebra $(V_M, (\ ,\ ), *)$, the stabiliser of a $3A$-axis u_ρ is the whole normaliser of the relevant order 3 subgroup:

$$N_M(\langle \rho \rangle) \cong 3 \cdot Fi_{24},$$

which is the non-split extension of the 3-group generated by ρ by the largest Fischer 3-transposition group Fi_{24} containing the simple sporadic Fisher group Fi'_{24} (projecting onto the centraliser of ρ) as an index 2 subgroup. The following is lemma 4 and theorem 3 in [N96], subject to Majorana rescaling.

Lemma 9.2 *Let u_ρ be a $3A$-axis in the Monster algebra, where ρ is the relevant element of order 3. Then the spectrum of $\mathrm{ad}(u_\rho)$ is*

$$1, \frac{1}{3}, \frac{1}{5}, \frac{1}{30}, 0$$

with the corresponding eigenspaces having dimension

$$1,\ 783 + 783,\ 8671,\ 64584 + 64584,\ 1 + 57477,$$

respectively, where the sums represent the decompositions of the eigenspaces into irreducibles of $N_M(\langle \rho \rangle) \cong 3 \cdot Fi_{24}$. The 1-eigenspace is spanned by u_ρ and the 1-dimensional component of the 0-eigenspace is spanned by the identity minus u_ρ. Furthermore, $C_M(\rho)$ has dimension $1 + 8671 + 1 + 57477$ and the ρ-fixed subalgebra possesses an involutory automorphism which negates the $\frac{1}{5}$-eigenspace and fixes the 1- and 0-eigenspaces. □

This lemma already puts considerable restriction on the fusion rules of the eigenspaces of $\mathrm{ad}(u_\rho)$. Further restrictions come in the VOA setting [Miy01] as follows.

Lemma 9.3 *In the Moonshine module $V^\#$, a $3A$-axis corresponds to a conformal vector of central charge $\frac{4}{5}$. The simple Virasoro module $L(\frac{4}{5}, 0)$, also known as the 3-state Pott model, has 10 irreducible modules. The restriction of the fusion rules to the irreducible modules $L(\frac{4}{5}, 0)$, $L(\frac{4}{5}, \frac{2}{5})$, $L(\frac{4}{5}, \frac{2}{3})$, and $L(\frac{4}{5}, \frac{1}{15})$ which intersect the Griess subalgebra V_2 of $V^\#$ are given (in terms of the Majorana eigenvalues) in Table 11.*

5.10 A_{12}

Majorana representations of A_{12} are of particular interest for several reasons (see [Iv18] and [FIM20]); in particular:

(a) A_{12} is the largest alternating subgroup of the Monster;

Table 11 Fusion rules for $3A$-axes

Sp	1	0	$\frac{1}{5}$	$\frac{1}{3}$	$\frac{1}{30}$
1	1	0	$\frac{1}{5}$	$\frac{1}{3}$	$\frac{1}{30}$
0	0	0	$\frac{1}{5}$	$\frac{1}{3}$	$\frac{1}{30}$
$\frac{1}{5}$	$\frac{1}{5}$	$\frac{1}{5}$	$1,0$	$\frac{1}{30}$	$\frac{1}{3},\frac{1}{30}$
$\frac{1}{3}$	$\frac{1}{3}$	$\frac{1}{3}$	$\frac{1}{30}$	$1,0,\frac{1}{3}$	$\frac{1}{5}\ \frac{1}{30}$
$\frac{1}{30}$	$\frac{1}{30}$	$\frac{1}{30}$	$\frac{1}{3},\frac{1}{30}$	$\frac{1}{5},\frac{1}{3}$	$1,0,\frac{1}{5},\frac{1}{3},\frac{1}{30}$

(b) most of the configurations, as in item II of Section 5.1, appear inside an A_{12}-subgroup, which is the centraliser of an A_5-subgroup of type $(2A, 3A, 5A)$ in the Monster;

(c) all $(2A, 3A)$ configurations except for the last four in Table 14 in the next section appear inside the A_{12}-subgroup in (b).

In the sequel, let $(V, (\ ,\ ),\ \cdot\ , A, A_{12})$ be a Majorana representation of A_{12} and T be the set of Majorana involutions associated to the axes in A; according to Section 5.4, the strategy for determining its structure is to accomplish the following steps:

(I) determine the possibilities for the set T;

(II) classify the possible shapes;

(III) identify the submodule of V spanned by the set A of axes;

(IV) calculate the submodule of V spanned by the sets of $3A$-, $4A$-, and $5A$-axes;

(V) determine the 2-closure of the linear span of A;

(VI) possibly show that the 2-closure of the linear span of A is the whole algebra.

This issue was first considered by A. Castillo-Ramirez and A. A. Ivanov who, in [CI14], showed that, under the hypothesis that the representation is based on the embedding in the Monster (so that T is the set of the involutions of cycle type 2^2 or 2^6; compare Table 13), the 2-closure has dimension 3960^{1} and gives a bound for the dimension of V. They also found several useful relations on $3A$, $4A$, and $5A$-axes which allow, in Step (III), to consider only the case of the $3A$-axes associated to elements of cycle type $(3,3)$ and of the $4A$-axes associated to elements of cycle type $(4,4)$.

In the general case, by [FIM22, lemma 3.1 and proposition 3.4] we have the following theorem which realised Step (I).

---

[1] Note that, due to a misprint, the dimension given in that paper is 3958.

Theorem 10.1 *The set T is either the class of involutions of cycle type 2^2 of A_{12}, or the union of classes of types 2^2 and 2^6.*

Step (II) was accomplished by [FIM22, proposition 3.4]: for each of the two possible choices of T there is a unique shape for the representation $(V, (\,,\,),\,\cdot\,, A, A_{12})$. In both cases, this is the same as the shape of the representation with respect to T induced by the embedding of A_{12} as the centraliser of an A_5-subgroup of type $(2A, 3A, 5A)$ of the Monster (see table 3 in [FIM22]).

To deal with Steps (III) and (IV), recall that, by Section 5.5.1, the submodules of V involved are factor modules of certain transitive permutation modules for A_{12}, such as the permutation modules over the set of its subgroups generated by permutations of cycle type 2^2, 2^6, 3^2, and 4^2. By corollary 6.3 in [FIM22], every $4A$-axis relative to a subgroup of A_{12} generated by a permutation of cycle type 4^2 is contained in the linear span of the Majorana axes and $3A$-axes. Thus we only need to deal with permutations of cycle type 2^2, 2^6, and 3^2. The structure of these permutation modules can be obtained by applying standard representation theory of the symmetric groups. We mention here that the permutation module of A_{12} over the set of its subgroups generated by cycles of type $(3, 3)$ was also obtained by applying a general method, which might be of independent interest, for studying the permutation modules of the symmetric groups which was developed in [FIM17].

Once the aforementioned permutation modules have been determined, one would ideally apply the machinery discussed in Section 5.5.1 to obtain the corresponding submodules of V. However, since the permutation modules arising in the cases 2^2 and 3^2 are not multiplicity-free, one needs to generalise the above machinery (see [FIM20]) to non-multiplicity-free actions. This was done in [FIM16b], getting Step (III) for T being the set of involutions of cycle type 2^2 and in [FIM22] if T also contains the set of involutions of cycle type 2^6. Here difficulties grow exponentially with the number of non-multiplicity-free components. In particular, to deal with the case of the $3A$-axes associated to elements of cycle type 3^2, more sophisticated arguments are needed (see [FIM22]).

Step (V) was accomplished by computing the intersections of the corresponding submodules of V, using the methods introduced in [FIM18] (see also [FIM20] for a brief description).

Lemma 10.2 *The decomposition into A_{12}-irreducible submodules of the 2-closure of the set A is given as*

(i) *in the sixth column of Table 12, if T is the set of the permutations of cycle type 2^2;*

Table 12   Irreducible $\mathbb{R}[A_n]$-submodules and their multiplicities in V° and in V

	V°					V
λ	$n=8$	$n=9$	$n=10$	$n=11$	$n=12$	
(n)	2	2	2	2	1	1
$(n-1,1)$	2	2	2	2	1	1
$(n-2,2)$	3	3	3	2	1	1
$(n-3,3)$	2	2	2	2	1	1
$(n-4,4)$	1	1	1	1	1	1
$(n-3,2,1)$	1	1	1	1	1	1
$(n-4,2,2)$	2	2	2	1	1	1
$(n-5,2,2,1)$	1	1	1	1	0	0
$(n-6,2,2,2)$	1	1	1	1	1	1
$(n-4,1,1,1,1)$	1	1+1	1	1	0	0
$(n-5,1,1,1,1,1)$	1	1	1+1	1	1	1
$(4,4,4)$						1
$\dim(V_n^\circ)$	462	1008	2052	3498	3498	3960

(ii) *in the seventh column of Table 12, if T is the union of permutations of cycle types 2^2 and 2^6.*

We now focus on saturated Majorana representations, which, in this case, means that T is the full set of the permutations of cycle types 2^2 and 2^6. In particular, we get that the 2-closure of A is the direct sum $U \oplus W$ of the linear span U of A with one irreducible A_{12}-submodule, W, isomorphic to the Specht module relative to the partition $(7, 1^5)$. In order to show that $U \oplus W$ is the whole algebra (and thus accomplish Step (VI)), we need to show that it contains all products between Majorana axes and $3A$-axes and between pairs of $3A$-axes. Using the projection of the $3A$-axes onto W (with respect to the above decomposition) and a recursive argument, in [FIM22, sections 10, 11] the computation of the products was reduced within a Majorana representation of A_7, which is 2-closed by proposition 2.7 in [Iv11b].

Theorem 10.3 *There is, up to equivalence, a unique saturated Majorana representation $(V, (\,,\,), \cdot, A, A_{12})$. In particular:*

(1) *it is based on the embedding of A_{12} in the Monster as the centraliser of a $(2A, 3A, 5A)$-subgroup isomorphic to A_5.*

(2) *V has dimension $3,960$ and its decomposition into irreducible submodules is given in the last column of Table 12,*

(3) *V is 2-closed and it is spanned by the Majorana axes and the $3A$-axes associated to permutations of cycle type 3^2.*

Table 13 Fusion of A_{12}-classes into the Monster

A_{12}	M	A_{12}	M	A_{12}	M	A_{12}	M
(1^{12})	$1A$	$(4,2)$	$4B$	$(6,2)$	$6C$	$(4,3,2)$	$12C$
(2^2)	$2A$	$(4,2^3)$	$4B$	(7)	$7A$	$(4,3^2,2)$	$12C$
(2^6)	$2A$	(5)	$5A$	$(8,2)$	$8B$	$(6,4)$	$12C$
(2^4)	$2B$	(5^2)	$5A$	$(8,4)$	$8B$	$(7,2^2)$	$14A$
(3)	$3A$	$(3,2^2)$	$6A$	(9)	$9A$	$(5,3)$	$15A$
(3^2)	$3A$	$(3^2,2^2)$	$6A$	$(9,3)$	$9A$	$(5,3^2)$	$15A$
(3^4)	$3A$	$(6,2^3)$	$6A$	$(5,2^2)$	$10A$	$(5,4,2)$	$20B$
(3^3)	$3B$	(6^2)	$6A$	$(10,2)$	$10A$	$(7,3)$	$21A$
(4^2)	$4A$	$(6,3,2)$	$6B$	(11)	$11A$	$(5,3,2^2)$	$30B$
$(4^2,2^2)$	$4A$	$(3,2^4)$	$6C$	$(4^2,3)$	$12A$	$(7,5)$	$35A$

The case of the non-saturated representation of A_{12}, that is when T, as the set of the permutations of cycle type 2^2, is still open. More generally, for the alternating groups A_n, with $8 \leq n \leq 12$, the following holds.

Theorem 10.4 *Let $n \in \{8, \ldots, 12\}$, let $G = A_n$, and let $(V, (\ , \ ), \ \cdot \ , A, G)$ be a Majorana representation of G where, in case $n = 12$, we assume that φ is not saturated. Denote by V° the 2-closure of A. Then*

(i) *the Majorana involutions are the bitranspositions of G;*

(ii) *the shape is unique for each G;*

(iii) *the restriction of the inner product on V° is unique for each G;*

(iv) *the irreducible $\mathbb{R}[G]$-submodules of V° and their multiplicities are the ones given in Table 12.*

The following conjecture was stated in [FIM22].

Conjecture 10.5 *Let $n \in \{8, \ldots, 12\}$, let $G := A_n$ and let $(V, (\ , \ ), \ \cdot \ , A, G)$ be a Majorana representation of G such that the set of Majorana involutions is the set of bitranspositions of G, and let V° be the 2-closure of A. Then $V \cong V^\circ \oplus S^{(4^2, n-8)}$.*

We conclude the section with the fusion table of conjugacy class of A_{12} (centralized by an A_5-subgroup of type $(2A, 3A, 5A)$) into the classes of the Monster M. This table is very useful, widely known, while the particular form in Table 13 is taken from [C17].

5.11 Perspectives in Majorana Theory

5.11.1 Atlas of Representations

By now, a considerable number of Majorana algebras have been constructed. Most of the original constructions rely on computer calculations [S12], [PW18], [MS21]. Some of the algebras are embedded in the Monster algebra; some are not. It would be helpful to organise this information in a convenient form including pairwise embeddings, dimension, nice generating sets, etc.

5.11.2 Representations of "Medium" Groups

By now, the Majorana representations of small simple groups are classified. When the group is large, the representations can be identified with that in the Monster, as was done for HN group. The difficult part we now face is representations of "medium" groups, a typical one being $Sp_6(2)$ (roughly from the middle of ATLAS [CCNPW85]). The problem seems to be with $3A$- and $4A$-axes: there are dozens of orbits of pairs, each with different inner products and so on. An extensive catalogue was produced by William Giuliano. Some progress can be made through a more detailed study of the properties of these axes and possibly their inclusion in the axiomatic. Recently identified representation of $U_3(5)$ [BI23] is a step in this direction.

5.11.3 $2 \cdot M_{22}.2$

The alternating group A_{12} is the centraliser in the Monster of an A_5-subgroup of shape $(2A, 3A, 5A)$. For an A_5 of shape $(2B, 3A, 5A)$ the centraliser is $2 \cdot M_{22}.2$ [N98]. The Majorana representation of this group is currently studied by Clara Franchi, Mario Mainardis, and the author.

5.11.4 Madeleine's Infinite Series and Representations of Extra-Special Groups

In [W21] Madeleine Whybrow constructed a one-parameter infinite series of Majorana algebras. These algebras are representations of the group $Z_4 * Q_8$. The Monster-embedded representation appears for the limit value of the parameter. It would be interesting to systematically classify the representations of extra-special groups and of their central products with Z_4. This project was started in a second-year project supervised by the author [CCSZ22].

5.11.5 3-Transposition Groups and their Majorana Representations

Let (G, D) be a 3-transposition group with transposition set D, and let B be the set of involutions in G which are products of commuting transpositions from D. Then $D \cup B$ is a 6-transposition set [N17], which can be taken as a starting point for the Majorana representation of G. The case $G \cong S_n$ is related to the bitransposition as in Section 5.10. The author has long been interested in which 3-transposition groups possess such Majorana representations and conjectured that only those embedded in the Monster do. Since 3-transposition groups from infinite series contain large symmetric subgroups, and since A_{13} [FIM22] does not possess a Majorana representation, we are left with a finite set of groups. Recently Albert Gevorgyan [G23] made a crucial progress in proving the conjecture.

5.11.6 Connections to Y-Groups

Another amazing feature of the Monster is in the Y-group conjecture proved by S. P. Norton and the author in 1990 during the Durham Symposium (cf. [N17]). It would be nice to bridge a connection between Y-groups and Majorana representations. Since in both cases A_{12} is the largest subgroup, this is one connection. Another approach could be constructing Majorana representations for the sequence of Y-groups reaching the Bi-Monster. Then one should start with the Weyl group of E_6, which is also known as $U_4(2)$ or $O_6^-(2)$. Majorana representations of this group are studied by Andrei Mamontov in Novosibirsk and the case clearly exhibits features of "medium" groups.

5.11.7 3A-Axes

The properties of 3A-axes in the Monster discussed in Section 5.10 can be axiomatised towards the relevant axes in 3A-subalgebras within the Majorana theory or more abstractly as axes associated to certain subgroups of order 3 in the target group. Some attempts in this direction were made in [C17] and they should be extended.

5.11.8 $(2A, 3A)$-Pairs

A very important project is related to the algebras generated by a 2A-axis a_t and a 3A-axis u_h. The corresponding pairs of axes in the Monster algebra have been classified by S. Norton up to conjugation and they correspond to rows in

Table 14 Pairs of $(2A, 3A)$-axes

$\langle t, h \rangle$	$(th)^M$	(a_t, u_h)	(u_h, u_h^t)
S_3	$2A$	$\frac{1}{4}$	$\frac{5}{8}$
6	$6A$	0	$\frac{5}{8}$
A_4	$3A$	$\frac{1}{9}$	$\frac{136}{405}$
S_4	$4B$	$\frac{1}{36}$	$\frac{136}{405}$
A_5	$5A$	$\frac{1}{18}$	$\frac{16}{405}$
$2 \times A_4$	$6C$	$\frac{1}{45}$	$\frac{136}{405}$
$3 \times S_3$	$6A$	$\frac{1}{20}$	0
$L_2(7)$	$7A$	$\frac{1}{24}$	$\frac{4}{81}$
$3 \times A_5$	$15A$	$\frac{11}{360}$	$\frac{188}{2025}$
S_4	$4A$	$\frac{13}{180}$	$\frac{168}{2025}$
$3 \times S_4$	$12C$	$\frac{1}{36}$	$\frac{64}{405}$
$2 \times L_2(7)$	$14A$	$\frac{11}{360}$	$\frac{228}{2025}$
$2 \times A_4$	$6A$	$\frac{2}{45}$	$\frac{168}{2025}$
$S_3 \times A_4$	$6B$	$\frac{17}{360}$	$\frac{68}{2025}$
$2 \times A_5$	$10A$	$\frac{1}{30}$	$\frac{208}{2025}$
$3^3.A_4$	$9A$	$\frac{13}{360}$	$\frac{148}{2025}$
$4^2.S_3$	$8B$	$\frac{1}{36}$	$\frac{72}{405}$
$L_2(11)$	$11A$	$\frac{1}{30}$	$\frac{168}{2025}$
$3 \times S_3$	$6D$	$\frac{1}{40}$	$\frac{4}{25}$
$GL_2(3)$	$8C$	$\frac{1}{36}$	$\frac{64}{675}$
$4.A_4$	$12A$	$\frac{1}{30}$	$\frac{56}{675}$
$2.4^2.S_3$	$8A$	$\frac{7}{180}$	$\frac{56}{675}$

Table 5.14, which is a rescaled extract from table 3 in [N96]. This project can be realised in three stages:

(I)   describe the $(2A, 3A)$-generated subalgebras in the Monster up to isomorphism;

(II) classify the $(2A, 3A)$-generated subalgebras within Majorana theory extended by axiomatisation of the $3A$-axes as in Section 5.9;

(III) classify the $(2A, 3A)$-generated subalgebras purely within Majorana theory.

Task (III) was attempted by the author with very little progress. Task (I) is under realisation and the intermediate task, (II), was never seriously attempted.

5.11.9 4A-Axes

The 4A-axes were considered in [N96] along with 2A- and 3A-axes. The fixed subalgebra of the relevant element of order 4 contains an extra automorphism of order 2. This should result in some fusion rule, similar to the fusion rules of the Majorana axes and of 3A-axes as in Section 5.10. A fragment of the fusion rule of 4A-axes can be found in [W21]. The abstract theory of 4A-axes is waiting for its development. The starting point could be lemma 5 in [N96]. Such a theory, when completed, would make it easier to understand the Majorana representations of A_8.

Acknowledgements

I am extremely thankful to Clara Franchi and Mario Mainardis for Section 5.10, to Albert Gevorgyan for the list of referencies on Majorana Theory, and to William Giuliano for a most careful proofreading of drafts.

References

[BI84] E. Bannai and T. Ito, *Algebraic Combinatorics: I. Association Schemes*, Benjamin, Menlo Park, California, 1984.

[BPZ84] A. A. Belavin, A. M. Polyakov and A. B. Zamolodchikov, Infinite conformal symmetry in two-dimensional quantum field theory, *Nuclear Physics* **B241** (1984), 333–380.

[C85] J. H. Conway, A simple construction for the Fischer–Griess monster group, *Invent. Math.* **79** (1985), 513–540.

[CCNPW85] J. H. Conway, R. T. Curtis, S. P. Norton, R. A. Parker and R. A. Wilson, *Atlas of Finite Groups*, Clarendon Press, Oxford, 1985.

[FLM88] I. B. Frenkel, J. Lepowsky and A. Meurman, *Vertex Operator Algebras and the Monster*, Academic Press, Boston, 1988.

[GAP4] The GAP Group, GAP – Groups, Algorithms, and Programming, Version 4.4.12, 2008. (http://www.gap-system.org)

[Gr82] R. L. Griess, The friendly giant, *Invent. Math.* **69** (1982), 1–102.

[GMS89] R. L. Griess, U. Meierfrankenfeld and Y. Segev, A uniqueness proof for the Monster, *Ann. Math. (2)* **130** (1989), 567–602.

[Iv92] A. A. Ivanov, A geometric characterization of the Monster, in *Groups, Combinatorics and Geometry*, Durham 1990, LMS Lecture Notes **165**, Cambridge Univ. Press, 1992, 46–62.

[Iv93] A. A. Ivanov, Constructing the Monster via its Y-presentation, in *Paul Erdös is Eighty*, Keszthely, Budapest, 1993, 253–269.

[Iv98]   A. A. Ivanov, *Geometries of Sporadic Groups: I. Petersen and Tilde Geometries*, Cambridge Univ. Press, 1998.

[ILLSS95] A. A. Ivanov, S. A. Linton, K. Lux, J. Saxl and L. H. Soicher, Distance transitive graphs of the sporadic groups, *Comm. Alg.* **23** (1995), 3379–3427.

[IS02]   A. A. Ivanov and S. V. Shpectorov, *Geometries of Sporadic Groups: II. Representations and Amalgams*, Cambridge Univ. Press, 2002.

[Iv05]   A. A. Ivanov, Constructing the Monster amalgam, *J. Algebra* **300** (2005), 571–589.

[Iv21]   A. A. Ivanov, Locally projective graphs and their densely embedded subgraphs, *Beitr. Algebra Geom.* **62** (2021), 363–374.

[Iv22]   A. A. Ivanov, A characterization of the Mathieu–Conway–Monster series of locally projective graphs, *J. Algebra* **607A** (2022), 426–453.

[Miy96]   M. Miyamoto, Griess algebras and conformal vectors in vertex operator algebras, *J. Algebra* **179** (1996), 523–548.

[Miy01]   M. Miyamoto, 3-state Pott model and automorphisms of vertex operator algebras of order 3, *J. Algebra* **239** (2001), 56–76.

[Miy03]   M. Miyamoto, Vertex operator algebras generated by two conformal vectors whose τ-involutions generate S_3, *J. Algebra* **268**, 653–671.

[Miy04]   M. Miyamoto, A new construction of the moonshine vertex operator algebra over the real number field, *Ann. of Math.* **159** (2004), 535–596.

[N82]   S. P. Norton, The uniqueness of the monster, in *Finite Simple Groups, Coming of Age* (J. McKay, ed.), Contemp. Math., **45**, pp. 271–285, AMS, Providence, RI, 1982.

[N85]   S. P. Norton, The uniqueness of the Fischer–Griess Monster, in *Finite Groups, Coming of Age* (J. McKay, ed.), Contemp. Math., **45**, pp. 271–285. AMS, Providence, RI, 1985.

[N96]   S. P. Norton, The Monster algebra: Some new formulae, in *Moonshine, the Monster and Related Topics*, Contemp. Math. **193**, pp. 297–306, AMS, Providence, RI, 1996.

[N98]   S. P. Norton, Anatomy of the Monster I, in *The Atlas of Finite Groups: Ten Years On*, LMS Lect. Notes Ser. 249, pp. 198–214, Cambridge Univ. Press, Cambridge, 1998.

[N17]   S. P. Norton, The Monster is fabulous, in *Finite Simple Groups: Thirty Years of the Atlas and Beyond*, Contemp. Math., **694**, pp.3–10, AMS, Providence, RI, 2017.

[RS84]   M. A. Ronan and G. Stroth, Minimal parabolic geometries for the sporadic groups, *Europ. J. Combin.* **5** (1984), 59–91.

[Sak07]   S. Sakuma, 6-Transposition property of τ-involutions of Vertex Operator Algebras, *International Math. Research Notes* **2007**, article rnm030.

[T73]   F. G. Timmesfeld, A characterization of the Chevalley and Steinberg groups over F_2, *Geom. Dedicata* **1** (1973), 269–321.

[Th79]   J. G. Thompson, Uniqueness of the Fischer–Griess Monster, *Bull. London Math. Soc.* **11** (1979), 340–346.

Publications on Majorana Theory in Chronological Order

[Iv09] A. A. Ivanov, *The Monster Group and Majorana Involutions*, Cambridge University Press, Cambridge 2009.

[IPSS10] A. A. Ivanov, D. V. Pasechnik, Á. Seress and S. Shpectorov, Majorana representations of the symmetric group of degree 4, *J. Algebra* **324** (2010), 2432–2463.

[Iv11a] A. A. Ivanov, On Majorana representations of A_6 and A_7, *Comm. Math. Phys.* **306** (2011), 1–16.

[Iv11b] A. A. Ivanov, Majorana representation of A_6 involving $3C$-algebras, *Bull. of Math. Sci.* **1** (2011), 365–378.

[IS12] A. A. Ivanov and Á. Seress, Majorana representations of A_5, *Math. Z.* **272** (2012), 269–295.

[IS10] A. A. Ivanov and S. Shpectorov, Majorana representations of $L_3(2)$, *Adv. Geom.* **14** (2012), 717–738.

[S12] Á. Seress, Construction of 2-closed M-representations, *Proc. of the 37th Int. Symp. Symb. and Alg. Comp.* (2012), 311–318.

[D12] S. Decelle, The $L_2(11)$-subalgebra of the Monster algebra, *Ars Math. Cont.* **7** (2012), 83–103.

[D13] S. Decelle, Majorana representations and the Coxeter groups $G^{(m,n,p)}$, PhD Thesis, Imperial College 2013.

[C13] A. Castillo-Ramirez, Idempotents of the Norton-Sakuma algebras, *J. Group Theory* **16** (2013), 419–444.

[K13] S. M. S. Khasraw, On Majorana algebras for the group S_4, Master Thesis, University of Birmingham, 2013.

[CR13] A. Castillo-Ramirez, Associative subalgebras of the Norton–Sakuma algebras, *ArXiv* (2013).

[FIM13] C. Franchi, A. A. Ivanov and M. Mainardis, Computing the dimension of a Majorana representation of the Harada–Norton group *Quad. del sem. matem.* (2013), 1–7.

[CI14] A. Castillo-Ramirez, A. A Ivanov, The axes of a Majorana representation of A_{12}, in *Groups of Exceptional Type, Coxeter Groups and Related Geometries: Springer Proceedings in Mathematics & Statistics*, **82**, New Delhi, 2014, 159–188.

[CL14] H. Y. Chen and C. Lam, An explicit Majorana representation of the group $3^2 : 2$ of $3C$-pure type, *Pacific J. Math.*, **271** (2014), 25–51.

[C14] A. Castillo-Ramirez, On Majorana algebras and representations, PhD Thesis, Imperial College, 2014.

[HRS15] J. I. Hall, F. Rehren and S. Shpectorov, Universal axial algebras and a theorem of Sakuma, *J. Algebra*, **421** (2015), 394–424.

[C15] A. Castillo-Ramirez, Associative subalgebras of low-dimensional Majorana algebras, *J. Algebra*, **421** (2015), 119–135.

[HRSh15] J. I. Hall, F. Rehren and S. Shpectorov, Primitive axial algebras of Jordan type, *J. Algebra*, **437** (2015), 79–115.

[R15] F. Rehren, Axial algebras, PhD Thesis, University of Birmingham, 2015.

[K15] S. M. S. Khasraw, M-axial algebras related to 4-transposition groups, PhD Thesis, University of Birmingham, 2015.

[FIM16a] C. Franchi, A. A. Ivanov, M. Mainardis, The $2A$-Majorana representation of the Harada–Norton group, *Ars Math. Contemp.*, **11** (2016), 175–187.

[FIM16b] C. Franchi, A. A. Ivanov and M. Mainardis, The $2A$-Majorana representations of the Harada–Norton group, *J. Algebraic Comb.*, **44** (2016), 265–292.

[FIM17] C. Franchi, A. A. Ivanov, M. Mainardis, Permutation modules for the symmetric group, *Proc. Amer. Math. Soc.*, **145**, Number 8 (2017), 3249–3262.

[Iv17] A. A. Ivanov, Majorana representation of the Monster group: *Finite Simple Groups: Thirty Years of the Atlas and Beyond*, Contemp. Math., **694**, Amer. Math. Soc., Providence, RI, (2017), 11–17.

[C17] C. S. Lim, From the Monster to Majorana: A study of $3A$-axes, PhD Thesis, Imperial College, 2017.

[PW18] M. Pfeiffer and M. Whybrow, Constructing Majorana reprsentations, *ArXiv* (2018).

[Iv18] Ivanov, A. A., The future of Majorana theory, in *Group Theory and Computation*, Indian Statistical Institute Series, Springer, Singapore, 2018, 107–118.

[W18a] M. Whybrow, Majorana algebras generated by a $2A$ algebra and one further axis, *J. Group Theory* (2018), 417–437.

[W18b] M. Whybrow, Majorana algebras and subgroups of the Monster, PhD Thesis, Imperial College London, 2018.

[FIM18] C. Franchi, A. A. Ivanov and M. Mainardis, Radicals of S_n-invariant positive definite hermitian forms, *Algebraic Comb.*, **4** (2018), 425–440.

[W21] M. Whybrow, An infinite family of axial algebras, *J. Algebra*, **577** (2021), 1–31.

[MSW19] A. Mamontov, A. Staroletov and M. Whybrow, Minimal 3-generated Majorana algebras, *J. Algebra*, **524** (2019), 367–394.

[MS20] J. McInroy and S. Shpectorov, Minimal 3-generated Majorana algebras, *J. Algebra*, **550** (2020), 379–409.

[FIM20] C. Franchi, A. A. Ivanov and M. Mainardis, Majorana representations of finite groups, *Alg. Coll.*, **27** (2020), 31–50.

[MS21] J. McInroy and S. Shpectorov, From forbidden configurations to a classification of some axial algebras of Monster type, *ArXiv* (2021).

[I21] A. A. Ivanov, A practical course in Majorana Theory, Lecture Notes, TGMC, Yichang, China, 2021.

[KMS22] S. M. S Khasraw, J. McInroy and S. Shpectorov, Enumerating 3-generated axial algebras of Monster type, *J. Pure App. Alg.*, **226** (2022).

[FIM22] C. Franchi, A. A. Ivanov and M. Mainardis, Saturated Majorana representations of A_{12}, *Trans. Amer. Math. Soc.* **375** (2022) 5753–5801. A. A. Ivanov, Closed Majorana representations of $3, 4^+$-transposition groups, *Adv. Geom.* **22** (2022), 487–494.

[BI23] A. E. Brouwer and A. A. Ivanov, Majorana algebra for the Hoffman–Singleton graph, *Geom. Dedic.* (2023) 217:4.

[G23] A. Gevorgyan, Towards the standard Majorana representations of 3-transposition groups, *Algebra Colloquium* **217**, 4 (2023).

[CCSZ22] D. Choudhury, D. Choudhury, A. Sheng and C. Zhou, Constructions of Majorana algebras, Project, Imperial College London, 2022.

Part II

Algebraic Combinatorics

6

The Geometry of the Freudenthal–Tits Magic Square

Hendrik Van Maldeghem

Dedicated to the memory of Jacques Tits

Abstract

We review some geometric properties of the Freudenthal–Tits Magic Square, considered as a square of buildings, incidence geometries and varieties, as partly originally defined by Jacques Tits. In particular, we establish new links between the vertical and horizontal layers.

CONTENTS

6.1 Introduction

On pages 141 and 142 of [28], Jacques Tits introduced two tables, each consisting of twelve geometries, more exactly varieties in a projective space, ordered in three rows of four entries (a 3×4 matrix). The last column of each table consists of geometries of exceptional type: the first table contains real forms of F_4, E_6 and E_7, respectively, while the second table is the complexification of the first one, which, in modern terms, just means the split form (in French "forme déployée", which rather means "unfolded form"). Each of the geometries in the last column is constructed using an octonion algebra; in the first table the algebra is a division algebra, in the second table it is the corresponding split algebra. The geometries in the preceding columns are the analogues of the ones in the last column, defined over the real field, the complex field or its split variant $\mathbb{R} \times \mathbb{R}$, and the quaternion skew field or its split variant, the algebra of real 2×2 matrices. The goal of displaying these tables is to put forward the question whether there exists a fourth row completing the tables to squares, such that the last entry is a form of F_8 involving an octonion algebra, and the other entries are analogues of that over the real, complex and quaternion algebras, respectively. Jacques Tits then went on and predicted how these geometries (before complexification) should look and which forms of E_8 they should represent, which dimensions they have, etc. Today we know exactly which geometries they are: they are so-called metasymplectic spaces. But still unsolved is a direct construction of the corresponding varieties. The ultimate goal of this geometric investigation of the Magic Square is to eventually fill that gap. However, the way to the solution of this problem is paved with diamonds and pearls of beautiful connections and geometrical delights. Some of these I want to share with the reader of this chapter, which aims to tell a coherent story rather than collect mathematical statements in a formal way.

What one cannot find in this chapter is the approach via the Lie algebras. Although also very interesting, we have to draw the line somewhere, and we prefer to do it between geometry and algebra. The "magic" in the Magic Square is the symmetry in the table despite the asymmetric construction of the Lie algebras. However, we will see that geometry provides enough magic in this sense.

6.2 Structure of the Chapter

This chapter does not intend to prove a main result, but just to describe the beauty hidden in the Square. We will introduce quite some different shades of the Magic Square. Traditionally, there is the split form of the Magic Square, and there is the non-split form. Since we will introduce some variants of each of them, we will give them different names to distinguish them. Each variant takes into account the form in which we consider a given geometric or combinatorial structure. Roughly, there are three levels, ranging from general to specific (in reverse chronological historical order of introduction in the literature). The first level is the level of the theory of buildings. At the time Jacques Tits wrote about the Magic Square, these were not explicitly around yet, but they provide a very suitable framework to start with. The non-split (called "Relative") and split form (called "Absolute"; terminology taken from algebraic groups) are defined in **Section 6.3** and **Section 6.4**, respectively. We assume the reader is somewhat familiar with building theory; we do not define buildings here but instead refer to the literature, in particular the seminal book of Jacques Tits [30]. It is best for the present chapter to view buildings as simplicial complexes rather than chamber systems.

In **Section 6.5**, we put the two previous forms in one Square by considering the corresponding indices (terminology taken from Tits [29]); nowadays sometimes called Tits indices or Tits diagrams. We interpret these diagrams in the combinatorial way à la Mühlherr, Petersson & Weiss [17] as fixed point diagrams, briefly called *fix diagrams*. This will enable us to define the *Delayed Magic Square* later on in Subsection 6.9.2. The Fix Magic Square bears the ultimate magic in its twisted symmetry, which is explained in Section 6.5.

Geometries related to buildings existed before the buildings were defined. In fact, "Incidence Geometry" as a research field found its origin exactly in Tits' thesis [28] where he introduced his Magic Square. This geometric point of view is slightly less general than the building point of view in that it favours a certain type of vertices, called the "points". Incidence geometry detaches the points from the subspaces they belong to by introducing an incidence relation. As a consequence, geometries do not necessarily have to live inside some projective space as a kind of variety. We define geometries for the non-split form in **Section 6.6**, yielding the Relative Geometric Magic Square, and also for the split form in **Section 6.7**, yielding the Complexified Geometric Magic Square (terminology from Tits [28]). We mention some remarkable characterization results singling out exactly the complexified geometries and their residues.

In **Section 6.8** we return to the origin and define the Magic Square as a 4×4 table of varieties, that is, point-line geometries embedded in projective space, and defined with algebraic formulas. Nevertheless, we also provide some purely geometric constructions. In this section, we look at the Square row by row.

Finally, in **Section 6.9**, we point out some connections between the different rows and columns of the Magic Square. We show that one can walk from South-East to North-West by taking residuals and equator geometries (**Subsection 6.9.1**). We introduce the Delayed Magic Square (**Subsection 6.9.2**) as explained before, and point out two instances in which this Delayed Magic Square puts itself in the frontline – by studying the minimal number of quadrics intersecting in a variety of the second row (**Subsection 6.9.3**), and by studying so-called domestic automorphisms (**Subsection 6.9.4**), giving rise to a second interpretation of the fix diagrams as opposition diagrams. In **Subsection 6.9.5** at last we bring together some conclusions about the global connectivity of the Square, made obvious in the present chapter.

6.3 The Relative Magic Square

The Magic Square has many different forms, and each one usually has different interpretations. Let us review some of these (and remember we restrict ourselves to geometric approaches).

On the diagram, level, the basic idea of the Magic Square is building up to the diagram of type F_4 via subdiagrams. This is established using the sequence $A_1 - A_2 - C_3 - F_4$. Notice that this is not in conformity with the Bourbaki labelling of the diagram of type F_4, as it starts with the vertex labelled 4 rather than with the vertex labelled 1.

Before we draw any Dynkin diagram, we would like to comment on the way these diagrams are usually drawn. Recall that each node of a Dynkin diagram represents a fundamental root of a crystallographic root system. Two nodes are not connected if the corresponding roots are perpendicular; otherwise we connect them with an edge, double edge, or triple edge according to whether the roots form an angle of 120, 135 or 150 degrees. Since Tits' thesis [28], one also furnishes every multiple edge with an arrow pointing to the smaller root. Dynkin diagrams have then be generalized to Coxeter diagrams, where nodes represent involutive generators, joined by an edge of weight n if the product of the generators has order n. In the diagram, one connects the nodes with an $(n-2)$-fold bond. Except for the case of $n = 6$, the Coxeter diagram of the Weyl group of a root system is exactly the Dynkin diagram where the arrows are removed. Now, Tits [27] already interpreted the Coxeter diagrams as diagrams belonging to the geometry of Lie groups corresponding to the given Dynkin diagrams. This idea was formalized in full generality by Buekenhout

[3], and so we obtain Buekenhout diagrams, constructed as follows: each node represents a class of objects in an incidence geometry, two nodes being joined by an $(n - 2)$-fold edge if the residue of each flag of corresponding cotype is a generalized n-gon. It turns out that the Buekenhout diagram of the natural geometry of a simple algebraic group is the same as the Coxeter diagram of its Weyl group. Hence, as one can see, in the geometry the arrows of the Dynkin diagrams are completely useless. However, there remains some geometric distinction between types C_n and B_n, which is a consequence of the commutation relations of the root groups. In the context of the Square, it is convenient to reintroduce this distinction for geometries as well. We do this by also considering the rank 1 residues, and distinguishing between "projective lines" and "polar lines". The idea is that a projective line will represent a line in a projective space, whereas a polar line will represent a polar space of rank 1 in a projective space. Examples of the latter are conics and Hermitian curves; hence point sets related to quadratic or Hermitian forms of Witt index 1. Heuristically, the points of a projective line are pairwise collinear, whereas every pair of points of a polar line is "symplectic" (at "distance 2" although there is no path of length 2 around; this "distance" must rather be thought of as a "grading", as for Lie algebras, and it does have a close connection with some 5-graded Lie algebras). The general rule is then that the rank 1 residues corresponding to short roots are polar lines, whereas the others are projective lines. Nodes corresponding to polar lines will be represented by white nodes, the other by black ones.

Viewed in this manner, we can build up to F_4 in the following way:

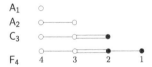

We have added the Bourbaki labelling for F_4 for clarity.

Now, these diagrams do not represent unique buildings, even if we specified a field. Let us focus for a moment on the F_4 diagram. It is well known and follows from the classification of buildings of that type, that the projective plane corresponding to the residues of type $\{1, 2\}$ (the plane represented by the black nodes) is always a plane over a field $\mathbb{K}$, whereas the projective plane corresponding to the residues of type $\{3, 4\}$ is defined over a quadratic alternative division algebra over $\mathbb{K}$. Discarding the exceptional case of an inseparable field extension in characteristic 2, there are exactly four types of such algebras. And here is where the magic starts: F_4 has rank 4 and the corresponding buildings come in 4 shades. Equality of two numbers in geometry is never a coincidence, or at least has an exceptional consequence! With this information, we can already make a perfect Square. Let us mention the corresponding algebra below the nodes. We let $\mathbb{K}$ be the ground field, and then we denote by $\mathbb{L}$ an arbitrary quadratic Galois extension of $\mathbb{K}$, by $\mathbb{H}$ an arbitrary quaternion division algebra over $\mathbb{K}$ and by $\mathbb{O}$ an arbitrary octonion division algebra (Cayley–Dickson algebra) over $\mathbb{K}$. We order the algebras in increasing complexity (or increasing dimension over $\mathbb{K}$):

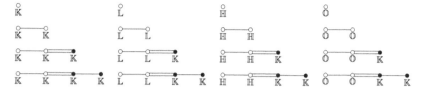

and call this table the *Relative Magic Square*. Its buildings are now all well defined and unique, given the algebraic structures, except that, in order to really have geometries and not "only" buildings, we need to decide which node we declare to correspond to points, and we also have to decide in the first row what we mean by a polar line over a quadratic alternative division algebra $\mathbb{A}$. Let us tackle the latter in the next section, the former being postponed to a later section.

6.4 The Absolute Magic Square

So what is a polar line over a quadratic alternative division algebra $\mathbb{A}$? In the spirit of our definition of "polar line", we should take a Hermitian form of Witt index 1 over $\mathbb{A}$. The corresponding field involution can be taken to be the natural involution in $\mathbb{A}$ defining the trace and norm (and given by the Cayley–Dickson process). Miraculously, if $\mathbb{A} = \mathbb{K}, \mathbb{L}, \mathbb{H}, \mathbb{O}$, these polar lines live naturally in split buildings of types $\ldots A_1, A_2, C_3, F_4$, respectively. In technical terms, using the vocabulary of Tits [29], these polar lines correspond to forms of algebraic groups of the respective types. Using the corresponding Tits diagrams (see [29]) together with our notation of black and white nodes to replace the arrows, the first row would become

Consequently, in this Square, both rows and columns are labelled by the sequence of diagrams $A_1 - A_2 - C_3 - F_4$. The latter are called the respective *absolute types* of the polar lines. Likewise, the projective planes over the quadratic alternative division rings as here have absolute types and corresponding

Tits diagram, as also the other buildings appearing in the Relative Square. Just listing the absolute types, we obtain the following with the aid of [29]:

A_1	A_2	C_3	F_4
A_2	$A_2{\times}A_2$	A_5	E_6
C_3	A_5	D_6	E_7
F_4	F_6	E_7	F_8

and call this the *Absolute Magic Square*. And here is again some magic: this table is symmetric! But this is not all. The symmetry goes further than simply the names of the types. It leads to another form of the Square.

6.5 The Fix Magic Square

Let us replace every absolute type in the Absolute Square with its Tits diagram, called *index* in [29]. Recall that a Tits diagram has the following geometric interpretation (see also [17]): the minimal flags fixed by the (Galois) descent group have the types given by the encircled nodes, and the diagram is drawn in a bent way if the descent group does not act type-preserving. In this context, and in the more general context of arbitrary automorphism groups fixing a flag opposite every fixed flag, this is sometimes referred to as the *fix diagram* of the automorphism group. Because of this, the following table will be called the *Fix Magic Square*.

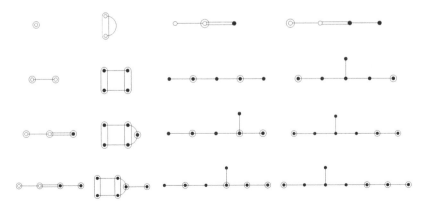

Clearly the symmetry is kind of broken since the same absolute types do not produce the same Tits diagrams. To re-establish the symmetry, in an even more glorious form, we need to get ahead of the facts for a moment. Firstly, it will appear that an important ingredient in the algebraic construction of most varieties is that of a Hermitian matrix, that is, a matrix which is symmetric up to a "twist", in this case a field automorphism. It would be a betrayal to itself if the Magic Square would also be "just symmetric", without an extra twist. Secondly, it will also appear that one can make a case to replace the forms in the first row by the anisotropic forms; that is, the empty Tits diagrams:

(This schizophrenic behaviour of the first row can be compared with the behaviour of "light" in physics – *particle* or *wave*. We will also always choose the interpretation that best fits our observations or our purposes.)

Now interpret the Tits diagrams in the Absolute Square (containing the row above as first row) as fix diagrams of certain involutions in Coxeter complexes. Let σ be an arbitrary involution of a Coxeter complex with a fix diagram in our Fix Square. Let ρ be the opposition map; that is, ρ maps each vertex of the Coxeter complex to its (unique) opposite. Then $\sigma\rho = \rho\sigma$ is an involution of the Coxeter complex with fix diagram exactly the one lying symmetrically (symmetry with respect to the main diagonal) in the Square. This way, one sees that the Fix Square itself is a kind of Hermitian matrix. Isn't that real magic?

For example, the involution corresponding to the first column is just the identity; it follows that the involution in the first row is then the opposition map. Note that the North-West cell has a conflict with itself: the involution cannot be at the same time the identity and the opposition map. So this is a little bug that does not bother us; on the contrary, it will be helpful when regarding this cell as the empty building.

Let us give two additional examples of this phenomenon.

Example 5.1   The Coxeter complex Σ of type D_6 can be defined using a set of twelve elements, say $\{-6, -5, \ldots, -1, 1, 2, \ldots, 6\}$. The vertices of Σ are all subsets of size distinct from 5 containing no pair of numbers with the same absolute value. The opposition map ρ is induced by the permutation of $\{-6, -5, \ldots, -1, 1, 2, \ldots, 6\}$ mapping each number x to its opposite $-x$. Let σ be induced by the permutation which interchanges the numbers i and j every time $|i + j| = 7$. Then σ stabilizes 2-cliques, 4-cliques and 6-cliques that contain an even number of positive numbers. These form a Coxeter complex of type C_3 on their own. The composition $\sigma\rho$ has exactly the same behaviour, although of course stabilizing different cliques. This explains why D_6 is on the diagonal of the Fix Square: It is self-conjugate.

Example 5.2 Recall the Bourbaki labelling of the nodes in the E_7 diagram:

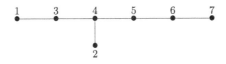

The Coxeter complex Σ of type E_7 can be modelled on the Gosset graph Γ, which is defined as follows: the 56 vertices of Γ are the pairs from the respective 8-sets $\{1, 2, \ldots, 8\}$ and $\{1', 2', \ldots, 8'\}$. Two pairs from the same set are adjacent if they intersect in precisely one element; two pairs $\{a, b\}$ and $\{c', d'\}$ from different sets are adjacent if $\{a, b\}$ and $\{c, d\}$ are disjoint. The elements (vertices) of type $2, 3, 4, 5, 6, 7$ of Σ are the maximal 7-cliques, the maximal 6-cliques, the 4-cliques, the 3-cliques, the edges and the vertices, respectively. The elements of type 1 are the cross-polytopes of size 12 (so-called *hexacrosses* or *6-orthoplexes*) contained in Γ (these are also the Coxeter subcomplexes of type D_6 as described in Example 5.1). There are 126 such, and 56 of these are determined by an ordered pair (i, j) with $i, j \in \{1, 2, 3, 4, 5, 6, 7, 8\}$, $i \neq j$, and induced on the vertices $\{i, k\}$ and $\{j', k'\}$, $k \notin \{i, j\}$, whereas the other 70 are determined by a 4-set $\{i, j, k, \ell\} \subseteq \{1, 2, 3, 4, 5, 6, 7, 8\}$ and are induced on the vertices $\{s, t\} \subseteq \{i, j, k, \ell\}$, $s \neq t$, and $\{u', v'\} \subseteq \{1', 2', 3', 4', 5', 6', 7', 8'\} \setminus \{i', j', k', \ell'\}$, $u \neq v$.

The opposition map ρ is induced by interchanging a with a', for all $a \in \{1, 2, \ldots, 8\}$. Let σ be the involution induced by the map sending $x \in \{1, 2, \ldots, 8\}$ to $(x + 4)'$, where $x + 4$ has to be read mod 8. Let us briefly write $x \pm 4$ for $x + 4$ mod 8. Clearly, σ does not fix any vertex of Γ, but it stabilizes precisely 24 edges, given by all pairs $\{\{a, b\}, \{(a \pm 4)', (b \pm 4)'\}\}$ with $a \neq b \pm 4$. It also stabilizes exactly 24 cross-polytopes of size 12, namely the eight determined by the ordered pairs $(i, i \pm 4)$, $i = 1, 2, \ldots, 8$, and the sixteen determined by all 4-subsets of $\{1, 2, \ldots, 8\}$ with the property that, for each $i \in \{1, 2, 3, 4\}$, it contains exactly one of i or $i + 4$. In each such cross-polytope, σ induces the map described in Example 5.1. It now easily follows that σ has fix diagram

●───◉───●───◉───◉───◉. Now $\sigma\rho = \rho\sigma$ fixes the eight vertices $\{i, i + 4\}$ and $\{i', (i + 4)'\}$, $i = 1, 2, 3, 4$, the twelve edges $\{\{i, i + 4\}, \{j', (j + 4)'\}\}$, with $i \in \{1, 2, 3, 4\}$ and $j \in \{1, 2, 3, 4\} \setminus \{i\}$, and the six cross-polytopes of size 12 determined by the 4-sets $\{i, j, i + 4, j + 4\}$, $i, j \in \{1, 2, 3, 4\}$, $i < j$, and nothing else. These form a cube (Coxeter complex of type C_3); the fix diagram of $\rho\sigma$ is now clearly ⊙———•———•———•———⊙———⊙ and we see the fix diagrams of σ and $\sigma\rho$ correspond under reflection about the main diagonal of the Fix Square.

The observation of this twisted symmetry will lead to rather interesting characterizations of certain automorphisms; see Section 6.9.4.

Digression The Coxeter complexes in the second row of the Magic Square look very similar: each can be modelled on the complement of the collinearity graph of a generalized quadrangle with three points per line (where we consider a single line as a degenerate generalized quadrangle; remember the rank 1 residues in the first cell are polar lines, so no collinear points exist, and hence the complement yields really a connected line with three points); they have order $(2,0)$, $(2,1)$, $(2,2)$, and $(2,4)$ and are uniquely determined by their order, for the columns 1, 2, 3, 4, respectively. One could ask whether something of that nature holds for the other rows. For the first row, we simply take away one line of the corresponding generalized quadrangle, and all edges joining vertices that correspond to points collinear to the same point of the line we took away; what remains is a graph that naturally models the Coxeter complex of the given geometry in the first row. The first cell becomes empty (compare with Subsection 6.9.5, where we argue why this cell could indeed also be seen as being empty; compare also with Subsection 6.8.2); the second cell becomes a hexagon, which models the Coxeter complex of type A_2 using the flags (chambers) as vertices; the third cell becomes a cuboctahedron; the fourth a 24-cell. Now what about the third row? We are looking for something that extends a generalized quadrangle. In the literature, an *extended generalized quadrangle of order* (s,t), $s,t \geq 1$, is a graph $\Gamma = (V(\Gamma), E(\Gamma))$ such that

(EGQ) For each vertex $v \in V(\Gamma)$, the point-line geometry with point set $\Gamma(v)$ (the set of vertices adjacent to v) and line set the maximal cliques of the graph induced on $\Gamma(v)$, is a generalized quadrangle of order (s,t).

It is shown in [7] that the diameter of an extended generalized quadrangle of order (s,t) is at most $s+1$. Moreover, it follows from theorem 2 of [8] that there are exactly three extended generalized quadrangles of order $(2,t)$ with diameter 3, and these have respective values $t = 1, 2, 4$. It can easily be checked using the description in [7] that the Gosset graph is the "2-complement" of the unique extended generalized quadrangle of diameter 3 and order $(2,4)$, where the 2-complement of a graph Γ is $\Gamma^2 - \Gamma$; that is, connect each pair of vertices at distance 2 and delete all other edges. The same is true for the ith cell of the third row: its Coxeter complex can be modelled on the 2-complement of the unique extended generalized quadrangle of diameter 3 and order $(2, 2^{i-2})$ in much the same way as the fourth cell and the Gosset graph.

In [8], a construction using quadrics is provided, making apparent the full automorphism group as a (subgroup of) an orthogonal classical group. But these groups must also be the Weyl groups of the corresponding Coxeter complex, hence Coxeter groups. This way, one sees that most Coxeter groups related to the Square are orthogonal classical groups. In fact, this even extends to the fourth row, but not entirely to the first one. We replace some cells of the Absolute Magic Square with their Coxeter group, using the notation of classical groups of the ATLAS [6], obtaining the following remarkable table:

$O_2^+(2)$	$O_3(2)$		
$O_3(2)$	$O_4^+(2)$	$O_5(2)$	$SO_6^-(2)$
	$O_5(2)$	$2^5.O_5(2)$	$2 \times O_7(2)$
	$SO_6^-(2)$	$2 \times O_7(2)$	$2.O_8^+(2).2$

6.6 The Relative Geometric Magic Square

We now turn to the geometries. We first briefly recall how to attach a geometry to a building by picking an arbitrary type of elements. We restrict ourselves to the irreducible thick case.

Everything that follows can be found at various places in [24], in particular in the chapters about parapolar spaces; see also [4].

Let Δ be an irreducible thick spherical building. Let n be its rank, let S be its type set, and let $s \in S$. Then we define a point-line geometry $\Gamma = (X, \mathscr{L}, *)$ as follows. The point set X is just the set of vertices of Δ of type s; the set $\mathscr{L}$ of lines are the flags of type $s^{\sim}$, where $s^{\sim}$ is the set of types adjacent to s in the Coxeter diagram of Δ. If x is a vertex of type s and F a flag of type $s^{\sim}$, then $x * F$ if $F \cup \{x\}$ is a flag. The geometry Γ is called a *Lie incidence geometry*. For instance, if Δ has type A_n, and $s = 1$ (remember we use Bourbaki labelling), then Γ is the point-line geometry of a projective space. If X_n is the Coxeter type of Δ and Γ is defined using $s \in S$ as noted, then we say that Γ has *type* $X_{n,s}$. In the diagram, we replace the corresponding node by x.

Lie incidence geometries of type $B_{n,1}$, $n \geq 2$, and $D_{n,1}$, $n \geq 3$, are polar spaces of rank n, and those of the latter type are sometimes called *hyperbolic*. All Lie incidence geometries as defined here and which are not projective

spaces or polar spaces, are *parapolar spaces*. Before recalling the definition of
a parapolar space, we need to review some basics about point-line geometries.

Let $\Gamma = (X,\mathscr{L})$ be a point-line geometry (if the incidence relation is not
mentioned, we assume it is induced by containment). Points $x, y \in X$ contained
in a common line are called *collinear*, denoted as $x \perp y$; the set of all points
collinear to x is denoted by $x^{\perp}$. We will always deal with situations where every
point is contained in at least one line, so $x \in x^{\perp}$. Also, for $S \subseteq X$, we denote
$S^{\perp} := \{x \in X \mid x \perp s \text{ for all } s \in S\}$.

The *point graph* of Γ is the graph on X with collinearity as adjacency relation.
The *distance* δ between two points $p, q \in X$ (denoted $\delta_{\Gamma}(p,q)$, or $\delta(p,q)$ if no
confusion is possible) is the distance between p and q in the collinearity graph,
where $\delta(p,q) = \infty$ if p and q are contained in distinct connected components
of the point graph; if $\delta := \delta(p,q)$ is finite, then a *geodesic path* or a *shortest
path* between p and q is a path between them in the point graph of length δ. The
diameter of Γ (denoted Diam Γ) is the diameter of the point graph. We say that
Γ is *connected* if every pair of vertices is at finite distance from one another.
The point-line geometry Γ is called a *partial linear space* if each pair of distinct
points is contained in at most one line.

A *subspace* of Γ is a subset A of X such that, if $x, y \in A$ are collinear and
distinct, then all lines containing both x and y are contained in A. A subspace
A is called *convex* if, for any pair of points $\{p,q\} \subseteq A$, every point occurring
in a geodesic between p and q is contained in A; it is *singular* if $\delta(p,q) \leq 1$ for
all $p, q \in A$. The intersection of all convex subspaces of Γ containing a given
subset $B \subseteq X$ is called the *convex subspace closure* of B. A proper subspace H
is called a *geometric hyperplane* if each line of Γ has either one or all its points
contained in H.

Now a parapolar space is a point-line geometry satisfying the following three
axioms:

(PPS1) There is line L and a point p such that no point of L is collinear to
p.

(PPS2) The geometry is connected.

(PPS3) Let x, y be two points at distance 2. Then either there is a unique point
collinear to both – and then the pair $\{x, y\}$ is called *special* – or the convex subspace closure of $\{x, y\}$ is a polar space – and then the pair $\{x, y\}$
is called a *symplectic pair*. Such polar spaces are called *symplecta*, or
symps for short.

(PPS4) Each line is contained in a symplecton.

The parapolar spaces we will encounter all have the rather peculiar property that all symps have the same rank, which is then called the (uniform) *symplectic rank* of the parapolar space. In contrast, the maximal singular subspaces (which will be projective spaces) will not all have the same dimension. The *singular ranks* of a parapolar space with only projective spaces as singular subspaces (which is automatic if the symplectic rank is at least 3) are the dimensions of the maximal singular subspaces.

We need two more notions. A parapolar space without special pairs is called *strong*. And a parapolar space of uniform symplectic rank ≥ 3 is called *locally connected* if for every point p, the graph on the lines passing through p, adjacent when contained in a common symp is connected. Equivalently, one can require that for every point p, the graph on the lines passing through p, adjacent when contained in a common singular plane, is connected.

A Lie incidence geometry is called a *long root geometry* if it has type $X_{n,s}$, with s the so-called *polar node* of the diagram X_n, $X \neq A$ (the long root geometry for type A needs a more general definition of Lie incidence geometry using flags as points instead of vertices of the corresponding building, but we will not bother the reader with this). The polar node in case $B, C, E,$ and F (we will not need G_2) is given by the (unique) fundamental root not perpendicular to the longest (highest) root. The types are $B_{n,2}$, $C_{n,1}$, $D_{n,2}$, $E_{6,2}$, $E_{7,1}$, $E_{8,8}$, and $F_{4,1}$. Long root geometries share a lot of interesting (even characteristic) properties across all types; however, for most types, they are not the "simplest" Lie incidence geometry in the sense that other choices for points produce geometries with smaller diameter or simpler structure. For E_8, it is the simplest one in that sense. For F_4, it is debatable whether or not the Lie incidence geometry of type $F_{4,4}$ is simpler than the one of type $F_{4,1}$. They both have the same diameter and global structure; in fact every characterization theorem of parapolar spaces trying to single out the long root geometries also embraces the Lie incidence geometry of type $F_{4,4}$. For an arbitrary building of type F_4, the Lie incidence geometry of type $F_{4,1}$ or $F_{4,4}$ is called a *metasymplectic space*, see [5] for an axiom system and more background information.

By the previous paragraph it is clear that the Square should contain long root geometries, and preferably in the fourth row, by the presence of type E_8 in the Absolute Square. So, carried over to the Relative Square, the geometries of the fourth row are the ones of type $F_{4,1}$. So we choose the last node as the type of vertices to play the role of the points, and we replace the node with an "×". We now do this for every row; however the diagrams in the second row are symmetric, so there we can choose the first vertex. Our Relative Geometric Square becomes:

$$
\begin{array}{llll}
\overset{\times}{\underset{\mathbb{K}}{\bullet}} & \overset{\times}{\underset{\mathbb{L}}{\bullet}} & \overset{\times}{\underset{\mathbb{H}}{\bullet}} & \overset{\times}{\underset{\mathbb{O}}{\bullet}}
\end{array}
$$

$$
\begin{array}{llll}
\overset{\times}{\underset{\mathbb{K}}{\bullet}}\!\!-\!\!\overset{\circ}{\underset{\mathbb{K}}{\bullet}} &
\overset{\times}{\underset{\mathbb{L}}{\bullet}}\!\!-\!\!\overset{\circ}{\underset{\mathbb{L}}{\bullet}} &
\overset{\times}{\underset{\mathbb{H}}{\bullet}}\!\!-\!\!\overset{\circ}{\underset{\mathbb{H}}{\bullet}} &
\overset{\times}{\underset{\mathbb{O}}{\bullet}}\!\!-\!\!\overset{\circ}{\underset{\mathbb{O}}{\bullet}}
\end{array}
$$

$$
\begin{array}{llll}
\overset{\circ}{\underset{\mathbb{K}}{\bullet}}\!-\!\overset{\circ}{\underset{\mathbb{K}}{\bullet}}\!\Rightarrow\!\overset{\times}{\underset{\mathbb{K}}{\bullet}} &
\overset{\circ}{\underset{\mathbb{L}}{\bullet}}\!-\!\overset{\circ}{\underset{\mathbb{L}}{\bullet}}\!-\!\overset{\times}{\underset{\mathbb{K}}{\bullet}} &
\overset{\circ}{\underset{\mathbb{H}}{\bullet}}\!-\!\overset{\circ}{\underset{\mathbb{H}}{\bullet}}\!-\!\overset{\times}{\underset{\mathbb{K}}{\bullet}} &
\overset{\circ}{\underset{\mathbb{O}}{\bullet}}\!-\!\overset{\circ}{\underset{\mathbb{O}}{\bullet}}\!-\!\overset{\times}{\underset{\mathbb{K}}{\bullet}}
\end{array}
$$

$$
\begin{array}{llll}
\overset{\circ}{\underset{\mathbb{K}}{\bullet}}\!-\!\overset{\circ}{\underset{\mathbb{K}}{\bullet}}\!-\!\overset{\bullet}{\underset{\mathbb{K}}{\bullet}}\!-\!\overset{\times}{\underset{\mathbb{K}}{\bullet}} &
\overset{\circ}{\underset{\mathbb{L}}{\bullet}}\!-\!\overset{\circ}{\underset{\mathbb{L}}{\bullet}}\!-\!\overset{\times}{\underset{\mathbb{K}}{\bullet}}\!-\!\overset{\times}{\underset{\mathbb{K}}{\bullet}} &
\overset{\circ}{\underset{\mathbb{H}}{\bullet}}\!-\!\overset{\circ}{\underset{\mathbb{H}}{\bullet}}\!-\!\overset{\times}{\underset{\mathbb{K}}{\bullet}}\!-\!\overset{\times}{\underset{\mathbb{K}}{\bullet}} &
\overset{\circ}{\underset{\mathbb{O}}{\bullet}}\!-\!\overset{\circ}{\underset{\mathbb{O}}{\bullet}}\!-\!\overset{\times}{\underset{\mathbb{K}}{\bullet}}\!-\!\overset{\times}{\underset{\mathbb{K}}{\bullet}}
\end{array}
$$

6.7 The Complexified Geometric Magic Square

We will now attach a geometry to each cell of the Absolute Square. This is done simply by taking the Lie incidence geometry with respect to the node of the absolute diagram that corresponds to the x-node in the Relative Geometric Square. We get the following Complexified Geometric Square:

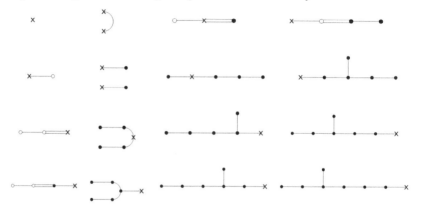

Using diagram types, this Square looks as follows:

$A_{1,1}$	$A_{2,\{1,2\}}$	$C_{3,2}$	$F_{4,4}$
$A_{2,1}$	$A_{2,1} \times A_{2,1}$	$A_{5,2}$	$E_{6,1}$
$C_{3,3}$	$A_{5,3}$	$D_{6,6}$	$E_{7,7}$
$F_{4,1}$	$E_{6,2}$	$E_{7,1}$	$E_{8,8}$

At this point, the squares represent 16 classes of Lie incidence geometries, with extraordinary properties and rather strong common characterizations. We mention two of them.

A parapolar space with uniform symplectic rank r possessing a singular subspace S of dimension $r-2$ with the property that $S^\perp$ is the union of two maximal singular subspaces of respective dimensions d_1, d_2 (possibly $d_1 = d_2$) is called $\{d_1, d_2\}$-*camel*. A parapolar space of uniform symplectic rank r with the property that no pair of symps intersects in a singular subspace of dimension k, $-1 \leq k \leq r-1$ is called k-*lacunary*. Table 1 shows some examples of $\{d_1, d_2\}$-camel and at the same time k-lacunary Lie incidence geometries with uniform symplectic rank r. The table shows the types; the fields or skew fields can be chosen arbitrarily (and in case of A_1 every set of at least three points qualifies; for A_2, every projective plane qualifies). All geometries have diameter 2, except the ones in the grey cells, which have diameter 3. All parapolar spaces are strong, except the ones written in white. All these examples have singular ranks d_1, d_2. Note that all geometries appearing in Table 1 either appear in the Complexified Magic Square, or are a residue of a geometry appearing in the second row of the Complexified Magic Square (and all geometries appearing in the 3×3 South-East corner of the Complexified Magic Square are in the table), with the understanding that types A_1 and A_2 stand for arbitrary projective lines and planes, respectively.

We now have the following two characterization theorems. The first one is taken from [13]; the second one combines the main results of [11] and [12].

Theorem 7.1 *Let Δ be a parapolar space of uniform symplectic rank $r \geq 2$ all symps of which are hyperbolic and all singular subspaces of which are projective spaces (remember this is automatic when $r \geq 3$). Assume Δ is locally connected if $r \geq 3$ and strong if $r = 2$. If Δ is $\{d_1, d_2\}$-camel, with $d_1 + d_2 \leq 2r$, then Δ is one of the geometries in* Table 1.

Theorem 7.2 *The k-lacunary parapolar spaces with uniform symplectic rank $r \geq k + 3$ are those in* Table 1 *for $r \in \{k+3, k+4, k+6\}$.*

Also the k-lacunary parapolar spaces with symplectic rank $k + 2$ are classified. Most other long root geometries appear there, along with dual polar spaces of rank 3 and half spin geometries, and some others; see [11]. Note also that the conditions in [11] are slightly weaker in that the rank of the parapolar space is not assumed to be uniform.

6.8 The Original Tits Magic Square

The point-line geometries displayed in the Relative and Complexified Magic Squares have natural embeddings in projective spaces (as is the case for all Lie incidence geometries related to split algebraic groups, by [16]). For the second and the third rows, these natural embeddings are even the absolutely universal embeddings, discarding the first cells. This is conjectured to be true for the last row, too. The corresponding point sets of these embedded geometries are rational varieties, as they can be described with polynomial equations. The latter will also be a feature in the connectivity of the square.

Table 1 Types of some k-lacunary Lie incidence geometries with symplectic rank r and singular ranks d_1, d_2.

r	d_1, d_2	$k = -1$	$k = 0$	$k = 1$	$k = 2$	$k = 3$	$k = 4$	$k = 5$
$k+2$	$k+1, k+2$		$A_{1,1} \times A_{2,1}$	$A_{4,2}$	$D_{5,5}$	$E_{6,1}$	$E_{7,7}$	
	$k+1, k+3$		$A_{1,1} \times A_{3,1}$	$A_{5,2}$	$D_{6,6}$			
$k+3$	$k+2, k+3$	$A_{1,1} \times A_{2,1}$	$A_{4,2}$	$D_{5,5}$	$E_{6,1}$	$E_{7,7}$		
	$k+3, k+3$	$A_{2,1} \times A_{2,1}$	$A_{5,3}$					
$k+4$	$k+3, k+4$	$A_{4,2}$	$D_{5,5}$	$E_{6,1}$	$E_{7,7}$			
	$k+3, k+5$	$A_{5,2}$	$D_{6,6}$					
$k+6$	$k+5, k+6$	$E_{6,1}$	$E_{7,7}$					

In this section, we describe these varieties, row by row, except for the fourth row, where the problem of finding an elementary common description of these varieties is still open. Note that in [28] Jacques Tits encircles the points of the dual geometry in order to have the smallest dimensions for the point sets.

6.8.1 The Second Row

We begin with the second row, which lies central with respect to the rows we are going to handle. The main ingredient of every construction is a non-degenerate quadratic alternative algebra $\mathbb{A}$ over a field $\mathbb{K}$. In the context of the Magic Square, $\mathbb{A}$ is not an inseparable field extension of $\mathbb{K}$ in characteristic 2. However, when we will mention characterization results, these will nevertheless turn up as extra examples, so we might as well retain that possibility. There is a natural involution σ in $\mathbb{A}$, usually denoted by $\sigma: x \mapsto \overline{x}$, and a natural inclusion $\mathbb{K} \subseteq \mathbb{A}$ with the properties $x\overline{x} \in \mathbb{K}$ and $x + \overline{x} \in \mathbb{K}$.

The possibilities for $\mathbb{A}$ are

(*i*) the field $\mathbb{K}$ itself; σ is the identity,
(*ii*) a separable quadratic field extension; σ is the non-trivial Galois automorphism,
(*iii*) the direct product $\mathbb{K} \times \mathbb{K}$ (with componentwise addition and multiplication); σ interchanges the two components and the natural inclusion of $\mathbb{K}$ in $\mathbb{A}$ is diagonal,
(*iv*) a quaternion division algebra over $\mathbb{K}$; σ is the natural involution,
(*v*) a split quaternion algebra over $\mathbb{K}$, which is isomorphic to the algebra of 2×2 matrices over $\mathbb{K}$; σ is taking the adjugate matrix and the natural inclusion of $\mathbb{K}$ in $\mathbb{A}$ is via the scalar matrices,
(*vi*) an octonion division algebra over $\mathbb{K}$; σ is the natural involution,
(*vii*) a split octonion algebra over $\mathbb{K}$ with natural involution σ,
(*viii*) an inseparable field extension of $\mathbb{K}$ in characteristic 2; σ is the identity.

So let $\mathbb{A}$ be one of these algebras. Set $d := \dim_{\mathbb{K}} \mathbb{A}$. ($d$ might possibly be infinite.) Now consider the *Veronese map*:

$$\rho_{\mathbb{A}}: \mathbb{A} \times \mathbb{A} \to \mathsf{PG}(3d + 2, \mathbb{K}) : (a, b) \mapsto (a\overline{a}, b\overline{b}, 1; b, \overline{a}, a\overline{b}),$$

where we view $\mathsf{PG}(3d + 2, \mathbb{K})$ as $\mathsf{PG}(V)$, with $V \cong \mathbb{K} \times \mathbb{K} \times \mathbb{K} \times \mathbb{A} \times \mathbb{A} \times \mathbb{A}$. Then the Zariski closure of the full image of $\rho_{\mathbb{A}}$ is a variety which we denote by $\mathscr{V}(\mathbb{K}, \mathbb{A})$.

Perhaps one word about the Zariski closure. The best way to think about that here is as follows: if $\mathbb{A}$ is an associative division algebra, then one can make the image homogeneous, and so the Zariski closure coincides with the set of points

$$\{(a\overline{a}, b\overline{b}, c\overline{c}; b\overline{c}, c\overline{a}, a\overline{b}) \mid a, b, c \in \mathbb{A}\}.$$

In the non-associative case we can alternatively take the following union:

$$\{(1, b\overline{b}, c\overline{c}; b\overline{c}, c, \overline{b}) \mid b, c \in \mathbb{A}\} \cup \{(a\overline{a}, 1, c\overline{c}; \overline{c}, c\overline{a}, a) \mid a, c \in \mathbb{A}\}$$
$$\cup \{(a\overline{a}, b\overline{b}, 1; b, \overline{a}, a\overline{b}) \mid a, b \in \mathbb{A}\}.$$

If $\mathbb{A}$ is not a division algebra, and $|\mathbb{K}| > 2$, then the Zariski closure of a point set P coincides with the *projective closure* of P; that is, the smallest set of points containing P and containing no affine line (an *affine line* is a projective line with one point deleted). If $|\mathbb{K}| = 2$ and $\mathbb{A}$ is not division, then first go to any extension field, take the projective closure and then restrict the field again to $\mathbb{K}$.

If $\mathbb{A}$ is division, then we obtain a geometry of the Relative Geometric Magic Square; otherwise the Complexified Geometric Magic Square. We now briefly discuss the different possibilities and provide the possibly different classical definition of construction.

The quadric Veronese surface $\mathcal{V}_2(\mathbb{K})$ – This is the image of the *Veronese map* $\mathrm{PG}(2, \mathbb{K}) \to \mathrm{PG}(5, \mathbb{K})$: $(x, y, z) \mapsto (x^2, y^2, z^2, yz, zx, xy)$. Here $\mathbb{A} = \mathbb{K}$.

We see that, although in essence a projective plane, the quadric Veronese surface behaves like a strong parapolar space of diameter 2 without lines: Every pair of points determines a unique symp, which is here a polar space of rank 1, in particular a quadric of Witt index 1. This is made clear by the diagram in the first column and second row of any form of the Magic square, which shows that the rank 1 residues are polar lines rather than projective lines.

The Segre variety $\mathcal{S}_{2,2}(\mathbb{K})$ – This is the image of the *Segre map* $\mathrm{PG}(2, \mathbb{K}) \times \mathrm{PG}(2, \mathbb{K}) \to \mathrm{PG}(8, \mathbb{K})$: $(x, y, z; u, v, w) \mapsto (xu, yu, zu, xv, yv, zv, xw, yw, zw)$. It is isomorphic to $\mathcal{V}(\mathbb{K}, \mathbb{K} \times \mathbb{K})$.

We may view the set of 3×3 matrices over $\mathbb{K}$ as a 9-dimensional vector space, and the set of symmetric 3×3 matrices as a 6-dimensional subspace. Then we may consider the corresponding projective spaces of (projective) dimension 8 and 5, respectively, in the classical way by considering the 1-spaces as the points. In this way, the Segre variety $\mathcal{S}_{2,2}(\mathbb{K})$ corresponds exactly to the rank 1 3×3 matrices; explicitly

$$\mathbb{K}(xu, yu, zu, xv, yv, zv, xw, yw, zw) \leftrightarrow \mathbb{K}\begin{pmatrix} xu & yu & zu \\ xv & yv & zv \\ xw & yw & zw \end{pmatrix}.$$

Similarly, the quadric Veronese surface $\mathcal{V}_2(\mathbb{K})$ corresponds exactly with the rank 1 symmetric 3×3 matrices; explicitly

$$\mathbb{K}(x^2, y^2, z^2, yz, zx, xy) \leftrightarrow \mathbb{K}\begin{pmatrix} x^2 & yx & zx \\ xy & y^2 & zy \\ xz & yz & z^2 \end{pmatrix}.$$

In particular, $\mathscr{V}_2(\mathbb{K})$ is a subvariety of $\mathscr{S}_{2,2}(\mathbb{K})$ obtained by intersecting with a 5-dimensional subspace.

We also encounter other Segre varieties; in general $\mathscr{S}_{n,m}(\mathbb{K})$ is defined as the image in $\mathsf{PG}(nm - 1, \mathbb{K})$ of the map $(x_i, y_j)_{1 \le i \le n, 1 \le j \le m} \mapsto (x_i y_j)_{1 \le i \le n, 1 \le j \le m}$. The images of the marginal maps defined by either fixing the x_i, $1 \le i \le n$, or the y_j, $1 \le j \le m$, are called the *generators* of the variety (in case of $\mathscr{S}_{2,2}(\mathbb{K})$ the generators are 2-dimensional projective subspaces, hence planes).

The line Grassmannian $\mathscr{G}_{2,6}(\mathbb{K})$ – Denote the set of lines of $\mathsf{PG}(5, \mathbb{K})$, or equivalently, the set of 2-spaces of $\mathbb{K}^6$ by $\binom{\mathbb{K}^6}{\mathbb{K}^2}$. Then $\mathscr{G}_{2,6}(\mathbb{K})$ is the image of the Plücker map $\binom{\mathbb{K}^6}{\mathbb{K}^2} \to \mathsf{PG}(14, \mathbb{K})$: $\langle (x_1, x_2, \ldots, x_6), (y_1, y_2, \ldots, y_6) \rangle \mapsto (x_i y_j - x_j y_i)_{1 \le i < j \le 6}$. Denote the coordinate of $\mathsf{PG}(14, \mathbb{K})$ corresponding to the entry $x_i y_j - x_j y_i$ by p_{ij}, $1 \le i < j \le 6$. By restricting to $y_1 = y_2 = y_3 = x_4 = x_5 = x_6 = 0$, we see that $\mathscr{S}_{2,2}(\mathbb{K})$ is a subvariety of $\mathscr{G}_{2,6}(\mathbb{K})$ obtained by intersecting with an 8-dimensional projective subspace with equation $p_{12} = p_{13} = p_{23} = p_{45} = p_{46} = p_{56} = 0$.

The line Grassmannian $\mathscr{G}_{2,6}(\mathbb{K})$ is isomorphic to $\mathscr{V}(\mathbb{K}, \mathbb{A})$, with $\mathbb{A}$ a split quaternion algebra over $\mathbb{K}$. The isomorphism can be seen by taking $a = \begin{pmatrix} x_1 & y_1 \\ x_2 & y_2 \end{pmatrix}$ and $b = \begin{pmatrix} x_3 & y_3 \\ x_4 & y_4 \end{pmatrix}$ in the definition of $\mathscr{V}(\mathbb{K}, \mathbb{A})$ above.

The Cartan variety $\mathscr{E}_{6,1}(\mathbb{K})$ – This variety is traditionally defined using a trilinear or cubic form. It is an exceptional variety in the sense that it cannot be defined, using classical notions like Plücker or Grassmann coordinates, from a projective space. We introduce the cubic form in the geometric way explored in [33].

Let $\Gamma = (X, \mathscr{L})$ be the generalized quadrangle of order $(2, 4)$, that is, the points and lines on the quadric in $\mathsf{PG}(5, 2)$ with equation $X_0 X_1 + X_2 X_3 = X_4^2 + X_4 X_5 + X_5^2$. Let $\mathscr{S}$ be a *regular spread*, that is, a set of nine lines of Γ partitioning the point set and enjoying the property that, if $L, M \in \mathscr{L}$, $L \ne M$, then the set of points off $L \cup M$ but contained in a line intersecting both L and M are the points of a member of $\mathscr{S}$. Label the standard basis of a 27-dimensional vector space over $\mathbb{K}$ with the points of Γ, and use x_p, $p \in X$, as the corresponding coordinate. Then we define the cubic form:

$$C\left((x_p)_{p \in X}\right) = \sum_{\{p,q,r\} \in \mathscr{S}} X_p X_q X_r - \sum_{\{p,q,r\} \in \mathscr{L} \backslash \mathscr{S}} X_p X_q X_r.$$

Then the Cartan variety $\mathscr{E}_{6,1}(\mathbb{K})$ consists of all projective points $\langle v \rangle$ such that $\nabla C(v) = \vec{o}$ (the gradient in the classical sense). We have $\mathscr{E}_{6,1}(\mathbb{K}) \cong \mathscr{V}(\mathbb{K}, \mathbb{A})$, with $\mathbb{A}$ a split octonion algebra over $\mathbb{K}$.

Veronesean representations of projective planes – This is the case where $\mathbb{A}$ is a division algebra. If it is also associative, then $\mathscr{V}(\mathbb{K}, \mathbb{A})$ is the image of the map:

$$\mathbb{A} \times \mathbb{A} \times \mathbb{A} \to \mathsf{PG}(3d+2, \mathbb{K}) : (a,b,c) \mapsto (a\overline{a}, b\overline{b}, c\overline{c}, b\overline{c}, c\overline{a}, a\overline{b}).$$

In general, it is the union of the three image sets:

$$\{(a\overline{a}, b\overline{b}, 1, b, \overline{a}, a\overline{b}) \mid a, b \in \mathbb{A}\} \cup \{(a\overline{a}, 1, c\overline{c}, \overline{c}, c\overline{a}, a) \mid a, c \in \mathbb{A}\}$$
$$\cup \{(1, b\overline{b}, c\overline{c}, b\overline{c}, c, \overline{b}) \mid b, c \in \mathbb{A}\}.$$

The variety $\mathscr{V}(\mathbb{K}, \mathbb{A})$ is a projective plane with obvious point set, and with a set of lines the d-dimensional quadrics with Witt index 1 entirely contained in it (if $|\mathbb{K}| = |\mathbb{A}| = 2$, we have to make an appropriate selection of those since every set of three non-collinear points is a conic, hence a 1-dimensional quadric of Witt index 1).

All the preceding varieties $\mathscr{V} = \mathscr{V}(\mathbb{K}, \mathbb{A})$ share the following properties, see [23]. There exists a unique set $\mathscr{H}$ of $(d+1)$-dimensional subspaces, called the *host spaces*, satisfying

(1) every pair of points of $\mathscr{V}$ is contained in at least one host space;
(2) the intersection of $\mathscr{V}$ with any host space is a non-empty non-degenerate d-dimensional quadric and the intersection of two distinct host spaces is always contained in $\mathscr{V}$;
(3) for each point p of $\mathscr{V}$, the space generated by the tangent spaces at p to the quadrics $\mathscr{V} \cap H$, with $p \in H \in \mathscr{H}$ has dimension $2d$.

These are precisely the properties, found by Zak [34], of the complex Severi varieties; that is, complex smooth $2d$-dimensional varieties in $\mathsf{PG}(3d+2, \mathbb{C})$ whose secant variety is not trivial (the secant variety of every complex smooth $2d$-dimensional variety in $\mathsf{PG}(n, \mathbb{C})$, with $n < 3d+2$ is trivial, that is, the whole of $\mathsf{PG}(n, \mathbb{C})$). Zak classified varieties satisfying the three properties (over the complex numbers) and only found the varieties $\mathscr{V}(\mathbb{C}, \mathbb{A})$, with $\mathbb{A}$ either $\mathbb{C}$, $\mathbb{C} \times \mathbb{C}$, the split quaternions over $\mathbb{C}$, or the split octonions over $\mathbb{C}$. One can ask whether that classification can be extended to general fields. In fact, we can do more. The following results come from [23], [15] and [10].

Theorem 8.1 *Assume we have a point set $\mathscr{V}$ in some (not necessarily finite-dimensional) projective space over some field $\mathbb{K}$, a set of $(d+1)$-dimensional host spaces $\mathscr{H}$ satisfying* (1), (2) *and* (3), *for some natural number $d \geq 1$. Then there exists a non-degenerate quadratic alternative algebra $\mathbb{A}$ over $\mathbb{K}$ such that $\mathscr{V} = \mathscr{V}(\mathbb{K}, \mathbb{A})$.*

We emphasize that the quadrics obtained as intersection of $\mathscr{V}$ with the members of $\mathscr{H}$ should not be assumed to be isomorphic, or necessarily have the same Witt index. No assumption whatsoever is made on these quadrics, besides the fact that they are d-dimensional; that is, they span their host space. Actually, the assumptions in the aforementioned papers are slightly weaker in that (3) is replaced with the following axiom.

(3′) For each point p of $\mathscr{V}$, the space generated by the tangent spaces at p to the quadrics $\mathscr{V} \cap H$, with $p \in H \in \mathscr{H}$ has dimension at most $2d$.

Under that weaker condition, the universal natural embeddings of the geometries in the conclusion of Theorem 7.1 also have to be added to the conclusion. So completely different properties lead to exactly the same geometries, in different forms (pure vs. embedded), restricting Theorem 8.1 to sets $\mathscr{V}$ containing lines. This is part of the Magic so characteristic for the geometries in the Magic Square.

Besides using the Veronese map, there is another way to construct the varieties related to the second row of the Magic Square. Let $\mathbb{K}$, $\mathbb{A}$, σ and V again be as before. Then we may view V as the set of all *Hermitian* 3×3 *matrices*; that is, 3×3 matrices M with on the diagonal scalars and off the diagonal members of $\mathbb{A}$, so that $M^{\mathrm{t}} = \overline{M}$ (with t the *transpose operator*). Explicitly,

$$M = \begin{pmatrix} x_1 & X_3 & \overline{X}_2 \\ \overline{X}_3 & x_2 & X_1 \\ X_2 & \overline{X}_1 & x_3 \end{pmatrix}, x_1, x_2, x_3 \in \mathbb{K}, \ X_1, X_2, X_3 \in \mathbb{A}.$$

These in fact form a Jordan algebra (under the Jordan multiplication $A * B = \frac{1}{2}(AB + BA)$, provided that char $\mathbb{K} \neq 2$). The projective points corresponding to all rank 1 such matrices constitute the variety $\mathscr{V}(\mathbb{K}, \mathbb{A})$. Writing down this condition explicitly by expressing that the columns of M are mutually proportional by right factors in $\mathbb{A}$, one obtains $\mathscr{V}(\mathbb{K}, \mathbb{A})$ as an intersection of $3d + 3$ quadrics with equations (where the x_i and the X_i refer to the coordinates as above in the matrix M):

$$\begin{cases} X_i \overline{X}_i &= x_{i+1} x_{i+2}, \\ x_i \overline{X}_i &= X_{i+1} X_{i+2}, \end{cases} \text{ for all } i \in \{1, 2, 3\} \pmod 3.$$

Now it is the right moment to look at the first row.

6.8.2 The First Row

The varieties of the first row of the Magic Square are obtained from the ones in the second row by a suitable hyperplane section. In the complexified case, the standard suitable hyperplane H is, with the previous notation, the one with equation $x_1 + x_2 + x_3 = 0$. Something peculiar happens now for the third and fourth cells. Indeed, the lines of the geometries are not all lines of the varieties. For instance, $\mathscr{E}_{6,1}(\mathbb{K})$ contains subspaces of dimension 5, so a hyperplane intersects this in a subspace of dimension 4 or 5; yet the geometry of type $\mathsf{F}_{4,4}$ does not have any singular subspaces of that dimension. It is even more extreme: the 5-spaces of $\mathscr{E}_{6,1}(\mathbb{K})$ that are contained in H become the symps of the new geometry. And a line of $\mathscr{E}_{6,1}(\mathbb{K})$ in H is a line of the new geometry if and only if it is contained in at least two 5-spaces that are contained in H and $\mathscr{E}_{6,1}(\mathbb{K})$. Similarly for $\mathscr{G}_{2,6}(\mathbb{K})$ and the variety corresponding to the geometry of type $\mathsf{C}_{3,2}$.

In general, we denote by $\mathscr{V}'(\mathbb{K}, \mathbb{A})$ the variety of the first row in $\mathsf{PG}(3d+1, \mathbb{K})$ obtained from $\mathscr{V}(\mathbb{K}, \mathbb{A})$ by intersecting with a suitable hyperplane of $\mathsf{PG}(3d + 2, \mathbb{K})$.

So this is a neat correspondence between the first two rows of the Complexified Geometric Magic Square. What about the Relative Geometric Magic Square? If we take the same equation $x_1 + x_2 + x_3 = 0$ for H, then it really depends on the field whether we get a non-empty variety. For instance, considering the real numbers, we see that we get a complete anisotropic variety, since the sum of squares is only zero if each square is zero. Hence here we see that it makes sense to take the empty Tits diagrams in the first row of the Relative Magic Square. But over a finite field, the second cell is a true Hermitian curve (the third and fourth cells do not exist over a finite field, in the Relative Square). In fact, for a finite field, one cannot find a hyperplane with empty intersection. This mixed behaviour implies different viewpoints and interpretations on the first row, and we already saw an example of how we can choose a certain interpretation to prove a point.

There is another, related correspondence between the first and second row of the Complexified Geometric Magic Square, recently proved by De Schepper & Victoor [14]. Before we can state it, we state a duality property of the varieties of the second row. This is kind of folklore. We use the terminology of the characterization in Theorem 8.1.

Theorem 8.2 *Let* $\mathbb{A}$ *be any non-degenerate quadratic alternative algebra over the field* $\mathbb{K}$. *Then each host space W of* $\mathscr{V}(\mathbb{K}, \mathbb{A})$ *is contained in a unique hyperplane H of* $\mathsf{PG}(3d+2, \mathbb{K})$ *which intersects* $\mathscr{V}(\mathbb{K}, \mathbb{A})$ *precisely in the points that belong to W and* $\mathscr{V}(\mathbb{K}, \mathbb{A})$. *The set of all such hyperplanes when varying the host space is in the dual projective space isomorphic to* $\mathscr{V}(\mathbb{K}, \mathbb{A})$.

We denote this dual variety by $\mathscr{V}^*(\mathbb{K}, \mathbb{A})$. Now we can state the beautiful result of De Schepper & Victoor.

Theorem 8.3 *Let Q be any non-degenerate quadric of* $\mathsf{PG}(3d + 2, \mathbb{K})$ *containing all points of* $\mathscr{V}(\mathbb{K}, \mathbb{A})$, *with* $\mathbb{A}$ *not division. Then the image in* $\mathscr{V}^*(\mathbb{K}, \mathbb{A})$

of the set of host spaces of $\mathcal{V}(\mathbb{K}, \mathbb{A})$ contained in a singular subspace of Q is isomorphic to a variety dual to $\mathcal{V}'(\mathbb{K}, \mathbb{A})$.

One expects the same to be true for the case of a division algebra $\mathbb{A}$ if one assumes that Q has maximal Witt index (this is proved in [14] for the smallest case, namely $\mathbb{A} = \mathbb{K}$).

Let us now have a look at the third row.

6.8.3 The Third Row

Again, we can construct all varieties corresponding to the geometries of the third row of the Relative and Complexified Geometric Magic Square in a uniform way using a Veronesean map. In the general case this goes as follows. For a 3×6 matrix M, we denote by $p_{ijk}(M)$, $1 \leq i < j < k \leq 6$, the determinant of the 3×3 matrix obtained from M by only keeping the columns labeled i, j and k. Now, for $\mathbb{K}$ and $\mathbb{A}$ as before, define the matrix:

$$M(x_1, x_2, x_3, X_1, X_2, X_3) = \begin{pmatrix} 1 & 0 & 0 & x_1 & X_3 & \overline{X}_2 \\ 0 & 1 & 0 & \overline{X}_3 & x_2 & X_1 \\ 0 & 0 & 1 & X_2 & \overline{X}_1 & x_3 \end{pmatrix},$$

$$x_1, x_2, x_3 \in \mathbb{K}, \quad X_1, X_2, X_3 \in \mathbb{A}.$$

Now let D be the set of triples $\{123, 125, 126, 134, 135, 145, 156, 234, 236, 246, 256, 345, 346, 456\}$. We may restrict to those p_{ijk} with $ijk \in D$ since clearly $p_{136} = -\overline{p}_{125}$, $p_{124} = \overline{p}_{236}$, $p_{235} = -\overline{p}_{134}$, $p_{356} = \overline{p}_{145}$, $p_{146} = -\overline{p}_{256}$, $p_{245} = -\overline{p}_{346}$. Then we define the *Veronese map* $\rho_{\mathbb{A}}^{\dagger}$ as follows:

$$\rho_{\mathbb{A}}^{\dagger} : \mathbb{K} \times \mathbb{K} \times \mathbb{K} \times \mathbb{A} \times \mathbb{A} \times \mathbb{A} \to \mathsf{PG}(6d + 7, \mathbb{K}) : (x_1, x_2, x_3, X_1, X_2, X_3)$$

$$\mapsto (p_{ijk})_{ijk \in D}.$$

Then we define the variety $\mathcal{V}^{\dagger}(\mathbb{K}, \mathbb{A})$ as the Zariski closure of the image of $\rho_{\mathbb{A}}^{\dagger}$.

Remark 8.4 One could wonder how exactly to calculate the 3×3 determinants when some entries belong to non-commutative and even non-associative structures. The answer is that it does not matter so much as long as the calculations are mutually consistent with cyclic permutations of the indices. For instance, once p_{145} is chosen as $x_2 X_2 - \overline{X}_3 \overline{X}_1$, we have to set p_{256} equal to $\overline{X}_1 \overline{X}_2 - x_3 X_3$. Concerning p_{456}, any way we place brackets to ensure the result belongs to $\mathbb{K}$, is fine. For instance,

$$p_{456} = x_1 x_2 x_3 + X_1(X_2 X_3) + (\overline{X}_3 \overline{X}_2)\overline{X}_1 - x_1 X_1 \overline{X}_1 - x_2 X_2 \overline{X}_2 - x_3 X_3 \overline{X}_3.$$

Another choice boils down to a simple recoordinatization. This freedom can be seen as another magic feature of the Square.

Each variety $\mathscr{V}^\dagger(\mathbb{K}, \mathbb{A})$ can also be described as the intersection of a number of quadrics (see [16]). Their equations in case of $\mathbb{A}$ being a split octonion algebra involves some nice combinatorics using the Gosset graph. However, the exact description would take us too far here.

Dual polar spaces – In the case of a division algebra $\mathbb{A}$, the geometry underlying $\mathscr{V}^\dagger(\mathbb{K}, \mathbb{A})$ is a dual polar space of rank 3. In this case, there is an efficient way to write the equations of the above mentioned quadrics down. The following is theorem 10.38 of [13].

Theorem 8.5 *Let $\mathbb{A}$ be a finite-dimensional alternative quadratic division algebra over $\mathbb{K}$ and set $d = \dim_\mathbb{K} \mathbb{A}$. Let V be the $(6d + 8)$-dimensional vector space over $\mathbb{K}$ consisting of the direct sum $\mathbb{K}^4 \oplus \mathbb{A}^3 \oplus \mathbb{K}^3 \oplus \mathbb{A}^3 \oplus \mathbb{K}$. We label the coordinates according to the generic point $(x, \ell_1, \ell_2, \ell_3, X_1, X_2, X_3, k_1, k_2, k_3, Y_1, Y_2, Y_3, y)$. Then the intersection of the $12d + 7$ quadrics in $\mathrm{PG}(V)$ with following equation is the point set of the variety $\mathscr{V}^\dagger(\mathbb{K}, \mathbb{A})$:*

$$
\begin{aligned}
&0 = xk_1 + \ell_2\ell_3 - X_1\overline{X}_1, \quad &&0 = xY_1 + X_2X_3 - \ell_1\overline{X}_1, \quad &&0 = k_2\overline{X}_1 + \ell_3Y_1 + X_2\overline{Y}_3, \\
&0 = xk_2 + \ell_3\ell_1 - X_2\overline{X}_2, \quad &&0 = xY_2 + X_3X_1 - \ell_2\overline{X}_2, \quad &&0 = k_3\overline{X}_1 + \ell_2Y_1 + \overline{Y}_2X_3, \\
&0 = xk_3 + \ell_1\ell_2 - X_3\overline{X}_3, \quad &&0 = xY_3 + X_1X_2 - \ell_3\overline{X}_3, \quad &&0 = k_3\overline{X}_2 + \ell_1Y_2 + X_3\overline{Y}_1, \\
&0 = y\ell_1 + k_2k_3 - Y_1\overline{Y}_1, \quad &&0 = yX_1 + Y_3Y_2 - k_1\overline{Y}_1, \quad &&0 = k_1\overline{X}_2 + \ell_3Y_2 + \overline{Y}_3X_1, \\
&0 = y\ell_2 + k_3k_1 - Y_2\overline{Y}_2, \quad &&0 = yX_2 + Y_1Y_3 - k_2\overline{Y}_2, \quad &&0 = k_1\overline{X}_3 + \ell_2Y_3 + X_1\overline{Y}_2, \\
&0 = y\ell_3 + k_1k_2 - Y_3\overline{Y}_3, \quad &&0 = yX_3 + Y_2Y_1 - k_3\overline{Y}_3, \quad &&0 = k_2\overline{X}_3 + \ell_1Y_3 + \overline{Y}_1X_2
\end{aligned}
$$

and $0 = xy + \ell_1k_1 - \ell_2k_2 - \ell_3k_3 - X_1Y_1 - \overline{Y}_1\overline{X}_1$.

Also, no quadric can be omited; that is, the intersection of each proper subset of this set of $12d + 7$ quadrics contains points off $\mathscr{V}^\dagger(\mathbb{K}, \mathbb{A})$.

Concerning the cases $\mathbb{A} = \mathbb{K} \times \mathbb{K}$ and $\mathbb{A}$ a quaternion algebra, their exist elegant geometric constructions of the varieties $\mathscr{V}^\dagger(\mathbb{K}, \mathbb{A})$ using varieties related to the North-East cell. In particular, we can construct the Grassmannian $\mathscr{G}_{3,6}(\mathbb{K})$ (the cell of type $\mathsf{A}_{5,3}$) using the smaller Grassmannian $\mathscr{G}_{2,5}(\mathbb{K})$ (related to the cell of type $\mathsf{A}_{5,2}$), and we can construct the half spin geometry $\mathscr{HS}_6(\mathbb{K})$ (the cell of type $\mathsf{D}_{6,6}$) using the half spin geometry $\mathscr{HS}_5(\mathbb{K})$ (related to the cell of type $\mathsf{E}_{6,1}$).

The plane Grassmannian $\mathscr{G}_{3,6}(\mathbb{K})$ – Let us construct $\mathscr{G}_{3,6}(\mathbb{K})$ in a completely geometric way. We proceed in three steps. (Proofs are easy and left to the reader.)

(Step 1) Let π_1 and π_2 be two disjoint planes in $\mathrm{PG}(5, \mathbb{K})$. Let α be a linear duality between π_1 and π_2; that is, an incidence preserving bijective map from the points and lines of π_1 to the lines and points, respectively, of π_2 preserving the cross ratio. Taking the union of all planes of $PG(5, \mathbb{K})$ joining a point $p \in \pi_1$ to its image $p^\alpha \subseteq \pi_2$, we obtain $\mathscr{G}_{2,4}(\mathbb{K})$.

(Step 2) It is well known that the points of $\mathscr{G}_{2,4}$ can be identified with the lines of $PG(3, \mathbb{K})$ in such a way that collinear points go to concurrent lines. Embed $\mathscr{G}_{2,4}(\mathbb{K})$ in a 5-space W of $PG(9, \mathbb{K})$ and choose a 3-space U disjoint from W. Let β be a bijective map from the point set of $\mathscr{G}_{2,4}(\mathbb{K})$ to the set of lines of U so that collinear points get mapped to concurrent lines, and so that β preserves the cross ratio. Taking the union of all planes of $PG(9, \mathbb{K})$ joining a point $p \in \mathscr{G}_{2,4}(\mathbb{K})$ to its image $p^\beta \subseteq U$, we obtain $\mathscr{G}_{2,5}(\mathbb{K})$.

(Step 3) Let Π_1 and Π_2 be two copies of $\mathscr{G}_{2,5}(\mathbb{K})$ in disjoint 9-spaces of $PG(19, \mathbb{K})$. We can consider Π_2 as a copy of $\mathscr{G}_{3,5}(\mathbb{K})$ and hence there is a bijective map γ from the set of points of Π_1 to the set of planes of Π_2 which are not contained in a 3-space contained in Π_2 mapping collinear points to planes sharing a point. Again we can choose γ such that it preserves the cross ratio. Taking the union of all solids of $PG(19, \mathbb{K})$ joining a point $p \in \Pi_1$ to its image $p^\gamma \subseteq \Pi_2$, we obtain $\mathscr{G}_{3,6}(\mathbb{K})$.

The half spin geometry $\mathscr{HS}_6(\mathbb{K})$ – Here, we also proceed in three steps.

(Step 1) Let Σ_1 and Σ_2 be two disjoint solids – that is, subspaces of dimension 3 – in $PG(7, \mathbb{K})$. Let α be a linear duality between Σ_1 and Σ_2, that is, an incidence preserving bijective map from the points, lines and planes of Σ_1 to the planes, lines and points, respectively, of Σ_2 preserving incidence and the cross ratio. Taking the union of all solids of $PG(7, \mathbb{K})$ joining a point $p \in \Sigma_1$ to its image $p^\alpha \subseteq \Sigma_2$, we obtain $\mathscr{HS}_4(\mathbb{K})$, which is isomorphic to a hyperbolic quadric (the triality quadric, or also called the quadric of Study).

(Step 2) It is well known that the points of $\mathscr{HS}_4(\mathbb{K})$ can be identified with one class of solids of $\mathscr{HS}_4(\mathbb{K})$ in such a way that collinear points go to solids sharing a line (this is due to triality). Embed two copies, say Ω_1 and Ω_2 of $\mathscr{HS}_4(\mathbb{K})$ in disjoint 7-spaces of $PG(15, \mathbb{K})$ and choose a bijective mapping β from the point set of Ω_1 to one class of solids of Ω_2 such that collinear points are mapped to solids sharing a line and so that the cross ration is preserved. Taking the union of all 4-spaces of $PG(15, \mathbb{K})$ joining a point $p \in \Omega_1$ to its image $p^\beta \subseteq \Omega_2$, we obtain $\mathscr{HS}_5(\mathbb{K})$.

(Step 3) Let Π_1 and Π_2 be two copies of $\mathscr{HS}_5(\mathbb{K})$ in disjoint 15-spaces of $PG(31, \mathbb{K})$. Considering the set of maximal 4-spaces of $\mathscr{HS}_5(\mathbb{K})$, there is a bijective map γ from the set of points of Π_1 to the set of maximal 4-spaces of Π_2 mapping collinear points to 4-spaces sharing a plane. Again we can choose γ such that it preserves the cross ratio. Taking the union of all 5-spaces of $\mathsf{PG}(31,\mathbb{K})$ joining a point $p \in \Pi_1$ to its image $p^\gamma \subseteq \Pi_2$, we obtain $\mathscr{HS}_6(\mathbb{K})$.

A similar construction of $\mathscr{V}^\dagger(\mathbb{K},\mathbb{A})$ for $\mathbb{A}$ the split octonion algebra over $\mathbb{K}$ is not available. There exists a more involved algebraic construction, but it falls beyond the scope of this chapter.

6.9 Connectivity of the Square

We now describe various results connecting in a systematic way cells of the Magic Square to other cells.

6.9.1 Global Connectivity: Equator Geometries and Residues

Diagrammatically, the first row and first column of the Absolute Magic Square seem to play an isolated role in that each cell is the type of a residue of a building of the type in any cell South, East and anywhere South-East of it, except if the original cell is in the first row or first column.

So we consider the South-East 3×3 corner of the Complexified Geometric Magic Square. The connectivity we would like to explain here is that of taking a point residual when going vertically North and taking an equator geometry when going horizontally West.

Going North: Point residual – Let $(X, \mathscr{L})$ be the Lie incidence geometry over the field $\mathbb{K}$ of type $\mathsf{A}_{5,3}$, $\mathsf{D}_{6,6}$, $\mathsf{E}_{7,7}$, $\mathsf{E}_{6,2}$, $\mathsf{E}_{7,1}$ or $\mathsf{E}_{8,8}$ (for type A, this may even be a skew field). Let $p \in X$ be arbitrary. Then the *point residual at* p is the geometry $(\mathscr{L}_p, \mathscr{P}_p)$, where $\mathscr{L}_p$ is the set of lines that contain p, and $\mathscr{P}_p$ the set of planes that contain p (and natural inclusion). This amounts to the same notion as the residue in the corresponding spherical building. Then $(\mathscr{L}_p, \mathscr{P}_p)$ is the Lie incidence geometry over $\mathbb{K}$ of respective type $\mathsf{A}_{2,1} \times \mathsf{A}_{2,1}$, $\mathsf{A}_{5,2}$, $\mathsf{E}_{6,1}$, $\mathsf{A}_{5,3}$, $\mathsf{D}_{6,6}$ and $\mathsf{E}_{7,7}$, that is, the type just North of the original type.

The same thing holds for the geometries in the fourth row of the Relative Geometric Square: the point residuals of the Lie incidence geometry for $\mathbb{A}$ a quadratic alternative division algebra, is

Going West: Equator geometry – Let $\Delta = (X, \mathscr{L})$ be the (split) Lie incidence geometry over the (skew) field $\mathbb{K}$ of type $\mathsf{C}_{3,2}$, $\mathsf{A}_{5,2}$, $\mathsf{D}_{6,6}$, $\mathsf{E}_{7,1}$, $\mathsf{F}_{4,4}$, $\mathsf{E}_{6,1}$, $\mathsf{E}_{7,7}$ or $\mathsf{E}_{8,8}$. Let t be the type of the corresponding building corresponding to the objects whose residues are buildings of the type just West in the Absolute Square. We observe that t is always a self-opposite type. We choose two opposite objects, Ω_1, Ω_2 of type t, and denote by $E(\Omega_1, \Omega_2)$ the set of points in X exactly in the middle on a geodesic from Ω_1 to Ω_2 in the full incidence graph of the corresponding building. Endow $E(\Omega_1, \Omega_2)$ with the set of lines contained in it, then $E(\Omega_1, \Omega_2)$ is the Lie incidence geometry over $\mathbb{K}$ of respective type $A_{2,\{1,2\}}$, $A_{2,1} \times A_{2,1}$, $A_{5,3}$, $E_{6,2}$, $C_{3,2}$, $A_{5,2}$, $D_{6,6}$ or $E_{7,1}$; that is, the type just West of the original type.

This definition of $E(\Omega_1, \Omega_2)$ is perhaps not very intuitive, and not completely geometric, rather combinatorial. For each particular type, there is an individual more geometric description. Let us review these quickly.

(i) **Case $C_{3,2}$.** Here, Ω_1 and Ω_2 are two opposite singular planes of Δ and $E(\Omega_1, \Omega_2)$ is the set of points of Δ collinear to a line of Ω_1 and one of Ω_2.

(ii) **Case $A_{5,2}$.** Here, Ω_1 and Ω_2 are two opposite planes of Δ that are maximal singular subspaces and $E(\Omega_1, \Omega_2)$ is the set of points of Δ collinear to a line of Ω_1 and one of Ω_2.

(iii) **Case $D_{6,6}$.** Here, Ω_1 and Ω_2 are two opposite singular 5-spaces of Δ and $E(\Omega_1, \Omega_2)$ is the set of points of Δ collinear to a plane of Ω_1 and one of Ω_2.

(iv) **Case $E_{7,1}$.** Here, Ω_1 and Ω_2 are two opposite subgeometries (called *paras* in [9]) of type $E_{6,1}$ of Δ and $E(\Omega_1, \Omega_2)$ is the set of points of Δ collinear to a 5-space of Ω_1 and one of Ω_2.

(v) **Case $F_{4,4}$.** Here, Ω_1 and Ω_2 are two opposite symplecta of Δ and $E(\Omega_1, \Omega_2)$ is the set of points of Δ collinear to a line of Ω_1 and one of Ω_2.

(vi) **Case $E_{6,1}$.** Here, Ω_1 and Ω_2 are two opposite singular 5-spaces of Δ and $E(\Omega_1, \Omega_2)$ is the set of points of Δ collinear to a solid of Ω_1 and one of Ω_2.

(vii) **Case $E_{7,7}$.** Here, Ω_1 and Ω_2 are two opposite symplecta of Δ and $E(\Omega_1, \Omega_2)$ is the set of points of Δ collinear to a singular 5-space of Ω_1 and one of Ω_2.

(viii) **Case $E_{8,8}$.** Here, Ω_1 and Ω_2 are two opposite points of Δ and $E(\Omega_1, \Omega_2)$ is the set of points x of Δ such that $\{x, \Omega_1\}$ and $\{x, \Omega_2\}$ are both symplectic pairs.

The Lie incidence geometries of the first column are not equator geometries of those of the second column. However, the results of Section 6.6 in [9] show that they are the intersection of two equator geometries of the geometries of the third column. More exactly, and with the above notation, by considering a(n arbitrary) third object Ω_3 of type t opposite both Ω_1 and Ω_2 and then intersecting $E(\Omega_1, \Omega_2) \cap E(\Omega_1, \Omega_3)$.

6.9.2 The Delayed Magic Square

Recall that the Relative Magic Square originates from the Absolute Magic Square by Galois descent. A characteristic feature of this Galois descent is that in the cases of the Magic Square, the Galois group had order 2. Hence the geometries of the Relative Geometric Magic Square (and also the corresponding varieties) are fixed point sets of the geometries of the Complexified Geometric Magic Square (and of the corresponding varieties, respectively) under the action of an involution. That involution is semi-linear in the sense that there exists a (unique) involutive field automorphism σ of the underlying field such that, if c is the cross-ratio of four points on a line, then c^σ is the cross-ratio of the image of the four points (which are usually also four points on a line, but could also be points in the dual geometry). Now, in each case, there also exists an involution that is linear (preserving cross-ratio rather than transforming it through a field automorphism) and has the same fix diagram as the Galois descent case. Magically, each fix geometry of such linear involution is isomorphic to the fix geometry of the Galois involution in the cell directly West to it, except for the ones in the first column, since there the linear involution is just the identity.

In the next table, we present the fix diagrams of the linear involutions together with the diagram of the fixed point set (which is a building, or a geometry depending on the point of view).

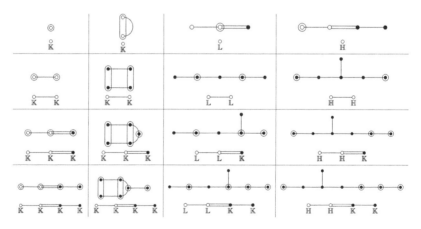

We call this the *Delayed Magic Square*. Some magic properties are mentioned in Section 6.9.4 below. For now we content ourselves with another surprising connection. In order to explain and motivate it, we need some preparation.

6.9.3 Minimum Number of Quadrics for the Varieties of the Second Row

Just as the Cartan variety was defined using a trilinear form, we can also define the other varieties $\mathscr{V}$ of the second row of the Complexified Geometric Magic Square using a trilinear form on a vector space V of dimension $n + 1 \in \{6, 9, 15, 27\}$. For the second and third cell, the definition is just the same:

$$C\left((x_p)_{p \in X}\right) = \sum_{\{p,q,r\} \in \mathscr{S}} X_p X_q X_r - \sum_{\{p,q,r\} \in \mathscr{L} \setminus \mathscr{S}} X_p X_q X_r,$$

where now for the second cell (the variety $\mathscr{S}_{2,2}(\mathbb{K})$) $(X, \mathscr{L})$ is a generalized quadrangle $\mathsf{GQ}(2, 1)$ of order $(2, 1)$; that is, a 3×3 grid, and $\mathscr{S}$ is a spread of $\mathsf{GQ}(2, 1)$; for the third cell (the variety $\mathscr{G}_{2,6}(\mathbb{K})$) $(X, \mathscr{L})$ is a generalized quadrangle $\mathsf{GQ}(2, 2)$ of order $(2, 2)$, and $\mathscr{S}$ is a spread of a subquadrangle of order $(2, 1)$. For $\mathscr{S}_{2,2}(\mathbb{K})$ this amounts to the determinant of a 3×3 matrix with entries the nine coordinates; for $\mathscr{V}_2(\mathbb{K})$, it is the determinant of a 3×3 symmetric matrix.

Now let ϕ be the operator from V to V mapping a vector with coordinates $(x_p)_{p \in X}$ to $\nabla C(v)$. We then have $\phi(\phi(v)) = C(v)v$. The points $\langle v \rangle$ with $\phi(v) \neq \vec{o}$ but $C(v) = 0$ are called *grey points* [2] with respect to $\mathscr{V}$. Their image under ϕ belongs to $\mathscr{V}$. They are precisely the points lying on a tangent or secant line of $\mathscr{V}$.

Each coordinate of $\nabla C\left((x_p)_{p \in X}\right)$ defines a quadratic form, and so $\mathscr{V}$, being the set of all projective points $\langle v \rangle$ with $\phi(v) = \vec{o}$, is the intersection of $n + 1$ quadrics. All these quadrics are linearly independent, and it is shown in [33] that (the equation of) every quadric containing $\mathscr{V}$ is a linear combination of (the equations of) these $n + 1$ quadrics. However, it is nevertheless possible to produce a smaller number of quadrics with common intersection exactly $\mathscr{V}$. Indeed, the following theorem is proved in [33].

Theorem 9.1 *Let $\mathscr{V}$ be either the quadratic Veronese surface $\mathscr{V}_2(\mathbb{K})$, the Segre variety $\mathscr{S}_{2,2}(\mathbb{K})$, the line Grassmannian $\mathscr{G}_{2,6}(\mathbb{K})$, or the Cartan variety $\mathscr{E}_6(\mathbb{K})$, in n-dimensional projective space $\mathsf{PG}(n, \mathbb{K})$ over $\mathbb{K}$, with $n = 5, 8, 14, 26$, respectively. Then $\mathscr{V}$ is the intersection of $n - d$ quadrics and no less, where d is the dimension of a maximum dimensional projective subspace of $\mathsf{PG}(n, \mathbb{K})$ consisting solely of grey points. More precisely, the equivalence classes of the systems of $n - d$ linearly independent quadrics intersecting precisely in $\mathscr{V}$ are in natural bijective correspondence with the d-dimensional projective subspaces of $\mathsf{PG}(n, \mathbb{K})$ consisting solely of grey points.*

One could ask what the largest subspaces of grey points are. And here we come to some unexpected magic with the Delayed Magic Square. Indeed, as noted, the image of each grey point under ϕ is a point of $\mathscr{V}$. Now it just happens so that, across all possible fields, the largest subspaces consisting entirely of grey points that are known have as image under ϕ precisely the subvariety given in the Delayed Magic Square obtained as fixed point set of a linear involution (this is also true for the first cell!)

As an example, taking as the cubic form for $\mathscr{S}_{2,2}(\mathbb{K})$ the determinant of the 3×3 matrix of coordinates x_{ij} (with self-explaining notation, $1 \leq i \leq 3$, $1 \leq j \leq 3$), a plane of grey points is determined by all anti-symmetric matrices. Noting that in this case $\nabla C\left((x_{ij})_{1 \leq i \leq 3, 1 \leq j \leq 3}\right)$ is the adjugate matrix of $(x_{ij})_{1 \leq i \leq 3, 1 \leq j \leq 3}$, one easily calculates that the global image of the set of anti-symmetric matrices under ϕ is exactly the set of symmetric matrices; that is, the image under ϕ of the corresponding plane of grey points is a subvariety of $\mathscr{S}_{2,2}(\mathbb{K})$ isomorphic to $\mathscr{V}_2(\mathbb{K})$.

6.9.4 Opposition Diagrams

The magical twisted symmetry of the Fix Magic Square is the basis of the notion of an *opposition diagram* (called *domesticity diagram* in [31] when it was introduced in the literature for the first time).

Let θ be an automorphism of a spherical building Δ. Then the *opposition diagram* of θ is the Coxeter diagram of Δ where a set of nodes is encircled if there exists a flag of that type mapped under θ to an opposite flag, and no subflag of it has the same property. It is drawn in a bent way if the action of θ on the types does not coincide with the opposition relation. It is shown in [18] that, as soon as Δ has rank at least 3 and does not contain projective planes of order 2 as some residue, then there exists a flag of type the union of all encircled nodes mapped to an opposite flag. Hence, for such buildings, θ maps some chamber to an opposite if and only if in the opposition diagram of θ, every node is in an encircled set of nodes (such diagrams are called *full*). If the opposition diagram has nodes that are not encircled, then we call θ *domestic*. Equivalently, θ is domestic if it does not map any chamber to an opposite. For instance, if not all nodes of the diagram of a spherical building are polar (recall that the *polar nodes* are the nodes corresponding to the fundamental roots that are not perpendicular to the highest root), then every central collineation (in other words, every long root elation) is domestic and in the opposition diagram exactly the polar nodes are encircled. Note that all nodes are polar if and only if the diagram is A_1 or A_2, and, indeed, in these cases, central collineations are not domestic. (All (irreducible) Dynkin diagrams admit exactly one polar node except A_n, $n \geq 2$, which admits two of them; namely, the two end nodes.)

Despite the immediate and easy example in the previous paragraph, domestic automorphisms seem to be rather rare, especially when we additionally assume that the automorphism also does not fix any chamber. In that case, it is also likely that the fix diagram is not full. More background and details about domestic collineations, including when the building does contain planes of order 2 as residues, can be found in the papers [18], [19], [20], where buildings containing no planes of order 2 as residue are called *large*.

The main conjecture on domestic automorphisms and the Magic Square is a thickening of the twisted symmetry of the Fix Magic Square (within the first row diagrams without encircled nodes to include the identity).

Conjecture 9.2   (*i*) *For each large spherical building with diagram in the Fix Magic Square, all domestic automorphisms with the given fix diagram are known and their opposition diagram is the fix diagram in the symmetric (with respect to the main diagonal) cell, except for the second half of the first column, where other possibilities for the opposition diagram than just those on the first row occur, if no subbuilding is fixed. Moreover, the fixed building in each case corresponds to the building of the Delayed Magic Square in the same cell.*

(*ii*) *Conversely, for each large spherical building with diagram in the Fix Magic Square, such that the corresponding fix diagram is not full, all automorphisms not fixing any chamber and with given opposition diagram are known and their fix diagram is the fix diagram in the symmetric cell. The fixed building is the corresponding building in the Delayed Magic Square.*

It is tacitly assumed that in the third column of the Delayed Magic Square the quadratic extension of the base field may also be inseparable in characteristic 2, and the quaternion division algebra in the fourth column may be a degree 4 inseparable extension in characteristic 2.

However, there is a counter example to (*ii*) of this conjecture for E_7. The renewed conjecture states that there are exactly two well-known counter examples to part (*ii*) of the old conjecture, one for E_7 and one for F_8. We explain this now in some more detail, reviewing the different cells of the Square. The (i, j)-cell will mean the cell in row i and column j, $1 \leq i \leq 4$, $1 \leq j \leq 4$.

The first column and the first row. Here, the conjecture is trivial in view of the fact that the opposition diagram of an automorphism has no encircled nodes if and only if the automorphism is the identity (this follows from [1], where it is proved that every nontrivial automorphism of a spherical building maps some flag to an opposite), and the fact that, in large split buildings of type C_3 and F_4 only the identity is a domestic collineation fixing a subbuilding (the

320   *Hendrik Van Maldeghem*

latter follows from [26] for type A_2, from the last section of [25] for type C_3, and from [20] for type F_4).

The cell with diagram $A_2 \times A_2$. This fix diagram is full and symmetric to itself, so nothing to prove here. The conjecture is trivially true.

The cells with diagram A_5. It is shown in [26] that every domestic collineation of a large projective space fixes at least one point. Hence the fix diagram •——⊚——•——⊚——• can never belong to a domestic collineation, showing in a trivial way part (*i*) of Conjecture 9.2 for the (2,3)-cell. Now let θ be an automorphism with fix diagram ⊞⊞⊕ . Since the diagram is bent, θ is a duality. Domestic dualities in large projective spaces are classified in [26], and exactly the symplectic polarities qualify. For 5-dimensional spaces they have opposition diagram exactly •——⊚——•——⊚——•. The fixed building is a symplectic polar space, which is exactly the (3,2)-cell in the Delayed Magic Square. Hence also part (*i*) of Conjecture 9.2 holds for the (3,2)-cell. Part (*ii*) of the conjecture follows immediately from part (*i*) and the uniqueness of domestic dualities in projective spaces.

The cell with diagram D_6. It is shown in [21] that a domestic automorphism of a large building of type D_6 either fixes at least one point, or has opposition diagram •——⊚——•——⊚——⊚. Moreover it is shown in [21] that θ being domestic with this opposition diagram is equivalent to θ being linear, the fixed line set being a partition of the point set and arising from a Hermitian polar space of rank 3 in projective 5-space over a quadratic extension $\mathbb{L}$ of the base field $\mathbb{K}$ by so-called "field reduction" (considering $\mathbb{L}$ as a vector space of dimension 2 over $\mathbb{K}$, the point set of the Hermitian polar space in 5-dimensional projective space over $\mathbb{L}$ becomes a set (spread) of lines of the hyperbolic quadric in 11-dimensional projective space over $\mathbb{K}$). This Hermitian polar space corresponds exactly to the Delayed Magic Square at position (3,3). Moreover, the fix diagram is the same as the opposition diagram. This shows the conjecture in this case. (But note that $\mathbb{L}$ may be inseparable, in which case "Hermitian" would rather be replaced with "mixed".)

We now come to the most interesting cells, those containing the exceptional diagrams of any type E.

The cells with diagram E_6. The situation here is completely analogous to the situation of projective spaces (the cells with diagram A_5). Indeed, if the fix diagram of an automorphism θ is ⊞⊞⊕——⊚, then θ is a duality, and it is shown in [31] that the only domestic dualities in large buildings of type E_6 are the symplectic dualities, which have as fix point building a split building of type F_4 (the $(4, 1)$-entry in the Delayed Magic Square). If θ is an automorphism having fix diagram ⊙———•——╎——•———⊙, then θ is a collineation. But from the table of opposition diagrams in [18], it follows that every domestic collineation, and hence also θ, fixes some object of type 2, a contradiction. Conjecture 9.2 follows for type E_6.

From now on, results are either not published yet, or even still in progress. So this serves rather as a survey of near-future results than of known results.

The cells with diagram E_7. This is the unique type for which the two distinct fix diagrams are both not full. So, Conjecture 9.2 hints at the existence of exactly two (classes of) domestic collineations fixing no chamber. But, according to the revised conjecture, there is a third. Here are the details (not published yet).

First we note that it is already shown in [20] that the two fix diagrams in the fix Magic Square of E_7 are the only two opposition diagrams for domestic collineations not fixing any chamber. Now first suppose that some collineation θ has opposition diagram ⊙——•——╎——•——⊙—⊙. This opposition diagram is realized for collineations which are the product of three perpendicular root elations, but if θ does not fix any chamber, then one can show that there is a quadratic extension $\mathbb{L}$ (may be inseparable in characteristic 2) such that the fixed building is ○——○══•——•, with labels $\mathbb{L}$ $\mathbb{L}$ $\mathbb{K}$ $\mathbb{K}$. The corresponding fix diagram is exactly •—⊙—•—⊙—⊙—⊙. Conversely, if the latter is the fix diagram of a domestic collineation θ, then θ obviously does not fix any chamber. It can be shown that the only possible opposition diagrams are the two fix diagrams under consideration. But a collineation with opposition diagram

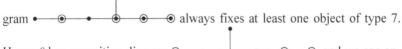

always fixes at least one object of type 7. Hence θ has opposition diagram ⊙———•——╎——•——⊙—⊙, and we can apply the first arguments again.

Now suppose that some collineation θ has opposition diagram •—⊙—•——╎——⊙—⊙. If θ does not fix any chamber, then one can show that there are two possibilities.

(1) Either there exists a quaternion division algebra $\mathbb{H}$ over the ground field $\mathbb{K}$ (possibly inseparable) such that the fixed building is ○——○══•, with labels $\mathbb{H}$ $\mathbb{H}$ $\mathbb{K}$, and the corresponding fix diagram is exactly ⊙———•——╎——•——⊙—⊙;

(2) or there exist two opposite objects Ω_1, Ω_2 of type 1 such that the equator geometry $E(\Omega_1, \Omega_2)$ in the Lie incidence geometry of type $\mathsf{E}_{7,1}$ is pointwise fixed, and θ acts fixed point freely on the set $\mathscr{I}$ of all objects Ω of type 1 such that Ω is opposite at least one of Ω_1 and Ω_2, and $E(\Omega, \Omega_i) = E(\Omega_1, \Omega_2)$, for $i \in \{1, 2\}$ such that Ω is opposite Ω_i. The set $\mathscr{I}$ is a kind of imaginary line; its stabilizer (in the adjoint group) acts on it like $\mathsf{PSL}_2(\mathbb{K})$, and it takes a quadratic extension of $\mathbb{K}$ to have a fixed point free action on it (a 2×2 matrix with imaginary eigenvalues). This is the class of counter examples appearing in the revised conjecture.

Conversely, suppose θ is a domestic collineation with fixed diagram

⊙———•——┴——•———⊙——⊙. Since no chamber is fixed, θ is domestic, and we already know that collineations with opposition diagram

⊙———•——┴——•———⊙——⊙ have different fixed point structure, the opposition

diagram of θ is •———⊙———•——┴——⊙———⊙. This would prove (the revision of) Conjecture 9.2 for the case of E_7.

The cell with diagram E_8. It is already shown in [20] that the only opposition diagram of any domestic collineation of a building of type E_8 fixing no

chamber is ⊙———•——┴——•———⊙——⊙——⊙. Hence in order to complete the conjecture, it suffices to show that any domestic collineation with that opposition diagram has the same diagram as fix diagram, and that the corresponding fixed building is $\underset{\mathbb{H}}{\circ}—\underset{\mathbb{H}}{\circ}—\underset{\mathbb{K}}{•}———•$. However, the revised conjecture states that there is a second possibility, which has a full fix diagram. This collineation is very similar to the counter example in the case of E_7; it is also defined using the equator geometries of objects of type 8. Proofs are in progress here.

6.9.5 Facts and Figures

To conclude, we emphasize the connectivity of the Square with some facts and figures, showing (1) increasing complexity going East and South, and (2) similarity between objects on the same row or column.

We start with a simple numerical observation. We saw that the first row of the Fix Magic Square leaves room for two interpretations. In the interpretation of involutions of Coxeter systems, the very first cell (at the utmost North-West position) is in contradiction with itself, and so for this moment we assume it corresponds to the empty cell, a Coxeter complex of rank 0. Writing down the ranks of the Coxeter systems for each cell, we obtain

0	2	3	4
2	4	5	6
3	5	6	7
4	6	7	8

and we observe that the rank in position (i, j) is the sum of the ranks in positions $(i, 1)$ and $(1, j)$, for all $\{i, j\} \subseteq \{1, 2, 3, 4\}$.

Common features of cells on the same columns are the following.

1. In the construction of the varieties in the Complexified Geometric Magic Square, and also in the definition of the Relative Magic Square, we saw that the first column uses a field, the second a field with a quadratic extension, the third a field with a quaternion algebra, and the fourth a field with an octonion algebra.

2. In the Fix Magic Square, interpreted as indices of semi-simple algebraic groups, the first column contains only split forms, the second quasi-split forms, the third non-split forms with small disconnected anisotropic kernels (direct products of A_1), and the fourth non-split with a rather large connected anisotropic kernel (of type D_4).

3. Overall, the Galois groups of the indices always have order 2; that is, the Galois descent is always with an involution. In the Delayed Magic Square, excluding the case of characteristic 2, the analogue of the Galois group is the group which pointwise fixes the subbuilding; in the first column it is the identity, in the second column it is always a group of order 2, in the third column it is the multiplicative group of norm 1 elements of a quadratic Galois extension, in the fourth column it is the multiplicative group of norm 1 elements of a quaternion algebra. This is a quite remarkable series, noting that the group of order 2 in the case of the second column can also be defined as the multiplicative group of elements of a field (in characteristic different from 2) with norm 1, where the norm here is just squaring (which is in conformity with the standard involution being the identity).

Common features of the cells on the same rows are the following.

1. The ranks of the buildings in the Relative Magic Square are the same as the row number. This is also equal to the number of encircled sets of vertices in the Fix Magic Square.

2. In the Relative Geometric Magic Square, the geometries in the first row are so-called Tits webs (coming from Moufang sets) and can be considered as

unitals or, in Tits' language [28], σ-conics, in the second row projective planes, in the third row dual polar spaces (of diameter 3), and in the fourth row metasymplectic spaces.

3. In the Complexified Geometric Magic Square, the second row contains strong parapolar spaces of diameter 2 (except for the first cell, where projective planes occur; although if we view each line of a projective plane as a polar line, as explained in Section 6.3, we obtain a parapolar space without lines. Compare with the one without symplecta we will meet in the symmetric cell below), the third row strong parapolar spaces of diameter 3 and the fourth row non-strong parapolar spaces of diameter 3 all of which are long root geometries. The first row, one would ask? This contains what we could call "would-be" long root parapolar spaces: the building in the first cell is really too small to be a sensible geometry; the second cell is the long root geometry for A_2, but this does not contain symplecta, so it can hardly be seen as a parapolar space; the third cell is the line Grassmannian of a symplectic polar space, which is non-strong of diameter 3, but not a "pure" long root geometry since in the symplectic case, the first node is the polar one, and not the second, as is the case for B_3. Similarly for the fourth cell: although there is seemingly symmetry here, it is the node with (Bourbaki) label 1 which is the polar node, and here we have the node with label 4 (but it also provides a metasymplectic space, a non-strong parapolar space of diameter 3 with thick symplecta of rank 3).

4. Going to the original Magic Square containing the varieties defined in Section 6.8, the first row contains a kind of Veronese representations of the corresponding unitals, the second row contains the Veronese representations of certain projective planes (over alternative quadratic algebras), and over the complex numbers, these are exactly the Severi varieties. The third row constitutes the (analogues of the) Lagrangean manifolds and the fourth row the varieties associated with the adjoint module.

This should convince the reader that the Geometric Magic Square is really an exceptionally beautiful object, well worth studying in some more detail.

References

[1] P. Abramenko & K. Brown, Automorphisms of non-spherical buildings have unbounded displacement, *Innov. Incid. Geom.* **10** (2009), 1–13.

[2] M. Aschbacher, The 27-dimensional module for E_6, I., *Invent. Math.* **89** (1987), 159–195.

[3] F. Buekenhout, Diagrams for geometries and groups, *J. Combin. Theory A* **27** (1979), 121–151.

[4]   F. Buekenhout and A. M. Cohen, Diagram Geometry, Related to Classical Groups and Buildings, *Ergebnisse der Mathematik und ihrer Grenzgebiete*, 3. Folge **57** (2013).

[5]   A. M. Cohen, An axiom system for metasymplectic spaces, *Geom. Ded.* **12** (1982), 417–433.

[6]   J. H. Conway, R. T. Curtis, S. P. Norton, R. A. Parker & R. A. Wilson, *Atlas of Finite Groups*, Clarendon Press, Oxford, 1985.

[7]   A. Del Fra & D. Ghinelli, Classification of extended generalized quadrangles with maximum diameter, *Discrete Math.* **105** (1992), 13–23.

[8]   A. Del Fra, D. Ghinelli, T. Meixner & A. Pasini, Flag-transitive extensons of C_n geometries, *Geom. Dedicata* **37** (1991), 253–273.

[9]   A. De Schepper, N. S. N. Sastry & H. Van Maldeghem, Buildings of exceptional type in buildings of type E_7, *Dissertationes Math.*, **573** (2022), 1–80.

[10] A. De Schepper, J. Schillewaert and H. Van Maldeghem, A uniform characterisation of the varieties of the second row of the Freudenthal-Tits Magic Square over arbitrary fields, *to appear in J. Comb. Alg.*

[11] A. De Schepper, J. Schillewaert, H. Van Maldeghem & M. Victoor, On exceptional Lie geometries, *Forum Math. Sigma* **9** (2021), paper No e2, 27pp.

[12] A. De Schepper, J. Schillewaert, H. Van Maldeghem & M. Victoor, A geometric characterization of Hjelmslev-Moufang planes, *Quart. J. Math.* **73** (2022), 369–394, https://doi.org/10.1093/qmath/haab043.

[13] A. De Schepper, J. Schillewaert, H. Van Maldeghem & M. Victoor, Construction and characterization of the varieties of the third row of the Freudenthal–Tits magic square, *submitted preprint*.

[14] A. De Schepper & M. Victoor, A geometric connection between the split first and second row of the Freudenthal–Tits magic square, *Innov. Incid. Geom.*, to appear.

[15] O. Krauss, J. Schillewaert and H. Van Maldeghem, Veronesean representations of Moufang planes. *Mich. Math. J.* **64** (2015), 819–847.

[16] W. Lichtenstein, A system of quadrics describing the orbit of the highest weight vector, *Proc. Amer. Math. Soc.* **84** (1982), 605–608.

[17] B. Mühlherr, H. Petersson & R. M. Weiss, *Descent in Buildings*, Annals of Mathematics Studies **190**, Princeton University Press, 2015.

[18] J. Parkinson & H. Van Maldeghem, Opposition diagrams for automorphisms of large spherical buildings, *J. Combin. Theory Ser.* A **162** (2019), 118–166.

[19] J. Parkinson & H. Van Maldeghem, Opposition diagrams for automorphisms of small spherical buildings, *Innov. Incid. Geom.* **17** (2019), 141–188.

[20] J. Parkinson & H. Van Maldeghem, Automorphisms and opposition in spherical buildings of exceptional type, I, *Canad. J. Math.* **74** (2022), 1517–1578, https://doi.org/10.4153/S0008414X21000341.

[21] J. Parkinson & H. Van Maldeghem, Automorphisms and opposition in spherical buildings of classical type, *submitted*.

[22] S. E. Payne & J. A. Thas, *Finite Generalized Quadrangles*, Research notes in Math. **110**, Pittman, 1984; second edition: Europ. Math. Soc. Series of Lectures in Mathematics, 2009.

[23] J. Schillewaert & H. Van Maldeghem, On the varieties of the second row of the split Freudenthal–Tits Magic Square, *Ann. Inst. Fourier* **67** (2017), 2265–2305.

[24] E. E. Shult, *Points and Lines, Characterizing the Classical Geometries*, Universitext, Springer, Heidelberg, 2011.

[25] B. Temmermans, J. A. Thas & H. Van Maldeghem, Collineations of polar spaces with restricted displacements, *Des. Codes Cryptogr.* **64** (2012), 61–80.

[26] B. Temmermans, J. A. Thas & H. Van Maldeghem, Domesticity in projective spaces, *Innov. Incid. Geom.* **12** (2011), 141–149.

[27] J. Tits, Espaces homogènes et groupes de Lie exceptionnels, in *Proc. Internat. Congr. Math.* 1954, vol. 1, Nordhoff, Groningen; North-Holland, Amsterdam (1954), 495–496.

[28] J. Tits, Sur certaines classes d'espaces homogènes de groupes de Lie, *Acad. Roy. Belg. Cl. Sci. Mém. Collect. 8º (2)* **29** (1955), 5–268.

[29] J. Tits, Classification of simple algebraic groups, in *Algebraic groups and discontinuous subgroups*, Proc. Summer Mathematical Inst., Boulder, July 5–August 6, 1965, *Proc. Symp. Pure Math.* **9**, Amer. Math. Soc., Providence, RI (1966), 33–62.

[30] J. Tits, *Buildings of Spherical Type and Finite BN-Pairs*, Springer Lecture Notes Series **386**, Springer-Verlag, 1974.

[31] H. Van Maldeghem, Symplectic polarities in buildings of type E6, *Des. Codes Cryptogr.* **65** (2012), 115–125.

[32] H. Van Maldeghem & M. Victoor, Combinatorial and geometric constructions of spherical buildings, *Surveys in Combinatorics 2019*, Cambridge University Press (ed. A. Lo et al.), *London Math. Soc. Lect. Notes Ser.* **456** (2019), 237–265.

[33] H. Van Maldeghem & M. Victoor, On Severi varieties as intersections of a minimum number of quadrics, *Cubo* **24** (2022), 307–331.

[34] F. L. Zak, *Tangents and Secants of Algebraic Varieties*, Translation of Mathematical Monographs **127**, American Mathematical Society, 1983, vii+164 pages.

7

On the Generation of Polar Grassmannians

Ilaria Cardinali, Luca Giuzzi and Antonio Pasini

Abstract

In this chapter we discuss grassmannians of classical polar spaces of finite rank defined over commutative division rings. We offer an up-to-date overview on the state of the art regarding their generation and the properties of their projective embeddings.

CONTENTS

7.1 Introduction

This chapter is an extended version of a talk given at the conference "Graphs and Groups, Geometries and GAP Summer school – External satellite conference of the 8ECM" in June 2021 by I. Cardinali.

This chapter is mainly about *polar grassmannians*; that is, we take a polar space Γ regarded as a point-line geometry and we construct another point-line geometry selecting as points the family of all singular subspaces of Γ of a given non-maximal dimension d, and as lines the sets of all d-dimensional singular subspaces of Γ through a given subspace of dimension $d - 1$ and all contained in a given singular subspace of dimension $d + 1$.

Polar grassmannians are very interesting objects which gained the attention of many researchers both from a theoretical point of view as well as for their applications. From a theoretical point of view, they are among the simplest gamma spaces we can consider which are not polar spaces and sport a very large automorphism group; from the point of view of applications, if a polar grassmannian is embedded into a projective space, it can be regarded as an algebraic set and used, for instance, to construct codes with good minimum distance and amenable to local decoding; see for example, [8, 9].

In this chapter we focus on theoretical aspects of polar grassmannians which regard the possibility of constructing a polar grassmannian starting from a small subset of its pointset. We shall discuss the effectiveness (achievements and failures) of what we call the *generation method*.

In general, by generation method of a given point-line geometry Γ, we mean the following. Given a subset X of its pointset, the following iterative construction yields the subspace $\langle X \rangle_\Gamma$ of Γ generated by X, namely the smallest subspace of Γ which contains X: we first take the points of X and then we add to them all the points of Γ which lie on the lines connecting any two points of X, thus obtaining a new set of points of Γ. We continue to iterate this construction as long as there are new points to add.

Ultimately, we end up with $\langle X \rangle_\Gamma$. If $\langle X \rangle_\Gamma = \Gamma$ we say that X is a generating set for Γ. The *generating rank* of Γ is the minimum cardinality of a generating set for it. Apart from the intrinsic interest in building up a geometry starting from a small set of points, the study of generating sets of a polar grassmannian Γ is very useful in investigating properties on the way Γ can be embedded in a projective space.

Among the projective embeddings of Γ, a special interest is reserved to the universal one. A natural question is about the existence and characterization of the universal embedding of a given point-line geometry (a polar grassmannian, for instance).

Indeed, suppose we know that Γ admits the universal embedding (as it is often the case) and we are aware of an embedding ε of Γ of finite dimension n and we conjecture that ε is universal. If we discover that Γ can be generated by a set of n points, then we are done: ε is indeed universal. This method works quite well in many cases, but not always. Sometimes it fails, as when all embeddings of Γ we know have dimension at most n but n points are never enough to generate Γ. In some cases, we can even prove that no embedding of Γ exists of dimension greater than n but every generating set of Γ contains more than n points. Failures like this also can occur when Γ is a polar grassmannian. In this chapter, we shall discuss these failures, trying to understand the reason why they occur.

In particular, we shall focus on the following two main topics: how to compute the generating rank of a polar grassmannian by reduction to easier cases where the generating rank is known and, when the polar grassmannian is embeddable, how to exploit the information thus achieved on generating sets in order to identify universal embeddings.

More geometries are involved in this investigation; these cannot be ignored, although they are not polar grassmannians, such as the point-hyperplane geometry of a projective space, which enters the stage when playing with orthogonal grassmannians. We shall briefly discuss them too.

Organization of the Chapter. Since the main topic of the chapter is polar grassmannians and polar grassmannians are defined by polar spaces, we have dedicated the first part of Section 7.2 to polar spaces. Polar grassmannians are also defined in Section 7.2 where we survey their basic properties. In Section 7.3, we report on known results on their generation and embeddings. Section 7.4 is a completion of Section 7.2. In Section 7.5, we discuss constructions of generating sets. In Section 7.6, we explain how the generating rank of an orthogonal grassmannian may depend on the field on which the underlying quadric is defined, and discuss some consequences of this fact. In Section 7.7, we write some comments on the techniques described in the rest of the chapter and we list some open problems.

7.2 Polar Spaces and Their Grassmannians

Our main sources for this section are Tits [45], Shult [42], Buekenhout and Cohen [7] and Pankov [36]. Additional references will be mentioned as needed.

7.2.1 Preliminaries on Partial Linear Spaces

A partial linear space is a point-line geometry $\Gamma = (\mathscr{P}, \mathscr{L})$, where $\mathscr{P}$ is the set of points and $\mathscr{L}$ is the set of lines, such that each line is incident with at least two points and for any two distinct points there is at most one line incident with them both. So, every line $\ell \in \mathscr{L}$ is uniquely determined by the set of the points incident with it; hence we can regard ℓ as a subset of $\mathscr{P}$ and a point p is incident with ℓ if and only if $p \in \ell$. Two points p and q of Γ are *collinear* (in symbols, $p \perp q$) if and only if there exists $\ell \in \mathscr{L}$ such that $p, q \in \ell$. The binary relation $\perp$ defines a graph on $\mathscr{P}$, called the *collinearity graph* of Γ. If this graph is connected, we say that Γ is *connected*. For $p \in \mathscr{P}$, we define $p^\perp := \{x \in \mathscr{P} : x \perp p\} \cup \{p\}$ and for $X \subseteq \mathscr{P}$, we set $X^\perp := \cap_{x \in X} x^\perp$.

7.2.1.1 Subspaces, Generating Sets and Subgeometries

Following Shult [42], a *subspace* of Γ is a subset $X \subseteq \mathscr{P}$ such that if a line $\ell \in \mathscr{L}$ meets X in at least two points, then $\ell \subseteq X$. If furthermore X is a *proper* subspace (namely $X \subset \mathscr{P}$) and every line of Γ meets X non-trivially, then X is said to be a *(geometric) hyperplane* of Γ. The poset of all subspaces of Γ is closed under intersection. Hence the intersection of all subspaces of Γ containing a given subset $X \subseteq \mathscr{P}$ is a subspace of Γ. We call it the *subspace generated* by X and we denote it by $\langle X \rangle_\Gamma$ (also $\langle X \rangle$ for short, when no ambiguity arises). In particular, if $\langle X \rangle_\Gamma = \mathscr{P}$, then X is a *generating set* for Γ.

The *generating rank* $\mathrm{grk}(\Gamma)$ of Γ is the minimum size of a generating set of Γ; that is, $\mathrm{grk}(\Gamma) := \min\{|X| : \langle X \rangle = \mathscr{P}\}$. A *smallest generating set* of Γ is a minimal generating set X of Γ of cardinality $|X| = \mathrm{grk}(\Gamma)$ (we warn that in general minimal generating sets exist of different cardinality; hence, in general, a minimal generating set is not a smallest one).

We have defined subspaces, but subgeometries will also be considered in the sequel. A partial linear space $\Gamma' = (\mathscr{P}', \mathscr{L}')$ is a *subgeometry* of $\Gamma = (\mathscr{P}, \mathscr{L})$ if $\mathscr{P}' \subseteq \mathscr{P}$ and every line $\ell' \in \mathscr{L}'$ is contained in a line of Γ. If moreover every line of Γ' is a line of Γ, namely $\mathscr{L}' \subseteq \mathscr{L}$, then we say that Γ' is a *full subgeometry* of Γ. Clearly, all subspaces of Γ are full subgeometries of Γ.

7.2.1.2 Embeddings

Let $\Gamma = (\mathscr{P}, \mathscr{L})$ be a partial linear space and V a vector space. A *projective embedding* of Γ (an *embedding* of Γ for short) is an injective map $\varepsilon \colon \mathscr{P} \to \mathrm{PG}(V)$ with the property that $\langle \varepsilon(\mathscr{P}) \rangle = \mathrm{PG}(V)$ and every line of Γ is mapped onto a projective line of PG(V). So, $\varepsilon(\mathscr{L}) := \{\varepsilon(\ell)\}_{\ell \in \mathscr{L}}$ is a set of lines of PG(V) and $\varepsilon(\Gamma) := (\varepsilon(\mathscr{P}), \varepsilon(\mathscr{L}))$ is a full subgeometry of (the point-line space of) PG(V).

The *dimension* dim(ε) of ε is defined as the dimension dim(V) of V. If $\mathbb{K}$ is the underlying division ring of V then ε is said to be *defined over* $\mathbb{K}$. If all embeddings of Γ are defined over the same division ring $\mathbb{K}$ then we say that Γ is *defined over* $\mathbb{K}$. Obviously, a partial linear space Γ admits an embedding (defined over $\mathbb{K}$) only if all lines of Γ have the same size $1 + \mathfrak{t} > 2$ (and $\mathfrak{t} = |\mathbb{K}|$).

If $\varepsilon_1 \colon \Gamma \to$ PG(V_1) and $\varepsilon_2 \colon \Gamma \to$ PG(V_2) are two projective embeddings of Γ we say that ε_1 *covers* ε_2 (in symbols, $\varepsilon_2 \leq \varepsilon_1$) if ε_1 and ε_2 are defined over the same division ring and there exists a semilinear mapping $f \colon V_1 \to V_2$ such that $\varepsilon_2 = f \circ \varepsilon_1$ (actually, writing $f \circ \varepsilon_2$ is an abuse, since morphisms of projective spaces are involved here rather than their underlying semilinear maps, but this is a harmless abuse). We say that ε_1 and ε_2 are *equivalent* (in symbols $\varepsilon_1 \simeq \varepsilon_2$) if $\varepsilon_1 \leq \varepsilon_2 \leq \varepsilon_1$. An embedding $\tilde{\varepsilon}$ is said to be *dominant* (also *relatively universal*) if $\tilde{\varepsilon} \leq \varepsilon$ implies $\tilde{\varepsilon} \simeq \varepsilon$. Equivalently, if $\tilde{\varepsilon} \geq \varepsilon$ and $\varepsilon' \geq \varepsilon$, then $\tilde{\varepsilon} \geq \varepsilon'$. Note that every embedding ε is covered by a dominant embedding (Ronan [41]), unique modulo equivalence. This dominant embedding is called the *hull* of ε.

An embedding $\tilde{\varepsilon}$ of Γ is *absolutely universal* (also just *universal*, for short) if it covers all projective embeddings of Γ. Clearly, when it exists, the universal embedding is unique up to equivalence and it is the hull of all embeddings of Γ. In this case, all embeddings of Γ are defined over the same division ring.

The *embedding rank* erk(Γ) of Γ is the least upper bound of the dimensions of the embeddings of Γ (the maximal dimension of an embedding of Γ when all these dimensions are finite and range in a finite set). Suppose that Γ admits the absolutely universal embedding, say $\tilde{\varepsilon}$. Then dim($\tilde{\varepsilon}$) = erk(Γ) and, for any embedding ε of Γ, if erk(Γ) $< \infty$, then $\varepsilon = \tilde{\varepsilon}$ if and only if dim($\tilde{\varepsilon}$) = erk(Γ).

7.2.1.3 The Standard Argument

Assuming Γ to be embeddable, let $\varepsilon \colon \Gamma \to$ PG(V) be a projective embedding of Γ and $X \subseteq \mathscr{P}$ a generating set of Γ. Then $\varepsilon(X)$ spans PG(V), hence $|X| \geq$ dim(ε). Therefore erk(Γ) $\leq$ grk(Γ). In particular, if $|X| =$ dim(ε) $< \infty$, then X is a smallest generating set, ε is dominant and grk(Γ) = erk(Γ) = $|X|$ = dim(ε). If moreover Γ admits the universal embedding, then ε is just the universal embedding of Γ. We call this way of arguing the *Standard Argument*.

7.2.1.4 Gamma Spaces

A *gamma space* is a partial linear space $\Gamma = (\mathscr{P}, \mathscr{L})$ where for any point-line pair $(p, \ell) \in \mathscr{P} \times \mathscr{L}$ the point p is collinear with either none, just one or all of the points of ℓ. Let Γ be a gamma space and X be a subspace of Γ. If all points of X are pairwise collinear, that is $X \subseteq X^\perp$, then X is called a *singular subspace*. By Zorn's argument, every singular subspace of Γ is contained in a maximal one. Note also that $X^\perp$ is always a subspace of Γ, for every subset $X \subseteq \mathscr{P}$. The equality $X = X^\perp$ characterizes maximal singular subspaces.

Polar spaces (to be defined in the next subsection) are examples of gamma spaces. Further examples are provided by projective and polar grassmannians, to be introduced later in this section.

7.2.2 Polar Spaces

A *polar space* is a gamma space $\Gamma = (\mathscr{P}, \mathscr{L})$ where $p^\perp \cap \ell \neq \emptyset$ for every point-line pair $(p, \ell) \in \mathscr{P} \times \mathscr{L}$. Note that this property forces Γ to be connected, except when $\mathscr{L} = \emptyset$. The subspace $\mathscr{P}^\perp = \{p \in \mathscr{P} : p^\perp = \mathscr{P}\}$ is called the *radical* of Γ and denoted by Rad(Γ). Obviously, Rad(Γ) is a singular subspace and is contained in all maximal singular subspaces of Γ. The polar space Γ is *degenerate* precisely when Rad$(\Gamma) \neq \emptyset$.

Suppose that all singular subspaces of Γ are projective spaces (as it is the case when Γ is non-degenerate; see e.g. Shult [42]). Recall that the rank of a projective space is its generating rank, namely its dimension augmented by 1. The *rank* rank(X) of a singular subspace X of Γ is its rank as a projective space. Singular subspaces of rank 3 (when they exist) are called *planes*.

Every subspace S of Γ can be naturally endowed with the structure of a polar space by taking as lines all the lines of Γ fully contained in S. In particular a singular subspace, regarded as a polar space, coincides with its own radical.

In the remainder of these notes, unless otherwise specified, we shall always assume that Γ is non-degenerate. Accordingly, the singular subspaces of Γ are projective spaces. If all maximal singular subspaces of Γ have finite rank, then they have the same rank. Their common rank, usually denoted by rank(Γ), is called the *rank* of Γ, not to be confused with the generating rank of Γ as a partial linear space, which is always greater than rank(Γ). For the sake of completeness, when Γ admits singular subspaces of infinite rank we put rank$(\Gamma) = \infty$. Polar spaces of rank 2 are called *generalized quadrangles*. Polar spaces of rank 1 are just raw sets.

Henceforth we assume that Γ has finite rank, say n. If S is a subspace of Γ, its *rank* is the maximal rank of the singular subspaces of Γ contained in S. In particular, the hyperplanes of Γ have rank either $n - 1$ or n.

Every non-maximal singular subspace of Γ is the intersection of the maximal singular subspaces which contain it. More explicitly, if M is a maximal singular subspace and $X \subset M$ is a proper subspace of M, then $X = M \cap N$ for a suitable maximal singular subspace $N \supset X$. When every singular subspace of rank

$n - 1$ is contained in exactly two maximal singular subspaces we say that Γ is *top-thin*. In particular, top-thin generalized quadrangles are called *grids*.

Given a singular subspace X of Γ of rank $m < n$, its *star* $X^\perp/X$ is the (non-degenerate) polar space of rank $n - m$ defined as follows: the singular subspaces of rank $m + 1$ which contain X are its points and (when $m < n - 1$) its lines are the families ℓ_Y of singular subspaces of rank $m + 1$ containing X and contained in a singular subspace $Y \supset X$ of rank $m + 2$.

In particular, if p is a point of Γ then $p^\perp/p$ is a non-degenerate polar space of rank $n-1$ while $p^\perp$ is a hyperplane of Γ, called the *singular hyperplane* with p as its *deepest point*. Of course, in spite of this terminology, a singular hyperplane is not a singular subspace, except possibly when rank$(\Gamma) = 1$. Indeed $p^\perp$ is a degenerate polar space of rank n with $\{p\} = \text{Rad}(p^\perp)$.

A *frame* of Γ is a set A of $2n$ points such that the collinearity graph of Γ induces on it is a complete n-partite graph with n classes of size 2. The singular subspaces of Γ spanned by the cliques of $(A, \perp)$ form a hyperoctahedron, called an *apartment* of Γ. Any two mutually disjoint maximal singular subspaces belong to a common apartment.

7.2.3 Sesquilinear and Pseudoquadratic Forms

Our next step is to discuss classical polar spaces. We shall do this in Section 7.2.4, but in view of that we need to recall some basics on reflexive sesquilinear forms and pseudoquadratic forms. Chapter 8 of Tits [45] is our source for this matter. Henceforth $\mathbb{K}$ is a division ring, σ is an antiautomorphism of $\mathbb{K}$ and V is a right $\mathbb{K}$-vector space.

Sesquilinear and pseudoquadratic forms are essential in order to describe the projective embeddings of polar spaces (and, in turn, to provide examples of the same). In particular, the universal embedding of a polar space (when it exists) can always be described by either a sesquilinear or a quadratic form; for many polar spaces, the only existing embedding is the universal one. However, in characteristic 2 there are more possibilities and pseudoquadratic forms are needed; see Section 7.2.4.1.

7.2.3.1 Reflexive Sesquilinear Forms

A σ-*sesquilinear form* is a function $f : V \times V \rightarrow \mathbb{K}$ such that $f(\sum_i x_i t_i, \sum_j y_j s_j) = \sum_{i,j} t_i^\sigma f(x_i, y_j) s_j$ for any choice of vectors $x_i, y_j \in V$ and scalars $t_i, s_j \in \mathbb{K}$. A sesquilinear form f is said to be *reflexive* when for any two vectors $x, y \in V$ we have $f(x, y) = 0$ if and only if $f(y, x) = 0$.

Let f be a non-trivial (i.e. not null) σ-sesquilinear form. Then f is reflexive if and only if there exists a scalar $\epsilon \in \mathbb{K}^*$ such that $f(y, x) = f(x, y)^\sigma \epsilon$ for any choice of $x, y \in V$; this condition forces $\epsilon^\sigma = \epsilon^{-1}$ and $t^{\sigma^2} = \epsilon t \epsilon^{-1}$ for any $t \in \mathbb{K}$ (see Bourbaki [6]). With ϵ as above, f is called a (σ, ϵ)-*sesquilinear form*. Clearly, $\epsilon \in \{1, -1\}$ if and only if $\sigma^2 = \mathrm{id}_\mathbb{K}$; also, $\sigma = \mathrm{id}_\mathbb{K}$ only if $\mathbb{K}$ is commutative. Let $\sigma = \mathrm{id}_\mathbb{K}$. If $\epsilon = 1$, then f is said to be *symmetric*; when $\mathrm{char}(\mathbb{K}) \neq 2$, then $\epsilon = -1$ if and only if $f(x, x) = 0$ for any $x \in V$. In this case, f is said to be *alternating*. When $\mathrm{char}(\mathbb{K}) = 2$, an *alternating* form is a $(\mathrm{id}_\mathbb{K}, 1)$-form f such that $f(x, x) = 0$ for every $x \in V$. A (σ, ϵ)-form with $\sigma \neq \mathrm{id}_\mathbb{K}$ and $\epsilon = 1$ (or $\epsilon = -1$) is called *hermitian* (respectively *antihermitian*).

Two vectors $v, w \in V$ are *orthogonal* with respect to f (in symbols $v \perp_f w$) if $f(v, w) = 0$. A vector $v \in V$ is *isotropic* if $v \perp_f v$. A subspace X of V is *totally isotropic* if $X \subset X^{\perp_f}$. In contrast, if 0 is the unique isotropic vector of a subspace X of V then we say that f is *anisotropic* over X. The same terminology is adopted for points and subspaces of $\mathrm{PG}(V)$, in an obvious way. The subspace $\mathrm{Rad}(f) = V^{\perp_f}$ is the *radical* of f. The form f is *degenerate* if $\mathrm{Rad}(f) \neq \{0\}$.

The isotropic points of $\mathrm{PG}(V)$ together with the totally isotropic lines of $\mathrm{PG}(V)$ form a polar space $\Gamma(f)$. The singular subspaces of $\Gamma(f)$ are precisely the totally isotropic subspaces of $\mathrm{PG}(V)$ and the radical of $\Gamma(f)$ is (the subspace of $\mathrm{PG}(V)$ corresponding to) $\mathrm{Rad}(f)$. In particular, $\Gamma(f)$ is non-degenerate if and only if f is non-degenerate.

Suppose that V is spanned by the isotropic vectors, as we can always assume, modulo replacing V with a suitable subspace of it. Then the inclusion mapping of $\Gamma(f)$ in $\mathrm{PG}(V)$ is a projective embedding.

Two non-trivial forms f and g define the same polar space $\Gamma = \Gamma(f) = \Gamma(g)$ if and only if they are proportional. Every reflexive sesquilinear form which is neither symmetric nor alternating is proportional to a hermitian form as well as to an antihermitian form. Explicitly, let f be (σ, ϵ)-sesquilinear and let $g = \kappa f$ for a scalar $\kappa \in \mathbb{K}^*$. Then g is (ρ, η)-sesquilinear with $\rho : t \to \kappa t^\sigma \kappa^{-1}$ and $\eta = \kappa \kappa^{-\sigma} \epsilon$. We claim that if $\sigma \neq \mathrm{id}_\mathbb{K}$ then we can always choose κ in such a way that $\eta = 1$. Indeed, if $\epsilon \neq -1$, then $\kappa = 1 + \epsilon^{-1}$ does the job. If $\epsilon = -1$, then pick $t \in \mathbb{K}$ such that $t \neq t^\sigma$ (such a t always exists because $\sigma \neq \mathrm{id}_\mathbb{K}$ by assumption) and put $\kappa = t - t^\sigma$. On the other hand, suppose that $\epsilon = 1 \neq -1$ (hence $\mathrm{char}(\mathbb{K}) \neq 2$) and let $t \in \mathbb{K}$ be such that $t \neq t^\sigma$. Again, put $\kappa = t - t^\sigma$. Then $\eta = -1$.

Thus, if we are interested only in the polar space $\Gamma(f)$ rather than in peculiar properties of the underlying form f, we can always assume that f is either alternating, symmetric or hermitian (or antihermitian, if we so prefer).

7.2.3.2 Pseudoquadratic Forms

Let σ and ϵ be as above but with $\epsilon \neq -1$ when $\sigma = \mathrm{id}_\mathbb{K}$ and $\mathrm{char}(\mathbb{K}) \neq 2$. Put $\mathbb{K}_{\sigma, \epsilon} := \{t - t^\sigma \epsilon : t \in \mathbb{K}\}$. The set $\mathbb{K}_{\sigma, \epsilon}$ is a subgroup of the additive group of

$\mathbb{K}$. Moreover $\mathbb{K}_{\sigma,\epsilon} \subset \mathbb{K}$, in view of the hypotheses we have assumed on σ and ϵ. Hence the quotient group $\mathbb{K}/\mathbb{K}_{\sigma,\epsilon}$ is non-trivial.

A function $\phi\colon V \to \mathbb{K}/\mathbb{K}_{\sigma,\epsilon}$ is a (σ,ϵ)-*pseudoquadratic form* if there exists a σ-sesquilinear form $g\colon V \times V \to \mathbb{K}$ such that $\phi(x) = g(x,x) + \mathbb{K}_{\sigma,\epsilon}$ for all $x \in V$.

The sesquilinear form g is not uniquely determined by ϕ and needs not to be reflexive. In contrast, the form $f_\phi\colon V \times V \to \mathbb{K}$ defined by the clause $f_\phi(x,y) := g(x,y) + g(y,x)^\sigma \epsilon$ only depends on ϕ and is (σ,ϵ)-sesquilinear (hence reflexive). Moreover $\phi(x+y) = \phi(x) + \phi(y) + (f_\phi(x,y) + \mathbb{K}_{\sigma,\epsilon})$ for any choice of $x,y \in V$. We call f_ϕ the *sesquilinearized* of ϕ.

In general, there is no way to recover ϕ from its sesquilinearized f_ϕ. However, suppose that the center of $\mathbb{K}$ contains an element μ such that $\mu + \mu^\sigma \neq 0$ (as is always the case when $\mathrm{char}(\mathbb{K}) \neq 2$: just choose $\mu = 1$). Then

$$\phi(x) = \frac{\mu}{\mu + \mu^\sigma} \cdot f_\phi(x,x) + \mathbb{K}_{\sigma,\epsilon}.$$

A vector $v \in V$ is *singular* for ϕ if $\phi(v) = 0$. A subspace X of V is *totally singular* if all of its nonzero vectors are singular while, if 0 is the unique singular vector of X, then we say that ϕ is *anisotropic* over X. The same terminology is used for points and subspaces of $\mathrm{PG}(V)$. The singular points of $\mathrm{PG}(V)$ together with the totally singular lines of $\mathrm{PG}(V)$ form a polar space $\Gamma(\phi)$. The singular subspaces of $\Gamma(\phi)$ are precisely the totally singular subspaces of $\mathrm{PG}(V)$. If the singular points of ϕ span V (as it is always the case when ϕ admits at least one singular vector not in $\mathrm{Rad}(\phi)$), then the inclusion mapping of $\Gamma(\phi)$ in $\mathrm{PG}(V)$ is a projective embedding.

The set of singular vectors of $\mathrm{Rad}(f_\phi)$ is a subspace of V. It is called the *radical* of ϕ and denoted by $\mathrm{Rad}(\phi)$. The form ϕ is *non-degenerate* if $\mathrm{Rad}(\phi) = \{0\}$, namely ϕ is anisotropic over $\mathrm{Rad}(f_\phi)$. The radical of the polar space $\Gamma(\phi)$ is (the subspace of $\mathrm{PG}(V)$ corresponding to) $\mathrm{Rad}(\phi)$. So, $\Gamma(\phi)$ is non-degenerate if and only if ϕ is non-degenerate.

All vectors that are singular for ϕ are isotropic for f_ϕ and the span $\langle v,w \rangle$ of two singular vectors $v,w \in V$ is totally singular if and only if $v \perp_{f_\phi} w$. It follows that $\Gamma(\phi)$ is a subspace of $\Gamma(f_\phi)$, but possibly different from $\Gamma(f_\phi)$. In particular, it can happen that $\Gamma(\phi)$ is non-degenerate and $\Gamma(f_\phi)$ is degenerate. However, when ϕ can be recovered from f_ϕ, then $\Gamma(\phi) = \Gamma(f_\phi)$. If this is the case, then all we can do with ϕ can be done with f_ϕ as well.

Proportionality can be defined for pseudoquadratic forms as for sesquilinear forms. Explicitly, let $\kappa \in \mathbb{K}^*$. Then $\kappa \mathbb{K}_{\sigma,\epsilon} = \mathbb{K}_{\rho,\eta}$ where $\rho\colon t \to \kappa t^\sigma \kappa^{-1}$ and $\eta = \kappa \kappa^{-\sigma} \epsilon$. If ϕ is a (σ,ϵ)-pseudoquadratic form represented as $\phi(x) = g(x,x) + \mathbb{K}_{\sigma,\epsilon}$ for a suitable σ-sesquilinear form g, then κg is ρ-sesquilinear and represents a (ρ, η)-pseudoquadratic form $\psi = \kappa\phi$ *proportional* to ϕ. The sesquilinearized of ϕ and ψ are proportional as well; explicitly, $f_\psi = \kappa f_\phi$. Likewise for sesquilinear forms, if ϕ and ψ are pseudoquadratic forms of V such that $\Gamma(\phi) \nsubseteq \mathrm{Rad}(f_\phi)$ and $\Gamma(\psi) \nsubseteq \mathrm{Rad}(f_\psi)$, then $\Gamma(\phi) = \Gamma(\psi)$ if and only if ϕ and ψ are proportional.

As shown in the previous subsection, with $\eta = \kappa\kappa^{-\sigma}\epsilon$ as here, we can always choose κ in such a way that $\eta = 1$. In other words, every (σ, ϵ)-pseudoquadratic form is proportional to a $(\rho, 1)$-pseudoquadratic form for a suitable antiautomorphism ρ of $\mathbb{K}$. So, when focusing on $\Gamma(\phi)$ regardless of peculiar properties of the form ϕ, we can always assume that ϕ is $(\sigma, 1)$-pseudoquadratic.

7.2.3.3 Some Terminology and Notation

We recall that the *Witt index* of a reflexive sesquilinear form $f: V \times V \to \mathbb{K}$ (a pseudoquadratic form $\phi: V \to \mathbb{K}/\mathbb{K}_{\sigma,1}$) is the common dimension of the maximal totally isotropic (totally singular) subspaces of V, provided that they have finite dimension; in other words, the Witt index of f (of ϕ) is the rank of $\Gamma(f)$ (respectively $\Gamma(\phi)$). Clearly, if the form is non-degenerate and n is its Witt index, then $\dim(V) \geq 2n$.

Let f be a non-degenerate alternating form of Witt index n (hence $\dim(V) = 2n$). Then $\Gamma(f)$ is called a *symplectic polar space* (also *symplectic variety*) and denoted by $\mathcal{W}_{2n-1}(\mathbb{K})$. In this chapter, if not otherwise specified, when dealing with symplectic varieties we always assume that $\mathrm{char}(\mathbb{K}) \neq 2$. Indeed, as we shall see later (Section 7.2.4.1), when $\mathrm{char}(\mathbb{K}) = 2$, the polar space called $\mathcal{W}_{2n-1}(\mathbb{K})$ is in fact a quadric *dressed up* as a symplectic variety.

The $(\mathrm{id}_\mathbb{K}, 1)$-pseudoquadratic forms are usually called *quadratic forms*. Let $\phi: V \to \mathbb{K}$ be a non-degenerate quadratic form of Witt index n. The polar space $\Gamma(\phi)$ is called a *quadric* (also an *orthogonal* polar space). When $\dim(V) \in \{2n, 2n+1, 2n+2\}$ (as when $\mathbb{K}$ is finite), the following terminology and notation is used. If $\dim(V) = 2n$, then $\Gamma(\phi)$ is called a *hyperbolic quadric* and denoted by $\mathcal{Q}_{2n-1}^+(\mathbb{K})$. When $\dim(V) = 2n+1$, then $\Gamma(\phi)$ is a *parabolic quadric*, denoted by $\mathcal{Q}_{2n}(\mathbb{K})$. When $\dim(V) = 2n + 2$ and either $\mathrm{char}(\mathbb{K}) \neq 2$ or $\mathrm{char}(\mathbb{K}) = 2$ and the sesquilinearized of ϕ is non-degenerate, then $\Gamma(\phi)$ is an *elliptic quadric* and denoted by $\mathcal{Q}_{2n+1}^-(\mathbb{K})$. We point out that while the classification presented here is exhaustive for quadrics over finite fields, it is far from so in general (e.g. over $\mathbb{Q}$ or $\mathbb{R}$).

Recall that hyperbolic quadrics define a top-thin polar spaces.[1] This property characterizes hyperbolic quadrics among embeddable polar spaces.

---

[1] The polar space associated to a hyperbolic quadric with $n \geq 3$ is top-thin of type C_n; there is also a building of type D_n which has the same points and lines and it is better behaved; see Section 7.6.2.2 for details.

Finally, let $\phi\colon V \to \mathbb{K}$ be a non-degenerate $(\sigma, 1)$-pseudoquadratic form of Witt index n with $\sigma \neq \mathrm{id}_\mathbb{K}$ and assume that $\mathbb{K}$ is commutative. Then $\mathbb{K}$ is a separable quadratic extension of its subfield $\mathbb{K}_0 = \{t \in \mathbb{K}\colon t^\sigma = t\}$ and σ is the unique non-trivial (involutory) element of the Galois group of $\mathbb{K}$ over $\mathbb{K}_0$. Moreover, ϕ and its sesquilinearized f_ϕ define the same polar space $\Gamma(\phi) = \Gamma(f_\phi)$. This polar space is called a *hermitian variety* (also *unitary polar space*). When $\dim(V) \in \{2n, 2n+1\}$ (as when $\mathbb{K}$ is finite), we denote $\Gamma(\phi)$ by $\mathscr{H}_{2n-1}(\mathbb{K})$ or $\mathscr{H}_{2n}(\mathbb{K})$, according to whether $\dim(V)$ is $2n$ or $2n + 1$. (Many authors prefer the symbols $\mathscr{H}_{2n-1}(\mathbb{K}_0)$ and $\mathscr{H}_{2n}(\mathbb{K}_0)$, but we shall stick to our convention).

7.2.4 Classical Polar Spaces

Obviously, embeddable point-line geometries are *thick-lined*, namely all of their lines have more than two points. As proved by Tits [45, chs. 7–9] (also Buekenhout and Cohen [7, chs. 7–11]), all thick-lined non-degenerate polar spaces of rank at least 3 are embeddable but for the following two exceptions, both of rank 3: the line-grassmannian of PG(3, $\mathbb{K}$) with $\mathbb{K}$ non-commutative and a family of polar spaces of rank 3 with non-desarguesian planes, described in [45, ch. 9] (also Freudenthal [30]), which we call *Freudenthal-Tits polar spaces*.

Let Γ be an embeddable non-degenerate polar space of finite rank $n \geq 2$. By Tits [45, ch. 8], the polar space Γ admits the universal embedding but for the following two exceptions of rank 2:

(1) Γ is a grid with lines of size $s + 1 > 4$, where s is a prime power if $s < \infty$. If $\varepsilon\colon \Gamma \to \mathrm{PG}(V)$ is an embedding of Γ, then $V \cong V(4, \mathbb{K})$ for a field $\mathbb{K}$ and $\varepsilon(\Gamma)$ is a hyperbolic quadric of $\mathrm{PG}(V)$. The field $\mathbb{K}$ is uniquely determined by Γ only if $s < \infty$.

(2) Γ is a generalized quadrangle admitting just two non-isomorphic embeddings $\varepsilon_1, \varepsilon_2\colon \Gamma \to \mathrm{PG}(3, \mathbb{K})$ for a quaternion division ring $\mathbb{K}$ (the same for ε_1 and ε_2). We refer to [45, 8.6] for more on these examples which we call *bi-embeddable quaternion quadrangles*.

We define a *classical* polar space as an embeddable non-degenerate polar space of finite rank at least 2 which does not belong to either of the above two cases.

7.2.4.1 The Universal Embedding of a Classical Polar Space

Henceforth Γ is a classical polar space of finite rank n and $\varepsilon\colon \Gamma \to \mathrm{PG}(V)$ is its universal embedding. So, $\Gamma \cong \varepsilon(\Gamma)$ and the latter is a full subgeometry of $\mathrm{PG}(V)$. Let $\mathbb{K}$ be the underlying division ring of V. As proved by Tits [45], either

(A) char($\mathbb{K}$) $\neq$ 2 and $\varepsilon(\Gamma)$ is a symplectic polar space, or

(B) $\varepsilon(\Gamma) = \Gamma(\phi)$ for a non-degenerate pseudoquadratic form $\phi \colon V \to \mathbb{K}/\mathbb{K}_{\sigma,1}$.

In case (A) we have dim(V) $= 2n$. In this case Γ is generated by each of its frames and ε is the unique embedding of Γ.

Assume case (B) and suppose that the sesquilinearized f_ϕ of ϕ is non-degenerate (as when char($\mathbb{K}$) $\neq$ 2 and, more generally, when $\Gamma(\phi) = \Gamma(f_\phi)$). Then ε is the unique embedding of Γ.

On the other hand, suppose that $R := \mathrm{Rad}(f_\phi) \neq \{0\}$ (hence char($\mathbb{K}$) $= 2$). No nonzero vector of R is singular for ϕ, since ϕ is non-degenerate. Hence R contains no point of $\Gamma(\phi)$. Moreover, every projective line of PG(V) containing at least two ϕ-singular points misses R. Therefore, for every subspace X of R, the projection $\pi_X \colon \mathrm{PG}(V) \to \mathrm{PG}(V/X)$ induces an injective mapping on $\Gamma(\phi)$. Accordingly, the composite $\varepsilon_X = \pi_X \circ \varepsilon$ is an embedding of Γ. All embeddings of Γ arise in this way, by factorizing V over subspaces of R. The embedding ε_R, obtained by factorizing V over R, is the minimal one.

All embeddings ε_X as here can be described by means of *generalized pseudoquadratic forms* (see [39]), except the minimal one when $\phi(R) = \mathbb{K}/\mathbb{K}_{\sigma,1}$. If $\phi(R) = \mathbb{K}/\mathbb{K}_{\sigma,1}$, then $\sigma = \mathrm{id}_\mathbb{K}$, the quadratic form ϕ induces the null form on $V' := V/R$ and $\varepsilon_R(\Gamma) = \Gamma(f)$ for a non-degenerate alternating form $f \colon V' \times V' \to \mathbb{K}$ (hence dim(V') $= 2n$). Accordingly, if char($\mathbb{K}$) $= 2$ and Γ admits an embedding $\varepsilon' \colon \Gamma \to \mathrm{PG}(V')$ such that $\varepsilon'(\Gamma)$ is a symplectic variety of PG(V'), then the universal embedding $\varepsilon \colon \Gamma \to \mathrm{PG}(V)$ embeds Γ as a quadric $\Gamma(\phi)$ of PG(V) and $V' = V/R$ with $R = \mathrm{Rad}(f_\phi)$. Moreover, dim(ε) $= 2n + d$ ($= d$ when d is infinite) where d is the degree of $\mathbb{K}$ over its subfield $\mathbb{K}^2 = \{t^2\}_{t \in \mathbb{K}}$ (see [28]). In particular, if $\mathbb{K}$ is perfect, then $d = 1$.

7.2.4.2 A Canonical Decomposition of V and the Anisotropic Gap of Γ

Let Γ be a classical polar space and $\varepsilon \colon \Gamma \to \mathrm{PG}(V)$ its universal embedding, as in the previous subsection. So, $\varepsilon(\Gamma) = \Gamma(f)$ with f a non-degenerate alternating form (case (A)) or $\varepsilon(\Gamma) = \Gamma(\phi)$ for a non-degenerate pseudoquadratic form ϕ (case (B)). It goes without saying that, when considering two orthogonal vectors or subspaces of V, we refer to orthogonality with respect to f (in case (A)) or with respect to the sesquilinearized of ϕ (case (B)). In the sequel, by a harmless abuse, we will freely shift from subspaces of PG(V) to the corresponding subspaces of V, even if correctness would impose us to stick to PG(V).

Given a frame A of Γ, its image $\varepsilon(A)$ spans a $2n$-dimensional subspace of V which splits as the direct sum $V_1 \oplus V_2 \oplus \cdots \oplus V_n$ of mutually orthogonal 2-dimensional subspaces $V_1, V_2, \ldots, V_n$. These subspaces bijectively correspond to the n pairs of non-collinear points of A and appear as secant lines to $\varepsilon(\Gamma)$

in PG(V). Their preimages $\varepsilon^{-1}(V_1),\ldots,\varepsilon^{-1}(V_n)$ are hyperbolic lines of Γ (a *hyperbolic line* of a polar space being the double perp $\{p,q\}^{\perp\perp}$ of two non-collinear points p and q).

The codimension of $V_1 \oplus V_2 \cdots \oplus V_n$ in V will be called the *anisotropic gap* of Γ (the *gap* of Γ for short) and denoted by gap(Γ). Clearly gap(Γ) = dim(V_0) for an orthogonal complement V_0 of $V_1 \oplus \cdots \oplus V_n$ in V. Trivially, $V_0 = \{0\}$ in case (A). In case (B), the form ϕ is anisotropic on V_0 and $V_0 \supseteq \mathrm{Rad}(f_\phi)$.

We will turn back to gap(Γ) in Section 7.4, where a geometric characterization of this number will be given, with no explicit mention of any embedding.

7.2.5 Polar Grassmannians

We assume that the reader is familiar with the definition of projective grassmannian. Polar grassmannians are defined similarly. Explicitly, let Γ be a nondegenerate polar space of finite rank $n \geq 2$. For $i \in \{0,1,\ldots,n\}$, let $\mathscr{P}_i$ be the set of singular subspaces of Γ of rank i. For instance, $\mathscr{P}_0 = \{\emptyset\}$, $\mathscr{P}_1 = \mathscr{P}$ (the point-set of Γ) and $\mathscr{P}_n$ is the family of maximal singular subspaces of Γ. With $k \in \{1,\ldots,n\}$, the *k–grassmannian* of Γ is the geometry $\mathrm{G}_k(\Gamma) := (\mathscr{P}_k,\mathscr{L}_k)$ with line-set $\mathscr{L}_k$ defined as follows. If $k < n$, then

$$\mathscr{L}_k = \{\ell_{X,Y}: (X,Y) \in \mathscr{P}_{k-1} \times \mathscr{P}_{k+1}, X \subset Y\}$$

where $\ell_{X,Y} := \{Z \in \mathscr{P}_k: X \subset Z \subset Y\}$. If $k = n$, then $\mathscr{L}_n = \{\ell_X\}_{X \in \mathscr{P}_{n-1}}$ with $\ell_X := \{Z \in \mathscr{P}_n: X \subset Z\}$. Obviously, $\mathrm{G}_1(\Gamma) = \Gamma$. The grassmannians $\mathrm{G}_2(\Gamma)$ and $\mathrm{G}_n(\Gamma)$ are called the *line-grassmannian* of Γ and the *dual polar space* of Γ, respectively.

For every $k = 1,2,\ldots,n$, the grassmannian $\mathrm{G}_k(\Gamma)$ is a gamma space. The dual polar space $\mathrm{G}_n(\Gamma)$ is also a *near polygon*: for every line ℓ and every point p, just one of the points of ℓ is at minimal distance from p in the collinearity graph of $\mathrm{G}_n(\Gamma)$.

7.2.5.1 Convex Subspaces and Frames

A subspace S of $\mathrm{G}_k(\Gamma)$ is *convex* if it is convex in the collinearity graph of $\mathrm{G}_k(\Gamma)$, namely every shortest path of that graph between any two points of S is fully contained in S. Obviously, singular subspaces are convex. Convex subspaces can be given a rank, the *convex rank* of a convex subspace S being the maximal length of the nested chains of convex subspaces properly contained in it. For instance, the convex rank of a singular subspace of $\mathrm{G}_k(\Gamma)$ is just its projective dimension. In particular, the empty subspace $\emptyset$, points and lines have convex rank equal to $-1, 0$ and 1 respectively. When $k > 1$, the maximal singular subspaces of $\mathrm{G}_k(\Gamma)$ have convex rank either k (if they correspond to singular subspaces of Γ of rank $k + 1$) or $n - k$ (when they correspond to maximal singular subspaces of $X^\perp/X$ with X a singular subspace of Γ of rank $k - 1$). The proper convex subspaces of $G_1(\Gamma) = \Gamma$ are just the singular subspaces of Γ.

Let $k > 1$. Then $G_k(\Gamma)$ admits convex subspaces isomorphic to non-degenerate polar spaces of rank at least 2, called *symps* in the literature (see e.g. [42]). Their convex ranks are their ranks as polar spaces. When $k < n$, the star $X^\perp/X$ of a singular subspace X of Γ of rank $k - 1$ yields a symp of $G_k(\Gamma)$ of rank $n - k + 1$. When $k < n - 1$, more symps exist, obtained as follows: let Y be a singular subspace of Γ of rank $k + 2$ and $X \subset Y$ a subspace of Y of rank $k - 2$. The quotient space Y/X is isomorphic to $PG(3, \mathbb{K})$, with $\mathbb{K}$ the underlying division ring of Γ. The line-grassmannian of Y/X is a symp of $G_k(\Gamma)$ of rank 3. We call those obtained in this way *symps of Grassmann type*.

When $k \in \{n - 1, n\}$, all symps have rank 2. When $k = n$, they are obtained as (the dual of) the stars $X^\perp/X$ of singular subspaces X of Γ. When $k = n - 1$, the symps are stars $X^\perp/X$ for X a singular subspace of Γ of rank $n - 2$. We call these *quads*. Observe however that in this case there might be also other families of convex subspaces, see [35, Theorem 7.4]. We refer in general to [35] for a classification of the convex subspaces of polar grassmannians.

A *frame* of $G_k(\Gamma)$ is the family $A_k := \mathscr{P}_k \cap \mathscr{A}$ for $\mathscr{A}$ an apartment of Γ. (A few authors, such as Pankov [36], [37], call the graph induced on A_k by the collinearity relation of Γ_k an *apartment* of Γ_k.) The size of A_k is easy to compute: we have $|A_k| = 2^k \cdot \binom{n}{k}$. So, if $G_k(\Gamma)$ is embeddable, then $G_k(\Gamma)$ is generated by a frame only if $\mathrm{erk}(G_k(\Gamma)) \leq 2^k \cdot \binom{n}{k}$. In particular, for $k = 1$ this inequality comes down to $\mathrm{erk}(\Gamma) \leq 2n$ and for $k = n$ it amounts to $\mathrm{erk}(G_n(\Gamma)) \leq 2^n$.

As we know from Section 7.2.4, if Γ is embeddable, then $\mathrm{erk}(\Gamma) \geq 2n$, with equality if and only if Γ is generated by its frames. A similar statement holds true for dual polar spaces: as proved by De Bruyn and Pasini [26], apart from pathological cases no example of which is known, if $G_n(\Gamma)$ is embeddable, then $\mathrm{erk}(G_k(\Gamma)) \geq 2^n$, with equality precisely when $G_n(\Gamma)$ is generated by its frames. Assume now $1 < k < n$. In all known examples where $G_k(\Gamma)$ is embeddable, we have $\mathrm{erk}(G_k(\Gamma)) \geq \binom{2n}{k} - \binom{2n}{k-2} > 2^k \binom{n}{k}$. In cases like these, $G_k(\Gamma)$ cannot be generated by any of its frames.

7.2.5.2 Embeddings

Suppose now that Γ is classical and let $\mathbb{K}$ be its underlying division ring. As $G_1(\Gamma) = \Gamma$, we may assume that $k > 1$.

If $\mathbb{K}$ is non-commutative and $1 < k < n - 1$, then $G_k(\Gamma)$ is not embeddable. Indeed the symps of $G_k(\Gamma)$ of Grassmann type are isomorphic to the line-grassmannian of $PG(3, \mathbb{K})$, which is not embeddable since $\mathbb{K}$ is not commutative. We guess that $G_n(\Gamma)$ and $G_{n-1}(\Gamma)$ (when $n > 2$) are also not embeddable. Indeed, as we shall see later, even when $\mathbb{K}$ is commutative, projective embeddings of $G_n(\Gamma)$ are known only in very special cases. However guessing can be risky in this matter; some surprises might lie in wait. For instance, Freudenthal–Tits polar spaces are not embeddable, neither are their line-grassmannians. Nevertheless their duals are embeddable. In fact they admit a (universal) 56-dimensional embedding (De Bruyn and Van Maldeghem [29]).

Before turning to the case where $\mathbb{K}$ is commutative, we fix some notation. Let $\varepsilon\colon \Gamma \to \mathrm{PG}(V)$ be an embedding of Γ. Then ε maps every singular subspace of Γ isomorphically onto a projective subspace of $\mathrm{PG}(V)$. In particular, $\varepsilon(\mathscr{P}_k)$ is a subset of the point-set of the k-grassmannian $G_k(V)$ of $\mathrm{PG}(V)$ and $\varepsilon(\ell)$ is a set of points of $G_k(V)$ for every line $\ell \in \mathscr{L}_k$ of $G_k(\Gamma)$. Put $\varepsilon(\mathscr{L}_k) := \{\varepsilon(\ell)\}_{\ell \subset \mathscr{L}_k}$ and $\varepsilon(G_k(\Gamma)) := (\varepsilon(\mathscr{P}_k), \varepsilon(\mathscr{L}_k))$. If $k < n$, then $\varepsilon(\ell)$ is a line of $G_k(V)$ for every $\ell \in \mathscr{L}_k$. In this case, $\varepsilon(G_k(\Gamma))$ is a full subgeometry of $G_k(V)$.

Suppose that $\mathbb{K}$ is commutative. Then $G_k(V)$ admits a projective embedding $\gamma_k\colon G_k(V) \to \mathrm{PG}(\wedge^k V)$, namely the Plücker embedding, which maps every k-dimensional subspace $\langle v_1, \ldots, v_k \rangle$ of V onto the point $\langle v_1 \wedge \cdots \wedge v_k \rangle$ of $\mathrm{PG}(\wedge^k V)$. The embedding γ_k induces an injective mapping on $\varepsilon(\mathscr{P}_k)$. Hence the composite $G_k(\varepsilon) := \gamma_k \circ \varepsilon$ is injective as well. We call it the *Plücker mapping* of $G_k(\Gamma)$ *induced by* ε (*Plücker embedding* of $G_k(\Gamma)$ if it is a projective embedding). If ε is universal (minimal), we call $G_k(\varepsilon)$ the *canonical* (*minimal*) Plücker mapping of $G_k(\Gamma)$. Clearly, if $\varepsilon \geq \varepsilon'$, then $G_k(\varepsilon) \geq G_k(\varepsilon')$ (the relation $\geq$ being defined for $G_k(\varepsilon)$ and $G_k(\varepsilon')$ just as if they were projective embeddings, even when they are not such). Also, $G_k(\varepsilon) \simeq G_k(\varepsilon')$ if and only if $\varepsilon \simeq \varepsilon'$.

Suppose that $k < n$. Then, as $\varepsilon(G_k(\Gamma))$ is a full subgeometry of $G_k(V)$, the restriction of γ_k to $\varepsilon(G_k(\Gamma))$ is a projective embedding of $\varepsilon(G_k(\Gamma))$ in a (possibly improper) subspace of $\mathrm{PG}(\wedge^k V)$. Accordingly, $\varepsilon_k := G_k(\varepsilon)$ is a projective embedding. In particular, let ε be universal. Let W be the subspace of $\wedge^k V$ corresponding to the subspace of $\mathrm{PG}(\wedge^k V)$ spanned by $\varepsilon_k(\mathscr{P}_k)$ and put $N := \dim(V)$. When N is finite, W has dimension either $\binom{N}{k} - \binom{N}{k-2}$ or $\binom{N}{k}$. Explicitly, $\dim(W) = \binom{N}{k}$ when either $\varepsilon(\Gamma)$ is a hermitian variety or $\mathrm{char}(\mathbb{K}) \neq 2$ and $\varepsilon(\Gamma)$ is a quadric, while, if either $\varepsilon(\Gamma)$ is symplectic or $\mathrm{char}(\mathbb{K}) = 2$ and $\varepsilon(\Gamma)$ is a quadric then $\dim(W) = \binom{N}{k} - \binom{N}{k-2}$ (see [10]). When N is infinite, then $\dim(W) = N$ in any case. The bound $\mathrm{erk}(G_k(\Gamma)) \geq \binom{2n}{k} - \binom{2n}{k-2}$ mentioned at the end of Section 7.2.5.1 follows from this (recall that $N \geq 2n$).

When $\mathrm{char}(\mathbb{K}) = 2$ and $\varepsilon(\Gamma) = \Gamma(\phi)$ is a quadric but it is neither hyperbolic nor elliptic, then non-canonical Plücker embeddings also exist. In particular, with $R = \mathrm{Rad}(f_\phi) \neq \{0\}$, let $\varepsilon' = \varepsilon_R$. If $\varepsilon_R(\Gamma)$ is symplectic then $\dim(G_k(\varepsilon')) = \binom{2n}{k} - \binom{2n}{k-2}$. Otherwise, Γ contains a non-full subgeometry Γ'

isomorphic to $\mathscr{Q}_{2n}(\mathbb{F}_2)$ and ε' embeds Γ' as a symplectic variety in the $\mathbb{F}_2$-span of a suitable subset of V/R. As $\varepsilon'(\Gamma')$ is symplectic, the $\mathbb{F}_2$-span of the $G_k(\varepsilon')$-image of $G_k(\Gamma'_k)$ has (vector) dimension equal to $\binom{2n}{k} - \binom{2n}{k-2}$. It follows that $\dim(G_k(\varepsilon')) \geq \binom{2n}{k} - \binom{2n}{k-2}$.

Let now $k = n$. Still assuming that $\mathbb{K}$ is commutative, let $\varepsilon \colon \Gamma \to \mathrm{PG}(V)$ be the universal embedding of Γ. As $\mathbb{K}$ is commutative, $\varepsilon(\Gamma)$ is either symplectic, orthogonal or hermitian. Clearly, $G_n(\Gamma)$ is embeddable only if the duals of the stars of the singular subspaces of rank $n - 2$ of Γ are classical. In view of this criterion, only the following possibilities survive:

(1) char$(\mathbb{K}) \neq 2$ and $\varepsilon(\Gamma)$ is symplectic;

(2) $\varepsilon(\Gamma) = \Gamma(\phi)$ is hermitian and $\dim(V) = 2n$;

(3) char$(\mathbb{K}) = 2$, $\varepsilon(\Gamma) = \Gamma(\phi)$ is a quadric and $\phi(\mathrm{Rad}(f_\phi)) = \mathbb{K}$;

(4) $\varepsilon(\Gamma)$ is a parabolic quadric;

(5) $\varepsilon(\Gamma)$ is an elliptic quadric.

Note that cases (3) and (4) overlap non-trivially but neither of the two includes the other one. In particular, if char$(\mathbb{K}) = 2$, let $\mathbb{K}^2 := \{x^2 \colon x \in \mathbb{K}\}$. Clearly, $\mathbb{K}^2$ is a subfield of $\mathbb{K}$. When $[\mathbb{K} \colon \mathbb{K}^2] = 1$, cases (3) and (4) are the same; however, if $[\mathbb{K} \colon \mathbb{K}^2] > 1$ the two possibilities are independent. Actually, following the terminology of Section 7.2.3.3, if char$(\mathbb{K}) = 2$ there can exist non-degenerate polar spaces Γ realized as quadrics such that the associated sesquilinearized is degenerate having anisotropic gap greater than 1. That is, $\varepsilon(\Gamma) = \Gamma(\phi)$ for $\phi \colon V \to \mathbb{K}$, a quadratic form with $\phi(\mathrm{Rad}(f_\phi)) = \mathbb{K}$ such that $\dim(V) > 2n + 1$ so that they are not parabolic quadrics. This happens when $\mathbb{K}$ is a non-perfect field of characteristic 2. We refer to [15] and [16] for more information on this regard.

Let $\varepsilon(\Gamma)$ be symplectic. Then $\varepsilon(G_n(\Gamma))$ is a full subgeometry of $G_n(V)$. In this case $G_n(\varepsilon)$ is a projective embedding of $G_n(\Gamma)$ in $\mathrm{PG}(W)$ for a subspace $W \subset \wedge^n V$ of dimension $\dim(W) = \binom{2n}{n} - \binom{2n}{n-2}$.

Assume case (2). Put $\varepsilon_n := G_n(\varepsilon)$ for short. Then $\varepsilon_n(G_n(\Gamma))$ is a non-full subgeometry of $G_n(V)$. The mapping ε_n is not a projective embedding. Nevertheless it can be regarded as a $\binom{2n}{n}$-dimensional projective embedding of $G_n(\Gamma)$ in a Baer subgeometry of $\mathrm{PG}(\wedge^n V)$ defined over the subfield $\mathbb{K}_0 = \{t \in \mathbb{K} \colon t = t^\sigma\}$ of $\mathbb{K}$, where σ is the involutory automorphism of $\mathbb{K}$ associated with the form ϕ (see Cooperstein and Shult [23]). We call this embedding the *Plücker-like embedding* of $G_n(\Gamma)$.

In case (3), the codimension of $R = \mathrm{Rad}(f_\phi)$ in V is equal to $2n$ and $\dim(R) = d = [\mathbb{K} \colon \mathbb{K}^2]$. Hence $\dim(V) = 2n + d > 2n$. The canonical Plücker mapping $G_n(\varepsilon)$ maps lines of $G_k(\Gamma)$ onto quadrics of $\mathrm{PG}(\wedge^n V)$. So, $G_n(\varepsilon)$ is not a projective embedding. However the minimal Plücker mapping is a projective embedding. Indeed, under the hypotheses of (3), if $R = \text{Rad}(f_\phi)$, $\varepsilon' = \varepsilon_R$ and $V' = V/R$, then $\varepsilon'(\Gamma)$ is a symplectic variety of $\text{PG}(V')$ and the minimal Plücker mapping $G_n(\varepsilon')$ is a projective embedding of $G_n(\Gamma)$ in $\text{PG}(W)$ for a subspace $W \subset \wedge^n V$ of dimension $\dim(W) = \binom{2n}{n} - \binom{2n}{n-2}$, just as in case (1).

In cases (4) and (5), the Plücker mapping $G_n(\varepsilon)$ utterly fails to work. Indeed in these cases, $\varepsilon(G_n(\Gamma))$ is not a subgeometry of $G_n(V)$, not even a non-full one, and $G_n(\varepsilon)$ maps lines of $G_n(\Gamma)$ onto conics (when Γ is parabolic) or elliptic quadrics (if Γ is elliptic). However, when Γ is parabolic, $G_n(\Gamma)$ admits a 2^n-dimensional projective embedding, called the *spin embedding* (see [23]; also [14]). We denote it by e_{spin}. When $\text{char}(\mathbb{K}) \neq 2$ then $G_n(\Gamma)$ is generated by its frames (Blok and Brouwer [2]). Hence e_{spin} is universal (since $G_n(\Gamma)$ admits the universal embedding; see Theorem 2.1). On the other hand, let $\mathbb{K}$ be a perfect field of characteristic 2. Then we are back to case (3) and the minimal Plücker mapping $G_n(\varepsilon')$ is a projective embedding with $\dim(G_n(\varepsilon')) = \binom{2n}{n} - \binom{2n}{n-2} > 2^n$. In this case, e_{spin} is not universal.

Finally, let Γ be elliptic. Then $G_n(\Gamma)$ admits a 2^n-dimensional embedding (see e.g. Cooperstein and Shult [23]). We denote it by e_{spin}^- and, following [10], we call it the *spin-like embedding* of $G_n(\Gamma)$. The dual polar space $G_n(\Gamma)$ is generated by a frame (Cooperstein and Shult [23, 4.1]). Hence e_{spin}^- is universal.

7.2.5.3 Weyl Embeddings

In order to complete the previous picture, Weyl embeddings should also be mentioned. Let Γ be either $\mathcal{W}_{2n-1}(\mathbb{K})$ (but with $\text{char}(\mathbb{K}) \neq 2$) or $\mathcal{Q}_{2n}(\mathbb{K})$ or $\mathcal{Q}_{2n-1}^+(\mathbb{K})$. For $1 \leq k \leq n$, but with $k < n$ when $\Gamma = \mathcal{Q}_{2n-1}^+(\mathbb{K})$, the k-grassmannian $G_k(\Gamma)$ admits a projective embedding in $\text{PG}(V(\lambda_k))$ where $V(\lambda_k)$ is the Weyl module for the commutator subgroup of $\text{Aut}(\Gamma)$, for a suitable weight λ_k of the root system of type C_n, B_n or D_n, respectively (see Humphreys [31] for this matter). More explicitly, λ_k is the kth fundamental dominant weight in all cases but when $\Gamma = \mathcal{Q}_{2n-1}^+(\mathbb{K})$ with $k = n - 1$; in the latter case λ_{n-1} is the sum of the last two fundamental dominant weights of the root system D_n. Following [14] and [10] we call this embedding the *Weyl embedding* of $G_k(\Gamma)$ and we denote it by e_k^W.

The Weyl embedding e_1^W is just the universal embedding of Γ. So, we can assume that $k > 1$. With $1 < k < n$, let $G_k(\varepsilon)$ be the canonical Plücker embedding of Γ_k. If $\text{char}(\mathbb{K}) \neq 2$, then $e_k^W = G_k(\varepsilon)$. Let $\text{char}(\mathbb{K}) = 2$ (hence Γ is $\mathcal{Q}_{2n}(\mathbb{K})$ or $\mathcal{Q}_{2n-1}^+(\mathbb{K})$). Then $e_k^W > G_k(\varepsilon)$ and $\dim(e_k^W) = \binom{N}{k}$ while $\dim(G_k(\varepsilon)) = \binom{N}{k} - \binom{N}{k-2}$, where $N = 2n + 1$ or $N = 2n$ according to whether Γ is $\mathcal{Q}_{2n}(\mathbb{K})$ or $\mathcal{Q}_{2n-1}^+(\mathbb{K})$ (see [14]). Finally, let $k = n$. If $\Gamma = \mathcal{W}_{2n-1}(\mathbb{K})$, then again $e_n^W = G_n(\varepsilon)$. If $\Gamma = \mathcal{Q}_{2n}(\mathbb{K})$, then $e_n^W = e_{\text{spin}}$.

Weyl-like embeddings are also considered in [10] for the grassmannians of a quadric $\mathcal{Q}$ of rank n in $\mathrm{PG}(V)$ with $\dim(V) > 2n + 1$. We are not going to recall their definition here. We only mention that if the underlying field $\mathbb{K}$ of $\mathcal{Q}$ has characteristic char$(\mathbb{K}) \neq 2$ and $k < n$, then the Weyl-like embedding of $G_k(\mathcal{Q})$ is just the same as the Plücker embedding. When char$(\mathbb{K}) = 2$, the Weyl-like embedding covers the canonical Plücker embedding but it is not known what its dimension is in general. When $\mathcal{Q} = \mathcal{Q}_{2n+1}^-(\mathbb{K})$ and $k = n$, then the Weyl-like embedding of $G_n(\mathcal{Q})$ is just the same as the spin-like embedding e_{spin}^-.

Only quadrics are considered in section 7 of [10], where Weyl-like embeddings are defined, but we believe that Weyl-like embeddings can also be defined for hermitian varieties, in nearly the same fashion as for quadrics, although we expect they will turn out to be the same as Plücker embeddings.

7.2.6 Universal Embeddings of Polar Grassmannians

Let Γ be a classical polar space of finite rank $n > 2$, defined over a commutative division ring $\mathbb{K}$. Recall that if $k < n$ then $G_k(\Gamma)$ is embeddable. The next statement follows from the main theorem of Kasikova and Shult [32].

Theorem 2.1   *With Γ as in the previous paragraph, all the following hold:*

(1) *If $k < n - 1$ then $G_k(\Gamma)$ admits the universal embedding.*
(2) *The grassmannian $G_{n-1}(\Gamma)$ admits the universal embedding except possibly when $\Gamma = \mathcal{Q}_{2n-1}^+(\mathbb{K})$ with $|\mathbb{K}| > 3$.*
(3) *The dual polar space $G_n(\Gamma)$ admits the universal embedding whenever it is embeddable.*

In claim (2.1) of this theorem, the hypothesis that $|\mathbb{K}| \geq 4$ when Γ is a hyperbolic quadric is justified as follows. In order to fulfill all hypotheses of Kasikova–Shult's theorem, we need that, if $G_k(\Gamma)$ admits quads (see Section 7.2.5.1), then its quads are embeddable and admit the universal embedding (in short, they are classical). As we have assumed that $\mathbb{K}$ is commutative, this requirement just rules out grids with lines of size $s > 4$. Quads exist in $G_k(\Gamma)$ only when $k = n - 1$ or $k = n$. As Γ is thick-lined, the quads of $G_n(\Gamma)$ are never grids. On the other hand, if $\Gamma = \mathcal{Q}_{2n-1}^+(\mathbb{K})$, then the quads of $G_{n-1}(\Gamma)$ are indeed grids. In this case, we need the hypothesis that $|\mathbb{K}| > 3$.

The next result, due to Blok and Pasini [5], is a little improvement of claim (2.1) of Theorem 2.1. Recall that a field $\mathbb{K}$ is *prime* if either $\mathbb{K} = \mathbb{F}_p$ for a prime p or $\mathbb{K} = \mathbb{Q}$ (the field of rational numbers).

Proposition 2.2   *If $\mathbb{K}$ is a prime field, then $G_{n-1}(\mathcal{Q}_{2n-1}^+(\mathbb{K}))$ admits the universal embedding.*

7.2.6.1 How to Recognize an Embedding as the Universal One

In view of Theorem 2.1 and Proposition 2.2, if we can prove that a given embedding of $G_k(\Gamma)$ is dominant (with $\mathbb{K}$ prime when $k = n - 1$ and $\Gamma = \mathscr{Q}^+_{2n-1}(\mathbb{K})$), then that embedding is the universal one.

The Standard Argument (7.2.1.3) is the trick most frequently used to prove that a given embedding is dominant. That argument works well when Γ is $W_{2n-1}(\mathbb{K})$ or it is hermitian but lives in a finite dimensional projective space. In these cases, by the Standard Argument (7.2.1.3), we can prove that for $1 < k < n$ the Plücker embedding is dominant (hence universal, by Theorem 2.1), except possibly when Γ is hermitian and defined over $\mathbb{F}_4$. A detailed account of these results will be given in Section 7.3. In contrast, when Γ is a quadric the Standard Argument (7.2.1.3) apparently gets stuck (see the final part of Section 7.5.2.3). So, we are facing the following obstruction, which we formulate as a pair of twin problems.

Problem 2.3 *Let $1 < k < n$ and let Γ be a quadric.*

(1) *Let* $\mathrm{char}(\mathbb{K}) \neq 2$. *Prove that the Plücker embedding of* $G_k(\Gamma)$ *is dominant.*

(2) *Prove that, if* $\Gamma = \mathscr{Q}_{2n}(\mathbb{K})$ *or* $\Gamma = \mathscr{Q}^+_{2n-1}(\mathbb{K})$, *then the Weyl embedding of* $G_k(\Gamma)$ *is dominant.*

The hypothesis $k < n$ in Problem 2.3 is motivated by the fact that, even if we do not know everything about the case $k = n$, our knowledge of that case is not so poor, as the results to be surveyed in Section 7.3 will show. In part (1) of Problem 2.3 we assume that $\mathrm{char}(\mathbb{K}) \neq 2$ because, as noticed in Section 7.2.5.3, when Γ is a hyperbolic or parabolic quadric and $\mathrm{char}(\mathbb{K}) = 2$, then the Weyl embedding properly covers the canonical Plücker embedding while if $\mathrm{char}(\mathbb{K}) \neq 2$, then $e_k^W = G_k(\varepsilon)$. So, when $\mathrm{char}(\mathbb{K}) \neq 2$, part (2) of Problem 2.3 is included in part (1).

A partial solution of Problem 2.3(2) is embodied in the main result of Völklein [46]. According to [46], the adjoint module of a Chevalley group G not of type C_n and defined over a field as described in Proposition 2.4, is presented by a certain set of relations, which we are not going to recall here. Therefore, if such a module affords a projective embedding of one of the geometries associated with G (as is the case for the modules considered in [46]), then that embedding is dominant. In particular, we obtain the following.

Proposition 2.4 *With Γ as in part (2) of Problem 2.3, suppose that $\mathbb{K}$ is either a perfect field of positive characteristic or a number field and let $n > 2$. Then the Weyl embedding of $G_2(\Gamma)$ is dominant.*

So, under the hypotheses of Proposition 2.4, the Weyl embedding e_2^W is universal, except possibly when $\Gamma = \mathcal{Q}_5^+(\mathbb{K})$ with $\mathbb{K}$ non-prime. Indeed we do not know if the universal embedding exists in this case.

Admittedly, Proposition 2.4 is not so far reaching: only the case $k = 2$ is considered in it and the hypotheses assumed on $\mathbb{K}$ rule out too many infinite fields. A more geometric proof of Proposition 2.4 is given in [14], where only the special case $\Gamma = \mathcal{Q}_5^+(\mathbb{K})$ of Völklein's result is exploited. By a similar proof, it is shown in [14] that the conclusions of Proposition 2.4 hold for $k = 3$ as well, but for adding the hypothesis that $\mathbb{K} \neq \mathbb{F}_2$.

Perhaps more can be done in this vein, but let's turn back to the Standard Argument (7.2.1.3): given an embedding η of $\mathrm{G}_k(\Gamma)$ with $\dim(\eta) < \infty$, which we guess is dominant, we should look for a generating set G of $\mathrm{G}_k(\Gamma)$ of size $|G| = \dim(\eta)$. If such a set exists, then η is dominant. However, as previously said, sometimes discovering generating sets of the right size seems to be impossible; in fact, in at least one case, to be discussed in the next subsection, no such set exists.

7.2.6.2 A Spiteful Geometry

Let $\mathbb{K} = \overline{\mathbb{F}}_p$ be the algebraic closure of the finite prime field $\mathbb{F}_p$ and $\mathcal{Q} = \mathcal{Q}_5^+(\mathbb{K})$. The field $\mathbb{K}$ is perfect of positive characteristic. Hence the Weyl embedding e_2^W of $\mathrm{G}_2(\mathcal{Q})$ (which is the same as the Plücker embedding when $p > 2$) is dominant, by Proposition 2.4. Note that $\dim(e_2^W) = 15$. This does not imply that $\mathrm{erk}(\mathrm{G}_2(\mathcal{Q})) = 15$. Indeed neither Theorem 2.1 nor Proposition 2.2 apply to $\mathrm{G}_2(\mathcal{Q})$, since $\mathrm{G}_2(\mathcal{Q})$ is one of the grassmannians excluded by the hypotheses of claim (2.1) of Theorem 2.1 and $\mathbb{K}$ is not a prime field. So, it might be that $\mathrm{G}_2(\mathcal{Q})$ admits no universal embeddings and $\mathrm{erk}(\mathrm{G}_2(\mathcal{Q})) > 15$. However, $\mathrm{erk}(\mathrm{G}_2(\mathcal{Q})) \leq 16$, as it follows from [13, theorem 1.5] (see Theorem 6.10 of the present chapter). We might hope for a generating set G of $\mathrm{G}_2(\mathcal{Q})$ such that $15 \leq |G| \leq 16$, but no such set exists. Indeed, according to [13, theorem 1.5] (see Theorem 6.6 of this chapter), no finite set of points of $\mathrm{G}_2(\mathcal{Q})$ can generate $\mathrm{G}_2(\mathcal{Q})$. We sum up as follows.

Result 2.5 *With $\mathcal{Q}$ as here, $\mathrm{G}_2(\mathcal{Q})$ admits a dominant embedding of dimension 15, the embedding rank of $\mathrm{G}_2(\mathcal{Q})$ is at most 16, but $\mathrm{grk}(\mathrm{G}_2(\mathcal{Q}))$ is infinite.*

More generally, if $\mathbb{K}$ is not finitely generated then the generating rank of $\mathrm{G}_k(\mathcal{Q}_{2k+1}^+(\mathbb{K}))$ is infinite (Section 7.6, Corollary 6.14). However the embedding rank of $\mathrm{G}_k(\mathcal{Q}_{2k+1}^+(\mathbb{K}))$ is likely to be finite, perhaps equal to $\binom{2k+2}{k}$. Such a discrepancy between $\mathrm{grk}(\mathrm{G}_k(\mathcal{Q}_{2k+1}^+(\mathbb{K})))$ and $\mathrm{erk}(\mathrm{G}_k(\mathcal{Q}_{2k+1}^+(\mathbb{K})))$ could undermine the significance of $\mathrm{grk}(\mathrm{G}_k(\mathcal{Q}))$ for any quadric $\mathcal{Q}$ of rank $n > k$, since the only way to compute $\mathrm{grk}(\mathrm{G}_k(\mathcal{Q}))$ which we can devise ultimately relies on the case $\mathcal{Q} = \mathcal{Q}^+_{2k+1}(\mathbb{K})$ (see Section 7.5.2.3, last paragraph).

7.3 A Survey of Known Results

We shall now offer a survey of known results on generation and embeddings of grassmannians of classical polar spaces. All polar spaces to be considered in this section are defined over a commutative division ring, say $\mathbb{K}$. Hence they are either symplectic or hermitian varieties or quadrics. We denote by n and d the rank and the anisotropic gap of the polar space under consideration and by ε its universal embedding; so $\dim(\varepsilon) = 2n + d$.

Most of the results to be mentioned in the sequel have been obtained via the Standard Argument (7.2.1.3). As this argument does not work when infinite dimensional embeddings occur, henceforth, if not otherwise specified, we assume that $\dim(\varepsilon) < \infty$, namely $d < \infty$.

7.3.1 The Symplectic Case

Let $\mathcal{W} = \mathcal{W}_{2n-1}(\mathbb{K})$. Recall that in this case $d = 0$; also $\mathrm{char}(\mathbb{K}) \neq 2$, according to the conventions stated in Section 7.2.3.3. For $1 \leq k \leq n$ we have

$$\mathrm{erk}(\mathrm{G}_k(\mathcal{W})) \;=\; \mathrm{grk}(\mathrm{G}_k(\mathcal{W})) \;=\; \binom{2n}{k} - \binom{2n}{k-2}. \tag{3.1}$$

(See [1, 3, 23, 27].) The Plücker embedding $\mathrm{G}_k(\varepsilon)$ is the universal embedding of $\mathrm{G}_k(\mathcal{W})$). Method of proof: Standard Argument (7.2.1.3).

7.3.2 The Hermitian Case

Let $\mathcal{H}$ be a hermitian variety. Let $1 \leq k < n$ and assume $\mathbb{K} \neq \mathbb{F}_4$ when $k > 1$. Then

$$\mathrm{grk}(\mathrm{G}_k(\mathcal{H})) \;=\; \mathrm{erk}(\mathrm{G}_k(\mathcal{H})) \;=\; \binom{2n+d}{k}. \tag{3.2}$$

(See [1] and [3] for $d = 0$ and [11] for $0 < d < \infty$; also Section 7.5.2 of this paper.) Method of proof: Standard Argument (7.2.1.3). The universal embedding of $\mathrm{G}_k(\mathcal{H})$ is the Plücker embedding $\mathrm{G}_k(\varepsilon)$.

Most likely, when $d = 0$, equalities (3.2) also hold with $\mathbb{K} = \mathbb{F}_4$, but the proof breaks down in this case. Indeed the construction of a smallest generating set for $\mathrm{G}_k(\mathcal{H})$ with $k < n$ ultimately relies on the knowledge of a smallest generating set of $\mathrm{G}_n(\mathcal{H})$ (see Section 7.5.2). However, as we shall see in a few lines, when $\mathbb{K} = \mathbb{F}_4$, all generating sets of $\mathrm{G}_n(\mathcal{H})$ have size greater than $\binom{2n}{n}$.

Let now $k = n$ and assume that $d = 0$. If $\mathbb{K} \neq \mathbb{F}_4$, then equalities like those of (3.2) still hold [1, 3, 23, 24, 27]. Explicitly:

$$\mathrm{grk}(\mathrm{G}_n(\mathscr{H})) \;=\; \mathrm{erk}(\mathrm{G}_n(\mathscr{H})) \;=\; \binom{2n}{n}. \qquad (3.3)$$

The universal embedding of $\mathrm{G}_n(\mathscr{H})$ is the Plücker-like embedding $\mathrm{G}_n(\varepsilon)$ (see Section 7.2.5.2). Equalities (3.3) are proved via the Standard Argument (7.2.1.3).

Still with $k = n$ and $d = 0$, suppose that $\mathbb{K} = \mathbb{F}_4$. Then $\mathrm{erk}(\mathrm{G}_n(\mathscr{H})) = (4^n + 2)/3$ (Li [34]). If $n = 2$, then $(4^n + 2)/3 = \binom{2n}{n} = 6 = \mathrm{grk}(\mathrm{G}_n(\mathscr{H}))$, but if $n > 2$, then $(4^n + 2)/3 > \binom{2n}{n}$ ($= \dim(\mathrm{G}_n(\varepsilon))$). We guess that the equality $\mathrm{grk}(\mathrm{G}_n(\mathscr{H})) = \mathrm{erk}(\mathrm{G}_n(\mathscr{H}))$ still holds true when $n > 2$ but a proof of this equality is known only for $n = 3$ (Cooperstein [20]).

When $d > 0$ (possibly infinite), the dual polar space $\mathrm{G}_n(\mathscr{H})$ is not embeddable but we can still consider the generating rank of $\mathrm{G}_n(\mathscr{H})$. It turns out that $\mathrm{grk}(\mathrm{G}_n(\mathscr{H})) \leq 2^n$ (see [23] for $d = 1$ and [11] for the general case; also Section 7.5.1 of this chapter).

7.3.3 The Orthogonal Case

Let $\mathscr{Q}$ be a quadric. Nothing is known on $\mathrm{grk}(\mathrm{G}_k(\mathscr{Q}))$ for $2 < k < n$ apart from lower bounds provided by the known embeddings of $\mathrm{G}_k(\mathscr{Q})$.

Let $k = n$. The case $d = 0$ is devoid of interest. In this case, $\mathscr{Q}$ is a hyperbolic quadric. Hence all lines of $\mathrm{G}_n(\mathscr{Q})$ have just two points, $\mathrm{G}_n(\mathscr{Q})$ is not embeddable and every set of points of $\mathrm{G}_n(\mathscr{Q})$ is a subspace of $\mathrm{G}_n(\mathscr{Q})$. Consequently, the full point-set of $\mathrm{G}_n(\mathscr{Q})$ is the unique generating set of $\mathrm{G}_n(\mathscr{Q})$.

Let $1 \leq d \leq 2$ and $\mathrm{char}(\mathbb{K}) \neq 2$. Then $\mathrm{G}_n(\mathscr{Q})$ is embeddable and

$$\mathrm{grk}(\mathrm{G}_n(\mathscr{Q})) \;=\; \mathrm{erk}(\mathrm{G}_n(\mathscr{Q})) \;=\; 2^n. \qquad (3.4)$$

(See [2, 23, 25, 48].) The universal embedding is the spin embedding (when $d = 1$, namely $\mathscr{Q}$ is parabolic) and the spin-like embedding ($d = 2$ and $\mathscr{Q}$ is elliptic). In either of these two cases, $\mathrm{G}_n(\mathscr{Q})$ is generated by its frames. Hence the spin embedding and the spin-like embedding are universal, by the Standard Argument (7.2.1.3).

Let $\mathrm{char}(\mathbb{K}) = 2$ and assume that $\mathscr{Q}$ is elliptic (hence $d = 2$). Then $\mathrm{G}_n(\mathscr{Q})$ is embeddable and $\mathrm{grk}(\mathrm{G}_n(\mathscr{Q})) = \mathrm{erk}(\mathrm{G}_n(\mathscr{Q})) = 2^n$, just as when $\mathrm{char}(\mathbb{K}) \neq 2$; see [23, 25]. The universal embedding is the spin-like embedding. Method of proof: Standard Argument (7.2.1.3).

Still with $\mathrm{char}(\mathbb{K}) = 2$ and $d > 0$, let ε_R be the minimal embedding of $\mathscr{Q}$ and suppose that $\varepsilon_R(\mathscr{Q}) \cong \mathscr{W}_{2n-1}(\mathbb{K})$. (We have considered this situation at the end of Section 7.2.4.1; here we only recall that this is indeed the case when $d = 1$ and $\mathbb{K}$ is perfect). Then $G_n(\mathcal{Q})$ is embeddable. Moreover, if $\mathbb{K} \neq \mathbb{F}_2$, then

$$\mathrm{grk}(G_n(\mathcal{Q})) = \mathrm{erk}(G_n(\mathcal{Q})) = \binom{2n}{n} - \binom{2n}{n-2}, \tag{3.5}$$

and the minimal Plücker embedding $G_n(\varepsilon_R)$ is the universal embedding of $G_n(\mathcal{Q})$ (see [27]). The proof exploits the Standard Argument (7.2.1.3). If $\mathbb{K} = \mathbb{F}_2$ and $d = 1$ then $\mathrm{erk}(G_n(\mathcal{Q})) = (2^n + 1)(2^{n-1} + 1)/3$ by Li [33]; it is likely that $\mathrm{grk}(G_n(\mathcal{Q})) = \mathrm{erk}(G_n(\mathcal{Q}))$ in this case too, but so far this equality has been proved only for $n \leq 5$; see [17].

Remark 3.1 The equalities in (3.5) hold true even if d is infinite. Indeed the embedding ε_R is $2n$-dimensional and no tracks of the cardinal number d are saved in it; that number is hidden in the properties of $\mathbb{K}$ as the degree of $\mathbb{K}$ over $\mathbb{K}^2$.

Finally, let $k = 2 < n$. Assume that $\mathbb{K} = \mathbb{F}_p$ for a prime p (hence $d \leq 2$). Then $\mathrm{grk}(G_2(\mathcal{Q})) = \mathrm{erk}(G_2(\mathcal{Q})) = \binom{2n+d}{2}$ (see [4, 19] for $d = 0$, [22] for $d = 1$ and [19] for $d = 2$; proofs rely on the Standard Argument (7.2.1.3)). In each of these three cases the universal embedding (which also exists for $n = 3$ and $d = 0$, by Proposition 2.2) is the Weyl embedding, which coincides with the Plücker embedding when $p > 2$.

Nothing is known on $\mathrm{grk}(G_2(\mathcal{Q}))$ for $\mathbb{K}$ an arbitrary field except that if $d \leq 1$ then $\mathrm{grk}(G_2(\mathcal{Q})) \leq \binom{2n+d}{2} + g$ where g is the minimal number of elements to be added to the prime subfield of $\mathbb{K}$ in order to generate the whole of $\mathbb{K}$; see [4].

Some more information is available about $\mathrm{erk}(G_k(\mathcal{Q}))$ when $n > k \in \{2,3\}$ and $d \leq 1$. As stated in Proposition 2.4, when $k = 2$, if $\mathbb{K}$ is either a perfect field of positive characteristic or a number field, then the Weyl embedding of $G_k(\mathcal{Q})$ is universal (hence $\mathrm{erk}(G_k(\mathcal{Q})) = \binom{2n+d}{k}$) except possibly when $d = 0$ and $n = k + 1$. The same statement holds for $k = 3$ under the additional hypothesis that $\mathbb{K} \neq \mathbb{F}_2$; see [14].

7.4 Another Way to Define the Anisotropic Gap

Let Γ be a non-degenerate polar space of finite rank $n \geq 2$. As in [11], we say that a subspace S of Γ is *nice* if it is non-degenerate of rank n, namely it contains a frame of Γ. Let $\mathfrak{N}(\Gamma)$ be the poset of all nice subspaces of Γ ordered with respect to inclusion. Clearly Γ is the maximum element of $\mathfrak{N}(\mathcal{P})$ while the minimal elements of $\mathfrak{N}(\mathcal{P})$ are the subspaces spanned by the frames of Γ.

Let $\omega(\Gamma)$ be the least upper bound of the lengths of the well-ordered chains of the poset $\mathfrak{R}(\Gamma)$, the length of such a chain being its cardinality, diminished by 1 when finite. Obviously, $\omega(\Gamma) = 0$ if and only if Γ is spanned by its frames. The following is proved in [11] (also [40] and [12, sections 3 and 4] for the non-commutative case; we warn that gap(Γ) is called *anisotropic defect* in [11]).

Theorem 4.1   *Let Γ be classical. Then* gap(Γ) $= \omega(\Gamma)$. *Moreover, every well-ordered chain of $\mathfrak{R}(\Gamma)$ is contained in a maximal well-ordered chain and all maximal well-ordered chains of $\mathfrak{R}(\Gamma)$ have the same length, namely* gap(Γ).

Remark 4.2   Under the Generalized Continuum Hypothesis, when gap(Γ) is infinite, non-well-ordered chains also exist in $\mathfrak{R}(\Gamma)$ of length $2^{\text{gap}(\Gamma)}$; see [40], also [11, Remark 2,12].

The proof of Theorem 4.1 relies on the next lemma, proved in [11] for nice subspaces and in [12] in full generality. Recall that, given a subspace S of Γ and a singular subspace X such that $X \subseteq S \subseteq X^{\perp}$, the points and lines of the star $X^{\perp}/X$ of X that correspond to singular subspaces contained in S form a subspace S/X of $X^{\perp}/X$. Recall also that $\text{Rad}(S) = S \cap S^{\perp}$ is a (possibly empty) singular subspace of Γ.

Lemma 4.3   *Let Γ be classical, let $\varepsilon \colon \Gamma \to \text{PG}(V)$ be its universal embedding and S a subspace of Γ. If* $\text{rank}(S/\text{Rad}(S)) > 1$, *then* $S = \varepsilon^{-1}(\langle \varepsilon(S) \rangle)$.

By Theorems 4.1 and Lemma 4.3, the following can be obtained (see [11]).

Corollary 4.4   *Let Γ be classical. Then* grk(Γ) $=$ erk(Γ) $= 2n + $ gap(Γ).

In view of Theorem 4.1, we can modify the definition of gap(Γ) as follows, so that it makes sense even if Γ is non-classical or non-embeddable: gap(Γ) $:= \omega(\Gamma)$ by definition. With gap(Γ) defined in this way, if Γ is a grid or a bi-embeddable quaternion quadrangle as in case (7.2.4) of Section 7.2.4 or the line-grassmannian of PG(3, $\mathbb{K}$) with $\mathbb{K}$ non-commutative or a Freudenthal–Tits polar space, then gap(Γ) $= 0$.

7.5 Searching for Smallest Generating Sets

Throughout this section, Γ is a classical polar space of rank n and (possibly infinite) gap $d := \text{gap}(\Gamma)$. We shall focus on the problem of reducing the search for smallest generating sets of k-grassmannians of Γ to cases where those sets are known or better understood. We refer to [11] for the proofs of all results to be stated in this section but one, namely Proposition 5.6, which follows from a result of Blok and Cooperstein [3]. We shall give a brief sketch of its proof.

The following fact will be freely used in the sequel: if $d > 0$, then Γ admits non-singular hyperplanes of rank n; moreover, every such hyperplane has gap $d - 1$ ($= d$ if d is infinite). Indeed let $\varepsilon \colon \Gamma \to \mathrm{PG}(V)$ be the universal embedding of Γ. Given a decomposition $V = V_1 \oplus \cdots \oplus V_n \oplus V_0$ of V as in Section 7.2.4.2, let X be a hyperplane of $\mathrm{PG}(V)$ containing the subspace of $\mathrm{PG}(V)$ corresponding to $V_1 \oplus \cdots \oplus V_n$. Then $\varepsilon^{-1}(X)$ is a hyperplane of Γ as required.

7.5.1 The Case $k = n$

Theorem 5.1 *Suppose that $d > 0$ and let H be a non-singular hyperplane of Γ of rank n. Then every generating set of $\mathrm{G}_n(H)$ generates $\mathrm{G}_n(\Gamma)$ as well. Accordingly, $\mathrm{grk}(\mathrm{G}_n(H)) \geq \mathrm{grk}(\mathrm{G}_n(\Gamma))$.*

Corollary 5.2 *Let $d < \infty$ and let S be a nice subspace of Γ as defined in Section 7.4. Then every generating set of $\mathrm{G}_n(S)$ also generates $\mathrm{G}_n(\Gamma)$.*

Note that the inequality $\mathrm{grk}(\mathrm{G}_n(H)) \geq \mathrm{grk}(\mathrm{G}_n(\Gamma))$ of Theorem 5.1 can be strict. This is the case when $\Gamma \cong \mathscr{Q}_{2n}(\mathbb{K})$ and $H \cong \mathscr{Q}^+_{2n-1}(\mathbb{K})$ or $\Gamma \cong \mathscr{H}_{2n}(\mathbb{K})$ and $H \cong \mathscr{H}_{2n-1}(\mathbb{K})$, for instance.

Corollary 5.2 can be used to reduce the problem of evaluating $\mathrm{grk}(\mathrm{G}_n(\Gamma))$ to a better understood case with smaller gap. For instance, let $1 < d < \infty$ and suppose that Γ is a quadric. Let $S \cong \mathscr{Q}_{2n}(\mathbb{K})$ be a nice subspace of Γ of gap 1, where $\mathbb{K}$ is the underlying field of Γ. Then $\mathrm{grk}(\mathrm{G}_n(\Gamma)) \leq \mathrm{grk}(\mathrm{G}_n(S))$ by Corollary 5.2. If moreover $\mathrm{char}(\mathbb{K}) \neq 2$, hence $\mathrm{grk}(\mathrm{G}_n(S)) = 2^n$, then $\mathrm{grk}(\mathrm{G}_n(\Gamma)) \leq 2^n$. When $\mathrm{char}(\mathbb{K}) = 2$, $d > 2$ and Γ admits at least one subspace $S \cong \mathscr{Q}^-_{2n+1}(\mathbb{K})$, we obtain the same conclusion. Similarly, when Γ is a hermitian variety with gap $d > 1$, by choosing $S \cong \mathscr{H}_{2n}(\mathbb{K})$ we get $\mathrm{grk}(\mathrm{G}_n(\Gamma)) \leq 2^n$.

As mentioned in Section 7.3, the generating rank of $\mathrm{G}_n(\Gamma)$ is known when Γ is $\mathscr{W}_{2n-1}(\mathbb{K})$, $\mathscr{H}_{2n-1}(\mathbb{K})$, $\mathscr{Q}_{2n}(\mathbb{K})$ (with $\mathbb{K} \neq \mathbb{F}_2$ when $n > 5$) and $\mathscr{Q}^-_{2n+1}(\mathbb{K})$.

7.5.2 The Case $k < n$

Throughout this subsection, we assume that $1 < k < n$ and $d < \infty$. We are interested in generating sets of size equal to the dimension of a (finite dimensional) embedding which we know or believe to be dominant; in short, generating sets of the right size, so to speak.

Suppose that Γ belongs to a certain class $\mathbf{C}$ of classical polar spaces closed under taking non-singular hyperplanes and stars of points of members of $\mathbf{C}$ of rank at least 3 and such that

($*$) for any integer $h > 1$ and any member Ξ of $\mathbf{C}$ of rank h and anisotropic gap 0, the dual polar space $\mathrm{G}_h(\Xi)$ is embeddable and generating sets of the right size are known to exist for it; in other words, $\mathrm{grk}(G_h(\Xi))$ and $\mathrm{erk}(G_h(\Xi))$ are known and $\mathrm{grk}(G_h(\Xi)) = \mathrm{erk}(G_h(\Xi))$.

In order to construct generating sets of $G_k(\Gamma)$ of hopefully the right size, we first produce a fairly large generating set of Γ (Section 7.5.2.1). Next we extract a subset from it, which still generates $G_k(\Gamma)$ and we hope that it is as small as possible (Section 7.5.2.2). In Section 7.5.2.3 we show how the latter construction can be embedded in a recursive procedure to produce a generating set G of $G_k(\Gamma)$. At each step of that procedure, the polar spaces involved all belong to **C**. Smallest generating sets of dual polar spaces as in ($*$) and frames of generalized quadrangles belonging to **C** are ultimately the only inputs which this procedure needs. If $|G| = \dim(\varepsilon)$, where ε is the embedding of $G_k(\Gamma)$ which we bet to be universal, then we are done: $|G| = \mathrm{grk}(G_k(\Gamma)) = \mathrm{erk}(G_k(\Gamma))$ and ε is indeed universal.

7.5.2.1 Large Generating Sets of $G_k(\Gamma)$

Given a hyperplane H of Γ, choose a point p of Γ exterior to H. Then $H \cap p^\perp$ is a non-degenerate subspace of Γ of rank $n-1$, isomorphic to the star $p^\perp/p$ of p. When $k > 2$, let $G_{H,p}$ be a generating set for $G_{k-2}(H \cap p^\perp)$. When $k = 2$ we put $G_{H,p} = \{\emptyset\}$. For any $Z \in G_{H,p}$, choose a singular subspace $\widehat{Z}$ of Γ of rank k containing Z and such that $\widehat{Z} \nsubseteq H \cup p^\perp$ and $\widehat{Z} \cap H \neq \widehat{Z} \cap p^\perp$. Put $\widehat{G}_{H,p} := \{\widehat{Z} : Z \in G_{H,p}\}$. In particular, when $k = 2$, the set $\widehat{G}_{H,p}$ consists of a single line of Γ. Also

$$\mathscr{P}_k(H) := \{X \in \mathscr{P}_k : X \subseteq H\}, \qquad \mathscr{P}_k(p) := \{X \in \mathscr{P}_k : p \in X\},$$

where $\mathscr{P}_k$ stands for the set of singular subspaces of Γ of rank k, as in Section 7.2.5 (note that $\mathscr{P}_k(H) \neq \emptyset$, as all hyperplanes of Γ have rank either n or $n-1$ and $k < n$ by assumption). Define:

$$\mathscr{P}_k(H, p, \widehat{G}_{H,p}) := \mathscr{P}_k(H) \cup \mathscr{P}_k(p) \cup \widehat{G}_{H,p}. \tag{5.6}$$

Lemma 5.3 *The set $\mathscr{P}_k(H, p, \widehat{G}_{H,p})$ generates $G_k(\Gamma)$.*

Let H be singular, say $H = q^\perp$ for a point $q \notin p^\perp$. So, $H \cap p^\perp = \{q, p\}^\perp$. Construction (5.6) can be refined as follows. As before, given a generating set $G_{q,p}$ of the $(k-2)$-grassmannian of $\{q,p\}^\perp$ (with $G_{q,p} = \{\emptyset\}$ when $k = 2$), for every $Z \in G_{q,p}$, we choose a singular subspace $\widehat{Z} \in \mathscr{P}_k$ containing Z and such that $\widehat{Z} \cap \{q,p\}^\perp = Z$ and put $\widehat{G}_{q,p} := \{\widehat{Z} : Z \in G_{q,p}\}$. Also $\mathscr{P}_k(q) := \{X \in \mathscr{P}_k : q \in X\}$ and $\mathscr{P}_k(\{q,p\}^\perp) := \{X \in \mathscr{P}_k : X \subseteq \{q,p\}^\perp\}$ ($\neq \emptyset$ since $\mathrm{rank}(\{q,p\}^\perp) = n - 2 \geq k$). Define:

$$\mathscr{P}_k(q, p, \widehat{G}_{q,p}) := \mathscr{P}_k(q) \cup \mathscr{P}_k(\{q,p\}^\perp) \cup \mathscr{P}_k(p) \cup \widehat{G}_{q,p}. \tag{5.7}$$

Lemma 5.4 *The set $\mathscr{P}_k(q, p, \widehat{G}_{q,p})$ generates $G_k(\Gamma)$.*

7.5.2.2 Smaller Generating Sets

Recall that $d = \text{gap}(\Gamma)$. Suppose that $d > 0$. Let H be a non-singular hyperplane of Γ of rank n. Then $\text{gap}(H) = d - 1$. With $p \notin H$ and $\widehat{G}_{H,p}$ defined as in Section 7.5.2.1, let $G(H)$ and $G(p)$ be generating sets of $G_k(H)$ and $G_{k-1}(p^\perp/p)$ respectively. Put

$$G_{k,n,d} := G(H) \cup G(p) \cup \widehat{G}_{H,p} \subseteq \mathscr{P}_k(H, p, \widehat{G}_{H,p}). \tag{5.8}$$

Lemma 5.3 implies the following.

Proposition 5.5 *The set $G_{n,k,d}$ generates $G_k(\Gamma)$.*

Dropping the hypothesis that $d > 0$, let p and q be non-collinear points. Define $\widehat{G}_{q,p}$ as in Section 7.5.2.1, choose generating sets $G(q)$, $G(p)$ and $G(q,p)$ of $G_{k-1}(q^\perp/q)$, $G_{k-1}(p^\perp/p)$ and $G_k(\{q,p\}^\perp)$ respectively and put

$$G'_{k,n,d} := G(q) \cup G(q,p) \cup G(p) \cup \widehat{G}_{q,p} \subseteq \mathscr{P}_k(q, p, \widehat{G}_{q,p}). \tag{5.9}$$

Proposition 5.6 *If $k = n - 1$, assume that $d = 0$; else $d \geq 0$. Then $G'_{k,n,d}$ generates $G_k(\Gamma)$.*

Proof When $k < n - 1$, the statement follows from Lemma 5.4. Suppose that $k = n - 1$. Then $\text{rank}(\{q,p\}^\perp) = n - 1 = k$. Hence $G(q,p)$ cannot span $G_k(\{q,p\}^\perp)$ in $G_k(\Gamma)$. Indeed now $G_k(\{q,p\}^\perp)$ is a dual polar space, hence its lines are not lines of $G_k(\Gamma)$. Consequently, we cannot claim that $G(q) \cup G(q,p)$ spans $\mathscr{P}_k(q^\perp)$ and we cannot apply Lemma 5.4.

However in this case, $d = 0$ by assumption. So, we can resort Blok and Cooperstein [3] for help. Indeed Proposition 5.3 of [3] deals just with the set $G'_{k,n,0}$, defined as in (5.9) but with $d = 0$. That proposition states that $G'_{k,n,0}$ generates $G_k(\Gamma)$. Actually Γ stands for $\mathscr{W}_{2n-1}(\mathbb{K})$ and $\mathscr{H}_{2n-1}(\mathbb{K})$ in [3], but the proof of Proposition 5.3 of [3] exploits only the fact that $d = 0$ for $\mathscr{W}_{2n-1}(\mathbb{K})$ and $\mathscr{H}_{2n-1}(\mathbb{K})$. So, the conclusions of that proposition hold true in general. $\square$

7.5.2.3 A Recursive Construction

Now we work by induction on the parameters n, d and k, assuming to know that generating sets of the right size exist when one of those parameters is given a lesser value. So, we decrease each of them step by step, firstly pulling d down to 0, next n down to $k + 1$ and finally k to 1. Explicitly, we do as follows.

1. Let $d > 0$. We choose the set $G_{k,n,d}$ defined in (5.8) as a candidate smallest generating set of Γ, with $G(H)$, $G(p)$ and $G_{H,p}$ of the right size. Choosing $G(H)$ of the right size is the same problem as finding a smallest generating set of $G_k(\Gamma)$, but with parameters $(k, n, d - 1)$ instead of (k, n, d). With $G(p)$

and $G_{H,p}$ we come down to $(k-1,n-1,d)$ and $(k-2,n-1,d)$, respectively, which we can assume to be well understood (by induction). By iterating this process, we eventually end up with $(k,n,0)$. If $k < n-1$, we continue as follows, otherwise we jump to the third phase of the process.

2. Let $d = 0$ but suppose $n > k+1$. Now we use $G'_{k,n,0}$. Choosing $G(q,p)$ of the right size is the same problem we want to solve for $G_k(\Gamma)$, but with $(k,n-1,0)$ in place of $(k,n,0)$. By iteration, we eventually end up with $(k,k+1,0)$.

3. Finally, assume $d = 0$ and $n = k+1$. If $k > 1$, we still use $G'_{k,k+1,0}$ (Proposition 5.6). When choosing $G(q)$, $G(p)$ and $G_{q,p}$ we shift from $(k,k+1,0)$ to $(k-1,k,0)$ and, in view of hypothesis (∗) on **C**, we are supposed to know how to choose $G(q,p)$ in the proper way. By iterating this process, we eventually end up with $(k,n,d) = (1,2,0)$. At this stage we are done: $G_k(\Gamma) = \Gamma$ is a generalized quadrangle of gap 0; so, a frame is enough to generate it.

Remark 5.7 Actually, we should use a letter different from Γ at the end of the above paragraph, since the geometry we are considering there is not the same Γ we started from, but it derives from it via a series of reductions; besides, writing k, n and d in the descriptions of phases 2 and 3 is incorrect as well, since k, n and d are the initial values of the corresponding parameters rather than the values they take at each step. For the sake of correctness, we should have introduced unknowns instead of Γ, k, n and d, which are data rather than unknowns, but the outcome would have been a bit too awkward. We have preferred to keep our notation, albeit wrong, hoping that this abuse will not confuse the reader.

The procedure we have described is exploited in [11] to improve the main result of Blok and Cooperstein [3] in the hermitian case, thus obtaining formula (3.2) of Section 7.3.2 for $d > 0$ and proving that the Plücker embedding is universal also in that case. Only part 1 of this procedure is used in [11], as to come down to the case $d = 0$, already fixed by Blok and Cooperstein in [3]. Parts 2 and 3 are basically the method used by Blok and Cooperstein in [3], except that they do the job in one phase, pulling $(k,n,0)$ down to $(1,n+1-k,0)$.

Our procedure jams when Γ is a quadric, since (∗) does not hold when **C** is the class of quadrics. Indeed the quadrics with anisotropic gap $d = 0$ are the hyperbolic ones; if $\Gamma = \mathcal{Q}^+_{2n-1}(\mathbb{K})$, then $G_n(\Gamma)$ is not embeddable and the unique generating set of $G_n(\Gamma)$ is the full point-set of $\mathcal{Q}^+_{2n-1}(\mathbb{K})$, which offers no insight into the generating ranks of other orthogonal grassmannians. However (∗) enters the game only in phase 3 of our procedure. The first two phases work for quadrics as well. They allow us to reduce the computation of $\mathrm{grk}(G_k(\mathcal{Q}))$ for $\mathcal{Q}$ a quadric of rank $n > k$ and finite gap $d \geq 0$ to the case $d = 0$ and $n = k+1$. However, once we reach that stage we are forced to stop. If we already know the generating rank of $G_k(\mathscr{Q}^+_{2k+1}(\mathbb{K}))$ (as when $k = 2$ and $\mathbb{K}$ is a finite prime field [19]), then we can infer what $\mathrm{grk}(G_k(\mathscr{Q}))$ is. Otherwise, reducing the problem to the computation of $\mathrm{grk}(G_k(\mathscr{Q}^+_{2k+1}(\mathbb{K})))$ is ultimately pointless.

7.6 Generation over Subfields

7.6.1 Subgeometries over Subrings

Given a division ring $\mathbb{K}$ and a subdivision ring $\mathbb{K}_0$ of $\mathbb{K}$, let V be a $\mathbb{K}$-vector space and E be a basis of V. Let V_0 be the $\mathbb{K}_0$-span of E. Then $\mathrm{PG}(V_0)$ can naturally be identified with a subgeometry of $\mathrm{PG}(V)$, which we still denote by $\mathrm{PG}(V_0)$ and call the $\mathbb{K}_0$-*subgeometry* of V (*based on E*).

Given an antiautomorphism σ of $\mathbb{K}$ such that $\sigma(\mathbb{K}_0) = \mathbb{K}_0$, let σ_0 be the antiautomorphism of $\mathbb{K}_0$ induced by σ. Every σ_0-sesquilinear form $f_0 \colon V_0 \times V_0 \to \mathbb{K}_0$ can be extended to a unique σ-sesquilinear form $f \colon V \times V \to \mathbb{K}$ and, conversely, if $f \colon V \times V \to \mathbb{K}$ is a σ-sesquilinear form such that $f(V_0 \times V_0) \subseteq \mathbb{K}_0$, then f induces a σ_0-sesquilinear form f_0 on $V_0 \times V_0$ and f is an extension of f_0 to $V \times V$. With f and f_0 as above, if $\epsilon \in \mathbb{K}_0^*$, then f is (σ, ϵ)-sesquilinear if and only if f_0 is (σ_0, ϵ)-sesquilinear. Suppose moreover that

$$\text{if } \mathbb{K}_{\sigma,\epsilon} \neq \mathbb{K}, \text{ then } \mathbb{K}_0 \neq \mathbb{K}_{\sigma,\epsilon} \cap \mathbb{K}_0 = (\mathbb{K}_0)_{\sigma_0,\epsilon} \; (= \{t_0 - t_0^{\sigma_0}\epsilon\}_{t_0 \in \mathbb{K}_0}). \quad (6.10)$$

Note that (6.10) trivially holds when $\sigma = \mathrm{id}_{\mathbb{K}}$. Assume that $\mathbb{K} \neq \mathbb{K}_{\sigma,\epsilon}$. Then (6.10) implies that

$$\{0\} \neq \frac{\mathbb{K}_0 + \mathbb{K}_{\sigma,\epsilon}}{\mathbb{K}_{\sigma,\epsilon}} \cong \frac{\mathbb{K}_0}{\mathbb{K}_{\sigma,\epsilon} \cap \mathbb{K}_0} = \frac{\mathbb{K}_0}{(\mathbb{K}_0)_{\sigma_0,\epsilon}}. \quad (6.11)$$

Recall that $\mathbb{K}_{\sigma,\epsilon} = \mathbb{K}$ if and only if $\mathrm{char}(\mathbb{K}) \neq 2$, $\sigma = \mathrm{id}_{\mathbb{K}}$ and $\epsilon = -1$. Assume that this is not the case and let $\phi \colon V \to \mathbb{K}/\mathbb{K}_{\sigma,\epsilon}$ be a (σ, ϵ)-pseudoquadratic form such that $\phi(V_0) \subseteq (\mathbb{K}_0 + \mathbb{K}_{\sigma,\epsilon})/\mathbb{K}_{\sigma,\epsilon}$. Then, in view of (6.11), the restriction ϕ_0 of ϕ to V_0 can be regarded as a (σ_0, ϵ)-pseudoquadratic form. The sesquilinearized f_{ϕ_0} of ϕ_0 is the restriction to $V_0 \times V_0$ of the sesquilinearized f_ϕ of ϕ. Conversely, every (σ_0, ϵ)-pseudoquadratic form ϕ_0 on V_0 can be extended to a unique (σ, ϵ)-pseudoquadratic form ϕ of V.

With ϕ and ϕ_0 as stated and $\Gamma = \Gamma(\phi)$ (the polar space associated to ϕ), the polar space $\Gamma_0 := \Gamma(\phi_0)$ is naturally identified with a subgeometry of Γ. Indeed, in view of (6.11), for a vector $v \in V_0$ we have $\phi_0(v) = 0$ (in $\mathbb{K}_0/(\mathbb{K}_0)_{\sigma_0,\epsilon}$) if and only if $\phi(v) = 0$ (in $\mathbb{K}/\mathbb{K}_{\sigma,\epsilon}$). In short,

$$\Gamma_0 = \Gamma \cap \mathrm{PG}(V_0), \quad (6.12)$$

where $\mathrm{PG}(V_0)$ is regarded as a subgeometry of $\mathrm{PG}(V)$. We call Γ_0 the $\mathbb{K}_0$-*subgeometry* of Γ (*based on E*). Similarly, still assuming (6.10), if f_0 is the

(σ_0, ϵ)-sesquilinear form induced on $V_0 \times V_0$ by a (σ, ϵ)-sesquilinear form $f :$ $V \times V \to \mathbb{K}$ such that $f(V_0 \times V_0) \subseteq \mathbb{K}_0$, then we call $\Gamma(f_0)$ the $\mathbb{K}_0$-*subgeometry* of $\Gamma(f)$ (*based on E*). Clearly, (6.12) also holds with $\Gamma = \Gamma(f)$ and $\Gamma_0 = \Gamma(f_0)$.

As E is a basis of V, if $v \in \mathrm{Rad}(\phi_0)$ (or $v \in \mathrm{Rad}(f_0)$), then $v \in \mathrm{Rad}(\phi)$ (respectively $v \in \mathrm{Rad}(f)$). So, $\mathrm{Rad}(f_{\phi_0}) \subseteq \mathrm{Rad}(f_\phi)$ and, if ϕ is non-degenerate, then ϕ_0 is non-degenerate. Similarly for f and f_0: if f is non-degenerate then f_0 is non-degenerate. It is also clear that $\mathrm{rank}(\Gamma_0) \leq \mathrm{rank}(\Gamma)$. By (6.12), the following holds.

Proposition 6.1 *Suppose that Γ has finite rank. Then* $\mathrm{rank}(\Gamma_0) = \mathrm{rank}(\Gamma)$ *if and only if* $\mathrm{PG}(V_0)$ *contains a frame of* Γ.

Remark 6.2 Hypothesis (6.10) is necessary for ϕ_0 to be a pseudoquadratic form. Without it, the restriction of ϕ to V_0 might turn to be a generalized pseudo-quadratic form in the sense of [39] or the null form. As we cannot treat reflexive sesquilinear forms differently from pseudoquadratic forms, we are forced to assume (6.10) also for reflexive sesquilinear forms. We warn that (6.10) does not prevent ϕ and ϕ_0 from being of different types, as when ϕ is hermitian and ϕ_0 is quadratic. If we like to avoid this, we need some additional hypotheses, like the following: $\sigma_0 = \mathrm{id}_{\mathbb{K}_0}$ only if $\sigma = \mathrm{id}_{\mathbb{K}}$.

7.6.2 Grassmannians of Buildings

In view of our purposes for this section, it is convenient to regard non-degenerate polar spaces of finite rank as buildings. We refer to Tits [45, chs. 6, 7] for details (also [38] or Shult [42] or Buekenhout and Cohen [7]). Here we just recall that the non-empty singular subspaces of a non-degenerate polar space Γ of finite rank n are the vertices of a simplicial complex $\mathscr{K}(\Gamma)$, the simplices of which are the flags of Γ, namely the nested chains of non-empty singular subspaces of Γ. The complex $\mathscr{K}(\Gamma)$, equipped with the family of subcomplexes $\mathscr{K}(\mathscr{A})$ for $\mathscr{A}$ an apartment of Γ, is a (weak) building as defined in [45, ch. 3] whose types correspond to the possible ranks of the singular subspaces.

Similarly, if Γ is a finite dimensional projective geometry, the non-empty proper subspaces of Γ are the vertices of a building $\mathscr{K}(\Gamma)$, the subspaces of Γ spanned by subsets of a basis of Γ form an apartment of $\mathscr{K}(\Gamma)$ and all apartments of $\mathscr{K}(\Gamma)$ arise in this way from bases of Γ. Of course, the simplices of the complex $\mathscr{K}(\Gamma)$ are the flags of Γ and the types correspond to the ranks of the subspaces appearing in a simplex.

7.6.2.1 The J-Grassmannian of a Building

Given a building $\mathscr{K}$ with set of types I, let $\mathscr{D} = (I, \sim)$ be the simple graph defined on I by the Coxeter diagram $D(\mathscr{K})$ of $\mathscr{K}$. Following Tits [45, ch. 12] (see also [38, ch. 5]), for $\emptyset \neq J \subseteq I$ the *J-grassmannian* of $\mathcal{H}$ is the point-line geometry $G_J(\mathcal{H})$ defined as follows. The points are the *J*-flags of $\mathcal{H}$ (namely the simplices of type J) and the lines are the flags of type $j^{\sim} \cup (J \setminus \{j\})$ for $j \in J$, with $j^{\sim}$ the neighborhood of j in $\mathcal{D}$ (where $j \notin j^{\sim}$ by convention). A point P and a line L of $G_J(\mathcal{H})$ are declared to be incident precisely when $P \cup L$ is a flag of $\mathcal{H}$. Points and lines of $G_J(\mathcal{H})$ will be called *J-points* and *J-lines*, for short. Properties of buildings force $G_J(\mathcal{H})$ to be a partial linear space (Tits [45, ch. 12], also Pasini [38, ch. 6]). Hence we can freely regard every *J*-line as the same as the set of its *J*-points, as usual.

A subset K of I is said to be *J-reduced* if no proper subset $T \subset K$ separates $K \setminus T$ from J in $\mathcal{D}$. If F is a flag of $\mathcal{H}$ with *J*-reduced type $t(F)$, then the *J-shadow* of F, namely the set of *J*-points P such that $P \cup F$ is a flag of $\mathcal{H}$, form a convex subspace of $G_J(\mathcal{H})$. We call these subspaces *shadow subspaces*. The *shadow rank* (*rank* for short) of a shadow subspace X of $G_J(\mathcal{H})$ is the cardinality of the union of the connected components of the graph $\mathcal{D} \setminus t(F)$ which meet $J \setminus t(F)$ non-trivially, where F is the *J*-reduced flag corresponding to X. The non-empty proper ordinary subspaces of $G_J(\mathcal{H})$, with their shadow ranks as types, form a geometry of rank n in the sense of diagram geometry [38, ch. 5].

Frames of $G_J(\mathcal{H})$ can also be defined. If $\mathscr{A}$ is an apartment of $\mathcal{H}$, the *J*-grassmannian $G_J(\mathscr{A})$ of $\mathscr{A}$ is a subgeometry of $G_J(\mathcal{H})$. The point-set of $G_J(\mathscr{A})$ is called a *frame* of $G_J(\mathcal{H})$.

It goes without saying that all these definitions agree with those stated in Section 7.2.5 when we start from the building of a polar space and J is a singleton. In fact, as we shall see in the next subsection, they generalize those of Section 7.2.5.

7.6.2.2 Special Cases

We have introduced grassmannians of buildings in view of two families of grassmannians which we cannot avoid discussing even if the elements of one of these two families are not polar grassmannians. The elements of the other family are polar grassmannians, but they are better understood in the setting of buildings. Before we turn to these geometries, we need to spend a few more words on projective and polar grassmannians.

For $n < \infty$, let Γ be an n-dimensional projective space or a non-degenerate polar space of rank n. Then the Coxeter diagram $D(\mathcal{H})$ of the building $\mathcal{H} = \mathcal{H}(\Gamma)$ is a string and it can be labeled in a standard way, by taking $I = \{1, 2, \dots, n\}$ as the set of types, where the type of a vertex of $\mathcal{H}$ is its rank as a subspace of Γ (if Γ is a projective space) or as a singular subspace of Γ (if Γ is a polar space). In Coxeter notation, the diagram $D(\mathcal{K})$ looks as follows, where A_n and C_n (or B_n, according to tastes) are the Coxeter names for these diagrams:

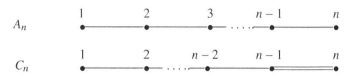

Remark 6.3 These diagrams indeed characterize projective and polar spaces (Tits [45, chs. 6, 7]): if $D(\mathcal{K}) = A_n$ (or $D(\mathcal{K}) = C_n$) then $\mathcal{K} = \mathcal{K}(\Gamma)$ for a projective (respectively, polar) space Γ.

With $\mathcal{K}$ as here, for $1 \le k \le n$ we obtain projective grassmannians and polar grassmannians as defined in Section 7.2.5. For instance, when $1 < k < n$, the neighbourhood of k in $D(\mathcal{K})$ is the pair $\{k-1, k+1\}$. The flags of $\mathcal{K}$ of type $\{k-1, k+1\}$ just correspond to the lines of $G_k(\Gamma)$ as defined in Section 7.2.5.

When Γ is a projective space, we shall also consider $G_{\{1,n\}}(\mathcal{K}(\Gamma))$, henceforth denoted $G_{1,n}(\Gamma)$ for short and called the *point-hyperplane geometry* of Γ. The points of $G_{1,n}(\Gamma)$ are the point-hyperplane flags of Γ and the lines are the flags of type $\{1, n-1\}$ or $\{2, n\}$. The $\{1, n\}$-points of a $\{1, n\}$-line $L = \{p, X\}$ of type $\{1, n-1\}$ are the point-hyperplane flags $\{p, H\}$ for H a hyperplane containing X and, if $L = \{\ell, H\}$ has type $\{2, n\}$, then its $\{1, n\}$-points are the flags $\{p, H\}$ with $p \in \ell$. When $n > 2$, the shadow subspaces of $G_{1,n}(\Gamma)$ of rank 2 correspond to flags of type $\{3, n\}$, $\{1, n-2\}$ or $\{2, n-1\}$ (vertices of type 3, 1 and 2, respectively, when $n = 3$). Flags of type $\{3, n\}$ or $\{1, n-2\}$ give rise to projective planes but flags of type $\{2, n-1\}$ yield grids.

When $\mathbb{K}$ is commutative, the symbol $A_n(\mathbb{K})$ is also used to denote the building $\mathcal{K}(\mathrm{PG}(n, \mathbb{K}))$. The $\{1, n\}$-grassmannian $G_{1,n}(A_n(\mathbb{K}))$ admits a (probably dominant) embedding in the adjoint module of $\mathrm{SL}_{n+1}(\mathbb{K})$. This module is associated with the longest root of the root system of type A_n. Accordingly, $G_{1,n}(A_n(\mathbb{K}))$ is often called the *long root geometry* for $\mathrm{SL}_{n+1}(\mathbb{K})$.

We now turn to $\mathcal{Q} = \mathcal{Q}_{2n-1}^+(\mathbb{K})$, with $n \ge 3$. It is well known [45, ch. 6] that maximal singular subspaces of this polar space split in two families, say $\mathcal{M}^+$ and $\mathcal{M}^-$, such that two maximal singular subspaces M and N belong to the same family if and only if $n - \mathrm{rank}(M \cap N)$ is even. It is also well known that $\mathcal{Q} \cong G_1(\mathcal{K})$ for a building $\mathcal{K}$ with diagram as follows:

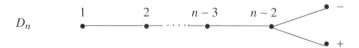

The elements of type $k < n-1$ are the singular subspaces of $\mathcal{Q}$ of rank k while the elements of type $+$ (or $-$) are the maximal singular subspaces of $\mathcal{Q}$ which belong to $\mathcal{M}^+$ (respectively $\mathcal{M}^-$). Two vertices of $\mathcal{K}$, one of which has type $k \in \{1, 2, \ldots, n-2\}$, form a flag of $\mathcal{K}$ if and only if one of the two contains the other one in $\mathcal{Q}$. Two vertices $M, N \in \mathcal{M}^+ \cup \mathcal{M}^-$ form a flag in $\mathcal{K}$ precisely when the singular subspace $M \cap N$ of $\mathcal{Q}$ has rank $n-1$ (hence one of M or N belongs to $\mathcal{M}^+$ and the other one to $\mathcal{M}^-$). This building is usually denoted by the symbol $D_n(\mathbb{K})$.

Note that $D_3(\mathbb{K})$ and $A_3(\mathbb{K})$ have the same diagram. Hence $D_3(\mathbb{K}) \cong A_3(\mathbb{K})$ by Remark 6.3. Accordingly, $G_{+,-}(D_3(\mathbb{K})) \cong G_{1,3}(A_3(\mathbb{K}))$. Also, $G_1(D_3(\mathbb{K})) \cong G_2(A_3(\mathbb{K})) \cong \mathcal{Q}_5^+(\mathbb{K})$ (Klein correspondence). Turning to the general case, let $\mathcal{Q} = \mathcal{Q}_{2n-1}^+(\mathbb{K})$. The equality $G_1(D_n(\mathbb{K})) = \mathcal{Q}$ is obvious. The flags of $D_n(\mathbb{K})$ of type $\{+, -\}$ naturally correspond to the singular subspaces of $\mathcal{Q}$ of rank $n-1$ and the lines of $G_{+,-}(D_n(\mathbb{K}))$ are provided by the flags of $D_n(\mathbb{K})$ of type $\{n-2, +\}$ and $\{n-2, -\}$, which together correspond to the flags of $\mathcal{Q}$ of type $\{n-2, n\}$. It is now clear that $G_{n-1}(\mathcal{Q}) \cong G_{+,-}(D_n(\mathbb{K}))$.

7.6.3 Sub-grassmannians over Subfields

We say that a building $\mathcal{K}_0$ is a *sub-building* of a building $\mathcal{K}$ if $\mathcal{K}_0$ is a subcomplex of $\mathcal{K}$ and all apartments of $\mathcal{K}_0$ are apartments of $\mathcal{K}$ as well. In view of the latter condition, $\mathcal{K}$ and $\mathcal{K}_0$ have the same Coxeter diagram. In particular, they have the same rank and we can assume that they also have the same set of types, say I. Clearly, $G_J(\mathcal{K}_0)$ is a subgeometry of $G_J(\mathcal{K})$ for any $\emptyset \neq J \subseteq I$.

In this section, we are interested in the following situation: $\mathcal{K}$ is defined over a division ring (in fact a field) $\mathbb{K}$ and $\mathcal{K}_0$ is defined over a subring (subfield) of $\mathbb{K}$; but we are not going to discuss this setting in full generality. We will only consider a few special cases where all relevant notions, like the meaning of the phrase, to be defined (over a given division ring) are easy to provide. Note however that, although many buildings are naturally associated with a division ring (usually a field), this is not the case for all of them (see e.g. Remark 6.4).

In the sequel we focus on the following four cases, all of which can rightly be said to be defined over the chosen field $\mathbb{K}$, namely: $A_n(\mathbb{K}) = \mathcal{K}(\mathrm{PG}(n, \mathbb{K}))$, $D_n(\mathbb{K})$ and the buildings $\mathcal{K}(\mathcal{Q}_{2n}(\mathbb{K}))$ and $\mathcal{K}(\mathcal{W}_{2n-1}(\mathbb{K}))$, usually denoted $B_n(\mathbb{K})$ and $C_n(\mathbb{K})$, respectively (Dynkin notation). Recall that $C_n(\mathbb{K})$ is defined only for char$(\mathbb{K}) \neq 2$.

Remark 6.4 The symbol $A_n(\mathbb{K})$ is used only when $\mathbb{K}$ is a field, namely it is commutative. When $\mathbb{K}$ is non-commutative, $\mathrm{PG}(n, \mathbb{K})$ and its dual $\mathrm{PG}(n, \mathbb{K}^{\mathrm{op}})$ give rise to the same building $\mathcal{K} = \mathcal{K}(\mathrm{PG}(n, \mathbb{K})) = \mathcal{K}(\mathrm{PG}(n, \mathbb{K}^{\mathrm{op}}))$. In this case, if a division ring should be attached to $\mathcal{K}$, both $\mathbb{K}$ and $\mathbb{K}^{\mathrm{op}}$ deserve this acknowledgment. However if $\mathbb{K} \not\cong \mathbb{K}^{\mathrm{op}}$, as it can happen when $\mathbb{K}$ is non-commutative, $\mathbb{K}$ and $\mathbb{K}^{\mathrm{op}}$ are different objects. Since there is no reason to prefer any of them, we renounce to match $\mathcal{K}$ with a single division ring. Note also that copies of both PG$(2, \mathbb{K})$ and PG$(2, \mathbb{K}^{\mathrm{op}})$ occur in $\mathrm{G}_J(\mathcal{K})$ for nearly all choices of J. When this is the case and $\mathbb{K}$ is non-commutative, no underlying division ring can be assigned to $\mathrm{G}_J(\mathcal{K})$.

Remark 6.5 The symbols $A_n(\mathbb{K})$, $B_n(\mathbb{K})$, $C_n(\mathbb{K})$ and $D_n(\mathbb{K})$ belong to the notation commonly used for Chevalley groups and Lie algebras, namely Dynkin notation, not to be confused with Coxeter notation. Note that in Dynkin notation, $B_n(\mathbb{K}) \neq C_n(\mathbb{K})$ while $B_n = C_n$ in Coxeter notation.

Given a field $\mathbb{K}$, a subfield $\mathbb{K}_0$ of $\mathbb{K}$, a finite dimensional $\mathbb{K}$-vector space V and a basis E of V, let V_0 be the $\mathbb{K}_0$-span of E, as in Section 7.6.1. With $n = \dim(V) - 1$, let $\mathcal{K} = A_n(\mathbb{K}) = \mathcal{K}(\mathrm{PG}(V))$ and $\mathcal{K}_0 = A_n(\mathbb{K}_0) = \mathcal{K}(\mathrm{PG}(V_0))$ with PG(V_0) regarded as a subgeometry of PG(V). The proper non-empty subsets of E span the simplices of an apartment $\mathcal{A}_0$ as well as of an apartment $\mathcal{A}$ of $\mathcal{K}$ and the simplices of $\mathcal{A}_0$ are naturally identified with those of $\mathcal{A}$. By a little abuse, we can identify $\mathcal{A}_0 = \mathcal{A}$. As E can be replaced by any basis of V_0, all apartments of $\mathcal{K}_0$ are apartments of $\mathcal{K}$ as well. Similarly, all vertices of $\mathcal{K}_0$ can be regarded as vertices of $\mathcal{K}$. In short, $\mathcal{K}_0$ is a sub-building of $\mathcal{K}$. Therefore $\mathrm{G}_J(\mathcal{K}_0)$ is a subgeometry of $\mathrm{G}_J(\mathcal{K})$, for every non-empty subset J of the common type-set $\{1, 2, \ldots, n\}$ of both $\mathcal{K}$ and $\mathcal{K}_0$.

Suppose now that $\dim(V) \in \{2n, 2n + 1\}$ and let $\phi \colon V \to \mathbb{K}$ be a non-degenerate quadratic form such that $\phi(V_0) \subseteq \mathbb{K}_0$. The quadric $\mathcal{Q} = \mathcal{Q}(\phi)$ associated to ϕ is non-degenerate of rank n and it is hyperbolic or parabolic according to whether $\dim(V) = 2n$ or $\dim(V) = 2n + 1$. Hypothesis (6.10) of Section 7.6.1 holds. Hence we can consider the quadratic form $\phi_0 \colon V_0 \to \mathbb{K}_0$ induced by ϕ on V_0 and, if the basis E of V has been chosen in such a way as to contain a frame of the quadric $\mathcal{Q} = \mathcal{Q}(\phi)$, then the quadric $\mathcal{Q}_0 = \mathcal{Q}(\phi_0)$ associated to ϕ_0 has the same rank n as $\mathcal{Q}$ (Proposition 6.1).

Suppose that E indeed contains a frame of $\mathcal{Q}$ and let $\dim(V) = 2n + 1$. Then $\mathcal{K}_0 := \mathcal{K}(\mathcal{Q}_0) = B_n(\mathbb{K}_0)$ is a sub-building of $\mathcal{K} := \mathcal{K}(\mathcal{Q}) = B_n(\mathbb{K})$ and $\mathrm{G}_J(\mathcal{K}_0)$ is a subgeometry of $\mathrm{G}_J(\mathcal{K})$ for every $\emptyset \neq J \subseteq \{1, 2, \ldots, n\}$.

Still assuming that E contains a frame of $\mathcal{Q}$, the same definitions as here can be stated when $\dim(V) = 2n$, both for $D_n(\mathbb{K})$ and $\mathcal{K}(\mathcal{Q}) = \mathrm{G}_1(D_n(\mathbb{K}))$. In particular, $D_n(\mathbb{K}_0)$ being obtained from $\mathcal{Q}_0$ in the same way as $D_n(\mathbb{K})$ from $\mathcal{Q}$, the building $D_n(\mathbb{K}_0)$ is contained in $D_n(\mathbb{K})$ as a sub-building and $\mathrm{G}_J(D_n(\mathbb{K}_0))$ is a subgeometry of $\mathrm{G}_J(D_n(\mathbb{K}))$ for every $\emptyset \neq J \subseteq \{1, \ldots, n - 2, +, -\}$.

Suppose char($\mathbb{K}$) $\neq$ 2. Still with dim(V) = $2n$, given a non-degenerate alternating form $f \colon V \times V \to \mathbb{K}$ such that $f(V_0 \times V_0) \subseteq \mathbb{K}_0$ let f_0 be the form induced by f on $V_0 \times V_0$. Then f_0 is alternating and non-degenerate and the symplectic variety $\mathscr{W}_0 = \mathscr{W}(f_0)$ associated to f_0 is a subgeometry of $\mathscr{W} = \mathscr{W}(f)$. Every frame of $\mathscr{W}_0$ is a frame of $\mathscr{W}$. Hence $\mathscr{K}_0 := \mathscr{K}(\mathscr{W}_0)$ is a sub-building of $\mathscr{K} := \mathscr{K}(\mathscr{W})$. Accordingly, $\mathrm{G}_J(\mathscr{K}_0)$ is a subgeometry of $\mathrm{G}_J(\mathscr{K})$ for every $\emptyset \neq J \subseteq \{1, 2, \ldots, n\}$.

7.6.4 Generating J-Grassmannians over Subfields

7.6.4.1 Definition and General Properties

Henceforth X_n stands for any of A_n, B_n, C_n or D_n. If (the point-set of) the geometry $\mathrm{G}_J(X_n(\mathbb{K}_0))$ generates $\mathrm{G}_J(X_n(\mathbb{K}))$, we say that $\mathrm{G}_J(X_n(\mathbb{K}))$ is *generated over $\mathbb{K}_0$*, also $\mathbb{K}_0$-*generated* for short.

If $\mathrm{G}_J(X_n(\mathbb{K}))$ is $\mathbb{K}_0$-generated, then every generating set of $\mathrm{G}_J(X_n(\mathbb{K}_0))$ also generates $\mathrm{G}_J(X_n(\mathbb{K}))$, hence $\mathrm{grk}(\mathrm{G}_J(X_n(\mathbb{K}_0))) \geq \mathrm{grk}(\mathrm{G}_J(X_n(\mathbb{K})))$. If $\mathrm{grk}(\mathrm{G}_J(X(\mathbb{K}_0)))$ is known and not too large (for instance, equal to the dimension of an embedding of $\mathrm{G}_J(X_n(\mathbb{K}))$), then we can infer that $\mathrm{grk}(\mathrm{G}_J(X_n(\mathbb{K})))$ = $\mathrm{grk}(\mathrm{G}_J(X_n(\mathbb{K}_0)))$. It is also clear that if $\mathrm{G}_J(X_n(\mathbb{K}))$ is $\mathbb{K}_0$-generated, then it is also $\mathbb{K}'$-generated for any subfield $\mathbb{K}'$ of $\mathbb{K}$ containing $\mathbb{K}_0$. In particular, suppose that $\mathrm{G}_J(X_n(\mathbb{K}))$ is generated by a frame (see Section 7.6.2.1, second to last paragraph), which we can always assume to be contained in $X_n(\mathbb{K}_0)$. Then $\mathrm{G}_J(X_n(\mathbb{K}))$ is generated over the prime subfield of $\mathbb{K}$, hence it is $\mathbb{K}'$-generated for every subfield $\mathbb{K}'$ of $\mathbb{K}$.

The next theorem is implicit in the proof of [13, lemma 1.4]. We make it explicit here.

Theorem 6.6 *Suppose that $\mathbb{K}$ is not finitely generated and $\mathrm{G}_J(X_n(\mathbb{K}))$ is not $\mathbb{K}_0$-generated for any finitely generated subfield $\mathbb{K}_0$ of $\mathbb{K}$. Then $\mathrm{grk}(\mathrm{G}_J(X_n(\mathbb{K})))$ is infinite.*

Proof It is not difficult to see that, for any choice of X_n among A_n, B_n, C_n or D_n and any non-empty subset J of the type-set of $X_n(\mathbb{K})$, every finite subset G of the point-set of $\mathrm{G}_J(X_n(\mathbb{K}))$ is contained in a subgeometry of $\mathrm{G}_J(X_n(\mathbb{K}))$ isomorphic to $\mathrm{G}_J(X_n(\mathbb{K}_0))$ for a finitely generated subfield $\mathbb{K}_0$ of $\mathbb{K}$. According to the hypotheses of the theorem, $\mathrm{G}_J(X_n(\mathbb{K}_0))$ spans a proper subspace of $\mathrm{G}_J(X_n(\mathbb{K}))$. Hence G cannot generate $\mathrm{G}_J(X_n(\mathbb{K}))$. $\square$

The hypotheses of Theorem 6.6 are satisfied in many cases, to be discussed later in this section. In each of them $\mathrm{grk}(\mathrm{G}_J(X_n(\mathbb{K}))$ is infinite by Theorem 6.6 but, nevertheless, when $\mathrm{G}_J(X_n(\mathbb{K}))$ is known to be embeddable, all of its known embeddings have finite dimension; in certain cases we even know that erk($G_J(X_n(\mathbb{K}))$) is finite. Noticeably, in each of these cases, grids occur among the shadow subspaces of $G_J(X_n(\mathbb{K}))$ of rank 2. Consequently, the main result of Kasikova and Shult [32] cannot be applied in any of them. So, we don't know if $G_J(X_n(\mathbb{K}))$ admits the absolutely universal embedding, even if some of its embeddings might be dominant.

7.6.4.2 Known Results

It is known (Cooperstein and Shult [22], Blok and Brouwer [2]) that the following grassmannians are generated by their frames: $G_k(A_n(\mathbb{K}))$ for $1 \le k \le n$, $G_k(D_n(\mathbb{K}))$ for $k \in \{1, +, -\}$, $G_n(B_n(\mathbb{K}))$ with char($\mathbb{K}$) $\ne 2$ and $G_1(C_n(\mathbb{K}))$. (Similar results are also obtained in [22] and [2] for buildings of type E_6 and E_7 but we are not going to report on them here.) Therefore, all the mentioned grassmannians are generated over the prime subfield of $\mathbb{K}$. It is easily seen that the same holds for $G_1(B_n(\mathbb{K}))$, even if this geometry is not spanned by any frame. Since grk($G_k(C_n(\mathbb{K}))$) does not depend on the particular choice of $\mathbb{K}$ (indeed grk($G_k(C_n(\mathbb{K}))$) = erk($G_k(C_n(\mathbb{K}))$) = $\binom{2n}{k} - \binom{2n}{k-2}$, as we know from [1] and [3]), the symplectic grassmannian $G_k(C_n(\mathbb{K}))$ is generated over the prime subfield of $\mathbb{K}$, for any $k \in \{1, 2, \ldots, n\}$.

We now turn to $G_{1,n}(A_n(\mathbb{K}))$, with $n > 2$. It is proved in [4] that $G_{1,n}(A_n(\mathbb{K}))$ is not $\mathbb{K}_0$-generated, for any proper subfield $\mathbb{K}_0$ of $\mathbb{K}$ (see also [11, theorem 5.10] for an alternative proof in the special case where $n = 3$). However, when $\mathbb{K}$ is generated by $\mathbb{K}_0 \cup \{a_1, \ldots, a_t\}$ for suitable elements $a_1, \ldots, a_t \in \mathbb{K} \setminus \mathbb{K}_0$, then $G_{1,n}(A_n(\mathbb{K}))$ can be generated by adding at most t elements to $G_{1,n}(A_n(\mathbb{K}_0))$ (Blok and Pasini [4]). In particular, when $\mathbb{K}$ is finite then $(n + 1)^2$ points are enough to generate $G_{1,n}(A_n(\mathbb{K}_0))$. Indeed, in this case, $\mathbb{K}$ is a simple extension of its prime subfield $\mathbb{K}_p$ and the generating rank of $G_{1,n}(A_n(\mathbb{K}_p))$ is equal to $(n + 1)^2 - 1$ (Cooperstein [19]).

Not so much is known about generation over subfields for $G_k(B_n(\mathbb{K}))$ for $1 < k < n$ and $G_k(D_n(\mathbb{K}))$ for $1 < k \le n - 2$. We have discussed these two cases in Section 7.3.3 and at the end of Section 7.5.2. Nevertheless, in [11] we have shown that for $\mathbb{K} = \mathbb{F}_4, \mathbb{F}_8$ or $\mathbb{F}_9$ the grassmannians $G_2(B_n(\mathbb{K}))$ ($n \ge 3$) and $G_2(D_n(\mathbb{K}))$ ($n > 3$) are generated over the corresponding prime subfields $\mathbb{F}_2$ or $\mathbb{F}_3$. The generating ranks of $G_2(B_n(\mathbb{K}_0))$ and $G_2(D_n(\mathbb{K}_0))$, for $\mathbb{K}_0$, a finite field of prime order, are known to be equal to $\binom{2n+1}{2}$ and $\binom{2n}{2}$, respectively (Cooperstein [19]). Hence $\binom{2n+1}{2}$ and $\binom{2n}{2}$ are the generating ranks of $G_2(B_n(\mathbb{K}))$ and $G_2(D_n(\mathbb{K}))$, respectively, when $\mathbb{K} = \mathbb{F}_4, \mathbb{F}_8$ or $\mathbb{F}_9$.

For the sake of completeness, we also mention a result on the 2-grassmannian of the quadric $\mathcal{Q}^- := \mathcal{Q}_{2n+1}^-(\mathbb{K})$, even if (the building of) this quadric does not belong to either of the families $B_n(\mathbb{K})$ or $D_n(\mathbb{K})$ considered in this subsection (in fact it belongs to the family of twisted type $^2D_{n+1}(\mathbb{K})$). It is proved in [11] that if $\mathbb{K} \in \{\mathbb{F}_4, \mathbb{F}_8, \mathbb{F}_9\}$ then $\mathrm{grk}(G_2(\mathcal{Q}^-)) = \binom{2n+2}{2}$.

7.6.4.3 Recent Results

We say that a subset J of the type-set of a building $\mathscr{K}$ is *disconnected* if the Coxeter diagram $D(\mathscr{K})$ of $\mathscr{K}$ induces a disconnected graph on J. The following is proved in [13].

Theorem 6.7 *Let X_n be either A_n or D_n and let J be a disconnected subset of the type-set of $X_n(\mathbb{K})$. Then $G_J(X_n(\mathbb{K}))$ is not $\mathbb{K}_0$-generated, for any proper subfield $\mathbb{K}_0$ of $\mathbb{K}$.*

Obviously $n \geq 3$ in Theorem 6.7, since we have defined D_n only for $n \geq 3$ and the Coxeter diagram A_2 admits no disconnected subsets. By combining this theorem with Theorem 6.6 we immediately obtain the following.

Corollary 6.8 *Under the hypotheses of Theorem 6.7, if the field $\mathbb{K}$ is not finitely generated over its prime subfield, then $\mathrm{grk}(G_J(X_n(\mathbb{K})))$ is infinite.*

The following is a special case of Theorem 6.7 and Corollary 6.8.

Corollary 6.9 *Let Γ be either $G_{1,n}(A_n(\mathbb{K}))$ (with $n \geq 3$) or $G_{+,-}(D_n(\mathbb{K}))$. Then Γ is not $\mathbb{K}_0$-generated for any proper subfield $\mathbb{K}_0$ of $\mathbb{K}$. In particular, if the field $\mathbb{K}$ is not finitely generated over its prime subfield, then $\mathrm{grk}(\Gamma)$ is infinite.*

Note that $\mathrm{erk}(G_{1,n}(A_n(\mathbb{K}))) \geq (n+1)^2 - 1$. Indeed $G_{1,n}(A_n(\mathbb{K}))$ admits a projective embedding in the adjoint module of $\mathrm{SL}_{n+1}(\mathbb{K})$, which has dimension equal to $(n+1)^2 - 1$. We call this embedding the *adjoint embedding* of $G_{1,n}(A_n(\mathbb{K}))$. The following is proved in [13].

Theorem 6.10 *Let $\overline{\mathbb{F}}_p$ be the algebraic closure of a finite prime field $\mathbb{F}_p$ and suppose that $n \geq 3$. Then $G_{1,n}(A_n(\overline{\mathbb{F}}_p))$ has infinite generating rank (by Corollary 6.9) but its embedding rank is at most $(n+1)^2$.*

Result 2.5 of Section 7.2.6.2 is a special case of Theorem 6.10 (recall that $G_2(\mathcal{Q}_5^+(\mathbb{K})) \cong G_{+,-}(D_3(\mathbb{K})) = G_{1,3}(A_3(\mathbb{K}))$).

Remark 6.11 In general, if $\mathbb{K}$ admits a non-trivial automorphism, then $\mathrm{Gr}_{1,n}(A_n(\mathbb{K}))$ admits an embedding of dimension one unit larger than the adjoint embedding (Van Maldeghem [47]). So, if the latter is dominant, then $\mathrm{Gr}_{1,n}(A_n(\mathbb{K}))$ admits no universal embedding. For instance, the field $\overline{\mathbb{F}}_p$ is perfect of positive characteristic and admits infinitely many non-trivial automorphisms. According to Völklein's theorem [46], when $\mathbb{K}$ is a perfect field of positive characteristic or a number field, the adjoint embedding of $G_{1,n}(A_n(\overline{F}_p))$ is dominant. Hence $G_{1,n}(A_n(\overline{F}_p))$ admits no universal embedding.

Remark 6.12 In Theorem 6.10, we have assumed that $n \geq 3$ but it is worth spending a few words also on the case $n = 2$. The geometry $G_{1,2}(A_2(\mathbb{K}))$ is a generalized hexagon with two lines on each point and it admits an 8-dimensional embedding, namely the adjoint embedding. If $\mathbb{K}$ satisfies the hypotheses of Theorem 6.10 then the adjoint embedding is dominant (Smith and Völklein [43]). However, the adjoint embedding is not the unique embedding of $G_{1,2}(A_2(\mathbb{K}))$. As proved by Thas and Van Maldeghem [44], when $\mathbb{K}$ is finite, $G_{1,2}(A_2(\mathbb{K}))$ admits dominant embeddings of dimension 8 and, when $\mathbb{K}$ is non-prime, also dominant embeddings of dimension 9. Consequently, when $\mathbb{K}$ is finite and not prime, $G_{1,2}(A_2(\mathbb{K}))$ admits no universal embedding when $\mathbb{K}$ is not prime. Most likely, the same holds true when $\mathbb{K}$ is infinite.

Remark 6.13 In contrast to what happens when $\mathbb{K}$ admits non-trivial automorphisms (Remark 6.11), if $\mathbb{K}$ is prime, $G_{1,n}(A_n(\mathbb{K}))$ admits the universal embedding (Blok and Pasini [5]).

Turning back to hyperbolic quadrics, recall that $\mathcal{Q}^+_{2n-1}(\mathbb{K}) = G_1(D_n(\mathbb{K}))$. The next corollary rephrases the part of Corollary 6.9 which concerns $D_n(\mathbb{K})$.

Corollary 6.14 *Let* $\mathcal{Q} = \mathcal{Q}^+_{2n-1}(\mathbb{K})$, *with* $n \geq 3$. *Then* $G_{n-1}(\mathcal{Q})$ *is never* $\mathbb{K}_0$-*generated, for any proper subfield* $\mathbb{K}_0$ *of* $\mathbb{K}$. *In particular, if* $\mathbb{K}$ *is not finitely generated, then* $\mathrm{grk}(G_{n-1}(\mathcal{Q}))$ *is infinite.*

7.7 Final Comments

As this survey shows, the Standard Argument (7.2.1.3) sports a rather poor performance when applied to grassmannians of quadrics, due to the difficulty of computing their generating ranks and the fact that in certain cases the generating rank turns out to be too large to give any information on the embedding rank of the grassmannian we are considering, as when the generating rank is infinite but we know or we can safely bet that the embedding rank is finite.

It is not clear to us which conclusions to draw from this unpleasant outcome. We could look for another way to organize the computation of generating ranks for most of orthogonal grassmannians, different from the one we have described in Section 7.5.2 of this chapter, so that we avoid bumping against $G_{n-1}(D_n(\mathbb{K}))$, which apparently is the bad guy in this play.

Otherwise, we could look for different ways to check if a given embedding $\varepsilon \colon \Gamma \to PG(V)$ of a given geometry Γ is dominant, where no use of the Standard Argument (7.2.1.3) is needed and, therefore, computations of generating ranks are not required. For instance, suppose that ε is G- *homogeneous* for a subgroup G of Aut(Γ), namely all elements of G are induced by collineations of $PG(V)$ stabilizing $\varepsilon(\Gamma)$. Suppose moreover that G is large enough that we can recover Γ inside G as a system of subgroups and their cosets. Then we could look for relations which present V as a G-module. Proving that ε is dominant is then the same as proving that those relations must be satisfied by any G-module which covers V. This is indeed the method used by Völklein in [16]; in the cases considered in [46], it works well. Maybe it is worth trying to implement this idea for a wider range of cases.

In certain cases, other ways can be tried to prove that a given embedding $\varepsilon \colon \Gamma \to PG(V)$ is universal, different from the one sketched here but still with no use of the Standard Argument (7.2.1.3). For instance, suppose that we can prove that all embeddings of Γ belong to a family $\mathcal{E}$ which contains an element $\tilde{\varepsilon} \colon \Gamma \to PG(\widetilde{V})$ which is initial with respect to $\mathcal{E}$ itself. Then all we have to do is to check if $\varepsilon \cong \tilde{\varepsilon}$. This is indeed the way by which one can prove that the universal embedding of a classical polar space is as described in Section 7.2.4.1 (see Tits [45, 8.2.9 and 8.10–8.19]). In this case, $\mathcal{E}$ consists of the embeddings $\varepsilon \colon \Gamma \to PG(V)$ such that $\varepsilon(\Gamma)$ is a subspace of $\Gamma(f)$ for a suitable reflexive sesquilinear form f of V (see also Buekenhout and Cohen [7, ch. 9]).

Open problems. We conclude our survey with some open problems we already addressed in the previous sections. For the convenience of the reader, we summarize them in the following list.

- Assume char($\mathbb{K}$) $\neq 2$. Prove that the Plücker embedding of $G_k(\Gamma)$ is dominant, for Γ a quadric.
- Prove that, if $\Gamma = \mathcal{Q}_{2n}(\mathbb{K})$ or $\Gamma = \mathcal{Q}_{2n-1}^+(\mathbb{K})$, then the Weyl-like embedding of $G_k(\Gamma)$, $1 < k < n$, is dominant, for any choice of the field $\mathbb{K}$.
- Determine the generating rank of $G_{n-1}(\mathcal{Q}_{2n-1}^+(\mathbb{K}))$.
- Determine the generating rank of the line-grassmannian of $\mathcal{Q}_{2n}(\mathbb{K})$ and $\mathcal{Q}_{2n-1}^+(\mathbb{K})$ for $n > 2$ and $\mathbb{K}$ a non-prime field of order at least 16.
- Determine the generating rank of k-orthogonal grassmannians for $k > 2$
- Determine (or prove it does not exist) the existence of the universal embedding for $G_{n-1}(\mathcal{Q}_{2n-1}(\mathbb{K}))$.

Acknowledgements

Both Ilaria Cardinali and Luca Giuzzi are affiliated with INDAM-GNSAGA whose support they here acknowledge.

References

[1] R. J. Blok. The generating rank of the unitary and symplectic grassmannians: hyperbolic and isotropic geometry. *European J. Combin.*, **28** (2007), 1368–1394.

[2] R. J. Blok and A. E. Brouwer. Spanning point-line geometries in buildings of spherical type, *J. Geom.* **62** (1998), 26–35.

[3] R. J. Blok and B. N. Cooperstein. The generating rank of the unitary and symplectic grassmannians, *J. Combin. Theorey Ser. A* **119** (2012), 1–13.

[4] R. J. Blok and A. Pasini. Point-line geometries with a generating set that depends on the underlying field, in *Finite Geometries* (eds. A. Blokhuis et al.), Kluwer, Dordrecth (2001), 1–25.

[5] R. J. Blok and A. Pasini. On absolutely universal embeddings, *Discr. Math.* **267** (2003), 45–62.

[6] N. Bourbaki. *Algébre, Chap. 9, Formes Sequilinéaires et Formes Quadratiques*, Actu. Sci. Ind. n. 1272, Hermann, Paris, 1959.

[7] F. Buekenhout and A. M. Cohen. *Diagram Geometry*, Springer, Berlin, 2013.

[8] I. Cardinali, L. Giuzzi, K. V. Kaipa and A. Pasini, Line polar grassmann codes of orthogonal type, *J. Pure Appl. Algebra* **220** (5) (2016), 1924–1934.

[9] I. Cardinali and L. Giuzzi, Enumerative coding for line polar grassmannians with applications to codes, *Finite Fields Appl.* **46** (2017) 107–138.

[10] I. Cardinali, L. Giuzzi and A. Pasini. Grassmann embeddings of polar grassmannians, *J. Combin. Th. A* **170** (2020), 105133.

[11] I. Cardinali, L. Giuzzi and A. Pasini. The generating rank of a polar grassmannian, *Adv. Geom.* **21**(4) (2021), 515–539.

[12] I. Cardinali, L. Giuzzi and A. Pasini. Nearly all subspaces of a classical polar space arise from its universal embedding, *Linear Algebra Appl.* **627** (2021), 287–307.

[13] I. Cardinali, L. Giuzzi and A. Pasini. On the generation of some Lie-type geometries, *J. Combin. Th. A.* **193** (2023) 105673. https://doi.org/10.1016/j.jcta .2022.105673.

[14] I. Cardinali and A. Pasini. Grassmann and Weyl embeddings of orthogonal grassmannians. *J. Alg. Combin.* **38** (2013), 863–888.

[15] I. Cardinali, H. Cuypers, L. Giuzzi and A. Pasini. Characterization of symplectic polar spaces, To appear in *Adv. Geom.* https://doi.org/10.1515/advgeom-2023-0006.

[16] I. Cardinali and L.Giuzzi, On orthogonal polar spaces, preprint (2022), arxiv:2301.05876.

[17] B. N. Cooperstein. On the generation of some dual polar spaces of symplectic type over *GF*(2), *European J. Combin.* **18** (1997), 741–749.

[18] B. N. Cooperstein. On the generation of dual polar spaces of unitary type over finite fields, *European J. Combin.* **18** (1997), 849–856.

[19] B. N. Cooperstein. Generating long root subgroup geometries of classical groups over finite prime fields, *Bull. Belg. Math. Soc.* **5** (1998), 531–548.

[20] B. N. Cooperstein. On the generation of some embeddable $GF(2)$ geometries, *J. Alg. Combin.* **13** (2001), 15–28.

[21] B. N. Cooperstein. Generation of embeddable incidence geometries: a survey, in *Topics in Diagram Geometry* (ed. A. Pasini), Vol. 12 of the series *Quaderni di Matematica*, II Univ. Naples (2003), 29–57.

[22] B. N. Cooperstein and E. E. Shult. Frames and bases of lie incidence geometries, *J. Geom.* **60** (1997), 17–46.

[23] B. N. Cooperstein and E. E. Shult. A note on embedding and generating dual polar spaces, *Adv. Geom.* **1** (2001), 37–48.

[24] B. De Bruyn. On the Grassmann modules for the unitary groups, *Linear Multilinear Algebra* **58** (7) (2010), 887–902.

[25] B. De Bruyn. A note on the spin-embedding of the dual polar space $DQ^-(2n+1,K)$, *Ars Combinatoria* **99** (2011), 365–375.

[26] B. De Bruyn and A. Pasini. Minimal scatterd sets and polarized embeddings of dual polar spaces, *European J. Combin.* **28** (2007), 1890–1909.

[27] B. De Bruyn and A. Pasini. Generating symplectic and hermitian dual polar spaces over arbitrary fields non-isomorphic to $\mathbb{F}_2$, *Elect. J. Combin.* **14** (1) (2007), Research Paper 54.

[28] B. De Bruyn and A. Pasini. On Symplectic Polar Spaces over non-Perfect Fields of Characteristic 2, *Lin. Multilinear Alg.* **57** (2009), 567–575.

[29] B. De Bruyn and H. Van Maldeghem. Universal and homogeneous embeddings of dual polar spaces of rank 3 defined over a quadratic alternative division ring, *J. Reine Agew. Math.* **715** (2016), 39–74.

[30] H. Freudenthal. Beziehung der E_7 un E_8 zur Oktavenebene, I-XI, *Proc. Kon. Ned. Akad. Wet. Ser. A*, **57** (1954), 218–230, 363–368; **57** (1955), 151–157, 277–285; **62** (1959), 165–201, 447–474; **66** (1963), 457–487.

[31] J. E. Humphreys. *Ordinary and modular representations of Chevalley groups*, Springer Lecture Notes in Math. Vol. 528 (1976).

[32] A. Kasikova and E. E. Shult. Absolute embeddings of point-line geometries, *J. Algebra* **238** (2001), 265–291.

[33] P. Li. On the universal embedding of the $Sp_{2n}(2)$ dual polar space, *J. Combin. Th. A* **94** (2001), 100–117.

[34] P. Li. On the universal embedding of the $U_{2n}(2)$ dual polar space, *J. Combin. Th. A* **98** (2002), 235–252.

[35] J. Meulewaeter and H. Van Maldegherm. Convex subspaces of Lie incidence geometries, Combinatorial Theory **2** (3) (2022), #13.

[36] M. Pankov. *Grassmannians of Classical Buildings*, Algebra and Discrete Mathematical Series 2, World Scientific, Singapore 2010.

[37] M. Pankov. Characterization of apartments in polar grassmannians, *Bull. Belg. Math. Soc. Simon Stevin* **19** (2012), 345–366.

[38] A. Pasini. *Diagram Geometries*, Oxford University Press (1994)

[39] A. Pasini. Embedded polar spaces revisited, *Inn. Inc. Geom.* **15** (2017), 31–72.

[40] A. Pasini. Sets of generators and chains of subspaces, *Contributions to Alg. and Geom.* **62** (2021), 457–474.

[41] M. A. Ronan. Embedding and hyperplanes of discrete geometries, *European J. Combin.* **8** (1987), 179–185.

[42] E. E. Shult. *Points and Lines,* Springer, Berlin, 2010.

[43] S. D. Smith and H. Völklein. A geometric presentation for the adjoint module of of $SL_3(k)$, *J. Algebra* **127** (1989), 127–138.

[44] J. A. Thas and H. Van Maldeghem. Classification of embeddings of the flag geometries of projective planes in finite projective planes, *J. Combin. Th. A* **90** (2000), 159–172 (Part 1); 241–256 (Part 2); 173–196 (Part 3).

[45] J. Tits. *Buildings of Spherical Type and Finite BN-Pairs*, Springer Lecture Notes in Math. Vol. 386 (1974).

[46] H. Völklein. On the geometry of the adjoint representation of a Chevalley group, *J. Algebra* **127** (1989), 139–154.

[47] H. Van Maldeghem, Private communication to A. Pasini.

[48] A. L. Wells. Universal projective embeddings of the grassmannian, half spinor and orthogonal geometries, *Quart. J. Math Oxford* (2), **34** (1983), 375–386.

8

Ovoidal Maximal Subspaces of Polar Spaces

Antonio Pasini and Hendrik Van Maldeghem

Abstract

Let $\mathscr{S}$ be a polar space of rank $n \geq 2$. A set of mutually non-collinear points of $\mathscr{S}$ is trivially a subspace of $\mathscr{S}$. We call it an ovoidal subspace. It is well known that when $n = 2$, all ovoids are maximal subspaces. However, as we shall see in this chapter, when $n > 2$, ovoids exist which are not maximal subspaces. Moreover, in the finite case, not all polar spaces admit ovoids. So, it is natural to ask whether ovoidal maximal subspaces exist in any polar space. In this chapter we provide a basically affirmative answer to this question, proving that ovoidal maximal subspaces exist in all polar spaces but the following ones: $Q_{2n}(2)$ with n even and greater than 2, $Q_{2n-1}^{+}(2)$ with $n \equiv 2, 3 \pmod 4$ and greater than 2 and $Q_{2n+1}^{-}(2)$ with $n \equiv 0, 3 \pmod 4$.

CONTENTS

8.1 Introduction

Let $\mathscr{S}$ be a non-degenerate thick-lined polar space of finite rank $n \geq 2$. According to Shult [17], a *subspace* of $\mathscr{S}$ is a set X of points of $\mathscr{S}$ such that if a line ℓ of $\mathscr{S}$ meets X in at least two points, then $X \supseteq \ell$. If furthermore X is a *proper* subspace (that is, not the full point set of $\mathscr{S}$) and every line of $\mathscr{S}$ meets X non-trivially, then X is said to be a *hyperplane*. All hyperplanes are *maximal subspaces* (Shult [17, 7.5.1]), namely they are maximal in the family of proper subspaces of $\mathscr{S}$.

Trivially, a non-empty set of mutually non-collinear points of $\mathscr{S}$ is a subspace of $\mathscr{S}$. We call it an *ovoidal subspace*, also a *partial ovoid* of $\mathscr{S}$. In other words, an ovoidal subspace of $\mathscr{S}$ is a set O of points such that every generator (i.e. maximal singular subspace) of $\mathscr{S}$ meets O in at most one point. If every generator of $\mathscr{S}$ meets O in exactly one point, then X is called an *ovoid*.

If O is an ovoid and $n = 2$, then O is a hyperplane, hence a maximal subspace of $\mathscr{S}$. On the other hand, when $n > 2$, all hyperplanes of $\mathscr{S}$ have rank at least $n - 1$, hence they cannot be ovoidal subspaces; moreover, if $\mathscr{S}$ is embeddable, then the hyperplanes of $\mathscr{S}$ are precisely the maximal subspaces of $\mathscr{S}$ of rank at least 2, namely the non-ovoidal ones (see [4, corollary 3]). Two questions arise quite naturally:

(A) Is it true that when $n = 2$ every ovoidal maximal subspace is an ovoid?
(B) Is it true that when $n > 2$ no ovoidal subspace is a maximal subspace?

The answer is NO for both questions. Question (A) is answered in the negative in [4, note 1]. Further counterexamples will be offered in the present chapter.

Question (B) is problem 4 of [4]. In this chapter, we shall prove that all polar spaces admit ovoidal maximal subspaces, namely maximal subspaces which are ovoidal (not to be confused with maximal ovoidal subspaces, which trivially exist in any case). Explicitly, we shall prove the following.

Theorem 1 *Let $\mathscr{S}$ be a (non-degenerate thick-lined) polar space of finite rank $n \geq 2$. Then $\mathscr{S}$ admits ovoidal maximal subspaces except precisely when $\mathscr{S}$ is one of the following:*

(1) $Q_{2n}(2) \cong W_{2n-1}(2)$ *with* $n \equiv 0 \pmod 2$, $n > 2$;
(2) $Q_{2n-1}^+(2)$ *with* $n \equiv 2, 3 \pmod 4$, $n > 2$;
(3) $Q_{2n+1}^-(2)$ *with* $n \equiv 0, 3 \pmod 4$.

Section 8.2 of this paper is entirely devoted to the proof of this theorem. An interesting property of matrices is implicit in Theorem 1. In order to state it, we need the following definition: we say that a square matrix $A = (a_{i,j})_{i,j=1}^N$

is *anti-diagonal* when $a_{i,j} = 0$ if and only if $i = j$. We also recall some terminology. Let $M_N(\mathbb{F})$ be the ring of square matrices of order N with entries in a given field $\mathbb{F}$. We say that two matrices $A, B \in M_N(\mathbb{F})$ are *T-equivalent* if $B = C^T A C$ for a non-singular matrix $C \in M_N(\mathbb{F})$, where C^T stands for the transpose of C. Suppose that $\mathbb{F}$ is a separable quadratic extension of a field $\mathbb{F}_0$ and let σ be the unique non-trivial element of the Galois group of $\mathbb{F}$ over $\mathbb{F}_0$. Recall that the adjoint of a matrix $A = (a_{i,j})_{i,j=1}^N$ (with respect to σ) is the matrix $A^* := (a'_{i,j})_{i,j=1}^N$ where $a'_{i,j} = a_{j,i}^\sigma$ for every choice of $i, j = 1, 2, \ldots, N$. We say that two matrices $A, B \in M_N(\mathbb{F})$ are $*$- *equivalent* if $B = C^* A C$ for a non-singular matrix $C \in M_N(\mathbb{F})$. We are not going to recall the definitions of symmetric, anti-symmetric, hermitian or anti-hermitian matrices. We only remind the reader of the fact that when char($\mathbb{F}$) $= 2$ all entries on the main diagonal of an anti-symmetric matrix are null, by definition of anti-symmetry. The following will be proved at the end of Section 8.2.

Corollary 2 *In the following claims, $\mathbb{F}$ is a field and we assume that $N > 1$.*

(1) *Let N be even. Then all non-singular anti-symmetric matrices of $M_N(\mathbb{F})$ are T-equivalent to anti-diagonal matrices.*

(2) *Let $N > 2$ and suppose that $\mathbb{F}$ is finite of odd order. When $N = 4$, suppose moreover that $\mathbb{F} \neq \mathbb{F}_3$. Then every non-singular symmetric matrix of $M_N(\mathbb{F})$ is T-equivalent to an anti-diagonal matrix.*

(3) *Let $\mathbb{F}$ be finite of square order. Then every non-singular hermitian or anti-hermitian matrix of $M_N(\mathbb{F})$ is $*$-equivalent to an anti-diagonal matrix.*

Turning back to ovoids, we know that in the rank 2 case, ovoids are maximal subspaces. Is the same true for polar spaces of arbitrary rank, possibly modulo a few exceptions? The next theorem, to be proved in Section 8.4, provides an answer to this question in the finite case.

Theorem 3 *Let $\mathscr{S}$ be a (non-degenerate thick-lined) finite polar space, different from both $Q_5^+(q)$ and $Q_7^+(q)$, with q odd in the latter case. Suppose that $\mathscr{S}$ admits ovoids. Then all ovoids of $\mathscr{S}$ are maximal subspaces.*

All quadrics $Q_5^+(q)$ are actually counterexamples to the conclusion of Theorem 3 (see Section 8.4, Remark 14) while only a few of the quadrics $Q_7^+(q)$ are known to admit ovoids. We might regard Theorem 3 as an affirmative answer to the above question, albeit limited to the finite case. However ovoids seem to be rare in finite polar spaces of arbitrary rank; so, we are not sure if the set of exceptions considered in Theorem 3 can rightly be regarded as a small one.

We end the chapter with a kind of counterpart to Theorem 3 for infinite polar spaces. It will imply examples of polar spaces containing ovoids that are not maximal subspaces, and examples of polar spaces all ovoids of which are automatically maximal subspaces.

Structure of the Chapter. Section 8.2 contains the proof of Theorem 1. The proof is divided into four parts. We prove first that every thick-lined generalized quadrangle admits ovoidal maximal subspaces. Next we consider embeddable polar spaces defined over division rings of order at least 3, proving that they also admit ovoidal maximal subspaces. After that, we turn to thick-lined non-embeddable polar spaces of rank 3, obtaining the conclusion with the help of a classification of their subspaces. Finally, we examine polar spaces defined over $\mathbb{F}_2$, thus completing the proof of Theorem 1.

The arguments exploited in Section 8.2 do not provide explicit descriptions of ovoidal maximal subspaces. However they show how to construct certain partial ovoids, called 'totally scattered' in Section 8.2, such that every maximal partial ovoid containing one of them is a maximal subspace.

In Section 8.3, we choose a more concrete approach. In the first part of Section 8.3, we offer an explicit construction of a family of ovoidal maximal subspaces in symplectic spaces. In the second part we consider embeddable polar spaces not of symplectic type, showing how totally scattered partial ovoids can be constructed for them. Section 8.4 is devoted to ovoids. It contains the proof of Theorem 3.

Notation. If X is a set of points of a polar space $\mathscr{S}$ we denote by $\langle X \rangle_{\mathscr{S}}$ the subspace of $\mathscr{S}$ spanned by X. Similarly, if X is a set of points of a projective space Σ then $\langle X \rangle_{\Sigma}$ is the subspace of Σ spanned by X. When no ambiguity will arise, we will freely omit the subscripts $\mathscr{S}$ or Σ from the symbols $\langle . \rangle_{\mathscr{S}}$ and $\langle . \rangle_{\Sigma}$, thus writing $\langle X \rangle$ instead of $\langle X \rangle_{\mathscr{S}}$, for instance, when $X \subseteq \mathscr{S}$.

For two points x and y of a polar space $\mathscr{S}$, if x and y are collinear we write $x \perp y$. Also, $x^{\perp}$ is the set of points of $\mathscr{S}$ collinear with x, with $x \in x^{\perp}$ by convention, and we put $X^{\perp} := \cap_{x \in X} x^{\perp}$, for X a set of points of $\mathscr{S}$.

Given a non-zero vector $\mathbf{v}$ of a vector space V, we denote by $[\mathbf{v}]$ the corresponding point of $PG(V)$. We also use the symbol $\perp$ to denote orthogonality between vectors or projective points. Thus, given a $\mathbb{K}$-vector space V and a reflexive sesquilinear form $f : V \times V \to \mathbb{K}$, when writing $\mathbf{v} \perp \mathbf{w}$ for two vectors $\mathbf{v}, \mathbf{w} \in V$ (or $[\mathbf{v}] \perp [\mathbf{w}]$ for two points $[\mathbf{v}], [\mathbf{w}] \in PG(V)$) we mean that $f(\mathbf{v}, \mathbf{w}) = 0$. Also, if x is a point of $PG(V)$, we denote by $x^{\perp}$ the subspace of $PG(V)$ formed by the points orthogonal to x and, for $X \subseteq PG(V)$, we put $X^{\perp} := \cap_{x \in X} x^{\perp}$.

Thus, we use the same notation for collinearity in polar spaces and orthogonality in vector or projective spaces, including symbols as $x^{\perp}$ and $X^{\perp}$. However the context will always avoid any ambiguity.

The symbols $W_{2n-1}(q)$, $Q_{2n-1}^+(q)$, $Q_{2n}(q)$, $Q_{2n+1}^-(q)$, $H_{2n-1}(q)$ and H_{2n} (q) have the usual meaning. Explicitly, $W_{2n-1}(q)$, $Q_{2n-1}^+(q)$ and $H_{2n-1}(q)$ are respectively the symplectic variety, the hyperbolic quadric and the hermitian variety of $PG(2n-1,q)$, $Q_{2n}(q)$ and $H_{2n}(q)$ are the quadric and the hermitian variety of $PG(2n,q)$ and $Q_{2n+1}^-(q)$ is the elliptic quadric of $PG(2n+1,q)$.

8.2 Proof of Theorem 1

8.2.1 The Rank 2 Case

Theorem 2.1 *All thick-lined generalized quadrangles admit ovoidal maximal subspaces.*

Proof Let $\mathscr{S}$ be a thick-lined generalized quadrangle, let p, q be two non-collinear points of $\mathscr{S}$ and let $r \in \{p,q\}^\perp$ be arbitrary. Choose $r_p \in \langle p, r \rangle \setminus \{p, r\}$ and $r_q \in \langle q, r \rangle \setminus \{q, r\}$. Put

$$\{p,q\}_{r|r_p,r_q}^\perp := (\{p,q\}^\perp \setminus \{r\}) \cup \{r_p, r_q\}. \tag{1}$$

Clearly, $\{p,q\}_{r|r_p,r_q}^\perp$ is a partial ovoid. Let O be a maximal partial ovoid containing $\{p,q\}_{r|r_p,r_q}^\perp$ and, for a point $a \notin O$, let $\mathscr{S}_a := \langle O \cup \{a\} \rangle$. By the maximality of O, the subspace $\mathscr{S}_a$ is a (possibly degenerate) full subquadrangle. We shall prove that $\mathscr{S}_a = \mathscr{S}$.

We firstly prove that $p, q \in \mathscr{S}_a$. As $\mathscr{S}_a$ is a subquadrangle, it contains at least one line ℓ through r_q and, since $\mathscr{S}_a$ is full, all points of ℓ belong to $\mathscr{S}_a$. If $\ell = \langle r,q \rangle$ then $q, r \in \mathscr{S}_a$, hence $\mathscr{S}_a$ also contains the line $\langle r, r_p \rangle = \langle r, p \rangle$. In this case we are done: $p, q \in \mathscr{S}_a$. So, suppose that $\ell \neq \langle r, q \rangle$. Hence $p \notin \ell$. The unique point x on ℓ collinear to p belongs to $\mathscr{S}_a$, and the unique point y on $\langle p, x \rangle$ collinear to q is clearly distinct from r and hence belongs to $\{p,q\}_{r|r_p,r_q}^\perp \subseteq \mathscr{S}_a$. Note that $x \neq y$. Hence $p \in \langle x, y \rangle \subseteq \mathscr{S}_a$. Similarly, $q \in \mathscr{S}_a$.

As $p \in \mathscr{S}_a$ and every line of $\mathscr{S}_a$ through p meets $\{p,q\}_{r|r_p,r_q}^\perp$ non-trivially, all lines of $\mathscr{S}$ through p belong to $\mathscr{S}_a$. This implies that $\mathscr{S}_a$ is an ideal subquadrangle of $\mathscr{S}$ as defined in section 1.8 of [20] (for every point $x \in \mathscr{S}_a$ all lines of $\mathscr{S}$ through x belong to $\mathscr{S}_a$). However $\mathscr{S}_a$ is a also a full subquadrangle of $\mathscr{S}$. Hence $\mathscr{S}_a = \mathscr{S}$ by 1.8.2 of [20]. $\square$

Remark 1 When the generalized quadrangle $\mathscr{S}$ has order (s,t) with s infinite and $s \geq t$ the conclusion of Theorem 2.1 also follows from a result of Cameron [3], according to which every generalized quadrangle of order (s,t) as here admits ovoids (in fact, a partition in ovoids).

8.2.2 The Embeddable Case

8.2.2.1 Preliminaries on Embeddings and Subspaces

We recall that a *projective embedding* of a polar space $\mathscr{S}$ (an *embedding* of $\mathscr{S}$ for short) is an injective mapping $e: \mathscr{S} \to \mathrm{PG}(V)$ from (the point set of) $\mathscr{S}$ to (the point set of) the projective geometry $\mathrm{PG}(V)$ of a vector space V, such that the set $e(\mathscr{S})$ spans $\mathrm{PG}(V)$ and e maps every line of $\mathscr{S}$ surjectively onto a projective line of $\mathrm{PG}(V)$. If $\mathbb{K}$ is the underlying division ring of V, then e is said to be *defined over* $\mathbb{K}$.

A polar space is *embeddable* if it admits a projective embedding. If all embeddings of $\mathscr{S}$ are defined over the same division ring, say $\mathbb{K}$, then $\mathscr{S}$ is said to be *defined over* $\mathbb{K}$.

Two embeddings $e: \mathscr{S} \to \mathrm{PG}(V)$ and $e': \mathscr{S} \to \mathrm{PG}(V')$ are *isomorphic* if $e' = \gamma \cdot e$ for an isomorphism γ from $\mathrm{PG}(V)$ to $\mathrm{PG}(V')$. An embedding $\tilde{e}: \mathscr{S} \to \mathrm{PG}(\widetilde{V})$ is *universal* if every embedding of $\mathscr{S}$ is isomorphic to $e_X := \pi_X \cdot \tilde{e}$ for a suitable (possibly trivial) subspace X of $\widetilde{V}$, where π_X stands for the projection of $\mathrm{PG}(\widetilde{V})$ onto $\mathrm{PG}(\widetilde{V}/X)$. The universal embedding, if it exists, is unique (modulo isomorphisms). Clearly, if $\mathscr{S}$ admits the universal embedding, then all of its embeddings are defined over the same division ring, namely the underlying division ring of its universal embedding.

We recall that all (non-degenerate thick-lined) polar spaces of rank $n > 3$ are embeddable and all those of rank 3 are embeddable but for the following two families: line-grassmannians of 3-dimensional projective spaces defined over non-commutative division rings and a family of thick polar spaces of rank 3 with non-desarguesian Moufang planes, which live inside buildings of type E_7 (see Tits [19, chapters 7–9]; also Buekenhout and Cohen [2, chapters 7–11]). We call the polar spaces of the latter family *Freudenthal-Tits polar spaces*; they are implicit in Freudenthal [9] and explicitly defined in Tits [19, chapter 9].

Remark 2   The approach chosen by Freudenthal [9] and Tits's definition in [19, chapter 9] are rather algebraic. The reader is referred to Mühlherr [13] for a more geometric approach and De Bruyn and Van Maldeghem [8] for an explicit concrete description of Freudenthal–Tits polar spaces.

As proved by Tits [19, 8.6], all embeddable polar spaces admit the universal embedding but for the following two families of rank 2: grids of order at least 4 (at least five points on each line) and certain generalized quadrangles defined over quaternion division rings (Tits [19, 8.6(II)(a)]). We also recall that, if $\mathscr{S}$ admits the universal embedding, say $\tilde{e}$, and is defined over a division ring of characteristic different from 2, then $\tilde{e}$ is the unique embedding of $\mathscr{S}$ (Tits [19, chapter 8]). In this case, the universal property of $\tilde{e}$ is ultimately vacuous.

Suppose that $\mathscr{S}$ is embeddable and let $e\colon \mathscr{S} \to \Sigma = \mathrm{PG}(V)$ be an embedding of $\mathscr{S}$. A subspace X of $\mathscr{S}$ *arises from* e if $e^{-1}(\langle e(X)\rangle_\Sigma) = X$. Clearly, if $\mathscr{S}$ admits the universal embedding and a subspace X of $\mathscr{S}$ arises from an embedding of $\mathscr{S}$, then X also arises from the universal embedding.

Every subspace of $\mathscr{S}$ is a possibly degenerate polar space. The *rank* of a subspace X of $\mathscr{S}$ is its rank as a polar space. If m is the rank of X and r the rank of the radical $X \cap X^\perp$ of X, then $m - r$ is the *reduced rank* of X. Clearly, the rank and the reduced rank of X are equal if and only if X is non-degenerate. The subspaces of rank 1 are precisely the ovoidal subspaces.

Let $n \geq 2$ be the rank of $\mathscr{S}$. Then all hyperplanes of $\mathscr{S}$ have reduced rank at least $n - 1$. For instance, let $H = x^\perp$ for a point x of $\mathscr{S}$. Then H is a hyperplane of $\mathscr{S}$, called a *singular* hyperplane; its rank is n and its reduced rank is $n - 1$.

Suppose that $\mathscr{S}$ is embeddable and let e be an embedding of $\mathscr{S}$. Then all singular hyperplanes of $\mathscr{S}$ arise from e but, if e is not universal, then hyperplanes of $\mathscr{S}$ exist which do not arise from e. However, let $n > 2$. Then $\mathscr{S}$ admits the universal embedding and all hyperplanes of $\mathscr{S}$ arise from it (as follows from a theorem of Ronan [15]). In fact this is a special case of a more general result, proved in [4]:

Proposition 2.2 *Let $n \geq 2$ and suppose that $\mathscr{S}$ admits the universal embedding, say $\tilde{e}$. Then all subspaces of $\mathscr{S}$ of reduced rank at least 2 arise from $\tilde{e}$.*

8.2.2.2 Preliminaries on Ovoidal Maximal Subspaces
As before, let $\mathscr{S}$ be an embeddable non-degenerate polar space of rank $n \geq 2$. As noticed in Section 8.2.2.1, if $n > 2$, then $\mathscr{S}$ admits the universal embedding. When $n = 2$, assume that the universal embedding of $\mathscr{S}$ exists. Let $\tilde{e}\colon \mathscr{S} \to \mathrm{PG}(\widetilde{V})$ be the universal embedding of $\mathscr{S}$.

Proposition 2.3 *Let $n > 2$. Then a partial ovoid O of $\mathscr{S}$ is a maximal subspace of $\mathscr{S}$ if and only if it is maximal as a partial ovoid and $\tilde{e}(O)$ spans $\mathrm{PG}(\widetilde{V})$.*

Proof Let O be an ovoidal maximal subspace of $\mathscr{S}$. Then O is a maximal ovoidal subspace. By way of contradiction, suppose that $\langle \tilde{e}(O)\rangle \subset \mathrm{PG}(\widetilde{V})$ and let X be a hyperplane of $\mathrm{PG}(\widetilde{V})$ containing $\tilde{e}(O)$. Then $H := \tilde{e}^{-1}(X)$ is a hyperplane of $\mathscr{S}$. However H contains at least a line of $\mathscr{S}$, since all hyperplanes of $\mathscr{S}$ have rank at least $n - 1$ and $n \geq 3$ by assumption. Hence $H \supset O$. This contradicts the hypothesis that O is a maximal subspace. Therefore $\tilde{e}(O)$ spans $\mathrm{PG}(\widetilde{V})$.

Conversely, suppose that $\tilde{e}(O)$ spans $\mathrm{PG}(V)$ and O is a maximal partial ovoid. Let X be a (possibly improper) subspace of $\mathscr{S}$ properly containing O and choose a point $x \in X \setminus O$. As O is maximal as a partial ovoid, if x is a point of $\mathscr{S}$ exterior to O, then $x \perp y$ for some point $y \in O$. Accordingly, X contains a line of $\mathscr{S}$, namely the line $\langle x, y \rangle$. Hence X has rank at least 2. If X is degenerate, then X is contained in a singular hyperplane $x^\perp$ of $\mathscr{S}$. Hence $\tilde{e}(O) \subseteq \tilde{e}(x)^\perp \subset \mathrm{PG}(\widetilde{V})$. However $\langle \tilde{e}(O) \rangle = \mathrm{PG}(\widetilde{V})$ by assumption. We get a contradiction, which forces us to conclude that X is non-degenerate. Accordingly, X has reduced rank at least 2. Therefore $X = \tilde{e}^{-1}(\langle \tilde{e}(X) \rangle)$, by Proposition 2.2. However $\langle \tilde{e}(X) \rangle \supseteq \langle \tilde{e}(O) \rangle$ and $\langle \tilde{e}(O) \rangle = \mathrm{PG}(\widetilde{V})$ by assumption. Hence $\langle \tilde{e}(X) \rangle = \mathrm{PG}(\widetilde{V})$. It follows that $X = \mathscr{S}$. We have proved that O is a maximal subspace of $\mathscr{S}$.   □

The next corollary trivially follows from Proposition 2.3.

Corollary 2.4   *If $n > 2$, then no ovoidal maximal subspace of $\mathscr{S}$ arises from $\tilde{e}$.*

Proposition 2.5   *Let $n = 2$. Then a partial ovoid O of $\mathscr{S}$ is a maximal subspace of $\mathscr{S}$ if and only if one of the following holds:*

(1) *O is an ovoid;*

(2) *O is a maximal partial ovoid and $\tilde{e}(O)$ spans $\mathrm{PG}(\widetilde{V})$.*

Proof   This statement can be proved by essentially the same arguments used to prove Proposition 2.3. We leave the details for the reader.   □

Remark 3   Conditions (1) and (2) of Proposition 2.5 are not mutually exclusive. Indeed ovoids are maximal partial ovoids and if an ovoid O does not arise from $\tilde{e}$ (in short, it is *non-classical*) then $\tilde{e}(O)$ spans $\mathrm{PG}(\widetilde{V})$.

The following is an obvious consequence of Propositions 2.3 and 2.5.

Corollary 2.6   *Suppose that $\mathscr{S}$ admits a partial ovoid O such that $\tilde{e}(O)$ spans $\mathrm{PG}(\widetilde{V})$. Then every maximal partial ovoid of $\mathscr{S}$ containing O is a maximal subspace of $\mathscr{S}$.*

8.2.2.3 Setting and More Notation and Terminology

Henceforth $\mathscr{S}$ is a (non-degenerate) embeddable polar space of rank $n \geq 2$ and, if $\mathscr{S}$ admits the universal embedding, then $e \colon \mathscr{S} \to \Sigma = \mathrm{PG}(V)$ is its universal embedding; otherwise, if $\mathscr{S}$ is a grid or a generalized quadrangle as in [19, 8.6(II)(a)], then e is any of the embeddings of $\mathscr{S}$. We denote by $\mathbb{K}$ the underlying division ring of V (which is also the underlying division ring of $\mathscr{S}$, except when $\mathscr{S}$ is a grid and $\mathbb{K}$ is infinite).

We shall keep the distinction between $\mathscr{S}$ and its e-image $e(\mathscr{S}) \subseteq \Sigma$. This distinction might look futile (especially when e is the unique embedding of $\mathscr{S}$), but it helps to avoid misunderstandings.

In the sequel we shall often deal with the subspace $\langle e(\{p,q\}^\perp)\rangle_\Sigma$ of Σ, for two non-collinear points p, q if $\mathscr{S}$. This subspace is the same as $\{e(p), e(q)\}^\perp$, where $\perp$ stands for the orthogonality relation of Σ associated to $e(\mathscr{S})$ rather than the collinearity relation in $\mathscr{S}$. Clearly, the subspace $\langle e(\{p,q\}^\perp)\rangle_\Sigma = \{e(p), e(q)\}^\perp$ has codimension 2 in Σ. In order to have a symbol not so clumsy as $\langle e(\{p,q\}^\perp)\rangle_\Sigma$, we put

$$\Sigma_{p,q} := \langle e(\{p,q\}^\perp)\rangle_\Sigma \; (= \{e(p), e(q)\}^\perp). \tag{2}$$

Finally, we say that a partial ovoid O of $\mathscr{S}$ is *totally scattered* (in Σ) if $e(O)$ spans Σ. In view of Proposition 2.3, when $n > 2$, proving that $\mathscr{S}$ admits an ovoidal maximal subspace is the same as proving that $\mathscr{S}$ admits a totally scattered partial ovoid.

8.2.2.4 Back to the Rank 2 Case

With $\mathscr{S}$ and Σ as before, suppose that $n = 2$. Let p, q be two non-collinear points of $\mathscr{S}$.

Lemma 2.7 *Suppose that $\mathscr{S}$ is not a grid and $\mathbb{K} \neq \mathbb{F}_2$. Then the set $e(\{p,q\}^\perp)$ properly contains a basis of $\Sigma_{p,q}$.*

Proof By way of contradiction, suppose that $e(\{p,q\}^\perp)$ is a basis of $\Sigma_{p,q}$. As this basis contains all singular points of $\Sigma_{p,q}$, the quadrangle $e(\mathscr{S})$ cannot be symplectic. Accordingly, $e(\mathscr{S})$ is defined by a $(\sigma, 1)$-quadratic form, say ϕ. Choose any three points p_1, p_2, p_3 of $\{p,q\}^\perp$ and let $\pi = \langle e(p_1), e(p_2), e(p_3)\rangle_\Sigma$ (recall that $\mathscr{S}$ is not a grid, by assumption). So, $\pi \cap e(\mathscr{S}) = \{e(p_1), e(p_2), e(p_3)\}$, since $e(\{p,q\}^\perp)$ is independent by assumption and contains all singular points of $\Sigma_{p,q}$. Accordingly, $e(p_1), e(p_2)$ and $e(p_3)$ are the unique points of π which are singular for ϕ and they span π. It follows that $\pi \cap e(\mathscr{S})$ is a conic defined over $\mathbb{F}_2$, namely σ is the identity and $\mathbb{K} = \mathbb{F}_2$. However $\mathbb{K} \neq \mathbb{F}_2$ by assumption. A contradiction has been reached. $\quad\square$

The set $\{p,q\}^\perp_{r|r_p, r_q}$ defined in (1) (see the proof of Theorem 2.1) is a partial ovoid. Moreover, by Lemma 2.7, when neither $\mathscr{S}$ is a grid nor $\mathbb{K} = \mathbb{F}_2$, then $e(\{p,q\}^\perp \setminus \{r\})$ spans $\Sigma_{p,q}$ for some $r \in \{p,q\}^\perp$, whence for any $r \in \{p,q\}^\perp$, since the stabilizer of p and q in $\mathrm{Aut}(\mathscr{S})$ acts transitively on $\{p,q\}^\perp$. However the span of $\Sigma_{p,q} \cup \{e(r_p), e(r_q)\}$ contains the line $\langle e(p), e(q)\rangle_\Sigma$ and $\langle e(p), e(q)\rangle_\Sigma \cup \Sigma_{p,q}$ spans Σ. Therefore $e(\{p,q\}^\perp_{r|r_p, r_q})$ spans Σ, consequently $\{p,q\}^\perp_{r|r_p, r_q}$ is totally scattered.

Remark 4   Actually, when $\mathscr{S}$ is as in [19, 8.6(II)(a)] the previous argument is incorrect, since the full automorphism group of $\mathscr{S}$ does not stabilize the chosen embedding e. However, a suitable subgroup of Aut($\mathscr{S}$) of index 2 does the job.

Lemma 2.8   *Suppose that* $\mathbb{K}$ *is infinite but* $\mathscr{S}$ *is not a grid. Then* $\{p,q\}^{\perp}_{r\,|r_p,r_q}$ *contains an infinite subset* C *such that the partial ovoid* $\{p,q\}^{\perp}_{r\,|r_p,r_q} \setminus C$ *is still totally scattered.*

Proof   As $\mathscr{S}$ is embeddable but not a grid, $\mathscr{S}$ cannot be semi-finite. Hence $\{p,q\}^{\perp}$ is infinite, since $\mathbb{K}$ is infinite by assumption. If $\dim(\Sigma_{p,q}) > 1$, then we can choose three points p_1, p_2, p_3 in $\{p,q\}^{\perp} \setminus \{r\}$, different from r and such that $e(p_1), e(p_2)$ and $e(p_3)$ span a plane π of Σ. This plane contains infinitely many points of $e(\mathscr{S})$, which obviously belong to $e(\{p,q\}^{\perp})$. The point $e(r)$ might be one of them. Put $C := \pi \setminus \{p_1, p_2, p_4, r\}$. Then C is an infinite subset of $\{p,q\}^{\perp}_{r\,|r_p,r_q}$ and $\{p,q\}^{\perp}_{r\,|r_p,r_q} \setminus C$ still spans Σ. When $\dim(\Sigma_{p,q}) = 1$, we argue in the same way, but considering two points $p_1, p_2 \in \{p,q\}^{\perp}$ instead of three.
□

The case where $\mathscr{S}$ is a grid has been put aside here. We deal with it in the following lemma.

Lemma 2.9   *Suppose that* $\mathscr{S}$ *is a grid and* $|\mathbb{K}| > 3$. *Then* $\mathscr{S}$ *admits an ovoid* O *containing a subset* C *of size* $|\mathbb{K}| - 3$ (= $|\mathbb{K}|$ *if* $\mathbb{K}$ *is infinite*) *such that* $O \setminus C$ *is totally scattered.*

Proof   As $|\mathbb{K}| \geq 4$ by assumption, $\mathscr{S}$ admits several non-equivalent embeddings and the embedding e which we have chosen for $\mathscr{S}$ is just one of them. The full automorphisms group of $\mathscr{S}$ acts transitively on the set of embeddings of $\mathscr{S}$ as well as on the set of ovoids of $\mathscr{S}$. None of the ovoids of $\mathscr{S}$ arises as a plane section from all embeddings of $\mathscr{S}$. So, there exists an ovoid O of $\mathscr{S}$ such that $e(O)$ is not a conic of Σ, namely $e(O)$ spans Σ. Hence $e(O)$ contains a basis $\{e(p_1), e(p_2), e(p_3), e(p_4)\}$ of Σ. The ovoid O and the set $C = O \setminus \{p_1, p_2, p_3, p_4\}$ have the required properties.
□

8.2.2.5 The General Embeddable Case

Let now $n \geq 2$. Note that, if e is universal, then, for any non-degenerate subspace $\mathscr{S}'$ of $\mathscr{S}$ of rank at least 2, the embedding $e': \mathscr{S}' \to \Sigma' := \langle e(\mathscr{S}') \rangle_{\Sigma}$ induced by e on $\mathscr{S}'$ is still universal except when $|\mathbb{K}| > 3$ and $\mathscr{S}'$ is a grid or $\mathscr{S}'$ is a generalized quadrangle as in [19, 8.6(II)(a)]. However, in the latter two cases we can always assume that e' is the embedding chosen for $\mathscr{S}'$ as in Section 8.2.2.3.

Lemma 2.10 *Let $\mathbb{K}$ be infinite. Then $\mathscr{S}$ admits a partial ovoid O containing an infinite subset C such that $O \setminus C$ is totally scattered.*

Proof We argue by induction on n. We firstly fix the inductive step. Suppose that the claim holds true for a given n, let $\mathrm{rank}(\mathscr{S}) = n + 1$ and let p, q be two non-collinear points of $\mathscr{S}$. By the induction hypothesis, $\mathscr{S}' := \{p, q\}^{\perp}$ admits a partial ovoid O' containing an infinite subset C' such that $e(O' \setminus C')$ spans $\Sigma_{p,q}$ $(= \langle e(\{p, q\}^{\perp}) \rangle_{\Sigma}$ according to definition (2) of Section 8.2.2.3). Choose $r \in C'$ and points $r_p \in \langle p, r \rangle \setminus \{p, r\}$ and $r_q \in \langle q, r \rangle \setminus \{q, r\}$. Then $O := (O' \setminus \{r\}) \cup \{r_p, r_q\}$ is a partial ovoid. Moreover, the set $C := C' \setminus \{r\} \subsetneq O$ is infinite and $e(O \setminus C)$ spans Σ, since $e(O' \setminus C') \subset e(O \setminus C)$ spans $\Sigma_{p,q}$ and $\{e(r_p), e(r_q)\} \cup \Sigma_{p,q}$ spans Σ. So, $O \setminus C$ is totally scattered.

The initial step remains to be fixed. Suppose $n = 2$. When $\mathscr{S}$ is thick, Lemma 2.8 does the job. When $\mathscr{S}$ is a grid, we can use Lemma 2.9. Note that, as $\mathbb{K}$ is infinite, the set C of lemma 2.9 is infinite as well. $\qquad\square$

We turn now to the finite case.

Lemma 2.11 *Let $\mathbb{K}$ be finite but different from $\mathbb{F}_2$ and let $U \subset e(\mathscr{S})$ be a basis of Σ formed by mutually non-orthogonal singular points. Then there exists a point $v \in \Sigma$ such that $v^{\perp} \cap U = \emptyset$.*

Proof Since $\mathbb{K}$ is finite, $\dim(\Sigma)$ is finite as well. Hence $H^{\perp} \neq \emptyset$ for every hyperplane H of Σ. If either $\mathrm{char}(\mathbb{K}) \neq 2$ or $e(\mathscr{S})$ is not a quadric then $H^{\perp}$ is a point, say v, and $v^{\perp} = H$. Clearly Σ admits hyperplanes disjoint from U. If H is such a hyperplane then $v^{\perp} \cap U = \emptyset$. In this case, we are done. When $\mathrm{char}(\mathbb{K}) = 2$, $e(\mathscr{S})$ is a quadric and $\dim(\Sigma)$ is odd (hence $\dim(\Sigma) = 2n \pm 1$ since $\mathbb{K}$ is finite), then $H^{\perp}$ is a point and, if v is that point, then $v^{\perp} = H$. As before, we can choose H disjoint from U thus obtaining that $v^{\perp} \cap U = \emptyset$.

Suppose now that $\mathrm{char}(\mathbb{K}) = 2$, $e(\mathscr{S})$ is a quadric and $\dim(\Sigma)$ is even. Then $\dim(\Sigma) = 2n$ and $e(\mathscr{S}) = Q_{2n+2}(\mathbb{K})$. Accordingly, Σ contains a unique point v_0 (the nucleus of the quadric $e(\mathscr{S})$) such that $v_0^{\perp} = \Sigma$. We shall prove that Σ admits a hyperplane H disjoint from U and such that $H^{\perp} \neq \{v_0\}$. With H chosen in that way, if $v \in H^{\perp} \setminus \{v_0\}$ then $v^{\perp} = H$, and we are done.

By assumption, $\Sigma = \mathrm{PG}(V)$ with $V = V(2n + 1, \mathbb{K})$, $e(\mathscr{S})$ is the quadric defined by a non-singular quadratic form $\phi \colon V \to \mathbb{K}$ and V admits a basis $(\mathbf{u}_i)_{i=1}^{2n+1}$ such that $\phi(\mathbf{u}_i) = 0$ for any $i = 1, 2, \ldots, 2n + 1$ and $f(\mathbf{u}_i, \mathbf{u}_j) \neq 0$ for $1 \leq i < j \leq 2n + 1$, where $f \colon V \times V \to \mathbb{K}$ is the bilinearization of ϕ. We shall prove that Σ admits a hyperplane H containing the nucleus v_0 of $e(\mathscr{S})$ and disjoint from $U = ([\mathbf{u}_i])_{i=1}^{2n+1}$.

Put $a_{i,j} = f(\mathbf{u}_i, \mathbf{u}_j)$. Then $A = (a_{i,j})_{i,j=1}^{2n+1}$ is the representative matrix of f with respect to $([\mathbf{u}_i])_{i=1}^{2n+1}$. By assumption, $a_{i,j} = 0$ if and only if $i = j$, namely

A is anti-diagonal. Moreover rank(A) = $2n$ and, if $\mathbf{v}_0 = \sum_{i=1}^{2n+1} \mathbf{u}_i \lambda_i$ represents v_0, then $A\mathbf{v}_0 = \mathbf{0}$. With no loss, we can assume that the first $2n$ rows of A are independent. Then necessarily $\lambda_{2n+1} \neq 0$. With no loss, $\lambda_{2n+1} = 1$, namely $\mathbf{v}_0 = \mathbf{u}_{2n+1} + \sum_{i=1}^{2n} \mathbf{u}_i \lambda_i$. Moreover, up to rescaling the vectors $\mathbf{u}_1, \ldots, \mathbf{u}_{2n}$, we can assume that $a_{i,2n+1} = 1$ for $i = 2, 3, \ldots, 2n$ and $a_{1,2n+1} = \mu \neq 1$ (recall that $|\mathbb{K}| > 2$ by assumption). Hence $\mu\lambda_1 + \sum_{i=2}^{2n} \lambda_i = a_{2n+1,2n+1} = 0$. Consequently $\sum_{i=1}^{2n} \lambda_i \neq 0$, because $\mu \neq 1$. Let H be the hyperplane of Σ defined by the following equation: $\sum_{i=1}^{2n} x_i + \sum_{i=1}^{2n} \lambda_i \cdot x_{2n+1} = 0$, where unknowns are taken with respect to the basis $(\mathbf{u}_i)_{i=1}^{2n+1}$ of V.

Clearly $v_0 \in H$ and $[\mathbf{u}_i] \notin H$ for every $i = 1, 2, \ldots, 2n$. Moreover $[\mathbf{u}_{2n+1}] \notin H$ because $\sum_{i=1}^{2n} \lambda_i \neq 0$. So, H contains the nucleus v_0 of $\mathscr{S}$ and is disjoint from U. As H contains v_0, $H^\perp$ is a line through v_0. If v is a point of that line different from v_0 then $v^\perp = H$. □

Lemma 2.12 *Let $\mathbb{K}$ be finite but different from $\mathbb{F}_2$. If $\mathscr{S} = Q_{2n-1}^+(3)$, assume moreover that $n > 2$. Then $\mathscr{S}$ admits a completely scattered partial ovoid.*

Proof As in the proof of Lemma 2.10, we argue by induction on n. We firstly fix the inductive step. Suppose that the claim holds true for n and let rank($\mathscr{S}$) = $n + 1$. Let p, q be non-collinear points of $\mathscr{S}$. By the inductive hypothesis, $\{p,q\}^\perp$ admits a partial ovoid O' such that $e(O')$ spans $\Sigma_{p,q}$. Clearly, $e(O')$ contains a basis U of $\Sigma_{p,q}$ formed by singular mutually non-orthogonal points. By Lemma 2.11 applied to $\Sigma_{p,q}$, the subspace $\Sigma_{p,q}$ contains a point v such that $v^\perp \cap U = \emptyset$. The triple $\{e(p), e(q), v\}$ spans a plane π of Σ and $\Sigma_{p,q} \cup \pi$ spans Σ. As $v^\perp \cap U = \emptyset$, we have $u^\perp \cap \pi = \langle e(p), e(q)\rangle$ for every point $u \in U$. It is clear that if $\mathbb{K} \neq \mathbb{F}_2$ then the plane π contains at least two singular points $e(a)$ and $e(b)$ exterior to the line $\langle e(p), e(q)\rangle_\Sigma$ and such that $v, e(a)$ and $e(b)$ are non-collinear in Σ. Accordingly, $O := \{a, b\} \cup e^{-1}(U)$ is a partial ovoid and $e(O) = \{e(a), e(b)\} \cup U$ spans Σ, namely O is totally scattered. The inductive step is performed.

The initial step remains to be done. When $\mathscr{S}$ is thick or $|\mathbb{K}| > 3$ then we can start the induction at $n = 2$. Indeed when $n = 2$ and $\mathscr{S}$ is not a grid, then $\{p,q\}^\perp_{r|r_p,r_q}$ is a totally scattered partial ovoid. When $|\mathbb{K}| > 3$ and $\mathscr{S}$ is a grid, then we can apply Lemma 2.9.

Suppose that $\mathbb{K} = \mathbb{F}_3$ and $\mathscr{S}$ is non-thick, namely $\mathscr{S} = Q_{2n-1}^+(3)$. In this case the claim of the lemma is false for $n = 2$. Thus, we are forced to start the induction at $n = 3$. Recall that, given a non-singular 6×6 symmetric matrix A with entries in $\mathbb{F}_3$, the matrix A represents a bilinear form f_A of $V(6,3)$ and the Witt index of f_A is 3 or 2 according to whether det(A) is equal to -1 or 1 respectively. Choose A anti-diagonal as follows:

$$A = \begin{pmatrix} 0 & 1 & 1 & 1 & 1 & 1 \\ 1 & 0 & 1 & 1 & 1 & 1 \\ 1 & 1 & 0 & 1 & 1 & 1 \\ 1 & 1 & 1 & 0 & 1 & 1 \\ 1 & 1 & 1 & 1 & 0 & 2 \\ 1 & 1 & 1 & 1 & 2 & 0 \end{pmatrix}.$$

It is easily seen that $\det(A) = -1$ (in $\mathbb{F}_3$). Hence the form f_A has Witt index 3, namely it defines $Q_5^+(3)$. The canonical basis of $V(6, 3)$ consists of mutually non-orthogonal singular vectors (with respect to f_A). Hence the corresponding points of PG$(5, 3)$ form a partial ovoid of $Q_5^+(3)$ with the required properties. □

Remark 5 The following is the main obstacle we face when trying to generalize Lemma 2.11 to the infinite case: if $\mathbb{K}$ is infinite, then dim(Σ) might be infinite; when dim(Σ) is infinite, it can happen that $H^\perp = \emptyset$. On the other hand, it is likely that we can safely replace the hypothesis that $|\mathbb{K}| < \infty$ with the weaker hypothesis that dim$(\Sigma) < \infty$.

By combining Lemmas 2.10 and 2.12 with Proposition 2.3 we immediately obtain the following.

Theorem 2.13 *Let $\mathscr{S}$ be embeddable of rank $n > 2$ and defined over a division ring different from $\mathbb{F}_2$. Then $\mathscr{S}$ admits ovoidal maximal subspaces.*

8.2.3 The Non-embeddable Case

Throughout this subsection, $\mathscr{S}$ is a non-embeddable thick-lined polar space of rank $n \geq 3$. As recalled in Section 8.2.2.1, we have $n = 3$ and the following are the only possibilities for $\mathscr{S}$:

(1) $\mathscr{S}$ is the line-grassmannian of $\Sigma = $ PG$(3, \mathbb{K})$ with $\mathbb{K}$ a non-commutative division ring; explicitly, the points and the lines of $\mathscr{S}$ are the lines and the full planar line pencils of Σ, respectively. In this case $\mathscr{S}$ is not thick; explicitly, every line of $\mathscr{S}$ belongs to just two planes (generators) of $\mathscr{S}$.

(2) $\mathscr{S}$ is a Freudenthal–Tits polar space, namely a polar space as defined in Tits [19, chapter 9] (see also Mühlherr [13] and De Bruyn and Van Maldeghem [8]). In this case $\mathscr{S}$ is thick. Its planes (generators) are Moufang but non-desarguesian.

We shall prove that $\mathscr{S}$ admits ovoidal maximal subspaces. We will obtain this result as a by-product of a classification of all subspaces of $\mathscr{S}$.

8.2.3.1 The Non-thick Case

Let $\mathscr{S}$ be the line-grassmannian of $\Sigma = \mathrm{PG}(3, \mathbb{K})$, with $\mathbb{K}$ non-commutative.

Lemma 2.14 *Let $\mathscr{S}'$ be a proper subspace of $\mathscr{S}$. Then one of the following occurs:*

(i) *$\mathscr{S}'$ is a partial ovoid;*

(ii) *$\mathscr{S}'$ consists of a set of lines through some point x which form a partial ovoid in the residue at x;*

(iii) *$\mathscr{S}' = \{p, q\}^{\perp}$ for two non-collinear points;*

(iv) *$\mathscr{S}'$ is either a plane or the union of two planes through a given line;*

(v) *$\mathscr{S}'$ is a singular hyperplane, namely $\mathscr{S}' = p^{\perp}$ for some point p.*

Proof Viewing $\mathscr{S}'$ as a set of lines of Σ, the property of being a subspace of $\mathscr{S}$ corresponds to the following: if $\mathscr{S}'$ contains two intersecting lines, then it contains all lines of the planar pencil containing those two lines. The cases (*i*)–(*v*) can be rephrased as follows:

(i) $\mathscr{S}'$ is a partial spread of Σ.

(ii) All members of $\mathscr{S}'$ meet a fixed line L of Σ; if a point $x \in L$ is contained in at least two members of $\mathscr{S}'$, then the members of $\mathscr{S}'$ through x form a full line pencil in a plane $\pi_x \supset L$ and $\pi_x \neq \pi_y$ for distinct points $x, y \in L$ for which π_x and π_y are defined; dually, if a plane $\pi \supset L$ contains at least two members of $\mathscr{S}'$ then the lines of $\mathscr{S}'$ in π form a pencil with center on L and no two such pencils have the same center.

(iii) There are two skew lines L_1, L_2 of Σ such that $\mathscr{S}'$ is the set of lines of Σ intersecting both L_1 and L_2.

(iv) The members of $\mathscr{S}'$ are all lines of a plane of Σ, all lines through a point of Σ or all lines of some plane π of Σ together with all lines through some point $x \in \pi$.

(v) The members of $\mathscr{S}'$ are the lines meeting a fixed line L (including L).

If $\mathscr{S}'$ does not contain intersecting members then clearly (*i*) holds. Suppose that $\mathscr{S}'$ does contain intersecting members, but not all lines of a plane of Σ and not all lines through a point of Σ. If all line pencils contained in $\mathscr{S}'$ contain the same line L, then $\mathscr{S}'$ is as in (*ii*).

So we may assume that $\mathscr{S}'$ contains two disjoint line pencils, say with vertex x_i and plane π_i, $i = 1, 2$. Our assumption implies that the line $L_1 := \langle x_1, x_2 \rangle$ is not contained in $\pi_1 \cup \pi_2$. Set $L_2 = \pi_1 \cap \pi_2$. Clearly, L_1 and L_2 are skew, for each point $y \in L_2$ the planar line pencil with vertex y and plane $\langle y, L_1 \rangle$ is contained in $\mathscr{S}'$ and for each point $x \in L_1$ the planar line pencil with vertex x and plane $\langle x, L_2 \rangle$ is contained in $\mathscr{S}'$. So $\mathscr{S}'$ contains all lines intersecting both L_1 and L_2. If $\mathscr{S}'$ does not contain any additional line, then we have case (*iii*).

Suppose now that $\mathscr{S}'$ contains a line L not intersecting both L_1 and L_2. If L intersects L_1 in a point, then by considering the plane $\langle L, L_1 \rangle$, we readily deduce that L_1 also belongs to $\mathscr{S}'$; if $L = L_1$, then similarly we readily deduce that every line intersecting L_1 belongs to $\mathscr{S}'$. But this contradicts our assumption that $\mathscr{S}'$ does not contain all lines of any plane of Σ. Hence L is disjoint from $L_1 \cup L_2$. We claim that every point x of Σ is the vertex of a unique planar line pencil contained in $\mathscr{S}'$. It is certainly contained in at most one such line pencil by assumption, so we only need to show it is contained in at least one.

Suppose first that $x \in L$. Hence $x \notin L_1 \cup L_2$ and there is a unique line L' through x meeting both L_1 and L_2. This line belongs to $\mathscr{S}'$. Consequently, the pencil containing L and L' is contained in $\mathscr{S}'$.

Assuming that $x \notin L$, set $\pi = \langle x, L \rangle$ and $x_i = \pi \cap L_i$, $i = 1, 2$. Suppose first that x, x_1, x_2 are not collinear. Then $\mathscr{S}'$ contains the planar line pencil with vertex $y := \langle x_1, x_2 \rangle \cap L$ and plane π, so $\langle x, y \rangle$ is a member of $\mathscr{S}'$. But also the unique line through x intersecting both L_1 and L_2 belongs to $\mathscr{S}'$ and the claim follows in this case. If x, x_1, x_2 are collinear, then we can find a line $M \neq \langle x_1, x_2 \rangle$ in π through x containing two points u_1, u_2 which are vertices of planar line pencils not containing M, with respective planes α_1 and α_2; a previous argument then shows that $\mathscr{S}'$ contains all lines intersecting both M and $\alpha_1 \cap \alpha_2$ and the claim follows also in this case.

Hence the members of $\mathscr{S}'$ form the line set of a generalized quadrangle $\mathscr{Q}$ fully embedded in Σ such that each point of Σ is also a point of $\mathscr{Q}$. This property forces $\mathscr{Q}$ to be a symplectic generalized quadrangle. Accordingly, the underlying division ring $\mathbb{K}$ of Σ is commutative, a contradiction.

Hence we may assume that $\mathscr{S}'$ contains all lines of a certain plane π or all lines through a certain point x. If $\mathscr{S}'$ is not as in (*iv*), then, up to duality of Σ, we may assume that $\mathscr{S}'$ contains all lines of π and additionally two lines L_1, L_2 intersecting π in distinct respective points x_1, x_2. Then clearly all lines of Σ through x_1 and all those through x_2 are contained in $\mathscr{S}'$. For an arbitrary line M intersecting $L := \langle x_1, x_2 \rangle$, we choose an arbitrary point $y \in M \setminus (L \cap M)$. Then $\langle y, x_i \rangle \in \mathscr{S}'$ for $i = 1, 2$ implies $M \in \mathscr{S}'$. Hence all lines meeting L belong to $\mathscr{S}'$. If no further lines belong to $\mathscr{S}'$, then (*v*) holds.

If $N \in \mathscr{S}'$ does not meet L, then we quickly deduce that all lines meeting N are contained in $\mathscr{S}'$. Now, since for an arbitrary point $z \notin L \cup N$, the line pencils with vertex z and respective planes $\langle z, L \rangle$ and $\langle z, N \rangle$ are contained in $\mathscr{S}'$, all lines through z are contained in $\mathscr{S}'$ and so $\mathscr{S}' = \mathscr{S}$. $\qquad\square$

We now define a partial ovoid O' of $\mathscr{S}$, which amounts to defining a partial spread of Σ, as follows. Let L_1, L_2 be two disjoint lines of Σ and $\beta\colon L_1 \to L_2$ an arbitrary bijection. Choose $r \in L_1$ arbitrarily. Choose a line M_1 through r in the plane $\langle L_1, \beta(r)\rangle$ distinct from both L_1 and $\langle r, \beta(r)\rangle$. Similarly, choose a line M_2 through $\beta(r)$ in the plane $\langle L_2, r\rangle$ distinct from both L_2 and $\langle r, \beta(r)\rangle$. Then O' consists of M_1, M_2 and every line $K_x := \langle x, \beta(x)\rangle$, for $x \in L_1 \setminus \{r\}$.

Let O be a maximal partial ovoid containing O'.

Theorem 2.15 *The maximal partial ovoid O is a maximal subspace of $\mathscr{S}$.*

Proof We show that O cannot be properly contained in any of the subspaces listed in Lemma 2.14.

(1) By definition of maximal partial ovoid, O is not contained in a strictly larger partial ovoid. This rules out (*i*).
(2) If a line L intersects M_1, M_2 and every line K_x, $x \in L_1 \setminus \{r\}$, then $L \notin \{L_1, L_2\}$. It is then readily deduced that L is disjoint from $L_1 \cup L_2$ and the line through r intersecting L and L_2 intersects L_2 in $\beta(r)$. Hence none of M_1 and M_2 intersects L, a contradiction. This implies that no line of Σ intersects every member of O. This rules out (*ii*), (*iii*) and (*v*). In case (*iv*), picking a line in π through x also shows that every member of O should intersect a fixed line. □

8.2.3.2 The Thick Case

Let $\mathscr{S}$ be a Freudenthal–Tits polar space. Then there is an octonion division algebra $\mathbb{O}$ over some field $\mathbb{K}$ coordinatizing the planes of $\mathscr{S}$ (see [19, chapter 9]). From the description given in [8], it easily follows that, for two opposite points p_1, p_2, the set of points $\{p_1, p_2\}^{\perp\perp}$ is an infinite set containing p_1, p_2 and determined by any pair of its points. We call such a set a *hyperbolic line*.

We could again give a complete classification of the subspaces of $\mathscr{S}$ (see Remark 6), but the following 'quasi-classification' is sufficient for our purposes.

Lemma 2.16 *Let $\mathscr{S}'$ be a proper subspace of $\mathscr{S}$. Then one of the following holds:*

 (*i*) $\mathscr{S}'$ *is a partial ovoid of $\mathscr{S}$;*
(*ii*) $\mathscr{S}'$ *consists of a set of lines through some point x which form a partial ovoid in the residue at x;*
(*iii*) $\mathscr{S}'$ *is a (non-degenerate) generalized quadrangle closed under taking hyperbolic lines;*
(*iv*) $\mathscr{S}'$ *is a set of planes through some fixed line;*
 (*v*) $\mathscr{S}' = p^\perp$ *for some point p.*

Proof If $\mathscr{S}'$ does not contain lines, then we clearly have (*i*). Suppose now $\mathscr{S}'$ contains lines but no planes. If all such lines contain a common point, then we have (*ii*). So suppose that $\mathscr{S}'$ contains two non-intersecting lines L_1, L_2. If these are not opposite, then $\mathscr{S}'$ contains a plane, a contradiction. Hence L_1 and L_2 are opposite. Let p_1 and p_2 be two opposite points with $(L_1 \cup L_2) \subseteq \{p_1, p_2\}^\perp =: U$ (we obtain p_i by choosing two planes through L_1 and projecting L_2 onto these planes). Now U is a generalized quadrangle which, according to proposition 5.9.4 of [20], does not contain proper full subquadrangles. Hence $U = \langle L_1, L_2 \rangle$ and therefore $U \subseteq \mathscr{S}'$. Also, since by definition $\{x_1, x_2\}^{\perp\perp}$ is contained in $\{p_1, p_2\}^\perp$ for each pair of opposite points $x_1, x_2 \in U$, the subspace U and hence $\mathscr{S}'$ is closed under taking hyperbolic lines. This is (*iii*).

So we may assume that $\mathscr{S}'$ contains planes. If all planes contain a common point x, and not a common line, then we have two opposite lines in the residue of x, which implies that $\mathscr{S}'$ is the full residue at x since the residual quadrangle does not have proper full subquadrangles. Hence (*iv*) or (*v*) arises.

Suppose that $\mathscr{S}'$ contains two opposite planes. The construction in [8] reveals the following property of $\mathscr{S}$. Restricting $\mathbb{O}$ in the construction to a quaternion subalgebra over $\mathbb{K}$, we obtain a sub polar space $\mathscr{S}_0$ of $\mathscr{S}$, which is not a subspace, but with the property that every plane of $\mathscr{S}$ through any line of $\mathscr{S}_0$ is also a plane of $\mathscr{S}_0$. We refer to the latter property as *idealness*. Now two opposite planes can always be included in such a sub polar space $\mathscr{S}_0$. Moreover, $\mathscr{S}_0$ lives in a projective 5-space. Therefore, by Proposition 2.2, every pair of opposite planes of $\mathscr{S}_0$ spans $\mathscr{S}_0$. Consequently $\mathscr{S}_0$ is a sub polar space of $\langle X_0 \cup Y_0 \rangle_\mathscr{S}$ for any two mutually disjoint planes X_0, Y_0 of $\mathscr{S}_0$. On the other hand, every pair of opposite planes of $\mathscr{S}$ can be regarded as a pair of opposite planes of a suitable copy of $\mathscr{S}_0$. Since we have assumed that $\mathscr{S}'$ contains two opposite planes, we can also assume that $\mathscr{S}_0$ is a sub polar space of $\mathscr{S}'$. Since $\mathscr{S}_0$ is ideal as a sub polar space, $\mathscr{S}'$ is ideal as a polar subspace. This forces $\mathscr{S}' = \mathscr{S}$. Indeed, for a point p of $\mathscr{S} \setminus \mathscr{S}'$, if any such point exists, select a plane π in $\mathscr{S}'$ and let $\pi' = \langle p, p^\perp \cap \pi \rangle$. The line $p^\perp \cap \pi$ belongs to $\mathscr{S}'$. Hence π' is a plane of $\mathscr{S}'$, by idealness of $\mathscr{S}'$. Accordingly, $p \in \mathscr{S}'$. $\qquad\square$

Now we choose two opposite points x_1, x_2 in $\mathscr{S}$ and an infinite ovoidal maximal subspace O' of $U := \{x_1, x_2\}^\perp$ as constructed in Theorem 2.1. Hence

$$O' \supseteq ((y_1^\perp \cap y_2^\perp \cap U) \setminus \{r\}) \cup \{u_1, u_2\},$$

where y_1, y_2 are opposite points of U, $r \in y_1^\perp \cap y_2^\perp \cap U$ and $u_i \in \langle y_i, r \rangle \setminus \{y_i, r\}$ for $i = 1, 2$. Notice that $y_1^\perp \cap y_2^\perp \cap U$ is a hyperbolic line. Indeed let z_1 and z_2 be any two distinct points of $\{y_1, y_2\}^\perp \cap U$, necessarily opposite since $\{y_1, y_2\}^{\perp} \cap U$ consists of mutually opposite points. Put $U' := \{z_1, z_2\}^{\perp}$. Then $U'^{\perp} = \{z_1, z_2\}^{\perp\perp}$ is a hyperbolic line. Moreover, any two opposite lines of U' span U'. In particular, U' is spanned by the lines $\langle x_1, y_1 \rangle$ and $\langle x_2, y_2 \rangle$. Therefore $U'^{\perp} = \{x_1, x_2, y_1, y_2\}^{\perp} = U \cap \{y_1, y_2\}^{\perp}$, namely $U \cap \{y_1, y_2\}^{\perp} = \{z_1, z_2\}^{\perp\perp}$.

Choose $r' \in (y_1^{\perp} \cap y_2^{\perp} \cap U) \setminus \{r\}$ and select points $w_i \in \langle x_i, r' \rangle \setminus \{x_i, r'\}$, $i = 1, 2$. Then define $O'' = (O' \setminus \{r'\}) \cup \{w_1, w_2\}$. Let O be a maximal partial ovoid of $\mathscr{S}$ containing O''.

Theorem 2.17 *The maximal partial ovoid O is a maximal subspace of $\mathscr{S}$.*

Proof We first show that no point of $\mathscr{S}$ is collinear to all points of O''. Indeed, all points collinear to $(y_1^{\perp} \cap y_2^{\perp} \cap U) \setminus \{r, r'\}$ constitute a generalized quadrangle $\mathscr{Q}$ contained in $\{v_1, v_2\}^{\perp}$, for two distinct points $v_1, v_2 \in (y_1^{\perp} \cap y_2^{\perp} \cap U) \setminus \{r, r'\}$. However $\{v_1, v_2\}^{\perp}$ admits no proper full subquadrangle. Hence $\mathscr{Q} = \{v_1, v_2\}^{\perp}$ and any two opposite lines of $\mathscr{Q}$ span $\mathscr{Q}$. In particular, $\mathscr{Q}$ is spanned by the opposite lines $\langle x_1, y_1 \rangle$ and $\langle x_2, y_2 \rangle$. Moreover, $\mathscr{Q} = \{r, r'\}^{\perp}$, since $r, r' \in \{x_1, x_2, y_1, y_2\}^{\perp}$ and $\mathscr{Q} = \langle x_1, y_1, x_2, y_2 \rangle$.

If a point z is collinear to all points of O'', then it belongs to $\mathscr{Q}$. However $\mathscr{Q} = \{r, r'\}^{\perp}$. Hence $z \perp r, r'$. On the other hand, z is also collinear to u_1, u_2, w_1, w_2, hence to x_1, x_2, y_1, y_2. This is impossible within $\mathscr{Q}$.

Hence a subspace containing lines and containing O cannot be of any of the types (i), (ii), (iv) or (v) of Lemma 2.16. So suppose O is contained in a subspace $\mathscr{S}'$ isomorphic to a generalized quadrangle and closed under taking hyperbolic lines, as in (iii) of Lemma 2.16. Then, for $z_1, z_2 \in O$, we deduce $y_1^{\perp} \cap y_2^{\perp} \cap U \subseteq \mathscr{S}'$. Hence $r \in \mathscr{S}'$, implying $y_i \in \langle r, u_i \rangle \subseteq \mathscr{S}'$, $i = 1, 2$. Also, $r' \in \mathscr{S}'$ implying $x_i \in \langle r', w_i \rangle \subseteq \mathscr{S}'$, $i = 1, 2$. Hence the two opposite planes $\langle x_1, y_1, z_1 \rangle$ and $\langle x_2, y_2, z_2 \rangle$ belong to $\mathscr{S}'$, and so $\mathscr{S}'$ coincides with $\mathscr{S}$. We have proved that O is indeed a maximal subspace. $\qquad\qquad \square$

Remark 6 Actually, in case (iii) of Lemma 2.16, we have $\mathscr{S}' = \{p, q\}^{\perp}$ for two opposite points p and q of $\mathscr{S}$. This can be proved by dimension arguments and a little more work, but we are not going into the details of that proof here.

With (iii) stated in this sharper way, Lemma 2.16 yields a complete classification of all non-ovoidal subspaces of $\mathscr{S}$, which can be summarized as follows: $\mathscr{S}$ admits only those subspaces that exist in any polar space, namely those of reduced rank at most 1 and intersections of singular hyperplanes. As a consequence, the maximal non-ovoidal subspaces of $\mathscr{S}$ are precisely its singular hyperplanes. Exactly the same holds in the non-embeddable non-thick case (Lemma 2.14). This also improves a result of Cohen and Shult [5], according to which all hyperplanes of a non-embeddable thick-lined polar space of rank 3 are singular.

8.2.4 The Case $\mathbb{K} = \mathbb{F}_2$

In order to finish the proof of Theorem 1, the case where $\mathscr{S}$ is a quadric defined over $\mathbb{F}_2$ remains to be examined.

Theorem 2.18 *Let $\mathscr{S}$ be a non-degenerate quadric of rank $n \geq 2$, defined over $\mathbb{F}_2$. Then $\mathscr{S}$ admits ovoidal maximal subspaces precisely in the following cases:*

(1) $\mathscr{S} = Q_{2n}(2)$ *with either $n = 2$ or n odd;*
(2) $\mathscr{S} = Q_{2n-1}^+(2)$ *with either $n = 2$ or $n \equiv 0, 1 \pmod 4$;*
(3) $\mathscr{S} = Q_{2n+1}^-(2)$ *with $n \equiv 1, 2 \pmod 4$.*

Proof The quadric $\mathscr{S}$ lives in PG$(N - 1, 2)$, with $N \in \{2n, 2n + 1, 2n + 2\}$. Since it is well known that $Q_4(2)$ and $Q_3^+(2)$ admit ovoids while $Q_5^-(2)$ admits no ovoids, we can assume that $N > 5$ and, in view of Propositions 2.3 and 2.5, the existence of an ovoidal maximal subspace of $\mathscr{S}$ is equivalent to the existence of a totally scattered partial ovoid of $\mathscr{S}$, which in turn is equivalent to the existence of a basis $E = (\mathbf{e}_1, \ldots, \mathbf{e}_N)$ of V such that, if ϕ is the quadratic form giving rise to $\mathscr{S}$ and f its bilinearization, we have $\phi(\mathbf{e}_i) = 0$ for $i = 1, 2, \ldots, N$ and $f(\mathbf{e}_i, \mathbf{e}_j) = 1$ for $1 \leq i < j \leq N$. Equivalently, ϕ admits the following expression with respect to E:

$$\phi(x_1 x_2, \ldots, x_N) = \sum_{i<j} x_i x_j. \tag{3}$$

So, the representative matrix of f with respect to E is $A = J + I$, where J is the $N \times N$ matrix with all entries equal to 1 and I is the identity matrix of order N.

Assume firstly $N = 2n + 1$. Then A has rank $2n = N - 1$ with kernel $\{\mathbf{0}, \mathbf{r}\}$, $\mathbf{r} = \sum_i \mathbf{e}_i$. Clearly, $\phi(\mathbf{r}) = \binom{N}{2} = n(2n + 1)$ (computed modulo 2). However, ϕ is non-degenerate if and only if $\phi(\mathbf{r}) = 1$; equivalently, n is odd. So, as ϕ is non-degenerate by assumption, ϕ can be expressed as in (3) if and only if n is odd. This proves the claim of Theorem 2.18 for N odd.

Suppose that N is even, say $N = 2m$. Hence A is non-singular. Define

$$\phi_m(x_1, \ldots, x_{2m}) := \sum_{1 \leq i < j \leq 2m} x_i x_j \quad \text{(see (3))};$$
$$\psi_m(x_1, \ldots, x_{2m}) := \phi_m(x_1, \ldots, x_{2m}) + \sum_{i=1}^{2m} x_i^2.$$

In these definitions, $m = 1$ is allowed. Both ϕ_m and ψ_m are non-degenerate quadratic forms. Let $\perp$ be the orthogonality relation associated with ϕ_m and suppose $m > 1$. Then ϕ_m induces ψ_{m-1} on $\{\mathbf{e}_{2m-1}, \mathbf{e}_{2m}\}^{\perp}$. As the form induced by ϕ_m on $\langle \mathbf{e}_{2n-1}, \mathbf{e}_{2n} \rangle$ is hyperbolic, ϕ_m and ψ_{m-1} have the same type, namely they are either both hyperbolic or both elliptic.

Suppose $m > 2$. Then ϕ_m induces ϕ_{m-2} on $\{\mathbf{e}_{2m-3}, \mathbf{e}_{2m-2}, \ \mathbf{e}_{2m-1}, \mathbf{e}_{2m}\}^{\perp}$. As ψ_{m-1} induces $\psi_1(x_{2m-3}, x_{2m-2}) = x_{2m-3}^2 + x_{2m-3}x_{2m-2} + x_{2m-2}^2$ on $\langle \mathbf{e}_{2m-3}, \mathbf{e}_{2m-2} \rangle$, which is elliptic, the types of ϕ_{m-2} and ψ_{m-1} are opposite. Hence ϕ_{m-2} and ϕ_m have opposite types. Clearly, ϕ_1 is hyperbolic. Therefore ϕ_{1+2k} is hyperbolic if k is even and elliptic if k is odd.

Consider now ϕ_2. We know that ϕ_2 and ψ_1 have the same type. However $\psi_1(x_1, x_2) = x_1^2 + x_1 x_2 + x_2^2$ is elliptic. Hence ϕ_2 is elliptic. By the aforementioned, ϕ_{2k} is elliptic if k is odd and hyperbolic if k is even. The theorem follows from the fact that the existence of an ovoidal maximal subspace of $\mathscr{S}$ is equivalent to ϕ admitting the expression (3), which is precisely the expression called ϕ_m here.   □

Remark 7   Suppose that the form ϕ defined in (3) is non-degenerate (hence either N is even or $N - 1 \equiv 0 \pmod{4}$) and let $\mathscr{S}$ be the quadric associated to ϕ. With $\mathbf{e}_1, \dots, \mathbf{e}_N$ as in the proof of Theorem 2.18, the set $O := \{[\mathbf{e}_1], \dots, [\mathbf{e}_N]\}$ is a partial ovoid of $\mathscr{S}$ and, if it is maximal as a partial ovoid, then it is a maximal subspace of $\mathscr{S}$.

As we shall see in a few lines, O is non-maximal if and only if $N \equiv 0 \pmod{4}$. Indeed, for a vector $\mathbf{u} \in V$, we have $\mathbf{e}_k \not\perp \mathbf{u}$ for every $k = 1, 2, \dots, N$ only if $\mathbf{u} = \sum_{i=1}^{N} \mathbf{e}_i$ and N is even. Consequently, when N is odd O is maximal. Let N be even, with $\mathbf{u}$ as above, and suppose that $\phi(\mathbf{u}) = 0$. Then $\binom{N}{2} = 0$ (in $\mathbb{F}_2$). Hence $N \equiv 0 \pmod{4}$. So, if $N \equiv 0 \pmod{4}$, then O is not maximal and $O \cup \{[\sum_{i=1}^{N} \mathbf{e}_i]\}$ is the unique maximal partial ovoid containing O (hence it is the unique ovoidal maximal subspace containing O); otherwise, O is maximal.

Note that neither O nor $O \cup \{\mathbf{u}\}$ (when $N \equiv 0 \pmod{4}$) are ovoids. Indeed, apart from the fact $Q_4(2), Q_3^+(2), Q_5^+(2)$ and $Q_7^+(2)$ are the only quadrics defined over $\mathbb{F}_2$ which admit ovoids (see Section 8.4.1), when $N > 5$ both O and $O \cup \{\mathbf{u}\}$ are far too small to be ovoids.

8.2.5 Conclusions

End of the Proof of Theorem 1. By combining Theorems 2.1, 2.13, 2.15, 2.17 and 2.18 we obtain Theorem 1.

Proof of Corollary 2. We could obtain part (1) of Corollary 2 from Lemmas 2.10 and 2.12, but the following elementary argument is enough. It is well known that all non-singular anti-symmetric matrices of $M_N(\mathbb{F})$ (N even) are mutually T-equivalent. So, all we have to do is find a non-singular, anti-diagonal and anti-symmetric matrix $A = (a_{i,j})_{i,j=1}^{N}$. Here is one: $a_{i,j} = 1$ if $i < j$, $a_{i,i} = 0$ and $a_{i,j} = -1$ if $i > j$. It is readily seen that, with A defined in this way, $\det(A) = 1$. Claim (1) of Corollary 2 is proved.

Turning to claims (2) and (3), let $\mathscr{S}_M$ be the polar space associated with the appropriate bilinear or hermitian form represented by M with respect to the canonical basis of $V = V(N, q)$. Claims (2) and (3) amount to the following: $\mathscr{S}_M$ contains a basis of $PG(N-1, q)$ formed by mutually non-orthogonal points. So, assuming that $N > 2$ and $(N, q) \neq (4, 3)$ when M is symmetric, we must prove the following: if $\text{rank}(\mathscr{S}_M) > 1$, then $\mathscr{S}_M$ admits a totally scattered partial ovoid; otherwise $\mathscr{S}_M$ has rank 1 and spans $PG(N - 1, q)$.

When $\text{rank}(\mathscr{S}_M) > 1$, the claim follows from Lemma 2.12. Note that $q = 2$ is forbidden here. Indeed q is odd in claim (2) while $q \geq 4$ in (3). Suppose that $\text{rank}(\mathscr{S}_M) \leq 1$. Note that $\mathscr{S}_M \neq \emptyset$. Indeed $N \geq 3$ in (2), which forces $|\mathscr{S}_M| \geq q + 1$, while if M is hermitian then $|\mathscr{S}_M| \geq 1 + \sqrt{q}$. Therefore $\mathscr{S}_M$ has rank 1 and spans $PG(N - 1, q)$. Explicitly, $\mathscr{S}_M$ is either a conic of $PG(2, q)$, an elliptic quadric of $PG(3, q)$, a set of $1 + \sqrt{q}$ points of $PG(1, q)$ or a hermitian unital of $PG(2, q)$.

Remark 8 The hypothesis $(N, \mathbb{F}) \neq (4, \mathbb{F}_3)$ cannot be dropped from claim (2) of Corollary 2. Indeed it is easily checked that every anti-diagonal symmetric matrix M of $M_4(\mathbb{F}_3)$ has determinant equal to -1. Therefore, if $M \in M_4(\mathbb{F}_3)$ is symmetric and anti-diagonal, then $\mathscr{S}_M$ is an elliptic quadric of $PG(3, 3)$. We miss hyperbolic quadrics. This is in conformity with the fact that each ovoid of $Q_3^+(3)$ arises from each embedding.

Remark 9 In claims (2) and (3) of Corollary 2, the hypothesis that $\mathbb{F}$ is finite is not strictly necessary, provided that those claims are rephrased as follows: every isotropic non-singular symmetric (hermitian) matrix of $M_N(\mathbb{F})$ is T-equivalent (∗-equivalent) to an anti-diagonal matrix, a symmetric or hermitian matrix $M \in M_N(\mathbb{F})$ being called *isotropic* if $\mathscr{S}_M \neq \emptyset$.

8.3 Constructions

8.3.1 A Construction in Symplectic Varieties

Let $\mathscr{S}$ be a symplectic polar space of rank $n \geq 2$, namely $\mathscr{S}$ admits an embedding $e \colon \mathscr{S} \to PG(V)$ such that $\dim(V) = 2n$ and $e(\mathscr{S})$ is the polar space associated to a non-degenerate alternating form $f \colon V \times V \to \mathbb{K}$, with $\mathbb{K}$ a commutative division ring.

Let A be a generator of $\mathscr{S}$ and $\{a_1, a_2, \ldots, a_n\}$ a basis of A. For every $k = 1, 2 \ldots, n$, let L_k be a hyperbolic line of $\mathscr{S}$ containing a_k and contained in

$\{a_1, \ldots, a_{k-1}, a_{k+1}, \ldots, a_n\}^\perp$. Let $\{\mathbf{e}_1, \mathbf{f}_2, \ldots, \mathbf{e}_n, \mathbf{f}_n\}$ be a basis of V such that $f(\mathbf{e}_i, \mathbf{f}_j) = \delta_{i,j}$ (Kronecker symbol) and $f(\mathbf{e}_i, \mathbf{e}_j) = f(\mathbf{f}_i, \mathbf{f}_j) = 0$ for any choice of $i, j = 1, \ldots, n$. We can assume that $[\mathbf{e}_i] = e(a_i)$ for $i = 1, 2, \ldots, n$. Thus, there exist scalars $\lambda_{i,j} \in \mathbb{K}$ such that $e(L_k) = \langle [\mathbf{e}_k], [\mathbf{f}_k + \sum_{i \neq k} \mathbf{e}_i \lambda_{i,k}] \rangle$. With the lines L_k defined as here, for $k \neq h$ we have $L_k \not\perp L_h$ (namely $L_k^\perp \cap L_h = \{a_h\}$) if and only if

$$\lambda_{i,j} \neq \lambda_{j,i}, \quad \text{for any choice of } i \neq j. \tag{4}$$

Condition (4) is very easy to satisfy. For instance, we can choose $\lambda_{i,j}$ arbitrarily for $i \leq j$ and put $\lambda_{j,i} = \lambda_{i,j} + 1$ for $i < j$. Suppose to have chosen $L_1, L_2, \ldots, L_n$ in such a way that (4) holds and let a be a point in $A \setminus \cup_{k=1}^n \langle a_i \rangle_{i \neq k}$, namely $e(a) = [\sum_{i=1}^n \mathbf{e}_i \mu_i]$ for $\mu_1, \mu_2, \ldots, \mu_n \in \mathbb{K}^*$. Put $O' := \cup_{k=1}^n (L_k \setminus \{a_k\})$ and $O := O' \cup \{a\}$.

Lemma 3.1 *The set O is a maximal partial ovoid of $\mathscr{S}$.*

Proof We firstly prove that O is a partial ovoid. No two points of the set $\cup_{k=1}^n (L_k \setminus \{a_k\})$ are orthogonal, since $L_1, \ldots, L_n$ are hyperbolic lines and if $k \neq h$ then $L_k^\perp \cap L_h = \{a_h\}$. It remains to prove that $a^\perp \cap L_k = \{a_k\}$. Suppose the contrary. Then $L_k \subseteq a^\perp$. However $L_k \perp a_i$ for every $i \neq k$. Moreover, the set $\{a_1, \ldots, a_{k-1}, a_{k+1}, \ldots, a_n, a\}$ spans A. It follows that $L_k \subseteq A^\perp = A$, a contradiction.

Maximality remains to be proved. Let b be a point exterior to O. We shall prove that $b^\perp \cap O \neq \emptyset$. We have $b^\perp \cap L_k \neq \emptyset$ for every k. If $b^\perp$ contains a point of $L_k \setminus \{a_k\}$, then we are done. So, suppose that $b^\perp \cap L_k = \{a_k\}$ for every k. Then $b \in A$. Accordingly, $b \perp a \in O$. $\qquad\square$

As $O' \subset O$, the set O' is also a partial ovoid, but not a maximal one. However,

Lemma 3.2 *Let $X \subseteq O'$ be such that $|X \cap L_k| \geq 2$ for every $k = 1, 2, \ldots, n$. Then $e(X)$ spans PG(V). In particular, $e(O')$ spans PG(V).*

Proof With X as in the hypotheses of the lemma, $\langle e(X) \rangle$ contains the line $e(L_k)$ for every $k = 1, 2, \ldots, n$. Accordingly, $\langle e(X) \rangle$ contains $[\mathbf{e}_k]$ and $[\mathbf{f}_k]$ for every k. However V is spanned by $\mathbf{e}_1, \ldots, \mathbf{e}_n, \mathbf{f}_1, \ldots, \mathbf{f}_n$. Therefore $\langle e(X) \rangle = $ PG(V). $\qquad\square$

Let $\tilde{e}\colon \mathscr{S} \to$ PG$(\widetilde{V})$ be the universal embedding of $\mathscr{S}$. If char$(\mathbb{K}) \neq 2$ then $\tilde{e} = e$ (hence $\widetilde{V} = V$). In this case $\tilde{e}(O')$ spans PG$(\widetilde{V})$, by Lemma 3.2 (with the terminology in Section 8.2, the partial ovoid O' is totally scattered). On the other hand, when char$(\mathbb{K}) = 2$ the embedding e is a proper quotient of $\tilde{e}$. In this case we cannot use Lemma 3.2 to obtain that $\langle \tilde{e}(O') \rangle = \mathrm{PG}(\widetilde{V})$. In fact $\langle \tilde{e}(O') \rangle \subset \mathrm{PG}(\widetilde{V})$ when $\mathbb{K} = \mathbb{F}_2$. Nevertheless:

Lemma 3.3 *Suppose that* $\mathrm{char}(\mathbb{K}) = 2$ *but* $\mathbb{K} \neq \mathbb{F}_2$. *Then* $\tilde{e}(O')$ *spans* $\mathrm{PG}(\widetilde{V})$.

Proof Under the hypotheses of the lemma, $\tilde{e}(\mathscr{S})$ is the quadric associated to a non-degenerate quadratic form $\phi\colon \widetilde{V} \to \mathbb{K}$ and $\langle \tilde{e}(O') \rangle = \mathrm{PG}(V')$ for a subspace V' of $\widetilde{V}$. We shall prove that $V' = \widetilde{V}$.

Let ϕ' be the form induced by ϕ on V' and $\mathscr{S}'$ the quadric defined by ϕ' in $\mathrm{PG}(V')$. As $|\mathbb{K}| > 2$, for every $k = 1, 2, \ldots, n$ the sets $\tilde{e}(L_k)$ and $\tilde{e}(L_k \setminus \{a_k\})$ span the same subspace X_k of $\mathrm{PG}(\widetilde{V})$. Hence $\mathrm{PG}(V') \supseteq \cup_{i=1}^n X_i$. Moreover $\tilde{e}(a_k) \in \mathrm{PG}(V')$ for every $k = 1, 2, \ldots, n$, as $a_k \in L_k$. It follows that $\mathscr{S}'$ contains $\tilde{e}(A)$, which is a generator of $\tilde{e}(\mathscr{S})$. However no point of Λ is collinear with all points of O, by construction. Therefore $\mathscr{S}'$ is non-degenerate of rank n. Turning back to X_k, its codimension in $\mathrm{PG}(\widetilde{V})$ is equal to $2n - 2$. Indeed $X_k = \langle \tilde{e}(L_k) \rangle$ and $\tilde{e}(L_k) = A_k^\perp \cap B_k^\perp$ (in $\tilde{e}(\mathscr{S})$) for any two submaximal singular subspaces A_k and B_k of $\tilde{e}(\mathscr{S})$ such that $A_k^\perp \cap B_k^\perp = \emptyset$ and $|A_k^\perp \cap B_k^\perp \cap \tilde{e}(L_k)| > 1$. Since $\mathscr{S}'$ is non-degenerate and has the same rank as $\tilde{e}(\mathscr{S})$, two singular subspaces A_k and B_k as required can always be chosen in $\mathscr{S}'$. Hence the equality $\tilde{e}(L_k) = A_k^\perp \cap B_k^\perp$ holds true in $\mathscr{S}'$, too. Consequently, X_k has codimension $2n - 2$ in $\mathrm{PG}(V')$ as well as in $\mathrm{PG}(\widetilde{V})$. This forces $V' = \widetilde{V}$. $\qquad\square$

Lemma 3.4 *If* $\mathbb{K} = \mathbb{F}_2$ *and* n *is odd, then* $\tilde{e}(O)$ *spans* $\mathrm{PG}(\widetilde{V})$.

Proof Suppose that $\mathbb{K} = \mathbb{F}_2$. Let ϕ be the quadratic form of $\widetilde{V}$ associated to $\tilde{e}(\mathscr{S})$. An ordered basis $(\mathbf{u}_1, \mathbf{v}_1, \ldots, \mathbf{u}_n, \mathbf{v}_n, \mathbf{w})$ can be chosen in $\widetilde{V}$ such that ϕ admits the following expression with respect to it:

$$\phi(x_1, y_1, \ldots, x_n, y_n, z) = \sum_{i=1}^n x_i y_i + z^2.$$

With no loss, we can assume to have chosen the vectors $\mathbf{u}_k$ in such a way that $[\mathbf{u}_k] = \tilde{e}(a_k)$ for every $k = 1, 2, \ldots, n$ and

$$\tilde{e}(L_k) = \{[\mathbf{u}_k], \, [\mathbf{v}_k + \sum_{i \neq k} \mathbf{u}_i \lambda_{i,k}], \, [\mathbf{u}_k + \mathbf{v}_k + \sum_{i \neq k} \mathbf{u}_i \lambda_{i,k} + \mathbf{w}]\}$$

for suitable scalars $\lambda_{i,j}$ such that $\lambda_{k,h} \neq \lambda_{k,h}$ when $k \neq h$ (see (4)). Since $[\mathbf{u}_k] = \tilde{e}(a_k)$ for every k, we also have $\tilde{e}(a) = [\sum_{k=1}^n \mathbf{u}_k]$.

Let X be the span of $\tilde{e}(O)$ in $\mathrm{PG}(\widetilde{V})$. Then X contains both $[\mathbf{v}_k + \sum_{i \neq k} \mathbf{u}_i \lambda_{i,k}]$ and $[\mathbf{u}_k + \mathbf{v}_k + \sum_{i \neq k} \mathbf{u}_i \lambda_{i,k} + \mathbf{w}]$, for every $k = 1, 2, \ldots, n$. Hence X also contains $[\mathbf{u}_k + \mathbf{w}]$. Consequently $[\mathbf{u}_k + \mathbf{u}_h] \in X$ for any choice of $k \neq h$. However X also contains $\tilde{e}(a) = [\sum_{k=1}^n \mathbf{u}_k]$ and n is odd, by assumption. Therefore

$[\mathbf{u}_k] \in X$ for any k. Hence $\mathbf{w} \in X$ and $\mathbf{v}_k \in X$ for any k. So, X contains all of $[\mathbf{u}_1], \ldots, [\mathbf{u}_n], [\mathbf{v}_1], \ldots, [\mathbf{v}_n]$ and $[\mathbf{w}]$. In short, $X = \mathrm{PG}(\widetilde{V})$. □

Theorem 3.5 *The set O is an ovoidal maximal subspace of $\mathscr{S}$, except precisely when $\mathbb{K} = \mathbb{F}_2$ and n is even and different from 2.*

Proof In view of Lemmas 3.3 and 3.4 and Propositions 2.3 and 2.5, if either $\mathbb{K} \neq \mathbb{F}_2$ or n is odd, then O is a maximal subspace of $\mathscr{S}$. If $\mathbb{K} = \mathbb{F}_2$ and $n = 2$, then $|O| = 5$, which is just the size of an ovoid of $\mathscr{S}$. Hence O is an ovoid. Finally, when $\mathbb{K} = \mathbb{F}_2$ and $n > 2$ is even, then $\mathscr{S}$ admits no ovoidal maximal subspace, by the isomorphism $W_{2n-1}(2) \cong O_{2n}(2)$ and Theorem 1. □

Remark 10 If $\mathbb{K} \neq \mathbb{F}_2$, then O is not an ovoid. Indeed, let $|\mathbb{K}| > 2$. Suppose firstly that for every $k = 1, \ldots, n$ there exists $i \neq k$ such that $\lambda_{i,k} \neq 0$. Then $B := \langle [\mathbf{f}_1], \ldots, [\mathbf{f}_n] \rangle$ is a generator of $\mathscr{S}$ but $O \cap B = \emptyset$. On the other hand, let $\lambda_{i,k} = 0$ for some k and every $i \neq k$, say $\lambda_{i,1} = 0$ for every $i > 1$. Then $\lambda_{1,2} \neq 0$ by (4). Given $t \in \mathbb{K} \setminus \{0, \lambda_{1,2}\}$ ($\neq \emptyset$ since $|\mathbb{K}| > 2$), put $\mathbf{f}'_1 := \mathbf{e}_2 t + \mathbf{f}_1$ and $\mathbf{f}'_2 := \mathbf{e}_1 t + \mathbf{f}_2$. Then $B' := \langle [\mathbf{f}'_1], [\mathbf{f}'_2], [\mathbf{f}_3], \ldots, [\mathbf{f}_n] \rangle$ is a generator of $\mathscr{S}$ and $B' \cap O = \emptyset$. In any case, O is not an ovoid.

Finally, let $\mathbb{K} = \mathbb{F}_2$. Then $|O| = 2n + 1$ while an ovoid of $W_{2n-1}(2)$, if it exists, has size $2^n + 1$. As $2n + 1 < 2^n + 1$ if $n > 2$, O is an ovoid if and only if $n = 2$. (Anyway, it is well known that no ovoids exist in $W_{2n-1}(2)$ when $n > 2$; see also Section 8.4.1.)

8.3.2 More Constructions

Throughout this subsection, $\mathscr{S}$ is the polar space associated to a non-degenerate $(\sigma, 1)$-pseudoquadratic form $\phi \colon V \to \mathbb{K}/\mathbb{K}_{\sigma,1}$ of finite Witt index $n \geq 2$, where $\mathbb{K}$ is a (possibly non-commutative) division ring, V is a $\mathbb{K}$-vector space, σ is an involutory anti-automorphism of $\mathbb{K}$ and $\mathbb{K}_{\sigma,1} := \{t - t^\sigma\}_{t \in \mathbb{K}}$. So, $\mathscr{S}$ is regarded as a subgeometry of $\Sigma := \mathrm{PG}(V)$ and the inclusion mapping of $\mathscr{S}$ into Σ provides the universal embedding of $\mathscr{S}$ except when $\dim(V) = 4$, $|\mathbb{K}| > 3$ and $\mathscr{S}$ is a hyperbolic quadric or a quadrangle as in [19, 8.6(II)(a)].

We shall show how to construct a totally scattered partial ovoid of $\mathscr{S}$, namely a partial ovoid O such that $\langle O \rangle_\Sigma = \Sigma$. We will only describe the constructions. The verifications that they indeed hit the target will be left to the reader.

We put $N := \dim(V) \geq 2n$. For ease of exposition, we assume that $N < \infty$, but everything we are going to say also holds when N is infinite.

8.3.2.1 The Orthogonal Case with $N = 2n$

Assume that $\sigma = \mathrm{id}_\mathbb{K}$ (hence $\mathbb{K}$ is commutative) and $N = 2n$. Then ϕ can be expressed as follows with respect to a suitable basis $(\mathbf{u}_1, \mathbf{v}_1, \ldots, \mathbf{u}_n, \mathbf{v}_n)$ of V:

$$\phi(x_1, y_1, \ldots, x_n, y_n) = \sum_{i=1}^{n} x_i y_i.$$

Suppose that char($\mathbb{K}$) $\neq$ 2. Let $L = (\lambda_{i,j})_{i,j=2}^{n}$ be an $(n-1) \times (n-1)$ anti-symmetric matrix. For every $k = 2, 3, \ldots, n$, put $I_k = \{2, 3, \ldots, n\} \setminus \{k\}$ and

$$\mathbf{x}_k(t) = -\mathbf{u}_k t^2 + \sum_{i \in I_k} \mathbf{u}_i \lambda_{i,k} + \mathbf{v}_k + (\mathbf{u}_1 + \mathbf{v}_1)t.$$

Put $X_k := \{\mathbf{x}_k(t)\}_{t \in \mathbb{K}^*}$. The set $O := \{\mathbf{u}_1\} \cup (\cup_{k=2}^{n} X_k)$ is a partial ovoid. Moreover, if $|\mathbb{K}| > 3$, then $\langle O \rangle_\Sigma$ contains $[\mathbf{u}_1], \ldots, [\mathbf{u}_n], [\mathbf{v}_1], \ldots, [\mathbf{v}_n]$. Hence $\langle O \rangle_\Sigma = \Sigma$.

The same construction applies when char($\mathbb{K}$) $= 2$ but now choosing $L = (\lambda_{i,j})_{i,j=2}^{n}$ in such a way that $\lambda_{i,j} + \lambda_{j,i} \neq 0$ for any choice of $i \neq j$ instead of $\lambda_i + \lambda_j = 0$.

Proposition 3.6 *If $|\mathbb{K}| > 3$, then the partial ovoid O is totally scattered.*

Remark 11   The partial ovoid O is not maximal. Indeed $O \cup \{\mathbf{v}_1\}$ is still a partial ovoid. However, if char($\mathbb{K}$) $\neq 2$ and $\mathbb{K}$ is quadratically closed, then $O \cup \{\mathbf{v}_1\}$ is maximal, whence it is a maximal subspace.

8.3.2.2 The Orthogonal Case with $N > 2n$ and char($\mathbb{K}$) $\neq 2$

Still with $\sigma = \mathrm{id}_\mathbb{K}$, suppose that $N > 2n$ and char($\mathbb{K}$) $\neq 2$. Put $m = N - 2n$. A basis $(\mathbf{u}_1, \mathbf{v}_1, \ldots, \mathbf{u}_n, \mathbf{v}_n, \mathbf{w}_1, \ldots, \mathbf{w}_m)$ of V exists such that ϕ admits the following expression with respect to it:

$$\phi(x_1, y_1, \ldots, x_n, y_n, z_1, \ldots, z_m) = \sum_{i=1}^{n} x_i y_i + \sum_{j=1}^{m} \mu_j z_j^2$$

and ϕ is anisotropic on $\langle \mathbf{w}_1, \ldots, \mathbf{w}_m \rangle$, namely $\sum_{j=1}^{m} \mu_j z_j^2 = 0$ only if $z_1 = z_2 = \ldots = z_m = 0$. We choose an anti-symmetric matrix $L = (\lambda_{i,j})_{i,j=1}^{n}$ and a non-singular matrix $A = (\alpha_{i,j})_{i,j=1}^{m}$ such that

$$\sum_{j=1}^{m} \mu_j \alpha_{j,r} \alpha_{j,s} \neq 0 \quad \text{for } 1 \leq r < s \leq m. \tag{5}$$

Example 3.7   Many non-singular matrices exist which satisfy condition (5). For instance, put $\alpha_{1,j} = 1$ for every j and $\alpha_{i,j} = \delta_{i,j}$ (Kronecker symbol) when $i > 1$. This is perhaps the easiest choice, but here is another one: choose m pairwise different square elements $\alpha_1, \ldots, \alpha_m \in \mathbb{K}^*$ and put $\alpha_{i,j} = \alpha_i^{j-1}$.

For $k = 1, 2, \ldots, n$ and $r = 1, 2, \ldots, m$ we put

$$\mathbf{x}_{k,r}(t) := -\mathbf{u}_k \cdot (\sum_{j=1}^{m} \mu_j \alpha_{j,r}^2) t^2 + \sum_{i \neq k} \mathbf{u}_i \lambda_{i,k} + \mathbf{v}_k + \sum_{j=1}^{m} \mathbf{w}_j \alpha_{j,r} t, \quad (t \in \mathbb{K}),$$

$X_{k,r} := \{[\mathbf{x}_{k,r}(t)]\}_{t \in \mathbb{K}^*}$ and $O := \cup_{k=1}^{n} \cup_{r=1}^{m} X_{k,r}$. The set O is a partial ovoid (but not a maximal one).

Proposition 3.8 *If $|\mathbb{K}| > 3$, then the partial ovoid O is totally scattered. If $\mathbb{K} = \mathbb{F}_3$ but $m > 1$, then the matrix A can be chosen in such a way that O is totally scattered.*

Remark 12 The hypothesis that $\mathbb{K} \neq \mathbb{F}_3$ when $m = 1$ cannot be removed from Lemma 3.8. Indeed, when $m = 1$ and $\mathbb{K} = \mathbb{F}_3$ our construction yields a partial ovoid of size $n + 1$, too small to span Σ.

8.3.2.3 The Orthogonal Case with $N > 2n$ and char($\mathbb{K}$) = 2
Still assuming that $\sigma = \mathrm{id}_{\mathbb{K}}$ and $N > 2n$, let now char($\mathbb{K}$) = 2. Then V admits a basis $(\mathbf{u}_1, \mathbf{u}_2, \ldots, \mathbf{u}_{2n-1}, \mathbf{u}_{2n}, \mathbf{v}_1, \mathbf{v}_2, \ldots, \mathbf{v}_{2\ell-1}, \mathbf{v}_{2\ell}, \mathbf{w}_1, \mathbf{w}_2, \ldots, \mathbf{w}_m)$, with $2\ell + m = N - 2n$, such that ϕ is expressed as follows with respect to it:

$$\phi(x_1, \ldots, x_{2n}, y_1, \ldots, y_{2\ell}, z_1, \ldots, z_m) = \sum_{i=1}^{n} x_{2i-1} x_{2i}$$
$$+ \psi(y_1, \ldots, y_{2\ell}, z_1, \ldots, z_m)$$

where $\psi(y_1, \ldots, y_{2\ell}, z_1, \ldots, z_m) = \sum_{i=1}^{\ell} (\kappa_i y_{2i-1}^2 + y_{2i-1} y_{2i} + y_{2i}^2 \chi_i) + \sum_{j=1}^{m} z_j^2 \mu_j$ for suitable scalars $\kappa_1, \chi_1, \ldots, \kappa_\ell, \chi_\ell, \mu_1, \ldots, \mu_m \in \mathbb{K}^*$ such that

$$\psi(y_1, \ldots, y_{2\ell}, z_1, \ldots, z_m) = 0 \text{ only if } y_1 = \ldots = y_{2\ell} = z_1 = \ldots = z_m = 0.$$

By assumption, at least one of ℓ or m is positive. To fix ideas, assume that both ℓ and m are positive. Let $L = (\lambda_{i,j})_{i,j=1}^{n}$ and $N = (\nu_{i,j})_{i,j=1}^{n}$ be $n \times n$ matrices such that $\lambda_{i,j} + \lambda_{j,i} \neq 0 \neq \nu_{i,j} + \nu_{j,i}$ and $\nu_{i,j} \neq \lambda_{i,j}, \lambda_{j,i}$ for any choice of $i \neq j$ and let $A = (\alpha_{i,j})_{i,j=1}^{\ell}$, $B = (\beta_{i,j})_{i,j=1}^{\ell}$ and $C = (\gamma_{i,j})_{i,j=1}^{m}$ be invertible matrices with A and B satisfying the following conditions:

$$\left.\begin{array}{l} \sum_{j=1}^{\ell} \kappa_j \alpha_{j,r} \alpha_{j,s} \neq 0 \quad \text{for } 1 \leq r < s \leq \ell, \\ \sum_{j=1}^{\ell} \chi_j \beta_{j,r} \beta_{j,s} \neq 0 \quad \text{for } 1 \leq r < s \leq \ell, \\ \sum_{j=1}^{\ell} \alpha_{j,r} \beta_{j,s} \neq 0 \quad \text{for any choice of } r, s \in \{1, 2, \ldots, \ell\}. \end{array}\right\} \quad (6)$$

We know from Example 3.7 that several ways exist to choose matrices A and B in such a way that the first two conditions of (6) are satisfied. Satisfying the third condition is not so difficult. For instance, the 'easiest' choice of Example 3.7 for A and a slight modification of that choice for B do the job.

With A, B and C as here, for every $k = 1, 2, \ldots, n$, every $r = 1, 2, \ldots, \ell$ and every $s = 1, 2, \ldots, m$ we put:

$$\mathbf{x}_{k,r}(t) := \mathbf{u}_{2k-1} \cdot \sum_{j=1}^{\ell} \kappa_j \alpha_{j,r}^2 t^2 + \sum_{i \neq k} \mathbf{u}_{2i-1} \lambda_{i,k} + \mathbf{u}_{2k} + \sum_{j=1}^{\ell} \mathbf{v}_{2j-1} \alpha_{j,r} t,$$

$$\mathbf{y}_{k,r}(t) := \mathbf{u}_{2k-1} \cdot \sum_{j=1}^{\ell} \chi_j \beta_{j,r}^2 t^2 + \sum_{i \neq k} \mathbf{u}_{2i-1} \lambda_{k,i} + \mathbf{u}_{2k} + \sum_{j=1}^{\ell} \mathbf{v}_{2j} \beta_{j,r} t,$$

$$\mathbf{z}_{k,s}(t) := \mathbf{u}_{2k-1} \cdot \sum_{j=1}^{m} \mu_j \gamma_{j,s}^2 t^2 + \sum_{i \neq k} \mathbf{u}_{2i-1} \nu_{i,k} + \mathbf{u}_{2k} + \sum_{j=1}^{m} \mathbf{w}_j \gamma_{j,s} t.$$

Put $X_{k,r} := \{\mathbf{x}_{k,r}(t)\}_{t \neq 0}$, $Y_{k,r} := \{\mathbf{y}_{k,r}(t)\}_{t \neq 0}$, $Z_{k,s} := \{\mathbf{z}_{k,s}(t)\}_{t \neq 0}$ and

$$O := \bigcup_{k,r,s=1}^{n,\ell,m} (X_{k,r} \cup Y_{k,r} \cup Z_{k,s}).$$

Then O is a partial ovoid and spans Σ. So far we have assumed that $\ell, m > 0$ (hence $\mathbb{K}$ is infinite). If $\ell = 0$, then we define only $\mathbf{z}_{k,s}(t)$ and put $O := \cup_{k,s} Z_{k,s}$. Similarly, when $m = 0$, then $O := \cup_{k,r}(X_{k,r} \cup Y_{k,r})$. Again, O is a partial ovoid; moreover, it spans Σ except when $(\ell, m) \in \{(1,0), (0,1)\}$ and $\mathbb{K} = \mathbb{F}_2$. In the end, the following holds.

Proposition 3.9 *If $\mathbb{K} \neq \mathbb{F}_2$ then the partial ovoid O is totally scattered.*

Remark 13 With $R := \langle \mathbf{w}_1 \ldots, \mathbf{w}_m \rangle$, let π_R be the canonical projection of $\Sigma = \mathrm{PG}(V)$ onto $\mathrm{PG}(V/R)$. Then π_R provides an embedding of $\mathscr{S}$ in $\mathrm{PG}(V/R)$. Suppose that $\ell = 0$ and $\phi(R) = \mathbb{K}$. Then $\pi_R(\mathscr{S})$ is symplectic and we are driven back to Section 8.3.1. With O as here, let O_R be the partial ovoid of $\pi_R(\mathscr{S})$ as constructed in Section 8.3.1 and let $\widetilde{O}_R = \pi_R^{-1}(O_R) \cap \mathscr{S}$ be its lifting to Σ. Then $O \neq \widetilde{O}_R$. Indeed $\widetilde{O}_R$ contains exactly one pointed conic for every $k = 1, 2, \ldots, n$ while O contains m bi-pointed conics for every k.

8.3.2.4 The Hermitian Case with $N = 2n$

Let $\sigma \neq \mathrm{id}_{\mathbb{K}}$ and $N = 2n$. A basis $(\mathbf{u}_1, \mathbf{v}_1, \ldots, \mathbf{u}_n, \mathbf{v}_n)$ of V exists such that ϕ admits the following expression with respect to it:

$$\phi(x_1, y_1, \ldots, x_n, y_n) = \sum_{i=1}^{n} x_i^{\sigma} y_i + \mathbb{K}/\mathbb{K}_{\sigma,1}.$$

Given a matrix $L = (\lambda_{i,j})_{i,j=1}^{n}$ such that $\lambda_{i,j} + \lambda_{j,i}^{\sigma} \neq 0$ for any $i \neq j$, we put $\mathbf{x}_k(t) := \mathbf{u}_k t + \sum_{i \neq k} \mathbf{u}_i \lambda_{i,k} + \mathbf{v}_k$ for $k = 1, 2, \ldots, n$ and $t \in \mathbb{K}_{\sigma,1}$, $X_k := \{[\mathbf{x}_k(t)]\}_{t \in \mathbb{K}_{\sigma,1} \setminus \{0\}}$ and $O := \cup_{k=1}^{n} X_k$.

Proposition 3.10 *The set O is a totally scattered partial ovoid.*

8.3.2.5 The Hermitian Case with $N > 2n$

Still with $\sigma \neq \mathrm{id}_{\mathbb{K}}$, let now $N > 2n$. Put $\mathbb{K}^{\sigma,1} := \{t \in \mathbb{K} \mid t + t^\sigma = 0\}$. Note that $\mathbb{K}_{\sigma,1} \subseteq \mathbb{K}^{\sigma,1}$. The vector space V admits a basis

$$(\mathbf{u}_1, \mathbf{u}_2, \ldots, \mathbf{u}_{2n-1}, \mathbf{u}_{2n}, \mathbf{v}_1, \mathbf{v}_2, \ldots, \mathbf{v}_{2\ell-1}, \mathbf{v}_{2\ell}, \mathbf{e}_1, \ldots, \mathbf{e}_{m_0}, \mathbf{f}_1, \ldots, \mathbf{f}_{m_1})$$

(possibly $\ell = 0$, $m_0 = 0$ or $m_1 = 0$) such that ϕ is expressed as follows with respect to it, with values taken modulo $\mathbb{K}_{\sigma,1}$:

$$\phi(x_1, \ldots, x_{2n}, y_1, \ldots, y_{2\ell}, t_1, \ldots, t_{m_0}, s_1, \ldots, s_{m_1}) =$$
$$= \sum_{i=1}^{n} x_{2i-1}^\sigma x_{2i} + \psi(y_1, \ldots, y_{2\ell}, t_1, \ldots, t_{m_0}, s_1, \ldots, s_{m_1})$$

where $\psi(y_1, \ldots, y_{2\ell}, t_1, \ldots, t_{m_0}, s_1, \ldots, s_{m_1})$ stands for the following:

$$\sum_{i=1}^{\ell} (y_{2i-1}^\sigma \kappa_i y_{2i-1} + y_{2i-1}^\sigma y_{2i} + y_{2i}^\sigma \chi_i y_{2i}) + \sum_{j=1}^{m_0} t_j^\sigma \mu_j t_j + \sum_{j=1}^{m_1} s^\sigma v_j s_j,$$

for suitable scalars $\kappa_1, \chi_1, \ldots, \kappa_\ell, \chi_\ell, \mu_1, \ldots, \mu_{m_0} \in \mathbb{K}^{\sigma,1} \setminus \mathbb{K}^{\sigma,1}$ and $v_1, \ldots, v_{m_1} \in \mathbb{K} \setminus \mathbb{K}^{\sigma,1}$ such that

$$\left. \begin{array}{l} \phi(y_1, \ldots, y_{2\ell}, t_1, \ldots, t_{m_0}, s_1, \ldots, s_{m_1}) \in \mathbb{K}_{\sigma,1} \text{ only if} \\ y_1 = y_2 = \ldots = y_{2\ell} = t_1 = \ldots = t_{m_0} = s_1 = \ldots = s_{m_1} = 0. \end{array} \right\}$$

Clearly, if $\mathbb{K}_{\sigma,1} = \mathbb{K}^{\sigma,1}$ (as it is the case when either $\mathrm{char}(\mathbb{K}) \neq 2$ or σ acts non-trivially on the center of $\mathbb{K}$), then $\ell = m_0 = 0$. In this case necessarily $m_1 > 0$, since $2\ell + m_0 + m_1 = N - 2n$ and $N > 2n$ by assumption.

To fix ideas, suppose that each of ℓ, m_0 and m_1 is positive. Choose matrices $L = (\lambda_{i,j})_{i,j=1}^n$, $M = (\mu_{i,j})_{i,j=1}^n$ and $N = (v_{i,j})_{i,j=1}^n$ such that $\lambda_{i,j} + \lambda_{j,i}^\sigma \neq 0$, $\mu_{i,j} + \mu_{j,i}^\sigma \neq 0$, $v_{i,j} + v_{j,i}^\sigma = 0$, $\mu_{i,j} \neq \lambda_{i,j} \lambda_{j,i}^\sigma$, $v_{i,j} \neq \lambda_{i,j} \lambda_{j,i}^\sigma$ and $v_{i,j} \neq \mu_{j,i}^\sigma$ for any choice of $i \neq j$. Moreover $A = (\alpha_{i,j})_{i,j=1}^\ell$, $B = (\beta_{i,j})_{i,j=1}^\ell$, $C = (\gamma_{i,j})_{i,i=1}^{m_0}$ and $D = (\delta_{i,j})_{i,j=1}^{m_1}$ are invertible matrices with A, B and D satisfying the following:

$$\begin{array}{ll} \sum_{j=1}^{\ell} \alpha_{j,r}^\sigma \kappa_j \alpha_{j,s} \neq 0 & \text{for } 1 \leq r < s \leq \ell, \\ \sum_{j=1}^{\ell} \beta_{j,r}^\sigma \chi_j \beta_{j,s} \neq 0 & \text{for } 1 \leq r < s \leq \ell, \\ \sum_{j=1}^{\ell} \alpha_{j,r}^\sigma \beta_{j,s} \neq 0 & \text{for } r, s = 1, \ldots, \ell. \\ \sum_{j=1}^{m_1} \delta_{j,r}^\sigma (v_j + v_j^\sigma) \delta_{j,s} \neq 0 & \text{for } 1 \leq r < s \leq m_1. \end{array}$$

For every $k = 1, 2, \ldots, n$, every $r = 1, 2, \ldots, \ell$ every $r_0 = 1, 2, \ldots, m_0$, every $r_1 = 1, 2, \ldots, m_1$ and any $s, t \in \mathbb{K}$ put:

$$\begin{aligned}
\mathbf{a}_{k,r}(s,t) &:= \mathbf{u}_{2k-1}s + \sum_{i\neq k}\mathbf{u}_{2i-1}\lambda_{i,k} + \mathbf{u}_{2k} + \sum_{j=1}^{\ell}\mathbf{v}_{2j-1}\alpha_{j,r}t, \\
\mathbf{b}_{k,r}(s,t) &:= \mathbf{u}_{2k-1}s + \sum_{i\neq k}\mathbf{u}_{2i-1}\lambda_{k,i} + \mathbf{u}_{2k} + \sum_{j=1}^{\ell}\mathbf{v}_{2j}\beta_{j,r}t, \\
\mathbf{c}_{k,r_0}(s,t) &:= \mathbf{u}_{2k-1}s + \sum_{i\neq k}\mathbf{u}_{2i-1}\mu_{i,k} + \mathbf{u}_{2k} + \sum_{j=1}^{m_0}\mathbf{e}_j\gamma_{j,r_0}t, \\
\mathbf{d}_{k,r_1}(s,t) &:= \mathbf{u}_{2k-1}s + \sum_{i\neq k}\mathbf{u}_{2i-1}\nu_{i,k} + \mathbf{u}_{2k} + \sum_{j=1}^{m_1}\mathbf{f}_j\delta_{j,r_1}t.
\end{aligned}$$

Next put

$$\begin{aligned}
A_{k,r} &:= \{\mathbf{a}_{k,r}(s,t) \mid s + \sum_{j=1}^{\ell}t^\sigma\alpha_{j,r}^\sigma\kappa_j\alpha_{j,r}t \in \mathbb{K}_{\sigma,1},\ t\neq 0\}, \\
B_{k,r} &:= \{\mathbf{b}_{k,r}(s,t) \mid s + \sum_{j=1}^{\ell}t^\sigma\beta_{j,r}^\sigma\chi_j\beta_{j,r}t \in \mathbb{K}_{\sigma,1},\ t\neq 0\}, \\
C_{k,r_0} &:= \{\mathbf{c}_{k,r_0}(s,t) \mid s + \sum_{j=1}^{m_0}t^\sigma\gamma_{j,r_0}^\sigma\mu_j\gamma_{j,r_0}t \in \mathbb{K}_{\sigma,1},\ t\neq 0\}, \\
D_{k,r_1} &:= \{\mathbf{d}_{k,r_1}(s,t) \mid s + \sum_{j=1}^{m_1}t^\sigma\delta_{j,r_1}^\sigma\nu_j\delta_{j,r_1}t \in \mathbb{K}_{\sigma,1},\ t\neq 0\}.
\end{aligned}$$

Finally, $O := \bigcup_{k,r,r_0,r_1=1}^{n,\ell,m_0,m_1}(A_{k,r}\cup B_{k,r}\cup C_{k,r_0}\cup D_{k,r_1})$.

We have assumed that neither ℓ nor m_0 is 0. Of course, when one of them is 0 we must accordingly modify the previous definition. For instance, when $\ell = 0 < m_0, m_1$ we form O as the union of the sets C_{k,r_0} and D_{k,r_1}. In this case, we omit to introduce the matrices L, A and B; we only need M, N, C and D. Similarly, if $m_0 = 0 < \ell, m_1$, then O is formed only by the sets $A_{k,r}$, $B_{k,r}$ and D_{k,r_1}. In this case, M and C are omitted. If $\ell = m_1 = 0$, then $O = \cup_{k,r_0}C_{k,r_1}$ and if $m_0 = m_1 = 0$, then $O = \cup_{k,r}(A_{k,r}\cup B_{k,r})$. Finally, if $\ell = m_0 = 0$, then only N and D are needed and $O = \cup_{k,r_1}D_{k,r_1}$. With O defined in this way,

Proposition 3.11 *The set O is a totally scattered partial ovoid.*

8.4 Ovoids and Maximal Subspaces

Throughout this section, but for the very last remark, $\mathscr{S}$ is a finite (non-degenerate, thick-lined) polar space of rank $n > 2$. Let q be the order of the underlying field of $\mathscr{S}$ and $t + 1$ the number of generators of $\mathscr{S}$ which contain a given singular subspace of rank $n - 1$. Recall that $t \in \{1, q, q^2, q^{1/2}, q^{3/2}\}$. The parameters q and t are the *orders* of $\mathscr{S}$.

We recall that $W_{2n-1}(q) \cong Q_{2n}(q)$ when q is even. Accordingly, the properties of $W_{2n-1}(q)$ with q even are not the same as when q is odd. This fact causes slight complications in the exposition, which we prefer to avoid. So, henceforth, when referring to $W_{2n-1}(q)$ we implicitly assume that q is odd.

We have defined ovoids in the Introduction of this chapter. We are not going to repeat that definition here. Instead we recall that, for a singular subspace X of $\mathscr{S}$ of rank $m < n$, the *star* $\mathscr{S}_X$ of X is the polar space of rank $n - m$ formed by the singular subspaces of $\mathscr{S}$ which properly contain X, those of rank $m + 1$ being taken as points of $\mathscr{S}_X$ and those of rank $m + 2$ as lines (if $m \leq n - 2$). Obviously, the generators of $\mathscr{S}_X$ are the generators of $\mathscr{S}$ which contain X.

8.4.1 Basics on Ovoids and Non-existence Results

Let q and t be the orders of $\mathscr{S}$. Since a partial ovoid meets every generator in at most one point, every partial ovoid O of $\mathscr{S}$ has size at most $q^{n-1}t + 1$ and it is an ovoid if and only if $|O| = q^{n-1}t + 1$.

Let O be an ovoid of $\mathscr{S}$ and let X be a non-maximal singular subspace of $\mathscr{S}$, disjoint from O. Let $O_X := \{\langle X, x \rangle \mid x \in O \cap X^\perp\}$. Then O_X is an ovoid of $\mathscr{S}_X$. In particular, if x is a point of $\mathscr{S}$ exterior to O, then O_x is the set of lines which join x to points of O and it is an ovoid of the star $\mathscr{S}_x$ of x. Accordingly, $|O_x| = |x^\perp \cap O| = q^{n-2}t + 1$.

The fact that O_X is an ovoid of $\mathscr{S}_X$ can be exploited to prove the non-existence of ovoids in certain polar spaces. Indeed, if we already know that $\mathscr{S}_X$ admits no ovoid, then we can conclude that no ovoids exist in $\mathscr{S}$. In this way, by reduction to the rank 2 case, we immediately see that $W_{2n-1}(q)$, $Q_{2n+1}^-(q)$ and $H_{2n}(q)$ admit no ovoid. Indeed it is well known that $W_3(q)$, $Q_5^-(q)$ and $H_4(q)$ admit no ovoids (see e.g. Payne and Thas [16]). Moreover, no ovoid exists in $Q_8(q)$, for any prime power q (Gunarwardena and Moorhouse [10]). Therefore $Q_{2n}(q)$ has no ovoid, for any $n \geq 4$.

As for $Q_6(q)$, the following is known. If q is even, then $Q_6(q)$ has no ovoids (Thas [18]) and no ovoids exist in $Q_6(p)$, for p prime, $p > 3$ (O'Keefe and Thas [14]). On the other hand, $Q_6(3^h)$ admits ovoids, for any positive integer h (Kantor [11]).

Not so much is known about $H_{2n-1}(q)$ for $n > 2$. It is known that $H_5(4)$ has no ovoid (De Beule and Metsch [6]), hence no ovoids exist in $H_{2n-1}(4)$ for any $n \geq 3$. It is also known that no ovoid exists in $H_{2n-1}(q)$ when the prime basis p of q satisfies the following inequality (Moorhouse [12]):

$$p^{2n-1} > \binom{2n+p-3}{p-1}^2 + 2 \cdot \binom{2n+p-3}{p-1} \cdot \binom{2n+p-3}{p-2}. \qquad (7)$$

Note that, for a given p, the second term of (7), say $f_p(n)$, is a polynomial of degree $2(p-1)$ in the unknown n. Therefore the ratio $p^{2n-1}/f_p(n)$ diverges as n diverges. So, (7) says that $H_{2n-1}(q)$ admits ovoids only if n is not too large compared to p. In other words, for every given prime p, if $n > 2$ then $H_{2n-1}(p^{2h})$ admits no ovoids except possibly for a finite number of choices of n. Inequality (7) embodies an upper bound for the number of those lucky choices, which only depends on p. We are not aware of any further existence or non-existence result for ovoids of $H_{2n-1}(q)$ when $n > 2$.

A few existence results are known for ovoids of $Q_{2n-1}^+(q)$, $n > 2$. For instance, $Q_5^+(q)$ admits ovoids for any q and $Q_7^+(q)$ admits ovoids for q a power of 2 or 3, for q an odd power of a prime $p \equiv 2 \pmod 3$ and for q prime (we refer to table 1 of [7] for this information). On the other hand, no ovoids exist in $Q^+_{2n-1}(q)$ if the prime basis p of q satisfies the following inequality (Blokhuis and Moorhouse [1]):

$$ p^{n-1} > \binom{2n+p-4}{p-1} + 2 \cdot \binom{2n+p-4}{p-2}. \tag{8} $$

Likewise (7), inequality (8) says that $Q^+_{2n-1}(p^h)$ admits ovoids only if n is not too large compared to p.

8.4.2 Proof of Theorem 3

Let $\mathscr{S}$ be finite with orders $q > 1$ and t and rank $n > 2$. Suppose that $\mathscr{S}$ is neither $Q^+_5(q)$ (for any q) nor $Q^+_7(q)$ (with q odd). Let O be an ovoid of $\mathscr{S}$. Given a point $x \notin O$, consider the subspace $\mathscr{S}(O, x) = \langle O \cup \{x\} \rangle$ of $\mathscr{S}$ generated by $O \cup \{x\}$. If $\mathscr{S}(O, x) = \mathscr{S}$ for every $x \notin O$, then O is a maximal subspace of $\mathscr{S}$.

By way of contradiction, suppose there exists a point $a \notin O$ such that $\mathscr{S}(O, a) \subset \mathscr{S}$. The subspace $\mathscr{S}(O, a)$ has rank at least 2, since it contains at least the $q^{n-2}t + 1$ lines which join a to points of O. The polar space $\mathscr{S}$ is embeddable and admits the universal embedding (indeed it is finite of rank $n > 2$). Hence the subspace $\mathscr{S}(O, a)$ arises from the universal embedding $\tilde{e}: \mathscr{S} \to \mathrm{PG}(\tilde{V})$ of $\mathscr{S}$ (Theorem 2.2), namely $\mathscr{S}(O, a) = \tilde{e}^{-1}(\langle \tilde{e}(O \cup \{a\}) \rangle)$. However $\mathscr{S}(O, a) \subset \mathscr{S}$ by assumption. Hence $\langle \tilde{e}(O \cup \{a\}) \rangle$ is contained in a hyperplane X of $\mathrm{PG}(\tilde{V})$. Accordingly, $O \cup \{a\}$ is contained in the hyperplane $H := \tilde{e}^{-1}(X)$ of $\mathscr{S}$.

If $H = b^\perp$ for some point b of $\mathscr{S}$, then b is joined with all points of $|O|$. However $|O| = q^{n-1}t + 1$ while $|b^\perp \cap O|$ is either 1 or $q^{n-2}t + 1$ according to whether $b \in O$ or $b \notin O$. In any case, we get a contradiction. Therefore H is non-singular. So, H is a non-degenerate polar space of rank $m \in \{n - 1, n\}$ and order (q, t'), where t' depends on m and the type of $\mathscr{S}$. However O, being contained in H, is also an ovoid of H. Consequently $q^{n-1}t + 1 = q^{m-1}t' + 1$. If $m = n$ then $t' = t$. This is impossible, as H cannot have the same rank and the same orders as $\mathscr{S}$. Therefore $m = n - 1$ and $t' = qt$.

On the other hand, if $t' > q$ then H is necessarily isomorphic to either $Q^-_{2m+1}(q)$ or $H_{2m}(q)$. However, as remarked in Section 8.4.1, neither $Q^-_{2m+1}(q)$ nor $H_{2m}(q)$ admit ovoids, while O is an ovoid of H. Therefore $t' \le q$ and the equality $t' = qt$ now forces $t = q$ and $t = 1$, namely $\mathscr{S} \cong Q^+_{2n-1}(q)$ and $H \cong Q_{2m}(q) = Q_{2n-2}(q)$.

As noticed in Section 8.4.1, the quadric $Q_{2m}(q)$ admits no ovoids when $m > 3$ and $Q_6(q)$ admits no ovoids if q is even. However O is an ovoid of

H. Therefore $m \leq 3$ and $m = 2$ if q is even, namely $n \leq 4$ and $n = 3$ if q is even. However, according to the hypotheses of the theorem, $t > 1$ when either $n = 3$ or $n = 4$ and q is even. We have reached a final contradiction. Consequently, $\mathscr{S}(O, x) = \mathscr{S}$ for every point $x \notin O$. □

Remark 14   The quadrics $Q_5^+(q)$ and $Q_7^+(3^h)$ are counterexamples to the conclusion of Theorem 3. Indeed $Q_5^+(q)$ contains $Q_3^-(q)$, which is indeed an ovoid of $O_5^+(q)$. However $Q_3^-(q)$ is not a maximal subspace of $Q_5^+(q)$, as it is contained in a hyperplane $Q_4(q)$ of $Q_5^+(q)$. Similarly, $Q_6(3^h)$ is a hyperplane of $Q_7^+(3^h)$ and contains ovoids (Section 8.4.1). The ovoids of $Q_6(3^h)$ are still ovoids in $Q_7^+(3^h)$, but they cannot be maximal subspaces of $Q_7^+(3^h)$.

Remark 15   The statement of Theorem 3 can be made slightly sharper by allowing $\mathscr{S} = Q_7^+(p)$ with p prime and different from 3. Indeed $Q_6(p)$ admits no ovoids for a prime $p \neq 3$.

Remark 16   Remark 14 suggests the following counterexample in the infinite case. Let $\mathscr{S} = Q_{2n+1}^+(\mathbb{K})$, with $\mathbb{K}$ an infinite field and $n \geq 2$. Then $\mathscr{S}$ contains hyperplanes isomorphic to $Q_{2n}(\mathbb{K})$. If $\mathscr{S}'$ is one of them, every generator of $\mathscr{S}'$ is contained in exactly two generators of $\mathscr{S}$ and every generator of $\mathscr{S}$ contains exactly one generator of $\mathscr{S}'$. The polar space $\mathscr{S}'$ admits ovoids, by Cameron [3] (in fact, it admits even a partition in ovoids). Let O be an ovoid of $\mathscr{S}'$. Since every generator of $\mathscr{S}$ contains exactly one generator of $\mathscr{S}'$, every generator of $\mathscr{S}$ meets O in exactly one point. Accordingly, O is also an ovoid of $\mathscr{S}$. However O is not a maximal subspace of $\mathscr{S}$, since it is contained in the hyperplane $\mathscr{S}'$ of $\mathscr{S}$. This is a special case of the following more general observation, which is the best we can do for the infinite case.

Proposition 4.1   *Let $\mathscr{S}$ be an infinite polar space of rank $r \geq 3$. If $\mathscr{S}$ is the line-grassmannian of $\mathrm{PG}(3, \mathbb{K})$, with $\mathbb{K}$ non-commutative, then every ovoid of $\mathscr{S}$ is a maximal subspace. If $\mathscr{S}$ is embeddable, and its universal embedding is finite-dimensional, then $\mathscr{S}$ possesses ovoids that are not maximal subspaces if and only if $\mathscr{S}$ admits non-singular hyperplanes of rank $r - 1$.*

Proof   We first note that no ovoid is contained in a singular hyperplane. Indeed, suppose for a contradiction that the ovoid O is contained in $p^\perp$ for some point p of $\mathscr{S}$. Then we can find a submaximal singular subspace W in $p^\perp$ not containing p and disjoint from O (since every projective space has a hyperplane avoiding two given points). Every generator containing W and distinct from $\langle W, p \rangle$ is disjoint from O, a contradiction.

   Suppose first that $\mathscr{S}$ is the line-grassmannian of $\mathrm{PG}(3, \mathbb{K})$, with $\mathbb{K}$ non-commutative, and suppose, for a contradiction, that an ovoid O is contained in a proper subspace X of $\mathscr{S}$. By Lemma 2.14, X is contained in the perp of a point, contradicting the previous paragraph.

Now suppose that $\mathscr{S}$ is embeddable, and its universal embedding is finite-dimensional, say in the projective space Σ. First suppose that $\mathscr{S}$ admits a non-singular hyperplane H of rank $r - 1$. By [3], H admits an ovoid O. Since H is a hyperplane, every generator of $\mathscr{S}$ intersects H in a submaximal singular subspace, that is, a generator of H, which by definition contains a member of O. Hence O is an ovoid of $\mathscr{S}$, which is not a maximal subspace as $O \subset H \subset \mathscr{S}$.

Now suppose that $\mathscr{S}$ possesses an ovoid O that is not a maximal subspace, and let $O \subset \mathscr{S}' \subset \mathscr{S}$, with $\mathscr{S}'$ a subspace. Then $\mathscr{S}'$ is a non-degenerate polar space by the first paragraph of this proof. Let S be the subspace of Σ corresponding to $\mathscr{S}'$. Suppose $\mathscr{S}'$ has rank r'. We claim that $r' < r$. Indeed, suppose for a contradiction that $r' = r$. There exists a submaximal singular subspace M' of $\mathscr{S}'$ disjoint from O. Since $\mathscr{S}'$ is not an ideal subspace (as otherwise $\mathscr{S}'$ would coincide with $\mathscr{S}$), there is a generator M not contained in $\mathscr{S}'$ but containing M'. Since also O is contained in $\mathscr{S}'$, $M \cap O = \emptyset$, a contradiction. The claim is proved. Now we can extend $\mathscr{S}'$ to a hyperplane of Σ and the proposition follows. $\square$

Note that the sharpened version of Lemma 2.16 referred to in Remark 6 implies that also each ovoid of the Freudenthal–Tits polar space is a maximal subspace, providing alternative evidence of the existence of ovoidal maximal subspaces in these polar spaces, using [3] once again.

Proposition 4.1 has many applications. We content ourselves by mentioning that each ovoid of every non-degenerate quadric in an even-dimensional projective space over an algebraically closed field is a maximal subspace, and that every quadric over $\mathbb{R}$ (and likewise every hermitian polar space over $\mathbb{C}$) admits ovoids which are not maximal as subspaces.

References

[1] A. Blokhuis and G. E. Moorhouse. Some p-ranks related to orthogonal spaces, *J. Algebraic Combin.* **4** (1995), 295–316.

[2] F. Buekenhout and A. M. Cohen. *Diagram Geometry*, Springer, Berlin 2013.

[3] P. J. Cameron, Ovoids in infinite incidence structures, *Arch. Math.* **62** (1994), 189–192.

[4] I. Cardinali, L. Giuzzi and A. Pasini. Nearly all subspaces of a classical polar space arise from its universal embedding. *Linear Algebra Appl.* **627** (2021), 287–307.

[5] A. M. Cohen and E. E. Shult. Affine polar spaces, *Geom. Dedicata* **35** (1990), 43–76.

[6] J. De Beule and K. Metsch, The hermitian variety $H(5,4)$ has no ovoid, *Bull. Belgian Math. Soc. Simon Stevin*, **12** (2005), 727–733.

[7] J. De Beule, A. Klein, K. Metsch and L. Storme. Partial ovoids and partial spreads of classical finite polar spaces, *Serdica Math. J.* **34** (2008), 689–414.

[8] B. De Bruyn and H. Van Maldeghem, Non-embeddable polar spaces, *Münster J. Math.* **7** (2014), 557–588.

[9] H. Freudenthal. Beziehung der E_7 un E_8 zur Oktavenebene, I-XI, *Proc. Kon. Ned. Akad. Wet. Ser. A*, **57** (1954), 218–230, 363–368; **57** (1955), 151–157, 277–285; **62** (1959), 165–201, 447–474; **66** (1963), 457–487.

[10] A. Gunawardena and G. E. Moorhouse. The non-existence of ovoids in $O_9(q)$, *Eropean J. Combin.* **18** (1997), 171–173.

[11] W. M. Kantor. Ovoids and translation planes, *Canadian J. Math.* **34** (1982), 153–157.

[12] G. E. Moorhouse, Som p-ranks related to hermitian varieties, *J. Satist. Plann. Inference* **56** (1996), 229–241.

[13] B. Mühlherr, A geometric approach to non-embeddable polar spaces of rank 3, *Bull. Belgian Math. Soc.* **42** (1990), 577–594.

[14] C. M. O'Keefe and J. A. Thas. Ovoids of the quadric $Q(2n, q)$, *European J. Combin.* **16** (1995), 87–92.

[15] M. A. Ronan. Embeddings and hyperplanes of ovoidal geometries, *European J. Combin.* **8** (1987), 179–185.

[16] S. E. Payne and J. A. Thas. Ovoids nd spreads in finite generalized quadrangles, *Geom. Dedicat,* **52** (1994), 227–253.

[17] E. E. Shult. *Points and Lines*, Springer, Berlin 2011.

[18] J. Thas. Ovoids and spreads in finite classical polar spaces, *Geom. Dedicata* **10** (1981), 135–143.

[19] J. Tits. *Buildings of Sperical Type and Finite BN-pairs*, Springer L.N. Vol. 386, 1974.

[20] H. Van Maldeghem, *Generalized Polygons*, Birkhäuser, 1998.

9

On the Behaviour of Regular Unipotent Elements from Subsystem Subgroups of Type A_3 in Irreducible Representations of Groups of Type A_n with Special Highest Weights

Tatsiana S. Busel, Irina D. Suprunenko

Abstract

Properties of the images of regular unipotent elements from subsystem subgroups of type A_3 in irreducible representations of an algebraic group of type A_n in positive characteristic are investigated. We prove that these images have Jordan blocks of all a priori possible sizes if some six consecutive highest weight coefficients satisfy a certain special condition.

CONTENTS

9.1 Introduction

The block structure of the images of regular unipotent elements from subsystem subgroups of type A_3 in irreducible representations of an algebraic group of type A_n in positive characteristic with certain local properties of highest weights is investigated. In this chapter, we consider the case where a certain linear combination of some six consecutive highest weight coefficients is sufficiently small with respect to the ground field characteristic.[*]

[*] This research has been supported by the Institute of Mathematics of the National Academy of Sciences of Belarus in the framework of the State Research Programme 'Convergence – 2025'. The authors thank the referee whose comments helped to improve the text.

In what follows, K is an algebraically closed field of characteristic $p \geq 11$, $G = A_n(K)$, $n \geq 6$, $\mathbb{N}_a$ is the set of all integers from 1 to a, and $\mathbb{N}_a^o$ and $\mathbb{N}_a^e$ are the sets of odd and even integers from 1 to a, respectively; ω_i, $1 \leq i \leq n$, are the fundamental weights of G. For a representation φ of a simple algebraic group Γ and an element $z \in \Gamma$, the symbol $J_\varphi(z)$ denotes the set of the dimensions of the Jordan blocks of the element $\varphi(z)$ (disregarding their multiplicities).

The following theorem is proved.

Theorem 1.1 *Let $G = A_n(K)$, $n \geq 6$, $x \in G$ be a regular unipotent element from a subsystem subgroup of type A_3, and φ be a p-restricted irreducible representation of G with highest weight $a_1\omega_1 + a_2\omega_2 + \cdots + a_n\omega_n$. Assume that for some $j \leq n - 5$, the following holds:*

(a) $6 + \sum\limits_{i=j}^{j+5} a_i \leq p,$

(b) $3a_j + 4(a_{j+1} + \cdots + a_{j+4}) + 3a_{j+5} \geq p,$

(c) $a_{j+1} + \cdots + a_{j+4} > 1$ or $a_j a_{j+5} \neq 0.$

Then $J_\varphi(x) = \mathbb{N}_p$.

The last assumption of Theorem 1.1 cannot be removed. It is proven in this chapter that for $n = 6$, the set $J_\varphi(x) \neq \mathbb{N}_p$ for the representation φ with highest weight $\frac{p-4}{3}\omega_1 + \omega_2$ (Proposition 3.2) and that $\{2, p - 2\} \not\subset J_\varphi(x)$ for the representations φ with highest weight $a_1\omega_1$, $a_1 > p/3$ (Lemma 3.3). One easily observes that representations satisfying the assumptions of Theorem 1.1 exist only for $p \geq 11$.

The images of irreducible representations of simple algebraic groups constitute an important family of subgroups in the classical algebraic groups. Such representations are parametrized by their highest weights, but in fact we know rather little about these images as linear groups, except for a few series of representations (mainly those connected with wedge or symmetric powers of the standard modules of the classical groups). So the investigations of the behaviour of individual elements in representations deepen our understanding of the subgroup structure of the classical groups.

Sometimes the presence of some special unipotent element in a semisimple (or even reductive) subgroup of a simple algebraic group yields a lot of information on this subgroup and allows one to put it into a certain well-defined list. For instance, semisimple subgroups of the simple algebraic groups containing regular unipotent elements of these groups are classified by Testerman and Zalesskii [23]; earlier maximal closed subgroups of these groups containing such elements were determined by Saxl and Seitz [15]. In [23] a quite unexpected result has been obtained: it has been proven there that connected reductive subgroups containing such elements do not lie in proper parabolic subgroups. In particular, relevant subgroups in the special linear groups are irreducible. Irreducible representations of the simple algebraic groups whose images contain unipotent matrices with a single nontrivial Jordan block have been classified in [22] for the classical groups and in [24] for exceptional ones. Results of [15] were applied in [9] to prove that for $p > 5$, the group $E_8(p)$ can be realized as the Galois group of an extension of the rational field.

Our experience shows that there are many other classes of unipotent elements that can be effectively used for recognizing representations and linear groups. Systematic investigations of properties of the images of unipotent elements in representations of algebraic groups allows us to distinguish typical and rare cases and to find such elements.

Obviously, one cannot determine the canonical Jordan form of such an image if the dimension of a representation being considered is not known. Notice that in positive characteristic one cannot expect that for arbitrary irreducible representations of simple algebraic groups, the dimension problem will be solved in the foreseeable future. Therefore it is worthwhile to consider a natural analogue of this form and to elaborate tools for determining the collection of Jordan block dimensions disregarding their multiplicities, that is, the set $J_\varphi(z)$, for a given unipotent element z and a fixed irreducible representation φ. At the current moment, all known results on this set that do not require finding the dimension of a representation concern elements lying in subsystem subgroups with simple components of small ranks.

P. H. Tiep and A. E. Zalesskii [25, theorem 2.20] described p-restricted irreducible representations φ of simple algebraic groups in characteristic $p > 3$ with $J_\varphi(z) \subset \{1, p - 1, p\}$ for all root elements z. This allowed them to show that the reduction mod p of certain complex representations of finite Chevalley groups is reducible.

In [10, 13, 26, 27, 28, 29] the set $J_\varphi(z)$ has been completely determined for root elements in simple algebraic groups of different types and for quadratic elements in groups of type A_n, in some cases. Now for root elements, the problem is solved in the following situations: for groups of rank > 2 with all roots of the same length, for groups of types C_n with $n > 3$ and F_4 under certain additional assumptions, and for long root elements in groups of type B_n with $n > 2$.

In [21] and [2] the images of regular unipotent elements from subsystem subgroups of types $A_2 \times A_1$ and of a product of commuting long and short root elements in p-restricted irreducible representations of groups of types A_n and B_n, respectively, with certain local properties of highest weights have been considered (here "local properties" means that certain assumptions are formulated only for some coefficients of the highest weights).

In terms of elements considered in Theorem 1.1, in [14] the set $J_\varphi(x)$ has been determined for p-restricted representations in characteristic $p \geq 11$ in the case where $3a_j + 4a_{j+1} + 3a_{j+2} < p$ for some j with $j \leq n-2$. The similar problem has been solved for p-restricted representations of groups of type C_n for $p \geq 11$ and regular unipotent elements from subsystem subgroups of type C_2 in two cases: (1) $3a_{n-1} + 4a_n < p$ [3]; (2) $3a_{n-1} + 4a_n \geq p$ and $4 + a_{n-3} + \cdots + a_n \leq p$ [4]. (For some representations φ of a special form, in [14] and [3] the set $J_\varphi(x)$ is found up to two dimensions.) In these papers and in the current chapter some approaches have been elaborated that are effective when the restrictions of relevant representations to subsystem subgroups being considered have many irreducible direct summands with not too large highest weights with respect to p. We expect that these approaches can be useful for investigating the behaviour of unipotent elements from subsystem subgroups of relatively small ranks in representations of classical algebraic groups in more general situations, therefore some preliminary results (Theorem 2.6, Propositions 2.15 and 2.22, and Lemma 2.20) are given in more generality than is required to achieve the direct aims of this chapter.

In all cases studied to date, in order to determine the set $J_\varphi(z)$ for an irreducible representation φ and an element z of order p, indecomposable components of the restriction of φ to a subgroup A of type A_1 containing z are considered. Usually it is easier to prove that $J_\varphi(z)$ contains an integer l if this is the case, than it is to show that $l \notin J_\varphi(z)$. Indeed, in the first case, it suffices to find a single indecomposable component ψ of $\varphi|A$ such that $l \in J_\psi(z)$, and in the second case it is necessary to analyze all indecomposable components which are unknown as a rule.

That is why now there are no results similar to Theorem 1.1 in those situations where one cannot expect to get Jordan blocks of all possible dimensions from 1 to the maximal one including the images of regular unipotent elements of subsystems subgroups of type A_2 in representations of groups of type A_n and those of short root elements in representations of groups of type B_n. Here for p-restricted representations with large enough highest weights with respect to p, only the odd dimensions of blocks are found under some assumptions on certain linear combinations of neighbouring coefficients of the highest weights [11, 13].

9.2 Notation and Preliminary Information

In what follows, $\mathbb{C}$ is the field of complex numbers, Γ is a simply connected simple algebraic group of a classical type over K, $T \subset \Gamma$ is a fixed maximal torus, $r(\Gamma)$ is the rank of Γ, α_i and ω_i $(1 \leq i \leq r)$ are the simple roots and the fundamental weights of Γ with respect to T, $\Lambda(\Gamma)$ and $\Lambda^+(\Gamma)$ are the set of weights and the set of dominant weights of Γ, ρ_Γ is the halfsum of the positive roots of Γ; α_Γ is the maximal short root of Γ if this group has roots of distinct lengths and the maximal root of this group otherwise; $W(\Gamma)$ is the Weyl group of Γ, $\langle \omega, \alpha \rangle$ is the value of a weight ω on a root α (in the sense of [19, §1]); $\omega(\varphi)$ ($\omega(M)$) is the highest weight of a representation φ (a module M), $\omega(m)$ is the weight of a weight vector m from some module, $\Gamma_{\mathbb{C}}$ is the simple simply connected algebraic group over $\mathbb{C}$ with the same root system as Γ; for an irreducible representation φ of Γ, we denote by $\varphi_{\mathbb{C}}$ the irreducible representation of $\Gamma_{\mathbb{C}}$ with highest weight $\omega(\varphi)$; similarly we define the $\Gamma_{\mathbb{C}}$-module $M_{\mathbb{C}}$ for an irreducible Γ-module M; dim M (dim φ) is the dimension of a Γ-module M (a representation φ), $\Lambda(M)$ ($\Lambda(\varphi)$) and $\Lambda^+(M)$ ($\Lambda^+(\varphi)$) are the sets of weights and dominant weights of a module M (a representation φ). If $\omega \in \Lambda^+(\Gamma)$, then $M(\omega)$, $V(\omega)$, and $T(\omega)$ are, respectively the irreducible module, the Weyl module, and the indecomposable tilting module of Γ with highest weight ω. It is always clear from the context which group is considered.

If H is a subgroup of Γ, then $M|H$ is the restriction of a Γ-module M to H. If $T \cap H$ is a maximal torus in H, then $\omega|H$ is the restriction of a weight ω to $T \cap H$. The symbol $d_\varphi(x)$ denotes the degree of the minimal polynomial of the image of an element $x \in \Gamma$ in a representation φ, and IrrM denotes the set of the composition factors of a module M (disregarding their multiplicities). The set of the weights of the group $A_1(K)$ is identified with the set $\mathbb{Z}$ of integers in the standard way: $a\omega_1 \mapsto a$.

Recall that an irreducible representation of a semisimple algebraic group in characteristic p is called p-restricted if its highest weight is a linear combination of the fundamental ones with all coefficients less than p. A subsystem subgroup is the subgroup generated by all root subgroups associated with the roots from a subsystem of the root system of Γ. Throughout, the text $\Gamma(i_1, \ldots, i_k)$ is a subsystem subgroup in Γ generated by the root subgroups associated with the roots $\pm\alpha_{i_1}, \ldots, \pm\alpha_{i_k}$.

The following facts will be essentially used in the proof of Theorem 1.1.

Theorem 2.1 *[18] Let $S = \Gamma(i_1, \ldots, i_k) \subseteq \Gamma$, M be an irreducible Γ-module with highest weight ω, and $v \in M$ be a nonzero highest weight vector. Then the subspace $KSv \subseteq M$ is an irreducible S-module with highest weight $\omega|S$ and a direct summand of the S-module M.*

It is well known that the image of a nontrivial unipotent element in a p-restricted representation of a group of type A_1 with highest weight a has a unique Jordan block of dimension $a + 1$ (see [13, lemma 10]). We often use this fact without an explicit reference.

Lemma 2.2 *[3, part I, corollary 2] Let U be a Γ-module. Assume that $V(\lambda)$ is irreducible for any composition factor $M(\lambda)$ of U. Then U is completely reducible.*

Lemma 2.3 *[13, lemma 12] Let Δ be a group and $U = U_1 \oplus \cdots \oplus U_k$ be a direct sum of Δ-modules. Then $J_U(z) = \bigcup_{j=1}^{k} J_{U_j}(z)$ for $z \in \Delta$.*

Lemma 2.4 *[3, part I, lemma 11] For $\langle \lambda + \rho_\Gamma, \alpha_\Gamma \rangle \leq p$, the Weyl module $V(\lambda)$ is irreducible.*

Corollary 2.5 *Let $\Gamma = A_3(K)$, $\omega = a_1\omega_1 + a_2\omega_2 + a_3\omega_3 \in \Lambda^+(\Gamma)$, and $a_1 + a_2 + a_3 \leq p - 3$. Then $V(\omega)$ is irreducible.*

This corollary follows directly from Lemma 2.4.

Theorem 2.6 *Let $S \subset \Gamma$ be a subsystem subgroup of type A_l, B_l, C_l, or D_l for $\Gamma = A_r(K)$, $B_r(K)$, $C_r(K)$, or $D_r(K)$, respectively, $l \geq 3$ for $\Gamma = B_r(K)$, $l \geq 2$ for $\Gamma = C_r(K)$, and $l \geq 4$ for $\Gamma = D_r(K)$. Assume that N is an irreducible Γ-module with highest weight λ and $\langle \lambda + \rho_\Gamma, \alpha_\Gamma \rangle \leq p$. Then for a unipotent element $u \in S$ and an irreducible S-module F, the set $J_F(u) \subset J_N(u)$ if $F_\mathbb{C}$ is a composition factor of the restriction $N_\mathbb{C}|S_\mathbb{C}$.*

Proof Obviously, it suffices to consider the case where S is a proper subgroup. By Lemma 2.4, the module $V(\lambda)$ is irreducible and isomorphic to N. Therefore the formal characters of the modules N and $N_\mathbb{C}$ are the same.

One easily observes that $\langle \rho_S, \alpha_S \rangle < \langle \rho_\Gamma, \alpha_\Gamma \rangle$. Let M be a composition factor of the restriction $N_\mathbb{C}|S_\mathbb{C}$ with highest weight μ. Then $\mu = \nu|S_\mathbb{C}$ for some weight $\nu \in \Lambda(N_\mathbb{C})$. As the roots α_S and α_Γ lie in the same $W(\Gamma)$-orbit, then $\langle \mu, \alpha_S \rangle = \langle \nu, \alpha_S \rangle \leq \langle \lambda, \alpha_\Gamma \rangle$. This implies that $\langle \mu + \rho_S, \alpha_S \rangle < \langle \lambda + \rho_\Gamma, \alpha_\Gamma \rangle \leq p$. Therefore the S-module $V(\mu)$ is irreducible by Lemma 2.4. This and the previously stated arguments on formal characters yield that an irreducible S-module F is a composition factor of the restriction $N|S$ if and only if $F_\mathbb{C}$ is a composition factor of $N_\mathbb{C}|S_\mathbb{C}$ and that all composition factors of the former restriction are isomorphic to corresponding Weyl modules. By Lemma 2.2, the restriction $N|S$ is completely reducible. It remains to apply Lemma 2.3. $\square$

Lemma 2.7 *[31, corollary 2.2.1] Let Δ be a simple algebraic group of a classical type over $\mathbb{C}$, $S \subset \Delta$ be a simple subsystem subgroup of the same series, and N_1, N_2, and N be irreducible Δ-modules with highest weights λ_1, λ_2, and $\lambda_1 + \lambda_2$, respectively. Then the set of the highest weights of the composition factors of the restriction $N|S$ consists of all sums of the form $\mu_1 + \mu_2$ where μ_i is the highest weight of a composition factor of $N_i|S$.*

This lemma follows from the classical branching rules [30].

The following lemma is well known.

Lemma 2.8 *Let τ be the irreducible representation of the group $\Delta = A_n(\mathbb{C})$ with highest weight ω_i, $1 \leq i \leq n$, $S \subset \Delta$ be a subsystem subgroup of type A_m, and $m < n$. Then the set of composition factors of the restriction $\tau|S$ consists of all irreducible representations with highest weights of the form ω_j where $0 \leq j \leq \min\{i, m + 1\}$ and $i - j \leq n - m$. Here ω_0 and ω_{m+1} are equal to 0.*

To prove Lemma 2.8, the realization of τ in the ith exterior power of the standard Δ-module is considered.

Proposition 2.9 *Let ψ be an irreducible representation of $\Delta = A_6(\mathbb{C})$ with highest weight $\sum_{i=1}^{6} a_i\omega_i$ and $S \subset \Delta$ be a subsystem subgroup of type A_3. Then the set of composition factors of $\psi|S$ consists of all irreducible representations with highest weights of the form $c_1\omega_1 + c_2\omega_2 + c_3\omega_3$ where*

$$\sum_{j=1}^{3} c_j \leq \sum_{i=1}^{6} a_i, \quad c_1 \leq \sum_{i=1}^{4} a_i, \quad c_2 \leq \sum_{i=2}^{5} a_i, \quad c_3 \leq \sum_{i=3}^{6} a_i,$$

$$c_1 + c_2 \leq \sum_{i=1}^{5} a_i, \quad c_2 + c_3 \leq \sum_{i=2}^{6} a_i. \tag{1}$$

Proof Set $\omega = \sum_{i=1}^{6} a_i\omega_i$. We use induction by the sum $\sum_{i=1}^{6} a_i$. If $\omega = \omega_i$, our conclusion follows from Lemma 2.8.

Let $\sum_{i=1}^{6} a_i > 1$ and the assertion of this proposition hold for the representations with highest weights with a smaller sum of these coefficients. Fix j with $a_j \neq 0$ and set $\lambda_1 = \omega - \omega_j$ and $\lambda_2 = \omega_j$. Using the induction hypothesis and Lemmas 2.7 and 2.8, one easily deduces that the highest weight of a composition factor of our restriction satisfies System (1).

Let $\mu = c_1\omega_1 + c_2\omega_2 + c_3\omega_3 \in \Lambda^+(S)$ and the triple (c_1, c_2, c_3) satisfy System (1). We shall show that there exist weights $\mu_1, \mu_2 \in \Lambda^+(S)$ and an index k such that $\mu = \mu_1 + \mu_2$, $a_k \neq 0$, and $M(\mu_1)$ and $M(\mu_2)$ are composition factors of $M(\omega - \omega_k)|S$ and $M(\omega_k)|S$, respectively. Then we shall apply Lemma 2.7 to complete the proof.

First assume that $0 < c_2 = \sum_{i=2}^{5} a_i$. Then $c_1 \leq a_1$ and $c_3 \leq a_6$. It is clear that $a_k \neq 0$ for some k with $2 \leq k \leq 5$. Fix such k and set $\mu_1 = \mu - \omega_2$ and $\mu_2 = \omega_2$.

Next, let $c_2 =, 0$ or $c_2 < \sum\limits_{i=2}^{5} a_i$. Suppose that $0 < c_1 = \sum\limits_{i=1}^{4} a_i$. Then $c_2 \le a_5$ and $c_2 + c_3 \le a_5 + a_6$. Obviously, in this case $a_k \ne 0$ for some k with $1 \le k \le 4$. We fix such k and put $\mu_1 = \mu - \omega_1$ and $\mu_2 = \omega_1$. If $0 < c_3 = \sum\limits_{i=3}^{6} a_i$, we conclude that $c_2 \le a_2$, $c_1 + c_2 \le a_1 + a_2$, and $a_k \ne 0$ for some k with $3 \le k \le 6$. Fix such k and set $\mu_1 = \mu - \omega_3$ and $\mu_2 = \omega_3$.

Now assume that for all j with $1 \le j \le 3$, the inequality for c_j in System (1) is strict or $c_j = 0$. Let $0 < c_1 + c_2 = \sum\limits_{i=1}^{5} a_i$. Then $c_3 \le a_6$. If $c_1 \ne 0$, it is clear that $a_k \ne 0$ for some k with $1 \le k \le 4$. We fix such k and put $\mu_1 = \mu - \omega_1$ and $\mu_2 = \omega_1$. Let $c_1 = 0$. Then $c_2 > 0$ and $a_k \ne 0$ for some k with $2 \le k \le 5$. Fix such k and set $\mu_1 = \mu - \omega_2$ and $\mu_2 = \omega_2$. Next, suppose that $0 < c_2 + c_3 = \sum\limits_{i=2}^{6} a_i$. Then $c_1 \le a_1$. If $c_3 > 0$, then $a_k \ne 0$ for some k with $3 \le k \le 6$. We fix such k and put $\mu_1 = \mu - \omega_3$ and $\mu_2 = \omega_3$. If $c_3 = 0$, we have $c_2 \ne 0$ and choose k, μ_1, and μ_2 as in the previous case.

Set $\Sigma = \{c_1, c_2, c_3, c_1 + c_2, c_2 + c_3\}$,

$$N(j) = \begin{cases} \{1,2,3,4\} & \text{for } j = 1, \\ \{2,3,4,5\} & \text{for } j = 2, \\ \{3,4,5,6\} & \text{for } j = 3. \end{cases}$$

Assume that for each nonzero $c \in \Sigma$ the inequality for c in System (1) is strict and $c_1 + c_2 + c_3 = \sum\limits_{i=1}^{6} a_i$. It is clear that in this situation there exist j with $1 \le j \le 3$ and $k \in N(j)$ such that c_j and $a_k \ne 0$. Fix such j and k and set $\mu_1 = \mu - \omega_j$ and $\mu_2 = \omega_j$.

In all remaining cases choose arbitrary k with $a_k \ne 0$ and put $\mu_1 = \mu$ and $\mu_2 = 0$.

Now k, μ_1, and μ_2 are chosen in all situations. Let $\omega - \omega_k = \sum\limits_{i=1}^{6} b_i$. The construction described then implies that the coefficients $b_i \ge 0$, $b_j = a_j - 1$ for some j, and $b_i = a_i$ for $i \ne j$. Furthermore, one can conclude that in all cases the weight μ_1 satisfies the inequalities of the system obtained from System (1) in the result of replacing the integers a_i by b_i. So by the induction hypothesis, $M(\mu_1)$ is a composition factor of the restriction $M(\omega - \omega_k)|S$. Lemma 2.8 implies that $M(\mu_2)$ is a composition factor of the restriction $M(\omega_k)|S$. Therefore we can apply Lemma 2.7. The proposition is proved. □

Definition 2.10 *A closed connected subgroup A of type A_1 in Γ is called good if the images of all roots under the homomorphism $\sigma \colon \Lambda(\Gamma) \to \Lambda(A)$ induced* by the restriction of weights from a maximal torus T of Γ to the maximal torus $T_A = T \cap A$ in A are at most $2p - 1$.

Theorem 2.11   *[17, theorem 1 and proposition 2.2] For any element $x \in \Gamma$ of order p, there exists a good subgroup A containing x and all such subgroups are conjugate; one can choose such a system of simple roots that $\sigma(\alpha_i) \in \{0, 1, 2\}$ and that coincides with the ith label on the labelled Dynkin diagram of x (here σ is the homomorphism from Definition 2.10).*

Recall that p is good for all classical simple algebraic groups since p is odd. Throughout the text, assume that $\sigma(\alpha_i)$ are such as in Theorem 2.11. Well-known facts of the representation theory for groups of type A_1 imply that for $\Gamma = A_r(K)$ and an arbitrary element of order p, a good A_1-subgroup is the image of a completely reducible representation of the group $A_1(K)$ with p-restricted components.

The sets $\Lambda(\Gamma)$ and $\Lambda(\Gamma_{\mathbb{C}})$ are identified in the standard way. For $\Gamma = A_r(K)$, denote by ε_i, $1 \le i \le r+1$, the weights of the standard $\Gamma_{\mathbb{C}}$-module defined in [1, §13].

Theorem 2.12   *Let $r \ge 3$, $\Gamma = A_r(K)$, $x \in \Gamma$ be a regular unipotent element from a subsystem subgroup of type A_3, A be a good A_1-subgroup containing x, and $\sigma : \Lambda(\Gamma) \to \Lambda(A)$ be a homomorphism from Theorem 2.11. Then $\sigma(\varepsilon_1) = 3$, $\sigma(\varepsilon_2) = 1$, $\sigma(\varepsilon_i) = 0$ for $3 \le i \le r-1$, $\sigma(\varepsilon_r) = -1$, $\sigma(\varepsilon_{r+1}) = -3$. Let φ be an irreducible p-restricted representation of Γ with highest weight $\sum\limits_{i=1}^{r} a_i \omega_i$.*
Then

$$\max_{\mu \in \Lambda(\varphi)} (\sigma(\mu)) = 3a_1 + 4(a_2 + \cdots + a_{r-1}) + 3a_r,$$

$$d_\varphi(x) = \min\{p, 1 + 3a_1 + 4(a_2 + \cdots + a_{r-1}) + 3a_r\}.$$

Proof   Let $x_{\mathbb{C}} \in \Gamma_{\mathbb{C}}$ be an element with the same Jordan form on the standard module as x. One easily concludes that in such a module, $x_{\mathbb{C}}$ (and x) has one Jordan block of size 4 and other blocks of size 1 (only a block of size 4 for $r = 3$). Set

$$N(x) = (3, 1, 0, \ldots, 0, -1, -3).$$

($N(x)$ contains $r + 1$ integers). By [20, proposition 2.12], the collections

$$(\sigma(\varepsilon_1), \sigma(\varepsilon_2), \ldots, \sigma(\varepsilon_r), \sigma(\varepsilon_{r+1}))$$

and $N(x)$ coincide taking into account multiplicities. Since $\sigma(\alpha_i) \ge 0$ for $i \le r$, this implies that $\sigma(\varepsilon_1) \ge \sigma(\varepsilon_2) \ge \cdots \ge \sigma(\varepsilon_{r+1})$. It is clear that $\max\limits_{\mu \in \Lambda(\varphi)} \{\sigma(\mu)\} =$

Table 1 Weights and blocks

$m_1(m_1 \geq m_3)$	m_2	m_3	missing block sizes
1	0	0	$3m_1 + 4m_2 + 3m_3 - 1 = 2$
1 (mod 2) > 1	0	0	$3m_1 + 4m_2 + 3m_3 - 1, 2$
2	0	0	$3m_1 + 4m_2 + 3m_3 - 1 = 5, 1$
2 (mod 4) > 2	0	0	$3m_1 + 4m_2 + 3m_3 - 1, 5, 1$
0 (mod 4)	0	0	$3m_1 + 4m_2 + 3m_3 - 1, 3$
0	1	0	$3m_1 + 4m_2 + 3m_3 - 1 = 3$
0	> 1	0	$3m_1 + 4m_2 + 3m_3 - 1, 3$
1 (mod 4)	0	1	1
0 (mod 4) > 0	0	2	1
2	> 0	0	1
0 (mod 2) > 2	1	0	1
2 (mod 4) > 2	2	0	1
1	1	1	1

$\sigma(\omega(\varphi))$. Since $\omega_i = \sum_{j=1}^{i} \varepsilon_j$, we get that $\sigma(\omega(\varphi)) = 3a_1 + 4(a_2 + \cdots + a_{r-1}) + 3a_r$. By [20, theorem 1.1],

$$d_\varphi(x) = min\{p, d_{\varphi_{\mathbb{C}}}(x_{\mathbb{C}})\}.$$

Let τ_j be the irreducible representation of $\Gamma_{\mathbb{C}}$ with highest weight ω_j, $1 \leq j \leq r$, and $m_j = d_{\tau_j}(x_{\mathbb{C}}) - 1$. According to [20, algorithm 1.4], m_j is equal to the sum of j maximal members of the collection $N(x)$. Therefore $m_1 = m_r = 3$ and $m_2 = \cdots = m_{r-1} = 4$. By [20, proposition 1.3],

$$d_{\varphi_{\mathbb{C}}}(x_{\mathbb{C}}) = 1 + \sum_{i=1}^{r} a_i m_i = 1 + 3a_1 + 4(a_2 + \cdots + a_{r-1}) + 3a_r.$$

The theorem is proved.   □

Theorem 2.13 *[12, theorem 5(iv), table 1] Let τ be an irreducible representation of $A_3(\mathbb{C})$ with highest weight $m_1\omega_1 + m_2\omega_2 + m_3\omega_3$ and $m_1 \geq m_3$. Then the image of a regular unipotent element in τ has blocks of all sizes:*

$i \equiv 3m_1 + 4m_2 + 3m_3 + 1 \pmod 2$ where $1 \leq i \leq 3m_1 + 4m_2 + 3m_3 + 1$,

except for the weights and blocks specified in Table 1.
In particular, for $\min\{m_1, m_2, m_3\} > 1$, there are blocks of all a priori possible sizes.

Remark 2.14 *Theorem 2.13 allows one to determine the dimensions of Jordan blocks of the elements being considered for arbitrary irreducible representations of $A_3(\mathbb{C})$ since it is well known that for mutually dual representations these dimensions coincide and the highest weight of the representation dual to τ is equal to $m_3\omega_1 + m_2\omega_2 + m_1\omega_3$.*

The following proposition is true for arbitrary odd p.

Proposition 2.15 *Let $z \in \Gamma$, $|z| = p$, and $z_{\mathbb{C}} \in \Gamma_{\mathbb{C}}$ be unipotent. Assume that the labelled Dynkin diagrams of the elements z and $z_{\mathbb{C}}$ are the same. Let ψ be an irreducible representation of Γ with highest weight λ, and σ be a homomorphism from Theorem 2.11 associated with a good A_1-subgroup containing z. Suppose that the Weyl module $V(\lambda)$ is irreducible and $\sigma(\lambda) < p$. Then $J_\psi(z) = J_{\psi_{\mathbb{C}}}(z_{\mathbb{C}})$.*

Proof Let $A \subset \Gamma$ be a good A_1-subgroup containing z. By [17, proposition 2.2 and the preceding remark], there exists a Zariski closed subgroup $A_{\mathbb{C}} \cong A_1(\mathbb{C})$ in $\Gamma_{\mathbb{C}}$ such that $x_{\mathbb{C}} \in A_{\mathbb{C}}$, $T_{\mathbb{C}} \cap A_{\mathbb{C}}$ is a maximal torus in $A_{\mathbb{C}}$ for some maximal torus $T_{\mathbb{C}} \subset \Gamma_{\mathbb{C}}$, and that the restriction of weights from $T_{\mathbb{C}}$ to $T_{\mathbb{C}} \cap A_{\mathbb{C}}$ yields the homomorphism σ. As $\sigma(\alpha_i) \geq 0$ and $\sigma(\lambda) < p$, it is clear that ψ is p-restricted and $\sigma(\mu) < p$ for all weights $\mu \in \Lambda(\psi)$. Now our proposition follows directly from [21, proposition 11]. $\qquad\square$

Corollary 2.16 *In the assumptions of Proposition 2.15, let $\Gamma = A_3(K)$ and $\lambda = c_1\omega_1 + c_2\omega_2 + c_3\omega_3$. Then the assertion of the proposition holds if $3c_1 + 4c_2 + 3c_3 < p$.*

Proof One easily observes that in the assumptions of the corollary, the module $V(\lambda)$ is irreducible by Lemma 2.4 and $\sigma(\lambda) < p$. It remains to apply Proposition 2.15. $\qquad\square$

Recall that a Γ-module is called a tilting module if it has both a filtration by Weyl modules and a filtration by dual Weyl modules.

Lemma 2.17 *[7, theorem 1.1 and proposition 1.2] and [6, corollary 1.3]*

(a) *For each dominant weight λ of Γ, there exists an indecomposable tilting module $T(\lambda)$, unique up to isomorphism, with highest weight λ.*

(b) *Any tilting module is a direct sum of tilting modules of form $T(\lambda)$.*

(c) *A direct summand of a tilting module is a tilting module.*

(d) *A tensor product of tilting modules is a tilting module.*

Lemma 2.18 *[17, lemmas 1.2 and 1.3] For $0 \leq c < p$, the module $T(c) \cong V(c) \cong M(c)$. If $p \leq c \leq 2p - 2$, write $c = r + p$. Then the maximal submodule M in $V(c)$ is isomorphic to $M(p - r - 2)$ and $V(c)/M \cong M(c)$. The module $T(c)$ has a filtration*

$$T(c) = M_1 \supset M_2 \supset M_3 \supset M_4 = 0$$

with $M_1/M_2 \cong M_3 \cong M(p - r - 2)$ and $M_2/M_3 \cong M(r + p)$;

$\dim(T(c)) = 2p$. *In this case, the module $T(c)$ is projective for the group $A_1(p)$, and therefore a nontrivial unipotent element has two blocks of size p in this module.*

Lemma 2.19 *Let $\Gamma = A_n(K)$, $A \subset \Gamma$ be the image of a completely reducible representation of the group $A_1(K)$ with p-restricted irreducible components, and $i < p$. Then $M(\omega_i)|A$ is a tilting module.*

Proof Recall that $M(\omega_i)$ is the ith exterior power of the standard Γ-module. Then we use the known fact that for $i < p$, this power is a direct summand of the ith tensor power of the standard module (a special case of the result [8, corollary 2.6e]) and Lemma 2.17.   □

Lemma 2.20 *[3, part 1, corollary 6] Let $A \subset \Gamma$ be a connected semisimple closed subgroup and N be an irreducible Γ-module with highest weight λ. Suppose that the restrictions to A of the irreducible Γ-modules with fundamental highest weights are tilting modules and that $\langle \lambda + \rho_\Gamma, \alpha_\Gamma \rangle \le p$. Then $N|A$ is a tilting module.*

Corollary 2.21 *Let $\Gamma = A_r(K)$, $p > r$, $z \in \Gamma$ be an element of order p, and A be a good A_1-subgroup containing z. Suppose that a module N and a weight λ satisfy the assertions of Lemma 2.20. Then $N|A$ is a tilting module.*

Proposition 2.22 *In the assumptions of Proposition 2.15, suppose that $\langle \lambda + \rho_\Gamma, \alpha_\Gamma \rangle \le p$ and $p \le \sigma(\lambda) = p + t \le 2p - 2$. Let a be a non-negative integer and $a < p - 2 - t$. In this case, $a + 1 \in J_\psi(z)$ if and only if $a + 1 \in J_{\psi_{\mathbb{C}}}(z_{\mathbb{C}})$.*

Proof The notation from the proof of Proposition 2.15 is used. Let M be an irreducible module with highest weight λ. By Corollary 2.5 and Lemma 2.20, the module $V(\lambda)$ is irreducible and $M|A$ is a tilting module. Therefore $\dim M_\mu = \dim(M_{\mathbb{C}})_\mu$ for any weight $\mu \in \Lambda(M)$ and the factors in the filtrations of the restrictions $M|A$ and $M_{\mathbb{C}}|A_{\mathbb{C}}$ by Weyl modules coincide (taking into account multiplicities). It is clear that if such filtration for the module $M_{\mathbb{C}}|A_{\mathbb{C}}$ has a factor $V(a)$, then $V(a)$ is a direct summand of this restriction. Obviously, $b \le p + t$ for any indecomposable component $T(b)$ of the restriction $M|A$. If $p \le b = p + u$, then by Lemma 2.18, any composition factor of $T(b)$ is isomorphic to $M(b)$ or $M(p - u - 2)$. Hence if $V(a)$ is among the factors of the filtration of $M|A$ by Weyl modules, then $V(a) \cong M(a)$ is a direct summand of this restriction. This implies that the modules $M_{\mathbb{C}}|A_{\mathbb{C}}$ and $M|A$ simultaneously have an indecomposable component isomorphic to $M(a)$ or have no such components. As $J_N(z) = \{p\}$ for an indecomposable component $N \cong T(b)$ with $b \ge p$, we get the assertion of the lemma.   □

Lemma 2.23 *Let* $\Gamma = A_3(K)$, N *be an irreducible* Γ-*module with highest weight* $\lambda = b_1\omega_1 + b_2\omega_2 + b_3\omega_3$, $3b_1 + 4b_2 + 3b_3 = p$, *and* $z \in \Gamma$ *be a regular unipotent element.*

Case 1. Assume that $b_1 b_2 b_3 \neq 0$. *Then* $p - 1 \in J_N(z)$.
Case 2. Let $b_1 b_2 b_3 = 0$. *Then* $p - 1 \notin J_N(z)$.

Proof Let A be a good A_1-subgroup containing z. One easily concludes that $\lambda \neq b_i\omega_i$. Obviously, $b_1 + b_2 + b_3 < p - 3$. Then by Corollary 2.21, $N|A$ is a tilting module.

Let σ be the homomorphism from Theorem 2.11 associated with z. It is well known that $\sigma(\alpha_i) = 2$ as z is a regular unipotent element [5, chapter 5]. By Theorem 2.12, $\sigma(\lambda) = p$. One easily deduces that $\sigma(\lambda - \alpha_i) = p - 2$ and $\sigma(\mu) < p - 2$ for other weights $\mu \in \Lambda(N)$.

Denote by N_a the weight subspace of weight a in the module $N|A$. It follows from [16, 1.5] that dim $N_{p-2} = 3$ or 2 in Case 1 or 2, respectively. Lemma 2.17 implies that $N|A = T(p) \oplus N_1$ where N_1 is a direct sum of irreducible modules $M(a)$ with $a \leq p - 2$. By Lemma 2.18, the multiplicity of the weight $p - 2$ in $T(p)$ is equal to 2. Now it is clear that $N|A$ has an indecomposable component isomorphic to $M(p - 2)$ in Case 1 and has no such components in Case 2. This implies the assertion of the lemma. □

In what follows, φ is a p-restricted irreducible representation of G, $\omega(\varphi) = \sum_{i=1}^{n} a_i\omega_i$, and M is a module affording φ.

9.3 Proof of Theorem 1.1

Suppose that the coefficients a_j, $a_{j+1}, \ldots, a_{j+5}$ satisfy the assertion of Theorem 1.1. Set $H = G(j, j + 1, \ldots, j + 5)$, $G_1 = G(j, j + 1, j + 2)$, and

$$I = \left\{ c_1\omega_1 + c_2\omega_2 + c_3\omega_3 \in \Lambda^+(G_1) \Big| \sum_{k=1}^{3} c_k \leq \sum_{i=j}^{j+5} a_i, \ c_1 \leq \sum_{i=j}^{j+3} a_i, \right.$$
$$\left. c_2 \leq \sum_{i=j+1}^{j+4} a_i, \ c_3 \leq \sum_{i=j+2}^{j+5} a_i, \ c_1 + c_2 \leq \sum_{i=j}^{j+4} a_i, \ c_2 + c_3 \leq \sum_{i=j+1}^{j+5} a_i \right\}.$$

It is clear that $H \cong A_6(K)$ and $G_1 \cong A_3(K)$. In what follows, $\omega' = \omega|H$ and $M' = M(\omega')$. We assume that x is a regular unipotent element of the subgroup G_1. To prove Theorem 1.1, we show that $J_{M'}(x) \subset J_M(x)$, find out that the restriction $M'|G_1$ is completely reducible, and determine its irreducible components.

Theorem 3.1 *Let $\mu \in I$. Then $J_{M(\mu)}(x) \subset J_M(x)$.*

Proof Since $\langle \omega' + \rho_H, \alpha_H \rangle = a_j + a_{j+1} + a_{j+2} + a_{j+3} + a_{j+4} + a_{j+5} + 6 \leq p$, one easily observes that M' satisfies the assertions of Theorem 2.6. By Proposition 2.9, $\mathrm{Irr}(M'_\mathbb{C}|G_{1,\mathbb{C}}) = \{M(\lambda)|\lambda \in I\}$. Theorem 2.6 implies that $J_{M(\mu)}(x) \subset J_{M'}(x)$. By Theorem 2.1, M' is a direct summand of $M|H$. Therefore $J_{M(\mu)}(x) \subset J_M(x)$. $\qquad\square$

Proof of Theorem 1.1. By Theorem 3.1, it suffices for every integer $k \in \mathbb{N}_p$ to find such a weight $\mu \in I$ that $k \in J_{M(\mu)}(x)$. It is clear that $0 \in I$. Therefore $1 \in J_M(x)$. Theorem 2.12 implies that $d_\varphi(x) = p$. Hence $p \in J_M(x)$. Recall that in the course of the proof of Theorem 2.6 we have shown that the Weyl module $V(\mu)$ is irreducible for $\mu \in I$. Let $x_\mathbb{C} \in A_3(\mathbb{C})$ be a regular unipotent element. We will use the symbol $M_{i,\mathbb{C}}$ instead of $(M_i)_\mathbb{C}$.

Set $Q = 3(a_j + a_{j+1} + a_{j+2} + a_{j+3} + a_{j+4} + a_{j+5})$. The proof is split into different subcases depending on the value of the parameter Q.

In all situations considered, our arguments follow the same scheme, but various weights from I are used. In every case, we choose three or four weights $\mu_i \in I$, consider the modules $M_i = M(\mu_i)$, and prove that $\mathbb{N}_p \setminus \{1,p\} \subset \cup_i J_{M_i}(x)$. Since $1, p \in J_\varphi(x)$, this yields that $J_\varphi(x) = \mathbb{N}_p$. As a rule, these weights μ_i are chosen such that $J_{M_1}(x)$ and $J_{M_2}(x)$ contain all integers from $\mathbb{N}_p^o$ and $\mathbb{N}_p^e$, respectively, except several explicitly indicated ones, and the sets $J_{M_3}(x)$ and $J_{M_4}(x)$ contain the remaining integers from $\mathbb{N}_p \setminus \{1,p\}$. We apply Propositions 2.15 and 2.22, Theorem 2.13, and Remark 2.14 to describe the set $J_{M_i}(x)$.

For a dominant weight $v = b_1\omega_1 + b_2\omega_2 + b_3\omega_3 \in \Lambda(G_1)$, set $s(v) = 3b_1 + 4b_2 + 3b_3$.

I. First suppose that $Q < p$. Set $b = p - 1 - Q$. Since $3a_j + 4(a_{j+1} + a_{j+2} + a_{j+3} + a_{j+4}) + 3a_{j+5} \geq p$, it is obvious that $0 \leq b < \sum_{i=j+1}^{j+4} a_i$.

(i) Let $b > 1$. One can conclude that there exist weights

$$\mu_1 = c_1^1\omega_1 + b\omega_2 + c_3^1\omega_3, \quad \mu_2 = c_1^2\omega_1 + (b-1)\omega_2 + c_3^2\omega_3 \in I$$

such that $c_1^1 + c_3^1 = \left(\sum_{i=j}^{j+5} a_i\right) - b$ and $c_1^2 + c_3^2 = 1 - b + \sum_{i=j}^{j+5} a_i$. We can check that $s(\mu_1) = p-1$ and $s(\mu_2) = p-2$. Therefore these weights satisfy the assumptions of Proposition 2.15. By this proposition, Theorem 2.13, and Remark 2.14,

$$J_{M_i}(x) = J_{M_{i,\mathbb{C}}}(x_\mathbb{C}) \text{ and } \mathbb{N}_p \setminus \{1\} \subset J_{M_1}(x) \cup J_{M_2}(x).$$

This implies that $J_M(x) = \mathbb{N}_p$ for $b > 1$.

(ii) Let $b = 1$. Then $\frac{Q}{3} = \frac{p-2}{3} \geq 3$ as $p \geq 11$. One easily observes that there exist μ_1, μ_2, and $\mu_3 \in I$ such that

$$
\begin{aligned}
\mu_1 &= c_1^1 \omega_1 + \omega_2 + c_3^1 \omega_3, & c_1^1 + c_3^1 &= \tfrac{p-5}{3}, \\
\mu_2 &= c_1^2 \omega_1 + c_3^2 \omega_3, & c_1^2 + c_3^2 &= \tfrac{p-2}{3}, \\
\mu_3 &= c_1^3 \omega_1 + \omega_2 + c_3^3 \omega_3, & c_1^3 + c_3^3 &= \tfrac{p-8}{3}.
\end{aligned}
$$

It is not difficult to check that $s(\mu_i) < p$ and that the weights μ_i satisfy the assumptions of Proposition 2.15. This proposition, Theorem 2.13, and Remark 2.14 yield that

$$J_{M_1}(x) \supset \mathbb{N}_p^o \setminus \{1\}, \ J_{M_2}(x) \supset \mathbb{N}_p^e \setminus \{p-3, 2\}, \text{ and } \{2, p-3\} \subset J_{M_3}(x).$$

Hence $J_M(x) = \mathbb{N}_p$.

(iii) Now let $b = 0$. Since $b < \sum\limits_{i=j+1}^{j+4} a_i$, the sum $\sum\limits_{i=j+1}^{j+4} a_i > 0$. It is clear that $Q = p - 1$. As $p \geq 11$ and $p \equiv 1 \pmod 3$, then $p \geq 13$ and $a_j + a_{j+1} + \cdots + a_{j+5} \geq 4$. One can conclude that there exist weights μ_1, μ_2, and $\mu_3 \in I$ with the following properties:

$$
\begin{aligned}
\mu_1 &= c_1^1 \omega_1 + c_3^1 \omega_3, & c_1^1 + c_3^1 &= \tfrac{p-1}{3}, \\
\mu_2 &= c_1^2 \omega_1 + \omega_2 + c_3^2 \omega_3, & c_1^2 + c_3^2 &= \tfrac{p-4}{3}, \\
\mu_3 &= c_1^3 \omega_1 + \omega_2 + c_3^3 \omega_3, & c_1^3 + c_3^3 &= \tfrac{p-7}{3},
\end{aligned}
$$

and $c_1^k c_3^k \neq 0$ for $1 \leq k \leq 3$ if $a_j a_{j+5} \neq 0$. If $\sum\limits_{i=j+1}^{j+4} a_i > 1$, one easily observes that there exists a weight $\mu_4 \in I$ such that

$$\mu_4 = c_1^4 \omega_1 + 2\omega_2 + c_3^4 \omega_4, \quad c_1^4 + c_3^4 = \frac{p-10}{3}.$$

We can directly verify that $s(\mu_i) < p$ for $i = 1, 3, 4$ and $s(\mu_2) = p$, therefore the weights μ_1, μ_3, and μ_4 satisfy the assumptions of Proposition 2.15. By this proposition, Theorem 2.13, and Remark 2.14,

$$\mathbb{N}_p^o \setminus \{1, 5, p-2\} \subset J_{M_1}(x) \text{ for } \frac{p-1}{3} \equiv 2 \pmod 4,$$

$$\mathbb{N}_p^o \setminus \{3, p-2\} \subset J_{M_1}(x) \text{ for } \frac{p-1}{3} \equiv 0 \pmod 4,$$

$$\{3, 5, p-2\} \subset J_{M_3}(x).$$

By Proposition 2.22, an integer $l \in \mathbb{N}_p^e \setminus \{p-1\}$ is simultaneously contained or not contained both in $J_{M_2}(x)$ and in $J_{M_{2,\mathbb{C}}}(x_{\mathbb{C}})$. Using Theorem 2.13 and Remark 2.14, we conclude that $\mathbb{N}_p^e \setminus \{p-1\} \subset J_{M_2}(x)$. If $\sum\limits_{i=j+1}^{j+4} a_i > 1$, then

$p - 1 \in J_{M_4}(x)$. Let $\sum\limits_{i=j+1}^{j+4} a_i = 1$. Then $a_j a_{j+5} \neq 0$. In this case by Lemma 2.23, $p - 1 \in J_{M_2}(x)$. Hence $J_\varphi(x) = \mathbb{N}_p$.

Now for $Q < p$, the theorem is proved.

II. Let $Q \geq p$. Then $\sum\limits_{i=j}^{j+5} a_i > 3$ as $p \geq 11$.

(i) Let $\sum\limits_{i=j+1}^{j+4} a_i > 1$. First suppose that $p \equiv 2 \pmod 3$. One easily observes that there exist weights μ_1, μ_2, and $\mu_3 \in I$ such that

$$\mu_1 = c_1^1 \omega_1 + \omega_2 + c_3^1 \omega_3, \qquad c_1^1 + c_3^1 = \frac{p-5}{3},$$

$$\mu_2 = c_1^2 \omega_1 + 2\omega_2 + c_3^2 \omega_3, \qquad c_1^2 + c_3^2 = \frac{p-8}{3},$$

$$\mu_3 = c_1^3 \omega_1 + c_3^3 \omega_3, \qquad c_1^3 + c_3^3 = \frac{p-2}{3}\omega_1.$$

Obviously, $s(\mu_1)$ and $s(\mu_3) < p$, and $s(\mu_2) = p$. Using Propositions 2.15 and 2.22, Theorem 2.13, and Remark 2.14, we conclude that $\mathbb{N}_p^o \setminus \{1\} \subset J_{M_1}(x)$, $\mathbb{N}_p^e \setminus \{p-1\} \subset J_{M_2}(x)$, and $p - 1 \in J_{M_3}(x)$. Therefore $J_M(x) = \mathbb{N}_p$.

Now let $p \equiv 1 \pmod 3$. Then $p \geq 13$. It is not difficult to show that there exist $\mu_1, \ldots, \mu_4 \in I$ such that

$$\mu_1 = c_1^1 \omega_1 + c_3^1 \omega_3, \qquad c_1^1 + c_3^1 = \frac{p-1}{3},$$

$$\mu_2 = c_1^2 \omega_1 + \omega_2 + c_3^2 \omega_3, \qquad c_1^2 + c_3^2 = \frac{p-4}{3},$$

$$\mu_3 = c_1^3 \omega_1 + \omega_2 + c_3^3 \omega_3, \qquad c_1^3 + c_3^3 = \frac{p-7}{3},$$

$$\mu_4 = c_1^4 \omega_1 + 2\omega_2 + c_3^4 \omega_3, \qquad c_1^4 + c_3^4 = \frac{p-10}{3}.$$

One can directly check that $s(\mu_i) < p$ for $i \neq 2$ and $s(\mu_2) = p$. Applying Proposition 2.15, Theorem 2.13, and Remark 2.14, we conclude that

$$\mathbb{N}_p^o \setminus \{p-2, 5, 1\} \subset J_{M_1}(x) \text{ if } \frac{p-1}{3} \equiv 2 \pmod 4,$$

$$\mathbb{N}_p^o \setminus \{p-2, 3\} \subset J_{M_1}(x) \text{ if } \frac{p-1}{3} \equiv 0 \pmod 4,$$

$$\mathbb{N}_p^e \setminus \{p-1\} \subset J_{M_2}(x), \ \{3, 5, p-2\} \subset J_{M_3}(x), \ p - 1 \in J_{M_4}(x).$$

Hence $J_\varphi(x) = \mathbb{N}_p$.

(ii) Let $\sum\limits_{i=j+1}^{j+4} a_i \leq 1$. Then $a_j a_{j+5} \neq 0$. Set $\Sigma = \frac{p-k}{3}$ for $p \equiv k \pmod 3$, $k = 1$ or 2. Since $p \geq 11$, we have $\Sigma > 2$. One easily observes that there exist weights μ_1 and $\mu_2 \in I$ such that $\mu_1 = c_1^1\omega_1 + c_3^1\omega_3$, $\mu_2 = c_1^2\omega_1 + c_3^2\omega_3$, $c_1^j c_3^j \neq 0$ for $j = 1, 2$, $c_1^1 + c_3^1 = \Sigma$, $c_1^2 + c_3^2 = \Sigma + 1$. Obviously, $s(\mu_1) < p$, and

$$s(\mu_2) = \begin{cases} p+2 & \text{for} \quad p \equiv 1 \pmod 3, \\ p+1 & \text{for} \quad p \equiv 2 \pmod 3. \end{cases}$$

By Proposition 2.15, Theorem 2.13, and Remark 2.14,

$$\mathbb{N}_p^o \setminus \{1\} \subset J_{M_1}(x) \text{ for } p \equiv 1 \pmod 3 \text{ and } J_{M_1}(x) = \mathbb{N}_p^e \text{ for } p \equiv 2 \pmod 3.$$

In what follows, A is a good A_1-subgroup containing x and σ is the homomorphism $\Lambda(G_1) \to \mathbb{Z}$ from Theorem 2.11; $M_{2,a} \subset M_2|A$ is the weight subspace of weight $a \in \mathbb{Z}$. Recall that $\sigma(\alpha_i) = 2$. By Corollary 2.21, $M_2|A$ is a tilting module. Lemma 2.17 implies that $M_2|A$ is a direct sum of modules $T(b)$ with $b \leq p+2$ for $p \equiv 1 \pmod 3$ and $b \leq p+1$ for $p \equiv 2 \pmod 3$. One easily concludes that at least one of the coefficients c_1^2 or $c_3^2 > 1$.

Let $p \equiv 1 \pmod 3$. One can check that $\dim M_{2,p+2} = 1$ and $\dim M_{2,p} = 2$. Set $\nu = \omega(M_2)$. By [16, 1.5], $\Lambda(M_2)$ contains the weights $\nu - \alpha_1 - \alpha_2$, $\nu - \alpha_3 - \alpha_2$, $\nu - \alpha_1 - \alpha_3$, and at least one of the weights $\nu - 2\alpha_1$ or $\nu - 2\alpha_3$. Therefore $\dim M_{2,p-2} \geq 4$. By Lemma 2.18, the multiplicities of the weights p and $p-2$ in $T(p+2)$ are zero, and the multiplicity of $p-2$ in $T(p)$ equals 2. Now one easily observes that $M_2|A = T(p+2) \oplus T(p) \oplus M(p-2) \oplus M_2'$ where M_2' is a direct sum of irreducible modules with highest weights $\leq p-2$. Hence $p-1 \in J_{M_2}(x)$. By Proposition 2.22, Theorem 2.13, and Remark 2.14, $\mathbb{N}_{p-5}^e \subset J_{M_2}(x)$. One can deduce that there exists a weight $\mu_3 = c_1^3\omega_1 + c_3^3\omega_3 \in I$ with $c_1^3 + c_3^3 = \frac{p-4}{3}$. It is clear that $p - 3 \in J_{M_3}(x)$. Therefore $J_M(x) = \mathbb{N}_p$.

Let $p \equiv 2 \pmod 3$. By Proposition 2.22, Theorem 2.13, and Remark 2.14, $\mathbb{N}_{p-4}^o \setminus \{1\} \subset J_{M_2}(x)$. Next, we shall show that $p-2 \in J_{M_2}(x)$. Using Lemma 2.18, we conclude that the weight $p-1$ has multiplicity 1 in $T(p+1)$, and the weight $p-3$ has multiplicity 2 in $T(p+1)$ and multiplicity 1 in $M(p-1)$. It is clear that $\dim M_{2,p+1} = 1$ and $\dim M_{2,p-1} = 2$. Arguing as for estimating the dimension of the subspace $M_{2,p-2}$ for $p \equiv 1 \pmod 3$, we can verify that $\dim M_{2,p-3} \geq 4$. This implies that $M_2|A = T(p+1) \oplus M(p-1) \oplus M(p-3) \oplus M_2'$ where M_2' is a direct sum of irreducible modules with highest weights $\leq p-3$. Therefore $p-2 \in J_{M_2}(x)$. Hence $J_M(x) = \mathbb{N}_p$.

All possibilities have been considered. The theorem is proved.

The following proposition and lemma imply that we cannot remove the last assumption in Theorem 1.1.

Proposition 3.2 *Let $n = 6$ and N be the irreducible G-module with highest weight $\lambda = \frac{p-4}{3}\omega_1 + \omega_2$. Then $J_N(x) = \mathbb{N}_p \setminus \{p-1\}$.*

Proof Set $I_1 = \left\{ a\omega_1,\ a \leq \frac{p-1}{3};\quad b\omega_1 + \omega_2,\ b \leq \frac{p-4}{3} \right\}$. By Proposition 2.9, the set of composition factors of the restriction $N_{\mathbb{C}}|_{G_1,\mathbb{C}}$ consists of irreducible modules with highest weights from I_1. One easily checks that $\langle \lambda + \rho_G, \alpha_G \rangle < p$. Arguing as in the proof of Theorem 2.6, we can show that $N|G_1$ is a direct sum of irreducible modules with highest weights from I_1 and all such modules are composition factors of this restriction.

Let $v_1 = \frac{p-7}{3}\omega_1 + \omega_2$ and $v_2 = \frac{p-4}{3}\omega_1 + \omega_2 \in \Lambda^+(G_1)$; $M_i = M(v_i)$. Obviously, $v_i \in I_1$, $s(v_1) = p - 3$, and $s(v_2) = p$. Propositions 2.15 and 2.22 and Theorem 2.13 imply that

$$\mathbb{N}^o_{p-2} \setminus \{1\} \subset J_{M_1}(x), \quad \mathbb{N}^e_{p-3} \subset J_{M_2}(x).$$

By Lemma 2.23, $p - 1 \notin J_{M_2}(x)$. Obviously, $p - 1 \notin J_F(x)$ for $F = M(\mu)$ with $\mu \in I_1 \setminus \{v_2\}$. The proposition is proved. $\square$

Lemma 3.3 *Let ψ be an irreducible representation of G with highest weight $a\omega_1$, $a > p/3$. Then $J_\psi(x) = \mathbb{N}_p \setminus \{2, p-2\}$ for $a \neq \frac{p+1}{3}$ and $J_\psi(x) = \mathbb{N}_p \setminus \{2, p-3, p-2\}$ for $a = \frac{p+1}{3}$.*

The proof follows directly from the results in [3, part II, subsection 2.1].

Acknowledgement To the memory of a remarkable mathematician and a nice person Igor Aleksandrovich Faradjev.

References

[1] N. Bourbaki, *Groupes et algebres de Lie*, chaps. VII–VIII, Hermann, Paris (1975).

[2] T. S. Busel, 'On the Jordan block structure of a product of long and short root elements in irreducible representations of algebraic groups of type B_r', *J. Math. Sci.*, **219**, No. 3, 346–354 (2016).

[3] T. S. Busel, I. D. Suprunenko, 'The block structure of the images of regular unipotent elements from subsystem symplectic subgroups of rank 2 in irreducible representations of symplectic groups': I *Siberian Advances in Mathematics*, **30**, No. 1, 1–20 (2020). II *Siberian Advances in Mathematics*, **30**, No. 4, 229–274 (2020). III *Siberian Advances in Mathematics*, **31**, No. 2, 79–97 (2021).

[4] T. S. Busel, I. D. Suprunenko, 'The block structure of the images of regular unipotent elements from subsystem subgroups of type C_2 in irreducible representations of groups of type C_n with certain highest weights', Submitted.

[5] R. W. Carter, *Finite Groups of Lie Type: Conjugacy Classes and Complex Characters*, Wiley, Chichester (1985).

[6] S. Donkin, 'A filtration for rational modules', *Math. Z.*, **177**, 1–8 (1981).

[7] S. Donkin, 'On tilting modules for algebraic groups', *Math. Z.*, **212**, 39–60 (1993).

[8] J. A. Green, *Polynomial Representations of GL_n*, Second corrected and augmented edition, Springer, **830** (2007).

[9] R. Guralnick, G. Malle, 'Rational rigidity for $E_8(p)$', *Compos. Math.*, **150**, No. 10, 1679–1702 (2014).

[10] A. A. Osinovskaya, 'Restrictions of representations of algebraic groups of types E_n and F_4 to naturally embedded A_1-subgroups and the behavior of root elements', *Comm. in Algebra*, **33**, 213–220 (2005).

[11] A. A. Osinovskaya, 'Regular unipotent elements from subsystem subgroups of type A_2 in representations of the special linear groups', *J. Math. Sci.*, **219**, No. 3, 473–483 (2016).

[12] A. A. Osinovskaya, 'Nilpotent elements in irreducible representations of simple Lie algebras of small rank', Preprint, No. 5 (554), Minsk: National Academy of Sciences of Belarus. Institute of Mathematics (1999).

[13] A. A. Osinovskaya, I. D. Suprunenko, 'On the Jordan block structure of images of some unipotent elements in modular irreducible representations of the classical algebraic groups', *J. Algebra*, **273**, 586–600 (2004).

[14] A. A. Osinovskaya, I. D. Suprunenko, 'Unipotent elements from subsystem subgroups of type A_3 in representations of the special linear group' [in Russian], *Doklady Natsional'noi academii nauk Belarus*, **56**, No. 4, 11–15 (2012).

[15] J. Saxl, G. M. Seitz, 'Subgroups of algebraic groups containing regular unipotent elements' *J. London Math. Soc.*, **55**, No. 2, 370–386 (1997).

[16] G. M. Seitz, 'The maximal subgroups of classical algebraic groups', *Memoirs Amer. Math. Soc.*, **67**, No. 365 (1987).

[17] G. M. Seitz, 'Unipotent elements, tilting modules, and saturation', *Invent. Math.*, **141**, No. 3, 467–502 (2000).

[18] S. Smith, 'Irreducible modules and parabolic subgroups', *J. Algebra*, **75**, 286–289 (1982).

[19] R. Steinberg, *Lectures on Chevalley Groups*, Yale University Mathematics Department, New Haven, CN (1968).

[20] I. D. Suprunenko, 'Minimal polynomials of elements of order p in irreducible representations of Chevalley groups over fields of characteristic p', *Siberian Advances in Mathematics*, **6**, No. 4, 97–150 (1996).

[21] I. D. Suprunenko, 'On the block structure of regular unipotent elements from subsystem subgroups of type $A_1 \times A_2$ in representations of the special linear group', *J. Math. Sci.*, **183**, No. 5, 715–726 (2012).

[22] I. D. Suprunenko, 'Unipotent elements of nonprime order in representations of the classical algebraic groups: two big Jordan blocks', *J. Math. Sci.*, **199**, No. 3, 350–374 (2014).

[23] D. M. Testerman, A. Zalesski, 'Irreducibility in algebraic groups and regular unipotent elements', *Proc. Amer. Math. Soc.*, **141**, No. 1, 13–28 (2013).

[24] D. M. Testerman, A. Zalesski, 'Irreducible representations of simple algebraic groups in which a unipotent element is represented by a matrix with a single non-trivial Jordan block', *Journal of Group Theory*, **21**, No. 1, 1–20 (2018).

[25] P. H. Tiep, A. E. Zalesskii, 'Mod p reducibility of unramified representations of finite groups of Lie type', *Proc. London Math. Soc.*, **84**, No. 2, 439–472 (2002).

[26] M. V. Velichko, 'On the behaviour of the root elements in irreducible representations of simple algebraic groups', *Trudy Instituta matematiki*, **13**, No. 2, 116–121 (2005).

[27] M. V. Velichko, 'On the behaviour of root elements in modular representations of symplectic groups' [in Russian], *Trudy Instituta matematiki*, **14**, No. 2, 28–34 (2006).

[28] M. V. Velichko, 'The Jordan block structure of images of long root elements in modular representations of algebraic groups of types B_n and F_4' [in Russian], *Trudy Instituta matematiki*, **19**, No. 2, 7–11 (2011).

[29] M. V. Velichko, I. D. Suprunenko, 'On the behaviour of small quadratic elements in representations of the special linear group with large highest weights', *J. Math. Sci.*, **147**, No. 5, 7021–7041 (2007).

[30] D. P. Zhelobenko, 'The classical groups. Spectral analysis of their finite-dimensional representations', *Russian Mathematical Surveys*, **17**, No. 1, 1–94 (1962).

[31] A. G. Zhilinskii, 'Coherent systems of representations of inductive families of simple complex Lie algebras' [in Russian], Preprint, No. 38 (438), Minsk: National Academy of Sciences of Belarus. Institute of Mathematics (1990).

10

Some Remarks on the Parameter c_2 for a Distance-Regular Graph with Classical Parameters

Jack H. Koolen, Jongyook Park and Qianqian Yang

Abstract

Let Γ be a distance-regular graph with classical parameters (D, b, α, β) and $b \geq 1$. It is known that Γ is Q-polynomial with respect to θ_1, where $\theta_1 = \frac{b_1}{b} - 1$ is the second-largest eigenvalue of Γ. And it was shown that for a distance-regular graph Γ with classical parameters (D, b, α, β), $D \geq 5$ and $b \geq 1$, if a_1 is large enough compared to b and Γ is thin, then the intersection number c_2 of Γ is bounded above by a function of b. In this chapter, we obtain a similar result without the assumption that the graph Γ is thin.

CONTENTS

10.1 Introduction

In this chapter, we study distance-regular graphs with classical parameters (D, b, α, β). It is known that the parameter b is an integer such that $b \neq 0, -1$ (Proposition 10.2). Note that C. Weng [7] classified the distance-regular graphs with classical parameters (D, b, α, β) and $b \leq -2$. So, we consider the distance-regular graphs with classical parameters (D, b, α, β) and $b \geq 1$.

Let Γ be a distance-regular graph with classical parameters (D, b, α, β) and $b \geq 1$. From [1, corollary 8.4.2], we know that Γ is Q-polynomial with respect to θ_1, where $\theta_1 = \frac{b_1}{b} - 1$ is the second largest eigenvalue of Γ. In [5], it was shown that for a distance-regular graph Γ with classical parameters (D, b, α, β), $D \geq 5$ and $b \geq 1$, if a_1 is large enough compared to b and Γ is thin, then the intersection number c_2 of Γ is bounded above by a function of b. Note that we mean thin in the sense of Terwilliger; that is, all the irreducible modules of any Terwilliger algebra of the distance-regular graph are thin. We refer to [6] for this notion. In Theorem 10.7, we obtain a similar result without the assumption that the graph Γ is thin. We note that the Grassmann graphs $J_q(n, D)$ have $\alpha = b$ and $c_2 = (b + 1)^2$, and all the known families of distance-regular graphs with classical parameters (D, b, α, β) and unbounded diameter D satisfy $\alpha \leq b + \sqrt{b}$ and $c_2 \leq (b + 1)(b + \sqrt{b} + 1)$ ([1, pp. 194–195]).

To prove Theorem 10.7, we find some properties of t-plexes of a regular graph by using Hoffman theory (Theorem 10.5). In order to justify Theorem 10.5, we give Lemmas 10.3 and 10.4. In Lemma 10.3, for a vertex x not in a clique C in a graph, we consider the number of neighbors of x in C using a result of [3]. And then we apply Lemma 10.3 to a t-plex in a graph, and this gives us Lemma 10.4. We also need Lemma 10.6, and in this lemma, we give an upper bound on the order of a t-plex of a distance-regular graph with classical parameters (D, b, α, β) under the assumptions that $\alpha > b \geq 1$ and the diameter $D \geq 6$ is large enough compared to b and t. Now, we combine those results, and we obtain Theorem 10.7. Note that we do not give the proofs for the results of this chapter. We only give a sketch of proof for Theorem 10.7. For the proofs we refer to the two manuscripts [4, 8].

10.2 Definitions and Preliminaries

All the graphs considered in this chapter are finite, undirected and simple. The reader is referred to [1, 2] for more information. Let G be a connected graph with vertex set $V(G)$. The *distance* $d(x, y)$ between two vertices $x, y \in V(G)$ is the length of a shortest path between x and y in G. The *diameter* $D = D(G)$ of G is the maximum distance between any two vertices of G. For each $x \in V(G)$, let $G_i(x)$ be the set of vertices in G at distance i from x ($0 \leq i \leq D$). In addition, define $G_{-1}(x) = G_{D+1}(x) = \emptyset$. For the sake of simplicity, we denote $G_1(x)$ by $G(x)$. For any vertex x of G, the subgraph induced on $G(x)$ is called the *local graph* of G at x, and we denote it by $\Delta(x)$. For a vertex x of G, the cardinality $|G(x)|$ of $G(x)$ is called the *valency* of x in G. In particular, G is *regular* with valency k if $k = |G(x)|$ holds for all $x \in V(G)$. The *adjacency matrix* $A = A(G)$

of G is the matrix whose rows and columns are indexed by vertices of G and the (x, y)-entry is 1 whenever x and y are adjacent and 0 otherwise. The *eigenvalues* of G are the eigenvalues of A.

An induced subgraph with n vertices in a graph is called a *clique* if each of its vertices has valency $n - 1$, and a *t-plex* if each of its vertices has valency at least $n - t$. Let G be a connected graph with diameter D. The graph G is called *distance-regular* if there exist integers b_i, c_i $(0 \leq i \leq D)$ such that for any two vertices $x, y \in V(G)$ with $d(x, y) = i$, there are precisely c_i neighbors of y in $G_{i-1}(x)$ and b_i neighbors of y in $G_{i+1}(x)$, where we define $b_D = c_0 = 0$. In particular, any distance-regular graph is regular with valency $k := b_0$. We define $a_i := k - b_i - c_i$ for notational convenience. Note that $a_i = |G(y) \cap G_i(x)|$ holds for any two vertices x, y with $d(x, y) = i$ $(0 \leq i \leq D)$ and that the numbers a_i, b_i and c_i $(0 \leq i \leq D)$ are called the *intersection numbers* of G. From now on, we denote a distance-regular graph by Γ.

The following lemma is a well-known result for smallest eigenvalues of local graphs of distance-regular graphs.

Lemma 10.1 (Cf. [1, theorem 4.4.3]) *Let Γ be a distance-regular graph with valency k, diameter $D \geq 3$ and distinct eigenvalues $k = \theta_0 > \theta_1 > \cdots > \theta_D$. For a vertex x of Γ, let $\lambda_{\min}(\Delta(x))$ be the smallest eigenvalue of the local graph $\Delta(x)$ of Γ at x. Let $b = \frac{b_1}{\theta_1 + 1}$. Then $\lambda_{\min}(\Delta(x)) \geq -b - 1$.*

A distance-regular graph Γ is said to have *classical parameters* (D, b, α, β) if the diameter of Γ is D and the intersection numbers of Γ can be expressed as follows:

$$b_i = \left(\begin{bmatrix} D \\ 1 \end{bmatrix} - \begin{bmatrix} i \\ 1 \end{bmatrix} \right)(\beta - \alpha \begin{bmatrix} i \\ 1 \end{bmatrix}), \quad 0 \leq i \leq D - 1,$$

$$c_i = \begin{bmatrix} i \\ 1 \end{bmatrix} \left(1 + \alpha \begin{bmatrix} i - 1 \\ 1 \end{bmatrix} \right), \quad 1 \leq i \leq D,$$

where

$$\begin{bmatrix} j \\ 1 \end{bmatrix} = \begin{cases} \frac{b^j - 1}{b - 1} & \text{if } b \neq 1, \\ \binom{j}{1} & \text{if } b = 1. \end{cases}$$

From [1, corollary 8.4.2], we know that the eigenvalues of Γ are

$$\begin{bmatrix} D - i \\ 1 \end{bmatrix} (\beta - \alpha \begin{bmatrix} i \\ 1 \end{bmatrix}) - \begin{bmatrix} i \\ 1 \end{bmatrix} = \frac{b_i}{b^i} - \begin{bmatrix} i \\ 1 \end{bmatrix}, \quad 0 \leq i \leq D.$$

We note that $c_2 = (b + 1)(\alpha + 1)$ and that if $b \geq 1$, then the eigenvalues $\theta_i = \frac{b_i}{b^i} - \begin{bmatrix} i \\ 1 \end{bmatrix} (0 \leq i \leq D)$ of Γ have natural ordering, i.e., $k = \theta_0 > \theta_1 > \cdots > \theta_D$.

From the following result we know that the parameter b of a distance-regular graph with parameters (D, b, α, β) is an integer.

Proposition 10.2 (Cf. [1, Proposition 6.2.1])   *Let Γ be a distance-regular graph with classical parameters (D, b, α, β) and diameter $D \geq 3$. Then b is an integer such that $b \neq 0, -1$.*

Remark 1   (i) *For $b \leq -2$, C. Weng [7] classified the distance-regular graphs with classical parameters (D, b, α, β) and $b \leq -2$. His result states: If $D \geq 4$, $a_1 \neq 0$ and $c_2 > 1$, then one of the following holds: (1) Γ is the dual polar graph $^2A_{2D-1}(-b)$, (2) Γ is the Hermitian forms graph $Her_{-b}(D)$, (3) $\alpha = \frac{b-1}{2}$, $\beta = -\frac{1+b^d}{2}$ and $-b$ is a power of an odd prime.*
(ii) *For $b \geq 1$, [1, corollary 8.4.2] says that a distance-regular graph Γ with classical parameters (D, b, α, β) and $b \geq 1$ is Q-polynomial with respect to θ_1, where $\theta_1 = \frac{b_1}{b} - 1$ is the second largest eigenvalue of Γ. In [5], it was shown that for a distance-regular graph Γ with classical parameters (D, b, α, β), $D \geq 5$ and $b \geq 1$, if a_1 is large enough compared to b and Γ is thin, then the intersection number c_2 of Γ is bounded above by a function of b.*

10.3 Hoffman Graphs

In this section, we give some properties of t-plexes in a graph (for a fixed t). To obtain those properties, we use Hoffman theory and the following two lemmas. We refer to [8] for the proofs of the results in this section.

In the following lemma, for a vertex x not in a clique C, we consider the number of neighbors of x in C. This result is a consequence of [3, lemma 3.1].

Lemma 10.3   *Let $\lambda \geq 1$ be an integer and let G be a graph with smallest eigenvalue $\lambda_{\min}(G) \geq -\lambda$. Let C be a clique in G with order c. If $c \geq \lambda^4 - 2\lambda^3 + 3\lambda^2 - 3\lambda + 2$, then a vertex outside C has either at most $\lambda(\lambda - 1)$ neighbors in C, or at least $c - (\lambda - 1)^2$ neighbors in C.*

We apply Lemma 10.3 to a t-plex, then we obtain the following lemma.

Lemma 10.4   *Let $\lambda \geq 1$ be an integer and let G be a graph with smallest eigenvalue $\lambda_{\min}(G) \geq -\lambda$. If P is a t-plex of G and x is not a vertex of P, then one of the following holds:*

(i) *x has at most $t\lambda(\lambda - 1)$ neighbors in P,*
(ii) *x has at least $|V(P)| - t(\lambda - 1)^2$ neighbors in P,*
(iii) $|V(P)| \leq t(\lambda^4 - 2\lambda^3 + 3\lambda^2 - 3\lambda + 2) - 1.$

The next result states that a k-regular graph with smallest eigenvalue at least $-\lambda$ has a nice family of $((\lambda - 1)^2 + 1)$-plexes if k is large compared to λ.

Theorem 10.5   *Let $\lambda \geq 1$ and $m \geq \lambda^2(\lambda^4 - 2\lambda^3 + 3\lambda^2 - 3\lambda + 2)$ be two integers. There exist constants $C(\lambda, m)$ and $K(\lambda, m)$ such that for any connected k-regular graph G with smallest eigenvalue $\lambda_{\min}(G)$, if $\lambda_{\min}(G) \geq -\lambda$ and $k \geq K(\lambda, m)$, then there exist (maximal) $((\lambda - 1)^2 + 1)$-plexes $P_1, P_2, \ldots, P_s$ in G satisfying the following five properties:*

(i) $|\{i = 1, 2, \ldots, s \mid x \in V(P_i)\}| \leq \lambda$ *for any vertex x of G,*
(ii) $|V(P_i) \cap V(P_j)| \leq \lambda - 1$ *for any $1 \leq i < j \leq s$,*
(iii) *for any vertex x of G, $|\{y \mid y \text{ is adjacent to } x\} - \{y \mid x, y \in V(P_i)$
     for some $i = 1, \ldots, s\}| \leq C(\lambda, m)$,*
(iv) $|V(P_i)| \geq m$ *for $i = 1, \ldots, s$,*
(v) *any vertex not in P_i ($i = 1, 2, \ldots, s$) has at most $\lambda(\lambda - 1)((\lambda - 1)^2 + 1)$
     neighbors in P_i.*

10.4 The Parameter c_2 is Bounded by a Function of b

In this section, we give our main result which shows that for a distance-regular graph with classical parameters (D, b, α, β), if $b \geq 1$ and the diameter D is large enough compared to b, then the intersection number c_2 is at most $(b + 2)^4$. This is a generalization of a result in [5], because we do not assume that the graph is thin. We refer to [4] for the proofs of the results in this section.

To prove our main result, we need the following lemma, and in the lemma, we give an upper bound of the order of a t-plex in a distance-regular graph with classical parameters (D, b, α, β) such that the diameter D is large enough compared to b and $\alpha > b \geq 1$. We note that the assumption $\alpha > b$ is natural as $\alpha \leq b$ implies that $c_2 = (b + 1)(\alpha + 1) \leq (b + 1)^2$.

Lemma 10.6   *Let Γ be a distance-regular graph with classical parameters (D, b, α, β) such that $D \geq 6$ and $\alpha > b \geq 1$. Let $t \geq 1$ be an integer. If the diameter D is large enough compared to t and b, then the order of any t-plex in Γ is smaller than $\beta + 2t + 1$.*

We combine all the results above in Sections 10.3 and 10.4, and then we obtain our main result as follows.

Theorem 10.7   *Let Γ be a distance-regular graph with classical parameters (D, b, α, β) such that $D \geq 6$ and $b \geq 1$. If the diameter D of Γ is large enough compared to b, then the intersection number c_2 of Γ is at most $(b + 2)^4$.*

Now, we give a sketch of the proof for Theorem 10.7.

Sketch of proof   If $\alpha \leq b$, then $c_2 = (b + 1)(\alpha + 1) \leq (b + 1)^2$. So, we may assume $\alpha > b \geq 1$. Also, we have $\beta > \alpha > 1$ as $b_1 = (\begin{bmatrix} D \\ 1 \end{bmatrix} - 1)(\beta - \alpha) > 0$.

Let x be a vertex of Γ and let $\Delta = \Delta(x)$ be the local graph of Γ at x. Note that Δ is an a_1-regular graph with its smallest eigenvalue $\lambda_{\min}(\Delta) \geq -b - 1$ (by Lemma 10.1). Since $a_1 = -1 + \beta + \alpha(\begin{bmatrix} D \\ 1 \end{bmatrix} - 1) > -1 + \beta + b(\begin{bmatrix} D \\ 1 \end{bmatrix} - 1)$ and D is large enough compared to b, we may assume that the valency a_1 of Δ is large enough compared to $b + 1$.

Set $\lambda = b + 1(\geq 2)$ and $m = (\lambda + 1)^6 \geq \lambda^2(\lambda^4 - 2\lambda^3 + 3\lambda^2 - 3\lambda + 2)$. Fix a vertex u of Δ. Then there exist a constant $C(\lambda, m)$ and a family of $((\lambda - 1)^2 + 1)$-plexes $P_1, P_2, \ldots, P_s$ in Δ satisfying the five conditions of Theorem 10.5 if a_1 is large enough compared to λ and m. Note that the constant $C(\lambda, m)$ only depends on λ as $m = (\lambda + 1)^6$.

We denote the set of neighbors of u in Δ by Z. Also, we denote the set of non-neighbors of u in Δ (excluding u itself) by W. We will count the number e of edges between Z and W in two different ways using $((\lambda - 1)^2 + 1)$-plexes among $P_1, P_2, \ldots, P_s$ containing u. And then we have that

$$e \geq (a_1 - \binom{\lambda}{2}(\lambda - 2) - C(\lambda, m))$$
$$\cdot (a_1 - 1 - (\beta + 2((\lambda - 1)^2 + 1) + \lambda(\lambda - 1)^2((\lambda - 1)^2 + 1) + C(\lambda, m)))$$

and

$$e \leq (k - 1 - a_1 - \lambda(\lambda - 1)^2)(\lambda^2(\lambda - 1)((\lambda - 1)^2 + 1) + C(\lambda, m)) + \lambda(\lambda - 1)^2 a_1.$$

By comparing these two inequalities, we obtain that $\alpha \leq C(\lambda, m) + \lambda^5 + 1$, and hence $c_2 = (\alpha + 1)(b + 1) \leq \lambda(C(\lambda, m) + \lambda^5 + 2)$.

Now, we set $m' := \lambda(C(\lambda, m) + \lambda^5 + 2) + 2(\lambda - 1)^2 + 3$. And we use Theorem 10.5 with the new m'. Then there exist a new constant $C(\lambda, m')$ and a new family of $((\lambda - 1)^2 + 1)$-plexes $P'_1, P'_2, \ldots, P'_{s'}$ in Δ satisfying the five conditions of Theorem 10.5 for this new m' if a_1 is large enough compared to λ and m'.

Next, we fix a vertex u of Δ and will show that all the $((\lambda - 1)^2 + 1)$-plexes $P'_1, P'_2, \ldots, P'_{s'}$ are cliques if a_1 is large enough compared to λ and m'. Let P be a $((\lambda - 1)^2 + 1)$-plex among $P'_1, P'_2, \ldots, P'_{s'}$. Assume that P is not a clique, that is, there exist two distinct vertices y_1 and y_2 in P that are not adjacent. Then from

Theorem 10.5 (iv), we know that y_1 and y_2 have at least $|V(P)| - 2 - 2(\lambda - 1)^2 > c_2$ common neighbors in P. This gives a contradiction, and thus P is a clique.

Now, we replace $((\lambda - 1)^2 + 1)$-plexes $P_1, P_2, \ldots, P_s$ by cliques $P'_1, P'_2, \ldots, P'_{s'}$. Following the same procedures, we obtain the desired result. Note that by using the fact that all these s' $((\lambda - 1)^2 + 1)$-plexes are cliques, we reduce the number of neighbors in a $((\lambda - 1)^2 + 1)$-plex for a vertex outside from $\lambda(\lambda - 1)((\lambda - 1)^2 + 1)$ to $\lambda(\lambda - 1)$ and that it is the essence why we can get an upper bound of α by $(\lambda + 1)^3$. □

Acknowledgments

Jack H. Koolen is partially supported by the National Key R. and D. Program of China (No. 2020YFA0713100), the National Natural Science Foundation of China (No. 12071454), and the Anhui Initiative in Quantum Information Technologies (No. AHY150000).

Jongyook Park is supported by the National Research Foundation of Korea (NRF) grant funded by the Korea government (MSIT) (NRF-2020R1A2C1A01 101838).

References

[1] A. E. Brouwer, A. M. Cohen and A. Neumaier, *Distance-Regular Graphs*, Springer-Verlag, Berlin, 1989.

[2] E. R. van Dam, J. H. Koolen and H. Tanaka, Distance-regular graphs, *Electron. J. Combin.* (2016), ♯DS22.

[3] G. R. W. Greaves, J. H. Koolen and J. Park, Augmenting the Delsarte bound: A forbidden interval for the order of maximal cliques in strongly regular graphs, *Eur. J. Combin.* 97 (2021), 103384.

[4] J. H. Koolen, J. Park and Q. Yang, *The Parameter c_2 for Distance-Regular Graphs with Classical Parameters*, manuscript (2022).

[5] Y.-Y. Tan, J. H. Koolen, M.-Y. Cao and J. Park, Thin Q-polynomial distance-regular graphs have bounded c_2, *Graphs Combin.* 38 (2022), 175.

[6] P. Terwilliger, The subconstituent algebra of an association scheme, (Part I), *J. Algebraic Combin.* 1 (1992), 363–388.

[7] C. Weng, Classical distance-regular graphs of negative type, *J. Combin. Theory Ser. B* 76 (1999), 93–116.

[8] Q. Yang and J. H. Koolen, *A Structure Theory for Regular Graphs with Fixed Smallest Eigenvalue*, manuscript (2022).

11

Distance-Regular Graphs, the Subconstituent Algebra, and the Q-Polynomial Property

Paul Terwilliger

Abstract

This survey chapter contains a tutorial introduction to distance-regular graphs, with an emphasis on the subconstituent algebra and the Q-polynomial property.

CONTENTS

11.1 Introduction

This survey chapter contains a tutorial introduction to distance-regular graphs, with an emphasis on the subconstituent algebra and the *Q*-polynomial property. Our treatment is roughly based on the unpublished lecture notes [61], along with two more recent versions [80]. The treatment is not comprehensive; instead we restrict out attention to those topics that seem most important. A proof is given for every main result in the chapter. We do not assume any prior knowledge about distance-regular graphs. We intend the present chapter to complement the excellent recent works [3, 19].

A hypercube is an elementary example of a *Q*-polynomial distance-regular graph. The subconstituent algebra of a hypercube is described in [27]. We advise the beginning reader to treat the hypercubes as a running example, using [27] as a guide.

As we go along, we will encounter a linear-algebraic object called a tridiagonal pair. As a warmup, we now define a tridiagonal pair.

Let $\mathbb{F}$ denote a field. Let V denote a nonzero, finite-dimensional vector space over $\mathbb{F}$. An $\mathbb{F}$-linear map $A: V \to V$ is called *diagonalizable* whenever V is spanned by the eigenspaces of A.

Definition 1.1 (See [33, definition 1.1])   A *tridiagonal pair* on V is an ordered pair of $\mathbb{F}$-linear maps $A : V \to V$ and $A^* : V \to V$ that satisfy the following four conditions.

(i) Each of A, A^* is diagonalizable.

(ii) There exists an ordering $\{V_i\}_{i=0}^d$ of the eigenspaces of A such that

$$A^* V_i \subseteq V_{i-1} + V_i + V_{i+1} \qquad (0 \leq i \leq d), \qquad (1)$$

where $V_{-1} = 0$ and $V_{d+1} = 0$.

(iii) There exists an ordering $\{V_i^*\}_{i=0}^{\delta}$ of the eigenspaces of A^* such that

$$AV_i^* \subseteq V_{i-1}^* + V_i^* + V_{i+1}^* \qquad (0 \le i \le \delta), \qquad (2)$$

where $V_{-1}^* = 0$ and $V_{\delta+1}^* = 0$.

(iv) There does not exist a subspace W of V such that $AW \subseteq W$, $A^*W \subseteq W$, $W \neq 0$, $W \neq V$.

Note 1.2   According to a common notational convention, A^* denotes the conjugate transpose of A. We are not using this convention. In a tridiagonal pair A, A^*, the maps A and A^* are arbitrary subject to (i)–(iv) above.

Referring to Definition 1.1, by [33, lemma 4.5] the integers d and δ from (1), (2) are equal; we call this common value the *diameter* of the pair.

The concept of a tridiagonal pair was formally introduced in [33]. However, the concept shows up earlier [67–69] in connection with the subconstituent algebra T of a Q-polynomial distance-regular graph. As we will see, for such a graph every irreducible T-module gives a tridiagonal pair in a natural way.

We refer the reader to [3, 32, 36, 37, 53] for background and historical remarks about tridiagonal pairs.

We will encounter tridiagonal pairs of the following sort.

Definition 1.3 (See [71, definition 1.1])   A *Leonard pair* on V is a tridiagonal pair A, A^* on V such that the eigenspaces $\{V_i\}_{i=0}^{d}$ and $\{V_i^*\}_{i=0}^{d}$ all have dimension one.

We refer the reader to [3, 71, 76, 78] for background and historical remarks about Leonard pairs.

11.2 Distance-Regular Graphs

We now turn our attention to distance-regular graphs. After a brief review of the basic definitions, we will describe the Bose–Mesner algebra, the dual Bose–Mesner algebra, and the subconstituent algebra. For more information we refer the reader to [2–4, 19, 27, 67–69].

Let $\mathbb{R}$ denote the field of real numbers. Let X denote a nonempty finite set. Let $\mathrm{Mat}_X(\mathbb{R})$ denote the $\mathbb{R}$-algebra consisting of the matrices with rows and columns indexed by X and all entries in $\mathbb{R}$. Let $I \in \mathrm{Mat}_X(\mathbb{R})$ denote the identity matrix. Let $V = \mathbb{R}^X$ denote the vector space over $\mathbb{R}$ consisting of the column vectors with coordinates indexed by X and all entries in $\mathbb{R}$. The algebra $\mathrm{Mat}_X(\mathbb{R})$ acts on V by left multiplication. We call V the *standard module*. We endow V with a bilinear form $\langle \, , \, \rangle$ that satisfies $\langle u, v \rangle = u^t v$ for $u, v \in V$, where t denotes transpose. This bilinear form is symmetric. For $u \in V$ we abbreviate

$\|u\|^2 = \langle u, u \rangle$. Note that $\|u\|^2 \geq 0$, with equality if and only if $u = 0$. For $u, v \in V$ and $B \in \mathrm{Mat}_X(\mathbb{R})$, we have $\langle Bu, v \rangle = \langle u, B^t v \rangle$. For all $y \in X$, define a vector $\hat{y} \in V$ that has y-coordinate 1 and all other coordinates 0. The vectors $\{\hat{y}\}_{y \in X}$ form an orthonormal basis for V. For later use, define a matrix $J \in \mathrm{Mat}_X(\mathbb{R})$ that has all entries 1.

Let $\Gamma = (X, \mathcal{R})$ denote a finite, undirected, connected graph, without loops or multiple edges, with vertex set X and edge set $\mathcal{R}$. Vertices y, z are *adjacent* whenever y, z form an edge. Let ∂ denote the path-length distance function for Γ, and define $D = \max\{\partial(y, z) | y, z \in X\}$. We call D the *diameter* of Γ. For $y \in X$ and an integer $i \geq 0$ define $\Gamma_i(y) = \{z \in X | \partial(y, z) = i\}$. We abbreviate $\Gamma(y) = \Gamma_1(y)$. For an integer $k \geq 0$, we say that Γ is *regular with valency k* whenever $|\Gamma(y)| = k$ for all $y \in X$. We say that Γ is *distance-regular* whenever for all integers h, i, j $(0 \leq h, i, j \leq D)$ and for all vertices $y, z \in X$ with $\partial(y, z) = h$, the number $p_{i,j}^h = |\Gamma_i(y) \cap \Gamma_j(z)|$ is independent of y and z. The $p_{i,j}^h$ are called the *intersection numbers* of Γ. From now until the end of Section 11.19, we assume that Γ is distance-regular with $D \geq 3$. By construction, $p_{i,j}^h = p_{j,i}^h$ for $0 \leq h, i, j \leq D$. We abbreviate

$$c_i = p_{1,i-1}^i \ (1 \leq i \leq D), \qquad a_i = p_{1,i}^i \ (0 \leq i \leq D),$$
$$b_i = p_{1,i+1}^i \ (0 \leq i \leq D-1).$$

Note that $a_0 = 0$ and $c_1 = 1$. Moreover

$$c_i > 0 \ (1 \leq i \leq D), \qquad b_i > 0 \ (0 \leq i \leq D-1).$$

The graph Γ is regular with valency $k = b_0$. Moreover,

$$c_i + a_i + b_i = k \qquad (0 \leq i \leq D),$$

where $c_0 = 0$ and $b_D = 0$. For $0 \leq i \leq D$, define $k_i = p_{i,i}^0$ and note that $k_i = |\Gamma_i(y)|$ for all $y \in X$. We have $k_0 = 1$ and $k_1 = k$. By a routine counting argument, $k_{i-1} b_{i-1} = k_i c_i$ for $1 \leq i \leq D$. Consequently,

$$k_i = \frac{b_0 b_1 \cdots b_{i-1}}{c_1 c_2 \cdots c_i} \qquad (0 \leq i \leq D). \tag{3}$$

By the triangle inequality, the following hold for $0 \leq h, i, j \leq D$:

(i) $p_{i,j}^h = 0$ if one of h, i, j is greater than the sum of the other two;
(ii) $p_{i,j}^h \neq 0$ if one of h, i, j is equal to the sum of the other two.

The following results are verified by routine counting arguments:

$$p_{0,j}^h = \delta_{h,j} \quad (0 \leq h, j \leq D); \qquad p_{i,0}^h = \delta_{h,i} \quad (0 \leq h, i \leq D);$$

$$p_{i,j}^0 = \delta_{i,j} k_i \quad (0 \leq i, j \leq D); \qquad \sum_{i=0}^{D} p_{i,j}^h = k_j \quad (0 \leq h, j \leq D).$$

We recall the Bose–Mesner algebra of Γ. For $0 \leq i \leq D$ define a matrix $A_i \in$ $\mathrm{Mat}_X(\mathbb{R})$ with (y, z)-entry:

$$(A_i)_{y,z} = \begin{cases} 1, & \text{if } \partial(y, z) = i; \\ 0, & \text{if } \partial(y, z) \neq i \end{cases} \qquad (y, z \in X).$$

We call A_i the ith *distance matrix* of Γ. For $y \in X$, we have $A_i \hat{y} = \sum_{z \in \Gamma_i(y)} \hat{z}$. We abbreviate $A = A_1$ and call this the *adjacency matrix* of Γ. We observe (i) $A_0 = I$; (ii) $\sum_{i=0}^{D} A_i = J$; (iii) $A_i^t = A_i$ $(0 \leq i \leq D)$; (iv) $A_i A_j = \sum_{h=0}^{D} p_{i,j}^h A_h$ $(0 \leq i, j \leq D)$. Consequently the matrices $\{A_i\}_{i=0}^{D}$ form a basis for a commutative subalgebra M of $\mathrm{Mat}_X(\mathbb{R})$, called the *Bose–Mesner algebra* of Γ. The distance matrices are symmetric and mutually commute. Therefore they can be simultaneously diagonalized. Consequently M has a second basis $\{E_i\}_{i=0}^{D}$ such that (i) $E_0 = |X|^{-1}J$; (ii) $\sum_{i=0}^{D} E_i = I$; (iii) $E_i^t = E_i$ $(0 \leq i \leq D)$; (iv) $E_i E_j = \delta_{i,j} E_i$ $(0 \leq i, j \leq D)$. We call $\{E_i\}_{i=0}^{D}$ the *primitive idempotents* of Γ. The primitive idempotent E_0 is called *trivial*. We have

$$V = \sum_{i=0}^{D} E_i V \qquad \text{(orthogonal direct sum)}.$$

For $0 \leq i \leq D$, the subspace $E_i V$ is a common eigenspace of M. Note that

$$E_0 V = \mathbb{R}\mathbf{1}, \qquad \mathbf{1} = \sum_{y \in X} \hat{y}.$$

For $0 \leq i \leq D$, let m_i denote the dimension of $E_i V$. We have $m_i = \mathrm{tr}(E_i)$, where tr denotes trace. Note that $m_0 = 1$.

We recall the dual Bose–Mesner algebras of Γ. For the rest of this section, fix a vertex $x \in X$. For $0 \leq i \leq D$, let $E_i^* = E_i^*(x)$ denote the diagonal matrix in $\mathrm{Mat}_X(\mathbb{R})$ with (y, y)-entry

$$(E_i^*)_{y,y} = \begin{cases} 1, & \text{if } \partial(x, y) = i; \\ 0, & \text{if } \partial(x, y) \neq i \end{cases} \qquad (y \in X). \qquad (4)$$

We call E_i^* the ith *dual primitive idempotent of* Γ *with respect to* x [67, p. 378]. For $y \in X$, we have $E_i^* \hat{y} = \hat{y}$ if $\partial(x, y) = i$, and $E_i^* \hat{y} = 0$ if $\partial(x, y) \neq i$. We observe (i) $\sum_{i=0}^{D} E_i^* = I$; (ii) $E_i^{*t} = E_i^*$ $(0 \leq i \leq D)$; (iii) $E_i^* E_j^* = \delta_{i,j} E_i^*$ $(0 \leq i, j \leq D)$. By these facts, $\{E_i^*\}_{i=0}^{D}$ form a basis for a commutative subalgebra $M^* = M^*(x)$ of $\mathrm{Mat}_X(\mathbb{R})$. We call M^* the *dual Bose–Mesner algebra of* Γ *with respect to* x [67, p. 378].

We recall the subconstituents of Γ with respect to x. From (4), we find

$$E_i^* V = \mathrm{Span}\{\hat{y} \mid y \in \Gamma_i(x)\} \qquad (0 \leq i \leq D). \qquad (5)$$

By (5) and since $\{\hat{y}\}_{y \in X}$ is an orthonormal basis for V, we find

$$V = \sum_{i=0}^{D} E_i^* V \qquad \text{(orthogonal direct sum)}.$$

For $0 \le i \le D$, the subspace $E_i^* V$ is a common eigenspace for M^*. Observe that the dimension of $E_i^* V$ is equal to k_i. Also $\text{tr}(E_i^*) = k_i$. We call $E_i^* V$ the *ith subconstituent of* Γ *with respect to* x. Note that $E_0^* V = \mathbb{R}\hat{x}$. Also note that

$$A E_i^* V \subseteq E_{i-1}^* V + E_i^* V + E_{i+1}^* V \qquad (0 \le i \le D),$$

where $E_{-1}^* = 0$ and $E_{D+1}^* = 0$.

We recall the subconstituent algebra of Γ with respect to x. Let $T = T(x)$ denote the subalgebra of $\text{Mat}_X(\mathbb{R})$ generated by M and M^*. Observe that T has finite dimension. The algebra T is closed under the transpose map, because M and M^* are closed under the transpose map. We call T the *subconstituent algebra* (or *Terwilliger algebra*) *of* Γ *with respect to* x [67, Definition 3.3]. See [13, 14, 18, 25, 27, 30, 63, 67–69] for more information on the subconstituent algebra.

We recall the T-modules. By a *T-module* we mean a subspace $W \subseteq V$ such that $BW \subseteq W$ for all $B \in T$. A T-module W is called *irreducible* whenever $W \ne 0$ and W does not contain a T-module besides 0 and W. Let W denote a T-module and let W' denote a T-module contained in W. Then the orthogonal complement of W' in W is a T-module. It follows that each T-module is an orthogonal direct sum of irreducible T-modules. In particular, V is an orthogonal direct sum of irreducible T-modules.

The following relations are verified by matrix multiplication:

$$|X| E_0^* E_0 E_0^* = E_0^*, \qquad\qquad |X| E_0 E_0^* E_0 = E_0.$$

11.3 Some Polynomials

Throughout this section, $\Gamma = (X, \mathcal{R})$ denotes a distance-regular graph with diameter $D \ge 3$. We will discuss two sequences of polynomials that are associated with the distance matrices of Γ.

Lemma 3.1 *We have*

$$A A_i = b_{i-1} A_{i-1} + a_i A_i + c_{i+1} A_{i+1} \qquad (1 \le i \le D - 1),$$
$$A A_D = b_{D-1} A_{D-1} + a_D A_D.$$

Proof This is $A_i A_j = \sum_{h=0}^{D} p_{i,j}^h A_h$ with $j = 1$. $\qquad\qquad\qquad \square$

Let λ denote an indeterminate. Let $\mathbb{R}[\lambda]$ denote the $\mathbb{R}$-algebra of polynomials in λ that have all coefficients in $\mathbb{R}$.

Definition 3.2 We define some polynomials $\{v_i\}_{i=0}^{D+1}$ in $\mathbb{R}[\lambda]$ such that

$$v_0 = 1, \qquad v_1 = \lambda,$$
$$\lambda v_i = b_{i-1}v_{i-1} + a_i v_i + c_{i+1}v_{i+1} \qquad (1 \leq i \leq D),$$

where $c_{D+1} = 1$.

Lemma 3.3 *The following* (i)–(iv) *hold:*

(i) *$\deg v_i = i$* $(0 \leq i \leq D+1)$;
(ii) *the coefficient of λ^i in v_i is $(c_1 c_2 \cdots c_i)^{-1}$* $(0 \leq i \leq D+1)$;
(iii) *$v_i(A) = A_i$* $(0 \leq i \leq D)$;
(iv) *$v_{D+1}(A) = 0$.*

Proof (i), (ii) By Definition 3.2.
(iii), (iv) Compare Lemma 3.1 and Definition 3.2. □

Corollary 3.4 *The following hold:*

(i) *the algebra M is generated by A;*
(ii) *the minimal polynomial of A is $c_1 c_2 \cdots c_D v_{D+1}$.*

Proof By Lemma 3.3 and since $\{A_i\}_{i=0}^D$ is a basis for M. □

Next we consider the eigenvalues of A. Since $\{E_i\}_{i=0}^D$ form a basis for M, there exist real numbers $\{\theta_i\}_{i=0}^D$ such that

$$A = \sum_{i=0}^D \theta_i E_i. \tag{6}$$

Lemma 3.5 *The following* (i)–(iii) *hold:*

(i) *the polynomial v_{D+1} has $D+1$ mutually distinct roots $\{\theta_i\}_{i=0}^D$;*
(ii) *the eigenspaces of A are $\{E_i V\}_{i=0}^D$;*
(iii) *for $0 \leq i \leq D$, θ_i is the eigenvalue of A for $E_i V$.*

Proof (i) The roots of v_{D+1} are mutually distinct by Corollary 3.4(ii) and since A is diagonalizable. These roots are $\{\theta_i\}_{i=0}^D$ by (6).
(ii), (iii) By (6). □

Definition 3.6 For $0 \leq i \leq D$ we call θ_i the *ith eigenvalue of* Γ (with respect to the given ordering of the primitive idempotents).

For convenience we adjust the normalization of the polynomials v_i.

Definition 3.7 Define the polynomial

$$u_i = \frac{v_i}{k_i} \qquad (0 \leq i \leq D). \tag{7}$$

Lemma 3.8 *We have*

$$u_0 = 1, \qquad u_1 = k^{-1}\lambda,$$

$$\lambda u_i = c_i u_{i-1} + a_i u_i + b_i u_{i+1} \qquad (1 \leq i \leq D-1),$$

$$\lambda u_D - c_D u_{D-1} - a_D u_D = k_D^{-1} v_{D+1}.$$

Proof Evaluate the recurrence in Definition 3.2 using $v_i = k_i u_i$ $(0 \leq i \leq D)$ and (3). $\qquad\square$

Recall that $\{A_i\}_{i=0}^D$ and $\{E_i\}_{i=0}^D$ are bases for the vector space M. Next we describe how these bases are related.

Lemma 3.9 *For $0 \leq j \leq D$, we have*

(i) $A_j = \sum_{i=0}^D v_j(\theta_i)E_i$;
(ii) $E_j = |X|^{-1}m_j \sum_{i=0}^D u_i(\theta_j)A_i$.

Proof   (i) We have

$$A_j = v_j(A) = v_j(A) \sum_{i=0}^D E_i = \sum_{i=0}^D v_j(\theta_i)E_i.$$

(ii) Define $S = |X|^{-1}m_j \sum_{i=0}^D u_i(\theta_j)A_i$. We show that $E_j = S$. Expanding AS using Lemma 3.1, we routinely obtain $AS = \theta_j S$. By this and since $S \in M$, we obtain $S = \alpha E_j$ for some $\alpha \in \mathbb{R}$. In the equation $S = \alpha E_j$, take the trace of each side. We have $\mathrm{tr}(S) = m_j$, because $\mathrm{tr}(A_\ell) = \delta_{0,\ell}|X|$ for $0 \leq \ell \leq D$. We have $\mathrm{tr}(E_j) = m_j$. By these comments $\alpha = 1$, so $E_j = S$. $\qquad\square$

Lemma 3.10 *For $0 \leq i, j \leq D$,*

$$E_i A_j = v_j(\theta_i)E_i = A_j E_i.$$

Proof To verify this equation, eliminate A_j using Lemma 3.9(i) and simplify the result. $\qquad\square$

Lemma 3.11 *For $0 \leq i \leq D$ we have*

(i) $v_i(\theta_0) = k_i$;
(ii) $u_i(\theta_0) = 1$.

Proof   (i) The matrix A_i has constant row sum k_i. Therefore $A_i J = k_i J$. We have $E_0 = |X|^{-1}J$, so $A_i E_0 = k_i E_0$. By Lemma 3.10, $A_i E_0 = v_i(\theta_0)E_0$. By these comments $v_i(\theta_0) = k_i$. $\qquad\square$
(ii) By (i) and $v_i = k_i u_i$.

It is often said that the polynomials $\{u_i\}_{i=0}^D$ and $\{v_i\}_{i=0}^D$ are orthogonal [2, p. 201]. Our next goal is to explain what this means. We will bring in a bilinear form, and explain what is orthogonal to what.

We endow the vector space $\text{Mat}_X(\mathbb{R})$ with a bilinear form $\langle \, , \, \rangle$ such that

$$\langle B, C \rangle = \text{tr}(BC^t) \qquad\qquad B, C \in \text{Mat}_X(\mathbb{R}).$$

This bilinear form is symmetric. For $B \in \text{Mat}_X(\mathbb{R})$ we abbreviate $\|B\|^2 = \langle B, B \rangle$. Note that $\|B\|^2 \geq 0$, with equality if and only if $B = 0$.

Lemma 3.12 *For $G, H, K \in \text{Mat}_X(\mathbb{R})$, we have*

$$\langle GH, K \rangle = \langle H, G^t K \rangle = \langle G, KH^t \rangle.$$

Proof Use $\text{tr}(BC) = \text{tr}(CB)$.   □

Lemma 3.13 *For $0 \leq i, j \leq D$, we have*

(i) $\langle E_i, E_j \rangle = \delta_{i,j} m_i$;
(ii) $\langle A_i, A_j \rangle = \delta_{i,j} k_i |X|$;
(iii) $\langle A_i, E_j \rangle = v_i(\theta_j) m_j$.

Proof (i) We have

$$\langle E_i, E_j \rangle = \text{tr}(E_i E_j^t) = \text{tr}(E_i E_j) = \delta_{i,j}\text{tr}(E_i) = \delta_{i,j} m_i.$$

(ii) We have

$$\langle A_i, A_j \rangle = \text{tr}(A_i A_j^t) = \text{tr}(A_i A_j) = \sum_{h=0}^{D} p_{i,j}^h \text{tr}(A_h) = p_{i,j}^0 |X|$$
$$= \delta_{i,j} k_i |X|.$$

(iii) Eliminate A_i using Lemma 3.9(i), and evaluate the result using (i).   □

Proposition 3.14 *We have*

$$\sum_{\ell=0}^{D} u_\ell(\theta_i) u_\ell(\theta_j) k_\ell = \delta_{i,j} m_i^{-1} |X| \qquad\qquad (0 \leq i, j \leq D),$$

$$\sum_{\ell=0}^{D} u_i(\theta_\ell) u_j(\theta_\ell) m_\ell = \delta_{i,j} k_i^{-1} |X| \qquad\qquad (0 \leq i, j \leq D).$$

Proof The first equation comes from $\langle E_i, E_j \rangle = \delta_{i,j} m_i$. In this equation, eliminate E_i and E_j using Lemma 3.9(ii), and evaluate the result using Lemma 3.13(ii). The second equation comes from $\langle A_i, A_j \rangle = \delta_{i,j} k_i |X|$. In this equation, eliminate A_i and A_j using Lemma 3.9(i), and evaluate the result using Lemma 3.13(i).   □

Proposition 3.15 *We have*

$$\sum_{\ell=0}^{D} v_\ell(\theta_i)v_\ell(\theta_j)k_\ell^{-1} = \delta_{i,j}m_i^{-1}|X| \qquad (0 \le i,j \le D),$$

$$\sum_{\ell=0}^{D} v_i(\theta_\ell)v_j(\theta_\ell)m_\ell = \delta_{i,j}k_i|X| \qquad (0 \le i,j \le D).$$

Proof Combine Definition 3.7 and Proposition 3.14. $\qquad\square$

Note 3.16 The relations in Proposition 3.14 and Proposition 3.15 are called the *orthogonality relations* for the polynomials $\{u_i\}_{i=0}^{D}$ and $\{v_i\}_{i=0}^{D}$, respectively.

Our next goal is to give some formulas for the intersection numbers $p_{i,j}^{h}$.

Lemma 3.17 *For* $0 \le h,i,j \le D$,

$$p_{i,j}^{h} = |X|^{-1}k_h^{-1}\langle A_iA_j, A_h\rangle = |X|^{-1}k_h^{-1}\langle A_h, A_iA_j\rangle.$$

Proof To verify these equations, expand A_iA_j using $A_iA_j = \sum_{\ell=0}^{D} p_{i,j}^{\ell}A_\ell$, and evaluate the results using Lemma 3.13(ii). $\qquad\square$

Lemma 3.18 *For* $0 \le h,i,j \le D$,

$$k_h p_{i,j}^{h} = k_i p_{j,h}^{i} = k_j p_{h,i}^{j} = |X|^{-1}tr(A_hA_iA_j).$$

Proof Routine application of Lemma 3.17. $\qquad\square$

Proposition 3.19 *For* $0 \le h,i,j \le D$,

$$p_{i,j}^{h} = |X|^{-1}k_ik_j \sum_{\ell=0}^{D} u_i(\theta_\ell)u_j(\theta_\ell)u_h(\theta_\ell)m_\ell.$$

Proof In the equation $p_{i,j}^{h} = |X|^{-1}k_h^{-1}\langle A_iA_j, A_h\rangle$, eliminate A_h, A_i, A_j using Lemma 3.9(i), and evaluate the result using Lemma 3.13(i). $\qquad\square$

11.4 The Geometry of the Eigenspaces

Throughout this section, $\Gamma = (X, \mathcal{R})$ denotes a distance-regular graph with diameter $D \ge 3$. Recall the standard module V and the adjacency matrix A. Recall that for $0 \le j \le D$, the subspace E_jV is an eigenspace of A with eigenvalue θ_j. This eigenspace is spanned by the vectors $\{E_j\hat{w}|w \in X\}$. Note that for $y, z \in X$, the following scalars are equal:

$$\langle E_j \hat{y}, E_j \hat{z} \rangle, \qquad\qquad (y,z)\text{-entry of } E_j,$$

$$\langle \hat{y}, E_j \hat{z} \rangle, \qquad\qquad y\text{-coordinate of } E_j \hat{z},$$

$$\langle E_j \hat{y}, \hat{z} \rangle, \qquad\qquad z\text{-coordinate of } E_j \hat{y}.$$

Next, we have some comments of a geometric nature.

Lemma 4.1 *For $0 \le i,j \le D$ and $y,z \in X$ with $\partial(y,z) = i$,*

(i) $\langle E_j \hat{y}, E_j \hat{z} \rangle = |X|^{-1} m_j u_i(\theta_j);$

(ii) $\|E_j \hat{y}\|^2 = \|E_j \hat{z}\|^2 = |X|^{-1} m_j;$

(iii) $u_i(\theta_j) = \dfrac{\langle E_j \hat{y}, E_j \hat{z} \rangle}{\|E_j \hat{y}\|\|E_j \hat{z}\|};$

(iv) $u_i(\theta_j)$ *is the cosine of the angle between $E_j \hat{y}$ and $E_j \hat{z}$.*

Proof (i) The (y,z)-entry of E_j is found in Lemma 3.9(ii).
(ii) Set $y = z$ and $i = 0$ in part (i).
(iii) Combine (i), (ii).
(iv) By (iii) and trigonometry.   ☐

Corollary 4.2 *We have*

$$-1 \le u_i(\theta_j) \le 1 \qquad\qquad (0 \le i,j \le D).$$

Proof By Lemma 4.1(iv) and trigonometry.   ☐

Corollary 4.3 *For $0 \le i,j \le D$, the following are equivalent:*

(i) $u_i(\theta_j) = 1;$

(ii) $E_j \hat{y} = E_j \hat{z}$ *for all $y,z \in X$ at $\partial(y,z) = i$;*

(iii) *there exists $y,z \in X$ such that $\partial(y,z) = i$ and $E_j \hat{y} = E_j \hat{z}$.*

Proof By Lemma 4.1(iv) and trigonometry.   ☐

Corollary 4.4 *For $0 \le i,j \le D$, the following are equivalent:*

(i) $u_i(\theta_j) = -1;$

(ii) $E_j \hat{y} = -E_j \hat{z}$ *for all $y,z \in X$ at $\partial(y,z) = i$;*

(iii) *there exists $y,z \in X$ such that $\partial(y,z) = i$ and $E_j \hat{y} = -E_j \hat{z}$.*

Proof By Lemma 4.1(iv) and trigonometry.   ☐

The following reformulation of Corollary 4.3 will be useful.

Lemma 4.5 *For $0 \le j \le D$, the following are equivalent:*

(i) $u_i(\theta_j) \ne 1$ *for $1 \le i \le D$;*

(ii) *the vectors $\{E_j \hat{y} | y \in X\}$ are mutually distinct.*

Assume that (i), (ii) *hold. Then $j \ne 0$.*

Proof The equivalence of (i), (ii) follows from Corollary 4.3. The last assertion is from Lemma 3.11(ii). □

Definition 4.6 For $0 \leq j \leq D$, we call E_j *nondegenerate* whenever the equivalent conditions (i), (ii) hold in Lemma 4.5. In this case $j \neq 0$.

The following definition is motivated by Lemma 4.1(iv).

Definition 4.7 For $0 \leq j \leq D$, we call the sequence $\{u_i(\theta_j)\}_{i=0}^{D}$ the *cosine sequence* of E_j (or θ_j).

Lemma 4.8 (See [4, section 4.1.B]) *For a real number θ and a sequence of real numbers $\{\sigma_i\}_{i=0}^{D}$, the following are equivalent:*

(i) *θ is an eigenvalue of Γ with cosine sequence $\{\sigma_i\}_{i=0}^{D}$;*
(ii) *$\sigma_0 = 1$, $\sigma_1 = k^{-1}\theta$, and*

$$c_i\sigma_{i-1} + a_i\sigma_i + b_i\sigma_{i+1} = \theta\sigma_i \qquad (1 \leq i \leq D-1),$$

$$c_D\sigma_{D-1} + a_D\sigma_D = \theta\sigma_D.$$

Proof Use Lemma 3.5(i) and Lemma 3.8. □

11.5 The Krein Parameters and the Dual Distance Matrices

Our next topic is the Krein parameters. Throughout this section, $\Gamma = (X, \mathcal{R})$ denotes a distance-regular graph with diameter $D \geq 3$.

For $B, C \in \mathrm{Mat}_X(\mathbb{R})$, define the matrix $B \circ C \in \mathrm{Mat}_X(\mathbb{R})$ with entries

$$(B \circ C)_{y,z} = B_{y,z}C_{y,z} \qquad (y, z \in X).$$

The operation $\circ$ is called entrywise multiplication, or Schur multiplication, or Hadamard multiplication. We have $A_i \circ A_j = \delta_{i,j}A_i$ for $0 \leq i, j \leq D$. Recall that $\{A_i\}_{i=0}^{D}$ is a basis for the Bose–Mesner algebra M. By these comments, M is closed under $\circ$. Recall that $\{E_i\}_{i=0}^{D}$ is a basis for M. Therefore, there exist real numbers $q_{i,j}^{h}$ ($0 \leq h, i, j \leq D$) such that

$$E_i \circ E_j = |X|^{-1} \sum_{h=0}^{D} q_{i,j}^{h}E_h \qquad (0 \leq i, j \leq D). \qquad (8)$$

The $q_{i,j}^{h}$ are called the *Krein parameters* of Γ. By construction $q_{i,j}^{h} = q_{j,i}^{h}$ for $0 \leq h, i, j \leq D$. Shortly we will show that $q_{i,j}^{h} \geq 0$ for $0 \leq h, i, j \leq D$.

In order to avoid dealing directly with entrywise multiplication, we bring in a certain map p. For the rest of this section, fix a vertex $x \in X$.

Definition 5.1 For $B \in \mathrm{Mat}_X(\mathbb{R})$, let B^p denote the diagonal matrix in $\mathrm{Mat}_X(\mathbb{R})$ with (y, y)-entry

$$(B^p)_{y,y} = B_{x,y} \qquad\qquad (y \in X).$$

Lemma 5.2 *For $B, C \in \mathrm{Mat}_X(\mathbb{R})$,*

$$(B \circ C)^p = B^p C^p.$$

Proof By Definition 5.1. □

Lemma 5.3 *We have*

(i) $A_i^p = E_i^*$ $\quad (0 \le i \le D)$;
(ii) $I^p = E_0^*$;
(iii) $J^p = I$.

Proof This is routinely checked using Definition 5.1. □

Recall the dual Bose–Mesner algebra $M^* = M^*(x)$.

Lemma 5.4 *The restriction $p|_M : M \to M^*$ is an isomorphism of vector spaces.*

Proof The map p is $\mathbb{R}$-linear, and sends the basis $\{A_i\}_{i=0}^D$ of M to the basis $\{E_i^*\}_{i=0}^D$ of M^*. □

We caution the reader that the map in Lemma 5.4 is not an algebra isomorphism in general.

Lemma 5.5 *For $B, C \in M$,*

$$\langle B^p, C^p \rangle = |X|^{-1} \langle B, C \rangle. \tag{9}$$

Proof Using Definition 5.1 one finds that each side of (9) is equal to the (x, x)-entry of BC^t. □

We mention some useful facts about the map p.

Lemma 5.6 *For $B \in M$, we have*

(i) $E_0 E_0^* B = E_0 B^p$;
(ii) $E_0^* E_0 B^p = |X|^{-1} E_0^* B$;
(iii) $B E_0^* E_0 = B^p E_0$;
(iv) $B^p E_0 E_0^* = |X|^{-1} B E_0^*$.

Proof (i) Recall that $E_0 = |X|^{-1} J$. Recall that for E_0^*, the (x, x)-entry is 1 and all other entries are 0. For $y, z \in X$, we compare the (y, z)-entry of each side of the given equation. We have

$$(E_0 E_0^* B)_{y,z} = (E_0)_{y,x} B_{x,z} = |X|^{-1} B_{x,z} = (E_0)_{y,z} (B^p)_{z,z}$$
$$= (E_0 B^p)_{y,z}.$$

(ii) In the equation (i), multiply each side on the left by E_0^* and evaluate the result using $|X| E_0^* E_0 E_0^* = E_0^*$.

(iii), (iv) Take the transpose of each side in (i), (ii) above. □

Definition 5.7 For $0 \leq i \leq D$, define the matrix $A_i^* = A_i^*(x)$ by

$$A_i^* = |X| (E_i)^p.$$

Thus A_i^* is diagonal with (y, y)-entry

$$(A_i^*)_{y,y} = |X| (E_i)_{x,y} \qquad (y \in X). \qquad (10)$$

We call A_i^* the ith *dual distance matrix of* Γ *with respect to* x.

Lemma 5.8 *With the notation in Definition 5.7,*

(i) *the matrices* $\{A_i^*\}_{i=0}^D$ *form a basis for* M^*;
(ii) *for* $0 \leq i \leq D$ *and* $y \in X$,

$$A_i^* \hat{y} = m_i u_j(\theta_i) \hat{y}, \qquad\qquad j = \partial(x, y).$$

Proof (i) By Lemma 5.4 and since $\{E_i\}_{i=0}^D$ form a basis for M.
(ii) By (10) and since $|X| (E_i)_{x,y} = m_i u_j(\theta_i)$. □

Lemma 5.9 *The following* (i)–(iv) *hold:*

(i) $A_0^* = I$;
(ii) $\sum_{i=0}^D A_i^* = |X| E_0^*$;
(iii) $(A_i^*)^t = A_i^*$ $(0 \leq i \leq D)$;
(iv) $A_i^* A_j^* = \sum_{h=0}^D q_{i,j}^h A_h^*$ $(0 \leq i, j \leq D)$.

Proof (i) Use $A_0^* = |X| E_0^p$ and $E_0 = |X|^{-1} J$.
(ii) Apply p to each side of $\sum_{i=0}^D E_i = I$.
(iii) A_i^* is diagonal.
(iv) Apply p to each side of (8), and evaluate the result using Lemma 5.2 along with Definition 5.7. □

We have seen that $\{E_i^*\}_{i=0}^D$ and $\{A_i^*\}_{i=0}^D$ are bases for the vector space M^*. Next we describe how these bases are related.

Lemma 5.10 *For* $0 \leq j \leq D$, *we have*

(i) $A_j^* = m_j \sum_{i=0}^D u_i(\theta_j) E_i^*$;
(ii) $E_j^* = |X|^{-1} \sum_{i=0}^D v_j(\theta_i) A_i^*$.

Proof   In Lemma 3.9 we displayed some equations that show how $\{E_i\}_{i=0}^{D}$ and $\{A_i\}_{i=0}^{D}$ are related. Apply p to each side of these equations.   □

Lemma 5.11   *For* $0 \le i, j \le D$,

$$E_i^* A_j^* = m_j u_i(\theta_j) E_i^* = A_j^* E_i^*.$$

Proof   To verify this equation, eliminate A_j^* using Lemma 5.10(i) and simplify the result.   □

Lemma 5.12   *For* $0 \le i, j \le D$, *we have*

(i) $\langle E_i^*, E_j^* \rangle = \delta_{i,j} k_i$;
(ii) $\langle A_i^*, A_j^* \rangle = \delta_{i,j} m_i |X|$;
(iii) $\langle A_i^*, E_j^* \rangle = m_i v_j(\theta_i)$.

Proof   (i) Routine.
(ii) Use Lemma 5.5 and Definition 5.7.
(iii) In the given equation, eliminate E_j^* using Lemma 5.10(ii), and simplify the result using (ii).   □

In the next few lemmas, we describe the Krein parameters in various ways.

Lemma 5.13   *For* $0 \le h, i, j \le D$,

$$q_{i,j}^h = |X|^{-1} m_h^{-1} \langle A_i^* A_j^*, A_h^* \rangle = |X|^{-1} m_h^{-1} \langle A_h^*, A_i^* A_j^* \rangle.$$

Proof   To verify these equations, expand $A_i^* A_j^*$ using $A_i^* A_j^* = \sum_{\ell=0}^{D} q_{i,j}^\ell A_\ell^*$, and evaluate the results using Lemma 5.12(ii).   □

Lemma 5.14   *For* $0 \le h, i, j \le D$,

$$m_h q_{i,j}^h = m_i q_{j,h}^i = m_j q_{h,i}^j = |X|^{-1} tr\left(A_h^* A_i^* A_j^*\right).$$

Proof   Routine application of Lemma 5.13.   □

Lemma 5.15   *The following* (i)–(iv) *hold.*

(i) $q_{0,j}^h = \delta_{h,j}$ $(0 \le h, j \le D)$;
(ii) $q_{i,0}^h = \delta_{h,i}$ $(0 \le h, i \le D)$;
(iii) $q_{i,j}^0 = \delta_{i,j} m_i$ $(0 \le i, j \le D)$;
(iv) $\sum_{i=0}^{D} q_{i,j}^h = m_j$ $(0 \le h, j \le D)$.

Proof   (i) By Lemmas 5.12, 5.13 we obtain

$$q_{0,j}^h = |X|^{-1} m_h^{-1} \langle A_0^* A_j^*, A_h^* \rangle = |X|^{-1} m_h^{-1} \langle A_j^*, A_h^* \rangle = \delta_{h,j}.$$

(ii) By (i) and $q_{i,0}^h = q_{0,i}^h$.

(iii) By (i) and Lemma 5.14.

(iv) We have

$$\sum_{i=0}^{D} q_{i,j}^{h} = |X|^{-1} m_h^{-1} \sum_{i=0}^{D} \langle A_i^* A_j^*, A_h^* \rangle = m_h^{-1} \langle E_0^* A_j^*, A_h^* \rangle$$

$$= m_h^{-1} m_j \langle E_0^*, A_h^* \rangle = m_j.$$ $\square$

Proposition 5.16   *For* $0 \le h, i, j \le D$,

$$q_{i,j}^{h} = |X|^{-1} m_i m_j \sum_{\ell=0}^{D} u_\ell(\theta_i) u_\ell(\theta_j) u_\ell(\theta_h) k_\ell.$$

Proof   In the equation $q_{i,j}^{h} = |X|^{-1} m_h^{-1} \langle A_i^* A_j^*, A_h^* \rangle$, eliminate A_h^*, A_i^*, A_j^* using Lemma 5.10(i), and evaluate the result using Lemma 5.12(i). $\square$

11.6 Reduction Rules

Throughout this section, $\Gamma = (X, \mathcal{R})$ denotes a distance-regular graph with diameter $D \ge 3$. Fix $x \in X$ and write $T = T(x)$. We will display a number of relations involving E_0 and E_0^*. These relations are informally known as reduction rules; see [25, section 7], [54, sections 9, 11, 13].

Lemma 6.1   *For* $0 \le i \le D$,

$$E_0 E_0^* A_i = E_0 E_i^*, \qquad\qquad E_0^* E_0 E_i^* = |X|^{-1} E_0^* A_i,$$
$$A_i E_0^* E_0 = E_i^* E_0, \qquad\qquad E_i^* E_0 E_0^* = |X|^{-1} A_i E_0^*.$$

Proof   Apply Lemma 5.6 with $B = A_i$. $\square$

Lemma 6.2   *For* $0 \le i \le D$,

$$E_0^* E_0 A_i^* = E_0^* E_i, \qquad\qquad E_0 E_0^* E_i = |X|^{-1} E_0 A_i^*,$$
$$A_i^* E_0 E_0^* = E_i E_0^*, \qquad\qquad E_i E_0^* E_0 = |X|^{-1} A_i^* E_0.$$

Proof   Apply Lemma 5.6 with $B = E_i$. $\square$

Lemma 6.3   *For* $0 \le i, j \le D$, *we have*

(i) $E_0 A_i^* E_j = \delta_{i,j} E_0 A_i^*$;
(ii) $E_0 E_i^* E_j = |X|^{-1} k_i u_i(\theta_j) E_0 A_j^*$;
(iii) $E_0 A_i^* A_j = k_j u_j(\theta_i) E_0 A_i^*$;
(iv) $E_0 E_i^* A_j = \sum_{h=0}^{D} p_{i,j}^{h} E_0 E_h^*$.

Proof   (i) Observe

$$E_0 A_i^* E_j = |X| E_0 E_0^* E_i E_j = \delta_{i,j} |X| E_0 E_0^* E_i = \delta_{i,j} E_0 A_i^*.$$

(ii) Observe

$$E_0 E_i^* E_j = E_0 E_0^* A_i E_j = k_i u_i(\theta_j) E_0 E_0^* E_j = |X|^{-1} k_i u_i(\theta_j) E_0 A_j^*.$$

(iii) Observe

$$E_0 A_i^* A_j = |X| E_0 E_0^* E_i A_j = |X| k_j u_j(\theta_i) E_0 E_0^* E_i = k_j u_j(\theta_i) E_0 A_i^*.$$

(iv) Observe

$$E_0 E_i^* A_j = E_0 E_0^* A_i A_j = \sum_{h=0}^{D} p_{i,j}^h E_0 E_0^* A_h$$
$$= \sum_{h=0}^{D} p_{i,j}^h E_0 E_h^*.$$

Lemma 6.4   *For $0 \le i, j \le D$, we have*

(i) $E_0^* A_i E_j^* = \delta_{i,j} E_0^* A_i$;
(ii) $E_0^* E_i E_j^* = |X|^{-1} m_i u_j(\theta_i) E_0^* A_j$;
(iii) $E_0^* A_i A_j^* = m_j u_i(\theta_j) E_0^* A_i$;
(iv) $E_0^* E_i A_j^* = \sum_{h=0}^{D} q_{i,j}^h E_0^* E_h$.

Proof   Similar to the proof of Lemma 6.3.

Lemma 6.5   *For $0 \le i, j \le D$, we have*

(i) $E_j A_i^* E_0 = \delta_{i,j} A_i^* E_0$;
(ii) $E_j E_i^* E_0 = |X|^{-1} k_i u_i(\theta_j) A_j^* E_0$;
(iii) $A_j A_i^* E_0 = k_j u_j(\theta_i) A_i^* E_0$;
(iv) $A_j E_i^* E_0 = \sum_{h=0}^{D} p_{i,j}^h E_h^* E_0$.

Proof   Take the transpose of everything in Lemma 6.3.

Lemma 6.6   *For $0 \le i, j \le D$, we have*

(i) $E_j^* A_i E_0^* = \delta_{i,j} A_i E_0^*$;
(ii) $E_j^* E_i E_0^* = |X|^{-1} m_i u_j(\theta_i) A_j E_0^*$;
(iii) $A_j^* A_i E_0^* = m_j u_i(\theta_j) A_i E_0^*$;
(iv) $A_j^* E_i E_0^* = \sum_{h=0}^{D} q_{i,j}^h E_h E_0^*$.

Proof   Take the transpose of everything in Lemma 6.4.

Lemma 6.7 *For $0 \le i \le D$, we have*

(i) $E_0 E_i^* E_0 = |X|^{-1} k_i E_0$;
(ii) $E_0^* E_i E_0^* = |X|^{-1} m_i E_0^*$.

Proof (i) Observe

$$E_0 E_i^* E_0 = E_0 E_0^* A_i E_0 = k_i E_0 E_0^* E_0 = |X|^{-1} k_i E_0.$$

(ii) Observe

$$E_0^* E_i E_0^* = E_0^* E_0 A_i^* E_0^* = m_i E_0^* E_0 E_0^* = |X|^{-1} m_i E_0^*. \qquad \square$$

Lemma 6.8 *For $0 \le i, j \le D$, we have*

(i) $A_i E_0^* A_j = |X| E_i^* E_0 E_j^*$;
(ii) $E_i E_0^* A_j = A_i^* E_0 E_j^*$;
(iii) $A_i E_0^* E_j = E_i^* E_0 A_j^*$;
(iv) $E_i E_0^* E_j = |X|^{-1} A_i^* E_0 A_j^*$.

Proof (i) Observe

$$A_i E_0^* A_j = |X| A_i E_0^* E_0 E_j^* = |X| E_i^* E_0 E_j^*.$$

(ii) Observe

$$E_i E_0^* A_j = |X| E_i E_0^* E_0 E_j^* = A_i^* E_0 E_j^*.$$

(iii) Observe

$$A_i E_0^* E_j = A_i E_0^* E_0 A_j^* = E_i^* E_0 A_j^*.$$

(iv) Observe

$$E_i E_0^* E_j = E_i E_0^* E_0 A_j^* = |X|^{-1} A_i^* E_0 A_j^*. \qquad \square$$

Corollary 6.9 *$M E_0^* M$ and $M^* E_0 M^*$ span the same subspace of Mat_X ($\mathbb{R}$).*

Proof By Lemma 6.8. $\qquad \square$

Lemma 6.10 (See [25, proposition 11.1]) *We have*

$$\sum_{i=0}^{D} k_i^{-1} E_i^* E_0 E_i^* = \sum_{j=0}^{D} m_j^{-1} E_j E_0^* E_j. \qquad (11)$$

Proof Observe

$$\sum_{i=0}^{D} k_i^{-1} E_i^* E_0 E_i^* = |X|^{-1} \sum_{i=0}^{D} k_i^{-1} A_i E_0^* A_i$$

$$= |X|^{-1} \sum_{i=0}^{D} k_i^{-1} \left(\sum_{r=0}^{D} v_i(\theta_r) E_r \right) E_0^* \left(\sum_{s=0}^{D} v_i(\theta_s) E_s \right)$$

$$= |X|^{-1} \sum_{r=0}^{D} \sum_{s=0}^{D} E_r E_0^* E_s \left(\sum_{i=0}^{D} k_i^{-1} v_i(\theta_r) v_i(\theta_s) \right)$$

$$= |X|^{-1} \sum_{r=0}^{D} \sum_{s=0}^{D} E_r E_0^* E_s \left(\delta_{r,s} m_r^{-1} |X| \right)$$

$$= \sum_{j=0}^{D} m_j^{-1} E_j E_0^* E_j. \qquad \square$$

Definition 6.11   Define the matrix $e_0 = e_0(x)$ to be $|X|$ times the common value of the matrices (11).

Referring to Definition 6.11, the matrix e_0 is symmetric. Moreover e_0 is contained in the common span of $M E_0^* M$ and $M^* E_0 M^*$. In the next result we describe another property of e_0.

Let $Z(T)$ denote the center of T.

Proposition 6.12 (See [25, corollary 11.3, proposition 11.4])   *For the matrix e_0 from Definition 6.11, we have*

   (i) $e_0 \in Z(T)$;
   (ii) $E_0 e_0 = E_0$;
   (iii) $E_0^* e_0 = E_0^*$;
   (iv) $e_0^2 = e_0$.

Proof (i) Using $e_0 = |X| \sum_{i=0}^{D} k_i^{-1} E_i^* E_0 E_i^*$, we see that e_0 commutes with E_ℓ^* for $0 \le \ell \le D$. Therefore e_0 commutes with everything in M^*. Using $e_0 = |X| \sum_{j=0}^{D} m_j^{-1} E_j E_0^* E_j$, we see that e_0 commutes with E_ℓ for $0 \le \ell \le D$. Therefore e_0 commutes with everything in M. The result follows since T is generated by M, M^*.
(ii) Observe

$$E_0 e_0 = E_0 |X| \sum_{j=0}^{D} m_j^{-1} E_j E_0^* E_j = |X| E_0 E_0^* E_0 = E_0.$$

(iii) Observe

$$E_0^* e_0 = E_0^* |X| \sum_{i=0}^{D} k_i^{-1} E_i^* E_0 E_i^* = |X| E_0^* E_0 E_0^* = E_0^*.$$

(iv) Observe

$$(I - e_0)e_0 \in (I - e_0)\mathrm{Span}(M E_0^* M) = \mathrm{Span}(M(I - e_0)E_0^* M)$$
$$= \mathrm{Span}(M 0 M) = 0. \qquad \square$$

We will say more about e_0 in the next section.

We finish this section with a comment.

Lemma 6.13   *For $B \in M$ we have the logical implications*

$$B = 0 \quad \Leftrightarrow \quad B E_0^* = 0 \quad \Leftrightarrow \quad E_0^* B = 0.$$

Moreover for $C \in M^*$ we have the logical implications

$$C = 0 \quad \Leftrightarrow \quad C E_0 = 0 \quad \Leftrightarrow \quad E_0 C = 0.$$

Proof   By Lemma 6.7 the following are nonzero for $0 \le i \le D$:

$$E_i E_0^*, \quad E_0^* E_i, \quad E_i^* E_0, \quad E_0 E_i^*.$$

The result is a routine consequence of this. $\square$

11.7 The Primary T-Module

Throughout this section, $\Gamma = (X, \mathcal{R})$ denotes a distance-regular graph with diameter $D \ge 3$. Fix $x \in X$ and write $T = T(x)$. Our next goal is to describe the primary T-module [13, section 5], [25, sections 8, 9], [54], [67, lemma 3.6]. Recall the vector $\mathbf{1} = \sum_{y \in X} \hat{y}$. For $0 \le i \le D$, define the vector $\mathbf{1}_i = \sum_{y \in \Gamma_i(x)} \hat{y}$. Observe that

$$A_i \hat{x} = \mathbf{1}_i = E_i^* \mathbf{1} \qquad (0 \le i \le D).$$

Consequently

$$M E_0^* V = M^* E_0 V. \qquad (12)$$

Lemma 7.1   *The vector space $M E_0^* V = M^* E_0 V$ is an irreducible T-module.*

Proof   Define $\mathcal{V} = M E_0^* V = M^* E_0 V$. We have $M \mathcal{V} \subseteq \mathcal{V}$ since $\mathcal{V} = M E_0^* V$. We have $M^* \mathcal{V} \subseteq \mathcal{V}$ since $\mathcal{V} = M^* E_0 V$. Therefore $T \mathcal{V} \subseteq \mathcal{V}$, so $\mathcal{V}$ is a T-module. We show that the T-module $\mathcal{V}$ is irreducible. The standard T-module V is a direct sum of irreducible T-modules. There exists an irreducible T-module that is not orthogonal to $\hat{x}$. This T-module contains $\hat{x}$, so it contains $M\hat{x} = \mathcal{V}$. This T-module must equal $\mathcal{V}$ by irreducibility. □

Definition 7.2   Define $\mathcal{V} = ME_0^*V = M^*E_0V$. The T-module $\mathcal{V}$ is called *primary*.

Lemma 7.3   *For $0 \leq i \leq D$, we have*

$$A_i^* \mathbf{1} = |X| E_i \hat{x}. \tag{13}$$

Proof   Both vectors in (13) have y-coordinate $|X|(E_i)_{x,y}$ for $y \in X$.   □

Definition 7.4   For $0 \leq i \leq D$, let $\mathbf{1}_i^*$ denote the common vector in (13).

We clarify the definitions. Note that $\mathbf{1}_0 = \hat{x}$ and $\mathbf{1}_0^* = \mathbf{1}$. Moreover

$$\mathbf{1}_0^* = \sum_{i=0}^{D} \mathbf{1}_i, \qquad \mathbf{1}_0 = |X|^{-1} \sum_{i=0}^{D} \mathbf{1}_i^*.$$

The following result is routinely verified.

Lemma 7.5   *For the primary T-module $\mathcal{V}$,*

   (i) *$\mathbf{1}_i$ is a basis for $E_i^* \mathcal{V}$ $(0 \leq i \leq D)$;*
   (ii) *$\{\mathbf{1}_i\}_{i=0}^{D}$ is a basis for $\mathcal{V}$;*
   (iii) *$\mathbf{1}_i^*$ is a basis for $E_i \mathcal{V}$ $(0 \leq i \leq D)$;*
   (iv) *$\{\mathbf{1}_i^*\}_{i=0}^{D}$ is a basis for $\mathcal{V}$.*

Next we explain how the primary T-module $\mathcal{V}$ is related to the matrix e_0 from Definition 6.11. Let $\mathcal{V}^\perp$ denote the orthogonal complement of $\mathcal{V}$ in V. We have an orthogonal direct sum $V = \mathcal{V} + \mathcal{V}^\perp$.

Lemma 7.6   *We have*

   (i) *$(e_0 - I)\mathcal{V} = 0$;*
   (ii) *$e_0 \mathcal{V}^\perp = 0$.*

In other words, e_0 acts on V as the orthogonal projection $V \to \mathcal{V}$.

Proof   (i) Observe

$$(e_0 - I)\mathcal{V} = (e_0 - I)ME_0^*V = M(e_0 - I)E_0^*V = 0.$$

(ii) Note that $\mathcal{V}^\perp$ is a T-module, so $e_0\mathcal{V}^\perp \subseteq \mathcal{V}^\perp$. Also,

$$e_0 \mathcal{V}^\perp \subseteq e_0 V \subseteq \operatorname{Span}\left(ME_0^*MV\right) \subseteq ME_0^*V = \mathcal{V}.$$

Therefore,

$$e_0 \mathcal{V}^\perp \subseteq \mathcal{V}^\perp \cap \mathcal{V} = 0. \qquad \square$$

We saw in Lemma 7.5 that $\{\mathbf{1}_i\}_{i=0}^{D}$ and $\{\mathbf{1}_i^*\}_{i=0}^{D}$ are bases for the primary T-module $\mathcal{V}$. Next we describe how these bases are related.

Lemma 7.7   *For $0 \leq j \leq D$, we have*

  (i) $\mathbf{1}_j = |X|^{-1}k_j \sum_{i=0}^{D} u_j(\theta_i)\mathbf{1}_i^*$;
  (ii) $\mathbf{1}_j^* = m_j \sum_{i=0}^{D} u_i(\theta_j)\mathbf{1}_i$.

Proof   (i) Observe

$$\mathbf{1}_j = A_j\hat{x} = k_j \sum_{i=0}^{D} u_j(\theta_i)E_i\hat{x} = |X|^{-1}k_j \sum_{i=0}^{D} u_j(\theta_i)\mathbf{1}_i^*.$$

(ii) Observe

$$\mathbf{1}_j^* = |X|E_j\hat{x} = m_j \sum_{i=0}^{D} u_i(\theta_j)A_i\hat{x} = m_j \sum_{i=0}^{D} u_i(\theta_j)\mathbf{1}_i. \qquad \square$$

Next we describe how the algebra T acts on the bases $\{\mathbf{1}_i\}_{i=0}^{D}$ and $\{\mathbf{1}_i^*\}_{i=0}^{D}$.

Lemma 7.8   *For $0 \leq i,j \leq D$ we have*

  (i) $E_i^*\mathbf{1}_j = \delta_{i,j}\mathbf{1}_j$;
  (ii) $A_i^*\mathbf{1}_j = m_i u_j(\theta_i)\mathbf{1}_j$;
  (iii) $E_i\mathbf{1}_j = |X|^{-1}m_i k_j u_j(\theta_i) \sum_{h=0}^{D} u_h(\theta_i)\mathbf{1}_h$;
  (iv) $A_i\mathbf{1}_j = \sum_{h=0}^{D} p_{i,j}^{h}\mathbf{1}_h$.

Proof   These are routinely checked using the reduction rules in Lemmas 6.5, 6.6 along with Lemma 7.7. $\square$

Lemma 7.9   *For $0 \leq i,j \leq D$, we have*

  (i) $E_i\mathbf{1}_j^* = \delta_{i,j}\mathbf{1}_j^*$;
  (ii) $A_i\mathbf{1}_j^* = k_i u_i(\theta_j)\mathbf{1}_j^*$;
  (iii) $E_i^*\mathbf{1}_j^* = |X|^{-1}k_i m_j u_i(\theta_j) \sum_{h=0}^{D} u_i(\theta_h)\mathbf{1}_h^*$;
  (iv) $A_i^*\mathbf{1}_j^* = \sum_{h=0}^{D} q_{i,j}^{h}\mathbf{1}_h^*$.

Proof   These are routinely checked using the reduction rules in Lemmas 6.5, 6.6 along with Lemma 7.7. $\square$

Next we bring in the bilinear form.

Lemma 7.10   *For $0 \leq i,j \leq D$, we have*

  (i) $\langle \mathbf{1}_i, \mathbf{1}_j \rangle = \delta_{i,j}k_i$;
  (ii) $\langle \mathbf{1}_i^*, \mathbf{1}_j^* \rangle = \delta_{i,j}|X|m_i$;
  (iii) $\langle \mathbf{1}_i, \mathbf{1}_j^* \rangle = k_i m_j u_i(\theta_j)$.

Proof   (i) Routine.

(ii) Observe

$$\langle \mathbf{1}_i^*, \mathbf{1}_j^* \rangle = |X|^2 \langle E_i \hat{x}, E_j \hat{x} \rangle = |X|^2 \langle \hat{x}, E_i E_j \hat{x} \rangle = \delta_{i,j} |X|^2 \langle \hat{x}, E_i \hat{x} \rangle$$
$$= \delta_{i,j} |X| m_i.$$

(iii) Observe

$$\langle \mathbf{1}_i, \mathbf{1}_j^* \rangle = |X| \langle A_i \hat{x}, E_j \hat{x} \rangle = |X| \langle \hat{x}, A_i E_j \hat{x} \rangle = |X| v_i(\theta_j) \langle \hat{x}, E_j \hat{x} \rangle$$
$$= k_i m_j u_i(\theta_j). \qquad \qquad \square$$

11.8 The Krein Condition and the Triple Product Relations

The Krein condition states that the Krein parameters are nonnegative. In this section, we will prove the Krein condition, and also derive some relations called the triple product relations.

Throughout this section, $\Gamma = (X, \mathcal{R})$ denotes a distance-regular graph with diameter $D \geq 3$. Fix $x \in X$ and write $T = T(x)$.

In the following result, the second item is a variation on [5, proposition 5.1].

Lemma 8.1   (See [5, proposition 5.1], [24, lemmas 3.1, 4.1].) *For* $0 \leq h, i, j, r, s, D$,

(i) $\langle E_h^* A_i E_j^*, E_r^* A_s E_t^* \rangle = \delta_{h,r} \delta_{i,s} \delta_{j,t} k_h p_{i,j}^h$;

(ii) $\langle E_h A_i^* E_j, E_r A_s^* E_t \rangle = \delta_{h,r} \delta_{i,s} \delta_{j,t} m_h q_{i,j}^h$.

Proof   (i) Using $\operatorname{tr}(BC) = \operatorname{tr}(CB)$,

$$\langle E_h^* A_i E_j^*, E_r^* A_s E_t^* \rangle = \operatorname{tr}\big(E_h^* A_i E_j^* (E_r^* A_s E_t^*)^t\big)$$
$$= \operatorname{tr}\big(E_h^* A_i E_j^* E_t^* A_s E_r^*\big)$$
$$= \delta_{h,r} \delta_{j,t} \operatorname{tr}\big(E_h^* A_i E_j^* A_s\big)$$

and

$$\operatorname{tr}\big(E_h^* A_i E_j^* A_s\big) = \sum_{y \in X} \sum_{z \in X} (E_h^*)_{y,y} (A_i)_{y,z} (E_j^*)_{z,z} (A_s)_{z,y}$$
$$= \sum_{y \in X} \sum_{z \in X} (E_h^*)_{y,y} (A_i \circ A_s)_{y,z} (E_j^*)_{z,z}$$
$$= \delta_{i,s} \sum_{y \in X} \sum_{z \in X} (E_h^*)_{y,y} (A_i)_{y,z} (E_j^*)_{z,z}$$
$$= \delta_{i,s} \sum_{\substack{y \in \Gamma_h(x), \\ z \in \Gamma_j(x), \\ \partial(y,z)=i}} 1$$
$$= \delta_{i,s} k_h p_{i,j}^h.$$

(ii) We have

$$
\begin{aligned}
\langle E_h A_i^* E_j, E_r A_s^* E_t \rangle &= \mathrm{tr}\big(E_h A_i^* E_j (E_r A_s^* E_t)^t\big)\\
&= \mathrm{tr}\big(E_h A_i^* E_j E_t A_s^* E_r\big)\\
&= \delta_{h,r}\delta_{j,t}\,\mathrm{tr}\big(E_h A_i^* E_j A_s^*\big)
\end{aligned}
$$

and

$$
\begin{aligned}
\mathrm{tr}\big(E_h A_i^* E_j A_s^*\big) &= \sum_{y\in X}\sum_{z\in X}(E_h)_{y,z}(A_i^*)_{z,z}(E_j)_{z,y}(A_s^*)_{y,y}\\
&= |X|^2 \sum_{y\in X}\sum_{z\in X}(E_h)_{y,z}(E_i)_{x,z}(E_j)_{z,y}(E_s)_{x,y}\\
&= |X|^2 \sum_{y\in X}\sum_{z\in X}(F_s)_{x,y}(E_h\circ E_j)_{y,z}(E_i)_{z,x}\\
&= |X|^2\Big((x,x)\text{-entry of } E_s(E_h\circ E_j)E_i\Big)\\
&= |X|\mathrm{tr}\big(E_s(E_h\circ E_j)E_i\big)\\
&= |X|\mathrm{tr}\big((E_h\circ E_j)E_i E_s\big)\\
&= \delta_{i,s}|X|\mathrm{tr}\big((E_h\circ E_j)E_i\big)\\
&= \delta_{i,s}\sum_{\ell=0}^{D} q_{h,j}^{\ell}\,\mathrm{tr}(E_\ell E_i)\\
&= \delta_{i,s}q_{h,j}^{i}\,\mathrm{tr}(E_i)\\
&= \delta_{i,s}q_{h,j}^{i}m_i\\
&= \delta_{i,s}m_h q_{i,j}^{h}.\qquad\qquad\qquad\square
\end{aligned}
$$

Corollary 8.2   *For $0 \le h,i,j \le D$ we have*

(i) $\|E_h^* A_i E_j^*\|^2 = k_h p_{i,j}^{h}$;

(ii) $\|E_h A_i^* E_j\|^2 = m_h q_{i,j}^{h}$.

Proof   Set $r=h$, $s=i$, $t=j$ in Lemma 8.1. $\square$

The following result is called the *Krein condition*. See [2, p. 69] for a discussion of the history.

Theorem 8.3   *We have $q_{i,j}^{h} \ge 0$ for $0 \le h,i,j \le D$.*

Proof   By Corollary 8.2(ii) and since $\|B\|^2 \ge 0$ for all $B \in \mathrm{Mat}_X(\mathbb{R})$. $\square$

Theorem 8.4 (See [67, lemma 3.2])   *For $0 \le h,i,j \le D$ we have*

(i) $E_h^* A_i E_j^* = 0$ *if and only if* $p_{i,j}^{h} = 0$;

(ii) $E_h A_i^* E_j = 0$ *if and only if* $q_{i,j}^h = 0$.

Proof By Corollary 8.2 and since $\|B\|^2 = 0$ implies $B = 0$ for all $B \in \mathrm{Mat}_X(\mathbb{R})$. □

The relations in Theorem 8.4 are called the *triple product relations*.
    We bring in some notation. For subspaces R, S of $\mathrm{Mat}_X(\mathbb{R})$, define

$$RS = \mathrm{Span}\{rs | r \in R, \ s \in S\}.$$

Theorem 8.5 (See [60, section 7]) *With the previous notation*

(i) *the vector space* $M^* M M^*$ *has an orthogonal basis*

$$\{E_h^* A_i E_j^* | 0 \le h, i, j \le D, \ p_{i,j}^h \ne 0\};$$

(ii) *the vector space* $M M^* M$ *has an orthogonal basis*

$$\{E_h A_i^* E_j | 0 \le h, i, j \le D, \ q_{i,j}^h \ne 0\}.$$

Proof By Lemma 8.1 and Theorem 8.4. □

We mention a consequence of Theorem 8.4.

Proposition 8.6 *For* $0 \le i, j \le D$ *we have*

$$A_i E_j^* V \subseteq \sum_{\substack{0 \le h \le D \\ p_{i,j}^h \ne 0}} E_h^* V, \qquad A_i^* E_j V \subseteq \sum_{\substack{0 \le h \le D \\ q_{i,j}^h \ne 0}} E_h V. \qquad (14)$$

Proof Concerning the containment on the left in (14),

$$A_i E_j^* V = I A_i E_j^* V = \sum_{h=0}^{D} E_h^* A_i E_j^* V = \sum_{\substack{0 \le h \le D \\ p_{i,j}^h \ne 0}} E_h^* A_i E_j^* V \subseteq \sum_{\substack{0 \le h \le D \\ p_{i,j}^h \ne 0}} E_h^* V.$$

The containment on the right in (14) is similarly obtained. □

We finish this section with some comments about the primary T-module.

Lemma 8.7 *For the primary T-module* V *and* $0 \le h, i, j \le D$,

(i) $E_h^* A_i E_j^* = 0$ *on* V *if and only if* $p_{i,j}^h = 0$;
(ii) $E_h A_i^* E_j = 0$ *on* V *if and only if* $q_{i,j}^h = 0$.

Proof Use parts (i), (iv) of Lemmas 7.8, 7.9. □

Lemma 8.8 *For* $0 \le i, j \le D$ *the following holds on the primary T-module* V:

$$E_i^* A_j E_i^* = p_{i,j}^i E_i^*, \qquad E_i A_j^* E_i = q_{i,j}^i E_i.$$

Proof Use parts (i), (iv) of Lemmas 7.8, 7.9. □

11.9 The Function Algebra and the Norton Algebra

Throughout this section, $\Gamma = (X, \mathcal{R})$ denotes a distance-regular graph with diameter $D \geq 3$. In Theorem 8.4, we saw that each vanishing Krein parameter gives a triple product relation. In this section, we consider the vanishing Krein parameters from another point of view, due to Cameron, Goethals, and Seidel [5, proposition 5.1]. We will also briefly mention Norton algebras.

Recall the basis $\{\hat{y}\}_{y \in X}$ for the standard module V.

Definition 9.1 We turn the vector space V into a commutative, associative, $\mathbb{R}$-algebra with multiplication $\circ$ defined as follows:

$$\hat{y} \circ \hat{z} = \delta_{y,z} \hat{y} \qquad y, z \in X. \tag{15}$$

The algebra V is isomorphic to the algebra of functions $X \to \mathbb{R}$. Motivated by this, we call the algebra V the *function algebra*.

In order to illustrate the multiplication $\circ$, let $v, w \in V$ and write

$$v = \sum_{y \in X} v_y \hat{y}, \qquad w = \sum_{y \in X} w_y \hat{y} \qquad v_y, w_y \in \mathbb{R}.$$

Then

$$v \circ w = \sum_{y \in X} v_y w_y \hat{y}.$$

Lemma 9.2 *For the function algebra V, the multiplicative identity is $\mathbf{1} = \sum_{y \in X} \hat{y}$.*

Proof Routine. □

For the rest of this section, fix $x \in X$ and write $T = T(x)$.

Lemma 9.3 *For $v \in V$ and $0 \leq i \leq D$,*

$$A_i^* v = |X| E_i \hat{x} \circ v. \tag{16}$$

Proof Write $v = \sum_{y \in X} v_y \hat{y}$. Pick $y \in X$. The y-coordinate of $A_i^* v$ is

$$(A_i^* v)_y = (A_i^*)_{y,y} v_y = |X| (E_i)_{x,y} v_y.$$

The y-coordinate of $E_i \hat{x} \circ v$ is

$$\big(E_i \hat{x} \circ v \big)_y = (E_i \hat{x})_y v_y = (E_i)_{y,x} v_y = (E_i)_{x,y} v_y.$$

The result follows. □

We bring in some notation. For subspaces R, S of V define

$$R \circ S = \mathrm{Span}\{r \circ s | r \in R, \; s \in S\}. \tag{17}$$

Theorem 9.4 (See [5, proposition 5.1])   *For* $0 \le i, j \le D$,

$$E_i V \circ E_j V = \sum_{\substack{0 \le h \le D \\ q_{i,j}^h \neq 0}} E_h V. \tag{18}$$

Proof   We first establish the inclusion $\subseteq$. By construction $E_i V = \mathrm{Span}\{E_i \hat{y} | y \in X\}$. We show that for $y \in X$,

$$E_i \hat{y} \circ E_j V \subseteq \sum_{\substack{0 \le h \le D \\ q_{i,j}^h \neq 0}} E_h V.$$

Since our base vertex x is arbitrary, we may assume without loss of generality that $x = y$. By Proposition 8.6 and Lemma 9.3,

$$E_i \hat{x} \circ E_j V = A_i^* E_j V \subseteq \sum_{\substack{0 \le h \le D \\ q_{i,j}^h \neq 0}} E_h V.$$

Next we establish the inclusion $\supseteq$. For $0 \le h \le D$ such that $q_{i,j}^h \neq 0$, we show that $E_i V \circ E_j V \supseteq E_h V$. We have

$$
\begin{aligned}
E_i V \circ E_j V &= \mathrm{Span}\{E_i \hat{y} \circ E_j \hat{z} | y, z \in X\} \\
&\supseteq \mathrm{Span}\{E_i \hat{y} \circ E_j \hat{y} | y \in X\} \\
&= \mathrm{Span}\{(E_i \circ E_j)\hat{y} | y \in X\} \\
&= (E_i \circ E_j) V \\
&\supseteq (E_i \circ E_j) E_h V \\
&= \left(|X|^{-1} \sum_{\ell=0}^{D} q_{i,j}^\ell E_\ell \right) E_h V \\
&= |X|^{-1} q_{i,j}^h E_h V \\
&= E_h V. \qquad\qquad\qquad\qquad \square
\end{aligned}
$$

Next we briefly review the Norton algebra.

Lemma 9.5 (See [5, proposition 5.2])   *For* $0 \le j \le D$ *we endow* $E_j V$ *with a binary operation* $\star$ *as follows:*

$$u \star v = E_j(u \circ v) \qquad\qquad u, v \in E_j V.$$

Then for $u, v, w \in E_j V$ *and* $\alpha \in \mathbb{R}$,

(i) $u \star v = v \star u$;

(ii) $u \star (v + w) = u \star v + u \star w$;

(iii) $(\alpha u) \star v = \alpha(u \star v)$.

Proof This is routinely checked.   □

Referring to Lemma 9.5, the vector space $E_j V$ together with the opertion $\star$ is called the *j*th *Norton algebra* for Γ; see [5]. This algebra is commutative and nonassociative. It has no multiplicative identity in general. See [31, 46, 50, 77] for recent results on the Norton algebra.

11.10 The Function Algebra and Nondegenerate Primitive Idempotents

Throughout this section, $\Gamma = (X, \mathcal{R})$ denotes a distance-regular graph with diameter $D \geq 3$. Recall the function algebra V from Definition 9.1. We will show that a primitive idempotent E of Γ is nondegenerate if and only if the eigenspace EV generates V in the function algebra.

For a subspace $U \subseteq V$, we describe the subalgebra of V generated by U. This subalgebra contains $\mathbf{1}$ by Lemma 9.2. To see what else is in the subalgebra, we define a binary relation on X called U-equivalence.

Definition 10.1   Vertices y, z in X are said to be *U-equivalent* whenever for all $u \in U$, the y-coordinate of u is equal to the z-coordinate of u. Observe that U-equivalence is an equivalence relation.

Definition 10.2   For a subset $Y \subseteq X$, define $\hat{Y} = \sum_{y \in Y} \hat{y}$.

Lemma 10.3   *For a subspace $U \subseteq V$, the following are equal:*

  (i) *the subalgebra of the function algebra V generated by U;*

 (ii) *Span$\{\hat{Y} | Y$ is a U-equivalence class$\}$.*

Proof   (i) $\subseteq$ (ii): Note that Span$\{\hat{Y} | Y$ is a U-equivalence class$\}$ is a subalgebra of V that contains U.

(i) $\supseteq$ (ii): Let Y denote a U-equivalence class. We show that $\hat{Y}$ is contained in the subalgebra of V generated by U. List the U-equivalence classes $Y = Y_0, Y_1, \ldots, Y_n$. For $u \in U$ write

$$u = \sum_{i=0}^{n} \alpha_i(u)\hat{Y}_i \qquad \alpha_i(u) \in \mathbb{R}.$$

For $1 \leq i \leq n$, there exists $u_i \in U$ such that $\alpha_0(u_i) \neq \alpha_i(u_i)$. We have

$$\hat{Y} = \prod_{i=1}^{n} \frac{u_i - \alpha_i(u_i)\mathbf{1}}{\alpha_0(u_i) - \alpha_i(u_i)},$$

where the product is with respect to $\circ$. Therefore $\hat{Y}$ is contained in the subalgebra of V generated by U. ◻

Corollary 10.4 *For a subspace $U \subseteq V$, the following are equivalent:*

  (i) *U generates the function algebra V;*
  (ii) *each U-equivalence class has cardinality one.*

Proof By Lemma 10.3. ◻

We have been discussing a subspace U of V. Next we consider the special case in which $U = EV$, where E is a primitive idempotent of Γ. Let us compute the EV-equivalence classes. We have $EV = \mathrm{Span}\{E\hat{w}|w \in X\}$. Before Lemma 4.1 we saw that for $y, z \in X$,

$$y\text{-coordinate of } E\hat{z} = z\text{-coordinate of } E\hat{y}. \tag{19}$$

Lemma 10.5 *Let E denote a primitive idempotent of Γ. Then for $y, z \in X$ the following are equivalent:*

  (i) *y, z are in the same EV-equivalence class;*
  (ii) *$E\hat{y} = E\hat{z}$.*

Proof Condition (i) holds if and only if the y-coordinate of $E\hat{w}$ is equal to the z-coordinate of $E\hat{w}$ for all $w \in X$. Condition (ii) holds if and only if the w-coordinate of $E\hat{y}$ is equal to the w-coordinate of $E\hat{z}$ for all $w \in X$. By these comments and (19), the conditions (i), (ii) are equivalent. ◻

Theorem 10.6 *Let E denote a primitive idempotent of Γ. Then the following are equivalent:*

  (i) *EV generates the function algebra V;*
  (ii) *E is nondegenerate.*

Proof By Corollary 10.4 and Lemma 10.5 we have the logical implications

EV generates the function algebra V

$\Leftrightarrow$ each EV-equivalence class has cardinality one

$\Leftrightarrow$ $\{E\hat{y}\}_{y \in X}$ are mutually distinct

$\Leftrightarrow$ E is nondegenerate. ◻

11.11 The Q-Polynomial Property and Askey–Wilson Duality

In this section, we discuss the Q-polynomial property and its connection to Askey–Wilson duality.

Throughout this section, $\Gamma = (X, \mathcal{R})$ denotes a distance-regular graph with diameter $D \geq 3$. Recall the primitive idempotents $\{E_i\}_{i=0}^{D}$ of Γ.

Definition 11.1   The ordering $\{E_i\}_{i=0}^{D}$ is called *Q-polynomial* whenever the following hold for $0 \leq h, i, j \leq D$:

(i) $q_{i,j}^{h} = 0$ if one of h, i, j is greater than the sum of the other two;

(ii) $q_{i,j}^{h} \neq 0$ if one of h, i, j is equal to the sum of the other two.

Definition 11.2   We say that Γ is *Q-polynomial* whenever there exists at least one Q-polynomial ordering of the primitive idempotents.

For the rest of this section, we assume that the ordering $\{E_i\}_{i=0}^{D}$ is Q-polynomial. Define

$$c_i^* = q_{1,i-1}^{i} \ (1 \leq i \leq D), \qquad a_i^* = q_{1,i}^{i} \ (0 \leq i \leq D),$$

$$b_i^* = q_{1,i+1}^{i} \ (0 \leq i \leq D - 1).$$

Note that $a_0^* = 0$ and $c_1^* = 1$. Moreover

$$c_i^* > 0 \ (1 \leq i \leq D), \qquad b_i^* > 0 \ (0 \leq i \leq D - 1).$$

From Lemma 5.15(iv), we obtain

$$c_i^* + a_i^* + b_i^* = m_1 \qquad (0 \leq i \leq D),$$

where $c_0^* = 0$ and $b_D^* = 0$. By Lemma 5.14, we have $m_i c_i^* = m_{i-1} b_{i-1}^*$ for $1 \leq i \leq D$. Consequently

$$m_i = \frac{b_0^* b_1^* \cdots b_{i-1}^*}{c_1^* c_2^* \cdots c_i^*} \qquad (0 \leq i \leq D). \qquad (20)$$

For the rest of this section, fix $x \in X$ and write $T = T(x)$. Recall the bases $\{A_i^*\}_{i=0}^{D}$ and $\{E_i^*\}_{i=0}^{D}$ for M^*. We abbreviate $A^* = A_1^*$ and call this the *dual adjacency matrix* (with respect to x and the given Q-polynomial structure).

Our next goal is to show that A_i^* is a polynomial of degree i in A^* for $0 \leq i \leq D$.

Lemma 11.3   *We have*

$$A^* A_i^* = b_{i-1}^* A_{i-1}^* + a_i^* A_i^* + c_{i+1}^* A_{i+1}^* \qquad (1 \leq i \leq D - 1),$$

$$A^* A_D^* = b_{D-1}^* A_{D-1}^* + a_D^* A_D^*.$$

Proof This is $A_i^* A_j^* = \sum_{h=0}^{D} q_{i,j}^h A_h^*$ with $j = 1$. □

Definition 11.4 We define some polynomials $\{v_i^*\}_{i=0}^{D+1}$ in $\mathbb{R}[\lambda]$ such that

$$v_0^* = 1, \qquad v_1^* = \lambda,$$
$$\lambda v_i^* = b_{i-1}^* v_{i-1}^* + a_i^* v_i^* + c_{i+1}^* v_{i+1}^* \qquad (1 \le i \le D),$$

where $c_{D+1}^* = 1$.

Lemma 11.5 *The following* (i)–(iv) *hold:*

(i) $\deg v_i^* = i \quad (0 \le i \le D + 1)$;
(ii) *the coefficient of* λ^i *in* v_i^* *is* $(c_1^* c_2^* \cdots c_i^*)^{-1} \quad (0 \le i \le D + 1)$;
(iii) $v_i^*(A^*) = A_i^* \quad (0 \le i \le D)$;
(iv) $v_{D+1}^*(A^*) = 0$.

Proof (i), (ii) By Definition 11.4.
(iii), (iv) Compare Lemma 11.3 and Definition 11.4. □

Corollary 11.6 *The following hold:*

(i) *the algebra* M^* *is generated by* A^*;
(ii) *the minimal polynomial of* A^* *is* $c_1^* c_2^* \cdots c_D^* v_{D+1}^*$.

Proof By Lemma 11.5 and since $\{A_i^*\}_{i=0}^{D}$ is a basis for M^*. □

Next we consider the eigenvalues of A^*. By Lemma 5.10(i),

$$A^* = m_1 \sum_{i=0}^{D} u_i(\theta_1) E_i^*.$$

Abbreviate

$$\theta_i^* = m_1 u_i(\theta_1) \qquad (0 \le i \le D), \qquad (21)$$

so that

$$A^* = \sum_{i=0}^{D} \theta_i^* E_i^*. \qquad (22)$$

Lemma 11.7 *The following* (i)–(iii) *hold:*

(i) *the polynomial* v_{D+1}^* *has* $D + 1$ *mutually distinct roots* $\{\theta_i^*\}_{i=0}^{D}$;
(ii) *the eigenspaces of* A^* *are* $\{E_i^* V\}_{i=0}^{D}$;
(iii) *for* $0 \le i \le D$, θ_i^* *is the eigenvalue of* A^* *for* $E_i^* V$.

Proof (i) The roots of v_{D+1}^* are mutually distinct by Corollary 11.6(ii) and
since A^* is diagonal. These roots are $\{\theta_i^*\}_{i=0}^{D}$ by (22).
(ii), (iii) By (22). □

Definition 11.8 For $0 \le i \le D$ we call θ_i^* the *ith dual eigenvalue of* Γ (with respect to the given Q-polynomial structure).

For convenience we adjust the normalization of the polynomials v_i^*.

Definition 11.9 Define the polynomial

$$u_i^* = \frac{v_i^*}{m_i} \qquad (0 \le i \le D). \tag{23}$$

Lemma 11.10 *We have*

$$u_0^* = 1, \qquad u_1^* = m_1^{-1}\lambda,$$
$$\lambda u_i^* = c_i^* u_{i-1}^* + a_i^* u_i^* + b_i^* u_{i+1}^* \qquad (1 \le i \le D-1),$$
$$\lambda u_D^* - c_D^* u_{D-1}^* - a_D^* u_D^* = m_D^{-1} v_{D+1}^*.$$

Proof Evaluate the recurrence in Definition 11.4 using $v_i^* = m_i u_i^*$ $(0 \le i \le D)$ and (20). □

We just defined the polynomials $\{u_i^*\}_{i=0}^D$. Next we explain how the polynomials $\{u_i\}_{i=0}^D$ and $\{u_i^*\}_{i=0}^D$ are related.

Theorem 11.11 (See [20, p. 14]) *We have*

$$u_i(\theta_j) = u_j^*(\theta_i^*) \qquad (0 \le i, j \le D). \tag{24}$$

Proof Using Lemma 5.11 and Lemma 11.5(iii),

$$u_i(\theta_j)E_i^* = m_j^{-1} A_j^* E_i^* = m_j^{-1} v_j^*(A^*)E_i^* = u_j^*(A^*)E_i^* = u_j^*(\theta_i^*)E_i^*.$$

The result follows. □

We will comment on Theorem 11.11 shortly.

Lemma 11.12 *For $0 \le i \le D$, we have*

    (i) $v_i^*(\theta_0^*) = m_i$;
    (ii) $u_i^*(\theta_0^*) = 1$.

Proof By Theorem 11.11, we obtain $u_i^*(\theta_0^*) = u_0(\theta_i) = 1$, giving (ii). By Definition 11.9, we get (i). □

In Propositions 3.14 and 3.15, we obtained some orthogonality relations for the polynomials $\{u_i\}_{i=0}^D$ and $\{v_i\}_{i=0}^D$. Next we give the analogous orthogonality relations for the polynomials $\{u_i^*\}_{i=0}^D$ and $\{v_i^*\}_{i=0}^D$.

Proposition 11.13 *We have*

$$\sum_{\ell=0}^{D} u_i^*(\theta_\ell^*)u_j^*(\theta_\ell^*)k_\ell = \delta_{i,j}m_i^{-1}|X| \qquad (0 \le i,j \le D),$$

$$\sum_{\ell=0}^{D} u_\ell^*(\theta_i^*)u_\ell^*(\theta_j^*)m_\ell = \delta_{i,j}k_i^{-1}|X| \qquad (0 \le i,j \le D).$$

Proof Combine Proposition 3.14 and Theorem 11.11. □

Proposition 11.14 *We have*

$$\sum_{\ell=0}^{D} v_i^*(\theta_\ell^*)v_j^*(\theta_\ell^*)k_\ell = \delta_{i,j}m_i|X| \qquad (0 \le i,j \le D),$$

$$\sum_{\ell=0}^{D} v_\ell^*(\theta_i^*)v_\ell^*(\theta_j^*)m_\ell^{-1} = \delta_{i,j}k_i^{-1}|X| \qquad (0 \le i,j \le D).$$

Proof Combine Definition 11.9 and Proposition 11.13. □

Equation (24) is called *Askey–Wilson duality* [73] or *Delsarte duality* [45]. In [45] D. Leonard classifies the orthogonal polynomial sequences that satisfy Askey–Wilson duality. See [2, p. 260] for a more comprehensive treatment. The classification shows that the orthogonal polynomial sequences that satisfy Askey–Wilson duality belong to the terminating branch of the Askey scheme; this branch consists of the q-Racah polynomials [1] along with their limiting cases [39]. The theory of Leonard pairs [29, 71, 73, 74, 76, 78] provides a modern approach to Askey–Wilson duality.

11.12 The Function Algebra Characterization of the Q-Polynomial Property

In this section, we characterize the Q-polynomial property in terms of the function algebra.

Throughout this section, $\Gamma = (X, \mathcal{R})$ denotes a distance-regular graph with diameter $D \ge 3$. Recall the function algebra V and the primitive idempotents $\{E_i\}_{i=0}^{D}$. Recall that

$$E_0V = \mathbb{R}\mathbf{1}, \qquad \mathbf{1} = \sum_{y \in X} \hat{y}.$$

We remind the reader that for subspaces R, S of V,

$$R \circ S = \mathrm{Span}\{r \circ s \mid r \in R,\ s \in S\}.$$

For an integer $n \geq 0$, define

$$R^{\circ n} = R \circ R \circ \cdots \circ R \qquad (n \text{ copies}). \qquad (25)$$

We interpret $R^{\circ 0} = \mathbb{R}\mathbf{1}$.

The following result appears in [61, lecture 23]. It is also mentioned in [19, p. 30].

Theorem 12.1   *The following are equivalent:*

(i) *the ordering $\{E_i\}_{i=0}^{D}$ is Q-polynomial;*
(ii) *E_1 is nondegenerate and*

$$E_1 V \circ E_i V \subseteq E_{i-1}V + E_i V + E_{i+1}V \qquad (0 \leq i \leq D), \qquad (26)$$

where $E_{-1} = 0$ and $E_{D+1} = 0$;
(iii) *for $0 \leq i \leq D$,*

$$\sum_{\ell=0}^{i} E_\ell V = \sum_{\ell=0}^{i} (E_1 V)^{\circ \ell}. \qquad (27)$$

Proof   (i) $\Rightarrow$ (ii) For the given Q-polynomial structure, the dual eigenvalues are $\theta_i^* = m_1 u_i(\theta_1)$ $(0 \leq i \leq D)$. The scalars $\{\theta_i^*\}_{i=0}^{D}$ are mutually distinct, so $\theta_i^* \neq \theta_0^*$ for $1 \leq i \leq D$. Therefore E_1 is nondegenerate. By Theorem 9.4, we obtain

$$E_1 V \circ E_i V = \sum_{\substack{0 \leq h \leq D \\ q_{1,i}^h \neq 0}} E_h V \qquad (0 \leq i \leq D).$$

The ordering $\{E_i\}_{i=0}^{D}$ is Q-polynomial, so $q_{1,i}^h = 0$ if $|h - i| > 1$ $(0 \leq h, i \leq D)$. By these comments we obtain (26).

(ii) $\Rightarrow$ (iii) For $0 \leq i \leq D$ define $P_i = \sum_{\ell=0}^{i}(E_1 V)^{\circ \ell}$. By Theorem 9.4, there exists a subset $S_i \subseteq \{0, 1, \ldots, D\}$ such that $P_i = \sum_{h \in S_i} E_h V$. We show that $S_i = \{0, 1, \ldots, i\}$ for $0 \leq i \leq D$. By construction, $S_0 = \{0\}$ and $S_1 = \{0, 1\}$. Using (26) we obtain $S_i \subseteq \{0, 1, \ldots, i\}$ for $0 \leq i \leq D$. By construction, $S_{i-1} \subseteq S_i$ for $1 \leq i \leq D$. For $1 \leq i \leq D$, we have $S_{i-1} \neq S_i$; otherwise $P_{i-1} = P_i$, which forces $i \geq 2$ and $E_1 V \circ P_{i-1} \subseteq P_{i-1}$, which forces $P_{i-1} = V$ by Theorem 10.6, which forces $S_{i-1} = \{0, 1, \ldots, D\}$, which contradicts $S_{i-1} \subseteq \{0, 1, \ldots, i - 1\} \subseteq \{0, 1, \ldots, D - 1\}$. By these comments, $S_i = \{0, 1, \ldots, i\}$ for $0 \leq i \leq D$.

(iii) $\Rightarrow$ (i) Let $0 \leq i \leq D$ and $0 \leq j \leq D - i$. We show that $q_{i,j}^{i+j} \neq 0$ and $q_{i,j}^h = 0$ for $i + j < h \leq D$. By (27),

$$(E_0 V + E_1 V + \cdots + E_i V) \circ (E_0 V + E_1 V + \cdots + E_j V)$$
$$= E_0 V + E_1 V + \cdots + E_{i+j}V.$$

By this and Theorem 9.4, we obtain $q_{i,j}^h = 0$ for $i + j < h \le D$. Also by Theorem 9.4, there exists $0 \le r \le i$ and $0 \le s \le j$ such that $q_{r,s}^{i+j} \ne 0$. By these comments, $i + j \le r + s$. By construction, $0 \le r \le i$ and $0 \le s \le j$, so $r = i$ and $s = j$. Therefore $q_{i,j}^{i+j} \ne 0$. □

11.13 Irreducible T-Modules and Tridiagonal Pairs

Throughout this section, $\Gamma = (X, \mathcal{R})$ denotes a distance-regular graph with diameter $D \ge 3$. We assume that Γ is Q-polynomial with respect to the ordering $\{E_i\}_{i=0}^D$ of the primitive idempotents. Fix $x \in X$ and write $T = T(x)$. We will show that A, A^* act on each irreducible T-module as a tridiagonal pair.

Lemma 13.1 *The algebra T is generated by A, A^*.*

Proof The algebra T is generated by M and M^*. Moreover M is generated by A, and M^* is generated by A^*. □

We review a few points:

- the eigenspaces of A are $\{E_i V\}_{i=0}^D$;
- for $0 \le i \le D$, θ_i is the eigenvalue of A for $E_i V$;
- the eigenspaces of A^* are $\{E_i^* V\}_{i=0}^D$;
- for $0 \le i \le D$, θ_i^* is the eigenvalue of A^* for $E_i^* V$;
- for $0 \le i \le D$, we have

$$A E_i^* V \subseteq E_{i-1}^* V + E_i^* V + E_{i+1}^* V,$$

where $E_{-1}^* = 0$ and $E_{D+1}^* = 0$.

Lemma 13.2 *For $0 \le i \le D$, we have*

$$A^* E_i V \subseteq E_{i-1} V + E_i V + E_{i+1} V,$$

where $E_{-1} = 0$ and $E_{D+1} = 0$.

Proof By the containment on the right in (14), together with Definition 11.1. □

In Section 11.2 we mentioned that the standard module V is an orthogonal direct sum of irreducible T-modules. Let W denote an irreducible T-module. Observe that W is an orthogonal direct sum of the nonzero subpaces among $\{E_i^* W\}_{i=0}^D$. Similarly W is an orthogonal direct sum of the nonzero subpaces among $\{E_i W\}_{i=0}^D$.

Lemma 13.3   *Let W denote an irreducible T-module. Then for $0 \le i \le D$, we have*

$$AE_i^*W \subseteq E_{i-1}^*W + E_i^*W + E_{i+1}^*W,$$
$$A^*E_iW \subseteq E_{i-1}W + E_iW + E_{i+1}W.$$

Proof   By Lemma 13.2 and the comment above it.   □

Let W denote an irreducible T-module. By the *endpoint* of W we mean $\min\{i|0 \le i \le D, E_i^*W \ne 0\}$. By the *diameter* of W we mean $|\{i|0 \le i \le D, E_i^*W \ne 0\}| - 1$. By the *dual endpoint* of W we mean $\min\{i|0 \le i \le D, E_iW \ne 0\}$. By the *dual diameter* of W we mean $|\{i|0 \le i \le D, E_iW \ne 0\}| - 1$.

Lemma 13.4   [67, lemma 3.4, lemma 3.9] *Let W denote an irreducible T-module with endpoint ρ and diameter d. Then ρ, d are nonnegative integers such that $\rho + d \le D$. Moreover the following* (i), (ii) *hold:*

  (i) $E_i^*W \ne 0$ *if and only if $\rho \le i \le \rho + d$* $(0 \le i \le D)$;
  (ii) $W = \sum_{i=\rho}^{\rho+d} E_i^*W$ *(orthogonal direct sum).*

Proof   (i) By construction $E_\rho^*W \ne 0$ and $E_i^*W = 0$ for $0 \le i < \rho$. Suppose there exists an integer i $(\rho < i \le \rho + d)$ such that $E_i^*W = 0$. Define $W' = E_\rho^*W + E_{\rho+1}^*W + \cdots + E_{i-1}^*W$. By construction $0 \ne W' \subseteq W$. Also by construction, $A^*W' \subseteq W'$. By Lemma 13.3 and $E_i^*W = 0$ we obtain $AW' \subseteq W'$. By these comments W' is a T-module. We have $W = W'$ since the T-module W is irreducible. This contradicts the fact that d is the diameter of W. We conclude that $E_i^*W \ne 0$ for $\rho \le i \le \rho + d$. By the definition of the diameter d we have $E_i^*W = 0$ for $\rho + d < i \le D$.

(ii) By (i) and the comments before Lemma 13.3.   □

Lemma 13.5   [67, lemma 3.4, lemma 3.12] *Let W denote an irreducible T-module with dual endpoint τ and dual diameter δ. Then τ, δ are nonnegative integers such that $\tau + \delta \le D$. Moreover the following* (i), (ii) *hold:*

  (i) $E_iW \ne 0$ *if and only if $\tau \le i \le \tau + \delta$* $(0 \le i \le D)$;
  (ii) $W = \sum_{i=\tau}^{\tau+\delta} E_iW$ *(orthogonal direct sum).*

Proof   Similar to the proof of Lemma 13.4.   □

Theorem 13.6 (See [67, lemmas 3.9, 3.12])   *The pair A, A^* acts on each irreducible T-module as a tridiagonal pair.*

Proof   By Definition 1.1 and Lemmas 13.1, 13.3, 13.4, 13.5.   □

Corollary 13.7   *Let W denote an irreducible T-module. The the diameter of W is equal to the dual diameter of W.*

Proof By the comment below Note 1.2.   □

It is an ongoing project to describe the irreducible T-modules for a Q-polynomial distance-regular graph Γ. Comprehensive treatments can be found in [3, 32, 36, 37, 67–69, 78]. In addition, there are papers about the thin condition [10, 21, 24, 64, 69, 82]; irreducible T-modules with endpoint one [30]; Γ being bipartite [6, 13, 14, 41]; Γ being almost-bipartite [9, 42]; Γ being dual bipartite [22]; Γ being almost dual bipartite [23]; Γ being 2-homogeneous [15, 17, 18, 53]; Γ being tight [56]; Γ being a hypercube [27]; Γ being a Doob graph [63]; Γ being a Johnson graph [49, 62]; Γ being a Grassmann graph [48]; Γ being a dual polar graph [84]; Γ having a spin model in the Bose–Mesner algebra [16, 52]. Some miscellaneous topics about irreducible T-modules can be found in [26, 34, 35, 40, 43, 44, 55, 59, 60, 75, 83].

11.14 Recurrent Sequences

In this section, we have some comments about finite sequences that satisfy a 3-term recurrence. In later sections we will apply these comments to Q-polynomial distance-regular graphs.

Throughout this section, fix an integer $D \geq 3$, and let $\{\theta_i\}_{i=0}^{D}$ denote scalars in $\mathbb{R}$.

Definition 14.1 Let β, γ, ϱ denote scalars in $\mathbb{R}$.

(i) The sequence $\{\theta_i\}_{i=0}^{D}$ is said to be *recurrent* whenever $\theta_{i-1} \neq \theta_i$ for $2 \leq i \leq D - 1$, and

$$\frac{\theta_{i-2} - \theta_{i+1}}{\theta_{i-1} - \theta_i} \tag{28}$$

is independent of i for $2 \leq i \leq D - 1$.

(ii) The sequence $\{\theta_i\}_{i=0}^{D}$ is said to be *β-recurrent* whenever

$$\theta_{i-2} - (\beta + 1)\theta_{i-1} + (\beta + 1)\theta_i - \theta_{i+1} \tag{29}$$

is zero for $2 \leq i \leq D - 1$.

(iii) The sequence $\{\theta_i\}_{i=0}^{D}$ is said to be *(β, γ)-recurrent* whenever

$$\theta_{i-1} - \beta\theta_i + \theta_{i+1} = \gamma \tag{30}$$

for $1 \leq i \leq D - 1$.

(iv) The sequence $\{\theta_i\}_{i=0}^{D}$ is said to be *(β, γ, ϱ)-recurrent* whenever

$$\theta_{i-1}^2 - \beta\theta_{i-1}\theta_i + \theta_i^2 - \gamma(\theta_{i-1} + \theta_i) = \varrho \tag{31}$$

for $1 \leq i \leq D$.

Lemma 14.2 *The following are equivalent:*

(i) *the sequence $\{\theta_i\}_{i=0}^D$ is recurrent;*

(ii) *the scalars $\theta_{i-1} \neq \theta_i$ for $2 \leq i \leq D-1$, and there exists $\beta \in \mathbb{R}$ such that $\{\theta_i\}_{i=0}^D$ is β-recurrent.*

Suppose (i), (ii) *hold. Then the common value of* (28) *is equal to $\beta + 1$.*

Proof Routine. □

Lemma 14.3 *For $\beta \in \mathbb{R}$ the following are equivalent:*

(i) *the sequence $\{\theta_i\}_{i=0}^D$ is β-recurrent;*

(ii) *there exists $\gamma \in \mathbb{R}$ such that $\{\theta_i\}_{i=0}^D$ is (β, γ)-recurrent.*

Proof Routine. □

Lemma 14.4 *The following* (i), (ii) *hold for all $\beta, \gamma \in \mathbb{R}$.*

(i) *Suppose $\{\theta_i\}_{i=0}^D$ is (β, γ)-recurrent. Then there exists $\varrho \in \mathbb{R}$ such that $\{\theta_i\}_{i=0}^D$ is (β, γ, ϱ)-recurrent.*

(ii) *Suppose $\{\theta_i\}_{i=0}^D$ is (β, γ, ϱ)-recurrent, and that $\theta_{i-1} \neq \theta_{i+1}$ for $1 \leq i \leq D-1$. Then $\{\theta_i\}_{i=0}^D$ is (β, γ)-recurrent.*

Proof Let p_i denote the expression on the left in (31), and observe

$$p_i - p_{i+1} = (\theta_{i-1} - \theta_{i+1})(\theta_{i-1} - \beta\theta_i + \theta_{i+1} - \gamma)$$

for $1 \leq i \leq D-1$. Assertions (i), (ii) are both routine consequences of this. □

The following result is handy.

Lemma 14.5 *Let $\beta, \gamma, \varrho \in \mathbb{R}$ and assume that $\{\theta_i\}_{i=0}^D$ is (β, γ, ϱ)-recurrent. Then*

$$(2 - \beta)\theta_i^2 - 2\gamma\theta_i - \varrho = (\theta_i - \theta_{i-1})(\theta_i - \theta_{i+1}) \qquad (0 \leq i \leq D), \quad (32)$$

where θ_{-1} and θ_{D+1} are defined by (30) *at $i = 0$ and $i = D$.*

Proof To verify (32) for $1 \leq i \leq D$, eliminate θ_{i+1} using (30), and evaluate the result using (31). To verify (32) for $0 \leq i \leq D-1$, eliminate θ_{i-1} using (30), and evaluate the result using (31). □

11.15 The Tridiagonal Relations

Throughout this section, $\Gamma = (X, \mathcal{R})$ denotes a distance-regular graph with diameter $D \geq 3$. Let $\{E_i\}_{i=0}^D$ denote a Q-polynomial ordering of the primitive idempotents of Γ. Fix $x \in X$ and write $T = T(x)$. We will show that A, A^* satisfy a pair of relations called the tridiagonal relations. We will also obtain a recurrence satisfied by the eigenvalues $\{\theta_i\}_{i=0}^D$ and the dual eigenvalues $\{\theta_i^*\}_{i=0}^D$. We now state our two main results.

Theorem 15.1 (See [69, lemma 5.4]) *With the above notation, there exist real numbers β, γ, γ^*, ϱ, ϱ^* such that*

$$0 = [A, A^2A^* - \beta AA^*A + A^*A^2 - \gamma(AA^* + A^*A) - \varrho A^*], \tag{33}$$

$$0 = [A^*, A^{*2}A - \beta A^*AA^* + AA^{*2} - \gamma^*(A^*A + AA^*) - \varrho^*A]. \tag{34}$$

Definition 15.2 The relations (33), (34) are called the *tridiagonal relations*; see [72].

Theorem 15.3 (See [45, proposition 3] and [2, theorem 5.1]) *With the above notation, the scalars*

$$\frac{\theta_{i-2} - \theta_{i+1}}{\theta_{i-1} - \theta_i}, \qquad \frac{\theta_{i-2}^* - \theta_{i+1}^*}{\theta_{i-1}^* - \theta_i^*}$$

are equal and independent of i for $2 \le i \le D - 1$.

We will prove Theorems 15.1 and 15.3 shortly.

Lemma 15.4 *For $0 \le i, j, r \le D$, we have*

(i) $E_i^* A^r E_j^* = \begin{cases} 0 & \text{if } r < |i - j|; \\ \ne 0 & \text{if } r = |i - j|, \end{cases}$

(ii) $E_i A^{*r} E_j = \begin{cases} 0 & \text{if } r < |i - j|; \\ \ne 0 & \text{if } r = |i - j|. \end{cases}$

Proof   (i) By Theorem 8.4(i), $E_i^* A_r E_j^* = 0$ if and only if $p_{r,j}^i = 0$. The matrix A_r is a polynomial in A with degree exactly r. The scalar $p_{r,j}^i$ is zero if $r < |i - j|$ and nonzero if $r = |i - j|$. The result follows.

(ii) Similar to the proof of (i).   □

We are going to prove a sequence of results. Each result has a 'dual' obtained by interchanging A, A^* and E_i, E_i^* for $0 \le i \le D$. We will not state each dual explicitly.

Lemma 15.5 *For $0 \le i, j, r, s \le D$, we have*

$$E_i^* A^r A^* A^s E_j^* = \begin{cases} \theta_{j+s}^* E_i^* A^{r+s} E_j^* & \text{if } r + s = i - j; \\ \theta_{j-s}^* E_i^* A^{r+s} E_j^* & \text{if } r + s = j - i; \\ 0 & \text{if } r + s < |i - j|. \end{cases}$$

Proof Using $I = \sum_{h=0}^{D} E_h^*$ and $A^* = \sum_{h=0}^{D} \theta_h^* E_h^*$, we find

$$E_i^* A^r I A^s E_j^* = \sum_{h=0}^{D} E_i^* A^r E_h^* A^s E_j^*,$$

$$E_i^* A^r A^* A^s E_j^* = \sum_{h=0}^{D} \theta_h^* E_i^* A^r E_h^* A^s E_j^*.$$

In these sums, evaluate each summand using Lemma 15.4(i). The result follows. □

Lemma 15.6 *For* $0 \le i \le D - 1$,

$$E_i A^* E_{i+1} - E_{i+1} A^* E_i = (E_0 + E_1 + \cdots + E_i) A^*$$
$$- A^* (E_0 + E_1 + \cdots + E_i).$$

Proof Recall the convention $E_{-1} = 0$. For $0 \le j \le D - 1$, we have

$$E_j A^* = E_j A^* (E_0 + E_1 + \cdots + E_D) = E_j A^* (E_{j-1} + E_j + E_{j+1}), \qquad (35)$$

and similarly

$$A^* E_j = (E_{j-1} + E_j + E_{j+1}) A^* E_j. \qquad (36)$$

Sum (35), (36) over $j = 0, 1, \ldots, i$. Take the difference between the two sums. □

Recall the Bose–Mesner algebra M of Γ.

Lemma 15.7 *We have*

$$\mathrm{Span}\{RA^* S - SA^* R | R, S \in M\} = \{YA^* - A^* Y | Y \in M\}.$$

Proof Recall that $\{E_i\}_{i=0}^{D}$ is a basis for M. Note that $\{E_0 + \cdots + E_i\}_{i=0}^{D}$ is a basis for M. Observe that

$$\mathrm{Span}\{RA^* S - SA^* R | R, S \in M\}$$
$$= \mathrm{Span}\{E_i A^* E_j - E_j A^* E_i | 0 \le i, j \le D\}$$
$$= \mathrm{Span}\{E_i A^* E_{i+1} - E_{i+1} A^* E_i | 0 \le i \le D - 1\}$$
$$= \mathrm{Span}\{(E_0 + \cdots + E_i) A^* - A^* (E_0 + \cdots + E_i) | 0 \le i \le D - 1\}$$
$$= \mathrm{Span}\{(E_0 + \cdots + E_i) A^* - A^* (E_0 + \cdots + E_i) | 0 \le i \le D\}$$
$$= \{YA^* - A^* Y | Y \in M\}. \qquad \square$$

Let λ, μ denote commuting indeterminates. We define some polynomials $P(\lambda, \mu)$ and $P^*(\lambda, \mu)$ as follows. Given real numbers β, γ, ϱ, define

$$P(\lambda, \mu) = \lambda^2 - \beta\lambda\mu + \mu^2 - \gamma(\lambda + \mu) - \varrho. \tag{37}$$

Given real numbers $\beta, \gamma^*, \varrho^*$, define

$$P^*(\lambda, \mu) = \lambda^2 - \beta\lambda\mu + \mu^2 - \gamma^*(\lambda + \mu) - \varrho^*. \tag{38}$$

Lemma 15.8   *For real numbers β, γ, ϱ, the following are equivalent:*

(i) $0 = [A, A^2A^* - \beta AA^*A + A^*A^2 - \gamma(AA^* + A^*A) - \varrho A^*]$;

(ii) $P(\theta_{i-1}, \theta_i) = 0$ *for* $1 \le i \le D$.

Proof   Let C denote the expression on the right in (i). We have

$$C = (E_0 + \cdots + E_D)C(E_0 + \cdots + E_D) = \sum_{i=0}^{D}\sum_{j=0}^{D} E_i C E_j.$$

For $0 \le i, j \le D$, use $E_i A = \theta_i E_i$ and $AE_j = \theta_j E_j$ to get

$$E_i C E_j = E_i A^* E_j(\theta_i - \theta_j)P(\theta_i, \theta_j). \tag{39}$$

(i) $\Rightarrow$ (ii): For $1 \le i \le D$, we show $P(\theta_{i-1}, \theta_i) = 0$. We have $C = 0$, so

$$0 = E_{i-1}C E_i = E_{i-1}A^*E_i(\theta_{i-1} - \theta_i)P(\theta_{i-1}, \theta_i).$$

We have $E_{i-1}A^*E_i \ne 0$ and $\theta_{i-1} - \theta_i \ne 0$, so $P(\theta_{i-1}, \theta_i) = 0$.

(ii) $\Rightarrow$ (i): We show that $C = 0$. The polynomial $P(\lambda, \mu)$ is symmetric, so $P(\theta_i, \theta_{i-1}) = 0$ for $1 \le i \le D$. To show that $C = 0$, it suffices to show that $E_i C E_j = 0$ for $0 \le i, j \le D$. Let i, j be given. We evaluate $E_i C E_j$ using (39). If $|i - j| > 1$ then $E_i A^* E_j = 0$, so $E_i C E_j = 0$. If $|i - j| = 1$ then $P(\theta_i, \theta_j) = 0$, so $E_i C E_j = 0$. If $i = j$ then $\theta_i - \theta_j = 0$, so $E_i C E_j = 0$. In all cases $E_i C E_j = 0$, so $C = 0$.   □

We are now ready to prove Theorem 15.1.

Proof of Theorem 15.1   By Lemma 15.7 (with $R = A^2$ and $S = A$), there exists $Z \in M$ such that

$$A^2A^*A - AA^*A^2 = ZA^* - A^*Z. \tag{40}$$

The matrices $\{A^i\}_{i=0}^{D}$ form a basis for M, so there exists a polynomial f with degree at most D such that $Z = f(A)$. Let d denote the degree of f. We show that $d = 3$. First assume that $d > 3$. Multiply each term in (40) on the left by E_d^* and the right by E_0^*. Evaluate the result using Lemmas 15.4, 15.5 to get

$$0 = c(\theta_0^* - \theta_d^*)E_d^*A^dE_0^*, \tag{41}$$

where c is the leading coefficient of f. By construction $c \ne 0$. We have $\theta_0^* - \theta_d^* \ne 0$ since $d \ne 0$. Also $E_d^*A^dE_0^* \ne 0$ by Lemma 15.4. Therefore (41) gives a contradiction. Next assume that $d < 3$. Multiply each term in (40) on the left by E_3^* and on the right by E_0^*. Evaluate the result using Lemmas 15.4, 15.5 to get

$$(\theta_1^* - \theta_2^*)E_3^* A^3 A_0^* = 0. \tag{42}$$

We have $\theta_1^* - \theta_2^* \neq 0$. Also $E_3^* A^3 E_0^* \neq 0$ by Lemma 15.4. Therefore (42) gives a contradiction. We have shown that $d = 3$. Define $\beta = c^{-1} - 1$. Divide both sides of (40) by c, and evaluate the result using $d = 3$ and $c^{-1} = \beta + 1$. We find that there exist $\gamma, \varrho \in \mathbb{R}$ such that

$$(\beta + 1)(A^2 A^* A - AA^* A^2) = A^3 A^* - A^* A^3 - \gamma(A^2 A^* - A^* A^2) \\ - \varrho(AA^* - A^* A).$$

In this equation, we rearrange the terms to get

$$0 = [A, A^2 A^* - \beta AA^* A + A^* A^2 - \gamma(AA^* + A^* A) - \varrho A^*]. \tag{43}$$

This is the first tridiagonal relation. To get the second tridiagonal relation, pick an integer i ($2 \leq i \leq D - 1$). Multiply each term in (43) on the left by E_{i-2}^* and on the right by E_{i+1}^*. Simplify the result using Lemma 15.5 to get

$$0 = E_{i-2}^* A^3 E_{i+1}^* \left(\theta_{i-2}^* - (\beta + 1)\theta_{i-1}^* + (\beta + 1)\theta_i^* - \theta_{i+1}^*\right). \tag{44}$$

We have $E_{i-2}^* A^3 E_{i+1}^* \neq 0$ by Lemma 15.4, so the coefficient in (44) must be zero. Therefore the sequence $\{\theta_i^*\}_{i=0}^D$ is β-recurrent. By Lemma 14.3, there exists $\gamma^* \in \mathbb{R}$ such that $\{\theta_i^*\}_{i=0}^D$ is (β, γ^*)-recurrent. By Lemma 14.4, there exists $\varrho^* \in \mathbb{R}$ such that $\{\theta_i^*\}_{i=0}^D$ is $(\beta, \gamma^*, \varrho^*)$-recurrent. Consequently $P^*(\theta_{i-1}^*, \theta_i^*) = 0$ for $1 \leq i \leq D$. Now $\beta, \gamma^*, \varrho^*$ satisfy (34) by Lemma 15.8. $\qquad\square$

Proposition 15.9 (See [72, theorem 4.3]) *We refer to the scalars $\beta, \gamma, \gamma^*, \varrho, \varrho^*$ from Theorem 15.1.*

(i) *The expressions*

$$\frac{\theta_{i-2} - \theta_{i+1}}{\theta_{i-1} - \theta_i}, \qquad\qquad \frac{\theta_{i-2}^* - \theta_{i+1}^*}{\theta_{i-1}^* - \theta_i^*}$$

are both equal to $\beta + 1$ for $2 \leq i \leq D - 1$;

(ii) $\gamma = \theta_{i-1} - \beta\theta_i + \theta_{i+1}$ $(1 \leq i \leq D - 1)$;

(iii) $\gamma^* = \theta_{i-1}^* - \beta\theta_i^* + \theta_{i+1}^*$ $(1 \leq i \leq D - 1)$;

(iv) $\varrho = \theta_{i-1}^2 - \beta\theta_{i-1}\theta_i + \theta_i^2 - \gamma(\theta_{i-1} + \theta_i)$ $(1 \leq i \leq D)$;

(v) $\varrho^* = \theta_{i-1}^{*2} - \beta\theta_{i-1}^*\theta_i^* + \theta_i^{*2} - \gamma^*(\theta_{i-1}^* + \theta_i^*)$ $(1 \leq i \leq D)$.

Proof   We start with item (iv).

(iv) By (33) and Lemma 15.8.

(v) By (34) and Lemma 15.8.

(ii) By Lemma 14.4 and (iv) of Proposition 15.9.

(iii) By Lemma 14.4 and (v) of Proposition 15.9.

(i) The sequence $\{\theta_i\}_{i=0}^{D}$ is (β, γ)-recurrent by (ii), so $\{\theta_i\}_{i=0}^{D}$ is β-recurrent. Similarly $\{\theta_i^*\}_{i=0}^{D}$ is β-recurrent. The result follows.   □

Theorem 15.3 is immediate from Proposition 15.9(i).

Remark 15.10   The tridiagonal relations (33), (34) are the defining relations for the tridiagonal algebra [72, definition 3.9]. Special cases of the tridiagonal algebra include the universal enveloping algebra of the Onsager Lie algebra O [76, remark 34.5], the q-Onsager algebra O_q [79], and the positive part U_q^+ [76, remark 34.7] of the q-deformed enveloping algebra $U_q(\widehat{\mathfrak{sl}}_2)$ [11].

11.16 The Primary T-Module and the Askey–Wilson Relations

Throughout this section, $\Gamma = (X, \mathcal{R})$ denotes a distance-regular graph with diameter $D \geq 3$. Fix $x \in X$ and write $T = T(x)$. In Section 11.7 we described the primary T-module. In this section, we give more information under the assumption that Γ is Q-polynomial.

Throughout this section, we assume that Γ is Q-polynomial with respect to the ordering $\{E_i\}_{i=0}^{D}$ of the primitive idempotents.

Lemma 16.1   *For the primary T-module* V, *the following hold.*

(i) *With respect to the basis* $\{1_i\}_{i=0}^{D}$, *the matrices representing* A *and* A^* *are*

$$
A: \begin{pmatrix}
a_0 & b_0 & & & & & \mathbf{0} \\
c_1 & a_1 & b_1 & & & & \\
& c_2 & \cdot & \cdot & & & \\
& & \cdot & \cdot & \cdot & & \\
& & & \cdot & \cdot & b_{D-1} \\
\mathbf{0} & & & & c_D & a_D
\end{pmatrix}, \ A^*: \ \mathrm{diag}(\theta_0^*, \theta_1^*, \ldots, \theta_D^*).
$$

(ii) *With respect to the basis* $\{1_i^*\}_{i=0}^{D}$, *the matrices representing* A *and* A^* *are*

$$A\colon \operatorname{diag}(\theta_0, \theta_1, \ldots, \theta_D), \ A^*\colon \begin{pmatrix} a_0^* & b_0^* & & & & \mathbf{0} \\ c_1^* & a_1^* & b_1^* & & & \\ & c_2^* & \cdot & \cdot & & \\ & & \cdot & \cdot & \cdot & \\ & & & \cdot & \cdot & b_{D-1}^* \\ \mathbf{0} & & & & c_D^* & a_D^* \end{pmatrix}.$$

Proof   By parts (ii), (iv) of Lemmas 7.8, 7.9.   □

Lemma 16.2   *The pair A, A^* acts on the primary T-module $\mathcal{V}$ as a Leonard pair.*

Proof   By Definition 1.3 and Lemma 16.1.   □

In Section 11.15, we saw that A, A^* satisfy the tridiagonal relations. In the literature, there is another pair of relations called the Askey–Wilson relations [85], that resemble the tridiagonal relations but are more elementary. It is shown in [81, theorem 1.5] that any Leonard pair satisfies the Askey–Wilson relations. This and Lemma 16.2 imply that A, A^* satisfy the Askey–Wilson relations on the primary T-module $\mathcal{V}$. We will show directly, that on the primary T-module $\mathcal{V}$ the A, A^* satisfy the Askey–Wilson relations.

We have a comment about Lemma 15.4. In that lemma there are some inequalities. We will need the fact that these inequalities hold on the primary T-module $\mathcal{V}$.

Lemma 16.3   *For $0 \le i, j \le D$ and $r = |i - j|$, the following hold on the primary T-module $\mathcal{V}$:*

$$E_i^* A^r E_j^* \ne 0, \qquad\qquad E_i A^{*r} E_j \ne 0.$$

Proof   Use the matrix representations in Lemma 16.1, together with the fact that $b_{i-1} c_i \ne 0$ and $b_{i-1}^* c_i^* \ne 0$ for $1 \le i \le D$.   □

We now display the Askey–Wilson relations. Recall the scalars $\beta, \gamma, \gamma^*, \varrho, \varrho^*$ from Theorem 15.1. The following result is a variation on [81, theorem 1.5].

Theorem 16.4 (See [81, theorem 1.5])   *There exist real numbers ω, η, η^* such that the following hold on the primary T-module $\mathcal{V}$:*

$$A^2 A^* - \beta A A^* A + A^* A^2 - \gamma(A A^* + A^* A) - \varrho A^* = \gamma^* A^2 + \omega A + \eta I, \quad (45)$$

$$A^{*2} A - \beta A^* A A^* + A A^{*2} - \gamma^*(A^* A + A A^*) - \varrho^* A = \gamma A^{*2} + \omega A^* + \eta^* I. \quad (46)$$

Proof   Let $\mathcal{L}$ denote the expression on the left in (45). By the tridiagonal relation (33), we see that $[A, \mathcal{L}] = 0$ on $\mathcal{V}$. The restriction of A to $\mathcal{V}$ is diagonalizable and has all eigenspaces of dimension one. Recall from linear algebra that for a diagonalizable linear transformation σ that has all eigenspaces of dimension one, any linear transformation that commutes with σ must be a polynomial in σ. Therefore there exists a polynomial f such that $\mathcal{L} = f(A)$ on $\mathcal{V}$. The minimal polynomial of A on $\mathcal{V}$ has degree $D + 1$, so we may choose f such that its degree is at most D. Let d denote the degree of f. We show that $d \leq 2$. Suppose that $d \geq 3$. In the equation $\mathcal{L} = f(A)$, multiply each term on the left by E_d^* and on the right by E_0^*. We have $E_d^* \mathcal{L} E_0^* = 0$ by Lemma 15.5, so $0 = E_d^* f(A) E_0^*$ on $\mathcal{V}$. By Lemma 15.4, we have $E_d^* f(A) E_0^* = \alpha E_d^* A^d E_0^*$, where α is the leading coefficient of f. By construction, $\alpha \neq 0$. By Lemma 16.3, we have $E_d^* A^d E_0^* \neq 0$ on $\mathcal{V}$. This is a contradiction, so $d \leq 2$. There exist real numbers $\varepsilon, \omega, \eta$ such that $\mathcal{L} = \varepsilon A^2 + \omega A + \eta I$ on $\mathcal{V}$. We show that $\gamma^* = \varepsilon$. Note that on $\mathcal{V}$,

$$
\begin{aligned}
\gamma^* E_2^* A^2 E_0^* &= (\theta_0^* - \beta \theta_1^* + \theta_2^*) E_2^* A^2 E_0^* = E_2^* \mathcal{L} E_0^* \\
&= E_2^* (\varepsilon A^2 + \omega A + \eta I) E_0^* = \varepsilon E_2^* A^2 E_0^*.
\end{aligned}
$$

By Lemma 16.3, we have $E_2^* A^2 E_0^* \neq 0$ on $\mathcal{V}$, so $\gamma^* = \varepsilon$. We have shown that (45) holds on $\mathcal{V}$. Interchanging the roles of A, A^* in the argument so far, we see that there exist real numbers ω^*, η^* such that on $\mathcal{V}$,

$$
A^{*2}A - \beta A^* A A^* + A A^{*2} - \gamma^* (A^* A + A A^*) - \varrho^* A = \gamma A^{*2} + \omega^* A^* + \eta^* I. \tag{47}
$$

We show that $\omega = \omega^*$. Take the commutator of (45) with A^*. This shows that on $\mathcal{V}$,

$$
\begin{aligned}
A^2 A^{*2} - &\beta A A^* A A^* + \beta A^* A A^* A - A^{*2} A^2 - \gamma (A A^{*2} - A^{*2} A) \\
&= \gamma^* (A^2 A^* - A^* A^2) + \omega (A A^* - A^* A).
\end{aligned}
$$

Next, take the commutator of (47) with A. This shows that on $\mathcal{V}$,

$$
\begin{aligned}
A^{*2} A^2 - &\beta A^* A A^* A + \beta A A^* A A^* - A^2 A^{*2} - \gamma^* (A^* A^2 - A^2 A^*) \\
&= \gamma (A^{*2} A - A A^{*2}) + \omega^* (A^* A - A A^*).
\end{aligned}
$$

Adding the above two equations, we find that on $\mathcal{V}$,

$$
0 = (\omega - \omega^*)(A A^* - A^* A).
$$

We have $A A^* \neq A^* A$ on $\mathcal{V}$, because the T-module $\mathcal{V}$ is irreducible. Therefore $\omega = \omega^*$. We have shown that (46) holds on $\mathcal{V}$. $\qquad\square$

Definition 16.5 (See [85])   The relations (45), (46) are called the *Askey–Wilson relations*.

Next we consider how to compute the scalars ω, η, η^* from Theorem 16.4. To facilitate this computation, we bring in some notation. Recall from Proposition 15.9 that

$$\gamma = \theta_{i-1} - \beta\theta_i + \theta_{i+1} \qquad (1 \le i \le D-1), \qquad (48)$$

$$\gamma^* = \theta^*_{i-1} - \beta\theta^*_i + \theta^*_{i+1} \qquad (1 \le i \le D-1). \qquad (49)$$

Definition 16.6 Define the real numbers

$$\theta_{-1}, \quad \theta_{D+1}, \quad \theta^*_{-1}, \quad \theta^*_{D+1}$$

such that (48), (49) hold at $i = 0$ and $i = D$.

Recall the polynomials P, P^* from (37), (38).

Lemma 16.7 *The following hold for $0 \le i \le D$:*

(i) $P(\theta_i, \theta_i) = (\theta_i - \theta_{i-1})(\theta_i - \theta_{i+1})$;

(ii) $P^*(\theta^*_i, \theta^*_i) = (\theta^*_i - \theta^*_{i-1})(\theta^*_i - \theta^*_{i+1})$.

Proof (i) By Lemma 14.5 and since $\{\theta_i\}_{i=0}^{D}$ is (β, γ, ϱ)-recurrent.

(ii) Similar to the proof of (i). $\square$

Proposition 16.8 (See [81, theorem 5.3]) *We have*

(i) $\omega = a^*_i(\theta_i - \theta_{i+1}) + a^*_{i-1}(\theta_{i-1} - \theta_{i-2}) - \gamma^*(\theta_{i-1} + \theta_i)$ $(1 \le i \le D)$,

(ii) $\omega = a_i(\theta^*_i - \theta^*_{i+1}) + a_{i-1}(\theta^*_{i-1} - \theta^*_{i-2}) - \gamma(\theta^*_{i-1} + \theta^*_i)$ $(1 \le i \le D)$,

(iii) $\eta = a^*_i(\theta_i - \theta_{i-1})(\theta_i - \theta_{i+1}) - \omega\theta_i - \gamma^*\theta^2_i$ $(0 \le i \le D)$,

(iv) $\eta^* = a_i(\theta^*_i - \theta^*_{i-1})(\theta^*_i - \theta^*_{i+1}) - \omega\theta^*_i - \gamma\theta^{*2}_i$ $(0 \le i \le D)$.

Proof We start with item (iii).

(iii) In the Askey–Wilson relation (45), multiply each term on the left by E_i and on the right by E_i. This shows that on $\mathcal{V}$,

$$E_i A^* E_i P(\theta_i, \theta_i) = E_i\left(\gamma^*\theta^2_i + \omega\theta_i + \eta\right).$$

By Lemma 8.8, we have $E_i A^* E_i = a^*_i E_i$ on $\mathcal{V}$. By Lemma 7.5(iii), $E_i \ne 0$ on $\mathcal{V}$. By these comments,

$$a^*_i P(\theta_i, \theta_i) = \gamma^*\theta^2_i + \omega\theta_i + \eta.$$

Solve this equation for η, and evaluate the result using Lemma 16.7(i).

(iv) Similar to the proof of (iii).

(i) Subtract (iii) (at i) from (iii) (at $i-1$).

(ii) Similar to the proof of (i). $\square$

11.17 The Pascasio Characterization of the Q-Polynomial Property

In [57], Pascasio characterized the Q-polynomial distance-regular graphs using the dual eigenvalues θ_i^* and the intersection numbers a_i. In this section, we give a proof of her result that uses some ideas of Hanson [28].

Throughout this section, $\Gamma = (X, \mathcal{R})$ denotes a distance-regular graph with diameter $D \geq 3$.

Definition 17.1 Let E denote a nontrivial primitive idempotent of Γ. We say that Γ is *Q-polynomial with respect to E* whenever there exists a Q-polynomial ordering $\{E_i\}_{i=0}^D$ of the primitive idempotents of Γ such that $E = E_1$.

Theorem 17.2 (See [57, theorem 1.2]) *Let $E = |X|^{-1} \sum_{i=0}^D \theta_i^* A_i$ denote a nontrivial primitive idempotent of Γ. Then Γ is Q-polynomial with respect to E if and only if the following conditions hold:*

(i) *$\theta_i^* \neq \theta_0^*$ ($1 \leq i \leq D$);*
(ii) *there exist real numbers β, γ^* such that*

$$\theta_{i-1}^* - \beta\theta_i^* + \theta_{i+1}^* = \gamma^* \qquad (1 \leq i \leq D - 1); \qquad (50)$$

(iii) *there exist real numbers γ, ω, η^* such that*

$$a_i(\theta_i^* - \theta_{i-1}^*)(\theta_i^* - \theta_{i+1}^*) = \gamma\theta_i^{*2} + \omega\theta_i^* + \eta^* \qquad (0 \leq i \leq D),$$

where θ_{-1}^*, θ_{D+1}^* are defined such that (50) holds at $i = 0$ and $i = D$.

We will prove Theorem 17.2 shortly. In the meantime, let $\{E_i\}_{i=0}^D$ denote an ordering of the primitive idempotents of Γ, and abbreviate $E = E_1$. Let $x \in X$ and write $T = T(x)$, $A^* = A_1^*$.

Definition 17.3 Define a graph Δ_E with vertex set $\{0, 1, \ldots, D\}$ such that for $0 \leq i, j \leq D$, the vertices i, j are adjacent whenever $i \neq j$ and $q_{i,j}^1 \neq 0$.

Lemma 17.4 *For the graph Δ_E in Definition 17.3, the vertex 0 is adjacent to vertex 1 and no other vertex.*

Proof We have $q_{0,j}^1 = \delta_{1,j}$ for $0 \leq j \leq D$.   □

Lemma 17.5 *For distinct vertices i, j of Δ_E, the following are equivalent:*

(i) *i, j are adjacent in Δ_E;*
(ii) *$E_i A^* E_j \neq 0$;*
(iii) *$E_i A^* E_j \neq 0$ on the primary T-module V.*

Proof By the triple product relations and Lemma 8.7.   □

Lemma 17.6 *For the graph Δ_E in Definition 17.3, assume that E is nondegenerate. Then Δ_E is connected.*

Proof Let $S \subseteq \{0, 1, \ldots, D\}$ denote the connected component of Δ_E that contains $0, 1$. We show that $S = \{0, 1, \ldots, D\}$. Define $U = \sum_{i \in S} E_i V$. By construction $E_0 V \subseteq U$ and $EV \subseteq U$. By Lemma 5.14 and Theorem 9.4, the following holds for $0 \le i \le D$:

$$EV \circ E_i V = \sum_{\substack{0 \le h \le D \\ q^1_{i,h} \ne 0}} E_h V.$$

Therefore $EV \circ U \subseteq U$. We assume that E is nondegenerate, so the function algebra V is generated by EV. By these comments, $U = V$, so $S = \{0, 1, \ldots, D\}$.   □

Proof of Theorem 17.2 First we assume that Γ is Q-polynomial with respect to E. We saw earlier that E satisfies (i)–(iii). Next we assume that E satisfies (i)–(iii), and show that Γ is Q-polynomial with respect to E. Fix $x \in X$ and write $T = T(x)$. Abbreviate $E = E_1$ and $A^* = A^*_1$. For the time being, let $\{E_i\}^D_{i=2}$ denote any ordering of the remaining nontrivial primitive idempotents of Γ. For $0 \le i \le D$, let θ_i denote the eigenvalue of Γ for E_i. Consider the graph Δ_E from Definition 17.3. By Lemma 17.4, in Δ_E the vertex 0 is adjacent to vertex 1 and no other vertices. Note that E is nondegenerate by condition (i) in the theorem statement, so the graph Δ_E is connected in view of Lemma 17.6. We will show that Δ_E is a path. By (50), the sequence $\{\theta^*_i\}^D_{i=0}$ is (β, γ^*)-recurrent. By Lemma 14.4, there exists $\varrho^* \in \mathbb{R}$ such that $\{\theta^*_i\}^D_{i=0}$ is $(\beta, \gamma^*, \varrho^*)$-recurrent. By this and Lemma 14.5, we obtain

$$P^*(\theta^*_i, \theta^*_i) = (\theta^*_i - \theta^*_{i-1})(\theta^*_i - \theta^*_{i+1}) \qquad (0 \le i \le D),$$

where the polynomial P^* is from (38).
Claim 1. On the primary T-module V,

$$A^{*2}A - \beta A^* A A^* + A A^{*2} - \gamma^*(A^* A + A A^*) - \varrho^* A = \gamma A^{*2} + \omega A^* + \eta^* I. \quad (51)$$

Proof of Claim 1. Let $\mathcal{L}$ denote the left-hand side of (51). On the T-module V,

$$\mathcal{L} = \sum_{i=0}^{D} \sum_{j=0}^{D} E_i^* \mathcal{L} E_j^*$$

$$= \sum_{i=0}^{D} \sum_{j=0}^{D} E_i^* A E_j^* P^*(\theta_i^*, \theta_j^*)$$

$$= \sum_{i=0}^{D} E_i^* A E_i^* P^*(\theta_i^*, \theta_i^*)$$

$$= \sum_{i=0}^{D} E_i^* a_i (\theta_i^* - \theta_{i-1}^*)(\theta_i^* - \theta_{i+1}^*)$$

$$= \sum_{i=0}^{D} E_i^* \left(\gamma \theta_i^{*2} + \omega \theta_i^* + \eta^* \right)$$

$$= \gamma A^{*2} + \omega A^* + \eta^* I.$$

We have shown (51).

Claim 2. Let i, j denote vertices in Δ_E that are at distance $\partial(i,j) = 2$. Assume that there exists a unique vertex h in Δ_E that is adjacent to both i and j. Then $\gamma = \theta_i - \beta \theta_h + \theta_j$.

Proof of Claim 2. In the equation (51), multiply each term on the left by E_i and on the right by E_j. Simplify the result using Lemma 17.5. To aid this simplification, note that

$$E_i A^{*2} E_j = E_i A^* \left(\sum_{r=0}^{D} E_r \right) A^* E_j = E_i A^* E_h A^* E_j$$

and

$$E_i A^* A A^* E_j = E_i A^* \left(\sum_{r=0}^{D} \theta_r E_r \right) A^* E_j = \theta_h E_i A^* E_h A^* E_j.$$

By these comments, the following holds on the T-module $\mathcal{V}$:

$$0 = E_i A^* E_h A^* E_j (\theta_i - \beta \theta_h + \theta_j - \gamma).$$

We show that $E_i A^* E_h A^* E_j \neq 0$ on $\mathcal{V}$. By Lemma 17.5 and the construction, $E_i A^* E_h$ and $E_h A^* E_j$ are nonzero on $\mathcal{V}$. The dimension of $E_h \mathcal{V}$ is one, so $E_h A^* E_j \mathcal{V} = E_h \mathcal{V}$. By these comments $E_i A^* E_h A^* E_j \mathcal{V} = E_i A^* E_h \mathcal{V} \neq 0$. We have shown that $E_i A^* E_h A^* E_j \neq 0$ on $\mathcal{V}$, so $\gamma = \theta_i - \beta \theta_h + \theta_j$. Claim 2 is proved.

We can now easily show that Δ_E is a path. Since Δ_E is connected, and since vertex 0 is adjacent only to vertex 1, it suffices to show that each vertex in Δ_E

is adjacent to at most two other vertices in Δ_E. Suppose there exists a vertex i of Δ_E that is adjacent to at least three vertices in Δ_E. Of all such vertices, pick i such that $\partial(0, i)$ is minimal. Without loss of generality, we may assume that the vertices of Δ_E are labelled such that $\partial(0, i) = i$, and vertices $0, 1, 2, \ldots, i$ form a path in Δ_E. By construction, $i \geq 1$. By assumption, there exist distinct vertices j, j' in Δ_E that are adjacent to i and not equal to $i - 1$. By construction, $\partial(i - 1, j) = 2$ and i is the unique vertex in Δ_E that is adjacent to both $i - 1, j$. By Claim 2, $\gamma = \theta_{i-1} - \beta\theta_i + \theta_j$. Repeating the argument with j replaced by j', we obtain $\gamma = \theta_{i-1} - \beta\theta_i + \theta_{j'}$. By these comments, $\theta_j = \theta_{j'}$ for a contradiction. We conclude that Δ_E is a path. Relabelling $\{E_i\}_{i=2}^D$ if necessary, we may assume without loss of generality that vertices $i - 1$ and i are adjacent in Δ_E for $1 \leq i \leq D$. The ordering $\{E_i\}_{i=0}^D$ is Q-polynomial by Theorem 12.1, because item (ii) of that theorem is satisfied by $E = E_1$. By these comments and Definition 17.1, the graph Γ is Q-polynomial with respect to E. $\qquad\square$

Note 17.7   Referring to Theorem 17.2, assume that Γ is Q-polynomial with respect to E. For this Q-polynomial structure, the eigenvalue sequence $\{\theta_i\}_{i=0}^D$ is obtained as follows:

- θ_0 is the valency k of Γ;
- $\theta_1 = k\theta_1^*/\theta_0^*$ by Lemma 4.8;
- $\theta_2, \theta_3, \ldots, \theta_D$ are recursively found using

$$\theta_{i-1} - \beta\theta_i + \theta_{i+1} = \gamma \qquad (1 \leq i \leq D - 1),$$

where β, γ are from Theorem 17.2.

Note 17.8   A variation on Theorem 17.2 is given in [38].

11.18 Distance-Regular Graphs with Classical Parameters

In [4, section 6.1], Brouwer, Cohen, and Neumaier introduce a type of distance-regular graph, said to have classical parameters. In [4, section 8.4], they show that these graphs are Q-polynomial. In this section, we give a short proof of this fact, using the Pascasio characterization from Theorem 17.2.

Throughout this section $\Gamma = (X, \mathcal{R})$ denotes a distance-regular graph with diameter $D \geq 3$. Let k denote the valency of Γ.

We now recall what it means for Γ to have classical parameters. We will use the following notation. For a nonzero integer b define

$$\begin{bmatrix} i \\ 1 \end{bmatrix} = \begin{bmatrix} i \\ 1 \end{bmatrix}_b = 1 + b + b^2 + \cdots + b^{i-1}.$$

Definition 18.1 (See [4, p. 193])   The graph Γ has *classical parameters* (D, b, α, σ) whenever the intersection numbers satisfy

$$c_i = \begin{bmatrix} i \\ 1 \end{bmatrix} \left(1 + \alpha \begin{bmatrix} i - 1 \\ 1 \end{bmatrix} \right) \qquad (0 \le i \le D),$$

$$b_i = \left(\begin{bmatrix} D \\ 1 \end{bmatrix} - \begin{bmatrix} i \\ 1 \end{bmatrix} \right) \left(\sigma - \alpha \begin{bmatrix} i \\ 1 \end{bmatrix} \right) \qquad (0 \le i \le D).$$

Theorem 18.2 (See [4, section 8.4])   *Assume that Γ has classical parameters* (D, b, α, σ). *Then the following (i)–(iv) hold.*

(i) $\theta = \frac{b_1}{b} - 1$ *is an eigenvalue of Γ.*

(ii) *Let* $E = |X|^{-1} \sum_{i=0}^{D} \theta_i^* A_i$ *denote the associated primitive idempotent. Then*

$$\frac{\theta_i^*}{\theta_0^*} = 1 + \left(\frac{\theta}{k} - 1 \right) \begin{bmatrix} i \\ 1 \end{bmatrix} b^{1-i} \qquad (0 \le i \le D).$$

(iii) $\theta \ne k$.

(iv) Γ *is Q-polynomial with respect to E.*

Proof   (i), (ii) Apply Lemma 4.8 with $\theta = \frac{b_1}{b} - 1$ and

$$\sigma_i = 1 + \left(\frac{\theta}{k} - 1 \right) \begin{bmatrix} i \\ 1 \end{bmatrix} b^{1-i} \qquad (0 \le i \le D).$$

(iii) Suppose $\theta = k$. We have $b_1 = b(k + 1)$, so $b > 0$. We have $b \ge 1$ since b is an integer. Therefore $b_1 \ge k + 1$, a contradiction. We have shown that $\theta \ne k$.
(iv) The conditions of Theorem 17.2 are satisfied using $\beta = b + b^{-1}$ and

$$\gamma = \frac{\alpha(b^D + 1) + \sigma(b - 1) + 1 - b}{b},$$

$$\gamma^* = \theta_0^* \frac{\alpha(b^D - b) + \sigma(b - 1) + b^2 - b}{kb},$$

$$\omega = \Psi(\theta_1^* - \theta_0^*) - 2\gamma\theta_0^*,$$

$$\eta^* = \gamma\theta_0^{*2} - \Psi\theta_0^*(\theta_1^* - \theta_0^*),$$

where

$$\Psi = 1 - \sigma - \frac{\alpha}{b} \begin{bmatrix} D + 1 \\ 1 \end{bmatrix}.$$

$\square$

Lemma 18.3 (See [4, corollary 8.4.2])   *Assume that Γ has classical parameters (D, b, α, σ). For the Q-polynomial structure in Theorem 18.2, the eigenvalue sequence is*

$$\theta_i = \frac{b_i}{b^i} - \begin{bmatrix} i \\ 1 \end{bmatrix} \qquad (0 \le i \le D).$$

Proof   Routine calculation using Note 17.7.   □

11.19 The Balanced Set Characterization of the Q-Polynomial Property

In this section, we give a characterization of the Q-polynomial property, known as the balanced set condition. The result is given in Theorem 19.2. The result first appeared in [66]. More recent versions can be found in [4, 70, 61, 3].

Throughout this section, $\Gamma = (X, \mathcal{R})$ denotes a distance-regular graph with diameter $D \ge 3$. Recall the valency k of Γ.

Lemma 19.1   *Fix $x \in X$ and write $T = T(x)$. Then for $0 \le i, j, \ell \le D$ and $y, z \in X$ the (y, z)-entry of $A_i A_\ell^* A_j$ is equal to*

$$|X| \sum_{w \in \Gamma_i(y) \cap \Gamma_j(z)} \langle E_\ell \hat{x}, E_\ell \hat{w} \rangle.$$

Proof   We have

$$(A_i A_\ell^* A_j)_{y,z} = \sum_{w \in X} (A_i)_{y,w} (A_\ell^*)_{w,w} (A_j)_{w,z} = \sum_{w \in \Gamma_i(y) \cap \Gamma_j(z)} (A_\ell^*)_{w,w}$$

$$= |X| \sum_{w \in \Gamma_i(y) \cap \Gamma_j(z)} \langle E_\ell \hat{x}, E_\ell \hat{w} \rangle.$$

□

For the rest of this section, let E denote a nontrivial primitive idempotent of Γ, and write $E = |X|^{-1} \sum_{n=0}^{D} \theta_n^* A_n$. Recall that E is nondegenerate if and only if $\theta_n^* \ne \theta_0^*$ for $1 \le n \le D$. Also recall that for $y, z \in X$,

$$\langle E \hat{y}, E \hat{z} \rangle = |X|^{-1} \theta_n^*, \qquad n = \partial(y, z). \tag{52}$$

Theorem 19.2 (See [66, theorem 1.1])   *The following (i)–(iii) are equivalent:*

(i) *Γ is Q-polynomial with respect to E;*
(ii) *E is nondegenerate and for all $0 \le i, j \le D$ and all distinct $y, z \in X$,*

$$\sum_{w \in \Gamma_i(y) \cap \Gamma_j(z)} E \hat{w} - \sum_{w \in \Gamma_j(y) \cap \Gamma_i(z)} E \hat{w} = p_{i,j}^h \frac{\theta_i^* - \theta_j^*}{\theta_0^* - \theta_h^*} (E \hat{y} - E \hat{z}),$$

where $h = \partial(y, z)$;

(iii) E *is nondegenerate and for all* $y, z \in X$,

$$\sum_{w \in \Gamma(y) \cap \Gamma_2(z)} E\hat{w} - \sum_{w \in \Gamma_2(y) \cap \Gamma(z)} E\hat{w} \in Span(E\hat{y} - E\hat{z}).$$

Proof Fix $x \in X$ and write $T = T(x)$. Write $E = E_1$ and $A^* = A_1^*$.
(i) $\Rightarrow$ (ii) E is nondegenerate since $\{\theta_n^*\}_{n=0}^D$ are mutually distinct. Recall the Bose–Mesner algebra M. By Lemma 15.7,

$$Span\{RA^*S - SA^*R | R, S \in M\} = \{YA^* - A^*Y | Y \in M\}.$$

Taking $R = A_i$ and $S = A_j$, we obtain

$$A_i A^* A_j - A_j A^* A_i = \sum_{n=1}^D r_{i,j}^n (A^* A_n - A_n A^*) \tag{53}$$

for some scalars $\{r_{i,j}^n\}_{n=1}^D$ in $\mathbb{R}$. We show that

$$r_{i,j}^h = p_{i,j}^h \frac{\theta_i^* - \theta_j^*}{\theta_0^* - \theta_h^*} \qquad (1 \le h \le D). \tag{54}$$

Let h be given, and pick $z \in X$ such that $\partial(x, z) = h$. We compute the (x, z)-entry of each term in (53). We do this using Lemma 19.1 (with $\ell = 1$ and $y = x$) along with (52). A brief calculation yields

$$p_{i,j}^h (\theta_i^* - \theta_j^*) = r_{i,j}^h (\theta_0^* - \theta_h^*),$$

and (54) follows. Pick distinct $y, z \in X$ and write $h = \partial(y, z)$. We show that

$$\sum_{w \in \Gamma_i(y) \cap \Gamma_j(z)} E\hat{w} - \sum_{w \in \Gamma_j(y) \cap \Gamma_i(z)} E\hat{w} = p_{i,j}^h \frac{\theta_i^* - \theta_j^*}{\theta_0^* - \theta_h^*} (E\hat{y} - E\hat{z}). \tag{55}$$

Since the base vertex x is arbitrary, without loss of generality it suffices to show that in (55), the left-hand side minus the right-hand side is orthogonal to $E\hat{x}$. This orthogonality is routinely obtained from (53) and (54) along with Lemma 19.1 (with $\ell = 1$).

(ii) $\Rightarrow$ (iii). Clear.

(iii) $\Rightarrow$ (i). We assume that E is nondegenerate, so $\theta_n^* \ne \theta_0^*$ for $1 \le n \le d$.

Claim 1. Pick an integer h ($1 \le h \le D$) and $y, z \in X$ such that $\partial(y, z) = h$. Then

$$\sum_{w \in \Gamma(y) \cap \Gamma_2(z)} E\hat{w} - \sum_{w \in \Gamma_2(y) \cap \Gamma(z)} E\hat{w} = r_{1,2}^h (E\hat{y} - E\hat{z}),$$

where

$$r_{1,2}^h = p_{1,2}^h \frac{\theta_1^* - \theta_2^*}{\theta_0^* - \theta_h^*}. \tag{56}$$

Proof of Claim 1. By assumption there exists $\alpha \in \mathbb{R}$ such that

$$\sum_{w \in \Gamma(y) \cap \Gamma_2(z)} E\hat{w} - \sum_{w \in \Gamma_2(y) \cap \Gamma(z)} E\hat{w} = \alpha(E\hat{y} - E\hat{z}).$$

For each term in this equation, take the inner product with $E\hat{y}$ using (52). A brief calculation yields

$$p_{1,2}^h(\theta_1^* - \theta_2^*) = \alpha(\theta_0^* - \theta_h^*).$$

Therefore

$$\alpha = p_{1,2}^h \frac{\theta_1^* - \theta_2^*}{\theta_0^* - \theta_h^*},$$

and Claim 1 is proved.

Claim 2. We have

$$AA^*A_2 - A_2A^*A = \sum_{n=1}^{D} r_{1,2}^n(A^*A_n - A_nA^*). \tag{57}$$

Proof of Claim 2. For $y, z \in X$ we compute the (y, z)-entry of the left-hand side of (57) minus the right-hand side of (57). We do this computation using Lemma 19.1 (with $\ell = 1$). For $y = z$, the (y, z)-entry is zero. For $y \neq z$, the (y, z)-entry is equal to $|X|$ times

$$\left\langle E\hat{x}, \sum_{w \in \Gamma(y) \cap \Gamma_2(z)} E\hat{w} - \sum_{w \in \Gamma_2(y) \cap \Gamma(z)} E\hat{w} - r_{1,2}^h(E\hat{y} - E\hat{z}) \right\rangle,$$

where $h = \partial(y, z)$. The above scalar is zero by Claim 1. Claim 2 is proved.

Conceivably $\theta_1^* = \theta_2^*$. In this case, $r_{1,2}^h = 0$ for $1 \leq h \leq D$. So by Claim 2, $AA^*A_2 = A_2A^*A$. In this equation, we eliminate A_2 using $A_2 = (A^2 - a_1A - kI)/c_2$ and get

$$A^2A^*A - AA^*A^2 = k(A^*A - AA^*). \tag{58}$$

We will return to this equation shortly.

Claim 3. Assume that $\theta_1^* \neq \theta_2^*$. Then there exist scalars $\beta, \gamma, \varrho \in \mathbb{R}$ such that

$$0 = [A, A^2A^* - \beta AA^*A + A^*A^2 - \gamma(AA^* + A^*A) - \varrho A^*]. \tag{59}$$

Proof of Claim 3. Referring to (56), the scalar $p_{1,2}^h$ is zero if $h > 3$ and nonzero if $h = 3$. Therefore $r_{1,2}^h$ is zero if $h > 3$ and nonzero if $h = 3$. The matrices A_2

and A_3 appear in (57). Recall that A_2 and A_3 are polynomials in A that have degrees 2 and 3, respectively. Evaluating (57) using this fact, we obtain

$$A^3A^* - A^*A^3 \in \text{Span}\left(A^2A^*A - AA^*A^2, A^2A^* - A^*A^2, AA^* - A^*A\right).$$

Therefore there exist $\beta, \gamma, \varrho \in \mathbb{R}$ such that

$$A^3A^* - A^*A^3 = (\beta + 1)(A^2A^*A - AA^*A^2) + \gamma(A^2A^* - A^*A^2)$$
$$+ \varrho(AA^* - A^*A).$$

In this equation, we rearrange the terms to obtain (59). Claim 3 is proved. Recall our notation $E = E_1$. For the time being, let $\{E_i\}_{i=2}^{D}$ denote any ordering of the remaining nontrivial primitive idempotents of Γ. For $0 \le i \le D$, let θ_i denote the eigenvalue of Γ for E_i. Recall the graph Δ_E from Definition 17.3. The graph Δ_E is connected since E is nondegenerate. Recall that in Δ_E, vertex 0 is adjacent to vertex 1 and no other vertex. We will show that Δ_E is a path. To do this, it suffices to show that each vertex i in Δ_E is adjacent to at most two vertices in Δ_E.

Claim 4. For distinct vertices i, j in Δ_E that are adjacent,

(i) if $\theta_1^* = \theta_2^*$ then $\theta_i\theta_j = -k$;
(ii) if $\theta_1^* \neq \theta_2^*$ then $P(\theta_i, \theta_j) = 0$, where we recall

$$P(\lambda, \mu) = \lambda^2 - \beta\lambda\mu + \mu^2 - \gamma(\lambda + \mu) - \varrho.$$

Proof of Claim 4. First assume that $\theta_1^* = \theta_2^*$. Then (58) holds. In (58), multiply each term on the left by E_i and on the right by E_j. Simplify the result to get

$$0 = E_iA^*E_j(\theta_i - \theta_j)(\theta_i\theta_j + k).$$

We have $E_iA^*E_j \neq 0$ since i, j are adjacent in Δ_E. The scalar $\theta_i - \theta_j$ is nonzero since $i \neq j$. Therefore $\theta_i\theta_j + k = 0$ so $\theta_i\theta_j = -k$. Next assume that $\theta_1^* \neq \theta_2^*$. Then (59) holds. In (59), multiply each term on the left by E_i and the right by E_j. Simplify the result to get

$$0 = E_iA^*E_j(\theta_i - \theta_j)P(\theta_i, \theta_j).$$

We have $E_iA^*E_j \neq 0$ since i, j are adjacent in Δ_E. The scalar $\theta_i - \theta_j$ is nonzero since $i \neq j$. Therefore $P(\theta_i, \theta_j) = 0$.

Claim 5. We have $\theta_1^* \neq \theta_2^*$.

Proof of Claim 5. Suppose that $\theta_1^* = \theta_2^*$. By Claim 4 and since vertex 0 is adjacent to vertex 1, we have $\theta_0\theta_1 = -k$. We have $\theta_0 = k$ so $\theta_1 = -1$. The graph Δ_E is connected, so vertex 1 is adjacent to some nonzero vertex j. By Claim 4 we have $\theta_1\theta_j = -k$. By this and $\theta_1 = -1$, we obtain $\theta_j = k$. This implies $j = 0$, for a contradiction. Claim 5 is proved.

Claim 6. Each vertex i in Δ_E is adjacent at most two vertices in Δ_E.

Proof of Claim 6. By Claims 4, 5 we see that for each vertex j in Δ_E that is adjacent vertex i, the eigenvalue θ_j is a root of the polynomial

$$P(\theta_i, \mu) = \theta_i^2 - \beta\theta_i\mu + \mu^2 - \gamma(\theta_i + \mu) - \varrho.$$

This polynomial is quadratic in μ, so it has at most two distinct roots. Claim 6 is proved.

We have shown that the graph Δ_E is a path. Consequently the graph Γ is Q-polynomial with respect to E. $\square$

The balanced set condition has subtle combinatorial implications; see [7, 8, 12, 47, 51, 65, 69].

11.20 Directions for Future Research

In this section, we extend the Q-polynomial property to graphs that are not necessarily distance-regular.

Throughout this section, let $\Gamma = (X, \mathcal{R})$ denote a finite, undirected, connected graph, without loops or multiple edges, with diameter $D \geq 1$. We do not assume that Γ is distance-regular. Let A denote the adjacency matrix of Γ.

Definition 20.1 An ordering $\{V_i\}_{i=0}^d$ of the eigenspaces of A is called Q-*polynomial* whenever

$$\sum_{\ell=0}^{i} V_\ell = \sum_{\ell=0}^{i} (V_1)^{\circ\ell} \qquad (0 \leq i \leq d). \tag{60}$$

We are using the notation (25).

Definition 20.2 The graph Γ is said to be Q-*polynomial* whenever there exists at least one Q-polynomial ordering of the eigenspaces of A.

Lemma 20.3 *Assume that Γ is Q-polynomial. Then Γ is regular.*

Proof Let $\{V_i\}_{i=0}^d$ denote a Q-polynomial ordering of the eigenspaces of A. Setting $i = 0$ in (60), we obtain $V_0 = \mathbb{R}\mathbf{1}$. Therefore, the vector $\mathbf{1}$ is an eigenvector for A. Consequently Γ is regular. $\square$

Assume for the moment that Γ is Q-polynomial. We do not expect that Γ is distance-regular. However, we do expect that the following conjecture is true. Let M denote the subalgebra of $\text{Mat}_X(\mathbb{R})$ generated by A.

Conjecture 20.4 If Γ is Q-polynomial, then $B \circ C \in M$ for all $B, C \in M$.

Note 20.5   Conjecture 20.4 asserts that if Γ is Q-polynomial, then M is the Bose–Mesner algebra of a symmetric association scheme [4, section 2.2].

For the rest of this section, fix $x \in X$. Define $D(x) = \max\{\partial(x, y)|y \in X\}$. For $0 \le i \le D(x)$ define the matrix $E_i^* = E_i^*(x)$ as in line (4). Note that $\{E_i^*\}_{i=0}^{D(x)}$ form a basis for a commutative subalgebra $M^* = M^*(x)$ of $\mathrm{Mat}_X(\mathbb{R})$.

Definition 20.6   Let $\{V_i\}_{i=0}^d$ denote an ordering of the eigenspaces of A. A matrix $A^* \in \mathrm{Mat}_X(\mathbb{R})$ is called a *dual adjacency matrix* (with respect to x and $\{V_i\}_{i=0}^d$) whenever

(i) A^* generates M^*;
(ii) for $0 \le i \le d$ we have

$$A^* V_i \subseteq V_{i-1} + V_i + V_{i+1},$$

where $V_{-1} = 0$ and $V_{d+1} = 0$.

Definition 20.7   An ordering $\{V_i\}_{i=0}^d$ of the eigenspaces of A is called Q-*polynomial with respect to* x whenever there exists a dual adjacency matrix with respect to x and $\{V_i\}_{i=0}^d$.

Definition 20.8   We say that Γ is Q-*polynomial with respect to* x whenever there exists an ordering of the eigenspaces of A that is Q-polynomial with respect to x.

Another generalization we could adopt, is to allow the adjacency matrix A to be weighted. A weighted adjacency matrix is obtained from the classical adjacency matrix by replacing each entry 1 by a nonzero real scalar. The only requirement on the scalars is that the weighted adjacency matrix is diagonalizable. A weighted adjacency matrix is used in [18].

Problem 20.9   Investigate how the above variations on the Q-polynomial property are related.

Remark 20.10   In [58], Sho Suda introduced the Q-polynomial property for coherent configurations. A coherent configuration is a combinatorial object more general than a graph.

Acknowledgment

The author thanks Edwin van Dam, Tatsuro Ito, Jack Koolen, and Hajime Tanaka for helpful comments about the chapter.

References

[1] R. Askey and J. Wilson. A set of orthogonal polynomials that generalize the Racah coefficients or 6–j symbols. *SIAM J. Math. Anal.* 10 (1979) 1008–1016.

[2] E. Bannai and T. Ito. *Algebraic Combinatorics I: Association Schemes.* Benjamin/Cummings, London, 1984.

[3] E. Bannai, E. Bannai, T. Ito, R. Tanaka. *Algebraic Combinatorics.* De Gruyter Series in Discrete Math and Applications, Vol. 5. De Gruyter, 2021. https://doi.org/10.1515/9783110630251.

[4] A. E. Brouwer, A. M. Cohen, and A. Neumaier. *Distance-Regular Graphs.* Springer-Verlag, Berlin, 1989.

[5] P. Cameron, J. Goethals, and J. Seidel. The Krein condition, spherical designs, Norton algebras, and permutation groups. *Indag. Math.* 40 (1978) 196–206.

[6] J. S. Caughman IV. The Terwilliger algebras of bipartite P- and Q-polynomial association schemes. *Discrete Math.* 196 (1999) 65–95.

[7] J. S. Caughman IV. The last subconstituent of a bipartite Q-polynomial distance-regular graph. *European J. Combin.* 24 (2003) 459–470.

[8] J. S. Caughman IV, E. Hart, and J. Ma. The last subconstituent of the Hemmeter graph. *Discrete Math.* 308 (2008) 3056–3060.

[9] J. S. Caughman IV, M. MacLean, and P. Terwilliger. The Terwilliger algebra of an almost-bipartite P- and Q-polynomial association scheme. *Discrete Math.* 292 (2005) 17–44.

[10] D. Cerzo. Structure of thin irreducible modules of a Q-polynomial distance-regular graph. *Linear Algebra Appl.* 433 (2010) 1573–1613.

[11] V. Chari and A. Pressley. Quantum affine algebras. *Comm. Math. Phys.* 142 (1991) 261–283.

[12] S. M. Cioabă, J. H. Koolen, and P. Terwilliger. Connectivity concerning the last two subconstituents of a Q-polynomial distance-regular graph. *J. Combin. Theory Ser. A* 177 (2021) Paper No. 105325, 6 pp.

[13] B. Curtin. Bipartite distance-regular graphs I. *Graphs Combin.* 15 (1999) 143–158.

[14] B. Curtin. Bipartite distance-regular graphs II. *Graphs Combin.* 15 (1999) 377–391.

[15] B. Curtin. 2-homogeneous bipartite distance-regular graphs. *Discrete Math.* 187 (1998) 39–70.

[16] B. Curtin. Distance-regular graphs which support a spin model are thin. *16th British Combinatorial Conference (London, 1997).* Discrete Math. 197/198 (1999) 205–216.

[17] B. Curtin. The Terwilliger algebra of a 2-homogeneous bipartite distance-regular graph. *J. Combin. Theory Ser. B* 81 (2001) 125–141.

[18] B. Curtin and K. Nomura. Distance-regular graphs related to the quantum enveloping algebra of sl(2). *J. Algebraic Combin.* 12 (2000) 25–36.

[19] E. R. van Dam, J. H. Koolen, H. Tanaka. Distance-regular graphs. *Electron. J. Combin.* (2016) DS22; arXiv:1410.6294.

[20] P. Delsarte. An algebraic approach to the association schemes of coding theory. *Philips Research Reports Suppl.* 10 (1973).

[21] G. Dickie. Twice Q-polynomial distance-regular graphs are thin. *European J. Combin.* 16 (1995) 555–560.

[22] G. Dickie and P. Terwilliger. Dual bipartite Q-polynomial distance-regular graphs. *European J. Combin.* 17 (1996) 613–623.

[23] G. Dickie. Q-polynomial structures for association schemes and distance-regular graphs. Thesis (Ph.D.), University of Wisconsin – Madison. ProQuest LLC, Ann Arbor, MI, 1995.

[24] G. Dickie and P. Terwilliger. A note on thin P-polynomial and dual-thin Q-polynomial symmetric association schemes. *J. Algebraic Combin.* 7 (1998) 5–15.

[25] E. Egge. A generalization of the Terwilliger algebra. *J. Algebra* 233 (2000) 213–252.

[26] A. Gavrilyuk and J. Koolen. The Terwilliger polynomial of a Q-polynomial distance-regular graph and its application to pseudo-partition graphs. *Linear Algebra Appl.* 466 (2015) 117–140.

[27] J. T. Go. The Terwilliger algebra of the hypercube. *European J. Combin.* 23 (2002) 399–429.

[28] E. Hanson. A characterization of Leonard pairs using the parameters $\{a_i\}_{i=0}^d$. *Linear Algebra Appl.* 438 (2013) 2289–2305.

[29] E. Hanson. How to recognize a Leonard pair. *Linear Multilinear Algebra* 69 (2021) 177–192.

[30] S. A. Hobart and T. Ito. The structure of nonthin irreducible T-modules: ladder bases and classical parameters. *J. Algebraic Combin.* 7 (1998) 53–75.

[31] J. Huang. Nonassociativity of the Norton algebras of some distance-regular graphs. *Electron. J. Combin.* 27 (2020) Paper No. 4.27; `arXiv:2001.05547`.

[32] T. Ito. TD-pairs and the q-Onsager algebra. *Sugaku Expositions* 32 (2019) 205–232.

[33] T. Ito, K. Tanabe, and P. Terwilliger. Some algebra related to P- and Q-polynomial association schemes. *Codes and Association Schemes (Piscataway NJ, 1999), 167–192, DIMACS Ser. Discrete Math. Theoret. Comput. Sci.* 56, Amer. Math. Soc., Providence, RI 2001; `arXiv:math.CO/0406556`.

[34] T. Ito and P. Terwilliger. Distance-regular graphs and the q-tetrahedron algebra. *European J. Combin.* 30 (2009) 682–697; `arXiv:math.CO/0608694`.

[35] T. Ito and P. Terwilliger. Distance-regular graphs of q-Racah type and the q-tetrahedron algebra. *Michigan Math. J.* 58 (2009) 241–254.

[36] T. Ito and P. Terwilliger. The augmented tridiagonal algebra. *Kyushu J. Math.* 64 (2010) 8–144; `arXiv:0904.2889`.

[37] T. Ito, K. Nomura, and P. Terwilliger. A classification of sharp tridiagonal pairs. *Linear Algebra Appl.* 435 (2011) 1857–1884.

[38] A. Jurišić, P. Terwilliger, and A. Žitnik. The Q-polynomial idempotents of a distance-regular graph. *J. Combin. Theory Ser. B* (2010) 683–690.

[39] R. Koekoek, P. A Lesky, and R. Swarttouw. *Hypergeometric Orthogonal Polynomials and Their q-Analogues.* With a foreword by Tom H. Koornwinder. Springer Monographs in Mathematics. Springer-Verlag, Berlin, 2010.

[40] M. S. Lang. Pseudo primitive idempotents and almost 2-homogeneous bipartite distance-regular graphs. *European J. Combin.* 29 (2008) 35–44.

[41] M. S. Lang. Bipartite distance-regular graphs: the Q-polynomial property and pseudo primitive idempotents. *Discrete Math.* 331 (2014) 27–35.

[42] M. S. Lang and P. Terwilliger. Almost-bipartite distance-regular graphs with the *Q*-polynomial property. *European J. Combin.* 28 (2007) 258–265.

[43] J. H. Lee. *Q*-polynomial distance-regular graphs and a double affine Hecke algebra of rank one. *Linear Algebra Appl.* 439 (2013) 3184–3240.

[44] J. H. Lee. Nonsymmetric Askey–Wilson polynomials and *Q*-polynomial distance-regular graphs. *J. Combin. Theory Ser. A* 147 (2017) 75–118.

[45] D. A. Leonard. Orthogonal polynomials, duality and association schemes. *SIAM J. Math. Anal.* 13 (1982) 656–663.

[46] F. Levstein, C. Maldonado, and D. Penazzi. Lattices, frames and Norton algebras of dual polar graphs. In *New Developments in Lie Theory and Its Applications*, 1–16, Contemp. Math., 544, Amer. Math. Soc., Providence, RI., 2011.

[47] H. A. Lewis. Homotopy in *Q*-polynomial distance-regular graphs. *Discrete Math.* 223 (2000) 189–206.

[48] X. Liang, T. Ito, and Y. Watanabe. The Terwilliger algebra of the Grassmann scheme $J_q(N,D)$ revisited from the viewpoint of the quantum affine algebra $U_q(\widehat{\mathfrak{sl}}_2)$. *Linear Algebra Appl.* 596 (2020) 117–144.

[49] X. Liang, Y. Tan, and T. Ito. An observation on Leonard system parameters for the Terwilliger algebra of the Johnson scheme $J(N,D)$. *Graphs Combin.* 33 (2017) 149–156.

[50] C. Maldonado and D. Penazzi. Lattices and Norton algebras of Johnson, Grassmann and Hamming graphs; arXiv:1204.1947v1.

[51] Š. Miklavič and P. Terwilliger. Bipartite *Q*-polynomial distance-regular graphs and uniform posets. *J. Algebraic Combin.* 38 (2013) 225–242.

[52] K. Nomura and P. Terwilliger. Leonard pairs, spin models, and distance-regular graphs. *J. Combin. Theory Ser. A* (2021) Paper No. 105312, 59 pp.

[53] K. Nomura and P. Terwilliger. Totally bipartite tridiagonal pairs. *Electron. J. Linear Algebra* 37 (2021) 434–491.

[54] K. Nomura and P. Terwilliger. Idempotent systems. *Algebr. Comb.* 4 (2021) 329–357.

[55] A. A. Pascasio. On the multiplicities of the primitive idempotents of a *Q*-polynomial distance-regular graph. *European J. Combin.* 23 (2002) 1073–1078.

[56] A. A. Pascasio. Tight distance-regular graphs and the *Q*-polynomial property. *Graphs Combin.* 17 (2001) 149–169.

[57] A. A. Pascasio. A characterization of *Q*-polynomial distance-regular graphs. *Discrete Math.* 308 (2008) 3090–3096.

[58] S. Suda. *Q*-polynomial coherent configurations. *Linear Algebra Appl.* 643 (2022) 166–195; arXiv:2104.04225.

[59] S. Sumalroj. A characterization of *Q*-polynomial distance-regular graphs using the intersection numbers. *Graphs Combin.* 34 (2018) 863–877.

[60] S. Sumalroj. A diagram associated with the subconstituent algebra of a distance-regular graph. *Ars Math. Contemp.* 17 (2019) 185–202.

[61] H. Suzuki and P. Terwilliger. Algebraic Graph Theory Lecture Notes. 1995; https://icu-hsuzuki.github.io/lecturenote/.

[62] Y. Tan, Y. Fan, T. Ito, and X. Liang. The Terwilliger algebra of the Johnson scheme $J(N,D)$ revisited from the viewpoint of group representations. *European J. Combin.* 80 (2019) 157–171.

[63] K. Tanabe. The irreducible modules of the Terwilliger algebras of Doob schemes. *J. Algebraic Combin.* 6 (1997) 173–195.

[64] H. Tanaka and T. Wang. The Terwilliger algebra of the twisted Grassmann graph: the thin case. *Electron. J. Combin.* 27 (2020) Paper No. 4.15.

[65] P. Terwilliger. Counting 4-vertex configurations in P- and Q-polynomial association schemes. Proceedings of the Conference on Groups and Geometry, Part B (Madison, Wis., 1985). *Algebras Groups Geom.* 2 (1985) 541–554.

[66] P. Terwilliger. A characterization of P- and Q-polynomial association schemes. *J. Combin. Theory Ser. A* 45 (1987) 8–26.

[67] P. Terwilliger. The subconstituent algebra of an association scheme I. *J. Algebraic Combin.* 1 (1992) 363–388.

[68] P. Terwilliger. The subconstituent algebra of an association scheme II. *J. Algebraic Combin.* 2 (1993) 73–103.

[69] P. Terwilliger. The subconstituent algebra of an association scheme III. *J. Algebraic Combin.* 2 (1993) 177–210.

[70] P. Terwilliger. A new inequality for distance-regular graphs. *Discrete Math.* 137 (1995) 319–332.

[71] P. Terwilliger. Two linear transformations each tridiagonal with respect to an eigenbasis of the other. *Linear Algebra Appl.* 330 (2001) 149–203; arXiv: math.RA/0406555.

[72] P. Terwilliger. Two relations that generalize the q-Serre relations and the Dolan-Grady relations. In A. N. Kirillov, A. Tsuchiya, and H. Umemura (eds.) *Physics and Combinatorics 1999 (Nagoya)*, 377–398, World Scientific Publishing, River Edge, NJ, 2001; arXiv:math.QA/0307016.

[73] P. Terwilliger. Leonard pairs and the q-Racah polynomials. *Linear Algebra Appl.* 387 (2004) 235–276. arXiv:math.QA/0306301.

[74] P. Terwilliger. Two linear transformations each tridiagonal with respect to an eigenbasis of the other; comments on the parameter array. *Des. Codes Cryptogr.* 34 (2005) 307–332. arXiv:math.RA/0306291.

[75] P. Terwilliger. The displacement and split decompositions for a Q-polynomial distance-regular graph. *Graphs Combin.* 21 (2005) 263–276. arXiv:math. CO/0306142.

[76] P. Terwilliger. An algebraic approach to the Askey scheme of orthogonal polynomials. Orthogonal polynomials and special functions, in *Orthogonal Polynomials and Special Functions*, 255–330, Lecture Notes in Math., 1883, Springer, Berlin, 2006.

[77] P. Terwilliger. The Norton algebra of a Q-polynomial distance-regular graph. *J. Combin. Theory Ser. A* 182 (2021) Paper No. 105477.

[78] P. Terwilliger. Notes on the Leonard system classification. *Graphs Combin.* 37 (2021) 1687–1748.

[79] P. Terwilliger. The q-Onsager algebra and the positive part of $U_q(\widehat{\mathfrak{sl}}_2)$. *Linear Algebra Appl.* 521 (2017) 19–56; arXiv:1506.08666.

[80] P. Terwilliger. Lecture Notes on Algebraic Graph Theory, 2009 and 2022; https://people.math.wisc.edu/~terwilli/teaching.html.

[81] P. Terwilliger and R. Vidunas. Leonard pairs and the Askey–Wilson relations *J. Algebra Appl.* 3 (2004) 411–426.

[82] P. Terwilliger and A. Žitnik. Distance-regular graphs of q-Racah type and the universal Askey–Wilson algebra. *J. Combin. Theory Ser. A* 125 (2014) 98–112.

[83] P. Terwilliger and A. Žitnik. The quantum adjacency algebra and subconstituent algebra of a graph. *J. Combin. Theory Ser. A* 166 (2019) 297–314.

[84] C. Worawannotai. Dual polar graphs, the quantum algebra $U_q(\mathfrak{sl}_2)$, and Leonard systems of dual q-Krawtchouk type. *Linear Algebra Appl.* 438 (2013) 443–497.

[85] A. Zhedanov. Hidden symmetry of Askey–Wilson polynomials. *Theoret. and Math. Phys.* 89 (1991) 1146–1157.

12

Terwilliger Algebras and the Weisfeiler–Leman Stabilization

Tatsuro Ito

In memory of Igor Faradjev

Abstract

This is an expository chapter about Terwilliger algebras revisited from the viewpoint of the Weisfeiler–Leman stabilization. We first discuss Terwilliger algebras in the framework of the Weisfeiler–Leman stabilization. Then for a tree, we use Terwilliger algebras to analyse the Weisfeiler–Leman stabilization, where the classification of irreducible representations of the Terwilliger algebra of the tree plays a crucial role. Finally we characterize distance-regular graphs in terms of the Weisfeiler–Leman stabilization and propose a new concept for a graph: the 'polynomial' property, which is weaker than 'distance-regular' and stronger than 'distance-polynomial' [22]. As the dual concept, we also propose the 'co-polynomial' property, which is weaker than 'Q-polynomial'. Jack Koolen showed that a P-polynomial association scheme is co-polynomial and a Q-polynomial association scheme is polynomial, and he conjectures that symmetric association schemes are co-polynomial if they do not have non-trivial fusion schemes.

CONTENTS

12.1 Introduction

This is an expository chapter about Terwilliger algebras revisited from the viewpoint of the Weisfeiler–Leman stabilization. It is inspired by the talk of Igor Faradjev at the Conference in Algebraic Graph Theory, Symmetry vs Regularity: The first 50 years since Weisfeiler–Leman stabilization, July 1–July 7, 2018, Pilsen Czech Republic.

I had not known anything about the Weisfeiler–Leman stabilization until I attended the lecture by Igor Faradjev [6], in which I learnt about the life of Boris Weisfeiler, the founder of the W-L stabilization, and the fact that the concept of coherent configurations [10, 11, 12] appeared earlier in the original paper of [23]. The W-L stabilization is in short an algorithm to get the coherent closure of a combinatorial object. For the original approach and its impact thereafter, readers are referred to the preface [17] by Ilia Ponomarenko to the original paper by Weisfeiler and Leman, which was published in 1968.

In this chapter, we first reformulate the W-L stabilization in terms of coherent configurations (Section 12.3) and then discuss Terwilliger algebras in its framework (Section 12.4), after preparing basics of coherent algebras and centralizer algebras (Section 12.2).

Terwilliger algebras were introduced in [20] for commutative association schemes and then in [21] for connected simple graphs, particularly for distance-regular graphs. Suppose that a finite group G acts on a set X and the action is generously transitive, that is, for any distinct $x, y \in X$, there exists an element $a \in G$ that interchanges x, y. Then a symmetric association scheme $\mathfrak{X}$ arises [1, page 54]. The Bose–Mesner algebra of $\mathfrak{X}$ coincides with the centralizer algebra of G acting on X. The Terwilliger algebra T of $\mathfrak{X}$ is defined with respect to a fixed base point $x_0 \in X : T = T(x_0)$. T turns out to be contained in the centralizer algebra of G_{x_0}, the stabilizer in G of x_0, and in many cases, T coincides with the centralizer algebra of G_{x_0}. In fact, Terwilliger introduced the subconstituent algebra, which we now call a Terwilliger algebra, expecting it would serve as a combinatorial analogue of the centralizer algebra of G_{x_0} even if $\mathfrak{X}$ does not come from a group action: he writes that we may view T as a 'combinatorial analog' of the centralizer algebra of G_{x_0}, where $G=\mathrm{Aut}(\mathfrak{X})$ [20, part I, p. 365].

One, however, may ask how and when T can be viewed as a combinatorial analogue of the centralizer algebra of G_{x_0}. We can formulate this question precisely by discussing T in the framework of the W-L stabilization. To do so, we prepare a set of notations as follows, which will be used throughout this chapter:

For a finite set X, $M_X(\mathbb{C})$ denotes the full matrix algebra consisting of all the matrices over the complex number field $\mathbb{C}$ whose rows and columns are indexed by the elements of X.

For $A \in M_X(\mathbb{C})$ and $x, y \in X$, $A(x, y)$ denotes the (x,y)-entry of A. $^t\overline{A}$ denotes the conjugate transpose of $A \in M_X(\mathbb{C})$, that is, $^t\overline{A}(x, y) = \overline{A(y, x)}$ for $x, y \in X$, where $\overline{A(y, x)}$ is the complex conjugate of $A(y, x)$. $A \circ B$ denotes the Hadamard product (entry-wise product) of $A, B \in M_X(\mathbb{C})$, i.e, $(A \circ B)(x, y) = A(x, y)B(x, y)$ for $x, y, \in X$.

$M_X(\mathbb{C})^\circ$ denotes the algebra with respect to the Hadamard product over the underlying vector space $M_X(\mathbb{C})$.

For a finite group G acting on a finite set X, $M_X^G(\mathbb{C})$ denotes the centralizer algebra of G, that is, the algebra consisting of all the matrices $A \in M_X(\mathbb{C})$ that commute with the action of G, or equivalently that satisfy $A(ax, ay) = A(x, y)$ for all $x, y \in X$ and all $a \in G$.

For a subset $\mathcal{F}$ of $M_X(\mathbb{C})$, $\mathrm{Aut}(\mathcal{F})$ denotes the automorphism group of $\mathcal{F}$, that is the group consisting of all the permutation matrices of $M_X(\mathbb{C})$ that commute with each member of $\mathcal{F}$.

We make one more preparation for a terminology, which will be also used throughout this chapter: a graph means a simple graph, that is, a finite undirected graph without loops or multiple edges, following [3].

We observe that $M_X(\mathbb{C})$ and $M_X^G(\mathbb{C})$ are coherent algebras, meaning they contain I, J (the identity matrix, the all-ones matrix, respectively) and they are closed both under the ordinary matrix product and under the Hadamard product, as well as under the conjugate transpose. For a subset $\mathcal{F}$ of $M_X(\mathbb{C})$ (resp. $M_X(\mathbb{C})^\circ$), the *closure* in $M_X(\mathbb{C})$ (resp. $M_X(\mathbb{C})^\circ$) of $\mathcal{F}$ means the smallest subalgebra of $M_X(\mathbb{C})$ (resp. $M_X(\mathbb{C})^\circ$) that contains $\mathcal{F}$ and I (resp. $\mathcal{F}$ and J), or equivalently the subalgebra of $M_X(\mathbb{C})$ (resp. $M_X(\mathbb{C})^\circ$) generated by $\mathcal{F}$ and I (resp. $\mathcal{F}$ and J).

The W-L stabilization proceeds as follows. We are given a subset $\mathcal{F}$ of $M_X(\mathbb{C})$ which is closed under the conjugate transpose: $^t\overline{A} \in \mathcal{F}$ for $A \in \mathcal{F}$. First take the closure in $M_X(\mathbb{C})^\circ$ of $\mathcal{F}$, and then the closure in $M_X(\mathbb{C})$ of the resulting subalgebra of $M_X(\mathbb{C})^\circ$. Continue these operations of taking the closures in turn in $M_X(\mathbb{C})^\circ$ and then in $M_X(\mathbb{C})$. Then the sequence stops in a finite number of steps and we get the coherent closure of $\mathcal{F}$, the smallest coherent algebra that contains $\mathcal{F}$. Let $\mathcal{M} = \mathcal{M}(\mathcal{F})$ denote the coherent closure of $\mathcal{F}$ and G the automorphism group of $\mathcal{F}$: $G = \mathrm{Aut}(\mathcal{F})$. Then it holds that $G = \mathrm{Aut}(\mathcal{M})$ and hence $\mathcal{M} \subseteq M_X^G(\mathbb{C})$. As is well-known, a combinatorial object, which is called a coherent configuration, corresponds with the coherent algebra $\mathcal{M}$. In this sense, we view $\mathcal{M}$ as a combinatorial analogue of the centralizer algebra $M_X^G(\mathbb{C})$. The coherent algebra $\mathcal{M}$ is called *Schurian* if $\mathcal{M} = M_X^G(\mathbb{C})$ holds.

For the sake of simplicity, we consider the Terwilliger algebra for a connected simple graph Γ: Terwilliger algebras for commutative association schemes can be treated in a similar way. Let X be the vertex set of Γ. Choose a base vertex $x_0 \in X$ and fix it. Define a diagonal matrix $E_i^* = E_i^*(x_0) \in M_X(\mathbb{C})$ by $E_i^*(x, y) = 1$ if $x = y$ and $\partial(x_0, x) = i$, 0 otherwise, where $\partial(x_0, x)$ is the length of a shortest path joining x_0 and x. The Terwilliger algebra $T = T(x_0)$ of Γ is defined to be the subalgebra of $M_X(\mathbb{C})$ generated by A and E_i^* $(i = 0, 1, \cdots, D)$, where A is the adjacency matrix of Γ and $D = \max\{\partial(x_0, x) | x \in X\}$.

Set $\mathcal{F} = \{A\}$ and $\mathcal{F}^{(x_0)} = \mathcal{F} \cup \{E_0^*\} = \{A, E_0^*\}$; we call $\mathcal{F}^{(x_0)}$ the *localization* at x_0 of $\mathcal{F}$. Let G (resp. H) denote the automorphism group of $\mathcal{F}$ (resp. $\mathcal{F}^{(x_0)}$): $G = \text{Aut}(\mathcal{F})$, $H = \text{Aut}(\mathcal{F}^{(x_0)})$. Then G coincides with the automorphism group of the graph Γ, and H with the stabilizer in G of x_0: $G = \text{Aut}(\Gamma)$, $H = G_{x_0}$.

Apply the W-L stabilization to $\mathcal{F}$. Then we have $\mathcal{M} \subseteq M_X^G(\mathbb{C})$, where $\mathcal{M} = \mathcal{M}(\mathcal{F})$, the coherent closure of $\mathcal{F}$, and $\mathcal{M}$ is viewed as a combinatorial analogue of the centralizer algebra $M_X^G(\mathbb{C})$. Apply the W-L stabilization to $\mathcal{F}^{(x_0)}$. Then the Terwilliger algebra T appears in the course of the stabilization process. In fact, T appears at the fifth step in the process of the stabilization (Proposition 4.3). In particular, T is contained in $\mathcal{M}(\mathcal{F}^{(x_0)})$, the coherent closure of $\mathcal{F}^{(x_0)}$. Let $\mathcal{M}^{(x_0)}$ denote the coherent closure of $\mathcal{F}^{(x_0)}$: $\mathcal{M}^{(x_0)} = \mathcal{M}(\mathcal{F}^{(x_0)})$. We have

$$T \subseteq \mathcal{M}^{(x_0)} \subseteq M_X^H(\mathbb{C}).$$

The coherent closure $\mathcal{M}^{(x_0)}$ is viewed as a combinatorial analogue of the centralizer algebra $M_X^H(\mathbb{C})$, but there may be a gap between T and $\mathcal{M}^{(x_0)}$ in general. The Terwilliger algebra T can be used in place of $\mathcal{M}^{(x_0)}$ when the gap is small or nil. The gap is measured by applying the W-L stabilization to T and counting how many steps we need to reach $\mathcal{M}^{(x_0)}$. For example, if the closure in $M_X(\mathbb{C})^\circ$ of T coincides with $\mathcal{M}^{(x_0)}$, T can be used as a good approximation of $\mathcal{M}^{(x_0)}$.

In Section 12.5, we treat the case where Γ is a tree. The W-L stabilization process is discussed in detail for a tree. We show that T is properly included in $\mathcal{M}^{(x_0)}$ in most cases, but it holds that

$$T \circ T = M_X^H(\mathbb{C}), \tag{1}$$

where $T \circ T = \{a \circ b \,|\, a, b \in T\} \subseteq M_X(\mathbb{C})^\circ$. In particular, $\mathcal{M}^{(x_0)} = M_X^H(\mathbb{C})$ holds: $\mathcal{M}^{(x_0)}$ is Schurian. For the proof of (1), the classification of irreducible representations of T and $M_X^H(\mathbb{C})$ [24] plays a crucial role. We may well say that in the case of a tree, the Terwilliger algebra is a bridge that connects regularity, which is local, to symmetry, which is global, through its representations.

Representations of Terwilliger algebras are deeply investigated for (P and Q)-polynomial association schemes [20], [2, chapter 6], which are also called Q-polynomial distance-regular graphs when considered as graphs. For this class of graphs, we may be able to show that T is a good approximation of $\mathrm{M}_X^H(\mathbb{C})$ in some sense, hopefully in the sense of (1). Hamming graphs and Johnson graphs are typical examples of Q-polynomial distance-regular graphs and it is known that $T = \mathrm{M}_X^H(\mathbb{C})$ holds for them [8, 19].

Finally in Section 12.6, we characterize distance-regular graphs in terms of the Weisfeiler–Leman stabilization and propose a new concept for a graph: the 'polynomial' property, which is weaker than 'distance-regular' and stronger than 'distance-polynomial' [22]. As the dual concept, we also propose the 'co-polynomial' property, which is weaker than 'Q-polynomial'. Jack Koolen showed that a P-polynomial association scheme is co-polynomial and a Q-polynomial association scheme is polynomial, and he conjectures that symmetric association schemes are co-polynomial if they do not have non-trivial fusion schemes (both are a personal communication).

12.2 Preliminaries

In this section, we summarize basics of coherent algebras and centralizer algebras for later use. The article [13] can serve as an excellent introduction to the representation theory of coherent algebras. The article [7] offers a general view of coherent (cellular) algebras and the Galois correspondence between them and permutation groups.

We keep the notations in the previous section. Recall that $\mathrm{M}_X(\mathbb{C})$ is the full matrix algebra consisting of all the matrices over the complex number field $\mathbb{C}$ whose rows and columns are indexed by the elements of a finite set X, and that $\mathrm{M}_X(\mathbb{C})^\circ$ is the algebra with respect to the Hadamard product over the underlying vector space $\mathrm{M}_X(\mathbb{C})$. Let $I \in \mathrm{M}_X(\mathbb{C})$ (resp. $J \in \mathrm{M}_X(\mathbb{C})^\circ$) denote the identity matrix (resp. the matrix with all entries 1). Note that I (resp. J) is the identity of the algebra $\mathrm{M}_X(\mathbb{C})$ (resp. the identity of the algebra $\mathrm{M}_X(\mathbb{C})^\circ$).

A linear subspace $\mathcal{M}$ of $\mathrm{M}_X(\mathbb{C})$ is said to be a *coherent algebra* if it satisfies the following three conditions:

(i) $\mathcal{M}$ is a subalgebra of $\mathrm{M}_X(\mathbb{C})^\circ$, containing J.
(ii) $\mathcal{M}$ is a subalgebra of $\mathrm{M}_X(\mathbb{C})$, containing I.
(iii) $\mathcal{M}$ is closed under the conjugate transpose: ${}^t\overline{A} \in \mathcal{M}, \forall A \in \mathcal{M}$.

The smallest coherent algebra is the linear space $\mathrm{Span}\{I, J\}$ and the largest one is $\mathrm{M}_X(\mathbb{C})$.

For any matrix A, we understand the 0th power of A in $\mathrm{M}_X(\mathbb{C})$ (resp $\mathrm{M}_X(\mathbb{C})^\circ$) is I (resp. J), and we require a subalgebra to be closed under the operation of taking any power of its elements. This is why a subalgebra of $M_X(\mathbb{C})$ (resp. $M_X(\mathbb{C})^\circ$) is required to contain I (resp. J).

Note that any subalgebra of $M_X(\mathbb{C})^\circ$ is commutative and semi-simple, because it does not contain nilpotent elements. So let $\{A_i \mid i \in \Lambda\}$ denote the primitive idempotents for a subalgebra $\mathcal{M}$ of $M_X(\mathbb{C})^\circ$. These three conditions (i)–(iii) for the linear subspace $\mathcal{M} = \mathrm{Span}\{A_i \mid i \in \Lambda\}$ are rephrased in terms of them as follows:

(i') Each A_i ($i \in \Lambda$) is a $(0,1)$ matrix and $J = \sum_{i \in \Lambda} A_i$.

(ii') For any $i, j \in \Lambda$, $A_i A_j$ is a linear combination of $A'_k s$, $k \in \Lambda$ and I is a linear combination of $A'_k s$, $k \in \Lambda$.

(iii') The set $\{A_i \mid i \in \Lambda\}$ is closed under conjugate transpose: $\{{}^t\overline{A_i} \mid i \in \Lambda\} = \{A_j \mid j \in \Lambda\}$.

Since A_i is a $(0,1)$ matrix, it corresponds with a relation R_i on X: $(x,y) \in R_i$ if and only if $A_i(x,y) = 1$. A combinatorial structure $\mathfrak{X} = (X, \{R_i\}_{i \in \Lambda})$, where $R_i \subset X \times X$, is called a *coherent configuration* if it comes from a coherent algebra, or equivalently if the set $\{A_i \mid i \in \Lambda\}$ of $(0,1)$ matrices, which corresponds with the set $\{R_i\}_{i \in \Lambda}$ of relations on X, satisfies the conditions (i')–(iii').

A coherent configuration $\mathfrak{X} = (X, \{R_i\}_{i \in \Lambda})$ is called an *association scheme* if one of R_i, $i \in \Lambda$, is the diagonal relation, in which case we usually choose R_0 to be the diagonal relation: $A_0 = I$. An association scheme is called *symmetric* if each relation R_i, $i \in \Lambda$, is symmetric, that is, each A_i, $i \in \Lambda$, is a symmetric matrix. It is well known and easy to show that if a coherent algebra $\mathcal{M}$ is symmetric, that is, if all the matrices of $\mathcal{M}$ are symmetric, then $\mathcal{M}$ is commutative (as a subalgebra of $M_X(\mathbb{C})$). It is also well known and easy to show that if a coherent algebra $\mathcal{M}$ is commutative, then the corresponding coherent configuration is an association scheme. An association scheme $\mathfrak{X} = (X, \{R_i\}_{i \in \Lambda})$ is said to be *primitive* if for all $i \neq 0$, the graph $\Gamma(X, R_i)$ with X the vertex set and R_i the edge set is connected.

Note that any subalgebra of $M_X(\mathbb{C})$ is semi-simple if it is closed under the conjugate transpose. So by (iii), a coherent algebra $\mathcal{M}$ is semi-simple as a subalgebra of $M_X(\mathbb{C})$, and we can 'use representations of $\mathcal{M}$ to analyse the coherent configurations $\mathfrak{X} = (X, \{R_i\}_{i \in \Lambda})$': the motto of algebraic combinatorics.

Let $\mathrm{Sym}(X)$ denote the symmetric group on X. We identify $\mathrm{Sym}(X)$ with the set of permutation matrices in $M_X(\mathbb{C})$. Precisely speaking, let $V = \mathbb{C}X$ denote the Hermitian space for which X is an orthonormal basis. Let $\mathrm{End}(V)$ denote the endomorphism algebra. Then $M_X(\mathbb{C})$ is the matrices of $\mathrm{End}(V)$ with respect to the basis X: $\mathrm{End}(V) \simeq M_X(\mathbb{C})$. Since $\mathrm{Sym}(X)$ permutes the basis X of V, it corresponds to the set of permutation matrices through the algebra isomorphism $\mathrm{End}(V) \simeq M_X(\mathbb{C})$.

For a subgroup G of $\mathrm{Sym}(X)$, the *centralizer algebra* $\mathrm{M}_X^G(\mathbb{C})$ of G is defined to be the set of matrices in $\mathrm{M}_X(\mathbb{C})$ that commute with all elements of G:

$$\mathrm{M}_X^G(\mathbb{C}) = \{A \in \mathrm{M}_X(\mathbb{C}) \mid AP = PA, \ \forall P \in G\}. \tag{2}$$

The following lemma is well known [1, p. 47].

Lemma 2.1   *A permutation matrix $P \in M_X(\mathbb{C})$ commutes with $A \in M_X(\mathbb{C})$ if and only if $A(x, y) = A(Px, Py)$, $\forall x, y \in X$.*

So we have

$$\mathrm{M}_X^G(\mathbb{C}) = \{A \in \mathrm{M}_X(\mathbb{C}) \mid A(x, y) = A(Px, Py), \ \forall x, y \in X, \forall P \in G\}. \tag{3}$$

Note that the centralizer algebra $\mathrm{M}_X^G(\mathbb{C})$ of G is usually denoted by $\mathrm{End}_G(V)$ through the algebra isomorphism $\mathrm{End}(V) \simeq \mathrm{M}_X(\mathbb{C})$, in which case $\mathrm{End}_G(V)$ consists of all the elements of $\mathrm{End}(V)$ that commute with the action of G on V, which is extended from that of G on X.

For a subset $\mathcal{F}$ of $\mathrm{M}_X(\mathbb{C})$, the *coherent closure* of $\mathcal{F}$ is defined to be the intersection of coherent algebras $\mathcal{M} \subseteq \mathrm{M}_X(\mathbb{C})$ that contain $\mathcal{F}$:

$$\mathcal{M}(\mathcal{F}) = \bigcap_{\mathcal{F} \subseteq \mathcal{M}:\, \text{coherent alg}} \mathcal{M}. \tag{4}$$

Note that this intersection is not empty, since there exists at least one coherent algebra that contains $\mathcal{F}$, for example $\mathrm{M}_X(\mathbb{C})$. The coherent closure $\mathcal{M}(\mathcal{F})$ turns out to be the smallest coherent algebra that contains $\mathcal{F}$, because the intersection of coherent algebras is again a coherent algebra.

For a non-empty subset $\mathcal{F}$ of $\mathrm{M}_X(\mathbb{C})$, the *automorphism group* of $\mathcal{F}$ is defined to be the group consisting of permutation matrices of $\mathrm{M}_X(\mathbb{C})$ that commute with all elements of $\mathcal{F}$:

$$\mathrm{Aut}(\mathcal{F}) = \{P \in \mathrm{Sym}(X) \mid PA = AP, \ \forall A \in \mathcal{F}\}. \tag{5}$$

If Γ is a graph and A is the adjacency matrix of Γ, then $\mathrm{Aut}(\mathcal{F})$ with $\mathcal{F} = \{A\}$ coincides with the automorphism group of Γ, because of Lemma 2.1.

For a non-empty subset $\mathcal{F}$ of $\mathrm{M}_X(\mathbb{C})$, let $\mathcal{M}$ be the coherent closure of $\mathcal{F}$ and G the automorphism group of $\mathcal{F}$: $\mathcal{M} = \mathcal{M}(\mathcal{F})$, $G = \mathrm{Aut}(\mathcal{F})$. Then $\mathcal{M}$ is contained in the centralizer algebra of G:

$$\mathcal{M} \subseteq \mathrm{M}_X^G(\mathbb{C}). \tag{6}$$

This is because $\mathrm{M}_X^G(\mathbb{C})$ is a coherent algebra that contains $\mathcal{F}$: it apparently contains I, J, and $\mathcal{F}$, and by the following lemma, it is a linear space which is closed both under the ordinary matrix product and under the Hadamard product, as well as under the conjugate transpose.

Lemma 2.2 *If $P \in \mathrm{Sym}(X)$ commutes with $A, B \in M_X(\mathbb{C})$, then P commutes with the following: linear combinations of A and B, the conjugate transpose ${}^t\overline{A}, {}^t\overline{B}$, the ordinary matrix product AB, the Hadamard product $A \circ B$.*

Remark 2.3   We have remarks relating to 2-closed groups and Schurian coherent algebras.

(*i*) For a subgroup G of $\mathrm{Sym}(X)$, the centralizer algebra $M_X^G(\mathbb{C})$ of G obviously satisfies

$$G \subseteq \mathrm{Aut}(M_X^G(\mathbb{C})), \tag{7}$$

but the equality does not hold in general. If it does, the group G is called *2-closed*.

(*ii*) For $G = \mathrm{Aut}(\mathcal{F})$ and $\mathcal{F}'$ with $\mathcal{F} \subseteq \mathcal{F}' \subseteq M_X^G(\mathbb{C})$, we have $G = \mathrm{Aut}(\mathcal{F}')$. This is because $\mathcal{F} \subseteq \mathcal{F}' \subseteq M_X^G(\mathbb{C})$ implies $G = \mathrm{Aut}(\mathcal{F}) \supseteq \mathrm{Aut}(\mathcal{F}') \supseteq \mathrm{Aut}(M_X^G(\mathbb{C})) \supseteq G$. On the other hand, if $\mathcal{F}' \subseteq M_X(\mathbb{C})$ satisfies $\mathrm{Aut}(\mathcal{F}') = G$, then obviously $\mathcal{F}' \subseteq M_X^G(\mathbb{C})$. Therefore $M_X^G(\mathbb{C})$ is the largest among subsets $\mathcal{F}'$ of $M_X(\mathbb{C})$ that satisfy $\mathrm{Aut}(\mathcal{F}') = G$. In particular, $G = \mathrm{Aut}(\mathcal{F})$ is 2-closed.

(*iii*) For a coherent algebra $\mathcal{M} \subseteq M_X(\mathbb{C})$, set $G = \mathrm{Aut}(\mathcal{M})$. Then we have

$$\mathcal{M} \subseteq M_X^G(\mathbb{C}).$$

The equality does not hold in general. If it does, the coherent algebra $\mathcal{M}$ is called *Schurian*.

12.3 The Weisfeiler–Leman Stabilization

In this section, we explain the Weisfeiler–Leman stabilization, which is an algorithm to obtain the coherent closure $\mathcal{M}(\mathcal{F})$ for a subset $\mathcal{F}$ of $M_X(\mathbb{C})$. A fairly complete picture of the Weisfeiler–Leman algorithm is available in the monograph [9].

Given a subset $\mathcal{F}$ of $M_X(\mathbb{C})$, the *closure* in $M_X(\mathbb{C})$ of $\mathcal{F}$ is defined to be the smallest subalgebra of $M_X(\mathbb{C})$ that contains $\mathcal{F}$ and I, or equivalently to be the subalgebra of $M_X(\mathbb{C})$ generated by $\mathcal{F}$ and I:

$$\langle \mathcal{F}, I \rangle \subseteq M_X(\mathbb{C}). \tag{8}$$

Similarly, given a subset $\mathcal{F}$ of $M_X(\mathbb{C})^\circ$, the *closure* in $M_X(\mathbb{C})^\circ$ of $\mathcal{F}$ is defined to be the smallest subalgebra of $M_X(\mathbb{C})^\circ$ that contains $\mathcal{F}$ and J, or equivalently to be the subalgebra of $M_X(\mathbb{C})^\circ$ generated by $\mathcal{F}$ and J:

$$\langle \mathcal{F}, J \rangle^\circ \subseteq M_X(\mathbb{C})^\circ. \tag{9}$$

For a subset $\mathcal{F}$ of $M_X(\mathbb{C})$, define ${}^t\overline{\mathcal{F}}$ by

$$ {}^t\overline{\mathcal{F}} = \{ {}^t\overline{A} \mid A \in \mathcal{F} \}. \tag{10} $$

The coherent closure $\mathcal{M}(\mathcal{F})$ contains ${}^t\overline{\mathcal{F}}$, since $\mathcal{M}(\mathcal{F})$ is closed under the conjugate transpose. In order to find $\mathcal{M}(\mathcal{F})$, we may replace $\mathcal{F}$ by $\mathcal{F} \cup {}^t\overline{\mathcal{F}}$ and assume $\mathcal{F}$ is closed under the conjugate transpose from the beginning: ${}^t\overline{\mathcal{F}} = \mathcal{F}$.

Define a sequence of subalgebras $\mathcal{A}_0 \subseteq \mathcal{A}_2 \subseteq \mathcal{A}_4 \subseteq \cdots$ of $M_X(\mathbb{C})^\circ$ and a sequence of subalgebras $\mathcal{A}_1 \subseteq \mathcal{A}_3 \subseteq \mathcal{A}_5 \subseteq \cdots$ of $M_X(\mathbb{C})$ by setting $\mathcal{A}_0 = \langle \mathcal{F}, J \rangle^\circ \subseteq M_X(\mathbb{C})^\circ$ and

$$ \mathcal{A}_{2i} = \langle \mathcal{A}_{2i-1}, J \rangle^\circ \subseteq M_X(\mathbb{C})^\circ, \quad i = 1,2,3,\ldots, \tag{11} $$
$$ \mathcal{A}_{2i+1} = \langle \mathcal{A}_{2i}, I \rangle \subseteq M_X(\mathbb{C}), \quad i = 0,1,2,\ldots, \tag{12} $$

using the closures (8), (9), inductively. Then we have a sequence of linear subspaces $\mathcal{A}_0 \subseteq \mathcal{A}_1 \subseteq \mathcal{A}_2 \subseteq \mathcal{A}_3 \subseteq \cdots$ of the coherent closure $\mathcal{M}(\mathcal{F})$ of $\mathcal{F}$. If $\mathcal{A}_i = \mathcal{A}_{i+1}$, it holds that $\mathcal{A}_i = \mathcal{A}_{i+1} = \cdots = \mathcal{M}(\mathcal{F})$. Since $\dim\mathcal{M}(\mathcal{F}) \leq \dim M_X(\mathbb{C}) < \infty$, there exists $r = r(\mathcal{F})$ such that

$$ \mathcal{A}_0 \subsetneq \mathcal{A}_1 \subsetneq \cdots \subsetneq \mathcal{A}_{r-1} \subsetneq \mathcal{A}_r = \mathcal{A}_{r+1} = \cdots = \mathcal{M}(\mathcal{F}). \tag{13} $$

We call $r = r(\mathcal{F})$ the *coherent length* of $\mathcal{F}$. Essentially $r = r(\mathcal{F})$ is what is called the iteration number in [16].

We close this section with remarks on the sequence of the linear subspaces $\mathcal{A}_0 \subseteq \mathcal{A}_1 \subseteq \mathcal{A}_2 \subseteq \mathcal{A}_3 \subseteq \cdots$ of the coherent closure $\mathcal{M}(\mathcal{F})$ of $\mathcal{F}$ that eventually reach $\mathcal{M}(\mathcal{F})$. Firstly, $\mathrm{Aut}(\mathcal{A}_i) = \mathrm{Aut}(\mathcal{F})$ holds for all i. Secondly, $\mathcal{A}_0 \subseteq \mathcal{A}_2 \subseteq \mathcal{A}_4 \subseteq \cdots$ is a sequence of commutative semi-simple algebras of $M_X(\mathbb{C})^\circ$, and so each of them induces a set of relations on X by means of its primitive idempotents. Thirdly, $\mathcal{A}_1 \subseteq \mathcal{A}_3 \subseteq \mathcal{A}_5 \subseteq \cdots$ is a sequence of semi-simple algebras of $M_X(\mathbb{C})$, and so their representations can be used to analyse the interactions between the relations on X that are induced by the sequence $\mathcal{A}_0 \subseteq \mathcal{A}_2 \subseteq \mathcal{A}_4 \subseteq \cdots$.

12.4 Terwilliger Algebras

In this section, we discuss Terwilliger algebras in the framework of the Weisfeiler Leman stabilization. For the sake of simplicity, we consider the Terwilliger algebra for a connected simple graph. Terwilliger algebras for commutative association schemes can be treated in a similar way.

Let Γ be a connected simple graph, X the vertex set of Γ, and A the adjacency matrix of Γ. The distance function of the graph Γ is denoted by ∂, that is, $\partial(x,y)$ is the length of a shortest path joining x and y ($x,y \in X$).

First, set $\mathcal{F} = \{A\}$ and apply the W-L stabilization to $\mathcal{F}$. By (11), (12), we have the sequence $\mathcal{A}_0 \subseteq \mathcal{A}_1 \subseteq \mathcal{A}_2 \subseteq \mathcal{A}_3 \subseteq \cdots$, which eventually reaches the coherent closure $\mathcal{M} = \mathcal{M}(\mathcal{F})$ of $\mathcal{F}$. The first three are

$$\mathcal{A}_0 = \mathrm{Span}\{A, J\} \subseteq \mathrm{M}_X(\mathbb{C})^\circ, \tag{14}$$

$$\mathcal{A}_1 = \langle\, A, I, J \,\rangle \subseteq \mathrm{M}_X(\mathbb{C}), \tag{15}$$

$$\mathcal{A}_2 = \langle\, \mathcal{A}_1 \,\rangle^\circ \subseteq \mathrm{M}_X(\mathbb{C})^\circ. \tag{16}$$

Define the *ith distance matrix* A_i by

$$A_i(x, y) = 1 \ \text{ if } \partial(x, y) = i, \ 0 \ \text{ otherwise.} \tag{17}$$

For $i = 0, 1$, we have $A_0 = I$, $A_1 = A$. The following lemma holds.

Lemma 4.1 *For $i = 2, 3, \ldots$, the distance matrix A_i is contained in $\mathcal{A}_2$.*

A proof is given in [25].

Next, choose a base vertex $x_0 \in X$ and fix it. Set

$$D = \max\{\partial(x_0, x) \,|\, x \in X\}, \tag{18}$$

and for $i = 0, 1, 2, \ldots, D$, define a diagonal matrix $E_i^* = E_i^*(x_0) \in \mathrm{M}_X(\mathbb{C})$ by

$$E_i^*(x, y) = 1 \ \text{ if } \ x = y \text{ and } \partial(x_0, x) = i, \ 0 \ \text{ otherwise.} \tag{19}$$

The Terwilliger algebra $T = T(x_0)$ of Γ is defined to be the subalgebra of $\mathrm{M}_X(\mathbb{C})$ generated by A and E_i^* $(i = 0, 1, \ldots, D)$:

$$T = T(x_0) = \langle\, A, E_i^* \,|\, i = 0, 1, \ldots, D \,\rangle \subseteq \mathrm{M}_X(\mathbb{C}). \tag{20}$$

Set $\mathcal{F}^{(x_0)} = \mathcal{F} \cup \{E_0^*\} = \{A, E_0^*\}$; we call $\mathcal{F}^{(x_0)}$ the *localization* at x_0 of $\mathcal{F}$. Apply the W-L stabilization to $\mathcal{F}^{(x_0)}$. By (11), (12), we have the sequence $\mathcal{A}_0^{(x_0)} \subseteq \mathcal{A}_1^{(x_0)} \subseteq \mathcal{A}_2^{(x_0)} \subseteq \mathcal{A}_3^{(x_0)} \subseteq \cdots$, which eventually reaches the coherent closure $\mathcal{M}^{(x_0)} = \mathcal{M}(\mathcal{F}^{(x_0)})$ of $\mathcal{F}^{(x_0)}$:

$$\mathcal{A}_0^{(x_0)} = \mathrm{Span}\{A, E_0^*, J\} \subseteq \mathrm{M}_X(\mathbb{C})^\circ, \tag{21}$$

$$\mathcal{A}_1^{(x_0)} = \langle\, A, E_0^*, I, J \,\rangle \subseteq \mathrm{M}_X(\mathbb{C}), \tag{22}$$

$$\mathcal{A}_2^{(x_0)} = \langle\, \mathcal{A}_1^{(x_0)} \,\rangle^\circ \subseteq \mathrm{M}_X(\mathbb{C})^\circ, \tag{23}$$

$$\mathcal{A}_{2i+1}^{(x_0)} = \langle\, \mathcal{A}_{2i}^{(x_0)} \,\rangle \subseteq \mathrm{M}_X(\mathbb{C}), \quad i = 1, 2, 3, \ldots, \tag{24}$$

$$\mathcal{A}_{2i}^{(x_0)} = \langle\, \mathcal{A}_{2i-1}^{(x_0)} \,\rangle^\circ \subseteq \mathrm{M}_X(\mathbb{C})^\circ, \quad i = 2, 3, 4, \ldots. \tag{25}$$

Note that

$$\mathcal{A}_i \subseteq \mathcal{A}_i^{(x_0)} \quad (i = 0, 1, 2, \ldots,). \tag{26}$$

So by Lemma 4.1, and (24), (25), we have

Lemma 4.2 $E_i^* = (E_0^* A_i E_0^*)^2 \circ (I - E_0^*) \in \mathcal{A}_4^{(x_0)}$ $(i = 1, 2, \ldots)$.

Therefore by (20), (24), we have

Proposition 4.3 $T = T(x_0) \subseteq \mathcal{A}_5^{(x_0)}$.

Let G (resp. H) denote the automorphism group of $\mathcal{F}$ (resp. $\mathcal{F}^{(x_0)}$): $G = \mathrm{Aut}(\mathcal{F})$, $H = \mathrm{Aut}(\mathcal{F}^{(x_0)})$. Then by Lemma 2.1, G coincides with the automorphism group of the graph Γ, and H with the stabilizer in G of x_0:

$$G = \mathrm{Aut}(\Gamma), \quad H = G_{x_0}. \tag{27}$$

Thus by (6), we have for the coherent closure $\mathcal{M} = \mathcal{M}(\mathcal{F})$ of $\mathcal{F}$

$$\mathcal{M} \subseteq \mathrm{M}_X^G(\mathbb{C}),$$

and for the coherent closure $\mathcal{M}^{(x_0)} = \mathcal{M}(\mathcal{F}^{(x_0)})$ of $\mathcal{F}^{(x_0)}$

$$T \subseteq \mathcal{M}^{(x_0)} \subseteq \mathrm{M}_X^H(\mathbb{C}). \tag{28}$$

If the coherent closure $\mathcal{M}$ of $\mathcal{F}$ is viewed as a combinatorial analogue of the centralizer algebra $\mathrm{M}_X^G(\mathbb{C})$ of $G = \mathrm{Aut}(\mathcal{F})$, the coherent closure $\mathcal{M}^{(x_0)}$ of $\mathcal{F}^{(x_0)}$ is viewed as a combinatorial analogue of the centralizer algebra $\mathrm{M}_X^H(\mathbb{C})$ of $H = \mathrm{Aut}(\mathcal{F}^{(x_0)})$. Suppose the gap between T and $\mathcal{M}^{(x_0)}$ is small or nil. Then Terwilliger algebra T can be used in place of $\mathcal{M}^{(x_0)}$. The problem is how to measure the gap. We propose to use the coherent length of T, namely to apply the W-L stabilization to T and count how many steps we need to reach $\mathcal{M}^{(x_0)}$. We ask:

Question 4.4   Which graph Γ satisfies the following properties $(i), (ii)$, $(iii), (iv)$?

(i) $T = \mathrm{M}_X^H(\mathbb{C})$.
(ii) $T = \mathcal{M}^{(x_0)}$.
(iii) $\langle T \rangle^\circ = \mathrm{M}_X^H(\mathbb{C})$, where $\langle T \rangle^\circ$ is the closure of T in $\mathrm{M}_X^H(\mathbb{C})^\circ$.
(iv) $\langle T \rangle^\circ = \mathcal{M}^{(x_0)}$.

Note that the property (i) (resp. (ii)) implies that T contains $\mathrm{M}_X^G(\mathbb{C})$ (resp. $\mathcal{M}$):

$$\mathrm{M}_X^G(\mathbb{C}) \subseteq T, \tag{29}$$

$$\mathcal{M} \subseteq T. \tag{30}$$

The properties (29), (30) are weaker than $(i), (ii)$, respectively, but they cannot be expected to hold for a Terwilliger algebra in general, and it is an interesting question whether (29), (30) hold or not for a given graph Γ.

When asking Question 4.4, we may assume that the automorphism group of Γ is generously transitive, which is perhaps the most important and interesting case. The property (i) holds for Hamming graphs and Johnson graphs [8, 19]. In the next section, the property (iii) will be proved for trees: actually the stronger property (1) is proved.

12.5 The Case of Trees

In this section, we analyse the Weisfeiler–Leman stabilization in more detail in the case of a tree, using the Terwilliger algebra.

Let Γ be a finite tree and X the vertex set of Γ. Choose a vertex $x_0 \in X$ and fix it. The rooted tree with x_0 designated the root of Γ is denoted by $\Gamma^{(x_0)}$. Allowing abuse of notations, we let Γ and $\Gamma^{(x_0)}$ also stand for the vertex set X: $X = \Gamma$, $X = \Gamma^{(x_0)}$.

We keep the notations in Section 12.4. So $T = T(x_0)$ is the Terwilliger algebra of Γ defined by (20), $G = \mathrm{Aut}(\Gamma)$ is the automorphism group of Γ, and so $H = G_{x_0}$, the stabilizer in G of x_0 as in (27), becomes the automorphism group of $\Gamma^{(x_0)}$: $H = \mathrm{Aut}(\Gamma^{(x_0)})$. For the centralizer algebra $M_X^H(\mathbb{C})$ of H, we have the inclusion (28)

$$T \subseteq \mathcal{M}^{(x_0)} \subseteq M_X^H(\mathbb{C}),$$

and we want to analyse it more closely. The Terwilliger algebra T and the centralizer algebra $M_X^H(\mathbb{C})$ of H are both semi-simple and their structures as semi-simple algebras are given in [24, theorem 7, theorem 8]. We will start with an explanation of these theorems, rewriting them in the language of matrices. We should note here that for the proof of [24, theorem 7], we essentially used the fact that the Terwilliger algebra $T = T(x_0)$ can recognize the isomorphism class of the rooted tree $\Gamma^{(x_0)}$ [15].

The vertex set X is partitioned according to the distance from x_0:

$$X = \bigcup_{0 \le i \le D} X_i, \tag{31}$$

where D is from (18) and

$$X_i = \{x \in X \mid \partial(x_0, x) = i\}, \tag{32}$$

in which ∂ is the distance function of the tree Γ. For $x \in X$, set

$$\Gamma^{(x)} = \{y \in X \mid \partial(x_0, y) = \partial(x_0, x) + \partial(x, y)\}. \tag{33}$$

We regard $\Gamma^{(x)}$ as the rooted tree with x the root that is induced on the subset $\Gamma^{(x)}$ of X. Using the rooted trees $\Gamma^{(x)}$, $x \in X_i$, we introduce an equivalence relation $\sim$ on X_i: $x \sim x'$ if and only if $\Gamma^{(x)} \simeq \Gamma^{(x')}$, that is, $\Gamma^{(x)}$ and $\Gamma^{(x')}$ are isomorphic as rooted trees. Let $X_i(\alpha)$, $\alpha \in \Lambda_i$, denote the equivalence classes on X_i:

$$X_i = \bigcup_{\alpha \in \Lambda_i} X_i(\alpha), \tag{34}$$

where Λ_i is the set of indices attached to the equivalence classes on X_i, and let $\{\Gamma^{(\alpha)}\}_{\alpha \in \Lambda_i}$ be a complete set of representatives for the isomorphism classes of $\Gamma^{(x)}$, $x \in X_i$:

$$X_i(\alpha) = \{x \in X_i \mid \Gamma^{(x)} \simeq \Gamma^{(\alpha)}\}. \tag{35}$$

The following subset $X_i'(\alpha)$ of $X_i(\alpha)$ plays a key role in describing irreducible representations of the Terwilliger algebra T of Γ:

$$X_i'(\alpha) = \{x \in X_i(\alpha) \mid \exists y \in X_i(\alpha), \ \partial(x,y) = 2\}, \ 1 \le i \le D. \tag{36}$$

For $i = 0$, $|\Lambda_0| = 1$ holds. With $\Lambda_0 = \{\alpha_0\}$, we define $X_0'(\alpha_0)$ by $X_0'(\alpha_0) = X_0 = \{x_0\}$. For $i = 1, 2, \ldots$, it may occur that $X_i'(\alpha)$ is an empty set.

For $0 \le i \le D$, $\alpha \in \Lambda_i$, let $\mathrm{Aut}(\Gamma^{(\alpha)})$ denote the automorphism group of the rooted tree $\Gamma^{(\alpha)}$ and define an equivalence relation $\approx$ on the vertex set of $\Gamma^{(\alpha)}$: $x \approx x'$ if and only if x, x' are in the same orbit of $\mathrm{Aut}(\Gamma^{(\alpha)})$. Regarding $\Gamma^{(\alpha)}$ as the vertex set of $\Gamma^{(\alpha)}$ by abuse of the notation, we define a subset $\mathrm{M}_{\Gamma^{(\alpha)}/\approx}(\mathbb{C})$ of $\mathrm{M}_{\Gamma^{(\alpha)}}(\mathbb{C})$ as follows:

$$\mathrm{M}_{\Gamma^{(\alpha)}/\approx}(\mathbb{C}) = \{B \in \mathrm{M}_{\Gamma^{(\alpha)}}(\mathbb{C}) \mid B(x,y) = B(x',y') \text{ if } x \approx x', y \approx y'\}. \tag{37}$$

Note that $\mathrm{M}_{\Gamma^{(\alpha)}/\approx}$ is a simple algebra isomorphic to the full matrix algebra $\mathrm{M}_n(\mathbb{C})$, where $n = n(\alpha)$ is the number of orbits of $\mathrm{Aut}(\Gamma^{(\alpha)})$ on $\Gamma^{(\alpha)}$. Note also that $|\Lambda_0| = 1$ for $i = 0$, and so with $\Lambda_0 = \{\alpha_0\}$, $\mathrm{M}_{\Gamma^{(\alpha_0)}}(\mathbb{C})$ is isomorphic to $\mathrm{M}_{\Gamma^{(x_0)}}(\mathbb{C})$ as algebras through the isomorphism $\Gamma^{(\alpha_0)} \simeq \Gamma^{(x_0)}$ as rooted trees. Therefore $\mathrm{M}_{\Gamma^{(\alpha_0)}/\approx}(\mathbb{C})$ is isomorphic to $\mathrm{M}_{\Gamma^{(x_0)}/\approx}(\mathbb{C})$ as algebras. We set

$$T_0 = \mathrm{M}_{\Gamma^{(x_0)}/\approx}(\mathbb{C}). \tag{38}$$

Since we regard $\Gamma^{(x_0)}$ as the vertex set X of $\Gamma^{(x_0)}$ by abuse of the notation, T_0 is a subalgebra of $M_X(\mathbb{C})$, which is isomorphic to $\mathrm{M}_n(\mathbb{C})$, where n is the number of orbits of $H = \mathrm{Aut}(\Gamma^{(x_0)})$ on X. Later we will see that T_0 turns out to be a subalgebra of T: we call T_0 the *principal component* of T.

Here we make general remarks on matrices, preparing notations further. Let Z be a non-empty subset of X. The identity matrix (resp. the all-ones matrix) in $\mathrm{M}_Z(\mathbb{C})$ is denoted by I_Z (resp. J_Z).

(1) We embed $\mathrm{M}_Z(\mathbb{C})$ in $\mathrm{M}_X(\mathbb{C})$, regarding matrices $A \in \mathrm{M}_Z(\mathbb{C})$ as $A \in \mathrm{M}_X(\mathbb{C})$ by setting $A(x,y) = 0$ unless $x, y \in Z$.

(2) If $Z = Z_1 \times Z_2$, then we have

$$\mathrm{M}_Z(\mathbb{C}) = \mathrm{M}_{Z_1}(\mathbb{C}) \otimes \mathrm{M}_{Z_2}(\mathbb{C}), \tag{39}$$

where $\otimes$ stands for the Kronecker product of matrices. Note that for $A_1, B_1 \in \mathrm{M}_{Z_1}(\mathbb{C})$ and $A_2, B_2 \in \mathrm{M}_{Z_2}(\mathbb{C})$,

$$(A_1 \otimes A_2)(B_1 \otimes B_2) = (A_1 B_1) \otimes (A_2 B_2), \tag{40}$$

$$(A_1 \otimes A_2) \circ (B_1 \otimes B_2) = (A_1 \circ B_1) \otimes (A_2 \circ B_2). \tag{41}$$

(3) Define $E_Z^{(0)}$ and $E_Z^{(1)}$ by

$$E_Z^{(0)} = \frac{1}{|Z|} J_Z, \quad E_Z^{(0)} + E_Z^{(1)} = I_Z, \tag{42}$$

unless $Z = \{x_0\}$. For $Z = \{x_0\}$, we define $E_{\{x_0\}}^{(0)} = 0$, $E_{\{x_0\}}^{(1)} = I_{\{x_0\}}$. Then it holds that

$$E_Z^{(0)} E_Z^{(0)} = E_Z^{(0)}, \quad E_Z^{(1)} E_Z^{(1)} = E_Z^{(1)}, \quad E_Z^{(0)} E_Z^{(1)} = E_Z^{(1)} E_Z^{(0)} = 0. \tag{43}$$

For $0 \le i \le D$, $\alpha \in \Lambda_i$, we identify $\Gamma^{(\alpha)} \times X_i'(\alpha)$ with $\cup_{x \in X_i'(\alpha)} \Gamma^{(x)}$ through the isomorphism $\Gamma^{(\alpha)} \simeq \Gamma^{(x)}$:

$$\Gamma^{(\alpha)} \times X_i'(\alpha) = \bigcup_{x \in X_i'(\alpha)} \Gamma^{(x)}. \tag{44}$$

Then we have $\mathrm{M}_{\Gamma^{(\alpha)} \times X_i'(\alpha)}(\mathbb{C}) = \mathrm{M}_Z(\mathbb{C})$ with $Z = \cup_{x \in X_i'(\alpha)} \Gamma^{(x)}$. So by (39),

$$\mathrm{M}_{\Gamma^{(\alpha)} \times X_i'(\alpha)}(\mathbb{C}) = \mathrm{M}_{\Gamma^{(\alpha)}}(\mathbb{C}) \otimes \mathrm{M}_{X_i'(\alpha)}(\mathbb{C}) \tag{45}$$

is embedded in $\mathrm{M}_X(\mathbb{C})$. If $X_i'(\alpha)$ is an empty set, we understand $\mathrm{M}_{X_i'(\alpha)}(\mathbb{C}) = 0$. Note that $\mathrm{M}_{\Gamma^{(\alpha)}/\approx}(\mathbb{C}) \otimes E_{X_i'(\alpha)}^{(1)}$ is contained in $\mathrm{M}_{\Gamma^{(\alpha)}}(\mathbb{C}) \otimes \mathrm{M}_{X_i'(\alpha)}(\mathbb{C})$, and we understand $\mathrm{M}_{\Gamma^{(\alpha)}/\approx}(\mathbb{C}) \otimes E_{X_i'(\alpha)}^{(1)} = 0$ if $X_i'(\alpha) = \emptyset$. For $i = 0$, $|\Lambda_0| = 1$ holds. With $\Lambda_0 = \{\alpha_0\}$, we have $X_0'(\alpha_0) = X_0 = \{x_0\}$. Recall $E_{X_0'(\alpha_0)}^{(1)} = I_{\{x_0\}}$ by definition. So $\mathrm{M}_{\Gamma^{(\alpha_0)}/\approx}(\mathbb{C}) \otimes E_{X_0'(\alpha_0)}^{(1)}$ is identified with $\mathrm{M}_{\Gamma^{(x_0)}/\approx}(\mathbb{C})$.

We are now prepared to rewrite [24, theorem 7] in the language of matrices.

Theorem 5.1 *The Terwilliger algebra $T = T(x_0)$ of a tree Γ is decomposed into the direct sum of simple algebras as follows:*

$$T = \bigoplus_{0 \le i \le D} \bigoplus_{\alpha \in \Lambda_i} \mathrm{M}_{\Gamma^{(\alpha)}/\approx}(\mathbb{C}) \otimes E_{X_i'(\alpha)}^{(1)},$$

where we understand T_0 appears as the summand for $i = 0$:

$$T_0 = \mathrm{M}_{\Gamma^{(x_0)}/\approx}(\mathbb{C}) = \mathrm{M}_{\Gamma^{(\alpha_0)}/\approx}(\mathbb{C}) \otimes E_{X_0'(\alpha_0)}^{(1)},$$

in which $\Lambda_0 = \{\alpha_0\}$, $X_0'(\alpha_0) = X_0 = \{x_0\}$ and $E_{X_0'(\alpha_0)}^{(1)} = I_{\{x_0\}}$.

Note that if $X_i'(\alpha) \neq \emptyset$, then $M_{\Gamma^{(\alpha)}/\approx}(\mathbb{C}) \otimes E_{X_i'(\alpha)}^{(1)}$ is by (40) isomorphic to $M_{\Gamma^{(\alpha)}/\approx}(\mathbb{C})$, which is a simple algebra, because $E_{X_i'(\alpha)}^{(1)}$ is an idempotent by (43).

Observe that for $0 \leq i \leq D$, $\alpha \in \Lambda_i$, the set $X_i'(\alpha)$ is invariant under the action of $H = \mathrm{Aut}(\Gamma^{(x_0)})$. Let $\mathrm{Orb}(H, X_i'(\alpha))$ denote the set of orbits of H on $X_i'(\alpha)$. Then for $Y \in \mathrm{Orb}(H, X_i'(\alpha))$, $M_{\Gamma^{(\alpha)}/\approx}(\mathbb{C}) \otimes E_Y^{(1)}$ is contained in $M_{\Gamma^{(\alpha)}}(\mathbb{C}) \otimes M_Y(\mathbb{C})$, particularly in $M_{\Gamma^{(\alpha)}}(\mathbb{C}) \otimes M_{X_i'(\alpha)}(\mathbb{C})$, and we understand $M_{\Gamma^{(\alpha)}/\approx}(\mathbb{C}) \otimes E_Y^{(1)} = 0$ if $X_i'(\alpha) = \emptyset$.

We are now prepared to rewrite [24, theorem 8] in the language of matrices.

Theorem 5.2   *The centralizer algebra $M_X^H(\mathbb{C})$ of $H = \mathrm{Aut}(\Gamma^{(x_0)})$ for a rooted tree $\Gamma^{(x_0)}$ is decomposed into the direct sum of simple algebras as follows:*

$$M_X^H(\mathbb{C}) = \bigoplus_{0 \leq i \leq D} \bigoplus_{\alpha \in \Lambda_i} \bigoplus_{Y \in \mathrm{Orb}(H, X_i'(\alpha))} M_{\Gamma^{(\alpha)}/\approx}(\mathbb{C}) \otimes E_Y^{(1)},$$

where we understand T_0 appears as the summand for $i = 0$:

$$T_0 = M_{\Gamma^{(x_0)}/\approx}(\mathbb{C}) = M_{\Gamma^{(\alpha_0)}/\approx}(\mathbb{C}) \otimes E_{Y_0}^{(1)},$$

in which $\Lambda_0 = \{\alpha_0\}$, $Y_0 = X_0'(\alpha_0) = \{x_0\}$ and $E_{Y_0}^{(1)} = I_{\{x_0\}}$.

Note that if $X_i'(\alpha) \neq \emptyset$, then $M_{\Gamma^{(\alpha)}/\approx}(\mathbb{C}) \otimes E_Y^{(1)}$ is by (40) isomorphic to $M_{\Gamma^{(\alpha)}/\approx}(\mathbb{C})$, which is a simple algebra, because $E_Y^{(1)}$ is an idempotent by (43).

For $0 \leq i \leq D$, $\alpha \in \Lambda_i$, $Y \in \mathrm{Orb}(H, X_i'(\alpha))$, $H = \mathrm{Aut}(\Gamma^{(x_0)})$ acts on

$$\Gamma^{(\alpha)} \times Y = \bigcup_{x \in Y} \Gamma^{(x)}$$

as the wreath product of $\mathrm{Aut}(\Gamma^{(\alpha)})$ with the symmetric group on Y:

$$H|_{\Gamma^{(\alpha)} \times Y} = \mathrm{Aut}(\Gamma^{(\alpha)}) \wr \mathrm{Sym}(Y). \tag{46}$$

So the principal component $T_0 = M_{\Gamma^{(x_0)}/\approx}(\mathbb{C})$ of T contains $B \otimes J_Y$ for all $B \in M_{\Gamma^{(\alpha)}/\approx}(\mathbb{C})$, $Y \in \mathrm{Orb}(H, X_i'(\alpha))$, particularly $J_{\Gamma^{(\alpha)}} \otimes J_Y$. Since $E_Y^{(1)} = I_Y - \frac{1}{|Y|}J_Y$, $E_{X_i'(\alpha)}^{(1)} = I_{X_i'(\alpha)} - \frac{1}{|X_i'(\alpha)|}J_{X_i'(\alpha)}$, we have

$$E_{X_i'(\alpha)}^{(1)} \circ J_Y = I_Y - \frac{1}{|X_i'(\alpha)|}J_Y$$

$$= E_Y^{(1)} + \left(\frac{1}{|Y|} - \frac{1}{|X_i'(\alpha)|}\right)J_Y.$$

Therefore for $B \in M_{\Gamma^{(\alpha)}/\approx}(\mathbb{C})$, $Y \in \mathrm{Orb}(H, X_i'(\alpha))$, we have by (41)

$$(B \otimes E^{(1)}_{X'_i(\alpha)}) \circ (J_{\Gamma(\alpha)} \otimes J_Y) = B \otimes E^{(1)}_Y + (\frac{1}{|Y|} - \frac{1}{|X'_i(\alpha)|})B \otimes J_Y,$$

and hence

$$(\mathrm{M}_{\Gamma(\alpha)/\approx}(\mathbb{C}) \otimes E^{(1)}_{X'_i(\alpha)}) \circ (J_{\Gamma(\alpha)} \otimes J_Y) + T_0 = \mathrm{M}_{\Gamma(\alpha)/\approx}(\mathbb{C}) \otimes E^{(1)}_Y + T_0.$$

So we have the following corollary to Theorem 5.1 and Theorem 5.2, which is stronger than (1).

Corollary 5.3   *It holds that*

$$T \circ T_0 = \mathrm{M}^H_X(\mathbb{C}),$$

where $T \circ T_0 = \{a \circ b \mid a \in T, \, b \in T_0\}$.

In the rest of this section, we discuss the W-L stabilization of a tree Γ for $\mathcal{F} = \{A\}$, where A is the adjacency matrix of Γ. We keep the notations in Section 12.3. So $\mathcal{M} = \mathcal{M}(\mathcal{F})$ is the coherent closure of $\mathcal{F}$ and $\mathrm{M}^G_X(\mathbb{C})$ is the centralizer algebra of G, where $G = \mathrm{Aut}(\mathcal{M}) = \mathrm{Aut}(\Gamma)$ is the automorphism group of Γ acting on the vertex set X of Γ. Let $\mathcal{A}_0 \subseteq \mathcal{A}_1 \subseteq \mathcal{A}_2 \subseteq \mathcal{A}_3 \subseteq \cdots$ be the sequence of linear subspaces of $\mathcal{M}$ in (13) and $r = r(\mathcal{F})$ the coherent length.

Theorem 5.4   *We have $r \leq 8$. More precisely*

$$\mathcal{A}_8 = \mathcal{M}(\mathcal{F}) = M^G_X(\mathbb{C}).$$

In particular, the coherent algebra $\mathcal{M}(\mathcal{F})$ is Schurian.

The fact that $\mathcal{M}(\mathcal{F})$ is Schurian for a tree is known already [5]. We give an outline of the proof of this theorem: full details are in [25].

Setting

$$D(x) = \mathrm{Max}\{\partial(x, y) \mid y \in X\}, \tag{47}$$

we define D as the minimum among $D(x), x \in X$:

$$D = \mathrm{Min}\{D(x) \mid x \in X\}. \tag{48}$$

Note that D is much smaller than the diameter of Γ. The *centre* X_0 is defined to be

$$X_0 = \{x \in X \mid D(x) = D\}. \tag{49}$$

It is well known that the centre X_0 of the tree Γ consists either of a single vertex x_0 or of adjacent two vertices x_0, x'_0:

Case 1: $X_0 = \{x_0\}$;
Case 2: $X_0 = \{x_0, x'_0\}$ and x_0, x'_0 are adjacent.

Suppose Case 1 occurs: the centre of Γ is $X_0 = \{x_0\}$. Then the automorphism group G of Γ stabilizes the centre x_0: $G_{x_0} = G$. So for $H = \mathrm{Aut}(\Gamma^{(x_0)})$, we have $H = G$ and $\mathrm{M}_X^H(\mathbb{C}) = \mathrm{M}_X^G(\mathbb{C})$.

Let A_i, $0 \le i \le D$, be the distance matrix from (17), $E_i^* = E_i^*(x_0)$, $0 \le i \le D$, the matrix from (19), and $T = T(x_0)$ the Terwilliger algebra from (20). By Lemma 4.1, we have

$$A_i \in \mathcal{A}_2,\ 0 \le i \le D,$$

and so the following lemma.

Lemma 5.5   *In the W-L stabilization for $\mathcal{F} = \{A\}$, we have*

(*i*) $E_0^* \in \mathcal{A}_5$,
(*ii*) $E_i^* \in \mathcal{A}_6$, $1 \le i \le D$,
(*iii*) $T \subseteq \mathcal{A}_7$.

The claims (*i*), (*ii*) follow from

$$E_0^* = \frac{1}{(N-1)!} \prod_{j=1}^{N-1} \{((A_0 + A_1 + \cdots + A_D)^2 \circ I) - jI\}, \qquad (50)$$

$$E_i^* = (E_0^* A_i E_0^*)^2 \circ I, \qquad (51)$$

where $N = |X|$. By Corollary 5.3, we obtain Theorem 5.4.

Suppose Case 2 occurs: the centre of Γ is $X_0 = \{x_0, x_0'\}$ and x_0, x_0' are adjacent. In place of $T(x_0)$ (resp. G_{x_0}) of Case 1, we use the Terwilliger algebra $T = T(X_0)$ with respect to the set $X_0 = \{x_0, x_0'\}$ [18] (resp. the global stabilizer $H = G_{X_0}$ in G of the set X_0). Apparently $H = G$ holds. The algebras T, $\mathrm{M}_X^H(\mathbb{C})$ are semi-simple, and as in Theorem 5.1, Theorem 5.2, the decomposition into the direct sum of simple algebras is determined for each of them. The rest of the arguments goes parallel to Case 1. We remark that Case 2 is essentially reduced to Case 1, if we consider the augmented tree $\Gamma^{(\infty)}$ of Γ which is obtained from Γ by inserting an extra vertex x_∞ on the edge between x_0, x_0': $\Gamma^{(\infty)}$ has the centre consisting of the single vertex x_∞.

12.6 Distance-Regular Graphs Revisited

In this section, we consider the class of regular graphs that have coherent length. Keeping the notations in Section 12.3 and Section 12.4, let Γ be a connected simple graph, X the vertex set of Γ and A the adjacency matrix of Γ. We apply the W-L stabilization to the graph Γ with $\mathcal{F} = \{A\}$. Then we have (14), (15), (16). Note that $\mathcal{A}_1 = \langle A, I, J \rangle$ is the generalized adjacency algebra of Γ.

We now assume in addition that our graph Γ is regular with valency k. Observe that the all-ones matrix J is a polynomial in A, since $\frac{1}{|X|}J$ is the projection onto the eigenspace belonging to the eigenvalue k of A. So instead of (15), (16), we have

$$\mathcal{A}_1 = \langle A \rangle = \mathrm{Span}\{A^i \,|\, 0 \le i \le D\} \subseteq \mathrm{M}_X(\mathbb{C}), \tag{52}$$

$$\mathcal{A}_2 = \langle A^i \,|\, 0 \le i \le D \rangle^\circ \subseteq \mathrm{M}_X(\mathbb{C})^\circ, \tag{53}$$

where $D + 1$ is the degree of the minimal polynomial of A. Note that $\mathcal{A}_1$ is the adjacency algebra of Γ. Let $\mathrm{diam}(\Gamma)$ denote the diameter of Γ. Then we have

$$\mathrm{diam}(\Gamma) \le D. \tag{54}$$

Let $r = r(\mathcal{F})$ be the coherent length from (13). Then $r = 0$ if and only if Γ is a complete graph or equivalently $\mathrm{diam}(\Gamma) = 1$. This is because by (14),

$$\mathcal{A}_0 = \mathrm{Span}\{A, J\} \subseteq \mathrm{M}_X(\mathbb{C})^\circ$$

is required to be the minimum coherent algebra of dimension 2, if $r = 0$.

We are interested in the class of regular connected simple graphs that have $r = 1$. Suppose $r = 1$, that is, $\mathcal{A}_1 = \mathcal{A}_2$ and $2 \le \mathrm{diam}(\Gamma)$. Let $\mathcal{M} = \mathcal{M}(\mathcal{F})$ be the coherent closure of $\mathcal{F} = \{A\}$. Then by (52) we have

$$\mathcal{M} = \mathrm{Span}\{A^i \,|\, 0 \le i \le D\}. \tag{55}$$

Let $\{A_\alpha \,|\, \alpha \in \Lambda\}$ denote the set of primitive idempotents of $\mathcal{M}$ with respect to the Hadamard product:

$$\mathcal{M} = \mathrm{Span}\{A_\alpha \,|\, \alpha \in \Lambda\}, \tag{56}$$

$$A_\alpha \circ A_\beta = \delta_{\alpha\beta} A_\alpha, \quad \sum_{\alpha \in \Lambda} A_\alpha = J, \tag{57}$$

where $\delta_{\alpha,\beta}$ is the Kronecker delta. Then there arises a symmetric association scheme

$$\mathfrak{X} = (X, \{R_\alpha\}_{\alpha \in \Lambda}) \tag{58}$$

for which $\mathcal{M}$ is the Bose–Mesner algebra, that is, R_α is the relation on X corresponding to the $(0, 1)$ matrix $A_\alpha, \alpha \in \Lambda$. Note that there is a special $\alpha_0 \in \Lambda$ for which

$$A_{\alpha_0} = I. \tag{59}$$

Note also

$$\dim \mathcal{M} = D + 1 = |\Lambda|. \tag{60}$$

Definition 6.1   *A symmetric association scheme* $\mathfrak{X} = (X, \{R_\alpha\}_{\alpha \in \Lambda})$ *is called polynomial if its Bose–Mesner algebra* $\mathcal{M}$ *satisfies* (55) *for some* $A \in \mathcal{M}$ *with the property that* $A \circ A = A$ *and* $A \circ I = 0$. *Note that* $A \circ A = A$ *holds if and only if* A *is a* $(0, 1)$ *matrix.*

So from a regular connected simple graph Γ with $r = 1$, a polynomial association scheme naturally arises. In this sense, a regular connected simple graph Γ with $r = 1$ is also called *polynomial*. Conversely start with a polynomial association scheme $\mathfrak{X} = (X, \{R_\alpha\}_{\alpha \in \Lambda})$. Then the Bose–Mesner algebra of $\mathfrak{X}$ is generated by a $(0, 1)$ matrix A with $A \circ I = 0$. It is easy to see that the graph Γ for which A is the adjacency matrix is regular, connected, and has $r = 1$, unless the Bose–Mesner algebra is spanned by I, J, in which case $r = 0$. So from a polynomial association scheme, a polynomial graph natually arises. Note that the $(0, 1)$ matrix A with $A \circ I = 0$ that generates the Bose-Mesner algebra of a polynomial association scheme may not be uniquely determined.

For a connected simple graph Γ, let A_i, $0 \le i \le \text{diam}(\Gamma)$, be the distance matrices from (17). By Lemma 4.1, it always holds that

$$\text{Span}\{A_i \mid 0 \le i \le \text{diam}(\Gamma)\} \subseteq \mathcal{A}_2. \tag{61}$$

If Γ is regular and a stronger condition

$$\text{Span}\{A_i \mid 0 \le i \le \text{diam}(\Gamma)\} \subseteq \mathcal{A}_1 \tag{62}$$

holds for Γ, we call Γ *distance-polynomial*. In other words, Γ is *distance-polynomial* if each distance matrix A_i is a polynomial in the adjacency matrix A of Γ: this is the original definition, in which Γ is not assumed to be regular, but since the super-regularity, particularly the regularity, is derived from the original definition [22], we can add the regularity to the defining condition from the beginning. A distance-polynomial graph Γ is called *distance-regular* if the equality holds in (62), that is,

$$\mathcal{A}_1 = \text{Span}\{A_i \mid 0 \le i \le \text{diam}(\Gamma)\}. \tag{63}$$

Observe that a distance-polynomial graph Γ is distance-regular if and only if the equality holds in (54), that is, $\text{diam}(\Gamma) = D$. This is because $\dim \mathcal{A}_1 = D + 1$ by (52). Also observe that the condition (63) implies that $\mathcal{A}_1$ is closed under the Hadamard product and hence $\mathcal{A}_1 = \mathcal{A}_2$.

Therefore if a graph Γ is distance-regular, then Γ is polynomial, and if a graph Γ is polynomial, then Γ is distance-polynomial. We are interested in the gap between the class of distance-regular graphs and that of polynomial graphs, since the gap is formulated in terms of symmetric association schemes.

We start with a symmetric association scheme $\mathfrak{X} = (X, \{R_\alpha\}_{\alpha \in \Lambda})$. The Bose–Mesner algebra $\mathcal{M}$ of $\mathfrak{X}$ is

$$\mathcal{M} = \mathrm{Span}\{A_\alpha \mid \alpha \in \Lambda\} \tag{64}$$

$$= \mathrm{Span}\{E_\lambda \mid \lambda \in \Lambda\}, \tag{65}$$

where $\{A_\alpha \mid \alpha \in \Lambda\}$ is the set of primitive idempotents of $\mathcal{M}$ with respect to the Hadamard product, which is characterized by (57), and $\{E_\lambda \mid \lambda \in \Lambda\}$ is the set of primitive idempotents of $\mathcal{M}$ with respect to the ordinary matrix product:

$$E_\lambda E_\mu = \delta_{\lambda\mu} E_\lambda, \quad \sum_{\lambda \in \Lambda} E_\lambda = I. \tag{66}$$

Note that there is a special $\lambda_0 \in \Lambda$ for which

$$E_{\lambda_0} = \frac{1}{|X|} J. \tag{67}$$

Compare (67) with (59). The first eigenmatrix $P = (p_\alpha(\lambda))_{\alpha,\lambda \in \Lambda}$ and the second eigenmatrix $Q = (q_\lambda(\alpha))_{\lambda,\alpha \in \Lambda}$ are the transition matrices between the two bases of $\mathcal{M}$:

$$A_\alpha = \sum_{\lambda \in \Lambda} p_\alpha(\lambda) E_\lambda, \tag{68}$$

$$E_\lambda = \frac{1}{|X|} \sum_{\alpha \in \Lambda} q_\lambda(\alpha) A_\alpha, \tag{69}$$

$$PQ = QP = |X| I. \tag{70}$$

For a subset Λ_1 of Λ, set

$$A_{\Lambda_1} = \sum_{\alpha \in \Lambda_1} A_\alpha, \tag{71}$$

$$p_{\Lambda_1}(\lambda) = \sum_{\alpha \in \Lambda_1} p_\alpha(\lambda). \tag{72}$$

By (65), (66) and $A_{\Lambda_1} = \sum_{\lambda \in \Lambda} p_{\Lambda_1}(\lambda) E_\lambda$, we immediately have the following lemma.

Lemma 6.2 *The $(0, 1)$ matrix A_{Λ_1} generates the subalgebra $\mathcal{M}$ of $M(\mathbb{C})$ if and only if $p_{\Lambda_1}(\lambda)$, $\lambda \in \Lambda$, are distinct.*

Given the first eigenmatrix $P = (p_\alpha(\lambda))_{\alpha,\lambda \in \Lambda}$, we can check in principle whether the symmetric association scheme is polynomial or not, using this lemma.

We now return to a polynomial association scheme $\mathfrak{X} = (X, \{R_\alpha\}_{\alpha \in \Lambda})$. The Bose–Mesner algebra of $\mathfrak{X}$ is $\mathcal{M} = \mathrm{Span}\{A_\alpha \mid \alpha \in \Lambda\}$, where A_α, $\alpha \in \Lambda$, is the relation matrix of R_α. By the definition of a polynomial association scheme, $\mathfrak{X}$ is a symmetric association scheme and its Bose–Mesner algebra $\mathcal{M}$ satisfies the following property: there exists a $(0, 1)$ matrix $A \in \mathcal{M}$ such that $A \circ I = 0$ and

$$A_\alpha = v_\alpha(A), \ \alpha \in \Lambda \tag{73}$$

for some polynomial $v_\alpha(x) \in \mathbb{C}[x]$. Let $\varphi(x)$ denote the minimal polynomial of A. Set $\deg \varphi(x) = D + 1$: the degree of $\varphi(x)$ is $D + 1$. We may assume

$$\deg v_\alpha(x) \le D, \ \alpha \in \Lambda. \tag{74}$$

Regarding $\mathcal{M}$ as a subalgebra of $M_X(\mathbb{C})$, we have

$$\mathcal{M} = \langle A \rangle = \operatorname{Span}\{A^i \,|\, 0 \le i \le D\}$$
$$\simeq \mathbb{C}[x]/(\varphi(x)) = \langle x \rangle = \operatorname{Span}\{x^i \,|\, 0 \le i \le D\},$$

where $(\varphi(x))$ is the ideal of the polynomial ring $\mathbb{C}[x]$ generated by $\varphi(x)$, and

$$\mathcal{M} = \operatorname{Span}\{v_\alpha(A) \,|\, \alpha \in \Lambda\} \tag{75}$$
$$\simeq \mathbb{C}[x]/(\varphi(x)) = \operatorname{Span}\{v_\alpha(x) \,|\, \alpha \in \Lambda\}. \tag{76}$$

Since $(A - kI)J = 0$ and $J = \sum_{\alpha \in \Lambda} A_\alpha = \sum_{\alpha \in \Lambda} v_\alpha(A)$, we have

$$\varphi(x) = \frac{1}{c}(x - k) \sum_{\alpha \in \Lambda} v_\alpha(x), \tag{77}$$

where k is the constant row sum of A and c is the leading coefficient of $\sum_{\alpha \in \Lambda} v_\alpha(x)$.

Let Γ be the polynomial graph for which the adjacency matrix is A. Define a partition

$$\Lambda = \bigcup_{0 \le i \le \operatorname{diam}(\Gamma)} \Lambda_i \tag{78}$$

of Λ by setting inductively as

$$\Lambda_0 = \{\alpha_0\}, A_{\alpha_0} = I, \tag{79}$$
$$\Lambda_i = \{\alpha \in \Lambda \,|\, A^i \circ A_\alpha \ne 0\} - (\Lambda_0 \cup \Lambda_1 \cup \cdots \cup \Lambda_{i-1}). \tag{80}$$

Then the ith distance matrix of Γ from (17) is given by

$$A_{\Lambda_i} := \sum_{\alpha \in \Lambda_i} A_\alpha, \quad 0 \le i \le \operatorname{diam}(\Gamma). \tag{81}$$

Note that the notation of the ith distance matrix is changed in (81) to be A_{Λ_i}.

The polynomial graph Γ becomes distance-regular if and only if $\operatorname{diam}(\Gamma) = D$, which is equivalent to the condition

$$|\Lambda_i| = 1, \quad 0 \le i \le \operatorname{diam}(\Gamma), \tag{82}$$

because $|\Lambda| = D + 1$ by (60). In this case, it is well known that

$$\Lambda_i = \{\alpha_i\}, \ \deg v_{\alpha_i}(x) = i, \quad 0 \le i \le D. \tag{83}$$

The polynomial association scheme $\mathfrak{X}$ is called *P-polynomial* if the polynomial graph Γ, from which $\mathfrak{X}$ arises, is distance-regular. In terms of the first eigenmatrix $P = (p_\alpha(\lambda))_{\alpha,\lambda\in\Lambda}$, a symmetric association scheme $\mathfrak{X}$ is *P-polynomial* if and only if there exists an ordering $\alpha_0, \alpha_1, \ldots, \alpha_D$ of the indexes in Λ ($D = |\Lambda| - 1$) such that $p_{\alpha_i}(\lambda) = v_{\alpha_i}(p_{\alpha_1}(\lambda))$ for some polynomial $v_{\alpha_i}(x)$ of degree $i, 0 \le \forall i \le D$. It seems that Lemma 6.2 suggests a huge gap between the class of polynomial association schemes and that of *P*-polynomial association schemes. However, when we check distribution diagrams of generously transitive groups, we are inclined to think the following conjecture plausibly holds for $N = 8$.

Conjecture 6.3 *There exists an absolute constant N such that any polynomial graph Γ that satisfies $|\Lambda_i| = 1$, $0 \le i \le N$, for Λ_i from (80) is distance-regular:*

$$|\Lambda_i| = 1, \ 0 \le i \le N \Longrightarrow |\Lambda_i| = 1, \ \forall i.$$

In the rest of this final section, we discuss co-polynomial association schemes as the dual objects of polynomial association schemes, tracing the previous arguments with $\{A_\alpha \mid \alpha \in \Lambda\}$ and $\{E_\lambda \mid \lambda \in \Lambda\}$ interchanged.

Definition 6.4 *A symmetric association scheme $\mathfrak{X} = (X, \{R_\alpha\}_{\alpha\in\Lambda})$ is called co-polynomial if its Bose–Mesner algebra $\mathcal{M}$ is generated as a subalgebra of $M_X(\mathbb{C})^\circ$ by some $E \in \mathcal{M}$ with the property that $EE = E$ and $EJ = 0$. An element $E \in \mathcal{M}$ is called an idempotent if $EE = E$ holds with respect to the ordinary matrix product.*

Let $\mathfrak{X} = (X, \{R_\alpha\}_{\alpha\in\Lambda})$ be a symmetric association scheme. Let $\mathcal{M}$ be the Bose–Mesner algebra of $\mathfrak{X}$. Then $\mathcal{M}$ is given by (64), (65) with the properties (57), (59), (66), (67):

$$\mathcal{M} = \mathrm{Span}\{A_\alpha \mid \alpha \in \Lambda\} = \mathrm{Span}\{E_\lambda \mid \lambda \in \Lambda\},$$

$$A_\alpha \circ A_\beta = \delta_{\alpha\beta} A_\alpha, \quad \sum_{\alpha\in\Lambda} A_\alpha = J, \quad A_{\alpha_0} = I,$$

$$E_\lambda E_\mu = \delta_{\lambda\mu} E_\lambda, \quad \sum_{\lambda\in\Lambda} E_\lambda = I, \quad E_{\lambda_0} = \frac{1}{|X|} J.$$

Let $P = (p_\alpha(\lambda))_{\alpha,\lambda\in\Lambda}$, and $Q = (q_\lambda(\alpha))_{\lambda,\alpha\in\Lambda}$ be the first eigenmatrix and the second eigenmatrix of $\mathfrak{X}$, respectively. So by (68), (69), (70)

$$A_\alpha = \sum_{\lambda\in\Lambda} p_\alpha(\lambda) E_\lambda, \quad E_\lambda = \frac{1}{|X|} \sum_{\alpha\in\Lambda} q_\lambda(\alpha) A_\alpha, \quad PQ = QP = |X|I.$$

For a subset Λ_1 of Λ, set

$$E_{\Lambda_1} = \sum_{\lambda \in \Lambda_1} E_\lambda, \quad q_{\Lambda_1}(\alpha) = \sum_{\lambda \in \Lambda_1} q_\lambda(\alpha).$$

By (64), (57) and $E_{\Lambda_1} = \frac{1}{|X|} \sum_{\alpha \in \Lambda} q_{\Lambda_1}(\alpha) A_\alpha$, we immediately have the following.

Lemma 6.5 *The idempotent matrix E_{Λ_1} generates the subalgebra $\mathcal{M}$ of $M(\mathbb{C})^\circ$ if and only if $q_{\Lambda_1}(\alpha)$, $\alpha \in \Lambda$, are distinct.*

Given the second eigenmatrix $Q = (q_\lambda(\alpha))_{\lambda, \alpha \in \Lambda}$, we can check in principle whether the symmetric association scheme is co-polynomial or not, using this lemma.

Now we assume that the symmetric association scheme $\mathfrak{X} = (X, \{R_\alpha\}_{\alpha \in \Lambda})$ is co-polynomial. By the definition of a co-polynomial association scheme, there exists an idempotent $E \in \mathcal{M}$ such that $EJ = 0$ and, in the subalgebra $\mathcal{M}$ of $M_X(\mathbb{C})^\circ$,

$$E_\lambda = \frac{1}{|X|} v_\lambda^*(|X|E), \ \lambda \in \Lambda \tag{84}$$

for some polynomial $v_\lambda^*(x) \in \mathbb{C}[x]$. Let $\varphi^*(x)$ denote the minimal polynomial of $|X|E$ in the subalgebra $\mathcal{M}$ of $M_X(\mathbb{C})^\circ$. Set $\deg \varphi^*(x) = D^* + 1$. So

$$\dim \mathcal{M} = D^* + 1 = |\Lambda|. \tag{85}$$

We may assume

$$\deg v_\lambda^*(x) \le D^*, \ \lambda \in \Lambda. \tag{86}$$

Regarding $\mathcal{M}$ as a subalgebra of $M_X(\mathbb{C})^\circ$, which is denoted by $\mathcal{M}^\circ$, we have

$$\mathcal{M}^\circ = \langle E \rangle^\circ = \mathrm{Span}\{(|X|E)^{\circ i} \mid 0 \le i \le D^*\}$$
$$\simeq \mathbb{C}[x]/(\varphi^*(x)) = \langle x \rangle = \mathrm{Span}\{x^i \mid 0 \le i \le D^*\},$$

where $(|X|E)^{\circ i}$ is the ith power of $|X|E$ with respect to the Hadamard product, and

$$\mathcal{M}^\circ = \mathrm{Span}\{v_\lambda^*(|X|E) \mid \lambda \in \Lambda\} \tag{87}$$
$$\simeq \mathbb{C}[x]/(\varphi^*(x)) = \mathrm{Span}\{v_\lambda^*(x) \mid \lambda \in \Lambda\}. \tag{88}$$

Since $(|X|E - mJ) \circ I = 0$ and $I = \sum_{\lambda \in \Lambda} E_\lambda = \frac{1}{|X|} \sum_{\lambda \in \Lambda} v_\lambda^*(|X|E)$, we have

$$\varphi^*(x) = \frac{1}{c^*}(x - m) \sum_{\lambda \in \Lambda} v_\lambda^*(x), \tag{89}$$

where m is the rank of E and c^* is the leading coefficient of $\sum_{\lambda \in \Lambda} v_\lambda^*(x)$.

Define the subset Λ_i^* of Λ inductively by

$$\Lambda_0^* = \{\lambda_0\}, \quad E_{\lambda_0} = \frac{1}{|X|}J, \tag{90}$$

$$\Lambda_i^* = \{\lambda \in \Lambda \mid E^{\circ i} E_\lambda \neq 0\} - (\Lambda_0^* \cup \Lambda_1^* \cup \cdots \cup \Lambda_{i-1}^*). \tag{91}$$

The *co-diameter* of $\mathfrak{X}$ is defined to be

$$\mathrm{diam}^*(\mathfrak{X}) = \max\{i \in \mathbb{Z}_{>0} \mid \Lambda_i^* \neq \emptyset\}. \tag{92}$$

Then we get a partition

$$\Lambda = \bigcup_{0 \leq i \leq \mathrm{diam}^*(\mathfrak{X})} \Lambda_i^* \tag{93}$$

and the *ith co-distance matrix*

$$E_{\Lambda_i^*} := \sum_{\lambda \in \Lambda_i^*} E_\lambda, \quad 0 \leq i \leq \mathrm{diam}^*(\mathfrak{X}). \tag{94}$$

Note that

$$\mathrm{diam}^*(\mathfrak{X}) \leq D^* \tag{95}$$

and compare it with (54).

The co-polynomial association scheme $\mathfrak{X}$ is called *Q-polynomial* if $\mathrm{diam}^*(\mathfrak{X}) = D^*$, which is equivalent to the condition

$$|\Lambda_i^*| = 1, \quad 0 \leq i \leq \mathrm{diam}^*(\mathfrak{X}), \tag{96}$$

because $|\Lambda| = D^* + 1$ by (85). In this case, it is well known that

$$\Lambda_i^* = \{\lambda_i\}, \quad \deg v_{\lambda_i}^*(x) = i, \quad 0 \leq i \leq D^*. \tag{97}$$

In terms of the second eigenmatrix $Q = (q_\lambda(\alpha))_{\lambda, \alpha \in \Lambda}$, a symmetric association scheme $\mathfrak{X}$ is Q-polynomial if and only if there exists an ordering $\lambda_0, \lambda_1, \ldots, \lambda_{D^*}$ of the indexes in Λ ($D^* = |\Lambda| - 1$) such that $q_{\lambda_i}(\alpha) = v_{\lambda_i}^*(q_{\lambda_1}(\alpha))$ for some polynomial $v_{\lambda_i}^*(x)$ of degree i, $0 \leq \forall i \leq D^*$. It seems that Lemma 6.5 suggests a huge gap between the class of co-polynomial association schemes and that of Q-polynomial association schemes. We are not sure if the dual version of Corollary 6.3 holds, since we have not yet started to collect examples of co-polynomial association schemes.

Question 6.6 *Does there exist an absolute constant N^* such that any co-polynomial association scheme $\mathfrak{X}$ that satisfies $|\Lambda_i^*| = 1$, $0 \leq i \leq N^*$, for Λ_i^* from (91) is Q-polynomial? In other words,*

$$|\Lambda_i^*| = 1, \; 0 \leq i \leq N^* \implies |\Lambda_i^*| = 1, \; \forall i \; ?$$

Recently, Jack Koolen proved the following ([14] personal communication).

Theorem 6.7   *A P-polynomial association scheme is co-polynomial and a Q-polynomial association scheme is polynomial.*

He conjectures (personal communication) as follows.

Conjecture 6.8   *Symmetric association schemes are co-polynomial if they do not have non-trivial fusion schemes.*

For fusion schemes, readers are referred to [2, section 2.7.2] and [4]. We close this expository chapter with the well-known conjecture [1]:

Conjecture 6.9   *In the class of primitive symmetric association schemes $\mathfrak{X} = (X, \{R_\alpha\}_{\alpha \in \Lambda})$ with sufficiently large $|\Lambda|$, $\mathfrak{X}$ is P-polynomial if and only if $\mathfrak{X}$ is Q-polynomial.*

Acknowledgements

I am grateful to the referee who read the manuscript carefully and gave many helpful comments and suggestions. His pointing out the reference [4] has resulted in a substantial improvement to Conjecture 6.8.

References

[1] E. Bannai and T. Ito, *Algebraic Combinatorics I: Association Schemes*, Benjamin/Cummings Lecture Note Series, Menlo Park, 1984.

[2] E. Bannai, E. Bannai, T. Ito and R. Tanaka, *Algebraic Combinatorics*, De Gruyter Series in Discrete Mathematics and Applications, Vol. 5, 2021.

[3] N. L. Biggs, *Algebraic Graph Theory*, Cambridge University Press, 1974.

[4] E. R. van Dam and M. Muzychuk, Some implications on amorphic association schemes, *J. Comb. Theory* (A) 117 (2010) 111–127.

[5] S. Evdokimov, I. Ponomarenko and G. Tinhofer, Forestal algebras and algebraic forests (on a new class of weakly compact graphs), *Discrete Math.* 225 (2000) 149–172.

[6] I. A. Faradjev, 'Symmetry vs Regularity': How it started and what it led to, www.iti.zcu.cz/wl2018/slides.html.

[7] I. A. Faradžev, M. H. Klin and M. E. Muzychuk, Cellular rings and groups of automorphisms of graphs, in *Investigations in Algebraic Theory of Combinatorial Objects*, I. A. Faradžev et al. (eds.), Kluwer Acad. Publ., Dordrecht, 1994, pp. 1–152.

[8] D. Gijswijt, A. Schrijver and H. Tanaka, New upper bounds for nonbinary codes based on the Terwilliger algebra and semidefinite programming, *J. Comb. Theory* (A) 113 (2006) 1719–1731.

[9] M. Grohe, *Descriptive Complexity, Canonisation, and Definable Graph Structure Theory*, Cambridge University Press, 2017.

[10] D. G. Higman, Coherent configurations, I, Rend. Semin. Mat. Univ. Padova 44 (1970) 1–25.

[11] D. G. Higman, Coherent configurations, I, Ordinary representation theory, *Geom. Dedic.* 4 (1975) 1–32.

[12] D. G. Higman, Coherent configurations, II, Weights, *Geom. Dedic.* 5 (1976) 413–424.

[13] D. G. Higman, Coherent algebras, *Linear Algebra Appl.* 93 (1987) 209–239.

[14] T.-T. Xia, Y.-Y. Tan, X. Liang and J. Koolen, On association schemes generated by a relation or an idempotent, preprint.

[15] S.-D. Li, Y.-Z. Fan, T. Ito, M. Karimi and J. Xu, The isomorphism problem of trees from the viewpoint of Terwilliger algebras, *J. Comb. Theory* (A) 177 (2021) 105328.

[16] M. Lichter, I. Ponomarenko and P. Schweitzer, Walk refinement, walk logic, and the iteration number of the Weisfeiler–Leman algorithm, *Proceedings of the 34th Annual ACM/IEEE Symposium on Logic in Computer Science*, June 2019, Article No. 33, pp. 1–13.

[17] I. Ponomarenko, The original paper by Weisfeiler and Leman: The preface, www.iti.zcu.cz/wl2018/wlpaper.html.

[18] H. Suzuki, The Terwilliger algebra associated with a set of vertices in a distance-regular graph, *J. Algebraic Comb.* 22 (2005) 5–38.

[19] Y.-Y. Tan, Y.-Z. Fan, T. Ito and X.-Y. Liang, The Terwilliger algebra of the Johnson scheme $J(N, D)$ revisited from the viewpoint of group representations, *Eur. J. Comb.* 80 (2019) 157–171.

[20] P. Terwilliger, The subconstituent algebra of an association scheme, I, II, III, *J. Algebraic Comb.* 1 (1992) 363–388; 2 (1993) 73–103; 2 (1993) 177–210.

[21] P. Terwilliger, Algebraic Graph Theory, unpublished lecture notes (1993).

[22] P. M. Weichsel, On distance-regularity in graphs, *J. Comb. Theory* (B) 32 (1982) 156–161.

[23] B. Weisfeiler, *On Construction and Identification of Graphs*, Lecture Notes in Mathematics, Vol. 558, Springer-Verlag, Berlin, 1976.

[24] J. Xu, T. Ito, and S.-D. Li, Irreducible representations of the Terwilliger algebra of a tree, *Graphs Comb.* 37 (2021) 1749–1773.

[25] J. Xu, T. Ito and S.-D. Li, The Weisfeiler–Leman stabilization of a tree, in preparation.

13

Extended Double Covers of Non-Symmetric Association Schemes of Class 2

Takuya Ikuta and Akihiro Munemasa

Dedicated to the memory of Igor Faradžev

Abstract

In this chapter, we give a method to construct non-symmetric association schemes of class 3 from non-symmetric association schemes of class 2. This construction is a non-symmetric analogue of the construction of Taylor graphs as an antipodal double cover of a complete graph. We also mention how our construction interacts with doubling introduced by Pasechnik.

CONTENTS

13.1 Introduction

The first eigenmatrix of a non-symmetric association scheme of class 2 on m points is given by

$$
\begin{bmatrix}
1 & \frac{m-1}{2} & \frac{m-1}{2} \\
1 & \frac{-1+\sqrt{-m}}{2} & \frac{-1-\sqrt{-m}}{2} \\
1 & \frac{-1-\sqrt{-m}}{2} & \frac{-1+\sqrt{-m}}{2}
\end{bmatrix},
\tag{1}
$$

* This work was supported by JSPS KAKENHI grant number 20K03527.

where $m \equiv 3 \pmod 4$ (see, for example [8]). The digraphs defined by a non-trivial relation of such an association scheme are known as doubly regular tournaments, and the existence of such a digraph is equivalent to that of a skew Hadamard matrix. See [7] for details.

In this chapter it is shown that a non-symmetric association scheme with the first eigenmatrix (1) gives rise to an association scheme of class 3 with the first eigenmatrix

$$
\begin{bmatrix}
1 & m & m & 1 \\
1 & \sqrt{-m} & -\sqrt{-m} & -1 \\
1 & -\sqrt{-m} & \sqrt{-m} & -1 \\
1 & -1 & -1 & 1
\end{bmatrix}.
\tag{2}
$$

We call the resulting association scheme with the first eigenmatrix (2), the *extended double cover* of the original association scheme of class 2. The construction is analogous to that of Taylor graphs (see [2, sect. 1.5]). In fact, the association scheme defined by a Taylor graph has the first eigenmatrix

$$
\begin{bmatrix}
1 & m & m & 1 \\
1 & \sqrt{m} & -\sqrt{m} & -1 \\
1 & -\sqrt{m} & \sqrt{m} & -1 \\
1 & -1 & -1 & 1
\end{bmatrix}.
\tag{3}
$$

The extended double cover is a non-symmetric fission of a cocktail party graph, and self-dual. According to S. Y. Song [9, (5.3) lemma], a self-dual non-symmetric fission of a complete multipartite graph has the first eigenmatrix

$$
\begin{bmatrix}
1 & k_1 & k_1 & k_2 \\
1 & \sqrt{-k_1} & -\sqrt{-k_1} & -1 \\
1 & -\sqrt{-k_1} & \sqrt{-k_1} & -1 \\
1 & -\frac{k_2+1}{2} & -\frac{k_2+1}{2} & k_2
\end{bmatrix},
$$

and our association scheme with the first eigenmatrix (2) is a special case where $k_2 = 1$.

According to I. A. Faradžev, M. H. Klin, and M. E. Muzichuk [4, theorem 2.6.6], D. Pasechnik invented a construction of a non-symmetric association scheme of class 2 on $2m + 1$ points with the first eigenmatrix

$$
\begin{bmatrix}
1 & m & m \\
1 & \frac{-1+\sqrt{-(2m+1)}}{2} & \frac{-1-\sqrt{-(2m+1)}}{2} \\
1 & \frac{-1-\sqrt{-(2m+1)}}{2} & \frac{-1+\sqrt{-(2m+1)}}{2}
\end{bmatrix},
\tag{4}
$$

provided that there exists an association scheme with first eigenmatrix (1). This construction leads to the doubling of skew-Hadamard matrices (see [7, theorem 14]). We will give a direct description of the extended double cover of the doubling in Theorem 4.5.

One may wonder if an association scheme with first eigenmatrix (2) is related to a skew-Hadamard matrix. However, by [6], such an association scheme does not contain a (complex) Hadamard matrix in its Bose–Mesner algebra.

The organization of this chapter is as follows. In Section 13.2 we introduce necessary notation and give useful properties of non-symmetric association schemes of class 2. In Section 13.3 we construct the adjacency matrices needed in our main theorem, and prove their multiplication formulas. In Section 13.4 we present our main results.

13.2 Preliminaries

For fundamentals of the theory of association schemes, we refer the reader to [1]. Let $\mathfrak{X} = (X, \{R_i\}_{i=0}^d)$ be a commutative association scheme of class d on n points. Let $A_0, A_1, \ldots, A_d$ be the adjacency matrices of $\mathfrak{X}$. The intersection numbers $p_{i,j}^\ell$ are defined by

$$A_i A_j = \sum_{\ell=0}^{d} p_{i,j}^\ell A_\ell,$$

and the intersection matrices $\{B_i\}_{i=0}^d$ are defined by $(B_i)_{j,\ell} = p_{i,j}^\ell$. The linear span $\mathcal{A} = \langle A_0, A_1, \ldots, A_d \rangle$ is called the *Bose–Mesner algebra* of $\mathfrak{X}$, and it has primitive idempotents $E_0 = \frac{1}{|X|}J, E_1, \ldots, E_d$. The first eigenmatrix $P = (P_{i,j})_{0 \leq i,j \leq d}$ is defined by

$$(A_0, A_1, \ldots, A_d) = (E_0, E_1, \ldots, E_d)P,$$

and $Q = |X|P^{-1}$ is called the second eigenmatrix of $\mathfrak{X}$. Then we have

$$|X|(E_0, E_1, \ldots, E_d) = (A_0, A_1, \ldots, A_d)Q.$$

Let k_i ($i = 0, 1, \ldots, d$) and m_i ($i = 0, 1, \ldots, d$) be the valencies and the multiplicities of $\mathfrak{X}$, respectively. Then the intersection numbers $p_{i,j}^\ell$ are given by

$$p_{i,j}^\ell = \frac{1}{nk_\ell} \sum_{v=0}^{d} m_v \overline{P_{v,i}} \; \overline{P_{v,j}} P_{v,\ell} \tag{5}$$

Now, assume that $\mathfrak{X}$ is a non-symmetric association scheme of class 2 on n points. Then $k_1 = k_2 = m_1 = m_2 = (m-1)/2$. By (1) and (5) we have

$$A_1^2 = \frac{m-3}{4} A_1 + \frac{m+1}{4} A_2, \tag{6}$$

$$A_2^2 = \frac{m+1}{4} A_1 + \frac{m-3}{4} A_2, \tag{7}$$

$$A_1 A_2 = \frac{m-1}{2} A_0 + \frac{m-3}{4} (A_1 + A_2). \tag{8}$$

Then by (6), (7), and (8) we have

$$A_1^2 + A_2^2 = \frac{m-1}{2} (A_1 + A_2), \tag{9}$$

$$J + 2A_1 A_2 = mA_0 + \frac{m-1}{2} (A_1 + A_2). \tag{10}$$

13.3 Construction of Adjacency Matrices

Definition 3.1   Let $\mathfrak{X}$ be a non-symmetric association scheme of class 2 on m points, and denote its adjacency matrices by $A_0 = I_m, A_1, A_2$. Define

$$C_0 = I_{2(m+1)}, \tag{11}$$

$$C_1 = \begin{bmatrix} 0 & 1 & 0 & 0 \\ 0 & A_1 & A_2 & 1^\top \\ 1^\top & A_2 & A_1 & 0 \\ 0 & 0 & 1 & 0 \end{bmatrix}, \tag{12}$$

$$C_2 = C_1^\top, \tag{13}$$

$$C_3 = J - C_0 - C_1 - C_2, \tag{14}$$

where 1 is the all-one row vector of length m. The association scheme defined by the set of adjacency matrices $\{C_0, C_1, C_2, C_3\}$ is called the *extended double cover* of $\mathfrak{X}$.

We will show in the next section that $\{C_0, C_1, C_2, C_3\}$ is indeed the set of adjacency matrices of an association scheme of class 3.

By (11)–(14), we have

$$C_2 = \begin{bmatrix} 0 & 0 & 1 & 0 \\ 1^\top & A_2 & A_1 & 0 \\ 0 & A_1 & A_2 & 1^\top \\ 0 & 1 & 0 & 0 \end{bmatrix}, \tag{15}$$

$$C_3 = \begin{bmatrix} 0 & 0 & 0 & 1 \\ 0 & 0 & I_m & 0 \\ 0 & I_m & 0 & 0 \\ 1 & 0 & 0 & 0 \end{bmatrix}. \tag{16}$$

Remark 3.2   In Definition 3.1, if A_1 and A_2 define isomorphic digraphs, then it is easy to see that C_1 and C_2 define isomorphic digraphs.

The next lemma is necessary in the next section in order to establish our main result.

Lemma 3.3   *We have the following:*

$$C_1^2 = C_2^2 = \frac{m-1}{2}(C_1 + C_2) + mC_3, \tag{17}$$

$$C_1C_2 = C_2C_1 = mC_0 + \frac{m-1}{2}(C_1 + C_2), \tag{18}$$

$$C_1C_3 = C_3C_1 = C_2, \tag{19}$$

$$C_2C_3 = C_3C_2 = C_1, \tag{20}$$

$$C_3^2 = C_0. \tag{21}$$

Proof   We can easily see that (19) and (21) hold. Then (20) follows immediately from (19) and (21).

Since

$$(C_1^2)_{1,1} = (C_2^2)_{4,4} = 0,$$

$$(C_1^2)_{1,4} = (C_2^2)_{4,1} = m,$$

$$(C_1^2)_{1,2} = (C_1^2)_{1,3} = (C_2^2)_{4,2} = (C_2^2)_{4,3} = \frac{m-1}{2}\mathbf{1},$$

$$(C_1^2)_{2,1} = (C_1^2)_{2,4} = (C_2^2)_{3,1} = (C_2^2)_{3,4} = \frac{m-1}{2}\mathbf{1}^\top,$$

$$(C_1^2)_{2,2} = (C_1^2)_{3,3} = A_1^2 + A_2^2 = \frac{m-1}{2}(A_1 + A_2) \qquad \text{(by (9))},$$

$$(C_1^2)_{3,2} = (C_2^2)_{2,3} = J + 2A_1A_2 = mI_m + \frac{m-1}{2}(A_1 + A_2) \quad \text{(by (10))},$$

we have $C_1^2 = \frac{m-1}{2}(C_1+C_2)+mC_3$. Similarly, we have $C_2^2 = \frac{m-1}{2}(C_1+C_2)+mC_3$. Finally, (18) follows from (17) and (19)–(21).   □

13.4 Main results

Theorem 4.1   *The extended double cover of a non-symmetric association scheme of class 2 on m points is an association scheme with the first eigenmatrix* (2).

Proof   Suppose $\mathfrak{X}$ is a non-symmetric association scheme of class 2 on m points. Let C_0, C_1, C_2, C_3 be the matrices given in Definition 3.1, and let $\mathcal{A} = \langle C_0, C_1, C_2, C_3 \rangle$ be their linear span over the field of complex numbers. First we observe that $\mathcal{A}$ is closed under multiplication by Lemma 3.3. Thus $\mathcal{A}$ is the Bose–Mesner algebra of an association scheme $\tilde{\mathfrak{X}}$ of class 3.

Secondly we compute the first eigenmatrix of $\tilde{\mathfrak{X}}$. By Lemma 3.3 the intersection matrices B_1, B_2, B_3 of $\tilde{\mathfrak{X}}$ are given by

$$B_1 = \begin{bmatrix} 0 & 1 & 0 & 0 \\ 0 & \frac{m-1}{2} & \frac{m-1}{2} & m \\ m & \frac{m-1}{2} & \frac{m-1}{2} & 0 \\ 0 & 0 & 1 & 0 \end{bmatrix}, \tag{22}$$

$$B_2 = \begin{bmatrix} 0 & 0 & 1 & 0 \\ m & \frac{m-1}{2} & \frac{m-1}{2} & 0 \\ 0 & \frac{m-1}{2} & \frac{m-1}{2} & m \\ 0 & 1 & 0 & 0 \end{bmatrix}, \tag{23}$$

$$B_3 = \begin{bmatrix} 0 & 0 & 0 & 1 \\ 0 & 0 & 1 & 0 \\ 0 & 1 & 0 & 0 \\ 1 & 0 & 0 & 0 \end{bmatrix}. \tag{24}$$

Let

$$E_0 = \frac{1}{2(m+1)}(C_0 + C_1 + C_2 + C_3), \tag{25}$$

$$E_1 = \frac{1}{4}(C_0 - \frac{1}{\sqrt{m}}(C_1 - C_2) - C_3), \tag{26}$$

$$E_2 = \frac{1}{4}(C_0 + \frac{1}{\sqrt{m}}(C_1 - C_2) - C_3) \tag{27}$$

$$E_3 = \frac{1}{2(m+1)}(mC_0 - (C_1 + C_2) + mC_3). \tag{28}$$

Then by (22)–(24) we have $E_i E_j = \delta_{i,j} E_i$. By (25)–(28) the second eigenmatrix of $\tilde{\mathfrak{X}}$ is given by

$$\begin{bmatrix} 1 & \frac{m+1}{2} & \frac{m+1}{2} & m \\ 1 & -\frac{m+1}{2\sqrt{m}} & \frac{m+1}{2\sqrt{m}} & -1 \\ 1 & \frac{m+1}{2\sqrt{m}} & -\frac{m+1}{2\sqrt{m}} & -1 \\ 1 & -\frac{m+1}{2} & -\frac{m+1}{2} & m \end{bmatrix}.$$

Then the first eigenmatrix of $\tilde{\mathfrak{X}}$ is given by (2). $\qquad\square$

Remark 4.2   In the database of [5], as08[6], as16[11], and as24[14] can be constructed by Theorem 4.1.

Remark 4.3   Let $\{A_0, A_1, A_2\}$ be the set of the adjacency matrices of a symmetric association scheme of class 2 with $k = 2\mu$ on m points, and

$$D_0 = I_{2(m+1)},$$

$$D_1 = \begin{bmatrix} 0 & 1 & 0 & 0 \\ 1^\top & A_1 & A_2 & 0 \\ 0^\top & A_2 & A_1 & 1^\top \\ 0 & 0 & 1 & 0 \end{bmatrix},$$

$$D_2 = \begin{bmatrix} 0 & 0 & 1 & 0 \\ 0 & A_2 & A_1 & 1^\top \\ 1^\top & A_1 & A_2 & 0 \\ 0 & 1 & 0 & 0 \end{bmatrix},$$

$$D_3 = J - D_0 - D_1 - D_2.$$

Then, D_1 is the adjacency matrix of a Taylor graph (see [2, sect. 1.5]), and its first eigenmatrix is given by (3). In this sense, the extended double cover can be regarded as a non-symmetric analogue of Taylor graphs.

The following construction is due to D. Pasechnik (announced in [3]; see [4, theorem 2.6.6]).

Definition 4.4 Let $\mathfrak{X}$ be a non-symmetric association scheme of class 2 on m points, and denote its adjacency matrices by $A_0 = I_m, A_1, A_2$. Define

$$\tilde{A}_0 = I_{2m+1}, \; \tilde{A}_1 = \begin{bmatrix} 0 & 1 & 0 \\ 0 & A_1 & A_2 + I_m \\ 1^\top & A_2 & A_2 \end{bmatrix}, \; \tilde{A}_2 = \begin{bmatrix} 0 & 0 & 1 \\ 1^\top & A_2 & A_1 \\ 0 & A_1 + I_m & A_1 \end{bmatrix},$$

where $A_2^\top = A_1$. The association scheme with adjacency matrices $\{\tilde{A}_0, \tilde{A}_1, \tilde{A}_2\}$ is called the *doubling* of $\mathfrak{X}$.

In Definition 4.4, even if A_1 and A_2 define isomorphic digraphs, the matrices $\tilde{A}_1$ and $\tilde{A}_2$ may define non-isomorphic digraphs.

Since the doubling of a non-symmetric association scheme of class 2 on m points is a non-symmetric association scheme of class 2 on $2m + 1$ points, we can construct the extended double cover on $4(m + 1)$ points by Theorem 4.1. Then we have the following.

Theorem 4.5 *Let $\{C_0, C_1, C_2, C_3\}$ be set of the adjacency matrices of the extended double cover of a non-symmetric association scheme $\mathfrak{X}$ of class 2 on m points. Then the set of adjacency matrices of the extended double cover of the doubling of $\mathfrak{X}$ is given by $\{C_0', C_1', C_2', C_3'\}$, where*

$$C'_0 = I_{4(m+1)}, \tag{29}$$

$$C'_1 = \begin{bmatrix} C_1 & C_2 + I_{2(m+1)} \\ C_2 + C_3 & C_2 \end{bmatrix}, \tag{30}$$

$$C'_2 = \begin{bmatrix} C_2 & C_1 + C_3 \\ C_1 + I_{2(m+1)} & C_1 \end{bmatrix}, \tag{31}$$

$$C'_3 = \begin{bmatrix} C_3 & 0 \\ 0 & C_3 \end{bmatrix}. \tag{32}$$

Proof Substituting the adjacency matrices $\tilde{A}_0, \tilde{A}_1, \tilde{A}_2$ in Definition 4.4 into (12) directly, we have

$$\left[\begin{array}{cccc|cccc} 0 & 1 & 1 & 1 & 0 & 0 & 0 & 0 \\ 0 & 0 & 1 & 0 & 0 & 0 & 1 & 1 \\ 0 & 0 & A_1 & A_2 + I_m & 1^\top & A_2 & A_1 & 1^\top \\ 0 & 1^\top & A_2 & A_2 & 0 & A_1 + I_m & A_1 & 1^\top \\ \hline 1 & 0 & 0 & 1 & 0 & 1 & 0 & 0 \\ 1^\top & 1^\top & A_2 & A_1 & 0 & A_1 & A_2 + I_m & 0 \\ 1^\top & 0 & A_1 + I_m & A_1 & 1^\top & A_2 & A_2 & 0 \\ 0 & 0 & 0 & 0 & 1 & 1 & 1 & 0 \end{array}\right].$$

Rearranging the rows and columns in the order $[5, 6, 3, 2, 1, 7, 4, 8]$, we obtain

$$\left[\begin{array}{cccc|cccc} 0 & 1 & 0 & 0 & 1 & 0 & 1 & 0 \\ 0 & A_1 & A_2 & 1^\top & 1^\top & A_2 + I_m & A_1 & 0 \\ 1^\top & A_2 & A_1 & 0 & 0 & A_1 & A_2 + I_m & 1^\top \\ 0 & 0 & 1 & 0 & 0 & 1 & 0 & 1 \\ \hline 0 & 0 & 1 & 1 & 0 & 0 & 1 & 0 \\ 1^\top & A_2 & A_1 + I_m & 0 & 1^\top & A_2 & A_1 & 0 \\ 0 & A_1 + I_m & A_2 & 1^\top & 0 & A_1 & A_2 & 1^\top \\ 1 & 1 & 0 & 0 & 0 & 1 & 0 & 0 \end{array}\right]$$

$$= \begin{bmatrix} C_1 & C_2 + I_{2(m+1)} \\ C_2 + C_3 & C_2 \end{bmatrix}$$

$$= C'_1.$$

Hence we have (30). By (15) and (30) we have (31). By (14), (29)–(31) we have (32). □

Acknowledgements

We would like to thank the anonymous referee for pointing out the connections to Taylor graphs.

References

[1] E. Bannai and T. Ito, *Algebraic combinatorics I: Association schemes*, Benjamin/Cummings, Menlo Park, 1984.

[2] A. E. Brouwer, A. M. Cohen and A. Neumaier, *Distance-regular graphs*, Springer-Verlag, Berlin, 1989.

[3] I. A. Faradžev, A. A. Ivanov, M. H. Klin and D. V. Pasechnik, Cellular subrings of Cartesian products of cellular rings, Proc. Int. Workshop "Algebraic and combinatorial coding theory," Sofia, Informa, 1988, 58–62.

[4] I. A. Faradžev, M. H. Klin and M. E. Muzichuk, *Cellular rings and groups of automorphisms of graphs*, in Investigations in algebraic theory of combinatorial objects, edited by I. A. Faradžev, A. A. Ivanov, M. H. Klin and A. J. Woldar, Math. Appl. (Soviet Ser.), vol. 84, Kluwer Acad. Publ., Dordrecht, 1994, pp. 1–152.

[5] A. Hanaki, *Classification of association schemes with small vertices*, http://math.shinshu-u.ac.jp/~hanaki/as/

[6] T. Ikuta and A. Munemasa, Complex Hadamard matrices attached to a 3-class non-symmetric association scheme, *Graphs Combin.*, 35 (6) (2019), 1293–1304.

[7] C. Koukouvinos, S. Stylianou, On skew-Hadamard matrices, *Discrete Math.* 308 (2008) 2723–2731.

[8] H. Nozaki and S. Suda, A characterization of skew Hadamard matrices and doubly regular tournaments, *Linear Algebra Appl.* 437 (2012), 1050–1056.

[9] S. Y. Song, *Class 3 association schemes whose symmetrizations have two classes*, J. Combin. Theory, Ser. A, 70 (1995) 1–29.

14

Using **GAP** Packages for Research in Graph Theory, Design Theory, and Finite Geometry

Leonard H. Soicher

Abstract

The **GAP** system is a freely available, open-source computer system for algebra and discrete mathematics, with an emphasis on computational group theory. This chapter provides a tutorial introduction to the GRAPE, DESIGN, and FinInG packages for GAP. The GRAPE package is used to construct and study graphs related to groups, designs, and finite geometries. The DESIGN package is used to construct, classify, partition, and study block designs. The FinInG package has comprehensive functionality for many types of finite incidence structures and their groups of automorphisms, and we focus on its application to projective spaces and coset geometries. Instructive examples are given throughout, including some research applications of the author.

CONTENTS

14.1 Introduction

This chapter is based on selected material from the author's (online) minicourse [36] given at the Graphs and Groups, Geometries and GAP (G2G2) Summer School, Rogla, Slovenia, 2021. Its purpose is to help researchers find out about and use certain packages of the GAP system [15] in their work on groups, graphs, designs, and finite geometries, and the fruitful interplay of these areas.

We will first focus on the GRAPE package [35], to construct and study graphs related to groups, designs, and finite geometries. Then we will look at the DESIGN package [34], to construct, classify, partition, and study block designs. After that, we will demonstrate certain functionality for projective spaces and coset geometries of the FinInG package [3] for finite incidence geometry. Groups play fundamental roles in the construction, classification, and study of many types of combinatorial structures, and we will see the heavy involvement of groups in all three packages presented.

We will also see many examples of the use of these packages, including some extended examples demonstrating research applications of the author. The given examples come from the log-file of a single continuous GAP computation (run from a file), which takes a total of about 13 minutes run-time on an i5 laptop computer running Linux. The research applications include showing that the Haemer's partial geometry [18] is uniquely determined, up to isomorphism as a partial linear space, by its point graph and parameters, a construction of the Moscow–Soicher graph [11, section 3.2.5], a classification and study of certain Sylvester designs [1], the determination that the Cohen–Tits near octagon [7, section 13.6] has no ovoid, but does have a resolution, classifications of maximal partial spreads in $PG(3,4)$, and the construction and study of a flag-transitive coset geometry for the Hall–Janko sporadic simple group J_2.

14.1.1 GAP and Its Packages

The GAP system is an internationally developed computer system for algebra and discrete mathematics, with an emphasis on computational group theory. GAP is used in research and teaching for studying groups and their representations, rings, vector spaces, algebras, graphs, designs, finite geometries, and more. The GAP system is open-source and freely available. It has a kernel written in the C language, but is mainly written in the higher-level GAP programming language. GAP has a library of thousands of functions written in the GAP language, and also provides large data libraries of mathematical objects. The packages described in this chapter make extensive use of GAP functionality, in particular its powerful group-theoretic machinery.

The main GAP website is www.gap-system.org, from which you can download the latest version of GAP to install on your computer. You will find extensive documentation, including the GAP reference manual, and learning resources at www.gap-system.org/Doc/doc.html.

GAP packages are structured open-source extensions to GAP, usually user-contributed. They provide extra functionality or data to GAP, usually have their own separate manual, integrate smoothly with GAP and its help system, and are distributed with GAP. Package authors get full credit and are usually responsible for the maintenance of their packages. Packages may also be formally refereed, which has been the case with the DESIGN and FinInG packages.

The reference manuals for the GAP packages described in this chapter are available from their entries in www.gap-system.org/Packages/packages.html, and via online help in GAP. The present author made significant use of these manuals in the preparation of this chapter. In particular, the descriptions of the data structures and functions of the GRAPE and DESIGN packages that are given here are largely derived from their respective manuals. The reader should consult the GRAPE, DESIGN, and FinInG manuals for thorough documentation of these packages, including further optional function parameters, and much further functionality not covered here. For even more in-depth understanding of GAP and its packages, see also their (open) source code.

14.1.2 Prerequisites

The author's G2G2 minicourse lectures and exercises [36] include an introduction to the use of GAP, but for this chapter, it is assumed that the reader is already familiar with the GAP system, including basic programming in GAP, and working with integers, lists, sets, records, user-defined functions, permutation groups, and group actions. The tutorial [16] included with the GAP distribution covers all the background in GAP that is needed here, and more. You may also find the introductory GAP course [22] helpful.

No particular familiarity with the GRAPE, DESIGN, and FinInG packages is assumed. However, it is assumed that the reader has some basic knowledge in graph theory and permutation group theory, including the theory of group actions. A good reference for (algebraic) graph theory is [17], and for permutation groups, [9] is recommended. Some familiarity with design theory and finite geometry is useful, but not essential. For background in design theory, see [10], and for finite geometry, see [2] and [26].

14.2 The GRAPE Package

The GRAPE package for GAP provides functionality for graphs, and is designed primarily for applications in algebraic graph theory, permutation group theory, design theory, and finite geometry.

Within GAP, a package is loaded using the LoadPackage command. For example:

```
gap> LoadPackage("grape",false);
true
```

The (optional) second parameter of the LoadPackage command was set to false to suppress the printing of the package banner. The return value of true indicates that the package was loaded correctly.

In general, GRAPE deals with finite directed graphs, which may have loops but have no multiple edges. For the mathematical purposes of GRAPE, a **graph** consists of a finite set of **vertices**, together with a set of ordered pairs of vertices called **edges**. An edge $[x, y]$ in a graph is a **loop** if $x = y$. A graph is **simple** if it has no loops and whenever $[x, y]$ is an edge, then so is $[y, x]$. A simple graph can thus be considered to be a finite undirected graph having no loops and no multiple edges. Some GRAPE functions only work for simple graphs, but these functions will check if an input graph is simple.

Graphs Γ and Δ are **isomorphic** if there is an **isomorphism** from Γ to Δ, that is, a bijection ϕ from the vertex-set of Γ to that of Δ, such that, if v, w are vertices of Γ then $[v, w]$ is an edge of Γ if and only if $[v^\phi, w^\phi]$ is an edge of Δ. An **automorphism** of a graph Γ is an isomorphism from Γ to itself. The set of all automorphisms of Γ forms a group of permutations of the vertices of Γ, called the **automorphism group** of Γ.

In GRAPE, a graph always comes together with an associated group of automorphisms. This is an important and fundamental feature of GRAPE. This group is set by GRAPE when the graph is constructed. It is used by GRAPE to store the graph compactly, to speed up computations with the graph, and can affect the output of certain GRAPE functions. Often, but not always, this group is the full automorphism group of the graph. To have a new group (of automorphisms) associated with a graph in GRAPE, the user should construct a new graph with the new group, using the function NewGroupGraph.

14.2.1 The Structure of a Graph in GRAPE

In GRAPE, a graph *gamma* is stored as a record, with mandatory components isGraph, order, group, schreierVector, representatives, and adjacencies. Usually, the user need not be aware of this record structure, and is strongly advised only to use GRAPE functions to construct and modify graphs.

The isGraph component is set to true to specify that the record is a GRAPE graph. The order component gives the number of vertices of *gamma*.

The vertices of *gamma* are always $\{1, 2, \ldots, gamma.\text{order}\}$, but they may also be given names, either by a user (using AssignVertexNames) or by a function constructing a graph (e.g. Graph, InducedSubgraph, CayleyGraph, QuotientGraph). The names component, if present, records these names, with *gamma*.names[i] being the name of vertex i. If the names component is not present, then the names are taken to be $1, 2, \ldots, gamma.\text{order}$.

The group component records the GAP permutation group associated with *gamma*. This group must be a subgroup of the automorphism group of *gamma*.

The `representatives` component records a set of orbit representatives for the action of *gamma*. group on the vertices of *gamma*, with *gamma* .`adjacencies`[*i*] being the set of vertices (out)adjacent to *gamma*.`representatives`[*i*].

`GeneratorsOfGroup`(*gamma*.`group`), together with the `schreierVector` and `adjacencies` components, are used to compute the (out)adjacency-set of a given vertex of *gamma* when needed. See [30] for details on how this is done, as well as other insights into the algorithms used by **GRAPE**.

The only mandatory component which may change once a graph is initially constructed is `adjacencies` (when an edge-orbit of *gamma*.group is added to, or removed from, *gamma*). A graph record may also have some additional optional components which record information about that graph.

Here is a very simple example of the use of **GRAPE**, in which we construct the famous Petersen graph and compute some of its properties. More explanation of the functions used will be given later.

```
gap> Petersen := Graph( SymmetricGroup([1..5]),
>      Combinations([1..5],2), OnSets,
>      function(x,y) return Intersection(x,y)=[]; end,
>      true );
rec( adjacencies := [ [ 8, 9, 10 ] ],
  group := Group([ (1,5,8,10,4)(2,6,9,3,7), (2,5)(3,6)(4,7) ]),
  isGraph := true,
  names := [ [ 1, 2 ], [ 1, 3 ], [ 1, 4 ], [ 1, 5 ], [ 2, 3 ],
      [ 2, 4 ], [ 2, 5 ], [ 3, 4 ], [ 3, 5 ], [ 4, 5 ] ],
  order := 10, representatives := [ 1 ],
  schreierVector := [ -1, 2, 2, 1, 1, 1, 2, 1, 1, 1 ] )
gap> Vertices(Petersen);
[ 1 .. 10 ]
gap> VertexNames(Petersen);
[ [ 1, 2 ], [ 1, 3 ], [ 1, 4 ], [ 1, 5 ], [ 2, 3 ], [ 2, 4 ],
  [ 2, 5 ], [ 3, 4 ], [ 3, 5 ], [ 4, 5 ] ]
gap> DirectedEdges(Petersen);
[ [ 1, 8 ], [ 1, 9 ], [ 1, 10 ], [ 2, 6 ], [ 2, 7 ], [ 2, 10 ],
  [ 3, 5 ], [ 3, 7 ], [ 3, 9 ], [ 4, 5 ], [ 4, 6 ], [ 4, 8 ],
  [ 5, 3 ], [ 5, 4 ], [ 5, 10 ], [ 6, 2 ], [ 6, 4 ], [ 6, 9 ],
  [ 7, 2 ], [ 7, 3 ], [ 7, 8 ], [ 8, 1 ], [ 8, 4 ], [ 8, 7 ],
  [ 9, 1 ], [ 9, 3 ], [ 9, 6 ], [ 10, 1 ], [ 10, 2 ], [ 10, 5 ] ]
gap> UndirectedEdges(Petersen);
[ [ 1, 8 ], [ 1, 9 ], [ 1, 10 ], [ 2, 6 ], [ 2, 7 ], [ 2, 10 ],
  [ 3, 5 ], [ 3, 7 ], [ 3, 9 ], [ 4, 5 ], [ 4, 6 ], [ 4, 8 ],
  [ 5, 10 ], [ 6, 9 ], [ 7, 8 ] ]
gap> Adjacency(Petersen,5);  # for example
[ 3, 4, 10 ]
gap> Diameter(Petersen);
2
```

```
gap> Girth(Petersen);
5
gap> autgrp:=AutomorphismGroup(Petersen);
Group([ (3,4)(6,7)(8,9), (2,3)(5,6)(9,10), (2,5)(3,6)(4,7), (1,2)
(6,8)
  (7,9) ])
gap> Size(autgrp);
120
gap> CliqueNumber(Petersen);
2
gap> MaximumClique(Petersen);
[ 1, 8 ]
gap> ChromaticNumber(Petersen);
3
gap> MinimumVertexColouring(Petersen);
[ 1, 1, 2, 3, 1, 2, 3, 2, 3, 2 ]
```

14.2.2 The Function Graph in GRAPE

This is the most general and useful way of constructing a graph in GRAPE. If you learn just one GRAPE function to construct graphs, this is it!

A basic call to this function has the form

Graph(G, L, *act*, *rel*).

The parameter L should be a list of elements of a set S on which the group G acts, with the action given by the function *act*. The parameter *rel* should be a boolean function defining a G-invariant relation on S (so that for g in G, x, y in S, $rel(x, y)$ if and only if $rel(act(x, g), act(y, g))$).

Then the function Graph returns a graph *gamma* which has as vertex-names (an immutable copy of)

Concatenation(Orbits(G, L, *act*)),

and for vertices v, w of *gamma*, $[v, w]$ is a (directed) edge if and only if

rel(VertexName(*gamma*, v), VertexName(*gamma*, w)).

There is an additional optional (boolean) parameter, *invt* (default: false), which if set to true, asserts that L is a duplicate-free list invariant (setwise) under G with respect to action *act*, and then the function Graph behaves as above, except that the vertex-names of *gamma* become (an immutable copy of) L. In particular, this allows a user to specify the ordering of the vertices in the returned graph, as GAP makes no guarantees about the order of the orbits returned by the function Orbits, nor about the order of the elements within these orbits.

The group associated with the graph *gamma* returned is the image of the action homomorphism for G acting via act on VertexNames(gamma).

We now present Peter Cameron's construction of the Hoffman–Singleton graph using the function Graph, closely following [9, section 3.6]. The vertices consist of the 35 3-subsets of $\{1, \ldots, 7\}$, together with one orbit of 15 projective planes of order 2 on the points $\{1, \ldots, 7\}$, under the action of the alternating group A_7. The adjacencies are described by the function HoffmanSingleton Adjacency as follows.

```
gap> projectiveplane:=Set(Orbit(Group((1,2,3,4,5,6,7)),[1,2,4],OnSets));
[ [ 1, 2, 4 ], [ 1, 3, 7 ], [ 1, 5, 6 ], [ 2, 3, 5 ], [ 2, 6, 7 ],
  [ 3, 4, 6 ], [ 4, 5, 7 ] ]
gap> #
gap> # Now  projectiveplane  is (the set of lines of) a projective plane
gap> # of order 2.
gap> #
gap> HoffmanSingletonAction:=function(x,g)
> #
> # This function gives the action of AlternatingGroup([1..7])
> # on the vertices of the Hoffman-Singleton graph, in
> # Peter Cameron's construction.
> #
> if Size(x)=3 then          # x is a 3-set
>    return OnSets(x,g);
> else                       # x is a projective plane
>    return OnSetsSets(x,g);
> fi;
> end;;
gap> HoffmanSingletonAdjacency:=function(x,y)
> #
> # This boolean function returns  true  iff vertices  x  and  y
> # are adjacent in the Hoffman-Singleton graph, in
> # Peter Cameron's construction.
> #
> if Size(x)=3 then               # x is a 3-set
>    if Size(y)=3 then            # y is a 3-set
>       return Intersection(x,y)=[]; # join iff x and y disjoint
>    else                         # y is a projective plane
>       return x in y;            # join iff x is a line of y
>    fi;
> else                            # x is a projective plane
>    if Size(y)=3 then            # y is a 3-set
>       return y in x;            # join iff y is a line of x
>    else                         # y is a projective plane
>       return false;             # don't join
>    fi;
> fi;
> end;;
gap> HoffmanSingleton:=Graph( AlternatingGroup([1..7]),
>      [[1,2,3], projectiveplane],
```

```
>     HoffmanSingletonAction,
>     HoffmanSingletonAdjacency));;
gap> IsSimpleGraph(HoffmanSingleton);
true
gap> OrderGraph(HoffmanSingleton);  # number of vertices
50
gap> VertexDegrees(HoffmanSingleton);  # set of vertex degrees
[ 7 ]
gap> Diameter(HoffmanSingleton);
2
gap> Girth(HoffmanSingleton);
5
gap> Size(HoffmanSingleton.group);
2520
gap> autgrp:=AutomorphismGroup(HoffmanSingleton);;
gap> Size(autgrp);
252000
gap> HoffmanSingleton:=NewGroupGraph(autgrp,HoffmanSingleton);;
gap> # So now the group associated with  HoffmanSingleton  is its
gap> # full automorphism group.
gap> Size(HoffmanSingleton.group);
252000
gap> DisplayCompositionSeries(HoffmanSingleton.group);
G (5 gens, size 252000)
 | C2
S (2 gens, size 126000)
 | U3(5)
1 (0 gens, size 1)
```

14.2.3 EdgeOrbitsGraph

A common way to construct a graph in GRAPE is via the function EdgeOrbits
Graph.

Where n is a non-negative integer, G is a permutation group on $\{1,\ldots,n\}$,
and *edges* is a list of ordered pairs of elements of $\{1,\ldots,n\}$, the function call

EdgeOrbitsGraph(G, *edges*, n)

returns the (directed) graph with vertex-set $\{1,\ldots,n\}$, and edge-set the union
of the G-orbits of the ordered pairs in *edges*. The group associated with the
returned graph is G. The parameter n may be omitted, in which case n is taken
to be the largest point moved by G.

Note that G may be the trivial permutation group, Group(()), in which
case the (directed) edges of the returned graph are precisely those in the list
edges (not including any repeats).

Here is an example.

```
gap> G:=AllPrimitiveGroups(NrMovedPoints,275,Size,2025*443520*2);
[ McL:2 ]
```

```
gap> G:=G[1];  # the automorphism group of the McLaughlin group
McL:2
gap> H:=Stabilizer(G,1);
<permutation group of size 6531840 with 3 generators>
gap> orbs:=List(Orbits(H,[1..275]),Set);;
gap> List(orbs,Length);
[ 1, 112, 162 ]
gap> y:=First(orbs,x->Length(x)=112)[1];
2
gap> McLaughlin:=EdgeOrbitsGraph(G,[[1,y]]);;  # the McLaughlin graph
gap> IsSimpleGraph(McLaughlin);
true
gap> OrderGraph(McLaughlin);
275
gap> VertexDegrees(McLaughlin);
[ 112 ]
```

Some other useful GRAPE functions to construct graphs are NullGraph, CompleteGraph, CayleyGraph, JohnsonGraph, and HammingGraph.

14.2.4 Automorphism Groups and Isomorphism Testing in GRAPE

GRAPE contains functionality to test graph isomorphism and to calculate the automorphism group of a graph. This functionality can also be applied to graphs with ordered vertex partitions (see Graphs with colour-classes in the GRAPE manual).

To do all this, by default GRAPE uses its included version (currently final patched version 2.2) of B. D. McKay's nauty software [25]. A Linux user may instead choose to have GRAPE use their own installed copy of T. Junttila's and P. Kaski's bliss software [20]. The nauty (and nauty/traces) and bliss software packages are powerful tools using 'partition backtrack' for computing the automorphism group of a graph and for canonically labelling a graph for the purpose of isomorphism testing. A basic introduction to these techniques is given in [30], with much more detailed expositions in [20, 25]. The GRAPE interfaces to nauty and bliss are seamless and transparent to the user.

As we have already seen, where *gamma* is a graph in GRAPE, the function call

AutomorphismGroup(*gamma*)

returns the automorphism group of *gamma*, which can then be studied using the permutation group functionality in GAP.

Where *gamma* and *delta* are graphs in GRAPE, the function call

GraphIsomorphism(*gamma*, *delta*)

returns a permutation giving an isomorphism from *gamma* to *delta* if these graphs are isomorphic, and returns the special value `fail` if the graphs are not isomorphic, whereas

`IsIsomorphicGraph( `*gamma*`, `*delta*` )`

returns `true` if the graphs are isomorphic, and `false` otherwise.

For example:

```
gap> J:=JohnsonGraph(7,3);;
gap> Size(AutomorphismGroup(J));
5040
gap> IsIsomorphicGraph(J,JohnsonGraph(7,4));
true
```

When comparing more than two graphs for pairwise isomorphism, you should use the function `GraphIsomorphismClassRepresentatives`. Where L is a list of graphs in GRAPE, the function call

`GraphIsomorphismClassRepresentatives( `L` )`

returns a list consisting of pairwise non-isomorphic elements of L (in some order), representing all the isomorphism classes of elements of L.

14.2.5 Local and Global Regularity Parameters

Let *gamma* be a simple connected graph, and let V be a singleton vertex or set of vertices of *gamma*.

We say that *gamma* has the **local parameter** $c_i(V)$ (respectively $a_i(V)$, $b_i(V)$), with respect to V, if the number of vertices at distance $i-1$ (respectively $i, i+1$) from V that are adjacent to a vertex w at distance i from V is the constant $c_i(V)$ (respectively $a_i(V)$, $b_i(V)$) depending only on i and V (and not the choice of w). See `Distance` in the GRAPE manual.

We say that *gamma* has the **global parameter** c_i (respectively a_i, b_i) if the number of vertices at distance $i-1$ (respectively $i, i+1$) from a vertex v that are adjacent to a vertex w at distance i from v is the constant c_i (respectively a_i, b_i) depending only on i (and not the choice of v or w).

In GRAPE, the function call

`LocalParameters( `*gamma*`, `V` )`

returns a list whose ith element is the list

$$[c_{i-1}(V), a_{i-1}(V), b_{i-1}(V)],$$

except that if some local parameter does not exist, then -1 is put in its place.

The function call

LocalParameters(*gamma*, V, G)

does the same, except that the additional parameter G is assumed to be a subgroup of the automorphism group of *gamma* fixing V setwise. Including such a G can result in a performance gain.

The function call

GlobalParameters(*gamma*)

returns a list of length one more than the diameter of *gamma* (see Diameter in the GRAPE manual), and whose ith element is the list

$$[c_{i-1}, a_{i-1}, b_{i-1}],$$

except that if some global parameter does not exist, then -1 is put in its place.

Note that (the simple connected graph) *gamma* is **distance-regular** if and only if this function returns no -1 in place of a global parameter (see [7]). See also IsDistanceRegular and CollapsedAdjacencyMat in the GRAPE manual.

Here are some examples.

```
gap> GlobalParameters(HoffmanSingleton);
[ [ 0, 0, 7 ], [ 1, 0, 6 ], [ 1, 6, 0 ] ]
gap> IsDistanceRegular(HoffmanSingleton);
true
gap> edge:=[1,Adjacency(HoffmanSingleton,1)[1]];
[ 1, 4 ]
gap> edge_stab:=Stabilizer(HoffmanSingleton.group,edge,OnSets);
<permutation group of size 1440 with 4 generators>
gap> LocalParameters(HoffmanSingleton,edge,edge_stab);
[ [ 0, 1, 6 ], [ 1, 0, 6 ], [ 2, 5, 0 ] ]
gap> GlobalParameters(McLaughlin);
[ [ 0, 0, 112 ], [ 1, 30, 81 ], [ 56, 56, 0 ] ]
gap> GlobalParameters(EdgeGraph(McLaughlin));
[ [ 0, 0, 222 ], [ 1, -1, -1 ], [ -1, -1, -1 ], [ 184, 38, 0 ] ]
```

14.2.6 The Sylvester Graph

The **Sylvester graph** is the unique distance-regular graph with intersection array $\{5, 4, 2; 1, 1, 4\}$, that is, with global parameters

$$[[0,0,5],[1,0,4],[1,2,2],[4,1,0]].$$

This graph can be constructed as the induced subgraph on the 36 vertices at distance 2 from a fixed edge in the Hoffman–Singleton graph (see [7, section 13.1.A]). We do this now using GRAPE.

```
gap> Sylvester:=DistanceSetInduced(HoffmanSingleton,2,edge,edge_stab);;
gap> GlobalParameters(Sylvester);
[ [ 0, 0, 5 ], [ 1, 0, 4 ], [ 1, 2, 2 ], [ 4, 1, 0 ] ]
```

14.2.7 Construction of a Distance-Regular Graph from a Transitive Group

Let G be a transitive permutation group on $V := \{1, \dots, n\}$. A **generalized orbital graph** for G is a simple graph with vertex-set V on which G acts naturally as a (vertex-transitive) group of automorphisms.

Here we construct the (non-null) generalized orbital graphs for the group $M_{22}{:}2$ (the automorphism group of the Mathieu group M_{22}) in its primitive action on 672 points. This includes the distance-regular so-called Moscow–Soicher graph (see [28] and [11, section 3.2.5]).

```
gap> n:=672;
672
gap> G:=AllPrimitiveGroups(NrMovedPoints,n,Size,443520*2);
[ M(22).2 ]
gap> G:=G[1];  # the automorphism group of the Mathieu group $M_{22}$
M(22).2
gap> RankAction(G,[1..n]);
6
gap> L:=GeneralizedOrbitalGraphs(G);;
gap> # the non-null generalized orbital graphs
gap> Length(L);
31
gap> List(L,GlobalParameters);
[ [ [ 0, 0, 55 ], [ 1, 8, 46 ], [ -1, -1, -1 ], [ 45, 10, 0 ] ],
  [ [ 0, 0, 385 ], [ 1, -1, -1 ], [ -1, -1, 0 ] ],
  [ [ 0, 0, 440 ], [ 1, -1, -1 ], [ -1, -1, 0 ] ],
  [ [ 0, 0, 605 ], [ 1, -1, -1 ], [ 540, 65, 0 ] ],
  [ [ 0, 0, 671 ], [ 1, 670, 0 ] ],
  [ [ 0, 0, 506 ], [ 1, -1, -1 ], [ 398, 108, 0 ] ],
  [ [ 0, 0, 550 ], [ 1, -1, -1 ], [ -1, -1, 0 ] ],
  [ [ 0, 0, 616 ], [ 1, -1, -1 ], [ 562, 54, 0 ] ],
  [ [ 0, 0, 451 ], [ 1, -1, -1 ], [ -1, -1, 0 ] ],
  [ [ 0, 0, 110 ], [ 1, 28, 81 ], [ 18, 80, 12 ], [ 90, 20, 0 ] ],
  [ [ 0, 0, 275 ], [ 1, -1, -1 ], [ -1, -1, 0 ] ],
  [ [ 0, 0, 341 ], [ 1, -1, -1 ], [ 170, 171, 0 ] ],
  [ [ 0, 0, 176 ], [ 1, 40, 135 ], [ 48, 128, 0 ] ],
  [ [ 0, 0, 220 ], [ 1, -1, -1 ], [ -1, -1, 0 ] ],
  [ [ 0, 0, 286 ], [ 1, -1, -1 ], [ -1, -1, 0 ] ],
  [ [ 0, 0, 121 ], [ 1, 20, 100 ], [ -1, -1, 0 ] ],
```

```
[ [ 0, 0, 330 ], [ 1, 158, 171 ], [ -1, -1, 0 ] ],
[ [ 0, 0, 385 ], [ 1, -1, -1 ], [ -1, -1, 0 ] ],
[ [ 0, 0, 550 ], [ 1, -1, -1 ], [ 450, 100, 0 ] ],
[ [ 0, 0, 616 ], [ 1, -1, -1 ], [ 570, 46, 0 ] ],
[ [ 0, 0, 451 ], [ 1, -1, -1 ], [ -1, -1, 0 ] ],
[ [ 0, 0, 495 ], [ 1, 366, 128 ], [ 360, 135, 0 ] ],
[ [ 0, 0, 561 ], [ 1, -1, -1 ], [ 480, 81, 0 ] ],
[ [ 0, 0, 396 ], [ 1, -1, -1 ], [ -1, -1, 0 ] ],
[ [ 0, 0, 55 ], [ 1, 0, 54 ], [ -1, -1, -1 ], [ 45, 10, 0 ] ],
[ [ 0, 0, 220 ], [ 1, -1, -1 ], [ -1, -1, 0 ] ],
[ [ 0, 0, 286 ], [ 1, -1, -1 ], [ -1, -1, 0 ] ],
[ [ 0, 0, 121 ], [ 1, -1, -1 ], [ -1, -1, 0 ] ],
[ [ 0, 0, 165 ], [ 1, 56, 108 ], [ -1, -1, 0 ] ],
[ [ 0, 0, 231 ], [ 1, -1, -1 ], [ -1, -1, 0 ] ],
[ [ 0, 0, 66 ], [ 1, 0, 65 ], [ -1, -1, 0 ] ] ]
gap> MoscowSoicher:=First(L,x->VertexDegrees(x)=[110]);;
gap> IsDistanceRegular(MoscowSoicher);
true
gap> GlobalParameters(MoscowSoicher);
[ [ 0, 0, 110 ], [ 1, 28, 81 ], [ 18, 80, 12 ], [ 90, 20, 0 ] ]
gap> AutomorphismGroup(MoscowSoicher) = G;
true
```

See also `VertexTransitiveDRGs` and `OrbitalGraphColadjMats` in the GRAPE manual for further study of vertex-transitive graphs and the (homogeneous) coherent configurations associated to transitive permutation groups.

14.2.8 Cliques

Let Γ be a simple graph. A **clique** of Γ is a set of pairwise adjacent vertices. A **maximal** clique of Γ is a clique which is not properly contained in any clique of Γ, while a **maximum** clique of Γ is a clique of largest size in Γ. The **clique number** of Γ is the size of a maximum clique of Γ. A **co-clique** (or **independent set**) of Γ is a set of pairwise non-adjacent vertices. Clearly, the co-cliques of Γ are precisely the cliques of the complement of Γ. The **independence number** of Γ is the size of a largest co-clique of Γ.

Now let *gamma* be a simple graph in GRAPE, with associated group $G :=$ *gamma*.group. Then GRAPE functions can compute the maximal cliques of *gamma*, or the cliques of given size in *gamma*, or the maximal cliques of given size in *gamma*, such that one such clique is determined or it is determined that no such cliques exist, or G-orbit generators of all such cliques are determined, or G-orbit representatives of all such cliques are determined. There is also the functionality to determine a maximum clique, and hence to determine the clique number of *gamma*. These are computationally HARD problems!

The main function for clique classification in GRAPE is `CompleteSubgraphsOfGivenSize`. We shall now study the use of this function for non-weighted cliques, but for a complete specification, see CompleteSubgraphsOfGiven Size in the GRAPE manual. Also see Complete Subgraphs, MaximumClique, and CliqueNumber.

In GRAPE, where *gamma* is a simple graph, k is a non-negative integer, *alls* is 0, 1, or 2, and *maxi* is true or false, the function call

CompleteSubgraphsOfGivenSize(*gamma*, k, *alls*, *maxi*)

returns a set K (possibly empty) of cliques of size k of *gamma*. These are all maximal cliques if *maxi*=true, and not necessarily maximal cliques if *maxi*=false. The parameter *maxi* is optional, and has default value false. The parameter *alls* controls how many cliques are returned.

First, suppose *alls* = 0. Then K will contain at most one element. If *maxi* = false, then K will contain a clique of size k if and only if *gamma* has such a clique, and if *maxi* = true, then K will contain a maximal clique of size k if and only if *gamma* has a maximal clique of that size.

In this function call with *alls* = 0, if *gamma*.group is not the full automorphism group of *gamma*, it may be (much) more efficient to replace *gamma* with a copy whose associated group is its full automorphism group.

If *alls* = 1 and *maxi* = false, then K will contain (perhaps properly) a set of *gamma*.group orbit-representatives of the size k cliques of *gamma*. If *alls* = 1 and *maxi* = true, then K will contain (perhaps properly) a set of *gamma*.group orbit-representatives of the size k maximal cliques of *gamma*.

If *alls* = 2 and *maxi* = false, then K will be a set of *gamma*.group orbit-representatives of all the size k cliques of *gamma*. If *alls* = 2 and *maxi* = true, then K will be a set of *gamma*.group orbit-representatives of the size k maximal cliques of *gamma*.

Here are some illustrative examples.

```
gap> gamma:=MoscowSoicher;;
gap> CompleteSubgraphsOfGivenSize(gamma,5,0,false);
[ [ 1, 2, 18, 35, 41 ] ]
gap> CompleteSubgraphsOfGivenSize(gamma,5,0,true);
[ [ 1, 2, 80, 124, 455 ] ]
gap> CompleteSubgraphsOfGivenSize(gamma,5,1,false);
[ [ 1, 2, 18, 35, 41 ], [ 1, 2, 18, 35, 377 ], [ 1, 2, 18, 119, 161 ],
  [ 1, 2, 18, 119, 277 ], [ 1, 2, 18, 161, 277 ],
  [ 1, 2, 18, 281, 287 ], [ 1, 2, 18, 281, 366 ],
  [ 1, 2, 80, 124, 455 ], [ 1, 2, 124, 183, 461 ] ]
gap> CompleteSubgraphsOfGivenSize(gamma,5,1,true);
[ [ 1, 2, 80, 124, 455 ] ]
gap> CompleteSubgraphsOfGivenSize(gamma,5,2,false);
[ [ 1, 2, 18, 35, 41 ], [ 1, 2, 18, 119, 161 ],
  [ 1, 2, 18, 281, 287 ], [ 1, 2, 80, 124, 455 ] ]
```

```
gap> CompleteSubgraphsOfGivenSize(gamma,5,2,true);
[ [ 1, 2, 80, 124, 455 ] ]
gap> CompleteSubgraphsOfGivenSize(gamma,7,0,false);
[ ]
gap> C:=CompleteSubgraphsOfGivenSize(gamma,6,2,true);
[ [ 1, 2, 18, 35, 41, 377 ], [ 1, 2, 18, 119, 161, 277 ],
  [ 1, 2, 18, 281, 287, 366 ] ]
gap> List(C,clique->Size(Stabilizer(gamma.group,clique,OnSets)));
[ 360, 144, 36 ]
gap> CliqueNumber(McLaughlin);  # the size of a largest clique
5
gap> MaximumClique(McLaughlin); # a clique of largest size
[ 1, 2, 17, 45, 193 ]
gap> CliqueNumber(ComplementGraph(McLaughlin));
22
gap> # This is the independence number of the McLaughlin graph,
gap> # the size of a largest co-clique.
```

The function `CompleteSubgraphsOfGivenSize` can also compute and classify cliques with given vertex-weight sum in a vertex-weighted graph, where the weights can be positive integers or non-zero d-vectors of non-negative integers (satisfying certain conditions with respect to the group associated with the graph). This type of clique functionality in GRAPE is used by the DESIGN package for GAP, for functionality to construct and classify block designs of many types (including those invariant under a specified group), as well as to construct and classify parallel classes and resolutions of block designs. We will see some of this DESIGN package functionality later.

14.2.9 Proper Vertex-Colourings

Let Γ be a simple graph. A **proper vertex-colouring** of Γ is a labelling of its vertices by elements from a set of **colours**, such that adjacent vertices are labelled with different colours. Where k is a non-negative integer, a **vertex k-colouring** of Γ is a proper vertex-colouring using at most k colours. A **minimum vertex-colouring** of Γ is a vertex k-colouring with k as small as possible, and the **chromatic number** $\chi(\Gamma)$ of Γ is the number of colours used in a minimum vertex-colouring of Γ.

In GRAPE, a proper vertex-colouring of a simple graph is given as a list of positive integers (the colours), indexed by the vertices of the graph, such that the ith list element is the colour of vertex i.

In GRAPE, where *gamma* is a simple graph and k is a non-negative integer, the function call

`VertexColouring( `*gamma*`, `k` )`

returns a vertex k-colouring of *gamma* if such a colouring exists; otherwise, the special value `fail` is returned. In general, this is a computationally HARD problem. See also `MinimumVertex Colouring` and `ChromaticNumber` in the GRAPE manual.

On Colouring the McLaughlin Graph

In the code which follows, we first make the induced subgraph in the McLaughlin graph on the vertices at distance 1 from a fixed vertex (here the vertex 1). This is the so-called first subconstituent of the McLaughlin graph. We then verify that its chromatic number is 8. After that, we make the second subconstituent of the McLaughlin graph, and determine that its chromatic number is 10.

```
gap> gamma:=DistanceSetInduced(McLaughlin,1,1);;
gap> # This is the induced subgraph on the vertices of the
gap> # McLaughlin graph at distance 1 from the vertex 1.
gap> GlobalParameters(gamma);
[ [ 0, 0, 30 ], [ 1, 2, 27 ], [ 10, 20, 0 ] ]
gap> VertexColouring(gamma,7);
fail
gap> VertexColouring(gamma,8)<>fail;
true
gap> delta:=DistanceSetInduced(McLaughlin,2,1);;
gap> # This is the induced subgraph on the vertices of the
gap> # McLaughlin graph at distance 2 from the vertex 1.
gap> GlobalParameters(delta);
[ [ 0, 0, 56 ], [ 1, 10, 45 ], [ 24, 32, 0 ] ]
gap> ChromaticNumber(delta);
10
```

Now here is a challenging problem. Let M be the McLaughlin graph. It follows from the preceding computation that $\chi(M) \leq 18$. In fact, using the 'ant colony optimization' program supplied as supplementary material to [23], the author found a vertex 16-colouring of M. On the other hand, the eigenvalue-based lower bound given in [8, corollary 3.6.4] shows that $\chi(M) \geq 15$. Thus, the chromatic number of the McLaughlin graph M is 15 or 16. The problem is to either find a vertex 15-colouring of M or to show there is no such colouring.

14.2.10 Partial Linear Spaces and Partial Geometries

Where s and t are positive integers, a **partial linear space** with **parameters** (s,t) consists of a finite set of **points**, together with a set of $(s + 1)$-subsets of the points, called **lines**, such that every point is on exactly $t + 1$ lines, and any two distinct points lie on (i.e. are contained in) at most one line (equivalently, any two distinct lines meet in at most one point).

Two partial linear spaces are **isomorphic** if there is a bijection from the point-set of the first to that of the second which, applied to the set of lines of the first, yields the set of lines of the second.

The **point graph** (or **collinearity graph**) of a partial linear space is the graph whose vertices are the points of the space, with two distinct points joined by an edge exactly when they lie on a common line.

Now, where *gamma* is a simple graph in GRAPE, and *s* and *t* are positive integers, the function call

PartialLinearSpaces(*gamma*, *s*, *t*)

returns a list of representatives of the distinct isomorphism classes of partial linear spaces with parameters (s, t) and point graph *gamma*. (This may take a (very) long time.) In the output of this function, a partial linear space S is given by its incidence graph *delta*, such that the group *delta*.group associated with *delta* is the automorphism group of S, acting on point-vertices and line-vertices, and preserving both sets. See the GRAPE manual entry for PartialLinearSpaces for more information, including details of optional parameters.

For example:

```
gap> K7:=CompleteGraph(SymmetricGroup([1..7]));;
gap> P:=PartialLinearSpaces(K7,2,2);
[ rec( adjacencies := [ [ 8, 9, 10 ], [ 1, 2, 3 ] ],
       group := Group([ (1,2)(5,6)(9,11)(10,12), (1,2,3)(5,6,7)
           (9,11,13)(10,12,14), (1,2,3)(4,7,6)(9,12,14)(10,11,13),
           (1,4,7,6,2,5,3)(8,9,13,10,11,12,14) ]), isGraph := true,
       isSimple := true,
       names := [ 1, 2, 3, 4, 5, 6, 7, [ 1, 2, 3 ], [ 1, 4, 5 ],
           [ 1, 6, 7 ], [ 2, 4, 6 ], [ 2, 5, 7 ], [ 3, 4, 7 ],
           [ 3, 5, 6 ] ], order := 14, representatives := [ 1, 8 ],
       schreierVector := [ -1, 1, 2, 4, 4, 1, 3, -2, 4, 1, 1, 3, 4, 2
           ] ) ]
gap> Length(P);
1
gap> GlobalParameters(P[1]);
[ [ 0, 0, 3 ], [ 1, 0, 2 ], [ 1, 0, 2 ], [ 3, 0, 0 ] ]
gap> Size(P[1].group);
168
gap> T:=ComplementGraph(JohnsonGraph(10,2));;
gap> P:=PartialLinearSpaces(T,4,6);;
gap> List(P,x->Size(x.group));
[ 216, 1512 ]
```

The Haemers Partial Geometry

The Haemers partial geometry [18] is a partial geometry with parameters $s = 4$, $t = 17$, $\alpha = 2$ and point graph the distance-2 graph of the edge graph of the Hoffman–Singleton graph. (The **distance-k graph** Γ_k of a simple graph Γ ha the same vertex set as Γ, but with two vertices joined by an edge in Γ_k if and only if they have distance k in Γ. The **edge graph** $L(\Gamma)$ of a simple graph Γ has as vertices the undirected edges of Γ, with two such vertices joined by an edge in $L(\Gamma)$ if and only if they have exactly one common vertex in Γ.)

As done in the GRAPE manual, we now use the GRAPE function Partial LinearSpaces to construct the Haemers partial geometry and its dual, prove that they are determined up to isomorphism (as partial linear spaces) by their respective point graphs and parameters, and determine their full automorphism groups.

```
gap> pointgraph:=DistanceGraph(EdgeGraph(HoffmanSingleton),2);;
gap> GlobalParameters(pointgraph);
[ [ 0, 0, 72 ], [ 1, 20, 51 ], [ 36, 36, 0 ] ]
gap> VertexName(pointgraph,1);  # example vertex-name
[ [ 1, 2, 3 ], [ 4, 5, 6 ] ]
gap> spaces:=PartialLinearSpaces(pointgraph,4,17);;
gap> Length(spaces);
1
gap> HaemersGeometry:=spaces[1];;
gap> #
gap> # Now  HaemersGeometry  is the incidence graph of the Haemers
gap> # partial geometry.
gap> #
gap> DisplayCompositionSeries(HaemersGeometry.group);
G (2 gens, size 2520)
 | A7
1 (0 gens, size 1)
gap> linevertex:=Adjacency(HaemersGeometry,1)[1];
176
gap> linegraph:=PointGraph(HaemersGeometry,linevertex);;
gap> GlobalParameters(linegraph);
[ [ 0, 0, 85 ], [ 1, 20, 64 ], [ 10, 75, 0 ] ]
gap> spaces:=PartialLinearSpaces(linegraph,17,4);;
gap> Length(spaces);
1
gap> DualHaemersGeometry:=spaces[1];;
gap> DisplayCompositionSeries(DualHaemersGeometry.group);
G (3 gens, size 2520)
 | A7
1 (0 gens, size 1)
```

14.2.11 Steve Linton's Function SmallestImageSet

An important behind-the-scenes workhorse included in GRAPE is Steve Linton's function SmallestImageSet. This function is used in GRAPE in the classification of cliques and the classification of partial linear spaces with given point graph and parameters, as well as in the DESIGN package in the classification of block designs. The function is of use in many other situations when classifying objects up to the action of a given permutation group G, when the objects can be represented as subsets of the permutation domain of G.

Let G be a permutation group on $X := \{1, \ldots, n\}$, and let $S \subseteq X$. Then the function call

```
SmallestImageSet( G, S )
```

returns the lexicographically least set in `Orbit(G,S,OnSets)`, without explicitly computing this (possibly huge) orbit. Thus, if T is any subset of X, then S and T are equivalent under the action of G if and only if

```
SmallestImageSet( G, S ) = SmallestImageSet( G, T ).
```

Typically, we set things up so that certain subsets of X represent the objects we are classifying, with two objects equivalent if and only if their representing sets are in the same G-orbit of sets. Then, if C is a list of subsets of X (say certain cliques of a given size in a graph), and we want to determine a set of (canonical) representatives for the distinct G-orbits of the elements of C, we can do this as:

```
Set( List( C, c->SmallestImageSet( G,c ) ) ).
```

Steve Linton's algorithm for `SmallestImageSet` is given in [24]. Further developments in the computation of minimal and canonical images with respect to a group action are given in [19].

14.3 The DESIGN Package

The DESIGN package for GAP provides functionality for constructing, classifying, partitioning, and studying block designs, including the computation of statistical efficiency measures for these designs. For the use of the DESIGN package for statistical designs, see [31] and [1]. Here we concentrate on the combinatorial aspects of block designs. DESIGN makes use of the GRAPE package.

14.3.1 Block Designs

A **block design** is an ordered pair (P, B), where P is a non-empty finite set whose elements are called **points**, and B is a non-empty finite multiset whose elements are called **blocks**, such that each block is a non-empty finite multiset of points. For our purposes, a **multiset** is a list, where order does not matter, but the number of times an element appears does. We represent a multiset as an ordered list in GAP.

A **parallel class** (or **spread**) of a block design $D = (P, B)$ is a sub(multi)set of B forming a partition of P, and a **resolution** of D is a partition of B into parallel classes. We say that a block design is **resolvable** if it has a resolution.

The DESIGN package deals with arbitrary block designs, but some DESIGN functions only work for **binary** block designs (i.e. those with no repeated element in any block of the design), but these functions will check if an input block design is binary.

An important class of binary block designs is t-designs. For t a non-negative integer and v, k, λ positive integers with $t \leq k \leq v$, a t-**design** with **parameters** t, v, k, λ, or a t-(v, k, λ) **design**, is a binary block design with exactly v points, such that each block has size k and each t-subset of the points is contained in exactly λ blocks.

A t-(v, k, λ) design is also an s-(v, k, λ_s) design for $0 \leq s \leq t$, where

$$\lambda_s = \lambda \binom{v - s}{t - s} \Big/ \binom{k - s}{t - s}.$$

For example, we compute these λ_s for a 5-$(24, 8, 1)$ design.

```
gap> LoadPackage("design",false);
true
gap> TDesignLambdas(5,24,8,1);
[ 759, 253, 77, 21, 5, 1 ]
```

The DESIGN package has extensive functionality for t-designs and their parameters. For example, you can compute an upper bound on the multiplicity of a block in any t-design with given parameters t, v, k, λ or in any resolvable such t-design. See the DESIGN manual (or online help in GAP) for specifications of the functions used in the following.

```
gap> TDesignLambdaMin(2,12,4);
3
gap> TDesignLambdas(2,12,4,3);
[ 33, 11, 3 ]
gap> TDesignBlockMultiplicityBound(2,12,4,3);
2
gap> ResolvableTDesignBlockMultiplicityBound(2,12,4,3);
1
gap> #
gap> # The DESIGN package knows the Bruck-Ryser-Chowla Theorem,
gap> # and so knows there is no projective plane of order 6.
gap> #
gap> TDesignBlockMultiplicityBound(2,43,7,1);
0
```

14.3.2 The Function `BlockDesign`

The DESIGN package function `BlockDesign` gives a straightforward way of constructing a block design.

Let v be a positive integer and B be a non-empty list of non-empty sorted lists of elements of $V := \{1, \ldots, v\}$. Then

`BlockDesign( v, B )`

returns the block design with point-set V and block multiset C, where C is `SortedList( B )`. Moreover, where G is a group of permutations on V, the function call

`BlockDesign( v, B, G )`

returns the block design with point-set V and block multiset C, where now C is the sorted list of the concatenation of each of the G-orbits of the elements in the list B. For example:

```
gap> G:=Group((1,2,3,4,5,6,7));;
gap> design:=BlockDesign(7,[[1,2,4],[1,2,4]],G);
rec( autSubgroup := Group([ (1,2,3,4,5,6,7) ]),
  blocks := [ [ 1, 2, 4 ], [ 1, 2, 4 ], [ 1, 3, 7 ], [ 1, 3, 7 ],
      [ 1, 5, 6 ], [ 1, 5, 6 ], [ 2, 3, 5 ], [ 2, 3, 5 ],
      [ 2, 6, 7 ], [ 2, 6, 7 ], [ 3, 4, 6 ], [ 3, 4, 6 ],
      [ 4, 5, 7 ], [ 4, 5, 7 ] ], isBlockDesign := true, v := 7 )
gap> IsSimpleBlockDesign(design);
false
gap> AllTDesignLambdas(design);
[ 14, 6, 2 ]
gap> Size(AutomorphismGroup(design));
168
```

We note that, given a block design D, the function call

`AllTDesignLambdas( D )`

returns the empty list if D is not a t-design. Otherwise, D is a binary block design with constant block size k, say, and this function returns an immutable list L of length $T + 1$, where T is the maximum $t \le k$ such that D is a t-design, and, for $i = 1, \ldots, T + 1$, the ith element of L is equal to the (constant) number of blocks of D containing an $(i - 1)$-subset of the point-set of D.

14.3.3 The Structure of a Block Design in DESIGN

In DESIGN, a block design D is stored as a record, with mandatory components `isBlockDesign`, v, and `blocks`.

The `isBlockDesign` component is set to `true` to specify that the record is a DESIGN package block design. The component `v` gives the number of points of D. The points of D are $\{1, 2, \ldots, D.v\}$, but they may also be given names in the optional component `pointNames`, with `D.pointNames[i]` the name of point i. The `blocks` component must be a sorted list of the blocks of D (including any repeats), with each block being a sorted list of points (including any repeats). A block design record may also have some optional components which store information about the design.

A non-expert user should only use functions in the DESIGN package to create block design records and their components.

14.3.4 Automorphism Groups and Isomorphism Testing in DESIGN

Let A and B be block designs. Then A and B are **isomorphic** if there is an **isomorphism** from A to B; that is, a bijection from the points of A to those of B which applied to the block multiset of A yields the block multiset of B. An **automorphism** of A is an isomorphism from A to itself. The set of all automorphisms of A forms a group, the **automorphism group** of A.

The DESIGN package contains functionality to test block design isomorphism and to calculate the automorphism group of a block design, but currently only for binary block designs (equivalently, block designs where every block is a set). To do this, the DESIGN package uses the graph automorphism group and isomorphism testing functionality in GRAPE.

Now suppose that A and B are binary block designs in the DESIGN package. Then the function call

```
AutomorphismGroup( A )
```

returns the automorphism group of A, which can then be studied using the permutation group functionality in GAP. The function call

```
IsIsomorphicBlockDesign( A, B )
```

returns `true` if the block designs are isomorphic, and `false` otherwise.

When comparing more than two binary block designs for pairwise isomorphism, you should use the function `BlockDesignIsomorphismClassRepresentatives`. Where L is a list of binary block designs in DESIGN package format, the function call

```
BlockDesignIsomorphismClassRepresentatives( L )
```

returns a list consisting of pairwise non-isomorphic elements of L, representing all the isomorphism classes of elements of L.

14.3.5 The Function `BlockDesigns`

The most important DESIGN package function is `BlockDesigns`, which can construct and classify block designs satisfying a wide range of user-specified properties (although this may require a very large amount of memory or time depending on the problem). The calling syntax is

`BlockDesigns( param )`

and the function returns a list of block designs whose properties are specified by the user in the record *param*, as described thoroughly in the full function documentation in the DESIGN manual. Only binary block designs with the given properties are generated if *param*.`blockDesign` is unbound or is a binary block design, which we assume to be the case in this chapter.

The required components of the record *param* are as follows:

- *param*.`v` should be a positive integer specifying the number of points (in each returned design)
- *param*.`blockSizes` should be a set of positive integers, specifying the allowable block size(s)
- *param*.`tSubsetStructure` should be a record, used to specify for one given $t \geq 0$, for each t-subset T of the points, the number of blocks containing T. This number may be a positive constant *lambda*, not depending on T, in which case, *param*.`tSubsetStructure` can simply be set to

  `rec( t := t, lambdas := [lambda] ).`

We will see an example using the more general form of *param*.`tSubsetStructure` later.

The default is to classify the required designs up to isomorphism. As an example, we now classify the 5-(12,6,1) designs, verifying the well-known result that the 'small Witt design' is the unique such design (up to isomorphism).

```
gap> designs := BlockDesigns( rec(
>     v:=12,  # there are exactly 12 points
>     blockSizes:=[6], # each block has size 6
>     tSubsetStructure:=rec(t:=5, lambdas:=[1])
>         # every 5-subset of the points is contained in exactly one block
>     ) );;
gap> Length(designs);
1
gap> AllTDesignLambdas(designs[1]); # as a check
[ 132, 66, 30, 12, 4, 1 ]
gap> Size(AutomorphismGroup(designs[1]));
95040
```

We next classify, up to isomorphism, the binary block designs having 11 points, such that each block has size 4 or 5, and every pair of distinct points is contained in exactly two blocks.

```
gap> designs:=BlockDesigns( rec( v:=11, # there are exactly 11 points
>     blockSizes:=[4,5],  # each block has size 4 or 5
>     tSubsetStructure:=rec(t:=2, lambdas:=[2])
>         # every 2-subset of the points is contained in exactly two blocks
> ) );;
gap> Length(designs);
5
gap> List(designs,BlockSizes);
[ [ 5 ], [ 4, 5 ], [ 4, 5 ], [ 4, 5 ], [ 4, 5 ] ]
gap> List(designs,BlockNumbers);
[ [ 11 ], [ 10, 5 ], [ 10, 5 ], [ 10, 5 ], [ 10, 5 ] ]
gap> List(designs,AllTDesignLambdas);
[ [ 11, 5, 2 ], [ ], [ ], [ ], [ ] ]
gap> List(designs,d->Size(AutomorphismGroup(d)));
[ 660, 6, 8, 12, 120 ]
```

Further properties may optionally be specified to the function BlockDesigns via the record *param*. How to do this is precisely detailed in the DESIGN package documentation for Block Designs. These further properties include:

- a total number b of blocks, specifying that every returned design has exactly b blocks

- a replication number r, specifying that in every returned design, every point is in exactly r blocks

- a list giving, for each specified possible block size, a corresponding maximum multiplicity of a block of that size

- a list giving, for each specified possible block size, a corresponding number of blocks of that size

- the possible sizes of intersections of pairs of blocks of given sizes

- a permutation group G on the point set $\{1, \ldots, v\}$, such that two designs are considered to be isomorphic if one is in the G-orbit of the other (if *param*.blockDesign is unbound, the default is a group G giving the usual notion of block design isomorphism)

- a subgroup H of G such that H is required to be a subgroup of the automorphism group of each returned design (the default is for H to be the trivial permutation group)

- whether the user wants a single design with the specified properties (if one exists), a list of G-orbit representatives of all such designs (i.e. isomorphism class representatives as determined by G; this is the default), or a list of such designs containing at least one representative from each G-orbit.

For example, we now construct one simple 2-(20,5,4) design invariant under a group of order 19 (a block design is **simple** if it has no repeated block).

```
gap> H:=CyclicGroup(IsPermGroup,19);
Group([ (1,2,3,4,5,6,7,8,9,10,11,12,13,14,15,16,17,18,19) ])
gap> D:=BlockDesigns( rec( v:=20,
>       blockSizes:=[5],
>       tSubsetStructure:=rec(t:=2, lambdas:=[4]),
>       blockMaxMultiplicities:=[1], # since we want a simple design
>       requiredAutSubgroup:=H,
>          # since we want a design whose automorphism group contains H
>       isoLevel:=0
>          # since we want just one design (if such a design exists)
>       ) );;
gap> Length(D);
1
gap> AllTDesignLambdas(D[1]); # a check
[ 76, 19, 4 ]
gap> Size(AutomorphismGroup(D[1]));
19
```

14.3.6 Computing Subdesigns Using the Function `BlockDesigns`

Here a **subdesign** of a block design D means a block design with the same point set as D and whose block multiset is a submultiset of the blocks of D.

Setting the optional parameter *param*.`blockDesign` to be a block design D asks `BlockDesigns(` *param* `)` to construct subdesigns of D with the other given properties. In this case, the default is to determine the subdesigns up to the action of the automorphism group of D.

We now give an example of the use of subdesigns, where we determine certain 'Sylvester designs', which are highly efficient block designs from a statistical design viewpoint. See [1]. A **Sylvester design** is a 1-(36,6,8) design, such that

- if a and b are distinct points, then $\{a, b\}$ is contained in just 1 or 2 blocks, and
- the graph on the 36 points whose edges are those $\{a, b\}$ contained in exactly 2 blocks is isomorphic to the Sylvester graph.

```
gap> Sylvester_3:=DistanceGraph(Sylvester,3);;
gap> GlobalParameters(Sylvester_3);
[ [ 0, 0, 10 ], [ 1, 4, 5 ], [ 2, 8, 0 ] ]
gap> IsIsomorphicGraph(Sylvester_3,HammingGraph(2,6));
true
gap> K:=CompleteSubgraphsOfGivenSize(Sylvester_3,6,2);
[ [ 1, 10, 13, 16, 22, 23 ] ]
gap> K:=Union(Orbits(Sylvester_3.group,K,OnSets));;
```

```
gap> Length(K);
12
gap> #
gap> # Now  K  is the set of cliques of size  6  of  Sylvester_3,
gap> # which is isomorphic to the Hamming graph  H(2,6).
gap> #
gap> complement_3:=ComplementGraph(Sylvester_3);;
gap> L:=CompleteSubgraphsOfGivenSize(complement_3,6,2);
[ [ 1, 2, 4, 8, 9, 35 ], [ 1, 2, 4, 8, 19, 28 ],
  [ 1, 2, 4, 9, 11, 30 ], [ 1, 2, 4, 11, 19, 36 ],
  [ 1, 2, 5, 19, 24, 36 ] ]
gap> L:=Union(Orbits(complement_3.group,L,OnSets));;
gap> Length(L);
720
gap> #
gap> # Now  L  is the set of independent sets of size  6  in Sylvester_3.
gap> #
gap> blocks:=Union(K,L);;
gap> BD:=BlockDesign(36,blocks);;
gap> #
gap> # We are here interested in those Sylvester designs that are
gap> # subdesigns of  BD.
gap> #
gap> edges:=UndirectedEdges(Sylvester);;
gap> nonedges:=UndirectedEdges(ComplementGraph(Sylvester));;
gap> Sdesigns := BlockDesigns(rec(v:=36, blockSizes:=[6],
>     tSubsetStructure:=
>        rec(t:=2, partition:=[edges,nonedges],lambdas:=[2,1]),
>     r:=8, # replication number
>     blockDesign:=BD, # to get subdesigns of  BD
>     isoGroup:=AutomorphismGroup(Sylvester)));;
gap> Length(Sdesigns);
4
gap> #
gap> # The elements of  Sdesigns  are (up to the action of the
gap> # automorphism group of the Sylvester graph, and so up to
gap> # isomorphism) the Sylvester designs whose blocks come from
gap> # the rows, columns, and transversals of the 6x6 array
gap> # defined by the distance-3 graph of the Sylvester graph.
gap> #
gap> List(Sdesigns,design->Size(AutomorphismGroup(design)));
[ 144, 16, 72, 1440 ]
```

Challenge problem: Classify *all* the Sylvester designs.

14.3.7 Partitioning Block Designs

The DESIGN package function `PartitionsIntoBlockDesigns` constructs partitions of (the block multiset of) a given block design *D*, such that the subdesigns of *D* whose block multisets are the parts of this partition each have the

same user-specified properties. See PartitionsInto BlockDesigns in the
DESIGN manual for full details.

For example, given a block design D having v points and each of whose
blocks has size k, we can classify the resolutions of D by classifying the partitions of D into 1-$(v, k, 1)$ designs, up to the action of the automorphism group
of D. We now do this for the Sylvester designs constucted in Section 14.3.6.

```
gap> L:=List(Sdesigns, design->
>       PartitionsIntoBlockDesigns(rec(v:=36, blockSizes:=[6],
>           blockDesign:=design,
>           tSubsetStructure:=rec(t:=1,lambdas:=[1]) ) ) );;
gap> List(L,Length);
[ 1, 0, 1, 1 ]
gap> # This gives the respective numbers of resolutions of the designs
gap> # in  Sdesigns,  up to the actions of the respective automorphism
gap> # groups of these designs.
```

14.4 Study of the Cohen–Tits Near Octagon

There is a unique distance-regular graph with intersection array $\{10, 8, 8, 2; 1, 1, 4, 5\}$; that is, with global parameters

```
[[0,0,10], [1,1,8], [1,1,8], [4,4,2], [5,5,0]].
```

The partial linear space CT with parameters $(2, 4)$ having this graph as point
graph is the **Cohen–Tits near octagon**. The lines of CT consist of all the 525
triangles (cliques of size 3) in this point graph. See [7, sections 6.4 and 13.6].

In the extended example of this section, we show that CT does not have an
ovoid (where an **ovoid** is a set of points such that every line lies on just one of
these points), but CT does have a resolution, both of which appear to be new
results. We start by constructing the point graph of CT on the conjugacy class
of size 315 of involutions in the Hall–Janko sporadic simple group J_2.

```
gap> n:=315;
315
gap> L:=AllPrimitiveGroups(NrMovedPoints,100,Size,604800,IsSimple,true);
[ J_2 ]
gap> J2:=L[1];
J_2
gap> CC:=Filtered(ConjugacyClasses(J2),x->Size(x)=n);;
gap> Length(CC);
1
gap> CC:=CC[1];;
gap> Order(Representative(CC));

2
gap> pointgraph:=Graph(J2,AsSet(CC),OnPoints,
>       function(x,y) return x*y=y*x and x<>y; end,
```

```
>       true);;
gap> GlobalParameters(pointgraph);
[ [ 0, 0, 10 ], [ 1, 1, 8 ], [ 1, 1, 8 ], [ 4, 4, 2 ], [ 5, 5, 0 ] ]
gap> #
gap> # Now  pointgraph  is the point graph of the Cohen-Tits near octagon.
gap> #
gap> A:=AutomorphismGroup(pointgraph);;
gap> Size(A);
1209600
gap> pointgraph:=NewGroupGraph(A,pointgraph);;
gap> K:=CompleteSubgraphsOfGivenSize(pointgraph,3,2);
[ [ 1, 2, 5 ] ]
gap> CT:=BlockDesign(n,K,A);;
gap> #
gap> # Now  CT  is the Cohen-Tits near octagon as a block design
gap> # in DESIGN package format.
gap> #
gap> NrBlockDesignPoints(CT);
315
gap> BlockSizes(CT);
[ 3 ]
gap> AllTDesignLambdas(CT);
[ 525, 5 ]
gap> dualCT:=DualBlockDesign(CT);;
gap> NrBlockDesignPoints(dualCT);
525
gap> BlockSizes(dualCT);
[ 5 ]
gap> AllTDesignLambdas(dualCT);
[ 315, 3 ]
gap> #
gap> # Does  dualCT  have a spread (parallel class)?
gap> #
gap> spreadsofdual:=BlockDesigns(rec(v:=NrBlockDesignPoints(dualCT),
>       blockSizes:=BlockSizes(dualCT),
>       blockDesign:=dualCT,
>       tSubsetStructure:=rec(t:=1, lambdas:=[1])));
[ ]
gap> #
gap> # So the dual of  CT  has no  1-(525,5,1)  subdesign,
gap> # i.e. no spread. Equivalently,  CT  has no ovoid.
gap> #
gap> # What about possible spreads and resolutions of  CT?
gap> #
gap> # We shall first consider the spreads of  CT  invariant under
gap> # a well-chosen subgroup  S  of its automorphism group  A.
gap> #
gap> CC:=Set(ConjugacyClasses(A),x->Group(Representative(x)));;
gap> S:=Filtered(CC,x->Size(x)=5 and Size(Centralizer(A,x))=300);;
gap> Length(S);
1
```

```
gap> S:=S[1];
<permutation group of size 5 with 1 generator>
gap> #
gap> # We now classify the  S-invariant  spreads of  CT,  up to the
gap> # action of the normalizer in  A  of  S.
gap> #
gap> spreads:=BlockDesigns(rec(v:=NrBlockDesignPoints(CT),
>       blockSizes:=BlockSizes(CT),
>       blockDesign:=CT,
>       requiredAutSubgroup:=S,
>       isoGroup:=Normalizer(A,S),
>       tSubsetStructure:=rec(t:=1, lambdas:=[1])));;
gap> Collected(List(spreads,AllTDesignLambdas));  # a check
[ [ [ 105, 1 ], 39 ] ]
gap> #
gap> # Now make each spread into the set of the indices of its blocks
gap> # in  CT.
gap> #
gap> spreads:=List(spreads,
>       spread->List(spread.blocks,B->PositionSorted(CT.blocks,B)));;
gap> #
gap> # We now make a graph whose vertices correspond to the spreads
gap> # invariant under a conjugate in  A  of  S,  with two such
gap> # spreads joined by an edge iff they are disjoint.
gap> #
gap> spreadsgraph:=Graph(Image(ActionHomomorphism(A,CT.blocks,OnSets)),
>       spreads, OnSets,
>       function(x,y) return Intersection(x,y)=[]; end);;
gap> OrderGraph(spreadsgraph);
2645160
gap> VertexDegrees(spreadsgraph);
[ 51, 53, 158, 163, 166, 167, 173, 176, 179, 188, 196, 201, 207, 208,
  210, 212, 223, 226, 248, 252, 269, 312, 327, 357, 383, 395, 438,
  832, 900, 1369, 1423, 1432, 1561, 1879, 4948 ]
gap> Size(spreadsgraph.group);
1209600
gap> #
gap> # Now a clique of size  5  in spreadsgraph would yield a resolution
gap> # of  CT.
gap> #
gap> CompleteSubgraphsOfGivenSize(spreadsgraph,5,0,true);
[ [ 309961, 458110, 501557, 828391, 1047808 ] ]
gap> #
gap> # Thus, the Cohen-Tits near octagon  CT  has a resolution.
gap> #
```

14.5 The FinInG Package

FinInG [3] is a large and powerful GAP package for finite incidence geometry. It requires the GAP packages GAPDoc, Forms, cvec, Orb, GenSS, and GRAPE, and can also make use of DESIGN.

A (finite) **incidence structure** consists of a finite set of **elements**, a reflexive symmetric **incidence** relation on the set of elements, and a function, called a **type** function, from the set of elements to a finite set of **types**, such that no two distinct elements having the same type are incident. An **incidence geometry** is an incidence structure such that every maximal set of pairwise incident elements contains an element of each type.

FinInG provides efficient functionality for constructing and studying the following incidence structures:

- projective and affine spaces
- classical polar spaces
- generalised polygons
- coset incidence structures.

FinInG also provides functionality for:

- groups of automorphisms of incidence structures
- specialised actions, orbits, and stabilizers
- morphisms of incidence geometries
- algebraic varieties over finite fields.

See [4] for a good overview of the FinInG package. The extensive FinInG manual includes much more detailed information, together with many useful examples. In this chapter, however, we look only into some FinInG functionality for projective spaces and coset incidence structures, concentrating on applications by the author.

14.5.1 The Projective Space PG(d, q)

Let V be a $(d + 1)$-dimensional vector space over the finite field GF(q).

In FinInG, the **projective space** PG(d, q) is the incidence geometry whose elements are the proper non-zero subspaces of V, with two elements being incident exactly when one is contained in the other, and having the same type if and only if they have the same dimension.

The **points** and **lines** of PG(d, q) are respectively the 1- and 2-dimensional subspaces of V. A **partial spread** (of lines) in a projective space is a set of lines whose pairwise intersection is the zero-subspace (projectively, the **empty subspace**). A **maximal** partial spread is one which is not contained in any larger partial spread of the projective space.

We now use FinInG and GRAPE to classify the maximal partial spreads of lines in the projective space PG(3,4), first, up to the action of the projective general linear group PGL(4,4) (which was done much more slowly in [29]), and then, up to the action of the full collineation group $P\Gamma L(4,4)$ of the projective space.

```
gap> LoadPackage("fining",false);
true
gap> d:=3;;
gap> q:=4;;
gap> pg:=PG(d,q);
ProjectiveSpace(3, 4)
gap> IsIncidenceStructure(pg);
true
gap> TypesOfElementsOfIncidenceStructure(pg);
[ "point", "line", "plane" ]
gap> lines:=Lines(pg);
<lines of ProjectiveSpace(3, 4)>
gap> Size(lines);
357
gap> points:=Points(pg);
<points of ProjectiveSpace(3, 4)>
gap> Size(points);
85
gap> lineset:=Set(List(lines));;
gap> G:=ProjectivityGroup(pg);
The FinInG projectivity group PGL(4,4)
gap> Size(G);
987033600
gap> #
gap> # Now construct the graph  gamma,  whose vertices are labelled
gap> # by the lines of  PG(d,q),  with two vertices joined by an edge
gap> # iff the corresponding lines have no point in common.
gap> #
gap> # The cliques of  gamma  give the partial spreads of  PG(d,q),
gap> # with the maximal cliques giving the maximal partial spreads.
gap> #
gap> gamma:=Graph(G, lineset, OnProjSubspaces,
>     function(x,y) return Meet(x,y)=EmptySubspace(pg); end,
>     true);;
gap> GlobalParameters(gamma);
[ [ 0, 0, 256 ], [ 1, 180, 75 ], [ 192, 64, 0 ] ]
gap> Size(gamma.group);
987033600
gap> maximalpartialspreads:=CompleteSubgraphs(gamma,-1,2);;
gap> # The maximal complete subgraphs of  gamma,  up to the action of
gap> # gamma.group.  See the GRAPE documentation on  CompleteSubgraphs.
gap> #
gap> L:=List(maximalpartialspreads,Length);
[ 13, 14, 11, 13, 11, 12, 11, 12, 13, 13, 14, 14, 14, 13, 11, 13, 11,
  11, 13, 12, 17, 13, 13, 13, 13, 17, 13, 13, 17 ]
gap> Collected(L);
[ [ 11, 6 ], [ 12, 3 ], [ 13, 13 ], [ 14, 4 ], [ 17, 3 ] ]
gap> #
gap> # Repeat the classification with  G  now being the collineation
gap> # group of  pg.
gap> #
```

```
gap> G:=CollineationGroup(pg);
The FinInG collineation group PGammaL(4,4)
gap> Size(G);
1974067200
gap> hom:=ActionHomomorphism(G,VertexNames(gamma),OnProjSubspaces);
<action homomorphism>
gap> gamma:=NewGroupGraph(Image(hom),gamma);;
gap> Size(gamma.group);
1974067200
gap> Size(AutomorphismGroup(gamma));
3948134400
gap> maximalpartialspreads:=CompleteSubgraphs(gamma,-1,2);;
gap> L:=List(maximalpartialspreads,Length);
[ 12, 13, 13, 13, 11, 17, 14, 14, 11, 12, 14, 11, 11, 13, 13, 13, 13,
  17, 13, 17 ]
gap> Collected(L);
[ [ 11, 4 ], [ 12, 2 ], [ 13, 8 ], [ 14, 3 ], [ 17, 3 ] ]
```

More sophisticated computations of maximal partial spreads using GRAPE are described in [33], where to illustrate certain theory, the maximal partial spreads in $PG(3, q)$ that are invariant under a group of order 5 are classified, for $q \in \{7, 8\}$.

14.5.2 Flag-Transitive Geometries and Coset Incidence Structures

Let Γ be an incidence structure. A **flag** of Γ is a set of pairwise incident elements, and the **type** of a flag is the set of the types of its elements. A **chamber** of Γ is a flag containing an element of each type. An **automorphism** of Γ is a permutation of the elements of Γ preserving the type of each element and preserving incidence between elements.

The incidence structure Γ is a **flag-transitive geometry** for a group G of automorphisms of Γ if Γ is an incidence geometry and G acts transitively on the set of chambers of Γ (and so G acts transitively on the flags of Γ of any given type).

Suppose now Γ is a flag-transitive geometry for a group G of automorphisms. Consider a chamber $\{c_1, c_2, \ldots, c_n\}$ of Γ, such that c_i has type i, and let G_i be the stabilizer in G of c_i. Now G acts transitively on the elements of type i, so these elements are in one-to-one correspondence with the right cosets of G_i in G. Furthermore, two elements of Γ are incident if and only if the corresponding cosets have a non-empty intersection.

Now let G be any finite group and let $G_1 \ldots, G_n$ be distinct subgroups of G. Then these **parabolic subgroups** determine an incidence structure called a **coset incidence structure** for G, with type set $\{1, \ldots, n\}$, the elements of type i being the right cosets of G_i in G, and with two elements being incident precisely when they have a non-empty intersection.

A coset incidence structure need not be an incidence geometry. However, in FinInG, if the function IsFlagTransitiveGeometry applied to a coset incidence structure returns true, this guarantees that the argument is a (flag-transitive) incidence geometry.

14.5.3 A Flag-Transitive Geometry for J_2 and J_2:2

We now use GRAPE and FinInG to construct and study a certain coset incidence structure which is a flag-transitive geometry for the Hall–Janko sporadic simple group J_2 and its automorphism group J_2:2. The elements of this geometry correspond to certain induced subgraphs in a vertex- and edge-transitive graph for J_2, having 280 vertices and vertex-degree 36. As far as the author is aware, this geometry has never before been published.

```
gap> n:=280;
280
gap> L:=AllPrimitiveGroups(NrMovedPoints,n,Size,604800,IsSimple,true);
[ J_2 ]
gap> J2:=L[1];
J_2
gap> H:=Stabilizer(J2,1);
<permutation group of size 2160 with 4 generators>
gap> orbs:=List(Orbits(H,[1..n]),Set);;
gap> List(orbs,Length);
[ 1, 108, 36, 135 ]
gap> orb36:=First(orbs,orb->Length(orb)=36);;
gap> edge:=[1,orb36[1]];
[ 1, 3 ]
gap> gamma:=EdgeOrbitsGraph(J2,[edge]);;
gap> GlobalParameters(gamma);
[ [ 0, 0, 36 ], [ 1, 8, 27 ], [ 4, 32, 0 ] ]
gap> AssignVertexNames(gamma,[1..n]);
gap> f:=First(orbs,orb->Length(orb)=135)[1];
4
gap> eps:=GeodesicsGraph(gamma,1,f);;
gap> # So  eps  is the induced subgraph on the vertices of the
gap> # geodesics from 1 to f, but not including 1 and f.
gap> #
gap> GlobalParameters(eps);
[ [ 0, 0, 2 ], [ 1, 0, 1 ], [ 2, 0, 0 ] ]
gap> # So eps is a 4-gon.
gap> #
gap> oct_vertices:=Union([1,f],VertexNames(eps));
[ 1, 3, 4, 118, 223, 265 ]
gap> # So the induced subgraph on  oct_vertices  is (the 1-skeleton of)
gap> # an octahedron.
gap> #
```

```
gap> oct_stab:=Stabilizer(J2,oct_vertices,OnSets);;
gap> Size(oct_stab);
96
gap> oct_graph:=InducedSubgraph(gamma,oct_vertices,oct_stab);;
gap> GlobalParameters(oct_graph);
[ [ 0, 0, 4 ], [ 1, 2, 1 ], [ 4, 0, 0 ] ]
gap> Size(oct_graph.group);
48
gap> #
gap> #  oct_graph  is a chosen special induced subgraph of  gamma,
gap> # and is isomorphic to (the 1-skeleton of) an octahedron.
gap> #
gap> K:=CompleteSubgraphsOfGivenSize(oct_graph,3,2);
[ [ 1, 2, 4 ] ]
gap> clique:=Set(K[1],x->VertexName(oct_graph,x));
[ 1, 3, 118 ]
gap> # Now  clique  is the set of vertices in  gamma  of a triangle
gap> # in  oct_graph.
gap> #
gap> parabolics:=[];
[  ]
gap> for i in [1..3] do
>     parabolics[i]:=Stabilizer(J2,clique{[1..i]},OnSets);
> od;
gap> Add(parabolics,oct_stab);
gap> List(parabolics,x->Size(J2)/Size(x));
[ 280, 5040, 10080, 6300 ]
gap> J2geom:=CosetGeometry(J2,parabolics);
CosetGeometry( J_2 )
gap> #
gap> # In FinInG,  CosetGeometry  makes a coset incidence structure
gap> # (which is not necessarily an incidence geometry).
gap> #
gap> IsFlagTransitiveGeometry(J2geom);
true
gap> #
gap> # So  J2geom  is a flag-transitive geometry for the group  J2.
gap> #
gap> # We now use FinInG to determine some properties of  J2geom.
gap> #
gap> TypesOfElementsOfIncidenceStructure(J2geom);
[ 1 .. 4 ]
gap> IsConnected(J2geom);
true
gap> IsResiduallyConnected(J2geom);
true
gap> IsThinGeometry(J2geom);
false
gap> IsThickGeometry(J2geom);
```

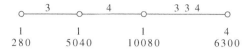

Figure 14.1 Diagram of the constructed flag-transitive geometry for J_2

```
false
gap> IsFirmGeometry(J2geom);
true
gap> Size(BorelSubgroup(J2geom));
2
gap> autgrp:=AutGroupIncidenceStructureWithNauty(J2geom);;
gap> #  the automorphism group of  J2geom
gap> #
gap> DisplayCompositionSeries(autgrp);
G (6 gens, size 1209600)
 | C2
S (3 gens, size 604800)
 | J2
1 (0 gens, size 1)
gap> #
gap> # FinInG can draw the "diagram" of a flag-transitive
gap> # coset geometry, making use of the GraphViz
gap> # software (available from https://graphviz.org).
gap> #
gap> DrawDiagram(DiagramOfGeometry(J2geom),"J2geomdiagram");
```

The diagram of the constructed flag-transitive geometry, as described in the FinInG manual and produced by the code here, is shown in Figure 14.1.

```
gap> incidencegraph:=IncidenceGraph(J2geom);;
gap> #
gap> # Now  incidencegraph  is the incidence graph of  J2geom,
gap> # in GRAPE format.
gap> #
gap> IsGraph(incidencegraph);
true
gap> OrderGraph(incidencegraph);  # number of vertices
21700
gap> Size(incidencegraph.group);
604800
gap> IsLoopy(incidencegraph);
true
```

We next read a file downloadable from [32], which provides GAP functions for the computation of fundamental groups, certain quotients of fundamental groups, and covers of finite simplicial complexes, making use of the theory and algorithms of [27] and the GRAPE package. We then compute the fundamental group of the clique complex of the incidence graph of the flag-transitive geometry for J_2 we have constructed. This clique complex is the so-called flag complex of the geometry.

```
gap> Read("fundamental_v2.g");
gap> # This loads functions for fundamental groups and covers of
gap> # finite simplicial complexes.
gap> #
gap> # We now remove the loops on the incidence graph of  J2geom
gap> # to make it compatible with the fundamental group software.
gap> #
gap> for rep in incidencegraph.representatives do
>     RemoveEdgeOrbit(incidencegraph,[rep,rep]);
> od;
gap> IsSimpleGraph(incidencegraph);
true
gap> VertexDegrees(incidencegraph);
[ 11, 23, 26, 279 ]
gap> F:=FundamentalRecordSimplicialComplex(incidencegraph);;
#I  now the presentation has 1 generators, the new generator is _x1
#I  now the presentation has 2 generators, the new generator is _x2
gap> G:=F.group;
<fp group on the generators [ _x1, _x2 ]>
gap> # Now  G  is the fundamental group of the clique complex
gap> # of  incidencegraph.
gap> Size(G);
12
gap> IsAbelian(G);
false
gap> StructureDescription(G);
"C3 : C4"
```

14.6 Other Software for Graphs, Designs, and Finite Geometries

In this chapter, we have focussed on three GAP packages, but the reader should be aware of other software available for research in graphs, designs, and finite geometries. We now briefly discuss some of this software. Much more detailed information is available in the references cited.

While GRAPE is designed for the construction and study of usually simple and sometimes very large graphs related to groups, designs, and finite geometries, the Digraphs package [12] for GAP focuses on directed graphs (digraphs), and provides a very extensive range of constructions and graph-theoretic functionality for simple graphs, digraphs, and (at present) multidigraphs. In particular, the Digraphs package can input digraphs in GRAPE and other formats, and includes many functions analogous to those in GRAPE, as well as providing efficient functionality for digraph homomorphisms, digraph drawing, I/O facilities and efficient storage for large collections of digraphs, and more. The Digraphs package also includes an improved version of the bliss software [20] for computing automorphism groups of digraphs and testing digraph isomorphism. Some aspects of the Digraphs package were covered in the author's G2G2 lectures [36].

The GAP package AGT [13] for algebraic graph theory provides functionality for the computation of spectral properties, various bounds, and regularity properties for simple graphs given in GRAPE format. The package also provides an extensive library of strongly regular graphs and lists of 'feasible' strongly regular graph parameters.

For computations with permutation groups and coherent configurations, the reader should consider the stand-alone COCO system [14], or one of its more modern descendents under development (see [21]). Andries Brouwer has made a Unix port of COCO, downloadable from [6].

Finally, Magma [5] is a large, powerful, comprehensive system for computations in algebra, number theory, algebraic geometry, and algebraic combinatorics. Magma's functionality has large overlaps with GAP and its packages. However, Magma is not open-source, and, with certain exceptions, is not free of charge.

References

[1] R. A. Bailey, P. J. Cameron, L. H. Soicher, and E. R. Williams, Substitutes for the non-existent square lattice designs for 36 varieties, *Journal of Agricultural, Biological and Environmental Statistics* **25** (2020), 487–499. (open-access) at https://doi.org/10.1007/s13253-020-00388-1

[2] S. Ball, *Finite Geometry and Combinatorial Applications*, Cambridge University Press, Cambridge, 2015.

[3] J. Bamberg, A. Betten, Ph. Cara, J. De Beule, M. Lavrauw, and M. Neunhoeffer, The FinInG package for GAP, Version 1.5, 2022, https://gap-packages.github.io/FinInG

[4] J. Bamberg, A. Betten, Ph. Cara, J. De Beule, M. Neunhöffer, and M. Lavrauw, FinInG: a package for Finite Incidence Geometry, 2016, https://arxiv.org/abs/1606.05530

[5] W. Bosma, J. Cannon, and C. Playout, The Magma algebra system. I: The user language, *Journal of Symbolic Computation* **24** (1997), 235–265.

[6] A. E. Brouwer, Unix port of the COCO computer algebra system, www.win.tue.nl/~aeb/ftpdocs/math/coco/coco-1.2b.tar.gz

[7] A. E. Brouwer, A. M. Cohen, and A. Neumaier, *Distance-Regular Graphs*, Springer-Verlag, Berlin, 1989.

[8] A. E. Brouwer and W. H. Haemers, *Spectra of Graphs*, Springer, New York, 2012.

[9] P. J. Cameron, *Permutation Groups*, Cambridge University Press, Cambridge, 1999.

[10] P. J. Cameron and J. H. van Lint, *Designs, Graphs, Codes and Their Links*, Cambridge University Press, Cambridge, 1991.

[11] E. R. van Dam, J. H. Koolen, and H. Tanaka, Distance-regular graphs, *Electronic Journal of Combinatorics* (2016), DS22.

[12] J. De Beule, J. Jonusas, J. Mitchell et al, The Digraphs package for GAP, Version 1.5.0, 2021, https://digraphs.github.io/Digraphs

[13] R. J. Evans, The AGT package for GAP, Version 0.2, 2020, https://gap-packages.github.io/agt

[14] I. A. Faradzev and M. H. Klin, Computer package for computations with coherent configurations, in *ISSAC '91: Proceedings of the 1991 International Symposium on Symbolic and Algebraic Computation*, S. M. Watt (ed.), ACM Press, New York, 1991, pp. 219–223.

[15] The GAP Group, GAP — Groups, Algorithms, and Programming, Version 4.12.0, 2022, www.gap-system.org

[16] The GAP Group, GAP — A Tutorial, Release 4.12.0, 2022, www.gap-system.org/Manuals/doc/tut/manual.pdf

[17] C. Godsil and G. Royle, *Algebraic Graph Theory*, Springer-Verlag, New York, 2001.

[18] W. Haemers, A new partial geometry constructed from the Hoffman–Singleton graph, in *Finite Geometries and Designs: Proceedings of the Second Isle of Thorns Conference 1980*, P. J. Cameron et al. (eds), LMS Lecture Note Series **49**, Cambridge University Press, Cambridge, 1981, pp. 119–127.

[19] C. Jefferson, E. Jonauskyte, M. Pfeiffer, and R. Waldecker, Minimal and canonical images, *Journal of Algebra* **521** (2019), 481–506.

[20] T. Junttila and P. Kaski, Engineering an efficient canonical labeling tool for large and sparse graphs, in *Proceedings of the Ninth Workshop on Algorithm Engineering and Experiments and the Fourth Workshop on Analytic Algorithmics and Combinatorics*, D. Applegate et al. (eds), SIAM, Philadelphia, 2007, pp. 135–149. www.tcs.hut.fi/Software/bliss/

[21] M. Klin, M. Muzychuk, and S. Reichard, Proper Jordan schemes exist. First examples, computer search, patterns of reasoning. An essay, https://arxiv.org/abs/1911.06160, 2019.

[22] A. Konovalov, Programming with GAP, https://alex-konovalov.github .io/gap-lesson/

[23] R. M. R. Lewis, *A Guide to Graph Colouring: Algorithms and Applications*, Springer International Publishing, Switzerland, 2016.

[24] S. Linton, Finding the smallest image of a set, in *ISSAC '04: Proceedings of the 2004 International Symposium on Symbolic and Algebraic Computation*, J. Gutierrez (ed.), ACM Press, New York, 2004, pp. 229–234.

[25] B. D. McKay and A. Piperno, Practical graph isomorphism, II, *Journal of Symbolic Computation* **60** (2014), 94–112. nauty and Traces: https://users .cecs.anu.edu.au/ bdm/nauty/

[26] A. Pasini, *Diagram Geometries*, Oxford University Press, Oxford, 1994.

[27] S. Rees and L. H. Soicher, An algorithmic approach to fundamental groups and covers of combinatorial cell complexes, *Journal of Symbolic Computation* **29** (2000), 59–77.

[28] L. H. Soicher, Yet another distance-regular graph related to a Golay code, *Electronic Journal of Combinatorics* **2** (1995), N1.

[29] L. H. Soicher, Computation of partial spreads, 2004, https://webspace .maths.qmul.ac.uk/l.h.soicher/partialspreads/

[30] L. H. Soicher, Computing with graphs and groups, in *Topics in Algebraic Graph Theory*, L. W. Beineke and R. J. Wilson (eds), Cambridge University Press, Cambridge, 2004, pp. 250–266.

[31] L. H. Soicher, Designs, groups and computing, in *Probabilistic Group Theory, Combinatorics, and Computing: Lectures from the Fifth de Brún Workshop*, A. Detinko et al. (eds), Lecture Notes in Mathematics **2070**, Springer, London, 2013, pp. 83–107.

[32] L. H. Soicher, Functions for computing fundamental groups, certain quotients of fundamental groups, and covers of finite simplicial complexes, Version 2.0, 2015, https://webspace.maths.qmul.ac.uk/l.h.soicher/ fundamental/fundamental_v2.g

[33] L. H. Soicher, On classifying objects with specified groups of automorphisms, friendly subgroups, and Sylow tower groups, *Portugaliae Mathematica* **74** (2017), 233–242.

[34] L. H. Soicher, The DESIGN package for GAP, Version 1.7, 2019, https: //gap-packages.github.io/design

[35] L. H. Soicher, The GRAPE package for GAP, Version 4.8.5, 2021, https: //gap-packages.github.io/grape

[36] L. H. Soicher, Using the GAP system and its packages for research in graph theory, design theory, and finite geometry, G2G2 minicourse lecture notes and exercises, 2021, https://webspace.maths.qmul.ac.uk/l.h.soicher/ g2g2/